A GLOSSARY OF FOOD SCIENCE AND TECHNOLOGY

食品科學詞彙

A Glossary of Food Science and Technology

食品科學詞彙

by Chi-fai Chau and Hoi-shan Kwan

周志輝 關海山 編譯

The Chinese University Press

A Glossary of Food Science and Technology
by Chi-fai Chau and Hoi-shan Kwan

ISBN: 962–996–092–3

Published by The Chinese University Press,
The Chinese University of Hong Kong,
Sha Tin, N.T., Hong Kong.
Fax: +852 2603 6692
+852 2603 7355
E-mail: cup@cuhk.edu.hk
Web-site: www.chineseupress.com

《食品科學詞彙》
周志輝、關海山 編譯

國際統一書號（ISBN）：962–996–092–3

出版：中文大學出版社
香港 新界 沙田・香港中文大學
圖文傳真：+852 2603 6692
+852 2603 7355
電子郵遞：cup@cuhk.edu.hk
網　　址：www.chineseupress.com

Printed in Hong Kong

Contents

Preface

(in English) ---------- vii

(in Chinese) ---------- ix

About the Editors

(in English) ---------- xi

(in Chinese) ---------- xii

Guide to the Glossary Entries

(in English) ---------- xiii

(in Chinese) ---------- xvi

English-Chinese Glossary ---------- 1

Chinese-English Glossary ---------- 375

目 錄

緒言
(英文) ---------- vii
(中文) ---------- ix
編者簡介
(英文) ---------- xi
(中文) ---------- xii
凡例
(英文) ---------- xiii
(中文) ---------- xvi
英中詞彙 ---------- 1
中英詞彙 ---------- 375

Preface

After exploring the fields of food science, nutrition and food technology for years, our interests and enthusiasms in these fields have not deteriorated but keep on growing. Sometimes when we are not sure about the Chinese meaning of some English technical terms, we will look for the Chinese translations in general dictionaries, scientific literature and various reference books. However, it is always difficult to locate appropriate reference materials. Because of this reason, we strongly believe that there is a need and urgency to have a glossary particularly compiled for food science and industry.

While reading various food-related reference materials, we often find that one English term may have varied Chinese translations. It is easy to understand why such a scenario appears in Mainland China and other Chinese communities because of the diversity in both culture and geographical distributions. Taking the terms "salad" and "food additive" as examples, there are different Chinese translations in the Mainland, Taiwan, Hong Kong and Macao. However, with the recent rapid and large-scale developments in the areas like food science and industry, food trading, investment, district cooperation and merging of markets in major Chinese communities, we discover that a number of Chinese terminologies have been adopted by Chinese in different geographical regions to describe the same food, such as the term "instant noodle."

Projecting on the long-term development of food science and industry in Mainland China and other Chinese communities, we believe that a glossary which has a wide coverage of food-related terms, with their Chinese translations in different geographical regions, will have much practical values. It will also be a powerful resource for people who are searching for specific information that cannot be obtained in general dictionaries. To compile this glossary, we have invested years of hard work with diligence and perseverance. We have conducted in-depth researches on a wealth of Chinese and foreign textbooks, reference books, glossaries, dictionaries and scientific literature. This glossary covers the terminologies and technical terms of areas including raw materials, processing, nutrition, chemistry, food hygiene, microorganism, food science analysis and additives, and so forth.

This glossary provides carefully selected defining translations of a quantity of essential food-related terms collected from Mainland China, Taiwan and Hong Kong, with a purpose to serve food science students, professionals in food-related business and general laymen. The first half of this book is an English-Chinese glossary which has 22,500 groups of English technical terms, and the latter half is a Chinese-English glossary which has 27,000 groups of Chinese technical terms.

Chi-Fai Chau
Hoi Shan Kwan
February 2005

緒　言

在食品科學、營養及科技這個學術領域探索多年，我們對食品行業的興趣和熱誠一直有增無減。然而，我們發現，每當自己翻查及閱讀各種飲食名著、中外參考書籍及科學文獻時，不時都會碰到一些與食品有關但又未能在一般詞彙或字典中找到的專業名詞，故此，我們經常需要翻查眾多參考書，來幫助自己閱讀及理解這些作品，可惜，當中很多中文名詞的譯名，仍然是不容易找到的。正因以上種種原因，我們深感一本專為食品科學和工業這兩個範疇而編寫的詞彙，實在非常重要和有價值。

在閱讀來自不同地方的資料時，我們經常發現同一個英文名詞可能會有多個中文譯名；中國地大物博，因地緣和生活文化等差異而出現這種情況，是可以理解的。就以“salad”一詞為例，在大陸、台灣及港澳地區，就分別譯作「色拉」、「沙拉」及「沙律」，甚至有人稱之為「涼拌菜」；又如 “food additive”，它在大陸和台灣分別譯作「食品添加劑」及「食品添加物」。隨着食品科學和工業的不斷發展，兩岸三地之間在食品業及飲食文化上的接觸、合作日趨頻繁，彼此融合之處漸多；時至今日，無論是在大陸、台灣或港澳地區，大家都可以隨便購買到由對方直接進口或於當地投資生產的食物製品。若我們細心比較一下產品包裝上的標籤用詞及食品成份表，便可發現一些以往只被台灣食品業所採用的譯名，如今已在大陸或香港等地漸漸被廣泛接納及採用；就以即食麵(instant noodle)為例，居住在香港的人以往只習慣叫它「即食麵」，時至今日，越來越多人知道「即食麵」、「速食麵」、「方便麵」及「泡麵」其實都是指同一類食品，很明顯，這是互相交流和互相影響的結果。

遠瞻全中國在食品科學和工業的長遠發展，一本詞彙若能同時收集中國多個地區所慣用的食品名詞及相對譯名，必定具有豐富的實用價值，並能迎合實際需要，這便是我們着手編譯這本詞彙的原因。本詞彙主要涵蓋的範圍包括：食品原料、加工、烹調、營養、化學、衛生、食物安全、微生物、食品科學分析及添加劑等。在編譯過程中，我們參考了大量與食品科學有關的中外教材、詞彙、字典、科學文獻及飲食叢書，並加以嚴謹的校對、比較、歸納與整合，以期提供既全面又準確的中、英文食品科學用詞及其相對的翻譯。本書除可方便讀者查閱一些普通英漢詞彙未有收錄的英文食品名詞外，亦可幫助他們翻查個別中文食品名詞的英文名稱。

本書的前半部份是英漢詞彙，共收22,500組英文用詞；而後半部份則屬漢英詞彙，共收中文用詞27,000組。本詞彙所收集的用詞以及譯名，乃來自大陸、台灣及香港等多個地區。我們嘗試把它們編排一起，是希望讓讀者同時了解個別食品在不同地方可能出現的中、英文名稱以及別名，從而增加本詞彙的實用性及參考價值。本詞彙之所以能夠順利完成，實有賴家人之全力支持和鼓勵。若有錯漏或不完備之處，還請各學者專家多多賜教指正。

周志輝

關海山

二零零五年二月

About the Editors

Chi-fai Chau graduated from the Food Science Department of National Chung Hsing University, Taiwan. He obtained his MSc degree in Food Science and Technology from the Food Protein Research and Development Center of Texas A&M University, USA. He then had his PhD in Biology from the Chinese University of Hong Kong, specializing in Food and Nutrition. He is now teaching at the Food Science Department of National Chung Hsing University. His research interests include Chinese medicinal plants, health foods and nanotechnology.

Hoi-shan Kwan graduated from the Biology Department of the Chinese University of Hong Kong and obtained his PhD in Microbiology from the University of California, Davis, USA. He is now the Associate Dean of the Science Faculty, Professor in Biology and Director of Food and Nutritional Sciences Programme in the Chinese University of Hong Kong. He has over twenty years' teaching and research experience in genetics, biotechnology and food microbiology.

編者簡介

周志輝：大學畢業於台灣國立中興大學食品科學系，在美國德州農工大學食物蛋白質研究及發展中心獲取食品科學及科技的碩士學位，及後於香港中文大學生物系獲取食品及營養科學的博士學位。現於台灣國立中興大學食品科學系任教，主要從事中草藥、保健食品及奈米技術等方面的研究及教學工作。

關海山：畢業於香港中文大學生物系，其後於美國加州大學戴維斯分校獲取微生物學哲學博士學位。現職香港中文大學理學院副院長、生物系教授和食品及營養科學課程主任，具二十多年在遺傳、生物技術與食品微生物學上的教學與研究經驗。

Guide to the Glossary Entries

1. In the English-Chinese section, English terms are listed in an alphabetic order and they are followed by their corresponding Chinese translations. Examples are given as follows:

apple	蘋果
dry weight	乾重
junk food	解饞食物

2. In the Chinese-English section, Chinese terms, which are arranged by the number of strokes of characters in increasing order, are followed by their corresponding English translations. Examples are given as follows:

千層糕	**stratiform cake**
品茗	**tea tasting**
蟹子	**crab roe**

3. If a term has more than one translation, all translated terms will be separated by the sign " ; ". Examples are given as follows:

field pea	普通豌豆；紫花豌豆；田豆
pizza	薄餅；比薩；意大利薄餅
乙醇	**alcohol；ethanol； ethyl alcohol； grain alcohol；spirit of wine**
素食主義	**grahamism；vegetarianism； veganism**

4. The abbreviation or full spelling of a particular term is placed in brackets "( )". Examples are given as follows:

FIFO (first-in first-out)	先進先出；先入先出
first-in first-out (FIFO)	先進先出；先入先出
sorbitan monolaurate (Span 20)	單月桂酸山梨醇酐酯；斯潘**20**；司盤**20**
危害分析重點控制	**hazard analysis critical control point (HACCP)**

5. If a term has some special meanings, the supplementary information will be placed in brackets. Examples are given as follows:

Å(Angstrom)	埃(**10^{-8}cm**)
brilliant black	黑色素；亮黑 (食用色素)
sackful	袋 (容量單位)；滿袋

6. In the English-Chinese section, vocabularies which have the same meanings as particular English entries are marked by the sign "[=]". If there are two or more vocabularies with the same meaning, they will be separated by semicolons " ; ". Examples are given as follows:

abattoir	屠宰場 **[= slaughterhouse]**
cargo rice	糙米 **[= brown rice; hulled rice; husked rice]**
safran	番紅花；藏紅花；藏花 **[= crocus; saffron]**

7. Chinese terms with square brackets "[ ]" are the terms of which their meanings remain unchanged even the characters inside the brackets "[ ]" are omitted. This arrangement can avoid unnecessary redundancy.

8. For English terms with different spellings, provided that the spelling difference is slight, brackets "( )" are used to indicate the repeated characters in different spellings. Without changing the alphabetical order of English terms, this arrangement can avoid unnecessary redundancy. Examples are given as follows:

farl(e)	燕麥薄餅；燕麥蛋糕
soybean	大豆；黃豆 **[=soya (bean); soja (bean)]**
tale odo(u)	陳腐氣味；霉臭；不新鮮氣味

凡　例

1. 在英漢詞彙部份，本書會先介紹英文名詞，再介紹其相對的中文名稱。英文名詞的排列次序是依照字母的排列順序，即如：

apple	蘋果
dry weight	乾重
junk food	解饞食物

2. 至於漢英詞彙部份，則是先列出中文名詞，再介紹其相對的英文名稱。各中文名詞的排列順序是依照筆畫之多寡，即如：

千層糕	**stratiform cake**
品茗	**tea tasting**
蟹子	**crab roe**

3. 若某食品用詞的譯名多於一個，則各譯名之間會以分號「；」分隔，即如：

field pea	普通豌豆；紫花豌豆；田豆
pizza	薄餅；比薩；意大利薄餅
乙醇	**alcohol; ethanol; ethyl alcohol; grain alcohol; spirit of wine**
素食主義	**grahamism; vegetarianism; veganism**

4. 個別食品用詞的簡稱、縮寫或全寫，本詞彙會以括弧「( )」把有關資料補充於該名詞之後，即如：

FIFO (first-in first-out)	**先進先出；先入先出**
first-in first-out (FIFO)	**先進先出；先入先出**
sorbitan monolaurate (Span 20)	**單月桂酸山梨醇酐酯；斯潘20；司盤20**
危害分析重點控制	**hazard analysis critical control point (HACCP)**

5. 若個別用詞具有某些特殊意義，本詞彙會按情況在有關譯名後以括弧「()」標示補充資料，即如：

Å(Angstrom)	**埃(10^{-8}cm)**
brilliant black	**黑色素；亮黑(食用色素)**
sackful	**袋 (容量單位)；滿袋**

6. 在英漢詞彙部份，若個別英文用詞與其他名詞意思相同，又或者擁有其他別名，則本書會在其相對的中文名稱後以符號「〔=〕」作資料補充。若補充資料多於一項，則各項資料之間會以分號「；」分隔，即如：

abattoir	屠宰場　**[= slaughterhouse]**
cargo rice	糙米　**[= brown rice; hulled rice; husked rice]**
safran	番紅花；藏紅花；藏花 **[= crocus; saffron]**

7. 若在中文用詞中出現括號「〔 〕」，即表示不論括號內的文字是否存在，該詞均被接受，即如：炭疽〔芽孢〕桿菌，既可作「炭疽桿菌」，亦可作「炭疽芽孢桿菌」。此編排方式既精簡，亦可避免過多重覆。

8. 若某英文用詞的拼寫方法多於一種，而當中只牽涉某些字母的增減，則本詞彙在不影響翻查次序的前提下，會以括弧「()」把有所區別的字母標示出來。此方式使本書在內容編排上更為精簡，能減少重覆，即如：

farl(e)	燕麥薄餅；燕麥蛋糕
soybean	大豆；黃豆 **[= soya (bean); soja (bean)]**
stale odo(u)r	陳腐氣味；霉臭；不新鮮氣味

A

Å(Angstrom) 埃 (10^{-8}cm)

A band A帶

A.V. (acid value) 酸價；酸值 [= acid number]

AACC (American Association of Cereal Chemists) 美國穀物化學師協會；美國穀類化學會

AAM (American Association of Microbiology) 美國微生物學會

AANN (American Academy of Applied Nutrition) 美國應用營養研究院

abacterial 無菌的

abaisse 千層餅；大蛋糕；壓薄的麵糰

abalone 鮑魚 [= ormer]

abalone mushroom 鮑魚菇

abalone poisoning 鮑魚中毒

abassina rigata 貝殼粉；蜆殼粉 [= conchiglie; shell pasta]

abat(t)is 家禽內臟

abattoir 屠宰場 [= slaughterhouse]

ABBA (American Board of Bioanalysis) 美國生物分析委員會

abbe refractometer 阿貝折射計；阿貝折光儀

abdomen 下腹

abequose 阿比可糖

Aberdeen Angus 黑安格斯牛

abernethy (biscuit) 葛縷子硬餅乾；艾伯內西餅乾

abgesottener cheese 奧地利牛奶軟乾酪；奧地利牛奶軟芝士

abietic acid 松脂酸；松香酸

abiotrophy 營養性衰竭

ablactation 斷乳

abnormal egg 異常蛋

abnormal fermentation 異常發酵

abnormal flavo(u)r 不正常風味；異常風味

abnormal milk 異常乳

abomasum 真胃；皺胃 [= true stomach]

aborted pink gill 角孢粉褶蕈

abrasion peeling 磨擦去皮

abrasive disc 磨盤

abrasive peeler 磨擦式去皮機

abrasive roller 磨擦式砂輥；碾米機的砂輥

abricotine 杏子酒；杏汁酒

abroach 開口；開孔；打開瓶蓋；打開罐頭

abrosia 禁食；飢餓；斷食

abscisic acid 脫落酸；離層酸

abscisin 脫落素；離層素

absinth(e) 苦艾酒；洋艾酒

absinthic oil 苦艾油

absinthin 苦艾素

absinthium 苦艾

absinthol 苦艾腦；苦艾醇

absolute (abs.) 絕對

absolute alcohol 無水酒精
absolute dry weight 絕乾重量 [= oven dry weight]
absolute humidity 絕對濕度
absolute juice 原汁；純汁
absolute pressure 絕對壓力
absolute sterility 絕對無菌性
absolute temperature 絕對溫度
absolute threshold 絕對低限；絕對閾值
absolute zero 絕對零度 [=－273.16℃；－459.69℉]
absorbance 吸光度；吸收比；吸收率；吸收能力 [= absorbancy]
absorbance index 吸光度指數；吸光系數；吸收指數
absorbancy 吸光度；吸收比；吸收率；吸收能力 [= absorbance]
absorbent 吸收劑
absorbent carbon 活性碳
absorbing agent 吸收劑
absorptiometer 吸收光度計；吸收計；液體吸氣計
absorption 吸收〔作用〕
absorption isobar 等壓吸收線
absorption isotherm 等溫吸收線
absorption refrigerating machine 吸收式製冷機；吸收式冷凍機
absorption refrigerator 吸收式冷凍櫃
absorption spectrophotometer 吸收分光光度計
absorptivity 吸收率；吸收性；吸收系數
abstergent 洗潔劑
abstinence 禁食；戒酒
abstract 摘要；提取；分離
abyssal algae 深水藻
abyssal fish 深海魚
abyssinian banana 埃塞俄比亞香蕉
acacia gum 阿拉伯樹膠 [= arabic gum; arabian gum; gum acacia; gum arabic]
acalcerosis 缺鈣症
acaricide 殺蟎劑
acarus 蟎
accelerated ag(e)ing 加速陳化；加速熟成
accelerated churning 加速攪拌
accelerated fermentation 快速發酵
accelerated freeze dryer 急速冷凍乾燥機；速凍乾燥機
accelerated freeze-drying (AFD) 急速冷凍乾燥
accelerato 促進劑；速濾劑
acceptability 接受性
acceptable daily intake (ADI) 每日容許攝取量；每日容許攝入量
acceptable quality level (AQL) 合格質量水平
acceptor 受體
accessory food 佐料；輔助食品；食品輔料
accessory fruit 假果；附果
accessory pigment 輔助色素
accredited milk 合格牛乳；特級牛乳
aceotin 乙酰甲基甲醇；乙偶姻 [= acetylmethylcarbinol]
acerbic fruit 酸澀水果
acerbity 苦味；酸澀味
acerola 西印度櫻桃
acescence 略帶酸味；微發酸；酒的酸敗
acescency 微酸味
acescent 略帶酸味的；微酸的
acesulfame 雙氧噁噻嗪
Acesulfame potassium (Acesulfame-K) 乙酰磺胺酸鉀
acetal 乙縮醛；縮醛
acetal phosphatide 縮醛磷脂

acetal resin 縮醛樹脂
acetaldehyde 乙醛；醋醛 [= ethanal]
acetaldehyde dehydrogenase 乙醛去氫酶；乙醛脱氫酶
acetaldehyde diethyl acetal 乙醛二乙縮醛
acetamidase 乙酰胺酶
acetanisole 對甲氧基苯乙酮；乙酰茴香醚
acetate 醋酸鹽；乙酸鹽；醋酸酯；乙酸酯
acetate film 醋酸纖維薄膜
acetator 釀醋裝置；醋化器
acetic 醋的；醋酸的；乙酸的
acetic acid 醋酸；乙酸 [= ethanoic acid]
acetic acid bacteria 醋酸菌
acetic acid fermentation 醋酸發酵
acetic acid preserves 醋〔酸浸〕漬食品
acetic and fatty acid esters of glycerol 醋酸和脂肪酸甘油酯
acetic anhydride 醋酐；乙酐
acetic ester 醋酸酯
acetic ether 乙酸乙酯；醋酸乙酯 [= ethyl acetate]
acetification 醋化作用
acetifier 醋化器
acetify 醋化
acetimeter 醋酸〔比重〕計 [= acetometer]
acetimetry 醋酸〔含量〕測定法
acetin 一乙酸甘油酯；醋精；甘油醋酸酯
acetoacetate 乙酰乙酸酯
acetoacetate decarboxylase 乙酰乙酸去羧酶
acetoacetic acid 乙酰乙酸
Acetobacter 醋酸桿菌屬
Acetobacter aceti 醋酸醋菌
Acetobacter melarogenum 黑醋菌
Acetobacter oxydans 氧化醋菌
Acetobacter peroxydans 過氧化醋菌
Acetobacter roseum 紅醋菌
Acetobacter suboxydans 次氧化醋菌；弱氧化醋酸桿菌
Acetobacter vini acetati 葡萄酒醋菌
acetoin 甲基乙酰甲醇
acetol 丙酮醇
acetolysis 乙酸解
acetometer 醋酸〔比重〕計 [= acetimeter]
acetometry 醋酸測定法
aceton(a)emia 丙酮血症
acetone 丙酮；二甲基酮
acetone body 丙酮體；酮體 [= ketone body]
acetone peroxide 過氧化丙酮
acetonuria 丙酮尿〔症〕；酮尿〔症〕 [= ketonuria]
acetophenone 苯乙酮；乙酰苯
acetous 醋酸的；酸的
acetyl benzene 苯乙酮；乙酰苯
acetyl cellulose 醋酸纖維素；乙酸纖維素 [= cellulose acetate]
acetyl CoA 乙酰輔酶A
3-acetyl-2, 5-dimethyl furan 3-乙酰基-2, 5-二甲基呋喃
acetyl number 乙酰價；乙酰值 [= acetyl value]
acetyl phosphate 乙酰磷酸酯
2-acetyl pyrazine 2-乙酰基吡嗪
2-acetyl thiazole 2-乙酰基噻唑
acetyl value 乙酰價；乙酰值 [= acetyl number]
acetylated distarch adipate 乙酰化二澱粉己二酸酯；乙酰化己二酸雙澱粉
acetylated distarch glycerol 乙酰化二澱粉甘油酯

acetylated distarch phosphate 乙酰化二澱粉磷酸酯；乙酰化磷酸雙澱粉
acetylated starch 乙酰化澱粉
acetylation 乙酰化〔作用〕
acetylcholine 乙酰膽鹼
acetylcholinesterase 乙酰膽鹼酯酶
acetylene 乙炔
2-acetylfuran 2-乙酰基呋喃
acetylmethylcarbinol 乙酰甲基甲醇；乙偶姻 [= acetoin]
2-acetylpyrrole 2-乙酰吡咯
3-acetylpyridine 3-乙酰吡啶
ACH index 臂胸臀指數；ACH 指數
achene 瘦果 [= akene]
achilleic acid 烏頭酸
Achilles tendon 跟腱
achiote 婀娜多樹；胭脂樹；婀娜多籽；胭脂樹籽
achlorhydria 胃酸缺乏症
Achromobacter 無色桿菌屬
achrondextrin 無色糊精
acid 酸
acid alcohol 酸性酒精
acid amide 酰胺
acid amino acid 酸性胺基酸
acid anhydride 酸酐
acid ash food 酸灰質食物
acid-base titration 酸鹼滴定
acid calcium phosphate 磷酸氫鈣；酸性磷酸鈣
acid canned food 酸性罐頭食品
acid clay 酸性白土；酸性黏土
acid cleaning 酸洗
acid coagulation 酸致凝結；酸凝〔作用〕
acid coagulation test 酸凝試驗
acid condiment 酸性調味料
acid-converted starch 酸轉化澱粉
acid-converted syrup 酸轉化糖漿
acid curd 酸凝乳〔塊〕
acid decomposition 酸水解；酸分解
acid degree 酸度
acid degumming 酸煉脱膠
acid denaturation 酸變性
acid digestion 酸煮解；酸消化
acid earth 酸性白土
acid endurance 耐酸性
acid-enzyme syrup 酸—酶法〔澱粉〕糖漿
acid-fast bacteria 耐酸細菌；抗酸細菌
acid-fast stain 耐酸染色；抗酸染色
acid flavo(u)r 酸味
acid fondant 酸味軟糖料；酸轉化微晶糖膏
acid food 酸性食品
acid-forming bacteria 產酸細菌
acid-free 無酸的
acid fuchsin 酸性品紅 [= acid magenta]
acid hydrolysis 酸水解
acid index 酸指數
acid insoluble ash 酸不溶灰分
acid magenta 酸性品紅 [= acid fuchsin]
acid medium 酸性培養基
acid modified starch 酸修飾澱粉；酸改性澱粉
acid number 酸價；酸值 [= acid value; A.V.]
acid peeling 酸去皮〔法〕
acid phosphatase 酸性磷酸〔酯〕酶
acid pickling 酸漬
acid production 產酸作用
acid-proof 耐酸的
acid proteinase 酸性蛋白酶
acid red 酸性紅

acid residue food 體內代謝中產酸的食品；生理酸性食品
acid resistance 耐酸性；抗酸性
acid resistant 耐酸的；抗酸的
acid rigor 酸僵直
acid saccharification 酸糖化
acid salt 酸式鹽
acid saponification 加酸皂化
acid smell 酸氣味；酸臭氣
acid stick 酸味水果棒糖
acid-sugar ratio 糖酸比；酸糖比
acid taste 酸味
acid test 酸度測定
acid treated starch 酸處理澱粉；酸變性澱粉
acid value (A.V.) 酸價；酸值 [= acid number]
acid violet 酸性紫
acid wash 酸洗
acid whey 酸〔性〕乳清
acid yellow 酸性黃
acidation 酸化〔作用〕
acidemia 酸血症
acidic 酸〔性〕的
acidic coal tar dye 酸性煤焦色素
acidic odor 酸臭
acidification 酸化；酸敗
acidified food 酸化食物
acidified juice 酸化果汁
acidified milk 酸化乳；發酵乳
acidifier 酸化器；酸化劑；酸味劑
acidimeter 酸度計；酸〔液〕比重計 [= acidometer]
acidimetry 酸量滴定法
acidity 酸度；酸性
acidity constant 酸度常數
acidity meter 酸度計
acidity regulator 酸度調節劑
acidity test 酸度檢驗
acidless 無酸的
acidoid 變酸性的；可能變酸的物質
acidolysis 酸解〔作用〕
acidometer 酸度計；酸〔液〕比重計 [= acidimeter]
acidophile 嗜酸細胞；嗜酸細菌
acidophilic 嗜酸的
acidophilic bacteria 嗜酸細菌
acidophilous 嗜酸的；適酸的
acidophilus milk 嗜酸菌乳；酸奶
acidosis 酸血症；酸毒症
acidulant 酸化劑
acidulating agent 酸化劑
acidulation 酸化；加酸
acidulous 帶微酸的
aciduric 耐酸的
aciduric bacteria 耐酸細菌
acinaceous 多籽的
acinose 細粒狀的
aconitic acid 烏頭酸
acorn 櫟實；槲實；橡樹果；橡子
acorn squash 橡實南瓜
acorn sugar 櫟實糖；槲實糖；槲皮醇；櫟醇 [= quercitol]
acridine orange 吖啶橙
acridity 辣味；辛辣
acriflavine 吖啶黃〔素〕
acrimony 辛辣性
acrolein 丙烯醛
acrose 消旋果糖；合成果糖；阿柯糖
acrylic acid 丙烯酸
act 法令；法規；條例；法案
actification 活性作用；再生作用
actin 肌動蛋白
actin filament 肌動蛋白纖維；肌動蛋白絲
actinator 光化器
actinidase 獼猴桃酶
actinidin(e) 獼猴桃鹼

Actinomyces 放線菌屬
actinomycosis 放線菌病
activated carbon 活性碳 [= activate carbon; activated charcoal]
activated carbon purifier 活性碳過濾器
activated charcoal 活性碳 [= active carbon; activated carbon]
activated clay 活性黏土
activated Fuller's earth 活性漂白土
activated yeast 活性酵母
activating agent 活化劑
activation 活性〔作用〕；激活
activation energy 活性能
activator 活化劑；活性劑
active calcium 活性鈣；活性離子鈣
active carbon 活性碳 [= activated carbon; activated charcoal]
active component 有效成分
active constituent 有效組分
active dry yeast 活性乾酵母
active oxygen method (AOM) 活性氧氣法；活性氧測定法（油脂穩定性試驗）
active site 活性部位
active surface 活性表面
activity 活性；活動性
actomyosin 肌動凝蛋白；肌動球蛋白
acute toxicity 急性毒性
acute toxin 急性毒素
acyl group 酰基
acylase 酰化酶；酰基轉移酶
acylated protein 酰化蛋白
acylated starch 酰化澱粉
acylation 酰化
Adamkiewiez reaction 阿丹奇維茲反應
adaptive enzyme 適應酶
added value 附加價值
added water judgement 加水判定（用於牛乳試驗）
addictive 嗜好；上癮；癖
additament 附加物；增加物
addition 添加〔作用〕；添加物
addition reaction 加成反應
additive 添加物；添加劑
additive action 相加作用
additive agent 添加劑；加成劑
adductor (muscle) 內收肌；閉殼肌
ade 果汁飲料
adenine 腺嘌呤
adenine nucleotide 腺嘌呤去胺酶
adenosine 腺苷
adenosine diphosphate (ADP) 腺苷二磷酸；二磷酸腺苷
adenosine monophosphate (AMP) 腺苷一磷酸；一磷酸腺苷；腺苷酸 [= 5'-adenylic acid; adenosine phosphate]
adenosine phosphate 腺苷一磷酸；腺苷酸 [= 5'-adenylic acid; adenosine monophosphate]
adenosine triphosphate (ATP) 腺苷三磷酸；三磷酸腺苷
adenylic acid 腺苷酸；腺苷一磷酸；一磷酸腺苷
adeps 動物脂；豚脂
adequate diet 適當膳食；足夠的飲食
adhering semolina 帶麩粗粒小麥粉
adhesion 附著；黏合；黏附力
adhesive 黏著劑；黏合劑；膠黏劑
adhesive depilating compound 膠黏脫毛化合物
adhumulone 附蛇麻葫酮；加葎草酮（啤酒花的苦味物質）
ADI (acceptable daily intake) 每日容許攝取量；每日容許攝入量

adiabatic calorimeter 絕熱量熱器；絕熱熱量計
adipic 脂肪的；多脂的
adipic acid 己二酸
adipocyte 脂肪細胞
adipogenesis 脂肪形成
adipolysis 脂解；脂肪分解
adipose 動物脂肪；脂肪的；肥胖的
adipose cell 脂肪細胞 [= fat cell]
adipose layer 皮下脂肪層
adipose tissue 脂肪組織
adiposis 肥胖病；脂肪過多症
adiposity 肥胖症；肥胖
adjunct 添加物；輔料；附加物
adjuvant 輔料；補助劑；佐藥
adlay 薏苡；薏仁；薏米 [= jobstears]
adlupulone 附蛇麻蘆酮；加蛇麻酮（啤酒花的苦味物質）
ADMI (American Dry Milk Institute) 美國乳粉學會
admix 混合；混合物；摻和物
admixture 混合；摻入；混合物；摻和物
adolescence 青春期
adonose 阿東〔戊〕糖；側金盞戊糖
ADP (adenosine diphosphate) 腺苷二磷酸；二磷酸腺苷
Adriance seamer 異形罐封罐機
ADSA (American Dairy Science Association) 美國乳業科學協會
adsorbate 被吸附物
adsorbent 吸附劑
adsorption 吸附〔作用〕
adsorption bleaching 吸附脱色
adsorption chromatography 吸附層析法；吸附色譜法
adsorption filtration 吸附過濾
adsorption isotherm 等溫吸附線
adsorption water 吸附水
adsuki bean 紅豆；赤豆；小豆 [= adzuki bean; azuki bean; red bean]
adulterant 摻雜物
adulterated flour 摻雜麵粉
adulterated food 摻雜食品；不符規格食品
adulteration 摻雜；摻假；摻假貨
advanced fry 魚秧；魚苗
adzuki bean 紅豆；赤豆；小豆 [= adsuki bean; azuki bean; red bean]
aequum 維持體重的飲食量；熱量所需量
aerated bread 充氣麵包
aerated candy 充氣糖果
aerated cream 充氣奶油
aerated water 充氣水；充二氧化碳水
aerated water machine 充二氧化碳機
aerated wine 充氣〔葡萄〕酒
aerating agent 充氣疏鬆劑
aeration 通氣；通風；混入氣體
aeration agitator 通氣攪拌器
aeration drying 通氣乾燥；通風乾燥
aeration tank 通氣槽；通風槽
aerator 通氣器；充氣器；暴氣槽
aerial hyphae 氣生菌絲
aerial infection 空氣感染；空氣傳染
aerial mycelium 氣生菌絲體
aerification 充氣；氣化；與空氣化合
Aerobacter 產氣桿菌屬；氣桿菌屬
aerobe 需氧菌；好氧菌；嗜氧菌
aerobic 需氧的；好氣的；依靠氧氣的
aerobic bacteria 嗜氧細菌；需氧細菌；好氣細菌
aerobic culture 嗜氧培養；好氧養；需氧培養；嗜氧菌種；好氧菌種；需氧菌種
aerobic fermentation 嗜氧發酵；好氧發酵；需氧發酵
aerobic glycolysis 需氧醣〔酵〕解

aerobic microorganism 嗜氧微生物；需氧微生物；好氧微生物
aerobic respiration 有氧呼吸
aerobics 有氧運動
aerobiosis 好氧生活
aerobium 需氧菌；嗜氧菌；好氧菌
aerogenic bacteria 產氣細菌
aeromicrobe 需氧菌；嗜氧菌；好氧菌
aerophilic bacteria 親氣細菌；好氣細菌；需氣細菌；嗜氣性細菌
aerophyte 氣生植物；附生植物
aerosol 氣霧劑；氣溶膠
aerosol can 噴霧罐；氣溶膠罐
aerosol food 噴射式食物
aerosporin 氣孢桿菌素；多黏菌素
aerotolerant anaerobe 耐氧厭氧菌
aerotolerant bacteria 耐氧細菌
AFD (accelerated freeze-drying) 急速冷凍乾燥
affection 疾病
AFFI (American Frozen Food Institute) 美國冷凍食品研究所
affination centrifuge 洗糖離心機；洗煉離心機
affination sugar 洗淨糖；洗煉糖
affination 精製；洗煉；洗糖
affinity 親和性；親合性
affinity absorber 親和吸收劑
affinity chromatography 親和層析法
aflagellar bacteria 無鞭毛細菌 [= aflagellate bacteria]
aflagellate bacteria 無鞭毛細菌 [= aflagellar bacteria]
aflatoxicosis 黃麴黴素中毒；黃麴毒素中毒
African lily 百子蓮
African spiderherb 非洲白花菜
after-bake 後〔期〕烘焙
after-cooling 後冷卻；二次冷卻
after-drying 後乾燥
after-fermentation 後發酵；二次發酵
after grinding 後研磨
after-massecuite 最後糖膏
after-maturation 後熟
after-milking 後擠乳
after-pressed juice 二次榨汁
after-ripening 後熟；催熟；追熟
after-saccharification 後糖化〔作用〕
after service 售後服務
after-taste 餘味；後味〔覺〕
after-treatment 後處理；再處理；病後療法
after wort 二次麥〔芽〕汁
afterdinner drink 餐後酒
afternoon milk 午後〔擠〕乳；夕奶；傍晚奶
afternoon tea 下午茶；午後茶點
afterproduct 後產品；副產品
afterrun 酒尾
afterwash 回洗
agalactosis 乳液分泌缺乏；無乳症
agar 洋菜；瓊脂；瓊膠 [= agar-agar]
agar diffusion method 洋菜擴散法
agar gel 洋菜凝膠；瓊脂凝膠
agar jelly 洋菜凍
agar medium 洋菜培養基
agar plate 洋菜平板；瓊脂平板
agar plate count 洋菜平板菌數；瓊脂平板菌數
agar plate method 洋菜平板〔培養〕法；瓊脂平板〔培養〕法
agar slant 洋菜斜面培養；瓊脂斜面培養
agar slope 洋菜斜面培養；瓊脂斜面培養
agar stab 洋菜穿刺培養；瓊脂穿刺培養

agar streak 洋菜畫線培養；瓊脂畫線培養

agar tube 洋菜培養試管；瓊脂培養試管

agar weed 洋菜原料藻；石花菜

agar-agar 洋菜；瓊脂；瓊膠 [= agar]

agaric 蘑菇木耳；傘菌；落葉松蕈

agaric(ic) acid 蘑菇酸；松蕈酸；瓊脂酸 [= agaricinic acid]

agaricin 蘑菇素；白木耳素；落葉松蕈素

agaricinic acid 蘑菇酸；松蕈酸；瓊脂酸 [= agaric(ic) acid]

Agaricus 蘑菇屬；傘菌屬；落葉松蕈

agaritine 傘菌氨酸

agarobiose 洋菜二糖；瓊脂二糖

agaroid 類洋菜；類瓊脂

agaropectin 洋菜硫〔酸〕糖；瓊脂膠

agarophyte 洋菜原料藻

agarose 洋菜糖；瓊脂糖

agave 龍舌蘭〔發酵飲料〕

age thickening 貯存〔期〕濃稠化；儲藏〔期〕濃稠化

aged beef 熟成牛肉

aged cheese 熟成乾酪；熟成芝士

aged flavo(u)r 陳釀風味；熟成風味

aged meat 熟成肉；柔化的肉

aged wine 陳酒；陳釀葡萄酒

ageing 熟成；陳化；柔化；老化 [= aging]

ageing cellar 陳釀酒窖；陳酒庫 [= aging cellar]

Agene 埃淨；埃京（漂白麵粉用的三氯化氮）

Agene process 埃淨作業法；埃京作業法

agenize 用埃淨漂白麵粉

agenized flour 埃淨漂白過的麵粉；三氯化氮漂白麵粉

ageusia 無味覺症；味覺喪失

agglomerant 黏聚劑；黏結劑；凝集劑

agglomeration 黏聚；凝聚；結塊；造粒

agglomerator 造粒機；結粒機

agglutinant 黏著劑；膠合物；凝集劑

agglutination 凝集〔作用〕；黏合〔作用〕

agglutinin 凝集素

aggregate 聚集體；集合體

aggregate fruit 聚合果

aggregation 聚集〔作用〕；聚合〔作用〕

aging 熟成；陳化；柔化；老化 [= ageing]

aging cellar 陳釀酒窖；陳酒庫 [= ageing cellar]

aging vat 熟化槽

aginomoto 味精；味之素；麩胺酸〔一〕鈉；谷胺酸〔一〕鈉 [= sodium glutamate; MSG; monosodium-glutamate]

agitated curing vat 攪拌醃漬槽

agitated dryer 攪動式乾燥機

agitating cooker 攪動式蒸煮器；迴轉式殺菌機

agitating crystallizer 攪拌式結晶罐

agitating device 攪拌裝置

agitating film evaporator 攪拌式薄膜蒸發器；攪拌式薄膜脱水器

agitating heater 攪動加熱器

agitating retort 攪動式殺菌斧；迴轉式殺菌鍋

agitating sterilizer 攪動式殺菌器；迴轉式殺菌器

agitating vat 攪拌桶

agitation 攪動；攪和

agitator 攪拌器；攪動器

agitator dryer 攪拌乾燥器

agitator flash evaporator 攪拌式瞬間蒸發器；攪拌式瞬間脱水器
agitator kettle 攪拌鍋
aglaia odorata flower oil 樹蘭〔花〕油
aglycon 糖苷配基；配糖基
agricultural 農業的；農藝的
agricultural chemicals 農用化學品；農藥
agricultural chemistry 農業化學
agricultural code 農業法規
agricultural engineering 農業工程學
agricultural product 農業產品；農業品
agriculture 農業；農藝
Agrobacterium 土壤桿菌屬
agronomy 農藝學；農學
aguacate 牛油果；鱷梨；油梨；酪梨 [= avocado; alligator pear; butter-fruit]
aguardiente 燒酒；劣質餾酒；南美蘭酒
aioli 蒜泥蛋黃醬
aipi 甜木薯
air bath 空氣浴；氣浴
air bladder 鰾；魚鰾；氣泡
air-blast cooling 強風冷卻
air-blast defrosting 強風解凍
air-blast freezer 強風冷凍機；鼓風凍結室
air-blast freezing 強風冷凍；鼓風凍結
air-blast thawer 強風解凍機
air blower 鼓風機
air catering 航空膳食
air classification 風選；氣流分級
air classifier 風選機；氣流分級機
air condenser 空氣冷凝器
air conditioner 空氣調節機；空調機
air conditioning 空氣調節；空調
air convection dryer 空氣對流乾燥機
air convection drying 空氣對流乾燥
air conveyor 氣流輸送器；氣動輸送機
air cooking 熱風烹煮；熱風烘烤
air cooled 空氣冷卻的；氣冷的
air cooler 空氣冷卻器
air cooling 空氣冷卻；氣冷
air cure 空氣中熟成
air curtain 氣簾；吹氣簾
air defrosting 空氣〔中〕解凍
air dehydration 風乾
air dried 風乾的
air dried meat 風乾肉
air dryer 空氣乾燥器；通風乾燥器
air drying 風乾
air filter 空氣過濾器
air flow 氣流
air freezing 氣冷式冷凍；空氣凍結
air heater 熱風機
air incinerator 空氣焚化爐
air incubator 有氧培養箱
air-jet agitation 空氣噴射攪拌
air-lift agitation 氣升攪拌；空氣升液攪拌
air-lift agitator 氣升攪拌器；空氣升液攪拌器；空氣鼓泡器
air-lift fermenter 氣升發酵槽；空氣升液發酵槽 [= gas-lift fermenter]
air meter 量氣計；風速計 [= airometer]
air oven 通氣烘箱；通氣烤箱；熱空氣烤箱
air oxidation 空氣氧化
air permeability 透氣性
air pollution 空氣污染
air pore 氣孔
air purifier 空氣淨化裝置；空氣濾淨器
air screen 空氣過濾網
air separator 風選機；空氣分離器
air slake 潮解

air slaking 潮解
air stream 氣流
air sweep 氣掃
air thawing 空氣解凍
airborne bacteria 空浮細菌；空氣傳播細菌
airborne contamination 空氣污染
airborne infection 空氣傳染
airometer 量氣計；風速計 [= air meter]
airproof 不透氣的；密封
airtight 氣密的；密封的；不透氣的
ajowan 阿鳩彎
ajowan oil 香旱芹油
akala 阿卡拉莓
akaryote 無核細胞
akebia 木通；通草
akee 阿奇樹
akene 瘦果 [= achene]
akola 阿拉伯辣椒
aku 鰹
akvavit 蒸餾酒；白蘭地〔烈性酒〕 [= aquavit]
ala 翼；翅
alamode 上面加冰淇淋的
alanine 丙氨酸
alantic acid 阿蘭酸；土木香酸
alantin 菊糖；菊粉 [= inulin]
Alaska blackfish 阿拉斯加黑魚
Alaska pollack 阿拉斯加青鱈；明泰魚；明太鱈魚
Alaska redfish 紅大麻哈魚
albacore (tuna) 長鰭金槍魚；長鰭鮪魚 [= long-finned tuna]
albedo 白軟皮；中果皮(柑橘類)
albumen 蛋白；卵白；胚乳
albumen whisk 蛋白攪泡器；卵白攪泡器；攪蛋器
albumin 白蛋白；清蛋白
albumin milk 蛋白乳
albuminoid 類白蛋白；硬蛋白
albuminolysis 白蛋白的分解〔作用〕
alcalase 鹼性蛋白酶；愛爾卡酶
Alcaligenes 產鹼桿菌屬
alcogel 醇凝膠
alcohol 酒精；乙醇 [= ethanol; ethyl alcohol]
alcohol coagulation test 酒精凝固試驗
alcohol dehydrogenase 乙醇去氫酶；醇脫氫酶
alcohol distillation 酒精蒸餾
alcohol fermentation 酒精發酵 [= alcoholic fermentation]
alcohol-free 無酒精的；不含酒精的
alcohol gauge 酒精〔比重〕計；酒精表
alcohol hydrometer 酒精比重計
alcohol-insoluble 酒精不可溶的；不醇溶的
alcohol plant 酒精廠
alcohol slop 酒精廢液；酒精發酵的廢液
alcohol-sol 醇溶膠 [= alcosol]
alcohol-soluble 酒精可溶的；醇溶的
alcohol strength 酒精濃度；酒精強度
alcohol yeast 酒精酵母
alcoholase 醇酶
alcoholemia 醇血症
alcoholic 含酒精的；酒精的；酗酒者
alcoholic acid 醇酸
alcoholic beverage 含酒精飲料；酒精性飲料
alcoholic carbonated drink 碳酸酒精飲料；汽酒
alcoholic cordial 含酒精的興奮性飲料；補酒；酒精性興奮劑
alcoholic drink 含酒精飲料；酒精性飲料

alcoholic extraction 酒精萃取；酒精抽提
alcoholic fermentation 酒精發酵 [= alcohol fermentation]
alcoholic liquor 蒸餾酒；燒酒
alcoholicity 酒精含量；酒精度
alcoholimeter 酒精比重計；酒精計 [= alcoholmeter; alcoholometer]
alcoholism 酒精中毒
alcoholization 醇化
alcoholmeter 酒精比重計；酒精計 [= alcoholimeter; alcoholometer]
alcoholomania 嗜酒狂
alcoholometer 酒精比重計；酒精計 [= alcoholmeter; alcoholimeter]
alcoholometry 酒精〔含量〕測定法
alcoholysis 醇解
alcomeal 粗製麵粉；小麥粗麵粉
alcosol 醇溶膠 [= alcohol sol]
aldabo 古巴燒酒
aldaric acid 醛醣二酸
aldehyde 醛
aldehyde dehydrogenase 醛去氫酶；醛脫氫酶
aldehyde oxidase 醛氧化酶
aldehydrase 醛酶
alditol 醛醣醇；醛醇
aldol resin 醛醇樹脂
aldolase 醛醇縮酶；醛縮酶
aldonic acid 醛糖酸；糖酸
aldose 醛糖
aldoside 醛糖苷
alduronic acid 糖醛酸
ale 艾爾啤酒；淡色啤酒；淺色麥酒
alegar 麥酒醋；啤酒醋
Alessandria sausage 亞歷山大香腸
aleurometer 麵糰脹力計；麵筋筋力儀；糊粉測定儀
aleuronat 高筋麵粉；麥蛋白
aleurone 糊粉
aleurone cell 糊粉細胞
aleurone grain 糊粉粒
aleurone layer 糊粉層
alewife 啤酒店女主人；北美洲大眼鯡魚
Alexandra kingpalm 亞歷山大椰子
alexipharmic 解毒藥
Alfa process 阿法〔白脫連續〕製法（連續分離奶油法）
alfabeto 字母〔意大利〕麵
alfalfa 苜蓿；紫花苜蓿 [= lucern(e)]
alfalfa sprout 苜蓿芽
alfatoxin 黃麴毒素；黃麴黴〔毒〕素
alfoncino 金眼鯛
alga(e) 海藻；藻類
algae extract power 海藻抽提粉
algaecide 殺藻劑 [= algicide]
algal 海藻的；藻類的
algal culture 藻類培養物；藻類養殖
algal polysaccharide 藻類多醣
algaroba 長角豆
algicide 殺藻劑 [= algaecide]
algin 褐藻膠；褐藻酸；海藻素
alginase 褐藻酸酶
alginate 褐藻酸鹽；海藻酸鹽
alginic acid 褐藻酸；海藻酸
algology 藻類學
alible food 滋養食物；有營養的食物
aliment 養料；滋養品；食物
alimental 有營養的；食物的
alimentary 營養的；食物的
alimentary anemia 營養性貧血
alimentary canal 消化道
alimentary crop 糧食〔農〕作物
alimentary glucosuria 營養性糖尿
alimentary infection 消化道感染
alimentary organ 消化器官
alimentary paste 麵食；麵條製品

alimentary toxicosis 食物中毒
alimentary tract 消化道
alimentation 營養；飼養
alimentology 營養學
alimentotherapy 營養療法；食物療法
aliphatic 脂肪族的；脂肪質的
aliphatic acid 脂〔肪〕族酸
aliphatic alcohol 脂〔肪〕族醇
aliquot 整等分液量；整分部分
alitame 阿力甜
alizarin 茜素
alizarin red 茜紅
alizarin yellow 茜黃
alkalescent 微鹼性的
alkali 鹼
alkali detergent 鹼性洗潔劑
alkali fountain 鹼性泉水
alkali metal 鹼金屬
alkali peeling 鹼去皮〔法〕
alkali producing bacteria 產鹼細菌
alkali refining 油脂鹼精煉；鹼煉 [= caustic refining]
alkali soluble 鹼溶性的
alkali solution 鹼性溶液
alkali starch 鹼澱粉
alkali wash 鹼洗
alkaligenous bacteria 產鹼細菌
alkalimeter 鹼度計；鹼濃度測定器；鹼量計 [= alkaline meter]
alkalimetric titration 鹼〔量〕滴定
alkaline 鹼的
alkaline-ash diet 鹼灰質膳食；鹼灰質飲食
alkaline-ashing 鹼灰化
alkaline-earth metal 鹼土金屬
alkaline flavo(u)r 鹼氣味
alkaline hydrolysis 鹼〔性〕水解
alkaline meter 鹼度計；鹼濃度測定器；鹼量計 [= alkalimeter]
alkaline rigor 鹼〔性〕僵直
alkaline saponification 鹼皂化
alkaline soapy odor 鹼臭；肥皂臭
alkaline taste 鹼味
alkaline treated starch 鹼處理澱粉
alkaline treatment 鹼處理
alkalinization 鹼化〔作用〕 [= alkalization]
alkalinous 鹼性的；含鹼的
alkalinty 鹼度；鹼性
alkalis 鹼性〔洗滌〕劑
alkalization 鹼化〔作用〕 [= alkalinization]
alkalize 鹼化
alkalized duck egg 皮蛋 [= alkalized egg; Chinese egg; thousand years' egg; pidan]
alkalized egg 皮蛋 [= alkalized duck egg]
alkalizer 鹼性劑
alkaloid 生物鹼；植物鹼基
alkalophilic bacteria 嗜鹼細菌
alkalosis 鹼毒症；鹼中毒
alkane 鏈烷
alkanet 朱草；紫草；紫草根(植物紅色染料)
alkannin 紫草素；紫草紅
alkanoic acid 鏈烷酸
alkekengi 酸漿果；冬季櫻
alkene 鏈烯
alkine 炔
alkograph 酒精儀
alkyd 醇酸樹脂
alkyd starch 醇酸澱粉
alkyl group 烷基
alkyl radical 烷自由基
alkylating agent 烷基化劑
alkyne 炔
all-day sucker 大棒棒糖
all purpose flour 通用麵粉；全用途麵粉

allantoic acid 尿囊酸
allantoin 尿囊素
allasch 阿拉希酒
allergen 過敏原
allergenic food 過敏〔原〕食物
allergy 過敏症；敏感症
alliaceous 蒜味的；蔥蒜味的
alliaceous condiment 蔥蒜〔性〕調味料
alliaceous odor 蔥蒜氣味
allicin 蒜素
alligator pear 牛油果；鱷梨；油梨；酪梨 [= aguacate; avocado; butter-fruit]
allin 蒜苷
allis 河鯡
allison tuna 黃鰭金槍魚；黃鰭鮪魚
allistatin 蒜抑菌素；蒜制菌素
Allium 蔥屬；蔥蒜
Allium cepa 洋蔥；玉蔥
Allium sativum 大蒜
Allium tuberosum 韭菜花
allocarpy 雜交果
allochlorophyll 別葉綠素；類似葉綠素
alloisoleucine 別異白胺酸
allolactose 別乳糖
allomerism 異質同晶〔現象〕
allomorphism 同質異晶〔現象〕
allomycin 別黴素
allophanic acid 脲基甲酸
allose 阿洛糖
allosteric effect 異位效應；變構效應
allotherm 變溫動物
allotriophagy 異食癖；嗜異癖
allowance 容許量；容限
allspice 多香果；眾香子；丁香辣椒；甘椒；牙買加辣椒；番辣椒 [= Jamaica pepper; pimento berry; pimento]
allspice leaf oil 多香果葉油；眾香葉油
allspice oil 眾香子油；多香果油
allulose 阿洛酮糖 [= psicose]
allura red AC 阿洛拉紅
allura red AC aluminium lake 阿洛拉紅鋁色澱；艷紅鋁色澱
allyl 3-cyclohexanepropionate 3-環己基丙酸烯丙酯；菠蘿酯
allyl alcohol 烯丙醇
allyl cinnamate 桂酸烯丙酯
allyl cyclohexaneacetate 環己基乙酸烯丙酯
allyl cyclohexanebutyrate 環己基丁酸烯丙酯
allyl heptanoate 庚酸烯丙酯
allyl hexanoate 己酸烯丙酯；菠蘿醛
allyl isothiocyanate 異硫氰酸烯丙酯
allyl isovalerate 異戊酸烯丙酯
allyl phenylacetate 苯乙酸烯丙酯
allyl propionate 丙酸烯丙酯
allyl starch 烯丙基澱粉
allyl tiglate 惕各酸烯丙酯
allyl α-ionone α-烯丙基紫羅蘭酮
allylamine 丙烯胺
allylic condiment 蒜味調味料
almond 杏仁；杏樹；扁桃
almond blanching machine 杏仁燙漂去皮機
almond cake 杏仁蛋糕；杏仁餅；杏仁粕
almond cream 杏仁霜；杏仁酪
almond decorticator 杏仁脫殼機
almond dragees 杏仁糖
almond flavo(u)r 杏仁香精
almond hardbake 杏仁脆餅；含杏仁的糖果
almond meal 杏仁粉
almond milk 杏仁乳；杏仁露

almond mill 杏仁磨粉機
almond oil 杏仁油
almond oil, bitter 苦杏仁油
almond oily flavo(u)r 杏仁油香精
almond rock 杏仁硬糖
almond syrup 杏仁糖漿
aloe 蘆薈
aloe vera 蘆薈
aloin 蘆薈素
alpha-amylase α澱粉酶；阿伐澱粉酶；澱粉液化酶 [= α-amylase]
alpha-cellulose α纖維素；阿伐纖維素
alpha-glucan α葡聚糖；阿伐葡聚糖
alpha-helix α螺旋；阿伐螺旋 [= α-helix]
alpha-starch α澱粉；阿伐澱粉 [= α-starch]
alpha-tocopherol α生育酚；阿伐生育酚 [= α-tocopherol]
alpine salami 高山薩拉米香腸；高山色拉米腸
alpine strawberry 高山草莓；高山士多啤梨
alsatian wine 法國阿爾薩斯白〔葡萄〕酒
alternate bearing 隔年結果果樹 [= biennial bearing]
alternative flour 代用麵粉
althea 蜀癸屬植物
altrose 阿卓糖
alum 硫酸鋁；明礬；礬 [= aluminium sulfate]
alumina 氧化鋁；礬土
aluminate 鋁酸鹽
aluminium 鋁 [= aluminum]
aluminium alum 硫酸銨鋁；銨明礬
aluminium ammonium sulfate 硫酸銨鋁；銨明礬
aluminium calcium silicate 硅酸鋁鈣
aluminium chloride 氯化鋁
aluminium lake 鋁色澱
aluminium potassium sulfate 硫酸鋁鉀；鉀明礬
aluminium potassium sulfate, anhydrous 無水硫酸鋁鉀；燒明礬
aluminium powder 鋁粉
aluminium sodium sulfate 硫酸鋁鈉；鈉明礬
aluminium sulfate 硫酸鋁 [= alum]
aluminum 鋁 [= aluminium]
aluminum ammonium sulfate 硫酸銨鋁
aluminum can 鋁罐
aluminum capping machine 鋁蓋封機
aluminum foil 鋁箔
aluminum lake 鋁色澱
aluminum ring pull end 環開鋁蓋；鋁製拉環式易開蓋
aluminum sulfate 硫酸鋁
alveogram 糰展性測定圖；麵糰吹泡張力曲線圖
alveograph 糰展性測定儀；麵筋拉力測定儀
Amadori rearrangement 安瑪多立重排〔反應〕
amanita 蠅虎蕈；毒傘蕈；傘形毒蕈
amanitatoxin 蠅虎蕈毒素；條蕈毒素
amaranth 莧菜；莧菜紅
amaranth alum lake 莧菜紅鋁色澱
amaranth aluminium lake 莧菜紅鋁色澱
amarelle 酸櫻桃
amaric acid 苦杏酸
amarine 苦杏素
amaroid 苦味質；苦杏質
amber 琥珀；琥珀色
amber acid 琥珀酸；丁二酸
amber malt 琥珀色〔焦〕麥芽
amberfish 圓鰺；萊蘭鰤
ambergris 龍涎香

amberjack 琥珀魚；紫鰤
ambient air 周圍空氣
ambient condition 周圍條件
ambient temperature 周圍溫度；環境溫度
ambrette 麝香梨
ambrette seed oil 麝香梨子油；麝葵籽油；黃葵籽油
ambrosia 佳餚；美味的食物
ambry 食品櫥；櫥櫃
amelioration 艾美立加糖改進發酵法
American Academy of Applied Nutrition (AANN) 美國應用營養研究院
American artichoke 美洲薊；菊薊
American Association of Cereal Chemists (AACC) 美國穀物化學師協會；美國穀類化學會
American Association of Microbiology (AAM) 美國微生物學會
American Board of Bioanalysis (ABBA) 美國生物分析委員會
American cranberry 大果蔓越橘；美洲酸果蔓果實
American Dairy Association (ADA) 美國乳品工業協會
American Dairy Science Association (ADSA) 美國乳業科學協會
American Dietetic Association (ADA) 美國膳食營養學會；美國飲食學會
American Dry Milk Institute (ADMI) 美國乳粉學會
American eel 美洲鰻鱺
American Frozen Food Institute (AFFI) 美國冷凍食品研究所
American ginseng 花旗參；西洋參；粉光參
American Institute of Nutrition (AIN) 美國營養學會
American Oil Chemists Society (AOCS) 美國油脂化學會
American oyster 美洲牡蠣
American plaice 擬庸鰈；歐蝶魚
American shad 美洲西鯡
American smalt 美洲胡瓜魚
American Society for Microbiology (ASM) 美國微生物學會
American Society of Brewing Chemists (ASBC) 美國釀造化學師學會
American Soybean Association (ASA) 美國大豆協會
American Spices Trade Association (ASTA) 美國香辛料協會
American Type Culture Collection (ATCC) 美國菌種保存中心；美國標準菌庫
americano 美式咖啡 [= caffe americano]
amerio method 艾美利接觸冷凍法；艾美利法
amethyst deceiver 紫晶蠟蘑(蕈類)
amicrobic 非微生物的
amidase 酰胺酶
amidated pectin 酰胺〔化〕果膠
amidation 酰胺化
amide 酰胺；氨基化合物
amide nitrogen 酰胺態氮
aminase 胺化酶；脱胺基酶
amination 胺化〔作用〕
amine 胺
amino acid 胺基酸；氨基酸
amino acid acylase 胺基酸酰化酶；胺基酸酰基轉移酶
amino acid analyzer 胺基酸分析儀
amino acid balance 胺基酸均衡
amino acid composition 胺基酸組成
amino acid decarboxylase 胺基酸去羧酶；胺基酸脱羧酶

amino acid fermentation 胺基酸發酵
amino acid fortification 胺基酸強化
amino acid nitrogen 胺基酸態氮
amino acid oxidase 胺基酸氧化酶
amino acid pattern 胺基酸範式
amino acid profile 胺基酸輪廓；胺基酸剖析
amino acid residue 胺基酸殘基
amino acid score 胺基酸價；胺基酸評分
amino acid seasoning 胺基酸調味料
amino acid sequenator 胺基酸定序儀；胺基酸順序分析儀 [= amino acid sequencer]
amino acid sequence 胺基酸定序；胺基酸順序
amino acid sequencer 胺基酸定序儀；胺基酸順序分析儀 [= amino acid sequenator]
amino acid side chain 胺基酸側鏈
amino acid solution 氨基酸溶液
amino acid transacetylase 胺基酸轉乙醯酶
amino acid transamination 胺基酸轉胺作用
amino alcohol 胺醇
amino compound 胺基化合物
amino group 胺基
amino nitrogen 胺基態氮
amino sugar 胺糖
amino-terminal 胺端
aminoacetic acid 甘胺酸；胺基醋酸；胺乙酸 [= collagen sugar; glycine]
aminoaciduria 胺基酸尿症
aminocarbonyl reaction 梅納反應；美拉德反應；羰胺反應；胺羰反應 [= carbonyl-amine reaction; Maillard reaction]
aminoethylated starch 氨基乙基化澱粉
aminolipid 胺脂質
aminolysis 氨解
aminopeptidase 胺〔基〕肽酶
aminopolypeptidase 胺〔基〕多肽酶
aminopterin 胺喋呤
aminopurine 胺〔基〕嘌呤
aminosaccharide 胺基糖類
aminotransferase 轉胺酶
amioca 糯玉米
ammonate 氨合物；氨合
ammonation 氨合〔作用〕
ammonia 氨
ammonia nitrogen 氨態氮
ammonia soda 臭粉
ammonia system baking powder 氨系發粉
ammoniacal odor 氨味；氨氣味
ammonialyase 解氨酶
ammoniate 氨化合物；加氨處理
ammoniated glycyrrhizin 氨化甘草甜素
ammonification 氨化〔作用〕
ammonifying bacteria 氨化細菌
ammonite 乾肉粉
ammonium acetate 乙酸銨
ammonium alginate 褐藻酸銨；海藻酸銨
ammonium bicarbonate 碳酸氫銨 [= ammonium hydrogen carbonate]
ammonium carbonate 碳酸銨
ammonium chloride 氯化銨
ammonium compound 銨化合物
ammonium dihydrogen phosphate 磷酸二氫銨
ammonium hydrogen carbonate 碳酸氫銨 [= ammonium bicarbonate]
ammonium hydroxide 氫氧化銨
ammonium magnesium phosphate 磷酸鎂銨

ammonium persulfate 過硫酸銨
ammonium phosphate 磷酸銨
ammonium phosphate, dibasic 硫酸〔氫〕二銨
ammonium phosphatide 磷脂酸銨鹽
ammonium polyphosphate 聚磷酸銨
ammonium saccharin 糖精銨
ammonium sulfate 硫酸銨
ammonium sulfite 亞硫酸銨
ammonolysis 氨解作用
Amomum 白豈蔻屬
amontillado 白雪莉酒；蒙蒂拉葡萄酒
amorphism 非晶形；無定形；無晶形
amorphous 非晶形的；無定形的
amorphous sugar 非晶形糖；無定形糖
AMP (adenosine monophosphate) 腺苷一磷酸；一磷酸腺苷；腺苷酸 [= 5'-adenylic acid; adenosine phosphate]
ampeliograph 葡萄圖譜
ampeliology 葡萄學
amphibian 兩棲類
amphiprotic solvent 兩性溶劑
ampholytic surfactant 兩性界面活性劑；兩性表面活性劑
amphoteric 兩性的
amphoteric emulsifier 兩性乳化劑
amphoteric ion-exchange resin 兩性離子交換樹脂
amphoteric starch 兩性澱粉
amphoteric surfactant 兩性表面活性劑
ampicillin 胺比西林；胺卡青黴素
ampoul 小玻璃瓶；細頸瓶
ampoule 安瓿；針劑(一次用量)
amygdala 杏仁；扁桃；扁桃形結構
amygdalate 苦杏仁酸鹽；苦杏仁酸酯；扁桃酸鹽；扁桃酸酯
amygdalin 苦杏仁苷
amygdalose 苦杏仁糖
amyl alcohol 戊醇 [= pentanol; pentyl alcohol]
amyl butyrate 丁酸戊酯
amyl cinnamate 桂皮酸戊酯
amyl formate 甲酸戊酯
amyl group 戊基 [= pentyl group]
amyl heptanoate 庚酸戊酯
amyl hexanoate 己酸戊酯
amyl mercaptan 戊硫醇
amyl octanoate 辛酸戊酯
amyl valerate 蘋果油；戊酸戊酯
amylaceum 葡萄糖
amylase 澱粉酶
α-amylase α澱粉酶；澱粉液化酶；阿伐澱粉酶 [= alpha-amylase]
β-amylase β澱粉酶；貝他澱粉酶；澱粉糖化酶 [beta-amylase]
α-amylcinnamaldehyde α-戊基桂醛；甲位戊基桂醛
α-amylcinnamaldehyde dimethylacetal α-戊基桂醛二甲縮醛
α-amylcinnamyl alcohol α-戊基桂醇；甲位戊基桂醇
amyliferous 含澱粉的；澱粉性的
amylo process 阿米諾法；阿米露法；黴菌糖化法；根黴糖化澱粉法
amylocellulose 澱粉纖維素
amylocoagulase 澱粉凝固酶
amylodextrin 澱粉糊精；可溶性澱粉
amylodyspepsia 澱粉消化不良
amyloglucosidase 澱粉葡萄糖苷酶
amylogram 連續黏度圖；澱粉黏焙力曲線圖；澱粉糊黏度圖譜
amylograph 連續黏度儀；澱粉黏焙力測量器；澱粉糊黏度圖示儀
amyloid 類澱粉；類澱粉黏膠性物質；澱粉質的；含澱粉的；澱粉食物
amyloin 麥芽糖糊精

amylolysis 澱粉水解；澱粉分解
amylolytic enzyme 澱粉水解酶；澱粉分解酶；澱粉糖化酶
amylolytic fermentation 澱粉分解發酵；澱粉糖化發酵
amylomaize 高直鏈澱粉玉米
amylomyces 澱粉酶
amylopectin 支鏈澱粉
amylopectin rice 糯米
amyloplast 澱粉〔質〕體
amylopsin 胰澱粉酶 [= pancreatic amylase]
amylorrhea 澱粉瀉
amylose 直鏈澱粉
amylose iodine complex 直鏈澱粉—碘的絡合物
amylum 澱粉
amyrin 芸香素；香樹素
amyris oil 芸香油；香樹油
anabiosis 復甦；復蘇
anabolic reaction 同化反應；合成代謝反應
anabolic steroid 同化類固醇
anabolism 合成代謝
anabolite 合成代謝產物
Anacardium occidentale 介壽果；檟如樹
anacidity 胃酸缺乏症
anaemia 貧血 [= anemia]
anaerobe 厭氧菌；嫌氣菌；厭氧微生物
anaerobia 厭氧菌；厭氣微生物
anaerobic 厭氧的；嫌氣〔性〕的 [= anerobic]
anaerobic bacteria 厭氧〔細〕菌；嫌氣〔細〕菌
anaerobic biochemical treatment 厭氧性生化處理；嫌氣性生化處理
anaerobic culture 厭氧培養；嫌氣培養
anaerobic dehydrogenase 厭氧去氫酶；厭氧脫氫酶
anaerobic fermentation 厭氧性發酵；嫌氣性發酵
anaerobic formers 厭氧菌；嫌氣菌
anaerobic microorganism 厭氧微生物；厭氣微生物
anaerobic respiration 厭氣呼吸；缺氧呼吸
anaerobion 厭氧菌；厭氣微生物
anaerobiosis 厭氣生活；厭氧生活；缺氧生活
anaerobium 厭氧菌；厭氣微生物
anaesthesia 麻醉
anaesthetic 麻醉劑
anal fin 臀鰭
analog protein 仿製蛋白
analog(ue) 類似物
analyser 分析器；分析員 [= analyzer]
analytical balance 分析天平；精密天平
analytical chemistry 分析化學
analytical column 分析柱
analytical reagent 分析試劑
analyzer 分析器；分析員 [= analyser]
ananas 鳳梨；菠蘿 [= pineapple]
anaphase 後期(細胞分裂)
anaphylactogen 過敏原；過敏元
anaphylatoxin 過敏毒素
anaphylaxis 過敏性反應
anastate 合成化謝產物
anatoxin 類毒素；變性毒素
anchor cap 安卡蓋；咬封蓋
anchoring agent 固著劑；固著促進劑
anchovy 鯷魚；鳳尾魚
anchovy oil 鯷魚油；鳳尾魚油
anchovy paste 鳳尾魚醬；鯷魚醬

Andropogon 高粱屬
anelli 環狀通心粉
anelli siciliani 指環粉
anemia 貧血 [= anaemia]
anerobic 厭氧的；嫌氣〔性〕的 [= anaerobic]
anesone 茴香利口酒
anethene 茴香烯；蒔蘿烯
anethole 茴香腦
aneurin 噻胺；硫胺；硫胺素；抗神經炎素；維生素B_1 [= thiamine; vitamin B_1]
aneurinase 抗神經炎素酶；硫胺素酶
anfora 意大利酒量單位
angel cake 安琪〔兒〕蛋糕；天使蛋糕
angel hair 天使髮絲 [= capellini]
angel's wings 突伸小側耳（蕈類）
angelfish 天使魚；刺蝶魚
angelica 當歸；白芷；羌活；安琪〔利〕卡白葡萄酒
angelica oil 當歸油
angelica root oil 圓當歸根油
angelica seed oil 圓當歸籽油
angiodystrophia 血管營養不良
anglerfish 琵琶魚；華臍魚
angola goat 安哥拉山羊
angostura 安古〔斯圖臘〕樹皮
angostura bitters 苦精
Angstrom (Å) 埃（10^{-8}cm）
Angus 安格斯牛
anhydration 去水；脱水
anhydrator 脱水器
anhydride 酐
anhydro sugar 脱水糖
anhydrous 無水的
anhydrous alcohol 無水酒精
anhydrous butter 脱水白脱；脱水乳酪
aniline dye 苯胺染料
aniline red 苯胺紅
animal casing 動物腸衣
animal farm 牧場；畜牧場
animal fat 動物脂肪
animal food 動物性食品
animal gelatin(e) 動物明膠
animal grease 動物油脂
animal husbandry 畜牧學；畜牧業
animal oil 動物油
animal pigment 動物色素
animal product 動物產品；畜產品
animal protein 動物〔性〕蛋白
animal starch 肝糖；糖原；動物澱粉 [= glycogen; liver starch]
animal tallow 動物脂肪
animal wax 動物蠟
anion 陰離子
anion exchange 陰離子交換
anion exchange resin 陰離子交換樹酯
anionic detergent 陰離子清潔劑
anionic starch 陰離子澱粉
anionic surface-active agent 陰離子界面活性劑；陰離子表面活性劑 [= anionic surfactant]
anionic surfactant 陰離子界面活性劑；陰離子表面活性劑 [= anionic surface-active agent]
anise 大茴香；茴香；茴芹；八角 [= Chinese aniseed; star anise]
anise oil 大茴香油；茴香油；茴芹油
anise pepper 大茴香椒粒
anise seed 大茴香；八角；茴香；茴芹 [= aniseed]
anise vodka 茴香伏特加酒
aniseed 大茴香；八角；茴香；茴芹 [= anise seed]
anisette 茴香酒
anisic acid 茴香酸
anisole 苯甲醚；茴香醚
anisomycin 大茴香黴菌

anisospore 異型孢子
anisotonic solution 異滲溶液；不等滲溶液
anisy formate 甲酸茴香酯
anisyl acetate 乙酸茴香酯
anisyl acetone 茴香基丙酮；茴香酮
anisyl alcohol 茴香醇；對甲氧基苯甲醇
anka 紅麴
ankaflavin 紅麴黃素
ankle 踝
annatto 胭脂籽；絳珠籽；玲瓏籽；蘿考籽；胭脂樹橙；胭脂樹橙〔紅〕色素；婀娜多紅色素 [= annotta; arnotta]
annealing 退火
annotta 胭脂樹橙〔紅〕色素；婀娜多紅色素 [= annatto; arnotta]
annulus 菌環；蕈環
anomalotrophy 營養異常
anorectic agent 食慾抑制劑
anorexia 厭食症；食慾缺乏症
anorexia nervosa 神經性厭食症；神經性食慾缺乏症
anorexigenic drug 食慾減退劑
anosmia 無嗅覺症；嗅覺缺失症
anoxia 缺氧症
anoxomer 阿諾克索默
anoxybiosis 厭氧生活
anoxybiotic bacteria 厭氧生活細菌
antacid 抗酸劑；抑酸劑 [= antiacid]
antagonism 拮抗〔作用〕
antagonist 拮抗物
antarctic krill 南極〔磷〕蝦
ante-mortem inspection 屠前檢驗
antelope 羚羊
antemortem inspection 屠前檢查；宰前檢查
anthocyanase 花色素酶
anthocyanin 花青素；花青苷；花色苷
anthocyanogen 白花色素
anthoxanthin 花黃質；黃酮
anthrax 炭疽熱；炭疽病
anthrax bacillus 炭疽熱桿菌
anti-yeast 抗酵母的
antiacid 抗酸劑；抑酸劑 [= antacid]
antiavidin 生物素；維生素H [= biotin]
antibacterial 抗菌的
antibacterial agent 抗菌劑
antibiosis 抗生
antibiotic 抗生素
antibody (Ab) 抗體
antibrowning agent 抗褐變劑
anticaking agent 抗結塊劑；抗結劑
anticoagulant 抗凝劑
antidote 解毒劑
antienzyme 抗酵素；抗酶
antiferment 抗發酵劑；抗酶
antifoam 消泡；防沫
antifoamer 消泡劑；消泡器
antifoaming agent 消泡劑；除沫劑；防沫劑 [= defoaming agent; kilfoam]
antifreeze 防凍劑
antifreezing agent 防凍劑
antifungal activity 抗真菌活性
antifungal agent 抗真菌劑 [= antimycotic agent]
antigen 抗原
antimicrobial 抗菌的；抗菌劑
antimicrobial activity 抗微生物活性；抗菌活性
antimicrobial agent 抗微生物劑；抗菌劑
antimildew 抗黴的
antimold 防黴劑
antimycin 抗黴素
antimycotic agent 抗真菌劑 [= antifungal agent]

antioxidant 抗氧化劑
antioxidative activity 抗氧化性
antisepsis 防腐；消毒
antiseptic 防腐的；消過毒的；殺菌的；防腐劑；消毒劑
antistatic agent 抗靜電劑；去靜電劑
antisticking 防黏
antisticking agent 防黏劑
antistickings 防黏劑
antitoxin 抗毒素
antitrypsin 抗胰蛋白酶
antiviral agent 抗病毒素
antivitamin 抗維生素
antizymotic agent 發酵抑制劑
anus 肛門
ANZFA (Australia New Zealand Food Authority) 澳洲及新西蘭食物管理局
AOAC (Association of Official Analytical Chemists) 美國公認分析化學〔師協〕會
AOCS (American Oil Chemists Society) 美國油脂化學會
AOM (active oxygen method) 活性氧氣法；活性氧測定法（油脂穩定性試驗）
apastia 禁食
aperient 通便藥
aperitif (wine) 飯前酒；開胃〔葡萄〕酒 [= appetizer wine]
aphtha 口瘡
aphthous fever 口蹄疫；鵝口瘡熱
apiarist 養蜂人；養蜂家
apiary 養蜂場；蜂房
apiculture 養蜂〔業〕
apiin 芹菜苷
apiole 芹菜腦
apiology 蜜蜂學
apiose 芹菜糖
Apium 芹菜屬；旱芹屬
Apium grareolens 芹菜；旱芹
aplasia 發育不全；先天萎縮
β-Apo-8' carotenal β-阿朴胡蘿蔔素醛
β-Apo-8' carotencic acid ethyl ester β-阿朴胡蘿蔔酸乙酯
apocholic acid 衍膽酸；原膽酸
apocyte 多核細胞
apoenzyme 缺輔基酶；脫輔基酶蛋白
apolipoprotein 缺脂肪蛋白；阿樸脂蛋白
apoprotein 缺輔基蛋白；脫輔基蛋白
aposia 不渴症
apparent digestibility 視消化率；表觀消化率
apparent viscosity 表觀黏度
Appennio sausage 意大利阿本尼香腸
appertization 商業滅菌法；阿佩爾滅菌法 [= commercial sterilization]
appertizing 商業滅菌；罐藏法
appetizer 開胃菜；開胃食品
appetizer wine 飯前酒；開胃〔葡萄〕酒 [= aperitif (wine)]
apple 蘋果
apple brandy 蘋果白蘭地酒
apple butter 蘋果醬；蘋果泥
apple cider 蘋果汁；蘋果淡酒
apple corer 蘋果去心器；蘋果挖核器
apple cream 蘋果酪；蘋果漿
apple essence 蘋果香精
apple flavo(u)r 蘋果香精
apple jack 法國蘋果白蘭地酒；卡巴度斯蘋果白蘭地酒；美國蘋果白蘭地酒 [= calvados]
apple jam 蘋果〔果〕醬
apple juice 蘋果汁
apple nugget 乾蘋果粒；蘋果塊
apple oily flavo(u)r 蘋果油香精
apple peeler 蘋果去皮機

apple pie 蘋果派；蘋果餡餅
apple pomace 蘋果漿；蘋果〔果〕渣
apple press 蘋果壓榨機
apple pulp 蘋果漿；碎蘋果肉
apple slicer 蘋果切片器
apple snail 福壽螺
apple strudel 焗蘋果酥
apple vinegar 蘋果醋
apple wine 蘋果酒
applesauce 蘋果醬；蘋果糊；蘋果沙司
applicator 電昏器(屠宰用)
applier 灌腸器；填充器
apricot 杏
apricot conserve 杏子罐頭
apricot jam 杏果醬；杏子醬
apricot kernel 杏仁
apricot oil 杏仁油
apricot plum 杏李；紅李
apricotine 杏子酒
apron 圍裙
AQL (acceptable quality level) 合格質量水平
aqua 水；溶液；液體
aqua fortis 硝酸
aquaculture 水產養殖〔業〕；水產業
aquatic animal 水生動物
aquatic fungi 水生真菌類
aquatic plant 水生植物
aquatic product 水產物；水產品
aquavit 蒸餾酒；白蘭地〔烈性酒〕 [= akvavit]
aqueous alcohol 含水酒精
aqueous solution 水溶液
araban 聚阿拉伯糖；阿拉伯聚糖
arabian coffee 阿拉伯咖啡樹
arabian gum 阿拉伯樹膠 [= acacia gum; arabic gum; gum acacia; gum arabic]
Arabian jasmine 茉莉花；水梨花
Arabian tea 阿拉伯茶
arabic acid 阿拉伯酸
arabic gum 阿拉伯〔樹〕膠 [= acacia gum; arabian gum; gum acacia; gum arabic]
arabica (coffee) 阿拉比卡咖啡；阿拉伯咖啡 [= coffee arabica]
arabinan 聚阿拉伯糖；阿拉伯聚糖
arabinose 阿拉伯糖；樹膠醛糖
arabitol 阿拉伯糖醇
arabonic acid 阿拉伯糖酸
arachain 花生仁蛋白酶
arachic acid 花生酸；二十烷酸 [= arachidic acid; eicosanoic acid]
arachidic acid 花生酸；二十烷酸 [= arachic acid; eicosanoic acid]
arachidonic acid 花生四烯酸；二十碳四烯酸 [= eicosatetraenoic acid]
arachin 花生球蛋白
Arachis 落花生屬；花生屬
arachis oil 花生油 [= peanut oil; earthnut oil; groundnut oil]
arak 阿拉克燒酒；甘蔗廢蜜伏特加
arame 黑海帶
arbinogalactan 阿拉伯半乳聚糖；松膠
arbutus 楊梅
arbutus jam 楊梅果醬
architectured food 造型食品
arctic shrimp 北極蝦
ardent drink 含酒精飲料
ardent spirit 烈酒；燒酒
area mixogram 麵糰調混性圖譜
areca nut 檳榔 [= betel nut]
areolar tissue 網狀結締組織；蜂窩組織
areometer 比重計；液體比重計
argentine 香魚
arginase 精胺酸酶

arginine 精胺酸；精氨酸
arginine monohydrochloride 精氨酸鹽酸鹽
argol 粗酒石
argon 氬
ariboflavinosis 維生素B_2缺乏症；核黃素缺乏症
aril 假種皮
ark shell 赤貝；毛蚶
arlac 強化嬰兒奶粉
arm mixer 臂式攪拌器；曲臂式攪拌器
armagnac 阿瑪白蘭地；阿馬尼亞克酒
Armillaria 松蕈屬；蜜環蕈屬
army bread 軍用麵包
Arnold sterilizer 阿諾滅菌器
arnotta 胭脂樹橙〔紅〕色素；婀娜多紅色素 [= annatto; annotta]
aroma 香氣；芳香；香味
aroma former 產香物；產香菌
aromatic 芳香的
aromatic amino acid 芳〔香〕族胺基酸
aromatic compound 芳香族化合物
aromatic condiment 芳香調味料
aromatic crop 芳香料作物
aromatic herb 芳香草
aromatic plant 芳香植物
aromatic seasoning 芳香調味料
aromatic substance 芳香物質
aromaticity 芳香性；芳香度
aromatics 芳〔香〕族化合物；芳香劑；香料
aromatized wine 香化酒；加香葡萄酒
arrack 亞力酒
arrow-toothed flounder 美洲箭齒鰈
arrow-toothed halibut 油鰈；亞洲箭齒鰈
arrowhead 慈菇
arrowroot 葛鬱金；竹芋
arrowroot starch 葛粉；竹芋澱粉
arsanilic acid 阿散酸；對胺苯基胂酸
arsenic 砷
Artemisia 艾屬
arteriosclerosis 動脈硬化〔症〕
artery 動脈
artery pumping 醃液動脈注射法；注射醃漬法
artichoke 朝鮮薊；雅芝竹；洋薊 [= globe artichoke]
artificial ag(e)ing 人工熟成；人工陳化
artificial butter 人造白脫；人造黃油；人造牛油；人造奶油
artificial casing 人造腸衣
artificial colo(u)r 人工色素；合成色素
artificial essence 人造香精；合成香精
artificial fat 人造脂肪
artificial flavo(u)r 人造〔食用〕香料；合成香料 [= synthetic flavo(u)r]
artificial flavo(u)ring substance 人造香料
artificial meat 人造肉
artificial milk 人造乳；合成乳 [= synthetic milk]
artificial ripening 人工催熟
artificial seasoning 人工調味料
artificial smoking 液燻；無煙燻製
artificial sweetener 人工甜味料
artificial vinegar 合成醋
arum 海芋植物
asafoetida 阿魏膠
asali 非洲蜂蜜酒
ascarid 蛔蟲
ascomycete 子囊菌〔類〕
ascorbic acid 抗壞血酸；維生素C [= vitamin C]
ascorbic acid oxidase 抗壞血酸氧化酶

ascorbyl palmitate 抗壞血酸〔基〕棕櫚酸酯；抗壞血酸-6-棕櫚酸鹽
ascorbyl stearate 抗壞血酸〔基〕硬脂酸酯
asepsis 無菌處理；無菌法；無菌；防腐處理
aseptic 無菌的；無菌性
aseptic box 無菌箱
aseptic canning 無菌裝罐
aseptic filling 無菌裝填；無菌灌裝
aseptic filtration 無菌過濾
aseptic method 無菌〔操作〕法
aseptic package 無菌包裝
aseptic paper 無菌紙；消毒紙
aseptic process 無菌製作；無菌製程；無菌加工
aseptic room 無菌室
aseptic tank 無菌桶；無菌槽；無菌罐
aseptic technic 無菌技術
asexual reproduction 無性生殖
asexual spore 無性孢子
ash 灰分
ashcake 灰焙玉米餅；灰爐烤餅
ashing 灰化
ashing crucible 灰化用坩堝
ashless 無灰的
ashless filter paper 無灰濾紙
asiatic ginseng 人蔘
asparaginase 天〔門〕冬酰胺酶
asparagine 天冬酰胺〔酸〕；天門冬素
asparagus 蘆筍；石刁柏；百刁柏
asparagus bean 豇豆；長豇豆；菜豆；豆角
asparagus broccoli 花莖甘藍；青花菜；莖椰菜 [= calabrese]
asparagus consomme 蘆筍清湯
asparagus lettuce 嫩莖萵苣；萵苣筍；香菜心
asparagus pea 四荚豆；四棱豆
aspartame 天門冬酰苯丙氨酸甲酯；甜味素；阿斯巴甜；天冬甜素；天冬苯丙二肽酯
aspartase 天〔門〕冬胺酸酶
aspartate 天〔門〕冬胺酸酯
aspartic acid 天〔門〕冬胺酸；氨基琥珀酸
aspergillic acid 麴菌酸；麴黴酸
aspergillin 麴菌素；麴黴素
aspergillosis 麴菌病；麴黴病
Aspergillus 麴菌屬；麴黴屬
Aspergillus awamori 泡盛麴菌
Aspergillus flavus 黃麴菌；黃麴黴
Aspergillus niger 黑麴菌
Aspergillus orchraceus 棕麴菌
Aspergillus oryzae 米麴菌
Aspergillus parasiticus 寄生麴菌
Aspergillus sojae 醬油麴菌
Aspergillus tamarii 溜麴菌
Aspergillus wentii 文氏麴菌
aspermous orange 無子柑橘
aspermous watermelon 無子西瓜
aspic 花式肉凍冷盤；肉凍；果凍裝飾配菜；寬葉薰衣草
aspic jelly 肉凍；魚凍
aspic oil 薰衣草油
aspirin 阿斯匹靈
asporous bacteria 無芽孢細菌
asporulate bacteria 不產芽孢細菌
ass' s-ear 驢耳朵鮑魚
assam tea 阿薩姆茶
assay 化驗
assayer 化驗員；分析專家；評味員
assimilable nitrogen 可同化氮
assimilation 同化〔作用〕；吸收〔作用〕
Association Internationalede Chimie Cerealiene (AICC) 國際穀物化學協會

Association of Official Analytical Chemists (AOAC) 美國公認分析化學〔師協〕會
assorted 什錦的
assorted biscuit 什錦餅乾
assorted chocolate 什錦巧克力；什錦朱古力
assorted pickle 什錦醃菜；什錦泡菜；什錦酸菜
assorted vegetable 什錦蔬菜
assorting cylinder 分級圓筒篩；滾筒式分級機；分級滾筒
astacin 蝦紅素
astaxanthin 還原蝦紅素；蝦青素；蝦黃質
asterin 紫苑苷；菊色素；葡糖苷
asterotoxin 海星毒素
astriction 收斂〔作用〕；澀味；便秘
astringency 收斂性；澀味
astringent 收斂劑；澀味劑；澀的
astringent flavo(u)r 澀風味
astringent taste 澀味
at-home meal 家用的膳食
ATCC (American Type Culture Collection) 美國菌種保存中心；美國標準菌庫
atherosclerosis 動脈〔粥狀〕硬化症
Athole brose 威士忌蜜飲料；阿索爾蜜酒
athrepsy 營養缺乏症
Atka fish 阿加鯖魚；多線魚；單鰭多線魚
Atka mackerel 阿加鯖魚；多線魚；遠東多線魚
Atlantic cod 大西洋鱈
Atlantic halibut 大西洋庸鰈
Atlantic herring 大西洋鯡
Atlantic mackerel 大西洋鯖
Atlantic salmon 大西洋鮭
Atlantic sardine 大西洋沙丁魚
Atlantic tuna 大西洋鮪魚；大西洋金槍魚
atmospheric cooking 常壓烹煮
atmospheric drying 風乾；晾乾
atmospheric evaporator 常壓蒸發器
atmospheric freeze drying 常壓冷凍乾燥
atmospheric humidity 大氣濕度
atmospheric pressure 大氣壓
atomic absorption spectrophotometer 原子吸〔收分〕光光度計
atomization 霧化；粉〔碎〕化
atomizer 霧化器；噴霧器；粉碎器
atomizing chamber 噴霧室；霧化室
atomizing damper 噴霧濕潤器
atomizing disk 噴霧盤；離心噴霧盤
atomizing nozzle 噴霧嘴
ATP (adenosine triphosphate) 腺苷三磷酸；三磷酸腺苷
atractylis concrete 蒼术〔硬〕脂
atractylis oil 蒼术油
atta 未篩麵粉；粗麵粉
attapulgite clay 凹凸棒石黏土
attar 玫瑰油；香精油；花油
attemperating tank 調溫槽
attemperation 調溫；溫度調節作用
attemperator 調溫器；調溫裝置
attenuation degree 發酵〔程〕度；耗糖度
attrition grinder 磨碎機
attrition mill 擦碎機；磨碎機
aubergine 茄子 [= eggplant; brinjal; garden-egg]
auffait 蜜餞夾心冰淇淋；水果夾心冰磚
auger feeder 螺旋喂送器；螺旋進料器
auger filler 螺旋充填機
auger packer 螺旋打包機

aujus 帶原汁的；原湯
aurantiamarin 橙皮苷；苦橙苷
aurantium 柑果
Auricularia 木耳屬
Auricularia auricula 木耳
aurore sauce 金黃調味醬；金黃醬；金黃色沙司
auroxanthin 金黃質
aurum 金黃烈酒；金酒
austere 澀味的
austerity 澀味〔性〕
Australia New Zealand Food Authority (ANZFA) 澳洲及新西蘭食物管理局
Australian nut 澳洲胡桃；昆士蘭〔堅〕果；夏威夷果 [= Queensland nut; macadamia nut]
auto-oxidation 自〔動〕氧化
auto-wrapper 自動包裝機
autocarp 自花受精果
autocatalysis 自身催化〔作用〕；自動催化〔作用〕
autocatalytic reaction 自觸媒反應；自〔身〕催化反應
autochthon 土生種；本生的動植物
autoclave 殺菌釜；高壓〔蒸汽〕滅菌器
autoclave sterilization 高壓蒸汽滅菌
autoclave sterilizer 殺菌釜；高壓蒸汽滅菌器
autoclaving 高壓蒸汽滅菌；高壓蒸汽處理
autocytolysis 細胞自溶〔作用〕
autodigestion 自〔體〕消化
autoenzyme 自溶酶
autofeeder 自動進料機；自動餵料機
autofermentation 自〔然〕發酵〔作用〕
autogamous crop 自花交配作物；自花受粉作物
autolysate 自溶產物
autolysin 自溶素
autolysis 自體消化；自溶；自家消化
autolytic deterioration 自然敗壞；自溶變質
autolytic enzyme 自溶酶；自解酶；自體消化酶
automat 自動售貨機；自動食品櫃
automatic bagging scale 自動裝袋稱量秤
automatic balance 自動天平；自動秤
automatic batch scale 自動分批配料秤
automatic bone meat separator 自動骨肉分離機
automatic deboning machine 自動去骨機
automatic feeder 自動進料器
automatic filler 自動裝填機；自動填充機
automatic filling machine 自動裝填機；自動灌裝機
automatic flanger 自動翻邊機
automatic marker 自動罐蓋打印機
automatic monitor 自動監測器
automatic peeler and corer 自動〔水果〕去皮去芯機
automatic peeling machine 自動〔水果〕去皮機
automatic round type fryer 自動迴轉式炸鍋
automatic sampler 自動取樣器
automatic thermostat 自動恆溫器
automonitor 自動監測器；自動監控程序
autoplast 葉綠體 [= chloroplast]
autotitrator 自動滴定器
autotroph 自營生物；自養生物
autotrophic bacteria 自營細菌；自養細菌
autumn crop 秋〔熟〕作物

autumn fattening 秋季長肥
autumn herring 秋鯡
autumn squash 印度南瓜；秋季南瓜
auxiliary material 佐料
available nutrient 有效營養素
Avena fatus 野燕麥
Avena sativa 燕麥
avenin 燕麥〔穀〕蛋白
avian flu 禽流感
avian influenza 禽流感
aviculture 野禽飼養；養禽業；養禽學
avidin 抗生物素蛋白
avirulence 無毒性
avitaminosis 維生素缺乏症
avocado 牛油果；鱷梨；油梨；酪梨 [= aguacate; alligator pear; butter-fruit]
Aw (water activity) 水〔分〕活性
awkeotsang 愛玉子
awn 芒；穀刺
awned wheat 有芒小麥
axenic 無菌的
axenic animal 無菌動物
axunge 油脂；藥用脂肪
Ayrshire (cow) 愛爾夏乳牛；埃爾夏乳牛
ayu (fish) 香魚 [= sweetfish]
azafrin 玄參紅酸
azeotropic distillation 共沸蒸餾
azeotropic drying 共沸乾燥
azeotropic freeze-drying 共沸冷凍乾燥
azo blue 偶氮藍
azo compound 偶氮化合物
azobacteria 固氮〔細〕菌 [= azotobacter (ia)]
azocarmine 偶氮胭脂紅
azodicarbonamide 偶氮〔二〕甲酰胺
azoprotein 偶氮蛋白
azorubin(e) 偶氮玉紅；藍光酸性紅；淡紅
azotobacter(ia) 固氮〔細〕菌 [= azobacteria]
azotoflavin 固氮黃素
azotometer 氮素計；氮量計；定氮儀 [= nitrometer]
azuki bean 紅豆；赤豆；小豆 [= adzuki bean; adsuki bean; red bean]
azyme 未膨發麵包；不經酵母發麵的麵包

B

B complex 複合維生素B [= vitamin B complex]
babaco 芭芭可果
babassu kernel oil 巴西棕櫚果仁油；巴西巴巴蘇棕櫚仁果
babassu nut 巴西棕櫚果；巴西巴巴蘇棕櫚果
Babcock bottle 巴布科克〔乳油〕瓶；貝氏〔乳脂〕瓶 [= Babcock flask]
Babcock flask 巴布科克〔乳脂〕瓶；貝氏〔乳脂〕瓶 [= Babcock bottle]
Babcock test 貝氏〔乳脂〕試驗；巴布科克〔乳脂〕試驗
baby abalone 九孔；小鮑魚
baby beef 小牛肉；犢牛肉(15個月以下)
baby cabbage 芽甘藍；抱子甘藍；球芽甘藍；北京芽菜
baby chick 小雞肉；秧雞
baby corn 玉米筍；珍珠筍
baby fish 小魚；幼魚
baby food 嬰兒食品
baby-neck clam 淺蜊
babysberry 楊梅
bacalao 鹹鱈魚乾；鱸科魚類
bacca 漿果
bacillemia 桿菌血症
bacilli 芽孢桿菌；桿菌 [= bacillus的複數]
bacillicide 殺桿菌劑
bacillin 枯草桿菌素；桿菌素
bacillomycin 芽孢菌黴素
bacillosis 桿菌病
bacillus 孢桿桿菌；桿菌
Bacillus aceticus 醋酸桿菌
Bacillus acidophilus 嗜酸乳桿菌
Bacillus anthracis 炭疽〔芽孢〕桿菌
Bacillus botulinus 肉毒〔芽孢〕桿菌
Bacillus cereus 仙人掌〔芽孢〕桿菌
Bacillus cereus food poisoning 仙人掌桿菌食物中毒
Bacillus coagulans 凝結〔芽孢〕桿菌
Bacillus coli 大腸桿菌
Bacillus coli anaerogenes 不產氣大腸桿菌
Bacillus natto 納豆〔芽孢〕桿菌
Bacillus panis 麵包〔芽孢〕桿菌
Bacillus polymyxa 多黏〔芽孢〕桿菌
Bacillus proteus 變形〔芽孢〕桿菌
Bacillus putrificus 腐敗〔芽孢〕桿菌
Bacillus stearothermophilus 脂嗜熱肪〔芽孢〕桿菌
Bacillus subtilis 枯草〔芽孢〕桿菌
Bacillus thermoamylolyticus 嗜熱澱粉分解〔芽孢〕桿菌
Bacillus tuberculosis 結核〔芽孢〕桿菌；結核桿菌
bacilysin 芽孢桿菌溶素；桿菌溶素
back fat 背脂；背膘
back fat skinner 背膘去皮機

back fillet 腰內肉；內里脊肉；鯨魚背條肉
back fin 背鰭
back flow 回流
back pork 背脊肉；通脊肉
back titration 反滴定
backboned animal 脊椎動物
backer 支持者；剝牲畜背皮的工人
backribs 背肋部
backside 胴體後背；背面
backstrap 羊背肌
bacon 培根；煙肉；醃燻豬五花肉；鹹肉
bacon hog 醃肉用閹豬
bacon pig 醃肉用小豬
bacon press(er) 醃肉壓型機；培根肉整形壓製機
bacon rind 脂肪皮
bacon slicer 醃肉切片機
bacon weed 醃肉雜碎
baconer 醃肉用豬
bacteremia 菌血症
bacteria 細菌
bacterial 細菌的
bacterial activity 細菌活性
bacterial amylase 細菌澱粉酶
bacterial colony 細菌菌落
bacterial contamination 細菌污染
bacterial content 細菌含量
bacterial count 細菌數
bacterial cultivation 細菌培養
bacterial decomposition 細菌分解
bacterial degradation 細菌降解
bacterial enzyme 細菌酵素
bacterial fermentation 細菌發酵
bacterial flora 細菌群落；細菌區系
bacterial food poisoning 細菌性食物中毒
bacterial infection 細菌感染
bacterial plate count 細菌平板菌落計數
bacterial population 細菌數〔目〕；細菌群體
bacterial spoilage 細菌性腐敗
bacterial spore 細菌孢子；細菌芽孢
bacterial taint 菌腐味；細菌腐壞味
bacterial toxin 細菌毒素
bactericidal 殺細菌的
bactericidal activity 殺細菌活性
bactericidal agent 殺〔細〕菌劑
bactericidal effect 殺細菌效應
bactericide 殺〔細〕菌劑
bacteriological 細菌學的
bacteriologist 細菌學家
bacteriology 細菌學
bacteriolysin 溶菌素
bacteriolysis 溶菌作用
bacteriophage 噬菌體；抗菌素 [= phage]
bacteriosis 細菌性疾病
bacteriostasis 靜菌作用；抑菌作用
bacteriostat 抑菌劑
bacteriostatic agent 抑〔細〕菌劑
bacteriotoxin 細菌毒素
bacterium 細菌
bactofugation 離心〔力〕除菌
bactofugeur 離心除菌機
bael 硬皮橘；印度枸杞
baffle (plate) 擋板；導流板；折流板；緩衝板
bag extractor 袋萃取器
bag filling machine 裝袋機
bag filter 袋濾器；袋濾機
bag holder 夾袋器
bag making machine 製袋機
bag sealer 封袋機
bag sealing machine 封袋機
bagasse 蔗渣；甘蔗渣 [= megasse]
bagasse board 蔗渣板

bagful 一袋量；滿口袋
bagger 裝袋工人；裝袋機
bagging 袋裝
bagging hopper 裝袋漏斗
bagging machine 裝袋機
bagoong 菲律賓發酵魚〔蝦〕醬；發酵魚醬；巴衮蝦醬
baguet(te) 法國〔長〕麵包
baikhovi tea 白毫茶
bailing press 打包機
bain-marie 水浴〔燉〕鍋；雙重蒸鍋；隔水燉鍋；保溫蒸盤
bake 烘焙；烘烤；焙烤食品
bake-in-bag (food) 袋烤食品
bake oil 糕點用油
baked bread 烤麵包
baked food 焙烤食品；烘焙食品
baked good 焙烤食品；烘焙食品
baked product 焙烤製品
bakehouse 麵包店；麵包烘房
bakemeat 肉餅；肉餡烤餅
baker 烘爐；烤爐；麵包師
baker's oil 焙烤食品用油
baker's salt 碳酸氫銨
baker's yeast 麵包酵母
bakers yeast glycan 焙烤酵母聚糖
bakery 烘烤食品廠；麵包店
bakery goods 烘烤食品
bakery product 烘烤製品
bakery refuse 麵包房的廢物渣屑
bakery shortening 烘烤食品用起酥油
bakery yeast 麵包酵母
bakeshop 麵包店；麵包烘房
baking 烘烤；烘焙；焗
baking agent 發粉
baking grease 烘焙用油脂
baking industry 烘焙食品工業；麵包工業
baking load 烤麵包
baking oven 烘箱；烤箱；烤爐
baking powder 發酵粉；焙粉；發粉
baking product 烘焙食品
baking soda 碳酸氫鈉；小蘇打 [= sodium hydrogen carbonate; bicarbonate of soda; sodium bicarbonate]
baking strength 麵糰發麵力
balance 天平；均衡
balanced diet 均衡膳食；均衡飲食
balanced growth 均衡生長
balata 巴拉他膠；巴拉塔樹膠
balbakwa 原條醃魚製品
Baldwin 鮑爾溫蘋果；赤龍蘋果
baled tea 壓塊茶磚
baleen whale 鬚鯨
baler 打包機；打包工人
ball cabbage 結球甘藍；結球高麗菜；結球椰菜
ball crusher 球磨碎機
ball machine 圓球糖成形機
ball mill 球磨機
baller 揉圓機；麵糰揉圓機
Balling (scale) 巴林度；巴林比重度
Balling hydrometer 巴林比重計；巴林計
ballotines 肉丸子
balm 香脂；香液；香油；香膏；香樹膠；蜜蜂花
balm-mint 香水薄荷
balsam 香脂；香液；香油；香膏；香樹膠；香脂木；香膠木
balsam fir oil 拔爾散冷杉油
balsam pear 苦瓜；涼瓜 [= bitter gourd; bitter melon]
Balsam Peru oil 秘魯香膏油
balsamweed 鈍葉鼠曲草
bamboo 竹
bamboo leaf 竹葉
bamboo shoot 冬筍；竹筍

bamboo skewer 竹串；竹籤
bamies 黃秋葵
banana 香蕉
banana fig 香蕉乾（去皮）
banana fish 金帶梅鯛
banana flavo(u)r 香蕉香精
banana fritter 油炸香蕉
banana oil 香蕉油
banana oily flavo(u)r 香蕉油香精
banana prawn 緬甸墨吉對蝦
banana squash 香蕉瓜；筍瓜
bananase 香蕉粉
banbury (biscuit) 葡萄乾夾心餅；班伯里餅乾
banbury cake 班伯里果料烤餅
banbury tart 班伯里果料塔餅
bancha (tea) 日本粗茶
band conveyor 帶式輸送機；輸送帶
band dryer 帶式乾燥機
band fish 背點棘赤刀魚
band oven 帶式烤爐
band-tailed sea perch 長尾海鱸魚
banded cavallal 六帶鰺
banded grunt 斑雞魚
banded grunter 鰾石鱸；星雞魚
banded pickerel 帶紋狗魚；美洲狗魚
banded shrimp 斑節蝦
banding machine 捆帶機
banjofish 壽魚
bank cod 大頭鱈
bank fishing 淺灘捕魚
banker 補鱈人；鱈魚船
banner oat 側穗燕麥
bannock 麥餅
bantam 矮腳雞；爪哇種矮雞
bantu beer 低醇黍啤酒
baobab 猢猻麵樹；猴麵包樹
bap 小麵包；早餐軟小麵包
bar 棒；條；酒吧
bar bread 長條〔大麥〕麵包
bar machine 冰棒冷凍機 [= ice bar freezer]
bar-spoon 調酒〔長柄〕匙
bar-type mill 橫杆式脫粒滾筒
Barbados cherry 西印度櫻桃
barbaloin 蘆薈苷
barbecue 烤肉；烤肉架；烤肉野餐；巴比丘 [= barbeque]
barbecue grill 燒烤爐
barbecue sauce 烤肉調味醬
barbecued meat 烤肉；燒肉；叉燒肉
barbel 魚觸鬚；白魚
barbeque 烤肉；烤肉架；烤肉野餐；巴比丘 [= barbecue]
barberry 伏牛花；伏牛花果
barbital 巴比妥
barbiturate 巴比妥酸鹽
barbotage 起泡作用；發泡作用
bard 焙根包；肥肉薄片；燻肉薄片
bardolino 意大利巴多利諾紅葡萄酒
bare toothed russule 菱紅菇
barfish 歐洲淡水鱸魚
Barfoed's test 巴式〔單醣〕試驗；巴弗特試驗
Baril wheat 阿根廷巴里爾小麥
barile 酒桶
barilla 蘇打灰
barium 鋇
bark 樹皮
barley 大麥 [= hordeum]
barley-bree 啤酒；威士忌 [= barley-broo; barley broth]
barley-broo 啤酒；威士忌 [= barley-bree; barley broth]
barley broth 啤酒；威士忌 [= barley-bree; barley-broo]
barley candy 硬糖果
barley extract 大麥浸出液

barley flour 大麥粉
barley grout 去殼大麥粒
barley koji 大麥麴
barley malt 大麥麥芽；麥芽
barley meal 大麥粉
barley sugar 粗粒糖；大麥糖；麥芽糖；硬糖果
barley water 大麥煎汁
barleycorn 大麥粒
barm 酵母；發酵劑
barmy 發酵的；發滿泡沫的；酵母的
barn 穀倉；畜舍
barnyard millet 稗子；湖南稗子
barolo 巴羅洛葡萄酒
barometer 氣壓計
barometric pressure 大氣壓力
barophile 嗜壓微生物；適壓微生物
barophilic bacteria 嗜壓細菌；適壓細菌
barracuda 金梭魚
barramunda 紅目鱸；澳洲肺魚 [= barramundi]
barramundi 紅目鱸；澳洲肺魚 [= barramunda]
barramundi cod 銳首擬石斑；老鼠斑
barred pickerel 帶紋狗魚
barrel 琵琶桶；桶
barrel churn 桶式乳油攪拌機
barrel filler 裝桶機
barrel heat exchanger 筒式熱交換器；筒式換熱器
barrel mixer 轉筒式混合器
barrelhouse 廉價小酒店
barrelled juice 桶裝果汁
barrelling 裝桶
barricas 大酒桶
barrier package 阻隔包裝；防潮包裝；防氣包裝
barril 小酒桶
barrow 閹公豬；手推車
barrow's bolete 丘形牛肝菌
bartender 調酒師
Bartlett (pear) 巴梨；洋梨
basal medium 基礎培養基
basal metabolic rate (BMR) 基礎代謝率
basal metabolism 基礎代謝
base 基礎；鹼
base forming food 鹼性食品；形成鹼基的食品；生理鹼性食品
base residue food 鹼性食品；生理鹼性食品
base stock 基〔本原〕料；基本組份
basella 白落葵
basella rubra red 落葵紅
basi 倍西飲料
basic 鹼〔性〕的；基礎的；基本農作物
basic amino acid 鹼性胺基酸
basic capacity 鹼度
basic coal tar dye 鹼性煤溚染料
basic colo(u)r 鹼性染料 [= basic dye]
basic crop 基本〔農〕作物；基本農產品
basic dye 鹼性染料 [= basic colo(u)r]
basic food 鹼性食品；生理鹼性食品；產鹼食品；主食
basic foodstuff 基本食物
basic ingredient 基本配料
basic protein 鹼性蛋白
basic salt 鹼式鹽
basicity 鹼性度；鹼度
basidiocarp 擔子果
basidiomycetes 擔子菌〔類〕
basidiospore 擔孢子
basification 鹼化〔作用〕
basil 丁香羅勒；羅勒；九層塔；洋紫蘇 [= sweet basil]
basil oil 羅勒油
basil thyme 羅勒百里香
basket centrifuge 籃式離心機

basket pasteurizer 籃式巴氏滅菌器
basmati rice 巴斯馬蒂米；喜馬拉雅山長形米
basophilic bacteria 嗜鹼細菌
bass 鱸魚；巴斯啤酒
bassora gum 黃蓍膠；刺槐樹膠
bassorin 黃蓍膠黏素；黃蓍膠糖
bastard cod 藍鱈魚
bastard mullet 大海鰱；五絲馬鱍
baste 烤肉塗醬；烤肉上塗油
basted meat 叉燒肉
bastimic rice 印度米
batata 番薯；甘薯；紅薯；白薯
batch 分批；一批
batch centrifuge 間歇式離心機
batch pasteurizer 間歇式巴氏滅菌器；分批巴氏滅菌器
batch sterilizer 間歇式殺菌器；分批殺菌器
batcher 計量送料機；分批送料機
batching table 計量配料臺
batfish 銀大眼鯧
Bath biscuit 葡萄乾比思克；巴思餅乾
Bath bun 葡萄乾甜捲麵包；巴思甜麵包
bath chap 醃豬頭肉
bathypelagic fish 深水魚
Batrachospermum 陟釐屬
batter 麵糊；麵漿
batter mix 麵漿混料；麵糊混料
battering 搥薄；麵拖；裹漿
batyl alcohol 鯊肝醇
Baumé (Bé) 波美；波美度 [= Beaume]
Baumé degree 波美度；波美比重度
Baumé gravity 波美比重
Baumé hydrometer 波美比重計
Baumé scale 波美度
Bavarian beer 巴伐利亞啤酒
Bavarian cream 巴伐利亞〔奶油〕甜點
Bavarian sausage 巴伐利亞香腸
bay 月桂
bay bark 桂皮
bay bolete 褐絨蓋牛肝菌
bay leaf 月桂葉；香葉 [= laurel leaf]
bay leaf oil 月桂油；月桂葉油 [= bay oil; laurel leaf oil; laurel oil]
bay oil 月桂油；月桂葉油 [= bay leaf oil; laurel leaf oil; laurel oil]
bay salt 曬鹽；粗海鹽
bayberry 月桂樹果實
baycherry 桂櫻
bazooka freezer 管道式速凍機
Bé (Baumé) 波美；波美度 [= Beaume]
beach conger 日本康吉鰻
beaded can 加強環罐；撐圈罐
beaked parsley 細葉香芹
beaker 燒杯；無柄大口杯
bean 豆
bean cake 豆餅；豆糟
bean curd 豆腐
bean curd cheese 豆腐乳；腐乳 [= fermented tofu; soy cheese; soybean cheese; sufu]
bean curd crouton 油炸豆腐
bean flour 豆粉
bean grader 豆類分級機；豆類篩選機
bean huller 豆類去莢機；豆類去皮機
bean noodle 粉絲；冬粉 [= bean vermicelli; bean thread (vermicelli); green bean vermicelli]
bean oil 大豆油
bean paste 豆醬；豆瓣醬；醬醪
bean pod 豆莢
bean puree 豆沙；豆泥
bean sprout 豆芽
bean thread (vermicelli) 粉絲；冬粉 [= bean noodle; bean vermicelli; green bean vermicelli]

bean vermicelli 粉絲；冬粉 [= bean noodle; bean thread (vermicelli); green bean vermicelli]
beanery 經濟餐館；廉價餐館
beany 大豆腥味的
beany flavo(u)r 豆臭；豆風味
bear prawn 短溝對蝦；熊蝦
bearded wheat 有芒小麥
beardie 泥鰍
beaten biscuit 鬆脆餅乾
beater 攪拌器；打漿機
beating 攪打；敲打
beating machine 攪打機
Beaume 波美；波美度 [= Baumé; Bé]
beauty berry 紫珠
bechamel (sauce) 白調味醬
bee 蜜蜂
beebread 蜜蜂食料
beech 山毛櫸樹
beech mast 山毛櫸堅果 [= beech nut]
beech nut 山毛櫸堅果 [= beech mast]
beechwood sugar 山毛櫸木糖
beef 牛肉
beef bacon 醃燻牛肉；牛肉培根；牛肉煙肉
beef ball 牛肉丸
beef breed 肉用牛種
beef broth 牛肉〔濃〕湯
beef bull 肉用公牛
beef casing 牛腸衣
beef cattle 肉用牛
beef extract 牛肉萃；牛肉浸汁
beef grade 牛肉等級
beef ham 牛腿肉；牛肉火腿
beef loin 牛腰背肉
beef roll 牛肉捲
beef royal 牛肉凍
beef rump butt 牛臀肉塊
beef sausage 牛肉香腸
beef steak 牛排
beef steer 閹牛；肉用閹牛
beef stew 燉牛肉
beef stock 原汁牛肉湯
beef suet 牛油；牛脂
beef tallow 牛脂
beef tapeworm 牛條蟲
beef tripe 牛肚
beefburger 牛肉漢堡；煎牛肉餅
beefsteak fungus 牛排蕈
beehive 蜂巢；蜂窩
beer 啤酒；麥芽汁飲料
beer barrel 啤酒桶
beer boot 啤酒〔靴形〕杯
beer can 啤酒罐
beer filter 啤酒過濾器
beer hall 啤酒店
Beer-Lambert law 貝耳定律 [= Beer's law]
beer mug 啤酒杯
beer must 麥芽汁
beer on draught 生啤；生啤酒 [= draft beer; draught beer]
beer tumbler 啤酒杯
beer wort 製啤酒的麥芽汁
beer yeast 啤酒酵母；釀造啤酒用之酵母 [= brewers' yeast]
Beer's law 貝耳定律 [= Beer-Lambert law]
beerhouse 啤酒店
beeriness 啤酒味
beestings 牛初乳
beeswax 蜂蠟
beeswing 陳年葡萄酒液面的薄膜；葡萄酒的酒石沉澱
beet (root) 甜菜；甜菜根；紅菜頭；紫菜頭
beet chip 甜菜絲；甜菜片 [= beet cossette]

beet cossette 甜菜絲；甜菜片 [= beet chip]
beet harvester 甜菜收獲機
beet juice 甜菜汁
beet marc 甜菜肉；甜菜榨渣
beet massecuite 甜菜糖膏
beet mill 甜菜糖廠
beet molasses 甜菜糖蜜
beet puller 甜菜挖掘機
beet pulp 甜菜渣；甜菜廢絲
beet red 甜菜紅；甜菜根紅 [= beet root red]
beet root red 甜菜紅；甜菜根紅 [= beet red]
beet sugar 甜菜糖
beet syrup 甜菜糖漿
beet topper 甜菜去頂器
beetroot 甜菜根
beggarweed 山綠豆；山蟻蝗(西印度)
begoon 茄子
beheading machine 去頭機
bell pasteurizer 鐘式巴氏滅菌器
bell pepper 燈籠椒；甜椒
bellayay 伯爾尼軟乾酪；伯爾尼軟芝士
bellows fish 魨魚
belly 腹脇肉；豬腹部肉；腩肉；腹肉；腹部
belly box 醃豬腹肉用箱
belly flap 腹壁
belly-full food 果腹食品；飽餐食品
belly meat 腹脇肉；豬腹部肉
belly nape 魚腹黑膜
belly roller 腹脇肉輥展機
belly wash 清涼飲料
belt conveyor 帶式輸送機；輸送帶
belt dryer 帶式乾燥機
belt fish 帶魚
belted bonito 東方狐鰹
beluga 白鯨；白色大鰉魚
beluga caviar 大粒魚子醬
bench 工作臺
bench mixer 臺上攪拌器；臺式混合器
bench scale 檯秤
Benedict's test 本尼迪特試驗；本氏試驗；別尼迪克還原糖試驗
Benedictine 本尼地甜酒；祝福酒
Bengal gram 鷹嘴豆；小黎豆
Bengal quince 硬皮桔
benne 胡麻；芝麻 [= sesame]
benne oil 芝麻油；香油 [= gingelly oil; jinjili oil; sesame oil]
beno 棕櫚燒酒
benthic fish 底棲魚；深海底的魚
benthon 水底生物；底棲生物
benthophyte 水底植物
bentonite 膨潤土
benzaldehyde 苯甲醛；安息香醛；合成苦杏仁油
benzaldehyde glyceryl acetal 苯甲醛甘油縮醛
benzaldehyde propylene glycol acetal 苯甲醛丙二醇縮醛
benzene 苯
benzidine test 聯苯胺試驗
benzoate 苯甲酸鹽；苯甲酸酯；安息香酸鹽
benzoic acid 苯甲酸；安息香酸
benzoic aldehyde 苯甲醛；安息香醛；合成苦杏仁油
benzoic sulfimide 糖精 [= benzosulfimide]
benzoin 安息香膠
benzoin gum 安息香樹膠
benzophenone 二苯甲酮
benzosulfimide 糖精 [= benzoic sulfimide]
benzothiazole 苯并噻唑
benzoyl peroxide 過氧化苯〔甲〕酰

benzoyl thiamin(e) disulfide 二硫化苯酰硫胺；二硫化苯甲酰噻胺

benzoylation 苯甲酰化

benzoylglycine 馬尿酸；苯甲酰甘胺酸 [= hippuric acid]

benzyl acetate 乙酸苄酯

benzyl alcohol 苯甲醇；苄醇

benzyl benzoate 苯甲酸苄酯

benzyl butyl ether 苄丁醚

benzyl butyrate 丁酸苄酯

benzyl cinnamate 桂酸苄酯；桂酸苯甲酯

benzyl formate 甲酸苄酯

benzyl isobutyl carbinol 苄基異丁基甲醇

benzyl isobutyrate 異丁酸苄酯

benzyl isoeugenyl ether 苄基異丁香醚；異丁香酚苄醚

benzyl isovalerate 異戊酸苄酯

benzyl phenyl acetate 苯乙酸苄酯

benzyl propionate 丙酸苄酯

benzyl salicylate 水楊酸苄酯；柳酸苄酯

benzyl tiglate 惕各酸苄酯

berberis 小檗屬植物；小檗果

bergamot 油橙；香檸檬；佛手柑

bergamot mint oil 香檸檬薄荷油

bergamot oil 香檸檬油；油橙油；佛手柑油

Bergey' s manual 伯杰氏細菌分類手冊

Bergundy 法國柏甘地葡萄酒；法國布爾戈尼葡萄酒；法國保根地葡萄酒

Berkshire 波克夏〔黑白〕豬；伯克夏〔黑白〕豬；巴克夏〔黑白〕豬

Berliner sausage 柏林香腸

berlinois 油炸圓餅

bermuda butter-cup 大黃花酢醬草

berry 漿果；小漿果

berry fruit 醬果類

berry pie 漿果派

berry sugar 細白砂糖

berry wine 醬果酒

beshow 裸蓋魚

besmoking 煙燻

bespicing 香辛料調味；添加香辛料

besugo 金線魚

beta-amylase β澱粉酶；澱粉糖化酶；貝他澱粉酶 [= β-amylase]

beta-carotene β胡蘿蔔素；貝他胡蘿蔔素 [= β-carotene]

beta-starch β澱粉；貝他澱粉 [= β-starch]

beta-tocopherol β生育酚；貝他生育酚 [= β-tocopherol]

betain(e) 甜菜鹼 [= oxyneurine]

betanin 甜菜苷

betel 荖藤；蒌葉

betel nut 檳榔 [= areca nut]

betel nut palm 檳榔樹

betty 水果烤布丁；貝蒂布丁

beverage 飲料

BHA (butylated hydroxyanisole) 丁基羥基茴香醚；丁基羥基甲氧苯；丁基大茴香醚

BHT (butylated hydroxytoluene) 二丁基羥基甲苯；2,6-二叔丁基對甲酚

biacetyl 雙乙酰；聯乙酰 [= diacetyl]

bibulosity 吸水性；酗酒

bicarbonate 碳酸氫鹽；重碳酸鹽

bicarbonate of soda 碳酸氫鈉；小蘇打 [= sodium hydrogen carbonate; sodium bicarbonate; baking soda]

biennial 二年生植物

biennial bearing 隔年結果果樹 [= alternate bearing]

biennial fruit 越年果；二年果

bierschinken 啤酒火腿腸

bierwurst 啤酒香腸

biffin 暗紅色的蘋果（英國）
biffin 蘋果乾
Bifidobacterium 雙叉乳酸桿菌屬
bifidus 雙叉乳酸桿菌
bifunctional catalyst 二功能觸媒
big-bud hickory 馬看山核桃
big-eye porgy 鋸大眼鯛
big-eye scad 白鰺；脂眼凹肩鰺
big-eye tuna 大目鮪；大眼擬金槍魚
big-eyed flathead 大眼桶
big-eyed mackerel 大眼鯖魚
big-mouthed flounder 大口鰈
big-mouthed sole 大口鰨
big shellbark hickory 光皮山核桃
big stem jersey 粉質甘薯
big stem mustard 大心芥菜
bigarade 酸橙 [= sour orange]
bigaradier 酸橙樹
bighead carp 黑鰱；大目鰱
bilberry 山桑；烏嘴莓；越橘莓
bile 膽汁
bile acid 膽汁酸
bile cyst 膽囊
bile medium 膽汁培養基
bile pigment 膽色素
bile salt 膽鹽
bilifuscin 膽褐素
bilirubin 膽紅素
biliverdin 膽綠素
bill of fare 菜單
biltong 乾肉條
bin 穀倉
bin-burned grain 倉燜穀物
bin dryer 箱型乾燥機
binder 黏結劑；黏合劑；包紮工具
binder flour 黏合用麵粉
binding agent 黏合劑
binding machine 包紮機；捆紮機
bino 棕櫚燒酒
bioassay 生物鑑定；生物試驗
biocatalysis 生物催化作用
biocatalyst 生物催化劑
biochemical oxygen demand 生化需氧量
biochemical oxygen demand loading 生化需氧負荷量
biochemical reaction 生化反應
biochemistry 生物化學
biochrome 生物色素
biocide 殺生物劑
biocolloid 生物膠體
biodegradability 生物降解性
biodegradable 生物可降解的；生物能分解的
biodegradation 生物降解作用
bioflavonoid 生物類黃酮
biohazard 生物危害
biolipid 生物脂
biological activity 生物活性
biological analysis 生物分析
biological assay 生物試驗
biological calcium carbonate 生物碳酸鈣
biological nitrogen fixation 生物固氮〔作用〕
biological oxygen demand 生物需氧量
biological oxygen demand loading 生物需氧負荷量
biological product 生物產品
biological value 生物價；生理價值
biology 生物學
biolysis 生物分解〔作用〕
biomass 生物量
bioreactor 生物反應器
biorisation 噴霧式巴氏殺菌器法
biorisator 噴霧殺菌器
bios 生物活素
biose 二〔碳〕醣；乙醣 [= diose]

biosynthesis 生物合成
biosystem 生物系統
biotechnology 生物技術；生物科技
biotin 生物素；維生素H [= antiavidin]
biotoxin 生物毒素
biphenyl 聯苯
birch beer 樺木汽水；樺樹汽水
birch seed 樺子
birch tar oil 樺焦油
birchleaf pear 堂梨
bird cherry 稠李
bird grape 烏葡萄
bird pepper 鳥椒；野辣椒 [= bush pepper]
bird's nest 燕窩
birlie 長形麵包；小麵包
biscuit 餅乾；比思吉；麵包乾
biscuit dough 餅乾麵糰
biscuit oven 餅乾烤爐
biscuit paste 餅乾麵糰
biscuit tin 餅乾罐
bishop 橘味葡萄酒；主教飲料
bisk 海鮮濃湯；濃湯 [= bisque]
bisque 海鮮濃湯；濃湯 [= bisk]
bisque ice-cream 果仁餅乾冰淇淋
bistort 拳參
biting taste 辛辣味
bitter 苦的；苦味
bitter acid 苦味酸
bitter almond 苦杏仁
bitter almond oil 苦杏仁油
bitter cassava 苦樹薯；苦根；參茨
bitter cherry 苦櫻桃
bitter chocolate 不含糖巧克力〔原料〕；可可漿
bitter cress 碎米薺
bitter gourd 苦瓜；涼瓜 [= bitter melon; balsam pear]
bitter herb 苦青草；苦草
bitter melon 苦瓜；涼瓜 [= bitter gourd; balsam pear]
bitter orange 苦橙；西班牙苦橙
bitter orange oil 苦橙油
bitter peptide 苦肽
bitter salt 瀉鹽 [= Epsom salt]
bitter wood extract 苦木抽提物
bittern 苦滷；鹽滷
bitterness 苦味
bitternut 山核桃
bitterroot 苦根馬齒莧
bitters 苦味物；苦啤酒；苦味酒
bittersweet chocolate 半甜巧克力
biuret 雙〔縮〕脲；縮二脲
biuret reaction 雙〔縮〕脲反應；縮二脲反應
bixin 胭脂樹紅色素；胭脂樹橙
blacan 蝦醬；蝦膏；發酵蝦醬 [= blachan; shrimp paste]
blachan 蝦醬；蝦膏；發酵蝦醬 [= blacan; shrimp paste]
black abalone 黑鮑
black aspergillus 黑麴菌
black bean 黑豆；烏豆；豇豆
black bean red 黑豆紅
black beer 黑啤酒
black berry 歐洲黑莓
black bream 棘鯛；黑鰭鯛
black butterfish 烏鯧
black carp 青魚
black caviar 黑魚子醬
black cod 裸蓋魚
black coffee 黑咖啡；純咖啡；齋啡
black conger eel 灰康吉鰻
black cowpea 黑豇豆
black croaker 黑姑魚
black currant red 黑加侖紅
black-eyed bean 豇豆；牛〔眼〕豆；黑眼豆

black-eyed pea 豇豆 [= cowpea]
black fish 黑巨頭鯨；產卵後鮭魚
black fungus 黑木耳
black glutinous rice 黑糯米
black gram 小豆；黑綠豆
black hickory 馬看山核桃
black house 煙燻室
black jack 焦糖
black koji 黑麴
black locust 洋槐
black malt 黑麥芽
black marlin 黑旗魚；黑槍魚
black meat 黑變肉
black molasses 廢糖蜜
black morel 高羊肚菌
black moss 髮菜
black mould 黑麴黴；黑黴
black mouth croaker 黑鰔；黑口
black mulberry 黑桑
black mullet 鯔；烏魚；烏仔魚；烏頭
black mustard 黑芥菜；黑芥子
black pepper 黑胡椒
black pepper oil 黑胡椒油
black perch 黑河鱸
black pomfret 黑鯧；烏鯧
black porgy 黑鯛 [= black sea bream]
black pudding 血香腸；黑布丁 [= black sausage]
black radish 黑蘿蔔
black raspberry 黑色覆盆子；美國黑莓；糙莓；黑樹莓 [= blackcap]
black salsify 黑皮波羅門參
black sausage 血香腸；黑布丁 [= black pudding]
black scraper 黑面魨；綠鰭馬面魨
black sea bream 黑鯛 [= black porgy]
black sesame (seed) 黑芝麻
black shell 珠母貝
black spearfish 黑槍魚；黑皮旗魚
black spotted grouper 鑲點石斑；小點石斑
black staining polypore 大部份蓋菌（蕈類）
black striped snapper 畫眉笛鯛；縱帶笛鯛；金雞魚
black sugar 紅糖；黑糖
black tea 紅茶
black treacle 廢糖蜜
black wrack 墨角藻
blackbass 黑鱸
blackberry 黑莓
blackcap 黑色覆盆子；美國黑莓；糙莓；黑樹莓 [= black raspberry]
blackcap raspberry 黑刺莓；糙莓；黑樹莓
blackcock 黑色雄松雞；黑雷雞
blackcurrant 黑醋栗；黑加侖子；黑茶藨子
blackening bolete 粉末牛肝菌
blackstrap 蘭蜜酒；廢蜜紅〔葡萄酒〕
blackstrap molasses 廢糖蜜；赤糖糊
blackthorn 黑刺李；金合歡
bladder 膀胱；氣泡
bladder wrack 泡葉藻
blade 上肩胛肉；牛肩肉
blade agitator 翼式攪拌器；槳式攪拌器
blade meat 肩部肉；前肩肉
blade mixer 槳式混合器
blade steak 肋骨牛排
blade stirrer 翼式攪拌器；槳式攪拌器
bladebone 肩胛骨
blanch(ing) 殺菁；燙漂；熱燙；漂白；預煮
blancher 殺菁劑；預煮機
blancmange 牛奶凍；布丁
bland diet 溫和膳食；無刺激膳食；清淡膳食

bland flavo(u)r 清淡風味
blank 空白；對照
blank taste 淡而無味
blank test 空白試驗
blanquillo 日本馬頭魚；日本方頭魚
blast drier 氣流乾燥器
blast freezer 氣流冷凍器
Blastomyces 芽生菌
blastomycosis 芽生菌病
blastospore 芽生孢子
Blaw-knox instantizer 布勞氏造粒機
Blaw-knox process 布勞氏〔即溶奶粉〕製造
bleached fat 脱色脂肪
bleached flour 漂白麵粉
bleached oil 脱色油
bleached starch 漂白澱粉
bleacher 漂白裝置；漂白劑
bleaching 漂白；脱色
bleaching agent 漂白劑
bleaching clay 脱色〔用黏〕土；漂白〔用黏〕土
bleaching earth 脱色土；漂白土
bleaching liquor 漂白液
bleaching powder 漂白粉
bleed 放血；放出；出血
bleeder 放氣閥；泄放閥；放血工
bleeding 放血；出血；麵糰泄氣
bleeding bread 泄氣的麵包
bleeding conveyor 放血用的傳送帶
blend 調合；混和；調合物
blended juice 調合果汁；混合果汁
blended seasoning 調合的調味料
blender 攪拌器；攪切器；搗碎器 [= blendor]
blending 混合；調合；混和
blendor 攪拌器；攪切器；搗碎器 [= blender]
blet 水果過熟腐爛；腐爛
bleu cheese 藍芝士；藍乾酪；青紋乾酪 [= blue cheese]
blewits 青紫色食用傘菌；藍紫洋菇
blind 無餡大餅；無孔眼的乾酪
blind roaster 閉式烤爐
blind test 盲檢試驗
blintze 薄烙餡餅；薄卷餅
blister 小牡蠣；水泡；起氣泡
blister pack 套蓋包裝；泡鼓包裝；鼓泡塑料襯墊包裝
blithemeat 歡樂食物 (兒童生日會)
bloat can 膨罐
bloater 醃燻鯡魚；醃燻鯖魚
bloch 狐鰹
block chocolate 巧克力塊；塊裝朱古力
block cutter 切塊機
block ice 冰塊；塊冰
block milk 磚乳；壓塊奶粉
block moulding machine 壓塊機
block quick freezing (BQF) 成塊急速冷凍法
block salt 塊鹽
block sugar 方糖 [= cube sugar; cut sugar]
blockwurst 方形香腸
blood agar 血液洋菜培養基；血瓊脂；血平板瓊脂
blood albumin 血清蛋白；血白蛋白
blood cholesterol 血膽固醇
blood cockle 灰蚶；血蚶
blood meal 血粉
blood medium 血液培養基
blood orange 血橙；紅橙
blood plasma powder 血清粉
blood powder 血粉
blood sausage 血腸
blood serum 血清
blood spot 血斑點
blood sugar 血糖

blood tellurite medium 血亞碲酸鹽培養基
blood vessel 血管
Bloody Mary 紅瑪莉酒；血腥瑪莉〔雞尾酒〕
bloody milk 血〔色〕乳
bloody stool 血便
blotter press 壓濾器
blow-up pan 蒸氣攪拌鍋
blowfish 東方魨屬；鯖兔河魨；栗色河魨
blowing agent 發泡劑
blown can 膨罐
blown tin 膨罐
blubber 鯨脂
blue cheese 藍芝士；藍乾酪；青紋乾酪 [= bleu cheese]
blue crab 藍蟹；青蟹
blue fish 鯥魚
blue footed scaly tooth 粗糙肉齒菌(蕈類)
blue-green algae 藍綠藻
blue green funnel cap 香杯傘(蕈類)
blue herring 美洲鯡
blue legs 偽裝香蘑(蕈類)
blue marlin 黑槍魚；箕作氏槍魚
blue mishima pufferfish 青縢
blue mo(u)ld 青黴
blue pea 藍豌豆
blue rimmed liberty cap 變暗藍裸蓋菇
blue shark 鋸峰齒鮫；藍錐齒鯊
blue-striped runner 紡錘鰤
blue tilapia 奧利亞吳郭魚
blue whale 藍鰮鯨；藍鯨
blueback salmon 紅鮭魚；紅大麻哈魚
blueberry 藍莓；紫黑醬果
bluebottle 青蠅；矢車菊
bluefin tuna 藍鰭鮪；金槍魚
bluefish 蘭勃舵魚；美洲青魚
bluegill sunfish 藍鰓魚；大鰭鱗鰓太陽魚
bluestem wheat 青杆小麥
bluewurst sausage 血香腸
BMR (basal metabolic rate) 基礎代謝率
boar 公豬；野豬
board 餐桌；膳食費用；伙食
boarding-out 外膳
boardinghouse 供膳宿舍
bobbio (sausage) 博多奧香腸；筒管式香腸
bobwhite 北美鶉
bock (beer) 巴克啤酒；博克啤酒
bockwurst sausage 博克香腸；牛犢肉小香腸；德國粗香腸
body flanged 罐身捲緣；罐緣
body icing 包冰衣；加細冰屑
body joint 罐身接縫
bodying 質感化
bodying agent 質感化劑；增稠劑
bogberry 蔓越桔；歐洲越桔；水越桔
bohea 武夷茶
Bohemian rye 波希米亞黑麥
Bohemian sausage 波希米亞香腸
boil 煮沸；沸騰
boil down 煮濃；蒸發濃縮
boil-in-bag food 蒸煮袋食品；袋煮食品
boilable pack 可〔蒸〕煮包裝
boiled chicken 白斬雞；白切雞
boiled dinner 水煮餐
boiled egg 煮蛋；帶殼水煮蛋
boiled food 水煮食品
boiled ham 熟火腿
boiled meat 白切肉；清燉肉
boiled oil 熟油
boiled sausage 水煮香腸
boiled sweet 硬糖〔果〕；熬煮硬糖 [= hard candy]

boiler 鍋爐；蒸煮器
boiler compound 鍋爐防垢劑
boiler room 鍋爐房
boilery 煮熬器；煮熬場；熬鹽場
boiling bag 蒸煮袋；可煮袋
boiling down 煮濃
boiling fermentation 劇烈發酵；沸騰狀發酵
boiling pan 蒸煮鍋；熬糖鍋
boiling point (b.p.) 沸點
boiling starch 煮熟澱粉
boiling water 開水；沸水
bolar blade steak 牛肩肉排
bolete 牛肝菌
boletus 牛肝菌
Bollman extractor 博爾門萃取器；博爾曼抽提器
bologna (sausage) 伯樂納腸；波羅納香腸；波洛尼亞香腸
bolter 篩粉機；篩子
bolting cloth 篩布
bolting machine 篩選機；篩粉機
bolting mill 篩磨機
bolting reel 轉筒篩；圓篩
bomb calorimeter 彈式熱量計
bombay duck 印度鐮齒魚；龍頭魚；水晶龍頭魚
bombe 邦比甜品；瓜形甜點心
bonavist bean 鵲豆；娥眉豆
bonbon 棒棒糖；小糖果；夾心糖
bonding agent 黏合劑
bone 骨
bone ash 骨灰
bone cleaner 去骨機；剔骨機
bone dry air 乾透空氣
bone dust 骨粉
bone-in 帶骨的；有骨的
bone-in ham 帶骨火腿
bone marrow 骨髓
bone meal 骨粉
bone phosphate 骨質磷酸鹽
bone tankage 骨肉粉；畜骨下腳料
boned 剔骨的；去骨的
boned meat 剔骨的肉；去骨的肉
boneless 無骨的；去骨的
boneless ham 去骨火腿；無骨火腿
boneless meat 去骨肉；剔骨肉
boning 去骨
bonito 鰹〔魚〕
bonito shark 灰鯖鯊
bonitonic acid 鰹酸
bony 多骨的
bony fish 多刺魚；多骨魚
borage 琉璃苣
borate 硼酸鹽
borax 硼砂
Bordeaux 波爾多葡萄酒
border bean 竹豆
borecole 羽衣甘藍
boregata 多綫魚
boric acid 硼酸
borlotto bean 博羅特豆
borneol 龍腦；冰片；莰醇 [= bornyl alcohol]
bornyl acetate 乙酸龍腦酯
bornyl alcohol 龍腦；冰片；莰醇 [= borneol]
boron 硼
boron-silicate glass 硼矽玻璃
borsch 羅宋湯
Boston bean 波士頓豆
Boston brown bread 波士頓褐饅頭；波士頓黑麵包
Boston butt 豬肩背肉；豬肩胛肉
boswellia 乳香
botany 植物學
bottle block 瓶塞
bottle cap 瓶蓋

bottle capping machine 瓶蓋機；封瓶蓋機
bottle corking machine 瓶塞機；封瓶塞機
bottle feeding 奶瓶餵養
bottle filler 裝瓶機
bottle filling machine 裝瓶機 [= bottling machine]
bottle food 瓶裝食物
bottle gourd 葫蘆；扁蒲
bottle gripper 夾瓶器
bottle opener 開瓶器；開瓶匙
bottle pasteurization 瓶內巴斯德殺菌
bottle pedestal 瓶托；托瓶盤
bottle washing machine 洗瓶機
bottled juice 瓶裝果汁
bottled milk 瓶裝乳；瓶裝牛奶
bottler 裝瓶機；灌瓶機
bottling 裝瓶；灌瓶
bottling machine 裝瓶機 [= bottle filling machine]
bottom-fed evaporator 底部供液蒸發器
bottom fermentation 底部發酵
bottom fish 底棲魚類 [= ground fish]
bottom seam 罐底捲封
bottom yeast 底部酵母；底層發酵酵母
botulin 肉毒桿菌毒素；肉毒梭菌毒素 [= botulinum toxin; botulinus toxin]
botulinal toxin 肉毒桿菌毒素；肉毒梭菌毒素
botulinogenic strain 肉毒桿菌毒素菌株；肉毒梭菌毒素菌株
botulinum 肉毒桿菌；肉毒梭菌
botulinum toxin 肉毒桿菌毒素；肉毒梭菌毒素 [= botulin]
botulinus 肉毒桿菌；肉毒梭菌
botulinus toxin 肉毒桿菌毒素；肉毒梭菌毒素 [= botulin]
botulism 肉毒中毒；肉毒桿菌中毒症；肉毒梭菌中毒症
bouchee 小肉餡餅；一口酥；一口食物
boudin 豬血腸
bouillabaisse 濃味魚羹；魚湯；濃味魚肉湯
bouillon 肉汁清湯；肉湯；羹
bouillon agar 肉汁洋菜〔培養基〕
bouillon medium 肉汁培養基
bouillon powder 肉湯粉；濃湯粉
bound water 結合水
boundary layer 界面層
boundary zone 界面區；界面帶
bouquet 酒香；花束
bouquet garni 香草紮；香辛料包
Bourbon (whisky) 波本威士忌；波旁威士忌
Bourbon tea 波本茶；波旁茶
bourbonal 波旁醛；乙基香蘭素；乙基香草醛 [= ethyl vanillin]
bovine 牛的
bovine serum albumin 牛血清白蛋白
bovine spongiform encephalopathy (BSE) 瘋牛症
bow tie pasta 大蝴蝶粉；蝴蝶形通心粉 [= farfalla；farfalle]
bowel 腸
bowfin 弓鰭魚
bowl 碗；缽
bowl cutter 轉盤斬拌機
bowl mill 碗磨
box curing 裝箱醃製
box dryer 箱型乾燥機
box in box 雙重紙盒
box press 箱型壓榨機
boxthorn 枸杞 [= matrimony vine]
boxty bread 馬鈴薯麵包
boysenberry 伯森莓；博伊森樹莓；波森莓

bozzoli 繭狀通心粉
Brabender unit 布氏單位；布拉班德爾單位
bracer 興奮性飲料
braciola 酒燴肉卷
bracken 蕨菜；過溝菜蕨
brackish 微鹹的；有鹽味的
bract 苞片；苞葉
bradyphagia 吞食徐緩〔症〕
braided ring 麻花狀麵包圈
brain 腦
braise 煎燉；炖 [= braize]
braised 紅燒的；煎燉的；炖熟的
braiser 鍋；鍋子
braising 煎燉；炖
braize 煎燉；炖 [= braise]
braizer 鍋；鍋子
bran 麩；糠；麥麩；米糠
bran bread 麩皮麵包
bran coat 粗糠；穀糠；麥皮
bran shorts 細麩皮；麩皮飼料
branched chain amino acid 支鏈胺基酸
branched chain reaction 支鏈反應
brand name 商標名稱；牌子
brandling 幼鮭〔魚〕
brandy 白蘭地酒
brandy eggnog 白蘭地蛋酒
brandy glass 白蘭地酒杯
brandy mint 薄荷；薄荷油；薄荷糖
brandy snap 薑餅；白蘭地酒小脆餅
brandyball 白蘭地糖果
brank 蕎麥
branny 有麩皮的；似糠的；似麩皮的
branquillos 紅馬頭魚
brasserie 啤酒屋；啤酒店
Brassica 芥菜類
brassicasterol 菜子固醇
bratwurst (sausage) 布拉德香腸；德國式油煎小香腸
braunschweiger 德國布氏肝腸；德國式燻肝香腸；德國布勞恩施魏克香腸
brause 碳酸飲料；汽水
brawn 醃製的野豬肉；豬頭皮凍；雜碎肉凍
brazil nut 巴西果；巴西栗
brazilian prawn 巴西對蝦
bread 麵包；麵包製品
bread and cheese 便餐
bread box 麵包箱
bread flour 麵包〔用〕麵粉
bread grain 麵包穀物
bread improver 麵包改良劑
bread loaf 整條麵包
bread mo(u)ld 麵包黴；麵包烤模
bread-nut tree 麵包樹；麵包果樹 [= breadfruit tree]
bread plate 麵包盤
bread shell 麵包皮〔殼〕
bread slicer 麵包切片機
bread stick 麵包棒；法式的棒形脆麵包
bread tin 麵包烤模
bread wheat 麵包〔用〕小麥
bread yeast 麵包酵母
breadcrumb 麵包糠；麵包屑
breaded 裹麵包屑後烹煮
breaded product 裹麵包屑製品
breadfruit 麵包果；麵包樹
breadfruit tree 麵包樹；麵包果樹 [= bread-nut tree]
breading 裹麵包屑
breadmaking 麵包製造；做麵包
breakfast 早餐；早點
breakfast cereal 早餐穀類食品
breakfast gruel 早餐稀粥；早餐粥品
breast 胸肉；胸部
breast and flap 羊胸肉
breast cut 胸肉；胸排；排骨

breast milk 母乳
breastfeeding 母乳育嬰
bree 湯；浸汁；清湯
breed 品種；繁殖；育種
breeding cow 種母牛
brei 漿；黏稠物
breve 貝菲咖啡
Brevibacterium 短桿菌屬
Brevibacterium acetylicum 醋酸短桿菌
Brevibacterium lactofermentus 乳酸發酵短桿菌
brew 釀造；釀酒；沖煮咖啡；沏茶
brewage 釀造；釀造酒
brewer 釀造業；釀造者；釀造師；啤酒師；沖煮用具
brewers' barley 啤酒用大麥
brewers' barley malt 啤酒用大麥芽
brewers' grain 啤酒粕；啤酒麥芽糟
brewers' grit 啤酒用碎穀；啤酒用粗碾穀粒
brewers' malt 啤酒用麥芽
brewers' wort 啤酒用麥芽汁
brewers' yeast 啤酒酵母；釀造啤酒用之酵母 [= beer yeast]
brewery 啤酒廠；釀酒廠 [= brewhouse]
brewhouse 啤酒廠；釀酒廠 [= brewery]
brewing 沏；釀造；沖煮〔咖啡〕
brewis 肉湯；泡湯的麵包
brewmaster 啤酒釀造師；釀酒技師
brewster 釀造者；釀造師
brick cheese 美國磚形乾酪；美國磚形芝士
brick sugar 方糖
brick tea 茶磚；磚茶 [= tea brick; tile tea]
bride cake 結婚蛋糕 [= wedding cake]
brie cheese 法國布里乾酪；法國布里芝士
bright beer 淺色啤酒
brill 菱平魚
brilliant alizarin blue 亮茜素藍
brilliant black 黑色素；亮黑（食用色素）
brilliant black BN 黑色素；亮黑BN（食用色素）
brilliant blue FCF 食用藍色一號；亮藍FCF
brilliant blue FCF aluminium lake 亮藍鋁色澱
brilliant blue FCF lake 食用藍色一號色澱
brilliant green lactose bile broth 亮綠乳糖膽汁培養液
brine 鹽水；食鹽水
brine cure 鹽水醃漬
brine cured 鹽漬的
brine flotation 鹽水浮選〔法〕
brine freezing 鹽浸冷凍；鹽水凍結
brine gauge 鹽水比重計
brine pack 鹽水充填包裝
brine pickle 鹽醃液
brine salting 鹽〔水〕漬；濕法醃漬
brined food 鹽漬食物
brined meat 鹽水醃肉；鹽漬肉類
brined vegetable 鹽漬蔬菜；鹹菜
briner 鹽漬桶
brininess 鹽度；鹹度
brining 鹽水漬
brinjal 茄子 [= eggplant; aubergine; garden-egg]
brinjaul 茄子
brinsen cheese 布林森乾酪；布林森芝士
brioche 乳酪蛋捲；奶油蛋捲；奶油糕點
briocke 奶油圓球蛋糕
brisk taste 濃〔厚〕味；時新風味
brisket 胸肉；牛胸肉；前胸肉

brisket end 胸肉尖
brisket navel end 胸腹肉
brisket point end, deckle-off 前胸肉；牛腩
brisling 黍鯡；小鯡魚
brit(t) 海洋小甲殼動物；小鯡魚；幼鯡魚；海中微生物
British compounds 含酒精的食物
British Thermal Unit (BTU) 英國熱量單位
brittle 酥糖；鬆脆薄片糖
brittleness 脆性；脆度
Brix 布里〔克氏〕糖度；白利糖度
Brix acid ratio 布里糖酸比；糖酸比
Brix degree 布里糖度；白利糖度
Brix hydrometer 布里比重計；布里糖度計；白利比重計；白利糖度計 [= Brix scale hydrometer]
Brix scale 布里糖度；白利糖度
Brix scale hydrometer 布里比重計；布里糖度計；白利比重計；白利糖度計 [= Brix hydrometer]
Brix spindle 布里糖度計；白利糖度計
broach 鐵叉；烤肉鐵叉
broad bean 蠶豆
broad bran 粗麩；粗糠
broad whitefish 寬鼻白鮭
broadbill swordfish 箭魚
broccoli 西蘭花；花椰菜；青花菜；椰菜；綠花椰菜
brochan 麥片粥；麥片糊
brochette 烤肉串；小串烤肉
broil 烤；焙
broiled eel 烤鰻
broiled meat 烤肉
broiler 小肉雞；肉雞；燒烤用具；烤焙用具
broiling 燒烤；烤 [= grilling]
broken tea 茶末；茶葉碎屑
brokens 破粒；碎米；碎屑
broma 固體食物；去油可可豆
bromate 溴酸鹽
bromatology 食品膳食學；食物學；食品學；飲食學
bromelain 鳳梨蛋白酶；菠蘿蛋白酶 [= bromelin]
bromelin 鳳梨蛋白酶；菠蘿蛋白酶 [= bromelain]
brominated flour 溴化麵粉
brominated vegetable oil 溴化〔植物〕油
bromocresol green 溴甲酚綠
bromocresol purple 溴甲酚紫
bromophenol blue (BPB) 溴酚藍
bromophenol red 溴酚紅
brook trout 美洲紅點鮭；溪紅點鮭
Brookfield viscometer 布氏黏度計；布魯克菲爾德回轉式黏度計
brose 燕麥粥；麥片湯；麥片粥
broth 肉湯；肉汁
broth agar 肉汁洋菜〔培養基〕；肉湯瓊脂〔培養基〕
broth culture 肉汁培養；肉湯培養
broth medium 肉汁培養基；肉湯培養基
brown algae 褐藻；褐色海藻
brown betty 蘋果烤布丁；褐色貝蒂布丁
brown birch scaber stalk 褐疣柄牛肝菌
brown bread 黑麵包；黑麥〔小麥〕麵包
brown colo(u)r 褐色著色劑；棕色著色劑
brown fish meal 褐魚粉
brown FK 棕色FK
brown grits 糙米碎屑
brown HT 棕色HT
brown malt 焦麥芽
brown matsutake 靴口蘑〔蕈類〕

brown rice 糙米 [= cargo rice; hulled rice；husked rice]
brown rot 褐腐病
brown roux 棕色油炒麵粉；褐色炒麵粉
brown sauce 棕色調味醬
brown seaweed 褐藻；褐色海草；褐色海藻
brown shorts 粗米糠；粗麥麩
brown slab sugar 片糖；黃片糖
brown sugar 黃糖；紅糖；赤〔砂〕糖；黑糖
brown sugar slab 片糖；黃片糖
brown trout 褐鱒；斑鱒
brownie 堅果巧克力糕；核仁巧克力餅；胡桃巧克力小方餅
browning 褐變
browning reaction 褐變反應
brucella 布魯氏桿菌
brucellosis 布〔魯〕氏桿菌病
brunch 早午餐
brush-toothed lizard fish 錦鱗蜥魚；狗母魚
brussel sprout 抱子甘藍；球芽甘藍；北京芽菜；芽甘藍
brut 低糖的酒；極乾香檳酒；原酒
BTU (British Thermal Unit) 英國熱量單位
bub 烈性麥芽酒；麥芽汁
bubble (chewing) gum 泡泡糖；吹波糖；泡泡口香糖
bubble and squeak 快餐菜；炒菜肉
bubble viscometer 氣泡黏度計
bubbly 泡泡酒；香檳酒；多泡的
bubbly-jock 雄火雞
buccan 烤肉
bucellas 布散拉斯葡萄酒
buchu leaf oil 布枯葉油
buck 公羊；公兔；雄鹿；灰水；鹼液
buckbean 睡菜
buckled can 凸罐
buckram 軟殼蟹
buckwheat 蕎麥
buckwheat flour 蕎麥粉
bud 芽；蓓蕾
buddeized milk 消毒的牛乳(添加過氧化氫)
budding 發芽
buff wax cap 草地濕傘(蕈類)
buffalo 水牛；美洲野牛
buffalo berry 水牛莓
buffer 緩衝器；緩衝劑；緩沖劑；緩衝液
buffer capacity 緩衝能力
buffer solution 緩衝液
buffet 自助餐；餐具櫃
buffet lunch 自助午餐；簡易午餐；小賣部便餐
buffet supper 自助晚餐
bugloss 牛舌草
buhr mill 石磨
built-up tin 三片罐；組合罐
bulb 球根；球形物；鱗莖
Bulgarian (sour) milk 保加利亞〔發酵〕酸乳
bulged can 凸罐
bulgood 酵母
bulgur 碎小麥
bulgur 布格麥食；碾碎的乾小麥；焦乾碎小麥
bulimia 貪食症；易餓病；食慾過盛
bulk goods 大包裝產品；散裝物品
bulk pack 散裝；大包裝
bulk packing 散裝
bulking agent 增積劑；增體劑；填充劑
bulking filler 增積劑
bull 公牛；未閹公牛

bull calf 公犢；小公牛
bull frog 牛蛙
bulletmackerel 平花鰹；扁頭鰹；圓舵鰹
bullhead coffee 帶莢咖啡豆
bullnose bass 鱸滑石斑
bullnose pepper 青椒；甜椒 [= sweet pepper]
bullock 閹牛
bullock' s heart 牛心梨
bully (beef) 醃牛肉；罐頭牛肉；鹹牛肉
bumbo 蘭酒；潘趣飲料；香糖酒
bun 小〔圓〕麵包；饅頭
buncuit 餡料夾層的甜點心
bunderfleish 瑞士牛肉乾
bung 豬直腸；牛盲腸；塞子
bung cap 直腸頭；大腸頭
bunghole 桶口；桶孔
burbot 淡水鱈；江鱈
burdock 牛蒡 [= lappa]
burger 漢堡；碎肉夾餅
burghul 碎小麥
Burgundy 柏根迪葡萄酒；布爾戈尼葡萄酒(法國)
burgundy slime head 皺環球蓋菇
Burgundy yeast 柏根迪酵母；布爾戈尼酵母
burmese 黃牛
burned taste 焦味
burnet 地榆
burnt ammonium alum 無水硫酸銨鋁；燒銨明礬
burnt flavo(u)r 焦〔風〕味；燒焦氣味
burnt sugar 焦糖；焦化糖
burnt sugar colo(u)r 焦糖色；醬色 [= caramel colo(u)r]
busa 土耳其發酵乳；土耳其奶酒
bush bean 矮生菜豆
bush fruit 灌木水果
bush pepper 鳥椒；野辣椒 [= bird pepper]
bush pumpkin 矮生南瓜；密生西葫蘆
butadiene-styrene rubber 丁苯橡膠
butan-1,3-diol 1,3-丁二醇
butan-3-one-2-yl butyrate 丁酸-3-丁酮-2-酯
butane 丁烷
butanoic acid 丁酸
butanol 丁醇 [= butyl alcohol]
2-butanol 仲丁醇
2-butanone 2-丁酮
butcher 屠宰工；屠夫；肉販
butcher bull 肉用公牛；肥豬
butcher cattle 肉用牛；育肥牛；肉用牲畜
butcher cow 肉用母牛；菜牛
butcher knife 屠刀；切肉刀
butcher' s breast 待宰家畜；肉用家畜
butcher' s cut 新鮮切割的肉塊
butcher' s meat 市販肉
butcher' s paper 包肉紙；防滲血紙
butchery 屠宰業；肉販；屠〔宰〕場
butoxide 農藥增效醚
butt 大酒桶；肩背肉；河鰈
butt square cut 方切後腿肉；後腿肉
butter 乳酪；黃油；白脫；奶油；牛油
butter bean 牛油豆；棉豆；利馬豆
butter bread 白脫麵包；乳酪麵包；奶油麵包
butter cake 白脫餅；乳酪餅；奶油蛋糕
butter churn 乳酪攪製器；奶油攪拌器
butter colo(u)r 白脫色素；乳酪色素；奶油色素
butter cookie 小甜酥餅；牛油曲奇餅
butter cream 白脫用鮮奶油；乳酪用鮮奶油；加糖的奶油漿

butter flavo(u)r 奶油香精
butter-fruit 牛油果；鱷梨；油梨；酪梨 [= avocado; alligator pear]
butter hardness tester 奶油硬度計
butter horn 奶油〔酥皮〕角
butter knife 白脱刀；乳酪刀
butter mushroom 黃牛肝菌
butter oil 白脱油；乳脂肪
butter paper 牛油紙；奶油包裝紙；白脱用羊皮紙
butter printer 白脱定型器；奶油定型器；奶油包裝機
butter sauce 乳酪醬；白脱醬；奶油醬
butter serum 乳酪乳清；奶油乳清；白脱乳清
butter starter 乳酪菌元；白脱菌元；奶油發酵劑
butter test bottle 乳脂計
butter whey 乳酪乳清；白脱乳清；奶油乳清
butter yellow 乳酪黃；白脱黃；奶油黃
butterfat 白脱脂；乳酪脂；乳脂肪
butterfish 銀大眼鯧；銀鯧；鯧魚；溜滑的魚
butterfly ray 日本鳶魟；燕仔魴；蝴蝶鯆
butterie 奶油麵包；奶油餅乾
butterine 瑪珈琳；人造乳酪；人造黃油；人造奶油；人造白脱；植物黃油；植物牛油 [= margarine; oleomargarine]
buttermilk 白脱乳；酪乳
butternut 油胡桃；灰胡桃
butternut squash 灰胡桃南瓜
butterscotch 乳酪硬糖；白脱硬糖；奶油硬糖(鹹味)
buttery 酒窖；備餐室；食物室；食品貯藏室
buttery tough shank 乳酪狀金錢菇
buttocks 臀部；臀部肉
button mushroom 蘑菇；鈕子洋菇
button onion 幼嫩洋葱；乳洋葱
butyl acetate 乙酸丁酯
butyl alcohol 丁醇 [= butanol]
butyl butyrate 丁酸丁酯
butyl butyryl lactate 丁酰乳酸丁酯
butyl formate 甲酸丁酯
butyl heptanoate 庚酸丁酯
butyl isobutyrate 異丁酸丁酯
butyl isovalerate 異戊酸丁酯
butyl lactate 乳酸丁酯
butyl phenylacetate 苯乙酸丁酯
butyl phthalyl butyl glycol(l)ate 丁基鄰苯二甲酰基乙醇酸丁酯
butyl *p*-hydroxybenzoate 對羥基苯甲酸丁酯；尼泊金丁酯
butyl rubber 丁基橡膠；異丁橡膠
butyl stearate 硬脂酸丁酯
butylated hydroxyanisole (BHA) 叔丁基對羥基茴香醚；丁基羥基茴香醚；丁基羥基甲氧苯；丁基大茴香醚
butylated hydroxytoluene 二丁基羥基甲苯
butylated hydroxytoluene (BHT) 二丁基羥基甲苯；2,6-二叔丁基對甲酚

butylparaben 對羥基苯甲酸丁酯；尼泊金丁酯
butyraldehyde 丁醛
butyrate 丁酸鹽
butyric acid 丁酸；酪酸
butyric acid bacteria 丁酸細菌
butyric acid fermentation 丁酸發酵
butyrin 三丁酸甘油酯
butyrinase 三丁酸甘油酯酶；乳脂酶
γ-butyrolactone γ-丁內酯；丙位丁內酯
butyrometer 乳脂計；測脂器
by-product 副產品
byne 麥芽
bynin(e) 麥芽醇溶蛋白

C

CA storage (controlled atmosphere storage) 控氣保藏；人工氣候保藏；氣調貯藏；控制大氣貯藏 [= gas storage]
cabardine 麝香
cabbage 甘藍；高麗菜；椰菜；卷心菜；包心菜
cabbage lettuce 結球萵苣；卷心萵苣 [= head lettuce]
cabbage mustard 芥藍〔菜〕；結球芥菜
cabbage roll 高麗菜捲
Cabernet (wine) 紅葡萄酒；解百衲葡萄酒
cabinet dryer 箱式乾燥機；乾燥箱
cabinet pasteurizer 櫃式巴氏滅菌器
caboose 船上的廚房
cacao 可可飲料；可可樹；可可豆；可可粉
cacao bean 可可豆
cacao brown 可可棕(色素)
cacao butter 可可脂；可可油 [= theobroma oil]
cacao husk 可可豆殼皮；可可殼
cacao mill 可可豆碾磨機
cacao nib 熟可可粒；可可豆肉；可可豆粉
cacao oil 可可脂
cacao pigment 可可豆色素
cacao powder 可可粉
cacao press 可可脂壓榨機
cacao red 可可紅(色素)
cacao shell 可可〔果〕殼
cachalot 抹香鯨
cachou 口香〔軟〕糖
cacochylia 消化不良
cacogeusia 劣味
cacotrophia 營養不良
cacotrophy 營養不良
cactus 仙人掌
caddy 茶葉罐；小罐；小盒
cade 杜松；小桶
cadenera orange 卡特尼拉甜橙
cadmium 鎘
caecum 盲腸 [= cecum]
Caesar's mushroom 白橙蓋鵝膏(蕈類)
caesio 褐梅鯛
café 咖啡館；餐館；酒館
café au lait 法國牛奶咖啡
café noir 法式黑咖啡
cafeteria 自助餐館
caffe americano 美式咖啡 [= americano]
caffe con montata 康培娜咖啡 [= caffe con panna]
caffe con panna 康培娜咖啡 [= caffe con montata]
caffe corretto 攙酒義式濃咖啡
caffe latte 拿鐵咖啡 [= latte]
caffe macchiato 馬奇雅圖咖啡 [= espresso macchiato; macchiato]

caffe mocha 摩卡咖啡；穆哈咖啡 [= mocha]
caffe ristretto 威士戚圖咖啡 [= espresso ristretto]
caffeic acid 咖啡酸
caffeidine 咖啡啶
caffeine 咖啡因；咖啡鹼；茶精
caffeine-free 無咖啡因的；無咖啡鹼的
caffeinism 咖啡鹼中毒
caffeol 咖啡醇；咖啡焦油；咖啡〔香〕油
caffeone 咖啡醇；咖啡焦油
cage mill 籠式磨；籠式破碎機
cage oil press 籠式榨油機
cage press 籠式壓榨機
caiman 凱門鱷
cainito 星蘋果
cajeput oil 白千層油
cajuput oil 白千層油
cake 蛋糕；餅；點心；塊狀物；粕
cake bread 蛋糕麵包；小白麵包
cake flour 糕餅用麵粉；細麵粉
cake hoop 活底烤蛋糕模子；活底烤蛋糕箍
cake mix 蛋糕用混料粉；糕餅用盒裝現成材料
cake mo(u)ld 糕餅模具
cake tea 茶磚；磚茶
caked sugar 塊糖；結塊蔗糖
caking 結塊；燒結
caking inhibitor 抗結塊劑
cal (calorie) 卡路里；卡 [= gram calorie]
Cal (Calorie) 大卡；仟卡 [= kilocalorie；kcal；great calorie]
calabash gourd 葫蘆；瓠瓜；扁蒲
calabazilla 南瓜；油瓜
calabrese 花莖甘藍；青花菜；莖椰菜 [= asparagus broccoli]
calamary 槍烏賊
calamite 蘆木
calamondin 四季桔；金橘
calamus oil 白菖油；菖蒲油
calandria evaporator 排管式蒸發器；列管式蒸發器
calciferol 麥角鈣化固醇；鈣化醇；骨化醇；維生素D_2 [= ergocalciferol; vitamin D_2]
calcification 鈣化；石灰化
calcium 鈣
calcium 5'-guanylate 5'-鳥苷酸鈣
calcium 5'-inosinate 5'-肌苷酸鈣
calcium 5'-ribonucleotide 5'-核糖核苷酸鈣
calcium acetate 乙酸鈣；醋酸鈣
calcium acid phosphate 酸性磷酸鈣
calcium alginate 褐藻酸鈣；海藻酸鈣
calcium aluminium silicate 硅酸鋁鈣
calcium ascorbate 抗壞血酸鈣
calcium benzoate 苯甲酸鈣
calcium bisulfite 亞硫酸氫鈣
calcium bromate 溴酸鈣
calcium carbonate 碳酸鈣
calcium carboxymethylcellulose 羧甲基纖維素鈣
calcium caseinate 酪蛋白鈣；酪朊酸鈣
calcium chloride 氯化鈣
calcium citrate 檸檬酸鈣
calcium cyclamate 環己基氨基磺酸鈣
calcium dihydrogen phosphate 磷酸二氫鈣；磷酸一鈣
calcium dihydrogen pyrophsophate 焦磷酸二氫鈣；酸性焦磷酸鈣
calcium disodium EDTA 乙二胺四乙酸二鈉鈣
calcium formate 甲酸鈣
calcium gluconate 葡〔萄〕糖酸鈣
calcium glutamate 谷氨酸鈣

calcium glycerophosphate 甘油磷酸鈣
calcium hydrogen sulfite 亞硫酸氫鈣
calcium hydroxide 氫氧化鈣；熟石灰；消石灰
calcium hypochlorite 次氯酸鈣；漂粉精
calcium iodate 碘酸鈣
calcium lactate 乳酸鈣
calcium lactobionate 乳糖醛酸鈣
calcium malate 蘋果酸鈣
calcium monophosphate 磷酸氫鈣
calcium oxide 氧化鈣；石灰
calcium pantothenate 泛酸鈣
calcium pectinate 果膠〔酯〕酸鈣
calcium peroxide 過氧化鈣
calcium phosphate 磷酸鈣
calcium phosphate, dibasic 磷酸氫鈣
calcium polyphosphate 聚磷酸鈣
calcium propionate 丙酸鈣
calcium pyrophsophate 焦磷酸鈣
calcium saccharate 蔗糖酸鈣
calcium saccharin 糖精鈣
calcium silicate 矽酸鈣；硅酸鈣
calcium sorbate 山梨酸鈣；己二烯酸鈣
calcium stearate 硬脂酸鈣
calcium stearoyl lactylate (CSL) 硬脂酰乳酸鈣
calcium sulfate 硫酸鈣；石膏
calcium tartrate 酒石酸鈣
caldron 大鍋 [= cauldron]
calefacient 發暖劑
calf 小牛；牛犢
calibration 校準
calico 白棉布
calico salmon 鮭魚；大麻哈魚
California anchovy 加利福尼亞鯷；加州鯷
California halibut 北美牙鮃
California sole 加利福尼亞鰨魚；加州鰨魚
calipash 龜背肉
calipee 龜腹肉
Calla 水芋屬
calmus oil 菖蒲油
Caloglossa 絹光紅藻屬；鷓鴣菜屬
calomel 甘汞；氯化亞汞
caloric unit 熱〔量〕單位
caloric value 熱值；卡值
calorie (cal) 卡路里；卡 [= gram calorie]
Calorie (Cal) 大卡；仟卡 [= kilocalorie; kcal; great calorie]
calorie value 熱值；卡值
calorific value 熱值；卡值
calorigenic food 產熱食物
calorimeter 卡計；熱量計
calorimetric bomb 熱量計彈筒；量熱彈
calorimetry 熱析法；量熱法；熱量測定
calorization 鋁化作用
caltrop 菱角；水栗子
caltrop starch 菱角澱粉；菱粉
calvados 法國蘋果白蘭地酒；卡巴度斯蘋果白蘭地酒；美國蘋果白蘭地 [= apple jack]
camel 駱駝
camellia oil 山茶油
camomile 黃金菊；春黃菊；甘菊
camomile oil 黃金菊油；春黃菊油；甘菊油
campesterol 油菜籽固醇；菜油固醇
camphene 莰烯
camphor 樟腦
can 罐頭；罐
can body 罐身
can body blank 罐身板；罐身鐵皮

can body forming machine 罐身成形機；罐身成圓機
can body making machine 罐身成形機；製罐機
can buckling 脹罐；膨罐
can clincher 預封機；假捲封機
can closing 封罐；封蓋
can cover 罐蓋；桶蓋
can dryer 罐〔頭〕乾燥機；空罐乾燥機
can enamel 罐頭塗料
can end 罐蓋
can end cement 罐蓋封口膠
can end curling machine 罐蓋捲邊機；蓋緣機
can end press 沖蓋機
can factory 製罐廠
can filler 裝罐機；罐頭充填機
can filling 裝罐
can flange 罐身捲緣；罐緣
can goods 罐頭食品
can-making 製罐
can opener 開罐器
can plate 馬口鐵皮
can poisoning 罐頭食品中毒
can sealer 封罐機
can sealing machine 封罐機
can seamer 封罐機
can seaming machine 封罐機
can seaming wheel 封罐滾輪
can soldering machine 空罐焊錫機
can straightener 空罐整形機
can stuffer 裝罐機
can washer 洗罐機
can washing machine 洗罐機
canada field 紫花豌豆
canakin 小罐；小杯
canal drier 隧道式乾燥機
canal dryer 隧道式乾燥機
cananga oil 香水樹油；依蘭油；卡南加油
canape 加拿薄點心；有魚肉的烤麵包
canary dextrin 黃〔色〕糊精
Canavalia 刀豆屬
canavalin 刀豆球蛋白
canavanine 刀豆胺酸
cancerogenesis 致癌
candelilla wax 堪帶蠟；小蠟燭樹蠟；小燭樹蠟；坎台里蠟
candicidin 殺念珠菌素；殺假絲菌素
Candida 念珠菌屬；假絲酵母菌屬
Candida albicans 白色念珠菌；白假絲酵母
Candida utilis 高蛋白假絲酵母
candidin 制念珠菌素；制假絲菌素
candied 甜蜜的；糖漬的；蜜餞的
candied fruit 水果蜜餞；蜜餞；果脯
candied peel 果皮蜜餞
candy 糖果；蜜餞；冰糖
candy center 糖果夾心
candy doctor 糖結晶抑制劑
candy floss 棉花糖
candy paper 糖果〔包裝〕紙
candy peel 蜜餞果皮；果皮蜜餞
candy press 糖果壓形機；糖果壓塊機
candy pulling 糖果拉製
candy stick 糖果棒；棒棒糖
candy wrapping 糖果包覆；糖果包裝
candying 包糖；用糖熬煮
cane 甘蔗
cane crusher 甘蔗榨汁機
cane field 甘蔗田；甘蔗園
cane joint 蔗節
cane juice 蔗汁；甘蔗汁
cane mill 蔗糖廠；甘蔗壓榨機；甘蔗輥壓機
cane molasses 甘蔗糖蜜；甘蔗廢蜜
cane plantation 甘蔗園；甘蔗養植場

cane stalk 甘蔗莖
cane sugar 蔗糖
cane syrup 蔗糖漿
cane trash 蔗渣
canelloni 包肉麵皮捲
canister 有蓋罐；小罐
canna 美人蕉
cannabic oil 大麻油
cannabinol 大麻酚；大麻醇
cannabis 大麻 [= hemp]
cannabis seed 大麻子；火麻仁 [= hemp seed]
canned beer 罐裝啤酒
canned fish 罐裝魚；魚罐頭
canned food 罐頭食品；罐裝食品；食品罐頭 [= tinned food]
canned fruit 罐裝水果；水果罐裝
canned fruit juice 罐裝果汁；果汁罐裝
canned goods 罐頭食品；罐裝食品；食品罐頭 [= tinned goods]
canned ham 火腿罐頭；罐裝火腿
canned meat 肉類罐頭；罐頭肉
canned product 罐頭製品
canned soup 罐裝湯；罐頭湯
canned vegetable 罐裝蔬菜；蔬菜罐頭
cannellini 長白菜豆
cannellino bean 坎尼里諾豆
cannelloni 意大利春捲
canner 罐頭製造商；罐頭殺菌鍋
cannery 罐頭工廠；罐頭食品廠
cannery retort 罐頭殺菌釜
cannikin 小罐；小杯；木水桶
canning 罐頭製造；罐藏〔法〕
canning industry 罐頭工業
cannon drying 爆化乾燥
canola oil 低芥酸菜籽油
cantaloup(e) 硬皮甜瓜；糙皮甜瓜；香瓜；哈密瓜；麝香瓜；羅馬甜瓜
canthaxanthin 杏菌紅素；斑蝥黃〔質〕
cap opener 開蓋器
cape gooseberry 燈籠果；好望角醋栗
cape hake 醃鱈魚；狗鱈
capelin 小海魚（香魚科）；毛鱗魚
capellini 意大利細麵；天使髮絲 [= angel hair]
caper 西洋白花菜；西洋白花菜蕾芽；酸豆
capercaillie 松雞；雷鳥 [= capercailzie]
capercailzie 松雞；雷鳥 [= capercaillie]
capillarity 毛細管現象
capillary 毛細管；微絲血管
capillary viscosimeter 毛細管黏度計
capon 閹雞（育肥供食用）
cappelletti 帽狀通心粉
capper 封蓋機；壓蓋機；封口機
cappicola sausage 意大利胡椒香腸；卡畢可拉豬肉香腸
capping 封蓋；封瓶
capping machine 封蓋機；壓蓋機
cappuccino 卡布奇諾咖啡；卡布基諾咖啡；熱牛奶咖啡
cappuccino chiaro 牛奶卡布奇諾咖啡 [= light cappuccino]
cappuccino scuro 黑卡布奇諾咖啡；黑卡布基諾咖啡 [= dark cappuccino]
capric acid 癸酸；羊脂酸 [= decanoic acid]
caprifig 野生無花果；卡必力無花果
caprin 羊蠟；甘油三癸酸酯；癸酸甘油酯
caproic acid 己酸 [= hexanoic acid]
caprylic acid 辛酸 [= octanoic acid]
capsaicin 辣椒素
capsanthin 辣椒紅素；辣椒質
capsanthol 辣椒醇
capsicine 辣椒鹼
capsicum 辣椒；辣椒屬
capsicum oil 辣椒油

capsicum oleoresin 辣椒油樹脂
capsorubin 辣椒紫紅素
capsulating process 膠囊〔包裹〕法
capsule 莢膜；膠囊；蒴果；瓶口套；瓶帽
captive-bolt gun 擊昏棒；擊昏槍
captive bolt pistol 擊昏槍
captive-bolt stunning 擊昏法
carambola 楊桃 [= star fruit]
caramel 焦糖；飴糖；醬色；卡拉蜜糖
caramel colo(u)r 焦糖色；醬色 [= burnt sugar colo(u)r]
caramel malt 焦麥芽；炒焦麥芽
caramelization 焦糖化〔作用〕
caramelized flavo(u)r 焦糖風味
caramelized sugar 焦糖；焦化糖
caraway 葛縷子；姬茴香
caraway oil 香芹油
caraway seed 香芹籽；藏茴香籽；葛縷籽
carbamate 胺基甲酸酯；胺基甲酸鹽
carbamide 尿素
carbohydrase 糖酶
carbohydrate 碳水化合物；糖類
carbolic acid 苯酚；石灰酸
carbon 碳
carbon black 碳黑
carbon dioxide 二氧化碳
carbon dioxide stunning 二氧化碳昏厥法；氣體殺法
carbon monoxide 一氧化碳
carbon tetrachloride 四氯化碳
carbonad 烤肉；烤炙的肉片
carbonatation 碳酸化作用；碳酸鹽法；碳酸飽和 [= carbonation]
carbonate 碳酸鹽；碳酸酯
carbonated beverage 碳酸飲料；充〔碳酸〕氣飲料
carbonated drink 碳酸飲料；充〔碳酸〕氣飲料
carbonated spring 碳酸泉水
carbonated water 蘇打水；碳酸水；汽水
carbonated wine 碳酸酒；汽酒
carbonating 充二氧化碳；碳酸化
carbonation 碳酸化作用；碳酸鹽法；碳酸飽和 [= carbonatation]
carbonator 碳酸化器
carbonic acid 碳酸
carbonic oxide 一氧化碳
carbonization 碳化〔作用〕
carbonizer 碳酸化器
carbonizing 碳化；焦化
carbonyl 羰基
carbonyl-amine reaction 梅納反應；美拉德反應；羰胺反應；胺羰反應 [= Maillard reaction; aminocarbonyl reaction]
carbonyl compound 羰基化合物
carboxyl group 羧基
carboxylase 羧〔基〕酶
carboxylation 羧化〔反應〕
carboxylic acid 羧酸
carboxymethyl cellulose (CMC) 羧甲基纖維素
carboxymethyl starch 羧甲基澱粉
carcass 屠體；胴體
carcass dressing 屠體剖洗；胴體修整
carcass meat 胴體肉
carcass setting 胴體死後僵直
carcass washer 胴體沖洗裝置
carcass weight 屠體重量
carcass yield 屠體收率
carcinogen 誘癌物質；致癌劑；誘癌劑
carcinogenesis 誘癌作用；致癌作用
carcinogenicity 誘癌性；致癌性
cardamom 小豆蔻 [= cardamon]

cardamon 小豆蔻　[= cardamom]
cardamon oil 小豆蔻油
cardiac muscle 心肌
cardiac muscle fiber 心肌纖維
cardiolipin 心磷脂
cardoon 刺菜薊；食用薊；刺棘薊；南歐朝鮮薊
cardy oil 紅花油
cargo rice 糙米　[= brown rice; hulled rice; husked rice]
Carica 番木瓜屬
carica papaya 番木瓜
carmine 洋紅；胭脂紅；胭脂蟲紅〔鋁〕
carminic acid 臙脂蟲酸；胭脂紅酸
carmoisine 偶氮玉紅；藍光酸性紅；淡紅
carnauba wax 巴西棕櫚酸
carniferrin 肉鐵質
carnitine 肉〔毒〕鹼
L-carnitine L-肉鹼；維生素B_t
carnivore 肉食動物；食蟲植物
carnosol 鼠尾草醇
carob 長角豆
carob bean 刺槐豆；角豆　[= locust bean]
carob bean gum 刺槐豆膠；角豆膠　[= locust bean gum]
Carolina rice 卡羅萊納米；巴特納米
carotenal 胡蘿蔔素醛
carotene 胡蘿蔔素；葉紅素；前維生素A　[= carotin]
β-carotene β胡蘿蔔素；貝他胡蘿蔔素　[= beta-carotene]
carotenoid 類胡蘿蔔素
carotenol 胡蘿蔔醇
carotin 胡蘿蔔素；葉紅素；前維生素A　[= carotene]
carotol 胡蘿蔔醇
carp 鯉魚
carpopodium 果柄
carpospore 果孢子；囊果孢子
carrabiose 鹿角菜雙糖；角叉藻二糖
carrageen 鹿角菜；角叉菜；鹿角菜苷　[= carragheen; Irish moss]
carrageenan 卡拉膠；角叉〔菜〕膠；紅藻膠　[= carrageenin; carragheenin]
carrageenin 卡拉膠；角叉〔菜〕膠；紅藻膠　[= carrageenan; carragheenin]
carragheen 鹿角菜；角叉菜；鹿角菜苷　[= carrageen; Irish moss]
carragheenin 卡拉膠；角叉〔菜〕膠；紅藻膠　[= carrageenan; carrageenin]
carrier cradle 殺菌籃〔吊架〕
carrier protein 載體蛋白
carrot 甘筍；胡蘿蔔；紅蘿蔔
carrot seed oil 胡蘿蔔籽油
carthaginian apple 石榴
carthamin 紅花素
carthamus oil 紅花油
carthamus red 紅花紅色素
carthamus yellow 紅花黃〔色素〕
cartilage 軟骨
cartilagin 軟骨蛋白
cartilaginous 軟骨的
carton 紙箱；硬紙盒
carton board 紙板；硬紙板
cartoning machine 紙盒製造機；製紙盒機
cartridge clarifier 砂糖精製器
cartridge filter 筒式過濾器
Carum carvi 葛縷子
carvacrol 香芹酚；香荊芥酚
carved meat 分切肉；分割肉
carveol 香芹醇；頁蒿萜醇
carver 切肉刀
carvier 魚子醬；鱘魚子醬　[= caviar]
carvone 香芹酮；貢蒿萜酮

carvyl acetate　乙酸葛縷酯；乙酸香芹酯
Carya 山核桃屬
caryophyllene　石竹烯；丁香油；丁子香烯
caryophyllin　丁香油素；石竹素
Caryophyllus aromaticus　丁香樹；丁子香
caryopsis　穎果
CAS (Chinese Agricultural Standard)　中國農業標準
casaba　冬甜瓜；加沙巴甜瓜；卡薩巴甜瓜　[= cassaba]
casareccia　卡薩瑞齊亞麵
cascarilla　西印度苦香樹
cascarilla bark oil　卡藜油；香苦木油
cascarilla oil　卡藜油；香苦木油
casease　酪蛋白酶
casein　酪蛋白；乾酪素
casein agar　酪蛋白洋菜〔培養基〕
casein hydrolyzate　酪蛋白水解物
casein whey　酪蛋白乳清
caseinate　酪蛋白鹽
caseinogen　酪蛋白元；酪蛋白原
cashew nut　腰果
casing　腸衣；外殼；裝箱；包裝
casing applier　灌腸嘴
casing fat　腸脂；腸油
casing flusher　洗腸器
casing packaging　腸衣包裝
casing stripper　腸衣剝刮機
casing stripping　腸衣剝刮
casing threading　腸衣穿串
cask　桶；木桶
cask beer　桶裝啤酒
cask fermentation　桶內發酵
cask hoop　木桶箍
cask smell　木桶味
cask stave　桶身木板
cask wine　桶裝葡萄酒
cassaba　冬甜瓜；加沙巴甜瓜；卡薩巴甜瓜　[= casaba]
cassava　木薯；樹薯　[= manioc; manihot]
cassava flour　木薯粉　[= tapioca]
cassava meal　木薯粉
cassava starch　木薯澱粉
cassava vermicelli　薯絲麵
casserole　燉鍋；燉雜燴；砂鍋菜
casserole dish　烘雜燴；砂鍋燉肉什錦
cassia　肉桂；中國肉桂；桂皮　[= Chinese cinnamon; cinnamon]
cassia bark　桂皮；肉桂；中國肉桂
cassia lignea　厚桂皮
cassia oil　肉桂油；桂皮油；中國肉桂油　[= Chinese cinnamon oil; cinnamon oil]
cassie concrete　金合歡浸膏；金合花浸膏
cassina　冬青茶
cassiri　南美的發酵玉米粉食品
cassowary　食火雞
caster　調味瓶；調味料瓶架
castilla　蛋糕
castor　蓖麻；調味〔料〕瓶
castor bean　蓖麻籽　[= castor oil bean]
castor oil　蓖麻油
castor oil bean　蓖麻籽　[= castor bean]
castor seed　蓖麻籽
castor sugar　糖粉；細白砂糖；小粒結晶砂糖
castrate　閹割；去勢
casual dining　休閒餐飲
cat shark　大喙篦鮫；寬紋虎鯊
catabolism　異化作用；分解代謝
catalana　卡特蘭納雞
catalase　過氧化氫酶；觸酶
catalysis　催化〔作用〕；觸酶〔作用〕
catalyst　催化劑；觸酶

catalytic agent 催化劑；觸酶
catalytic reaction 催化反應；觸酶反應
catalyze 催化
catchup 番茄醬 [= catsup; ketchup]
catechin 兒茶素；兒茶酸
catechol 兒茶酚
catecholamine 兒茶酚胺
catecholase 兒茶酚酶
catechu 兒茶；兒茶黑色素
catering 外燴；膳食製備；供應餐食
catering department 餐飲部
caterpillar juice extractor 履帶果汁萃取器
catfish 鯰魚；土刹
catgut 腸線
cathartic 瀉藥
cathepsin 細胞自溶酶；組織蛋白酶
cation 陽離子
cation exchange 陽離子交換
cationic detergent 陽離子清潔劑
cationic surfactant 陽離子表面活性劑
catjang 豇豆；眉豆
catlap 淡飲料；淡茶；稀薄飲料
catostomid fish 胭脂魚
catsup 番茄醬 [= catchup; ketchup]
cattle 牛；家畜
cattle farm 養牛場
cattle raising 牧牛
cattle range 牧場
cattley guava 草莓番石榴
catty 斤
Caucasian milk 高加索酸凝乳
caudal fin 尾鰭
caudle 流質滋補食物；粥湯；麥糊
caul fat 腹脂；網油；網膜脂；花油 [= caul web fat]
caul web fat 腹脂；網油；網膜脂；花油 [= caul fat]
cauldron 大鍋 [= caldron]
cauliflower 椰菜花；花椰菜；花菜
cauliflower fungus 繡球菌(蕈類)
caustic lye 苛性鹼液
caustic peeling 鹼液去皮〔法〕
caustic refining 油脂鹼精煉；鹼煉 [= alkali refining]
caustic soda 苛性鈉；氫氧化鉀
caustic washing 鹼洗
cavatappi 螺絲形通心粉
caviar 魚子醬；鱘魚子醬 [= carvier]
cayenne 辣椒
cayenne pepper 番椒；辣椒；番椒粉；辣椒粉
cecum 盲腸 [= caecum]
cedar leaf oil 雪松葉油；側柏葉油；金鐘柏油 [= thuja oil]
cedar nut 雪松〔堅〕果；松子
cedar oil 洋杉油；雪松油
cedar wood oil 柏木油
cedra 橘皮蜜餞
cedrat 枸櫞；香櫞
cedryl acetate 乙酸柏木酯
celandine 白屈菜
celeriac 根芹菜；塊根芹
celery 西芹；芹菜
celery cabbage 結球白菜；大白菜 [= Chinese cabbage]
celery oil 芹菜籽油
celery seed 芹菜籽
celery seed oil 芹菜籽油
cell culture 細胞培養〔物〕
cell plasm 細胞質
cell tissue 細胞組織
cell wall 細胞壁
cellar 地窖；酒窖；藏酒量；窖藏酒
cellar storage 地窖儲藏；窖藏
cellaret 酒瓶架；酒櫥
cellobiase 纖維二糖酶
cellobiose 纖維二糖

cellodextrin 纖維糊精
cellophane 玻璃紙；賽璐凡；賽璐玢
cellotriose 纖維三糖
celluflour 纖維素粉；纖維粉末
cellular fluid 細胞液
cellulase 纖維素酶
celluloid 賽璐珞
cellulolytic enzyme 纖維素分解酵素
cellulosan 纖維聚糖
cellulose 纖維素
cellulose acetate 醋酸纖維素；乙酸纖維素 [= acetyl cellulose]
cellulose casing 纖維素腸衣
cellulose gel 纖維素凝膠；微晶纖維素；結晶纖維素
cellulose gum 纖維素膠；羧甲基纖維素 [= carboxymethyl cellulose; CMC]
cellulose thimble 濾紙筒
cellulosic material 纖維素物質
celotex 甘蔗板；蔗渣板
Celsius thermometer 攝氏溫度計
cementation 黏固〔作用〕
centaury 矢車菊
centipoise (cp) 公泊；厘泊(黏度單位)
centistoke 厘史；厘司(黏度單位)
central exhaust 中央排氣
central kitchen 中央廚房
centralized slaughtering 中央屠宰
centralized slaughtering facility 中央屠宰設施
centre sweets 夾心糖果
centrifugal atomization 離心噴霧
centrifugal atomizer 離心噴霧器
centrifugal ball mill 離心球磨機
centrifugal clarifier 離心淨化器
centrifugal dish atomizer 離心盤式噴霧器
centrifugal disintegrator 離心粉碎機
centrifugal dryer 離心脫水機
centrifugal evaporator 離心蒸發機
centrifugal film evaporator 離心薄膜蒸發機
centrifugal honey 離心分離蜂蜜
centrifugal mill 離心粉碎機
centrifugal purification 離心澄清；離心淨化
centrifugal rendering 離心提煉油脂
centrifugal spray dryer 離心噴霧乾燥機
centrifugal syrup 離心分離糖漿
centrifugal washing 離心洗滌
centrifugate 離心分離物；離心液
centrifugation 離心〔作用〕
centrifuge 離心機
centrifuge basket 離心籃
centrifuge bottle 離心瓶
centrifuge tube 離心管
cep 石蕈
cephalin 腦磷脂 [= kephalin]
cera 蜂蠟
ceramic 陶器；陶器的
ceramic ware 陶瓷器皿
cere 蠟；蠟膜
cereal 穀類；禾穀類；穀類食品
cereal crops 穀類作物
cereal food 穀類食品
cereal grain 穀粒
cereal mash 麥芽漿；麥芽汁
cereal milk 穀漿；穀乳
cereal product 穀類製品；穀物加工品
cereal starch 穀物澱粉
cerebose 腦糖
cerelose 工業葡萄糖
ceride 脂蠟
ceriman 蓬萊蕉
cerin 蠟素
cerotic 蜂蠟的
cerotic acid 蠟酸

certified colo(u)r 法定食用色素
cervelas 燻臘腸；塞〔爾〕維拉〔特〕燻香腸（德式） [= cervelat]
cervelat 燻臘腸；塞〔爾〕維拉〔特〕燻香腸（德式） [= cervelas]
cesium 銫
cesspool 污水池
cetacean 鯨；鯨魚類動物
cetanol 鯨蠟醇；十六烷醇 [= cetyl alcohol]
cetyl alcohol 鯨蠟醇；十六〔烷〕醇 [= cetanol]
cevitamic acid 抗壞血酸；維生素C
Ceylon cinnamon 錫蘭肉桂；錫蘭桂皮
Ceylon spinach 落葵；龍宮菜；非洲菠菜
Ceylon tea 錫蘭茶；錫蘭紅茶
CFU (colony forming unit) 菌落形成單位
Chablis 查布利白葡萄酒；夏布里白葡萄酒（法國）
chaff 穀殼；穀糠；麩皮
chaffer 篩選機；穀殼機
chafing dish 火鍋
chain initiation 連鎖開始反應
chain propagation 連鎖繁衍反應
chain reaction 連鎖反應
chain store 連鎖商店
chain termination 連鎖終止反應
chakka 茶卡凝乳（印度）
chalice 杯；高腳酒杯；聖餐杯
challah 白麵包（猶太教的安息日用） [= hallah]
chamber dryer 箱型乾燥機
chamomile oil 春黃菊油
champagne 香檳〔酒〕
champagne cider 香檳蘋果酒
champagne cocktail 香檳雞尾酒
champagne glass 香檳酒杯
champagnization 香檳酒釀造
champagnize 釀製成香檳酒
champignon 洋菇；雙孢蘑菇；食用香蕈
chanterelle 雞油菌；杏菇；鬱金菌
chaps 豬顎
char 紅頭鮭；湖紅點鮭；炭
char filtration 炭過濾
char grilling 碳烤
charcoal 木炭；炭
charcoal burner 藍黃紅菇
chard 火餤菜；紅頭菜；莙蓬菜；牛皮菜
charentais 沙藍泰斯瓜
charlock 田芥菜；野芥菜
charlotte 水果夾心奶油蛋糕；水果〔奶油〕布丁
charolais 夏拉萊牛；夏洛來牛；查羅萊斯牛（法國）
charqui 巴西肉乾
charr 嘉魚；紅點鮭
charring 炭化；焦化
chartreuse 夏爾特甜酒；查爾特勒士甜酒（法國）
chaser 酒後的無酒精清淡飲料
chaser mill 滾碾機；碾碎機
chateau 酒堡；葡萄種植園
chauche 白玉〔葡萄酒〕
chavicol 佳味酚；蒌葉酚
chayote 佛手瓜；梨瓜；香櫞瓜 [= vegetable pear]
cheat 雀麥
cheddar cheese 車打芝士；切達乾酪
cheek 頰肉
cheek bone 頰骨
cheek meat 頰肉
cheese 芝士；起司；乾酪
cheese and meat loaf 芝士肉羅夫；乾酪肉羅夫；乾酪肉糕

cheese cake 芝士蛋糕；起司糕；乾酪餅
cheese cloth 紗布；乾酪包布
cheese coagulum 芝士凝塊；乾酪凝塊
cheese colo(u)r 起司色素；乾酪色素
cheese curd 芝士凝塊；乾酪凝塊
cheese curing 芝士熟成；乾酪熟成
cheese eyeing 芝士孔；乾酪孔
cheese flavo(u)r 乾酪風味；芝士風味 [= cheesy flavo(u)r]
cheese food 芝士食品；乾酪食品
cheese grinder 芝士刨；乾酪刨；乾酪研磨機
cheese hoop 乾酪模子；乾酪箍
cheese mould 乾酪模；乾酪壓模機；乾酪上的黴
cheese powder 芝士粉；乾酪粉
cheese press 乾酪榨乾機；乾酪壓榨機
cheese rind 芝士皮；乾酪皮
cheese ripening 乾酪成熟；乾酪熟成
cheese starter 乾酪發酵劑；乾酪菌元
cheese trier 芝士取樣器；乾酪取樣器
cheese vat 乾酪槽；乾酪桶
cheese wax 乾酪蠟
cheese whey 乾酪乳皮；乾酪乳清
cheesy 乾酪的；乾酪味的
cheesy flavo(u)r 乾酪風味；芝士風味 [= cheese flavo(u)r]
chef 主廚；廚師
chelate 螯合物；螯合的
chelating agent 鉗合劑；螯合劑；多價螯合劑 [= sequestering agent; sequestrant]
chelation 螯合〔作用〕；鉗合〔作用〕
chemical 化學藥品；化學的
chemical additive 化學添加物；化學添加劑
chemical analysis 化學分析
chemical composition 化學成分
chemical food poisoning 化學性食物中毒
chemical leavening 化學發麵劑
chemical peeling 化學去皮
chemical preservative 化學防腐劑
chemical reagent 化學試劑
chemical sanitizer 化學消毒劑
chemical seasoning 化學調味料
chemical stabilizer 化學穩定劑
chemical sterilization 化學滅菌
chemical sterilizer 化學滅菌劑
chemistry 化學
chemosterilant 化學滅菌劑
chemosynthesis 化學合成
chemotrophic bacteria 化能營養細菌
cherimoya 冷子番荔枝；南美番荔枝
Cherokee rose 金櫻子
cherry 櫻桃；車厘子
cherry flavo(u)r 櫻桃香精
cherry gum 櫻桃膠
cherry oily flavo(u)r 櫻桃油香精
cherry plum 櫻桃李
cherry red 櫻桃紅
cherry salmon 櫻桃鮭；馬蘇大麻哈魚；櫻花鉤吻鮭魚
cherry stone 櫻桃核
cherry tomato 櫻桃蕃茄
chervil 細葉香芹；山蘿蔔；茴芹
chessel 芝士桶；乾酪桶；乾酪模子
chesset 乾酪桶；芝士桶；凝酪鍋；凝酪槽 [= chisset]
chest drying machine 箱式乾燥機；多層乾燥機
chester white 切斯特白豬；吉士白豬（美國）
chestnut 栗子；板栗
chestnut bolete 褐圓孔牛肝菌
chestnut paste 栗泥
chetomin 黑毛黴菌

chew 咀嚼
chewiness 咀嚼度；咀嚼性
chewing gum 口香糖；香口膠
chewing gum base components 膠姆糖基礎劑
chewing tobacco 口嚼煙〔草〕
chewy 耐嚼的；有咬勁的
chia seed 野鼠尾草籽
chianina 契安尼娜牛；克安尼娜牛（意大利）
Chianti 意大利〔西昂蒂〕紅葡萄酒
chick 小雞；雛雞
chicken 小雞；雞肉
chicken broth 雞湯
chicken cutlet 雞排
chicken drumstick 雞小腿
chicken fat 雞油
chicken feet 雞腳
chicken giblet 雞雜
chicken gizzard 雞肫
chicken halibut 小庸鰈
chicken nugget 去骨雞塊
chicken of the woods 硫磺菌（蕈類）
chicken paw 雞爪
chicken soup 雞湯
chicken stew 燉雞濃湯
chicken thigh 雞〔上〕腿；雞大腿；雞髀
chicken wing 雞翅膀
chickling 野豌豆；小雞；雛雞
chickpea 鷹嘴豆；藜豆
chicle (gum) 糖膠樹膠；人心果樹膠 [= gum chicle]
chicory 菊苣
chicory concrete 菊苣浸膏
chifferi 通心粉
chiffon cake 戚風蛋糕；泡沫蛋白鬆糕
chile 辣椒；紅辣椒；智利辣椒 [= chili; chilli]
chile pepper 辣椒；智利辣椒
chili 辣椒；紅辣椒；智利辣椒 [= chile; chilli]
chili pepper 辣椒；智利辣椒
chili powder 辣椒粉
chili sauce 辣椒醬；調味辣椒醬
chill 冷卻
chill haze 冷卻混濁
chilled food 冷卻食品
chilled juice 冷卻果汁
chilled meat 冷卻肉
chilled room 冷卻室
chilled storage 冷藏
chiller 冷卻室；冷卻器
chilli 辣椒；紅辣椒；智利辣椒 [= chile; chili]
chilli orange 辣椒橙
chilli pepper 辣椒；智利辣椒
chilli powder 辣椒粉
chilli sauce 辣椒醬；調味辣椒醬
chilling 冷卻；冷卻的
chilling injury 冷傷
chilling room 冷卻室；冷藏室
chilling shrink 冷縮
chillproofing 冷卻檢驗
chilver 哺乳的母羊；雌綿羊；母羊肉
chimyl alcohol 鮫肝醇
China bean 豇豆
China white wine 中國白酒；白乾
chine (bone) 背脊肉；脊椎骨
Chinese Agricultural Standard (CAS) 中國農業標準
Chinese aniseed 大茴香；八角；茴芹 [= anise; star anise]
Chinese artichoke 中國薊
Chinese cabbage 結球白菜；大白菜；津白 [= celery cabbage]
Chinese cheese 豆腐乳；腐乳
Chinese chestnut 板栗；中國栗

Chinese chive 韮菜
Chinese cinnamon 肉桂；中國肉桂；桂皮 [= cassia; cinnamon]
Chinese cinnamon oil 肉桂油；桂皮油；中國肉桂油 [= cassia oil; cinnamon oil]
Chinese cookery 中式烹調法
Chinese cooking 中式烹調
Chinese date 棗 [= jujube]
Chinese dumpling 餛飩；餃子
Chinese egg 皮蛋 [= alkalized duck egg; alkalized egg; pidan; thousand years' egg]
Chinese food 中式食品
Chinese gooseberry 奇異果；中國鵝莓；獼猴桃；藤梨 [= kiwi (fruit)]
Chinese heading cabbage 結球白菜
Chinese herring 鰳魚；鰽白
Chinese ide 草魚；鯇魚
Chinese kale 芥藍〔菜〕；芥蘭菜
Chinese leek 韮菜
Chinese mackerel 中華鯖；中國馬鮫
Chinese mushroom 草菇
Chinese noodle 中式麵條
Chinese pear 沙梨；白梨
Chinese petit fours 中式甜點；中式美點
Chinese prickly ash 花椒；山椒 [= flatspine prickly ash; Szechuan peppercorn]
Chinese quince 木瓜
Chinese restaurant syndrome 中國餐館症候群
Chinese sausage 中國臘腸；中式香腸
Chinese sturgeon 中國鱘
Chinese swine 中國豬
Chinese tea 中國茶
Chinese turnip 沙葛 [= jicama]
Chinese walnut 野核桃；臺灣胡桃
Chinese wolfberry 枸杞
Chinese yam 山藥；薯蕷
chinook salmon 大鱗鮭魚；大鱗大麻哈魚
chip 薄片；炸〔馬鈴〕薯片
chip basket 小果籃
chipolate sausage 契普拉塔豬肉香腸(西班牙)
chippy bread 碎麵包皮
chisel 粗麵粉；粗磨全麥麵粉；麩；糠
chisset 乾酪桶；芝士桶；凝酪鍋；凝酪槽 [= chesset]
chitin 甲殼素；甲殼質；幾丁質
chitinase 幾丁質酶；甲殼質酶
chitodextrin 幾丁糊精；殼糊精
chitosamine 幾丁胺糖；葡萄胺糖；殼糖胺
chitosan 幾丁聚糖；聚葡萄胺糖；殼聚糖；聚甲殼糖
chitose 幾丁糖；殼糖
chive 細香葱；蝦夷葱
chlorate 氯酸鹽
chlorella 綠藻；小球藻
chloride 氯化物
chlorinated lime 漂白粉
chlorination 氯化〔處理〕
chlorine 氯
chlorine dioxide 二氧化氯
chlorinity 氯含量
chloroform 三氯甲烷；氯仿 [= trichloromethane]
chlorogenic acid 漂木酸；綠原酸
chlorophyceae 綠藻類
chlorophyll 葉綠素
chlorophyllin 葉綠酸
chloroplast 葉綠體 [= autoplast]
chocolate 巧克力；朱古力
chocolate bar 巧克力塊；朱古力塊

chocolate brittle 巧克力酥糖；脆心巧克力
chocolate candy 巧克力糖；朱古力糖
chocolate chip 巧克力屑
chocolate coat(ing) 巧克力糖衣；巧克力糖皮
chocolate flavo(u)r 巧克力香精；朱古力香精
chocolate fudge 巧克力焦軟糖；巧克力糖漿
chocolate goods 巧克力製品；朱古力製品
chocolate liquor 巧克力漿
chocolate milk 巧克力奶；朱古力奶
chocolate powder 巧克力粉；朱古力粉
chocolate sundae 巧克力聖代；朱古力新地
chocolate tree 可可樹
chocolate wafer 巧克力薄鬆餅；朱古力威化餅
chokeberry 山楸梅；阿龍尼亞苦味果；花楸果
cholecalciferol 維生素D_3；膽鈣化固醇 [= vitamin D_3]
cholera 霍亂
Cholera vibrio 霍亂弧菌
cholesterase 膽固醇酶；膽甾醇酶
cholesterin 膽甾醇；膽固醇
cholesterol 膽固醇；膽甾醇
cholesterol ester 膽固醇酯
cholic acid 膽酸
choline acetate 乙酸膽鹼
choline bitartrate 酒石酸氫膽鹼
choline chloride 氯化膽鹼
choline hydrochloride 鹽酸膽鹼
choline phosphate 磷酸膽鹼
choline 膽鹼
chondrin 軟骨膠
chondroitin 軟骨素
chondroprotein 軟骨蛋白
Chondrus crispus 鹿角菜；角叉菜
chop 排骨；肉排
chop suey 炒雜碎；炒雜燴菜
chophouse 小餐館
chopped ham 碎火腿
chopped meat 碎肉；肉餡
chopper 切碎機；斬拌機
chopping board 剁肉板；砧板
chopping machine 切碎機；斬拌機
chopsticks 筷子
chord(a) 腱；帶；索
chorizo (sausage) 西班牙口利左豬肉香腸
chow-mein 炒麵
chowchow 中國蜜餞；糖漬食品；醃什錦菜
chowder 海鮮濃湯；雜燴
chromate 鉻酸鹽
chromatometer 色覺計，比色計；色度計
chromium 鉻
chromogenesis 色素生成；產色作用
chromogenic bacteria 產色細菌；發色細菌
chromogenic reagent 發色劑
chromosome 染色體
chronic gastritis 慢性胃炎
chronic intoxication 慢性中毒
chronic toxicity 慢性毒性
chrysanthemaxanthin 菊黃質
chrysanthemin 菊黃酮；菊胺
chrysanthemum 菊花
chrysanthemum concrete 菊花浸膏
chrysanthemum tea 菊花茶；菊花晶
chrysin 金黃素
chrysoidine 橘紅；柯衣定
chub 蘭勃舵魚；白鮭；圓鰭雅羅魚
chub mackerel 日本鯖

chuck 牛肩頸肉
chuck and blade 肩胛肉
chuck eye log 下肩胛肋眼心；沙朗心
chuck eye roll 下肩胛眼肉捲；前腿心
chuck rib meat 肩胛小排肉
chuck roll 去骨牛肩頸肉
chuck tender 肩胛里肌；黃瓜條
chufa 油莎草；荸薺
chufa nut 杏仁核；土杏紅
chump 大肉片；羊後腰脊肉
churn 奶油攪拌機；攪動器；攪乳器
churn dasher 攪動翼；攪乳擋板
churning 攪動；攪乳
chutnee 印度的酸辣調味果醬；印度的醃漬果蔬 [= chutney]
chutney 印度的酸辣調味果醬；印度的醃漬果蔬 [= chutnee]
chyle 乳糜
chylomicron 乳糜微粒
chymase 凝乳酶
chyme 食糜 [= chymus]
chymosin 凝乳酶
chymosinogen 凝乳酶元；凝乳酶原
chymotrypsin 胰凝乳蛋白酶
chymotrypsinogen 胰凝乳蛋白酶元；胰凝乳蛋白酶原
chymus 食糜 [= chyme]
cibo 葱
cibo(u)l 葱；青葱；大葱 [= spring onion; Welsh onion]
cibophobia 拒食症
cider 蘋果汁；蘋果酒 [= cyder]
cider brandy 蘋果白蘭地
cider press 蘋果榨汁器
cider vinegar 蘋果醋
ciguatera poisoning 魚肉中毒；肉毒魚類中毒
ciguatera toxin 魚肉毒素；雪茄毒素；肉毒魚毒素 [= ciguatoxin]
ciguatoxin 魚肉毒素；雪茄毒素；肉毒魚毒素 [= ciguatera toxin]
cineole 桉葉素；桉樹腦
cinnabar chanterelle 紅雞油菌（蕈類）
cinnamal 桂醛
cinnamaldehyde 桂醛；肉桂醛
cinnamaldehyde ethylene glycol acetal 乙二醇縮桂醛；桂醛乙二醇縮醛
cinnamic acid 桂皮酸；肉桂酸
cinnamic alcohol 桂皮醇；肉桂醇
cinnamic aldehyde 桂皮醛；肉桂醛
cinnamic oil 桂皮油；肉桂油
cinnamon 肉桂；中國肉桂；桂皮 [= cassia; Chinese cinnamon]
cinnamon bark 肉桂；桂皮
cinnamon bark oil 肉桂油；桂皮油
cinnamon leaf oil 肉桂葉油；桂葉油
cinnamon oil 肉桂油；桂皮油；中國肉桂油 [= cassia oil; Chinese cinnamon oil]
cinnamon vine 山藥；薯蕷
cinnamyl acetate 乙酸桂酯
cinnamyl alcohol 桂醇；苯丙烯醇
cinnamyl anthranilate 鄰氨基苯甲酸桂酯
cinnamyl butyrate 丁酸桂酯
cinnamyl cinnamate 桂酸桂酯
cinnamyl formate 甲酸桂酯
cinnamyl isobutyrate 異丁酸桂酯
cinnamyl isovalerate 異戊酸桂酯
cinnamyl propionate 丙酸桂酯
CIP (cleaning in place) 定位洗滌〔法〕；就地清洗〔法〕
circular colony 環狀菌落
circular muscle 環肌
circulating water 循環水
cis-3-hexenyl 2-methylbutyrate 2-甲基丁酸葉醇酯

cis-3-hexenyl isovalerate 順式-3-己烯異戊酸酯
cis-4-decen-1-al 順式-4-癸烯醛
cis-4-hepten-1-al 順式-4-庚烯醛；4-庚烯醛
cisco 加拿大白鮭
cissa 嗜異食症；嗜異癖
cistern 蓄水池；水塔
cisticercus 囊蟲
citral 檸檬醛
citral diethyl acetal 檸檬醛二乙縮醛
citranaxanthin 柑橘皮類胡蘿蔔素
citrate 檸檬酸鹽
citric acid 檸檬酸
citrin(e) 柑橘苷；檸檬素；維生素P
citrinin 橘黴素
citriodora oil 檸檬桉油
citro-press 柑橘壓榨機
Citrobacter freundii 弗氏檸檬酸桿菌
citroglyceride 檸檬酸和脂肪酸甘油脂
citromycetin 檸檬酸菌素
citron 枸橼；香橼；香木橼
citron day-lily 黃花；金針
citron oil 檸檬油
citronella (grass) 香茅；檸檬草 [= lemon grass]
citronella oil 香茅油
citronellal 香茅醛
citronellol 香茅醇
citronellyl acetate 乙酸香茅酯
citronellyl butyrate 丁酸香茅酯
citronellyl formate 甲酸香茅酯
citronellyl isobutyrate 異丁酸香茅酯
citronellyl propionate 丙酸香茅酯
citroptin 萊姆素
citroxanthin 橘黃質；檸黃質
Citrullus 西瓜屬
Citrus 柑橘類植物
Citrus aurantium 酸橙
citrus brandy 柑橘白蘭地
citrus flavo(u)ring 柑橘香料；柑橘香精
Citrus foonensis 蘆柑
citrus fruit 柑橘類水果
Citrus grandis 柚
citrus juice 柑橘枝
Citrus limon 檸檬
citrus molasses 柑橘糖蜜
citrus oil 橘皮油
citrus paradisi 葡萄柚
citrus peel 柑橘果皮
citrus pulp 柑橘渣
citrus red No.2 橘紅2號
Citrus sinensis 甜橙
Citrus tangerina 紅橘
citrus wine 柑橘酒
city milk 鮮乳；市乳；消毒牛乳 [= market milk]
clabber 凝結酸乳；酸牛奶
Cladosporium 分枝孢子菌屬；芽枝黴屬
clairet 淡紅葡萄酒
clam 蛤；蚌蛤；蛤蜊；蜆肉
clam chowder 蛤肉雜燴；雜燴文蛤
clam meat 蛤肉
clam nectar 蛤濃汁
clam patty 蛤餡餅
clammy bread 發黏麵包
clap bread 燕麥薄餅
claret 紅葡萄酒
clarificant 澄清劑
clarification 淨化；澄清
clarified juice 澄清果汁
clarified molasses 淨化糖蜜；澄清糖蜜
clarifier 澄清器；澄清劑
clarifying agent 澄清劑
clarifying tank 澄清槽；沉澱槽
clary 鼠尾草 [= sage]
clary (sage) oil 香紫蘇油

clary oil 鼠尾草油；香丹參油；香紫蘇油 [= clary sage oil]
clary sage 鼠尾草；香丹參；香紫蘇
clary sage oil 鼠尾草油；香丹參油；香紫蘇油 [= clary oil]
classificator 分級器；篩分機；分粒機
classifier 選別機；分級機
claw 爪；蟹螯；鉗
clayed sugar 白土脱色糖；漂白蔗糖
clean flour 純麵粉
clean in table 原台位清洗〔法〕
cleaned coffee 去莢咖啡豆
cleaning agent 清潔劑；洗滌劑
cleaning in place (CIP) 定位洗滌〔法〕；就地清洗〔法〕
cleaning in place system (CIP system) 定位洗滌系統；就地清洗系統
cleanser 清潔劑；淨化器
clear bone 無骨；去骨
clear dough 麵筋；光潔麵糰
clear egg 無精蛋
clear flour 二等麵粉
clear ice 透明冰衣
clear juice 澄清果汁；清果汁
clear mint 透明薄荷硬糖
clear pork 淨豬肉；去骨豬肉
clear soup 清湯
clearing agent 澄清劑
clementine 克萊門氏小柑橘；克門提柑
climbing film evaporator 升膜〔式〕蒸發器
climbing perch 攀鱸
climbing-prickly ash 藤山椒；藤崖椒
clincher 預封機；置蓋機；假封機
clincher machine 假封機；預封機
clingstone 果肉緊黏於果核的
clingstone peach 核黏桃；黏核桃
clinical nutrition 臨床營養〔學〕
clipper 結紮機
clod 上肩胛肉；牛肩肉
close milk filter 封閉式濾乳機
closed circuit dryer 密閉循環乾燥機
closed fermenter 密閉式發酵罐
closed panel method 隔離品評法
closing machine 封灌機；封口機；封蓋機
clostridial spore 梭菌芽孢
Clostridium 梭〔孢桿〕菌屬；梭狀芽孢桿菌屬
Clostridium botulinum 肉毒〔梭孢〕桿菌；肉毒〔梭狀芽孢〕桿菌
Clostridium perfringens 產氣莢膜梭〔孢桿〕菌；產氣莢膜桿菌
clot 凝塊；凝結
cloth filter 布濾器；濾布
clotted milk 凝乳塊；酸凝乳
clotting 凝結的
clotting enzyme 凝結酵素；凝結酶
cloud test 混濁溫度試驗；霧點試驗
cloudberry 雲莓；野生黃〔草〕莓
clouded funnel cap 煙雲杯傘 (蕈類)
cloudiness 濁度；混濁度
clouding agent 混濁劑
cloudy juice 混濁果汁
clove 丁香；小鱗莖；蒜瓣
clove (bud) oil 丁香〔花蕾〕油
clove leaf oil 丁香葉油
clove oil 丁香油
clove stem oil 丁香莖油
clove wine 丁香〔葡萄〕酒
clover 三葉草；苜蓿
club cheese 總匯軟芝士；總匯軟乾酪
club footed funnel cap 棒柄杯傘 (蕈類)
club gourd 蛇瓜
club sandwich 總匯三明治；什錦三明治；公司三明治
club soda 蘇打水；汽水
club steak 前腰牛排

club wheat 密穗小麥
clump 菌塊；菌落
clupanodonic acid 魚酸；鯡魚酸
Clupea 鯡魚類
clupeoid 鯡魚
cluster 串；叢；群；簇
cluster pepper 簇生辣椒
cluster wheat 密穗小麥
clustered grey gill 荷葉離褶傘(蕈類)
CMC (carboxymethyl cellulose) 羧甲基纖維素
coacervate 凝聚層
coacervation 凝聚
coagel 凝聚膠
coagulant 凝結劑
coagulase 凝結酶；凝固酶
coagulate 凝結；凝結物；凝塊
coagulated milk 凝固乳
coagulating agent 凝結劑
coagulation 凝結；凝聚
coagulator 固化劑；凝聚劑；凝結劑
coagulin 凝聚素；凝固素
coagulum 凝結物；凝塊；凝結塊
coal gas 煤氣
coal tar 煤苔；煤焦油
coal tar colo(u)r 煤苔色素；煤焦油色素
coalescence 凝結；聚結
coarse bran 粗麩；粗糠
coarse break(ing) 粗壓碎；粗軋碎
coarse chopped sausage 粗絞〔餡〕香腸
coarse cut 粗切；粗絞
coarse cut sausage 粗絞〔餡〕香腸
coarse fish 下雜魚；低檔魚類
coarse flour 未篩麵粉；粗磨製麵粉
coarse grain 粗粒
coarse granular sugar 粗粒砂糖；大晶粒砂糖
coarse granulated sugar 粗粒砂糖
coarse grinding 粗磨；粗碾
coarse grit 粗粒〔穀〕粉
coarse meal 未篩粗磨粉
coarse mesh 粗〔孔〕篩
coarse salt 粗〔粒〕鹽
coarse screen 粗篩
coarse screenings 粗篩餘物；粗篩屑
coarse semolina 粗粒小麥粉；麥渣
coarse sieve 粗篩
coarse sizings 粗粒麵粉；小麥粗粒
coarseness 粗糙度；粗粒度
coated mill rice 亮滑米；上光精白米
coating agent 被膜劑；塗層劑
cob 果穗軸；玉米穗軸；玉米芯；小圓麵包
cob corn 帶穗玉米
cob nut 歐洲榛子
cobalamin 鈷維生素；維生素B_{12}
cobalt 鈷
cobamic acid 鈷胺酸
cobbler 水果派；水果餡餅；高杯拉雞尾酒
coburg 十字花烤包
coca 古柯
Coca Cola 可口可樂 [= Coke]
coca leaf extract 古柯葉抽提液
cocci 球菌
coccus 球菌 [= coccus的複數]
cochin 九斤〔黃〕雞
cochineal 胭脂紅；胭脂蟲紅
cochineal extract 胭脂蟲紅萃取液
cochineal red A 新胭脂紅；胭脂紅A；食用紅色六號
cock 公雞；旋塞；薑汁麵包
cockerel 小公雞
cockle 海扇；鳥蛤；麥仙翁
cockscomb 雞冠
cocktail 雞尾酒；開胃食品
cocktail glass 雞尾酒杯

cocktail juice 什錦果汁；混合果汁
cocktail lounge 酒廊；雞尾酒會
cocktail party 雞尾酒會
cocktail punch 什錦果汁；什果賓治
coco 椰子樹
coco plam 椰子樹；可可椰子
coco plum 椰李
cocoa 可可飲料；可可樹；可可豆；可可粉
cocoa bean 可可豆
cocoa butter 可可脂；可可油 [= theobroma oil]
cocoa butter substitute 可可脂代用品；代可可脂
cocoa cake 可可蛋糕；可可油餅
cocoa husk 可可豆殼皮；可可殼
cocoa nib 熟可可粒；可可豆肉；可可豆粉
cocoa powder 可可粉
cocoa shell 可可〔果〕殼
cocoa tincture 可可酊
cocoanut oily flavo(u)r 椰子油香精
coconut 椰子
coconut aldehyde 椰子醛
coconut butter 椰子脂
coconut cake 椰子油餅
coconut crab 椰子蟹
coconut cream 椰子漿
coconut meal 椰子粕
coconut milk 椰〔子〕汁 [= coconut water]
coconut oil 椰子油
coconut tree 椰子樹
coconut water 椰〔子〕汁 [= coconut milk]
cocurrent dryer 並流式乾燥機；順流式乾燥機
cocurrent spray dryer 並流式噴霧乾燥機；順流式噴霧乾燥機
cod 鱈；莢；殼
cod liver oil 鱈魚肝油
cod roe 鱈卵
coda cheese 卡達乾酪；卡達芝士
coddle 文火燉煮
Codex Alimentarius 國際食品法典；食品法規
Codex Alimentarius Commission 國際食品法典委員會；聯合國食品法典委員會
Codex Committee on Food Additive 食品添加劑法規委員會；國際食品添加物法典委員會
Codex Committee on Food Hygiene 食品衛生法規委員會；國際食品衛生法典委員會
Codex of Food Chemicals 食品用化學品藥典
codfish 鱈；鱈魚；鱈魚肉
codfish brick 醃鱈〔肉〕塊
codlin 未成熟小蘋果；尖頭長蘋果
codling 未成熟小蘋果；野生蘋果；幼鱈
coenzyme 輔酶
coferment 輔酶
coffee bar 咖啡館；咖啡吧
coffee 咖啡
coffee arabica 阿拉比卡咖啡；阿拉伯咖啡 [= arabica; arabica coffee]
coffee bean 咖啡豆
coffee bean decorticator 咖啡豆去殼機
coffee berry 咖啡果 [= coffee cherry]
coffee break 咖啡時間
coffee brew 煮咖啡
coffee cake 花式甜麵包；早餐點心
coffee cherry 咖啡果 [= coffee berry]
coffee cocktail 蛋雞尾酒
coffee cream 咖啡用鮮奶油；咖啡稀奶油

coffee creamer 咖啡用非乳製白油
coffee cup 咖啡杯
coffee drink 咖啡飲料
coffee dripper 咖啡過濾漏斗；咖啡沖濾壺
coffee filter bag 咖啡過濾紙；咖啡過濾袋
coffee filter paper 咖啡過濾紙；過濾咖啡用紙
coffee fondant 咖啡軟糖
coffee grinder 咖啡研磨機；咖啡磨
coffee grounds 咖啡渣
coffee maker 煮咖啡機
coffee mate 咖啡伴侶；咖啡用白油
coffee mill 咖啡磨
coffee palace 咖啡館；咖啡廳
coffee percolator 咖啡滲濾壺；煮咖啡漏壺
coffee pot 咖啡壺
coffee robusta 羅巴斯達咖啡；濃咖啡 [= Robusta]
coffee room 咖啡室
coffee siphon 咖啡泡濾壺
coffee table 茶几
coffee tincture 咖啡酊
coffee tree 咖啡樹
coffee urn 煮咖啡壺
coffee whitener 咖啡奶精；咖啡用白油
cognac 康雅〔白蘭地〕酒；科涅克〔白蘭地〕酒；干邑〔白蘭地〕酒；法國白蘭地酒
cognac oil 康乃克油
cognac oil, green 天然康乃克油
cognate amino acid 同源胺基酸
coherence 黏著；黏合；內聚現象
cohesion 黏結；黏合；內聚現象；內聚力
cohesive 內聚性；黏合劑
cohesive force 內聚力
coho (salmon) 銀鮭；銀大麻哈魚
cohune 巴西棕櫚；羽狀葉棕櫚樹
cohune nut 羽狀葉棕櫚堅果
cohune palm 羽狀葉棕櫚
cohurmulone 合葎草酮
coil condenser 旋管冷凝器；盤管冷凝器
coil evaporator 盤管蒸發器；旋管蒸發器
coil-type heat exchanger 旋管式熱交換器；盤管式熱交換器
Coke 可口可樂 [= Coca Cola]
coke 焦炭；可樂
cola 可樂樹；可樂豆；可樂果；可樂類飲料
cola (type) beverage 可樂型飲料
cola nut 可樂豆；可樂果
cola nut extract 可樂果抽提液
cola seed 可樂豆；可樂果
cola type flavo(u)r 可樂型香精
colander 濾鍋；濾器；篩
colation 過濾
colature 濾液；粗濾產物
colcannon 愛爾蘭馬鈴薯菜糊
colchicine 秋水仙鹼
cold air dryer 冷風乾燥機
cold beverage 冷飲料；凍飲；冷飲
cold break 冷碎〔法〕；冷打漿
cold closet 冷藏箱；冷藏室
cold damage 冷傷害；凍傷
cold denaturation 冷變性
cold dish 冷盤；冷盤菜〔餚〕
cold distillation 低溫蒸餾
cold drawn oil 冷榨油
cold drink 冷飲；凍飲；清涼飲料
cold drying 低溫乾燥；冷風乾燥
cold endurance 耐冷性；耐寒性
cold fermentation 低溫發酵
cold food pantry 冷盤室

cold injury 冷傷
cold liming 冷灰〔澄清〕法；冷汁加灰
cold meat 冷肉；冷盤肉
cold pack(ing) 冷裝〔法〕；冷充填
cold pasteurization 低溫巴氏殺菌法；低溫巴斯德殺菌法
cold plate 冷卻台
cold preservation 冷藏
cold press 冷壓〔榨〕；冷榨；冷壓〔榨〕機
cold pressed oil 冷壓油；冷榨油
cold pulping 冷碎〔法〕
cold rendering 低溫熬油
cold resistance 抗冷性；耐寒性
cold rigor 冷僵直
cold room 冷卻室；冷藏房
cold shortening 冷收縮
cold smoking 冷燻〔法〕
cold sterilization 冷式滅菌；照射滅菌
cold stocker 低溫儲藏庫
cold storage 冷藏
cold storage room 冷藏室
cold storage warehouse 冷藏倉庫；冷凍倉庫
cold store 冷藏庫；冰箱
cold test 低溫試驗；耐冷試驗；油脂凝固點試驗
cold treatment 冷處理
coldpressed oil 冷榨油
cole 甘藍；海甘藍；油菜；蕓薹
coleorhiza 胚根鞘
coleseed 油菜子
coleseed oil 菜籽油；菜油
coleslaw 椰菜絲沙律；甘藍沙拉；卷心菜絲色拉；涼拌卷心菜絲 [= slaw]
colewort 甘藍；海甘藍；蕓薹
coley 科萊鱈
coli-aerogenes organisms 大腸〔桿〕菌群
colibacillosis 大腸桿菌病
coliform 大腸〔桿〕菌群；似大腸〔埃希氏〕桿菌
coliform bacteria 大腸〔桿〕菌群
coliform count 大腸〔桿〕菌數
coliform group 大腸〔桿〕菌群
coliform mastitis 大腸桿菌群乳房炎
colitis 結腸炎；大腸炎
colitoxin 大腸桿菌毒素
collagen 膠原〔質〕；膠原蛋白
collagen casing 膠原腸衣
collagen fiber 膠原纖維
collagen sugar 甘胺酸；胺基醋酸；胺基乙酸 [= aminoacetic acid; glycine]
collagenase 膠原〔蛋白〕酶
collapsed can 凹罐；擠扁罐
collapsible box 可折疊紙盒
collapsible tube 可擠軟管；牙膏管
collar 肉捲
collar butt 上肩肉；梅頭肉
collard 羽衣甘藍
collins 柯林斯雞尾酒
colloid 膠體
colloid mill 膠磨機；膠體磨
colloidal 膠體的；膠質的
colloidal solution 膠體溶液
colloidal suspension 膠體懸浮液
collop 小肉片；薄肉片；肉排
colo(u)r 顏色；色素；著色劑
colo(u)r densitometer 色密度計
colo(u)r density 色密度；顏色強度
colo(u)r difference meter 色差計
colo(u)r disc 比色板；色板；比色盤 [= colo(u)r disk]
colo(u)r disk 比色板；色板；比色盤 [= colo(u)r disc]
colo(u)r fading 退色
colo(u)r fixation 顏色固定
colo(u)r fixative 發色劑；顏色固定劑；保色劑；護色劑

colo(u)r intensity 色度；顏色強度
colo(u)r reaction 呈色反應；顯色反應
colo(u)r retention agent 保色劑；固色劑
colo(u)r test 呈色試驗；顯色試驗
colo(u)r thermometer 色溫度劑
colo(u)rant 著色劑 [= colo(u)ring agent]
colo(u)rimeter 比色計
colo(u)rimetric analysis 比色分析〔法〕
colo(u)rimetric determination 比色測定〔法〕
colo(u)rimetry 比色法
colo(u)ring 著色；著色劑
colo(u)ring adjunct 增色劑
colo(u)ring agent 著色劑 [= colo(u)rant]
colo(u)rless 無色的
colon 結腸
colon bacillus 大腸桿菌群
colonial morphology 菌落形態
colonial sugar 蔗糖
colony 菌落
colony count 菌落數
colony counter 菌落計數計
colony forming unit (CFU) 菌落形成單位
colophony 松香
colostrum 初乳
coloured malt 炒麥芽
column chromatography 柱層析法；管柱層析法
colupulone 合蛇麻酮；輔蛇麻蘆酮
colza 菜籽；油菜籽 [= rapeseed]
colza oil 菜籽油 [= rapeseed oil]
comb honey 巢蜜
comb shell 蝦夷扇貝
combat food 軍糧
combined water 化合水
comestible 可吃的
comestibles 食物
comet fish 康氏烟管魚
comfit 糖果；蜜餞
comfrey 康富力；紫草科植物
Comma bacillus 霍亂弧菌
commercial fishes 商業魚類；經濟魚類
commercial goods 商品
commercial sterilization 商業滅菌法；阿佩爾滅菌法 [= appertization]
commercial strain 商用菌株
comminuted cheese 粉碎芝士；粉碎乾酪
comminuted juice 帶肉果汁；含果粒果汁
comminuted meat 碎肉；肉餡
comminution 粉碎〔作用〕
comminutor 粉碎機
common bean 菜豆；刀豆
common chanterelle 雞油菌 (蕈類)
common deceiver 漆蠟蘑 (蕈類)
common eel 鱔魚；鰻鱺
common funnel cap 中凸蓋杯傘 (蕈類)
common head cabbage 甘藍；結球甘藍；捲心菜 [= head cabbage; headed cabbage; heading cabbage]
common hedgehog fungus 捲緣齒菌 (蕈類)
common Japanese conger 繁星康吉鰻；星鰻
common jujube 棗
common minnow 寬鰭鱲魚
common morel 羊肚菌
common oyster mushroom 蠔菇；糙皮側耳 (蕈類)
common perch 河鱸
common plum 西洋李；歐洲李；洋李
common pomegranate 石榴
common puffball 網紋馬勃 (蕈類)

common salt 食鹽
common sea bass 真鱸
common sole 鰨
common stinkhorn 白鬼筆(蕈類)
common wheel mummy cap 輪枝小皮傘(蕈類)
common white saddle 皺馬鞍菌(蕈類)
common yellow daylily 金針菜；食用萱草；黃花菜
Commonwealth Scientific and Industrial Research Organization (CSIRO) 聯邦科學與工業研究組織
compact package 緊縮包裝；緊密包裝
compact ration 壓縮乾糧；壓縮軍糧
compartment dryer 隔間乾燥器；廂式乾燥器
complector 瑪珈琳製造器；人造奶油製造器
complementary colo(u)r 補色
complete fermentation 完全發酵
complete food 完全食品
complete medium 完全培養基
complete protein 完全蛋白質
complete sterilization 完全滅菌
completely fermented tea 全發酵茶
completely fermented wine 發酵成熟的酒
complex 複合物；錯合物；絡合物；複體
complex lipid 複合脂質
complex medium 複合培養基
complex seasoning 複合調味料
complexing agent 錯合劑
composite can 紙身罐；金屬底蓋的紙罐
composite flour 複合〔麵〕粉
composite food 複合食品
composite paper can 紙身罐；金屬底蓋的紙罐
compote 糖煮水果；果盤；糖漬水果片
compound 化合物；封口膠
compound barm 混合酵母
compound fat 混合油脂
compressed dehydrated food 壓縮乾燥食品；壓縮脫水食品
compressed food 壓製食品；壓成塊的食品
compressed salt 鹽磚
compressed tea 茶磚；磚茶
compressed yeast 壓榨酵母
compression filtration 壓濾
compression granulation 壓縮造粒〔法〕
compression packing 壓縮包裝
compressometer 壓縮計；壓力計
compressor 壓縮機
conalbumin 伴白蛋白；伴清蛋白
concentrate 濃縮物；濃縮
concentrated cream 濃縮鮮奶油；高脂稀奶油
concentrated juice 濃縮果汁
concentrated milk 濃縮乳；煉乳(無蔗糖)
concentrated solution 濃縮液
concentrated whey 濃縮乳清
concentration 濃縮；濃度
concentrator 濃縮器；集中器
conch 貝殼；海螺
conche 巧克力精磨機；巧克力精煉機；巧克力攪拌揉捏機
conchiglie 貝殼粉；蜆殼粉
[= abassina rigata; shell pasta]
conchiglie rigate 稜紋貝通心粉
conching machine 巧克力捏磨機；巧克力精煉機
concoction 混合；調製；調合；混合物
Concord grape 康考特葡萄；康科特葡萄
concurrent flow dryer 順流式乾燥機

condemned 被判定為不適宜食用的
condemned meat 不適宜食用的肉類
condemned product 不適宜食用的產品
condensate 冷凝物；冷凝液
condensation 冷凝；凝聚；縮合
condensed buttermilk 濃縮白脫乳；濃縮酪乳
condensed cream 濃縮奶油(加糖)
condensed juice 濃縮果汁
condensed milk 濃縮乳；煉乳；加糖煉乳
condensed seasoning 濃縮調味料
condensed skim milk 脱脂煉乳
condensed soup 濃縮湯
condensed soy 醬油膏
condenser 冷凝器；電容器
condensery 煉乳廠
condiment 調味料；調味品
conditioning 調節；水份調節；熟成；回潮
conduction 傳導
conduction heating 傳導加熱
cone 球果；錐形冰淇淋杯；圓錐形
cone cap 鼠色小孔菌(蕈類)
cone crusher 錐形壓碎機；錐形榨汁機
cone mill 錐形磨；錐形壓碎機
cone shell 雞心螺
cone wheat 圓錐小麥
confection 蜜餞；糖食；糖果
confectionary 糖果店；糖果；糖食的
confectioner 糖果店；糖果製造者；蛋糕裝飾師
confectioner's coating 糖果的包衣；糖飾
confectioner's sugar 粉糖；糖粉
confectionery 糖果；糖食；糖果業；糖果店
confetti 糖果
confiture 果醬；蜜餞
congee 粥；米粥 [= conjee; rice congee]
congelation 凍結；凍凝；凝固
conger 康吉鰻；穴鰻
conger eel 海鰻；康吉鰻；海白鰻；繁星糯鰻；花點糯鰻
Congo coffee 剛果咖啡樹
congou 功夫茶(中國紅茶)
conical ball mill 錐形球磨
conical blender 錐形混合機
conical breaker 錐形破碎機
conical dryer 錐形乾燥機
conical mixer 錐形混合機
conidendrin 松柏抗氧化酚；鐵杉內酯；降二氫愈瘡木酸
conifer tuft 煙色垂幕菇
conjee 粥；米粥 [= congee; rice congee]
conjugate acid 共軛酸
conjugate base 共軛鹼
conjugated protein 複合蛋白質；綴合蛋白質
conjugation 接合；共軛；結合
conjugative lipid 複合脂質
connective tissue 結締組織
conpoy 干貝；瑤柱
conservation 保藏；保存；防腐
conservative 防腐劑
conservatory 貯藏室；溫室
conserved fruit 蜜餞
conserving agent 防腐劑；保藏劑 [= preservative; preserving agent]
consistence 黏稠度；濃度；一致性
consistency 黏稠度；濃度；一致性
consistometer 黏稠度計；稠度計
consomme 清肉湯
constant weight 恆重
constipation 便秘
constituent 組分；成分

constitution water 化合水
consumer 消費者
consumer flour 家用麵粉；二等小麥粉
consumer goods 零售商品
consumer pack(age) 零售包裝
contact icing 接觸冰藏；雪封保藏
contaminant 污染物〔質〕
contamination 沾污；污染；污染物
content 含量；內含物
Continental breakfast 歐〔陸〕式早餐；簡便早餐
continuous centrifuge 連續離心分離機
continuous extraction 連續萃取
continuous filtration 連續過濾
continuous flow centrifugation 連續流動離心分離
continuous pasteurizer 連續式巴氏殺菌器
continuous retort 連續〔高壓〕殺菌釜
continuous sterilizer 連續式滅菌器
contractile protein 收縮性蛋白質
contraflow dryer 對流式乾燥機；逆流式乾燥機
control 對照；控制
control experiment 對照實驗
controlled atmosphere 控制氣體；人工氣體〔環境〕；調控大氣
controlled atmosphere packaging 控氣包裝；氣控包裝
controlled atmosphere storage (CA storage) 控氣保藏；人工氣候保藏；氣調貯藏；控制大氣貯藏 [= gas storage]
convection 對流
convection dryer 對流式乾燥機
convection oven 對流〔式〕烘箱
convenience food 簡便食品；方便食品
convenience goods 簡便商品
convenience package 簡易包裝
convenience store 簡便商店
conventional curing 傳統醃製
conventional dry curing 傳統乾醃法
conventional pickle curing 傳統濕醃法
conversion factor 換算系數
converted rice 改良米
converted starch 糖化澱粉；轉化澱粉
conveyor 傳送機；輸送帶
conveyor belt 輸送帶；傳送帶
conveyor dryer 帶式乾燥機
conveyor rail 傳送吊軌
conveyor table 輸送台
cony 家兔
cook 廚師；烹調
cook-before-eating 食用前需煮熟
cook flavo(u)r 烹煮風味；煮熟風味
cook-in-pouch food 連袋蒸煮食品
cookbook 食譜
cooked ham 熟火腿
cooked meat 熟肉
cooked meat product 熟肉製品
cooked rice 米飯
cooked sausage 熟香腸
cooked shrinkage rate 烹煮收縮率
cooked taste 烹煮味；煮熟味
cooker 蒸煮器；烹調器；炊具
cookery 烹飪法；烹調術
cookhouse 船上廚房；屋外廚房
cookie 曲奇餅；甜酥餅；小甜餅；脆餅
cooking 烹調；烹飪；蒸煮
cooking equipment 烹調設備
cooking fat 烹調用脂肪
cooking loss 烹調損失；烹煮損耗
cooking oil 烹調用油
cooking pan 蒸煮鍋；烹調鍋
cooking stove 廚灶
cooking time 蒸煮時間；烹調時間
cooking utensil 炊具
cooking wine 烹調酒；料酒

cookroom 廚房
cookshop 小餐館；小菜館
cooky 小甜餅；餅乾
cool fermentation 低溫發酵；冷發酵
cool storage 冷藏
cool wash 冷洗
coolant 冷卻劑；冷煤
cooled meat 冷卻肉
cooler 冰箱；冷卻器；冷卻劑；冷卻室；冷飲
cooling 冷卻；冷卻的
cooling agent 冷卻劑
cooling bath 冷卻浴
cooling chamber 冷卻室
cooling grinder 冷卻研磨機
cooling medium 冷卻介質
cooling pond 冷卻池
cooling room 冷卻室
cooling tank 冷卻槽
cooling unit 冷卻裝置
cooling water 冷卻水
coolness 冷卻程度；涼爽
cooper 箍桶；製桶工人；品酒員；零售酒商
cooperage 製桶業；製桶工場
copa 乾椰子瓤
copaiba oil 苦配巴油
copied food 仿冒食品
copolymer 共聚物
copper 銅
copper chlorophyll 銅葉綠素；葉綠素銅
copper gluconate 葡〔萄〕糖酸銅
copper poisoning 銅中毒
copper sulfate 硫酸銅
copra 乾椰子肉
copra meal 椰子粕
copra oil 椰子油
coprecipitation 共〔同〕沉澱
coproduct 副產品；副產物
coral fish 珊瑚礁魚類
coral tooth fungus 珊瑚狀猴頭菇
corb shell 蜆
corbicula 蜆；蜊仔；台灣蜆
cord 腱；帶；索
cordial 興奮劑；興奮性飲料；可蒂雞尾酒；露酒；加香甜酒
cordial punch 可蒂潘趣飲料；酒味果汁
cordyceps sinensis 冬蟲夏草
core 果心；果核
coreopsis 金雞菊
coreopsis yellow extract 菊花黃浸膏
corer 去果心器；挖核器
coriander 芫荽；胡荽；芫茜；香菜
coriander oil 胡荽籽油；芫荽籽油
coriander seed 芫荽籽；胡荽籽
coring 去核
coring knife 去核刀；挖〔果〕心刀
corium 真皮
cork 軟木；軟木塞
cork liner 軟木墊
cork oak 軟木橡樹
cork press 壓軟木塞機
cork stopper 軟木塞
cork washer 軟木墊圈
corkage 開瓶費；拔瓶塞
corker 壓塞機；封塞機；塞瓶工人
corking machine 封塞機
corkscrew 螺旋拔塞器；開塞鑽
corky 軟木塞的
corm 球莖；地下球莖
corn 玉米；玉蜀黍；粟米 [= maize]
corn-and-cob 整穗玉米 [= corn on the cob]
corn-and-cob crusher 玉米脫粒機
corn belt 玉米產區
corn bread 玉米麵包

corn chip 玉米脆片；玉米片
corn cutter 玉米脫粒機
corn distillery 玉米釀酒廠；玉米酒精廠
corn ear 玉米穗
corn flake 玉米〔雪〕片
corn flour 玉米粉
corn gluten 玉米蛋白；玉米麩質；玉米麵筋
corn gluten meal 玉米蛋白粗粉；玉米麩質粗粉
corn grit 碎玉米；玉米渣
corn gruel 玉米粥；玉米糜
corn husk 玉米殼
corn husker 玉米去殼機；玉米去殼工人
corn kernel 玉米粒
corn mash 玉米醪
corn meal 玉米粉；玉米粗粉；玉蜀黍粉
corn mill 玉米碾磨機；玉米磨
corn oil 玉米油；粟米油
corn on the cob 整穗玉米 [= corn-and-cob]
corn popper 爆玉米花機
corn puff 爆玉米花
corn rice 脫皮玉米粒
corn sheller 玉米脫粒機
corn silk 玉米穗絲
corn steep liquor 玉米浸液
corn sugar 玉米糖
corn sweetener 澱粉糖
corn syrup 玉米糖漿；澱粉糖漿 [= maize syrup]
corn whisky 玉米威士忌
cornbeef 碎鹹牛肉
corncob 玉米穗軸；玉米芯
corned 鹽醃的
corned beef 鹹牛肉；醃牛肉
cornetfish 鱗烟管魚；康氏烟管魚 [= flutemouth]
cornflower bolete 藍圓孔牛肝菌
corning 輕醃的；鹽醃的
Cornish 可尼西雞；康尼希雞
cornstarch 玉米澱粉；粟粉
cornucopia 三角形包裝；錐形紙袋
Corriedale 考利台爾羊；考力代羊
corroborant 強壯劑；補藥
corroborant food 滋補食品
corrugated slice 波形切片
corticosterone 皮質酮；腎上腺酮
cortisol 皮質固醇
cortisone 可體松；可的松；皮質酮
Corynebacterium 棒狀桿菌屬
coryneform 棒狀桿菌的
cos lettuce 立生萵苣；直立萵苣
cossette 甜菜片；甜菜絲
costa rib 肋骨；排肋
costal cartilage 肋軟骨
costard 可斯達蘋果（英國）
costmary 脂香菊
costus root oil 雲木香油；廣木香根油
cotplum 杏李
cottage bread 農家自製麵包
cottage cheese 酪農乾酪；酪農芝士；卡達乾酪；卡達芝士；鄉村乾酪 [= pot cheese]
cotton candy 棉花糖
cotton plug 棉花塞
cotton wool 棉花；脫脂棉；原棉
cottonseed 棉籽
cottonseed hull 棉花殼
cottonseed meal 棉籽粕
cottonseed oil 棉籽油
cotyledon 子葉
cough drop 止咳糖；止咳藥片
cough sweets 止咳糖
cough syrup 止咳糖漿

coumaric acid 香豆酸
coumarin 香豆素；薰草素
coumarone 香豆酮；薰草酮
counter balance 托盤天平；平衡
counter-current 逆流
counter-current extraction 逆流萃取；逆流提取
counter-current flow 逆流流動
counter-current multistage extraction 逆流多段萃取
counter-current spray dryer 逆流式噴霧乾燥機
counter diffusion 相互擴散；逆擴散
counterflow 逆流
counterflow condenser 逆流冷凝器
counterflow drying 逆流乾燥
counterflow heat exchanger 逆流式熱交換機
counterflow spray dryer 逆流式噴霧乾燥機
counterflow tunnel dryer 逆流式隧道乾燥機
counterstain 對比染色；復染色
country cure 乾鹽醃製
country style sausage 鄉村式香腸；土製香腸；家庭自製型香腸
coupling sugar 配聯糖；偶合糖
courgette 小胡瓜；筍瓜；小西葫蘆 [= zucchini]
course 一道菜；課程
court bouillon 什錦魚湯；酸辣芹菜魚湯
couscous 庫斯庫斯
couverture 巧克力糖衣；朱古力糖衣
cover remover 去蓋器
cover sterilizer 罐蓋滅菌器
covered dish 燉鍋；烘雜燴
covering chocolate 包衣巧克力
covering machine 掛糖衣機；包封機
coverplate 蓋板
coverplate oven 附蓋烤爐；升降門室烤爐
cow 母牛；乳牛
cow milk 牛乳
cow tree 乳樹；牛奶樹
cowberry 越橘；越桔
cowberry red 越橘紅
cowheel 牛蹄凍；燉牛蹄筋
cowpea 豇豆 [= black-eyed pea]
cowy flavo(u)r 乳牛臭；牛膻味
cox's orange pippin 舵手橘蘋果
crab 蟹；蟳；螃蟹
crab apple 小酸蘋果；山楂子；海棠〔蘋果〕
crab floating cannery 蟹罐頭〔加工〕船
crab meat 蟹肉
crab roe 蟹子；蟹黃
crab russule 黃孢紅菇
crabapple 小蘋果
cracked egg 裂紋蛋
cracked grain 裂紋殼粒；破碎殼粒
cracked ice 碎冰
cracked wheat 破碎小麥粒
cracker 蘇打餅乾；脆餅乾；胡桃夾
cracking mill 破碎機
crackling 豬油渣；烤豬的脆皮
cracknel 硬脆餅乾；豬油渣；脆煎肉餅
crambe 海甘藍
crambe cabbage 包心菜
crammed duck 填鴨
crammed peking duck 北京填鴨
cran 鯡斗
cranberry 紅莓；蔓越橘；酸果蔓；苔桃
crate 條板箱；板條箱
crawfish 小龍蝦
crayfish 小龍蝦(淡水)

cream 忌廉；乳酪；乳油；奶油；稀奶油
cream beater 乳油攪拌器；奶油攪打器
cream bun 乳油小麵包
cream butyrometer 乳油用乳脂計
cream cheese 奶油芝士；乳油乾酪；稀奶油乾酪
cream cracker 奶油蘇打餅；忌廉梳打餅
cream filling 乳油夾心；奶油糖膏餡
cream fondant 夾心糖餡用的糖料
cream ice 冰淇淋
cream layer 乳油層；稀奶油層；奶油層
cream nut 巴西果
cream of tartar 酒石酸氫鉀；塔塔粉；酒石 [= potassium bitartrate; potassium acid tartrate; potassium hydrogen tartrate]
cream off 撇除奶油
cream pie 鮮乳油派；忌廉批
cream powder 乳油粉；脱水奶油
cream puff 乳油小麵包；奶油泡芙
cream roll 乳油蛋捲；奶油蛋捲
cream roux 乳油炒麵粉；奶油麵粉糊（湯料用）
cream sauce 乳油白醬料；奶油沙司；白汁
cream separator 乳油分離器；奶油分離器
cream soda 冰淇淋蘇打〔汽水〕；忌廉梳打〔汽水〕
cream soup 奶油濃湯；忌廉湯
cream-soup bowl 附耳湯碗
cream style corn 碎玉米羹；玉米糊
cream style soup 濃湯
creamer 乳油分離器；奶油攪打器；乳酪分離器
creamery 乳油〔乾酪〕製造廠；乳品商店；奶油製造廠
creaming 乳油化；乳油分離；奶油分離；乳油上浮；奶油上浮；形成乳狀物
creaming agent 乳油分層劑；起泡劑；乳脂上浮促進劑
creaming property 乳油分層性；浮油起泡性；乳脂凝聚析出現象
creamy 乳脂狀的；含乳脂的；含奶油的
creamy chocolate 乳油狀巧克力；鮮乳油巧克力；巧克力乳漿
creamy rock 奶油硬糖；乳油硬糖
creamy white 乳白色
creatine 肌酸；肌氨酸
creatine phosphate 磷酸肌酸；肌酸磷酸 [= phosphocreatine]
creatinine 肌酐；肌酸酐
crème 甜〔烈〕酒；乳酪；奶油；乳油
creosote 雜酚油
crescent 新月形小麵包；牛角麵包
cresol 甲酚
cress 水芹
cresti di gallo 雞冠形通心粉
p-cresyl acetate 乙酸對甲酚酯
p-cresyl phenylacetate 苯乙酸對甲酚酯
crewet 調味品瓶；調味品樽
cribble 粗篩；粗〔穀〕粉
crimson crab 旭蟹
crimson sea bream 鯛；血鯛；真鯛
crinoid 海百合
cripple 殘廢牲畜
crisp 脆物；炸馬鈴薯片；酥的；鬆脆的
crisp fat 豬油渣
crisphead lettuce 結球萵苣；卷心萵苣
crisping agent 脆化劑
crispness 脆性；脆度

crisppock 脆燕麥餅
critical control point (CCP) 重要控制點
critical point 臨界點
critical temperature 臨界溫度
croaker 黃姑魚；石首魚
crocetin 藏花酸
crocin 藏花素；番紅花苷
crocodile 鱷魚
crocus 番紅花；藏紅花；藏花 [= saffron; safran]
croissant 牛角麵包
crone 老母綿羊
cronton 煎麵包
crookneck (pumpkin) 長頸南瓜；彎頸南瓜
crookneck squash 長頸葫蘆；彎頸葫蘆；長頸南瓜
crop 作物；農作物
crop husbandry 耕種栽培
crop rotation 輪作；輪耕
cropping season 耕作期；耕作季
croquette 油炸菜肉丸；炸丸子；炸肉餅
cross bonded starch 交鏈澱粉；交聯澱粉 [= cross linked starch]
cross breeding 雜交育種；雜交繁殖
cross contamination 交叉污染
cross flow dryer 錯流式乾燥機
cross linked starch 交鏈澱粉；交聯澱粉 [= cross bonded starch]
cross linking gel 交鏈凝膠；交聯凝膠
cross striated muscle 骨骼肌；橫紋肌
croustade 脆麵包片；酥炸麵包塊；脆皮酥盒
crouton 脆麵包丁；油炸碎麵包片
crown roast 皇冠羊肋排烤肉
crowning machine 壓蓋機
crream meal 白玉米粉
crucian (carp) 鯽魚；歐洲鯽魚
crucible 坩堝
crude extract 粗萃
crude fat 粗脂肪
crude fiber 粗纖維
crude flour 粗磨麵粉
crude fruit 未成熟水果
crude oil 原油；粗油；毛油
crude protein 粗蛋白質
crude salt 粗鹽；原鹽；非精製鹽
crude sugar 粗製砂糖
cruet 調味品瓶；調味品罇
crule 淡味小烤餅
cruller 油煎餅；油炸麻花；油炸圓圈
crumb 碎麵包；麵包屑；搗碎
crumb softener 麵包保軟劑；麵包柔軟劑
crumble 弄碎；粉碎；麵包屑
crumbly 易碎的；脆的；沾上麵包屑的
crumbly bread 多碎屑的麵包；麵包屑
crumby bread 柔軟麵包
crumpet 烤餅；脆烤圓餅
crumpling resistance 耐皺性；耐揉性
crush 破碎；壓碎；碾碎；粉碎；碎塊
crushed ice 碎冰
crushed juice 含果肉的果汁；壓碎汁
crushed wheat 全麥粗粉
crusher 壓碎機；破碎機
crushing 壓碎；破碎
crushing mill 壓碎機；碾碎機
crust 麵包皮；餅皮；硬殼；酒垢
crusta 甲殼
crustacean 甲殼類動物
crustless cheese 無硬皮芝士；無皮乾酪；融製乾酪
crusty 有硬殼的
crusty bread 硬皮麵包
crutcher 螺旋攪拌器；垂直攪拌鍋
cryo-milling 冷凍粉碎
cryochemistry 極冷化學；低溫化學

cryogen 極冷劑；冷卻劑
cryogenic 低溫學的
cryogenic freezing 極低溫冷凍；低溫凍結
cryogenic temperature 極冷溫度
cryogenics 低溫學；低溫實驗法
cryohydrate 冰鹽
cryometer 低溫計
cryophile 嗜寒微生物；嗜冷微生物
cryophilic bacteria 嗜寒細菌；嗜冷細菌
cryopreservation 極冷保藏；低溫保藏
cryosel 冷鹽；冰鹽
cryptoxanthin 玉米黃質；隱黃質
crystal 晶體；結晶；水晶
crystal cellulose 微晶纖維素；結晶纖維素
crystal malt 焦化麥芽
crystal sugar 冰糖；晶糖；砂糖
crystal vinegar 脫色醋
crystallin 晶體蛋白
crystalline cellulose 結晶纖維素
crystalline dextrose 結晶葡萄糖
crystalline granule 半結晶體
crystalline tea 速溶茶晶
crystallite 微晶
crystallization 結晶
crystallized fruit 糖晶水果蜜餞；蜜餞水果
crystallized sugar 晶糖；結晶砂糖
crystallizing 結晶；沾上砂糖
crystallizing agent 結晶〔助〕劑
crystallizing evaporator 結晶蒸發器；蒸發結晶器
crystallizing point 結晶點；起晶點；結晶溫度
crystalloid 晶體
crystallox 糖晶；糖精鈉
crytal fruit 拌有砂糖的蜜餞
csipetke 西陪特克餃子(匈牙利)
Cuban suagr 古巴糖
cube 方塊；立方體
cube cutter 切塊機；切丁機
cube ice 小冰塊；方塊冰
cube roll steak 沙朗牛排
cube sugar 方糖 [= cut sugar; block sugar]
cubeb oil 華澄茄油
cubed meat 肉丁
cubelet 小方塊；丁
cuber 切丁機；壓方塊機
cucumber 黃瓜；胡瓜；青瓜
cucumber alcohol 黃瓜醇
cucumber oil 黃瓜子油
cucumber pickle 酸漬黃瓜；醃黃瓜
cucumber seed oil 黃瓜子油
cucurbit 南瓜；葫蘆
cucurbitol 南瓜子醇
cud 反芻的食物
cuisine 烹飪；菜餚；烹調風格
cuke 小黃瓜；黃瓜
culinary art 烹調技術
culinary value 營養值
cull eliminator 揀選台；分級台；分選機
cull fruit 淘汰果；剔出的殘果
cullage 淘汰品；剔除〔物品〕
cullender 粗濾器；濾鍋；爪籬
culm 莖；桿
cultivar 栽培〔品〕種
cultivated mushroom 洋菇
cultivation 耕種；耕作；培養；栽培
culture 菌種；種菌；培養；培養物
culture bottle 培養瓶
culture broth 培養液
culture tube 培養試管
culture yeast 培養酵母

cultured buttermilk 發酵酪乳；酸性酪乳
cultured dish 培養皿
cultured medium 培養基
cultured milk 發酵乳；酸乳
cumaldehyde 枯茗醛；蒔蘿醛
cumin 枯茗；歐蒔蘿；小茴香；孜然芹；孜然芹籽 [= cummin]
cumin aldehyde 小茴香醛；枯茗醛
cumin oil 小茴香油；枯茗油；姬茴香油；孜然油
cumin seed 枯茗籽；孜然籽；小茴香籽；歐蒔蘿籽 [= cummin seed]
cuminic acid 小茴香酸；枯茗酸
cummin 枯茗；歐蒔蘿；小茴香；孜然芹；孜然芹籽 [= cumin]
cummin seed 枯茗籽；小茴香籽；歐蒔蘿籽 [= cumin seed]
cumquat 金橘；金桔 [= kumquat]
cup 杯子；酒杯
cup and saucer 附碟杯
cup cake 小蛋糕；紙杯蛋糕
cup fungus 杯子洋菇；碗菌 [= cup mushroom]
cup like morel 肋狀皺盤菌（蕈類）
cup mushroom 杯子洋菇；碗菌 [=cup fungus]
cup noodle 杯裝速食麵；杯麵
cup rice 杯裝速食飯
cup test 品嚐試驗
cupboard 碗櫥；食櫥；碗櫃
cupful 一杯之量；滿杯
cupping room 品嚐室
cupric sulfate 硫酸銅
cuprous iodide 碘化亞銅
curacao 克利沙酒；橘味烈酒；柑香酒；庫拉索酒
curative food 療效食品
curcuma 薑黃
curcuma oil 薑黃油
curcumin 薑黃；薑黃素；酸性黃 [= turmeric yellow]
curd 凝乳；凝塊
curd mill 乳塊切碎機；乾酪凝乳切碎機
curd particle 凝乳粒
curd tart 凝乳餡餅
curd vat 凝乳槽
curdled 凝固的；凝結的
curdled cream 凝塊鮮乳油；結塊奶油
curdy 凝結成塊的；凝乳狀的
curdy cheese 未成熟的乾酪
cure 保藏；醃製；熟成
cured cheese 熟成芝士；熟成乾酪；發酵成熟的芝士
cured fish 鹹乾魚；鹹魚；醃乾魚
cured flavo(u)r 醃肉香味
cured ham 醃製火腿
cured meat 醃肉；鹹肉；醃製肉食
cured meat colo(u)r 醃肉色
cured product 醃藏食品；醃製產品
cured raisin 風乾葡萄乾
curing 醃漬；醃製；保藏；熟成
curing agent 醃料；醃製劑
curing frame 煙燻架
curing mix 混合醃劑；醃製用混合鹽
curing period 醃製期；熟成期
curing pickle 醃漬用鹽汁
curing room 醃製室；熟成室
curing salt 醃製用〔混合〕鹽
curing tank 醃〔製〕槽；醃缸
curl 罐蓋捲曲部；蓋邊
curled lettuce 皺葉萵苣
curled mustard 皺葉芥菜；鬈葉芥
curling machine 捲邊機
currant 無核小葡萄乾；穗醋栗
currant bread 葡萄乾麵包
currant tomato 醋栗蕃茄

curried 加入咖哩〔粉〕的；用咖哩〔粉〕調味的；咖哩味的
curried beef 咖哩牛肉 [= curry beef]
curried chicken 咖哩雞肉 [= curry chicken]
curry 咖哩；咖哩粉；咖哩食品；用咖哩粉調味
curry beef 咖哩牛肉 [= curried beef]
curry chicken 咖哩雞肉 [= curried chicken]
curry powder 咖哩粉
curry rice 咖哩飯
cushaw 南瓜；倭瓜
cushcush 蔗渣屑
cushion 臀部肉；襯墊
cusk 紅鱈；單鰭鱈
custard 卡士達〔點心〕；奶油蛋羹；奶油凍；蛋奶甜羹
custard apple 牛心梨；釋迦〔果〕；蕃荔枝；番荔枝
custard cream 卡士達糊；蛋奶羹
custard pie 卡士達派
custard powder 吉士粉；卡士達粉；調色玉米澱粉
custard pudding 卡士達布丁；牛奶布丁
custard sauce 卡士達醬；蛋黃沙司
custard squash 扇貝南瓜
cut 肉塊；分割肉
cut back 原果汁回添〔法〕；稀釋；削減
cut corn 玉米片
cut mixer 斬拌機
cut-out Brix 開罐糖度
cut-out test 開罐試驗
cut sugar 方糖 [= cube sugar; block sugar]
cut up 分割；切碎；按部位分割；按量分割
cut-wrap machine 切割包裝機
cutch 兒茶
cutlass fish 隱足魚；帶魚
cutlery 刀叉餐具；餐具
cutlet 小肉片；一人份量的肉排；肉片；炸肉排
cuts chart 肉類分切圖；分切圖
cutter 切割機；切刀；刀具
cutting 分割；切割；切肉；絞肉
cutting board 砧板；切肉板
cuttle 墨魚；花枝；墨鰂；烏賊 [= cuttlefish]
cuttle bone 墨魚骨
cuttlefish 墨魚；花枝；墨鰂；烏賊 [= cuttle]
cuvee 桶內葡萄酒；酒桶內容物
cuvette 小透明玻璃管；光析管
cyanide 氰化物
cyanocobalamin 氰鈷胺素；氰鈷維生素；維生素B_{12}
cyclamate 賽克拉美；環己烷基磺酸鹽（人工甜味劑）
cyclamen aldehyde 兔耳草醛；仙客來醛
cyclic hydrocarbon 環烴
cyclic peptide 環肽
cyclitol 環糖醇；環醇；環多醇
cyclodextrin 環糊精
β-cyclodextrin β-環糊精
cycloglucan 環糊精；環葡聚糖
cyclohexane 環己烷
cyclohexanol 環己醇
cyclohexyl acetate 醋酸環己酯
cyclohexyl butyrate 丁酸環己酯
cyclohexyl sulfamic acid 環己基氨基磺酸
cyclone 旋風器；旋風分離器
cyclone collector 旋風收集器
cyclone scrubber 旋風除塵器

cyclone separator 旋風分離器
cyder 蘋果汁；蘋果酒 [= cider]
cylinder dryer 筒式乾燥機；滚筒乾燥機
cylinder drying machine 筒式乾燥機；滚筒乾燥機
cylindrical heat exchanger 圓筒形熱交換器；圓筒形換熱器
cylindrical retort 圓〔筒形〕殺菌釜
p-cymene 對百里香素
cyperus oil 莎草油；香附〔子〕油
cyprinid 鯉科的魚類；鯉科的
cyst 胞囊；孢囊；休眠孢子
cysteic acid 磺基丙胺酸
cysteine (Cys) 半胱胺酸
L-cysteine (mono)hydrochloride L-半胱氨酸鹽酸鹽
cysticercosis 囊蟲病
cysticercus 囊蟲
cystine 胱胺酸；胱氨酸
cytase 溶胞酶；細胞溶解酶
cytidine 胞〔嘧啶核〕苷
cytobiology 細胞生物學
cytochemistry 細胞化學
cytochrome 細胞色素 [= histohaematin]
cytochrome oxidase 細胞色素氧化酶
cytochrome reductase 細胞色素還原酶
cytoflavin 細胞黃素
cytolipin 胞糖脂；細胞糖苷脂
cytolysin 細胞溶素
cytolysis 細胞溶解
cytolytic enzyme 細胞溶解酶
cytoplasm 細胞質
cytosine 胞嘧啶
cytotoxin 細胞毒素
Czapek bouillon 察氏肉汁〔培養基〕
Czapek Dox agar 察氏培養基 [= Czapek's medium]
Czapek's medium 察氏培養基 [= Czapek Dox agar]

D

D-Rump 修清臀肉；尾龍肉；冧肉
D value 九成滅菌時間；D值
[= decimal reduction value]
dab 黃蓋鰈
dace 鯪魚
dai-dai flower oil 玳玳花油
daikon (radish) 日本蘿蔔
daily food 日常食品
dainty 可口的；美味的；可口食品
daiquiri 黛可莉〔雞尾〕酒
dairy 製乳場；牛奶場；乳品業
dairy breed 乳用牛種
dairy cattle 乳牛
dairy farm 乳牛場；牧場
dairy food 乳製品
dairy industry 乳品〔加工〕業
dairy product 乳製品
dairyman 乳牛場場主；乳製品加工工人
daisy 黛西雞尾酒；雛菊
daisy ham 去骨燻火腿
damaged sugar 結晶不好的糖
damascone 突厥酮
dammar resin 達馬樹脂；達馬樹膠
damp off 濕潤腐爛；枯萎
damping machine 調濕機
damping vat 汽蒸〔增濕〕桶
dampness 潮濕；濕氣
dampproof 防潮的；防濕的
damson (plant) 西洋李子；大母松李；蜜李
dandelion 蒲公英
dandy cart 鮮乳車；送牛奶車
danger zone 危險區域
Danish agar 紅藻膠 [= furcellaran]
Danish bread 丹麥酥餅
Danish landrace 丹麥藍瑞斯豬；丹麥蘭德瑞斯豬；丹麥長白豬
Danish pastry 丹麥餡餅
Danish sausage 丹麥香腸
daphne oil 瑞香油
daphnetin 瑞香素
daren flour 達蘭麵粉
darjeeling 紅茶
dark beer 黑啤酒
dark cappuccino 黑卡布奇諾咖啡；黑卡布基諾咖啡 [= cappuccino scuro]
dark chocolate 黑巧克力；黑朱古力
dark crystal 原糖；紅糖
dark flour 深色麵粉
dark malt 深色麥芽
dark meat 深色肉；暗色肉；血合肉；紅肉
dark muscle 深色肌肉；血合肉；暗紅色肉
dark reaction 暗反應
dark scaled knight cap 暗鱗口蘑（蕈類）
dark soy (sauce) 老抽；深色醬油

dark, firm, dry muscle (DFD muscle) 深暗緊密和乾燥肉；色深質硬乾燥肉；暗乾肉；DFD肉
darkening 發黑；變黑
darnel 黑麥草；毒麥
dash churn 沖擊攪動器；撞擊攪乳器
dasheen 芋；芋頭 [= taro]
dasher 攪動翼；攪拌裝置
date 棗；棗椰果；海棗；椰棗
date kernel oil 海棗油
date palm 棗椰樹；棗椰子；海棗
date paste 棗泥
dateplum 柿子；君遷子
david peach 山桃
daylily 金針
DE (dextrose equivalent) 葡萄糖當量；葡萄糖值 [= dextrose equivalent (value)]
deacidification 去酸〔作用〕；脫酸〔作用〕
deactivation 去活化；鈍化；失活
deactivator 去活化劑；鈍化劑；失活劑
dead colo(u)r 肉色蒼白；肉色暗淡
dead meal 廢棄〔麵粉〕原料
dead water 靜水；死水
deaerated juice 脫氣果汁
deaeration 脫氣〔作用〕
deaerator 脫氣裝置
deamidization 去醯胺〔基〕作用；脫醯胺〔基〕作用
deaminase 去胺〔基〕酶；脫胺〔基〕酶 [= desaminase]
deamination 去胺〔基〕作用；脫胺〔基〕作用 [= desamination]
deaquation 脫水作用
dearth 缺乏；饑荒
deastringent 去澀味；脫澀味
death phase 死亡期
death stiffening 死後僵直
debittering 去苦味；脫苦味
debittering yeast 去苦味酵母；脫苦味酵母
debitterize 去苦味；脫苦味
debone 去骨
deboner 去骨機；去骨工人
deboning machine 去骨機
debranching enzyme 去支酶；剪支酶；脫支酶
debris 餘渣；碎屑；殘渣
decadienoic acid 癸二烯酸
decaffeinated coffee 去咖啡因咖啡；無咖啡因咖啡
decaffeination 去咖啡因；除去咖啡因
γ-decalactone γ-癸內酯；丙位癸內酯
δ-decalactone δ-癸內酯；丁位癸內酯
decalcification 去鈣作用；脫鈣作用
decalcify 去鈣；脫鈣
decanal 癸醛
decanal dimethyl acetal 癸醛二甲縮醛
decanoic acid 癸酸；羊脂酸 [= capric acid]
decanol 癸醇
decantation 傾析
decanter 傾析器；傾析用的玻璃瓶；傾析〔式〕離心機
decanting tank 傾析槽
decap 開瓶；去瓶蓋
decapitation 去頭
decapper 開瓶器
decapsulation 去莢
decarbonation 去二氧化碳；除碳酸
decarbonization 去碳〔作用〕；脫碳〔作用〕
decarboxylase 去羧酶；脫羧酶
decarboxylation 去羧基〔作用〕；脫羧基〔作用〕
decay 腐爛；腐朽；衰變
deceiving knight cap 絲蓋口蘑（蕈類）

dechlorination 去氯〔作用〕；脱去氯〔作用〕
deciduous tooth 乳齒
deciliter 公合；十分之一公升
decimal reduction value D值；九成滅菌時間 [= D value]
deck oven 櫃式烤爐
decoct 熬；煎煮；煎藥
decoction 熬；煎煮；煎熬；熬出物；煎熬出的藥汁；煎劑
decolo(u)rant 脱色劑；漂白劑
decolo(u)ration 脱色〔作用〕；漂白〔作用〕
decolo(u)rization 脱色〔作用〕；漂白〔作用〕
decolo(u)rizer 脱色劑；漂白劑
decolo(u)rizing agent 脱色劑；漂白劑
decomposability 可分解性
decomposer 分解體；分解槽
decomposition 分解〔作用〕
decomposition product 分解產物
deconcentrator 減濃器
decontaminant 去污染劑；淨化劑
decontaminating agent 去污染劑；淨化劑
decontamination 去污染；淨化
decorated cake 裝飾蛋糕；花飾蛋糕
decortication 剝皮；脱皮；去殼
decorticator 剝皮機；脱皮機；去殼機
decosahexaenoic acid 二十碳六烯酸
decrowner 開瓶器；去瓶蓋器
decyl acetate 醋酸癸酯；乙酸癸酯
decyl alcohol 癸醇
dedust 除塵；去灰
deemulsification 浮濁澄清；去乳化；反乳化
deemulsifying agent 破乳劑；浮濁澄清劑；反乳化劑
deep agar method 深層洋菜〔培養〕法；厚層瓊脂〔培養〕法
deep colony 深層菌落
deep culture 深層培養
deep drawn can 深沖壓罐；深拉拔罐
deep fat frying 油炸 [= deep frying]
deep freezer 超低溫冷凍櫃；超低溫冷凍庫；深度冷凍櫃
deep freezing 超低溫冷凍；深度冷凍
deep fried product 油炸食品
deep frozen meat 超低溫冷凍肉
deep frozen product 超低温凍結食品；深度凍結食品
deep fryer 油炸鍋
deep frying 油炸 [= deep fat frying]
deep frying pan 油炸鍋
deep milk medium 深層牛乳培養基
deep sea fish 深海魚
deep sea fishery 遠洋魚業；深海魚業
deer 鹿
deesterification 去酯化〔作用〕；脱酯〔作用〕
defatted food 脱脂食品
defatted meal 脱酯粕
defatted peanut 脱脂花生
defatted soy flour 脱脂大豆粉
defeather 去毛；脱毛；拔毛；打毛
defeatherer 去毛機；打毛機
defeathering machine 去毛機；打毛機
defecation 澄清〔作用〕；淨化〔作用〕；加灰澄清
defecator 澄清器；清淨器；澄清槽
defeco-carbonation process 石灰碳酸澄清法
defective absorption 吸收不良；吸收障礙
deficiency 缺乏；不足
deficiency disease 營養缺乏症

defined medium 合成培養基 [= synthetic medium]
deflation 放氣；脱氣
deflavour 去雜味
deflocculating agent 去絮凝劑；反絮凝劑
deflocculation 去絮凝〔作用〕；反絮凝〔作用〕
deflocculator 去絮凝離心機；反絮凝離心機
defluxion 溢流
defoam 消泡；去除泡沫
defoamer 消泡劑；消泡裝置；去沫劑
defoaming agent 消泡劑；除沫劑；防沫劑 [= antifoaming agent; kilfoam]
defoliant 落葉劑；脱葉劑
defoliation 落葉；打葉
deformation 變形
defreeze 解凍
defrost 解凍；除霜；融霜；沖霜
defrostation 解凍；除霜；融霜；沖霜
defrosted meat 解凍肉
defroster 除霜裝置；解凍裝置
defrosting machine 除霜機；解凍機
defrosting room 解凍室
defuselation 去除雜醇油
degas 脱氣
degasification 脱氣〔作用〕；去氣〔作用〕；除氣〔作用〕
degassing 脱氣；除氣；放氣
degelatinization 去膠〔作用〕
degeneration 退化
degerm 除菌；去胚芽；脱胚
degerminating machine 去胚芽機
degerminating mill 去胚芽磨
degerminator 去胚芽機；去胚器
deglutition 吞嚥
degradation 降解〔作用〕
degradation product 降解產物
degras 氧化魚油
degrease 脱脂；去除油污
degreaser 去油污劑；脱脂劑
degree of doneness 熟嫩程度
degree of spoilage 腐敗程度
degreening 去綠〔作用〕
degum 脱膠；水洗
degumming 脱膠；水洗
degumming agent 脱膠劑
degust(ate) 嘗〔味〕
degustation 嘗味；品嚐
dehair 去毛；脱毛；刮毛
dehairer 去毛機；打毛機；脱毛機
dehairing machine 去毛機；脱毛機
dehide 剝皮
dehiding 剝皮
dehiscent fruit 裂果
dehull 去殼；脱殼；去皮；脱皮
dehuller 去皮機；脱皮機；去殼機；脱殼機
dehumidification 除濕〔作用〕；減濕〔作用〕
dehumidifier 除濕機；抽濕機
dehusk 去皮；脱殼
dehydrant 脱水劑；乾燥劑
dehydrase 脱水酶
dehydrated food 脱水食品
dehydrated meat 脱水肉；肉乾
dehydrated soup 脱水湯粉
dehydrated vegetable 脱水蔬菜
dehydrater 脱水劑；乾燥劑；脱水器 [= dehydrator]
dehydration 脱水〔作用〕
dehydrator 脱水劑；乾燥劑；脱水器 [= dehydrater]
dehydroacetic acid (DHA) 脱氫醋酸；去氫醋酸
dehydroascorbic acid 去氫抗壞血酸；脱氫抗壞血酸

dehydrocanning 脫水製罐
dehydrocholesterol 去氫膽固醇；脫氫膽固醇
dehydrofreezing 脫水冷凍〔法〕；冷凍脫水〔法〕
dehydrofrozen 脫水冷凍的；冷凍脫水的
dehydrogenase 去氫酶；脫氫酶
dehydrogenation 脫氫〔作用〕；去氫〔作用〕
deicer 防凍劑；防冰器；除冰裝置
deicing fluid 防凍液
deionization 去離子〔化〕作用；消電離作用
deionized water 去離子水
deionizer 去離子器；去離子劑
dejeuner 早餐；午餐
delactosed milk 去乳糖乳；脫〔乳〕糖乳
delactosed whey 去乳糖乳清；脫〔乳〕糖乳清
deleterious 有害的；有毒的
delicacy 美味食品；精美食品
delicate flavo(u)r 美味；鮮味
delicate food 精緻食物；美味食品
delicatessen 熟食店；熟食；現成菜餚；現成食物
delicatessen shop 熟食店
Delicious 五爪甜蘋果；元帥蘋果
delicious milk cap 松乳菇
deliquate 沖淡
deliquescence 潮解；液化
demargarination 去固態油〔作用〕；去硬脂〔作用〕
demerara sugar 紅蔗糖（含6—9%蔗糖漿）；淡褐色粗糖
demersal fish 底層魚類；底棲魚類
demethylation 去甲基〔作用〕
demifine 中等的葡萄酒
demineralization 去礦物質〔作用〕；脫礦質〔作用〕
demineralized water 去礦物質水；軟化水
demineralized whey 去鹽乳清
demineralizer 去礦物質裝置
demitasse 小杯黑咖啡；小杯清咖啡；小咖啡杯
demulsification 去乳化〔作用〕；反乳化〔作用〕
demulsifier 去乳化劑；反乳化劑
demulsify 去乳化；反乳化
demulsifying agent 去乳化劑；反乳化劑
denaturant 變性劑
denaturation 變性〔作用〕
denatured protein 變性蛋白質
denaturing agent 變性劑
dendritic salt 枝晶食鹽；易溶食鹽；星狀結晶〔易溶〕食鹽
denitration 去硝〔作用〕
denitrification 去氮〔作用〕；脫氮〔作用〕
denitrifying bacteria 去氮細菌；脫氮細菌
densimeter 密度計；比重計
densitometer 光密度計；比重計
density 密度
dent corn 馬齒玉米；臼齒形玉米
dented can 凹罐
denting 罐頭凹陷
denutrition 營養不足；營養缺乏
deodorant 除臭劑；脫臭劑
deodorization 除臭；脫臭
deodorizer 脫臭器；除臭劑；脫臭劑
deoiler 脫油裝置
deoxidant 去氧劑；脫氧劑；除氧劑
deoxidation 去氧〔作用〕；脫氧〔作用〕
deoxidization 去氧〔作用〕；脫氧〔作用〕

deoxidizer 去氧劑
deoxycholic acid 脱氧膽酸
deoxygenase 去氧酶；脱氧酶
deoxyribonuclease 去氧核糖核酸酶；脱氧核糖核酸酶
deoxyribonucleic acid (DNA) 脱氧核糖核酸；去氧核糖核酸
deoxyribose 脱氧核糖；去氧核糖
deoxysugar 脱氧糖；去氧糖
depectinization 去果膠；脱果膠
depectinized juice 去果膠果汁；脱〔果〕膠果汁
dephlegmation 分餾；分凝
dephlegmator 分餾器；分餾塔
depigmentation 去色素；脱色
depilate 去毛；脱毛
depilating compound 脱毛膠料；去毛膠料
depilation 去毛；脱毛
depilator 去毛機；脱毛機
depilatory 脱毛的；去毛劑；脱毛劑
deplumation 去〔羽〕毛；拔〔羽〕毛
depolymerase 解聚酶
depolymerization 解聚〔合〕作用
deposited fat 儲存脂肪；沉積脂肪；儲藏脂肪
deposition 沉積物；沉積作用
depository 倉庫；儲藏室
depot 倉庫
depot fat 儲存脂肪；沉積脂肪；儲藏脂肪
depraved appetite 食慾不振
depressant 抑制劑；鎮靜劑
deproteinization 去蛋白質〔作用〕；脱蛋白質〔作用〕
deproteinize 去蛋白質；脱蛋白質
depurant 淨化劑；淨化器
depuration 淨化；純化；精煉；清潔
depurator 淨化器；淨化劑
derind 去皮
derinder 去皮機
derivative 衍生物
derivatized starch 衍生澱粉
derived lipid 衍生脂質
desaccharification 去糖〔作用〕；提糖〔作用〕
desaccharify 去糖；提糖
desalination 脱鹽；去鹽；化淡
desalted 去鹽的；脱鹽的
desalting 去鹽；脱鹽
desaminase 去胺〔基〕酶；脱胺〔基〕酶 [= deaminase]
desamination 去胺〔基〕作用；脱胺〔基〕作用 [= deamination]
desaturation 去飽和〔作用〕
descale 去鱗；去鍋垢
descaler 去鱗器；去〔鍋〕垢劑
desiccant 乾燥劑 [= desiccating agent]
desiccate 乾燥物；乾燥
desiccating agent 乾燥劑 [= desiccant]
desiccation 乾燥〔作用〕
desiccative 乾燥的；乾燥劑
desiccator 乾燥器；防潮箱；保乾器
designer food 特製食品
desintegrator 磨碎機；粉碎機
desolventization 去溶劑；除溶劑
desolventize 去溶劑；除溶劑
desolventizer 脱溶劑器；除溶劑器
desorption 解吸附〔作用〕；脱附〔作用〕
desoxycholic acid 脱氧膽酸
desoxyribonucleic acid 脱氧核糖核酸；去氧核糖核酸
desoxyribose 脱氧核糖；去氧核糖
desoxysugar 脱氧糖；去氧糖
despumation 去浮渣；除泡沫
dessert 甜點；甜品
dessert service 甜點餐具
dessert wine 餐後〔甜〕酒；餐後葡萄酒

destabilization 去穩定〔作用〕；破壞穩定〔作用〕
destabilizer 去穩定劑；破壞穩定劑
destainer 去染劑
destarch 脫漿；脫澱粉
destearination 去固態油〔作用〕；去硬脂〔作用〕
destruction curve 破壞曲線；致死曲線
desugar 去糖；脫糖；提糖
desulfurization 去硫〔作用〕；脫硫〔作用〕
deswelling 消脹〔作用〕
detacher 鬆粉器
detector 檢波器；偵察器；探測器
detergent 清潔劑；去垢劑
deterioration 變質；敗壞
determination 測定
detersive 清潔劑
detinning 脫錫
detoxi(fi)cation 解毒
detoxicant 解毒劑
devein 去腸〔線〕
deveined shrimp 去腸〔對〕蝦
developer 揉麵機；和麵機
developing agent 顯影劑
development 顯影；發育；發展
deviation 偏差；誤差
device 裝置；設備
devil 辛辣菜餚
devilfish 玫瑰毒鮋；灰鯨；大鰩魚；章魚
devilled 辣味的
dew point 露點
dew point hygrometer 露點濕度計
dewater 去水；脫水
dewatering screen 瀝水篩
dewax 脫蠟
dewberry 露珠莓；黑莓
dewlap 垂肉(喉部)

dexpanthenol D-泛醇
dextran 1,6-聚葡萄糖；葡聚糖
dextranase 1,6-聚葡萄糖酶；葡聚糖酶
dextrin(e) 糊精
dextrin(e) gum 糊精膠
dextrinase 糊精酶
dextrinization 糊精化〔作用〕
dextrorotation 右旋〔作用〕
dextrose 葡萄糖；右旋糖 [= glucose; grape sugar]
dextrose equivalent (DE) 葡萄糖當量；葡萄糖值 [= dextrose equivalent (value)]
dextrose equivalent (value) 葡萄糖當量；葡萄糖值
dextrose syrup 葡萄糖漿 [= glucose syrup]
DFD muscle (dark, firm, dry muscle) 深暗緊密和乾燥肉；色深質硬乾燥肉；暗乾肉；DFD肉
dhal 豆
diabetes 多尿症；糖尿病
diabetes mellitus 糖尿病
diabetic food 糖尿病人食品
diacetin 二乙酸甘油酯；二醋酸甘油酯
diacetyl 雙乙酰；丁二酮；聯乙酰 [= biacetyl]
diacetyl tartaric acid esters of mono and diglycerides 雙甘油酯
dialdehyde starch 二醛澱粉；雙醛澱粉
dialysate 透析液；滲析液
dialysis 透析；滲析
dialysis tube 透析管；滲析管
dialyzer 透析器；滲析器；滲析膜
diamino acid 二胺基酸
diaphragm 橫膈膜 [= midriff]
diarrhea 腹瀉
diastase 澱粉酶；澱粉糖化酶
diastatic activity 澱粉〔糖化〕酶活性

diastatic enzyme 澱粉分解酶；糖化酶
diathermic heating 高頻加熱
diatom 矽藻；硅藻
diatom earth 矽藻土；硅藻土
diatomaceous earth 矽藻土；硅藻土
diatomite 矽藻土；硅藻土
diazotroph 固氮生物
dibenzoyl thiamine 二苯酰硫胺素
dibenzoyl thiamine hydrochloride 二苯酰硫胺鹽酸鹽
dibenzyl ether 二苄醚
dibutyl sebacate 癸二酸二丁酯
dibutyl sulfide 二丁基硫醚
dicalcium phosphate 磷酸二鈣
dicalcium pyrophosphate 焦磷酸鈣
dicarboxylic acid 二元酸；二羧酸
diced 切粒；切成方塊
diced meat 肉丁
dicer 切丁機
dichlorinated sodium isocyanurate 二氯異氰尿酸鈉
dichlorodifluoromethane 二氯二氟甲烷
dichloromethane 二氯甲烷
dichromate 二鉻酸鹽；重鉻酸鹽
dicing machine 切丁機
dicoumarin 雙香豆素
dicoumarol 雙羥薰草素；雙羥香豆素
dielectric constant 介電常數
dielectric defrosting 介電解凍；高頻加熱解凍
dielectric drying 介電乾燥；高頻加熱乾燥
dielectric heating 介電加熱；高頻加熱
dielectric thawing 介電解凍；高頻加熱解凍
dienoic acid 雙烯酸；二烯酸
diet 飲食；膳食
diet food 減肥食物；膳食療養食品；醫療食品
dietary 規定食物；膳食規定；飲食的；膳食表
dietary allowance 膳食營養攝取量
dietary cure 膳食療養法；食療法
dietary deficiency 膳食營養缺乏
dietary factor 膳食要素；食物要素
dietary fiber 膳食纖維
dietary supplement 膳食增補劑；營養增補劑
dietary therapy 膳食治療；飲食治療
dietetic 飲食的；食療的
dietetic beverage 食療飲料；療效飲料
dietetic food 食療食物；療效食品
dietetics 膳食學；營養學；食療法
diethyl acetaldehyde 二乙基乙醛
diethyl ether 乙醚；二乙基醚 [= ether]
diethyl malonate 丙二酸二乙酯
diethyl pyrocarbonate 焦碳酸二乙酯
diethyl sebacate 癸二酸二乙酯
diethyl succinate 丁二酸二乙酯；琥珀酸二乙酯
diethyl tartrate 酒石酸二乙酯
diethylacetic acid 二乙基乙酸
diethylene glycol monoethyl ether 二甘醇一乙醚
dietician 營養師；飲食學家；營養學家 [= dietitian]
dietics 飲食學；營養學
dietitian 營養師；飲食學家；營養學家 [= dietician]
dietotherapy 膳食療法；飲食療法
differential medium 鑑別培養基
differential thawing 分段解凍
diffluence 分流；溶解
diffusate 滲出液
diffuser 擴散器；浸提器；滲出器 [= diffusor]
diffusibility 擴散性
diffusion 擴散；浸提；滲出

diffusion coefficient 擴散系數
diffusion gradient 擴散梯度
diffusion juice 滲出汁
diffusion water 浸提用水
diffusivity 擴散率
diffusor 擴散器；浸提器；滲出器 [= diffuser]
difurfuryl disulfide 二糠基二硫醚
difurfuryl sulfide 二糠基硫醚
digenic acid 海人〔草〕酸
digest 消化
digestant 助消化劑；胃藥
digester 消化槽；煮解器；浸煮器；高壓蒸煮鍋
digestibility 消化率；消化性
digestibility coefficient 消化〔率〕系數
digestible 可消化的
digestible nutrient 可消化營養素
digestion 消化；煮解
digestive 消化的；助消化的；助消化劑
digestive enzyme 消化酶
digestive fluid 消化液
digestive gland 消化腺
digestive juice 消化液
digestive problem 消化不良
digestive tract 消化道
digitalin 毛地黃苷
digitonin 毛地黃皂苷
digitoxin 毛地黃毒苷
diglyceride 甘油二〔酸〕酯
dihydroanhydrovitamin 二氫去水維生素
dihydrocarveol 二氫香芹酚；二氫黃蒿萜醇
dihydrocarvone 二氫香芹酮；二氫黃蒿萜酮
dihydrocarvyl acetate 乙酸二氫葛縷酯
dihydrochalcone (DHC) 二氫查耳酮
dihydrocholesterol 二氫膽固醇
dihydrocoumarin 二氫香豆素
dihydrofolic acid 二氫葉酸
dilatometer 膨張計
dilaurin 甘油二月桂酸酯；二月桂精
dilauryl thiodipropionate 硫代二丙酸二月桂酯
dilinolein 甘油二亞麻油酸酯；二亞油精
dill 蒔蘿；土茴香
dill seed 蒔蘿籽
dill seed oil 蒔蘿籽油
dill weed 蒔蘿葉
diluent 稀釋液；稀釋劑；沖淡劑
dilute 稀釋〔的〕；沖淡〔的〕
dilution 稀釋；沖淡；稀釋物
dimer 二聚物
dimerization 二聚〔合〕作用
dimethoxybenzene 二甲基對苯二酚；對二甲氧〔基〕苯
3,4-dimethoxybenzenecarbonal 3,4-二甲氧基苯甲醛
dimethyl anthranilate 鄰甲氨基苯甲酸甲酯
dimethyl benzyl carbinol 二甲基苄基原醇
dimethyl benzyl carbinyl acetate 乙酸二甲基苄基甲醇酯
dimethyl benzyl carbinyl butyrate 丁酸二甲基苄基甲醇酯
dimethyl dicarbonate (DMDC) 焦碳酸二甲酯；二碳酸二甲酯
dimethyl disulfide 二甲基二硫醚
dimethyl ketone 丙酮；二甲基酮 [= acetone]
dimethyl sulfide 二甲硫；甲硫醚；二甲基硫醚
dimethyl trisulfide 二甲基三硫醚
dimethylamine 二甲胺
dimethylpyrazine 二甲基吡嗪

dimethylpyrrole 二甲基吡咯
3,7-dimethylxanthine 3,7-二甲基黃嘌呤；可可鹼
dimorphism 雙晶性；二態性
dimorphous 雙晶的；二態的
dine 正餐；進餐
diner 用餐者；餐車
diner-out 外食者
dinette 小餐室；小餐車
dining 用餐；吃飯
dining car 餐車
dining hall 餐廳；食堂
dining room 餐廳
dining table 餐桌
dinkel (wheat) 斯佩爾特小麥
dinner 正餐；宴會
dinner biscuit 餐用比司克；餐用餅乾
dinner bucket 餐盒；飯桶
dinner pail 餐盒
dinner party 宴會
dinner roll 餐用麵包
dinner service 成套的餐具；整套的餐具
dinner set 成套的餐具；整套的餐具
dinner wine 餐酒
dioctyl sodium sulfosuccinate 磺基鈉代丁二酸二辛酯
diol 二〔元〕醇
diolein 甘油二油酸酯；二油酯
diose 二〔碳〕醣；乙醣 [= biose]
dioxide 二氧化物
dioxin 戴奧辛；二噁甲；二氧芑
dip 浸；蘸；沾；調味汁
dip freezing 浸〔漬冷〕凍
dipeptidase 二肽酶
dipeptide 二肽；縮二胺酸
diphenol 二酚；聯苯酚
diphenol oxidase 二酚氧化酶
diphenyl 聯苯
diphenyl oxide 二苯醚
diphenylamine 二苯胺
diplococcin 雙球菌素
diplococcus 雙球菌
diploid 二倍體
dipotassium (hydrogen) phosphate 磷酸氫二鉀
dipotassium 5’-guanylate 5’-鳥苷酸二鉀
dipotassium 5’-inosinate 5’-肌苷酸二鉀
dipotassium tartrate 酒石酸二甲
dipper 長柄杓
dipping 浸漬
dipsomania 嗜酒狂；嗜酒癖
direct fermentation 直接發酵〔法〕
direct fire 直接火；活火
direct flame sterilization 直接火燄滅菌
direct flash pasteurization 直接蒸汽瞬間巴斯德殺菌
direct flowing coffee cup 咖啡直流杯
direct freezing 直接冷凍
direct heating 直接加熱
direct immersion freezing 直接浸漬冷凍；直接浸漬式速凍
direct microscopic count 直接鏡檢數
direct microscopy 直接鏡檢法
direct steam 直接蒸汽
direct steam injection heating 直接蒸汽噴射加熱
dirt 污物；污垢
disaccharidase 雙糖酶；二糖酶
disaccharide 雙糖；二糖 [= disaccharose]
disaccharose 雙糖；二糖 [= disaccharide]
disaggregation 解聚〔作用〕
disc 圓盤；圓片狀通心粉 [= disk]
disc atomizer 盤式噴霧器

disc bowl centrifuge 盤碗離心機；碟式離心機
disc colorimeter 盤式比色計
disc dryer 盤式乾燥器
disc evaporator 盤式離心蒸發器
disc mill 盤式磨
disc pulverizer 盤式粉碎機
disc shredder 盤式切碎機
disc slicer 圓刀切片機；圓盤切片機
discer 瓶口加環機
discharge atomizer 排放噴霧器
discharge fan 排氣風扇
discharge gas 廢氣
discolo(u)ration 變色；褪色；脱色
discolo(u)ring 褪色；脱色
discontinuous agitating retort 不連續攪動殺菌釜；迴轉式間歇殺菌釜
discontinuous pasteurization 間歇巴斯德殺菌法
discontinuous phase 不連續相
discontinuous sterilization 間歇滅菌法
disembowel 開膛取臟；除去內臟
disgerminator 去芽機；除胚芽機
disgorgement 除酒渣；除酵母泥渣
disgorger 除〔香檳〕酒泥機
dish 盤；碟；菜餚；盤〔裝〕菜
dishcloth 抹布；碗布
dishcloth gourd 絲瓜；菜瓜 [= sponge cucumber; sponge gourd]
dishwasher 洗碗機；洗碗工人
dishwater 洗碗水
disinfectant 消毒劑；殺菌劑
disinfecting agent 消毒劑；殺菌劑
disinfection 消毒；殺菌
disinfector 消毒器
disinfestant 滅蟲劑；殺蟲劑；除蟲劑
disinfestation 滅蟲；除蟲
disintegration 分解；粉碎
disintegrator 粉碎機
disk 圓盤；圓片狀通心粉 [= disc]
disodium (dihydrogen) pyrophosphate 焦磷酸二鈉；焦磷酸二氫二鈉
disodium 5'-cytidylate 5'-胞苷酸〔二〕鈉
disodium 5'-guanylate 5'-鳥苷酸〔二〕鈉
disodium 5'-inosinate 5'-肌苷酸〔二〕鈉
disodium 5'-ribonucleotide 5'-核糖核苷酸〔二〕鈉
disodium 5'-uridilate 5'-尿苷酸二鈉
disodium citrate 檸檬酸二鈉
disodium dihydrogen pyrophosphate 焦磷酸二氫二鈉；酸性焦磷酸鈉
disodium EDTA 乙二胺四乙酸二鈉
disodium glycyrrhizinate 甘草甜素二鈉；甘草酸二鈉
disodium guanylate 鳥〔核〕苷酸二鈉
disodium hydrogen phosphate 磷酸氫二鈉
disodium inosinate 肌苷酸二鈉；肉苷酸二鈉
disodium phosphate 磷酸二鈉
disodium pyrophosphate 焦磷酸二鈉
disodium stannous citrate 檸檬酸亞錫二鈉
disodium succinate 琥珀酸二鈉
disoxidation 去氧；脱氧
dispensable amino acid 非必需胺基酸 [= nonessential amino acid]
dispenser 分配器；自動販賣機
dispersant 分散劑
dispersed phase 分散相
dispersibility 可分散性
dispersing 分離劑
dispersing agent 分散劑
dispersion 分散〔作用〕；分散液
dispersion medium 分散媒介；分散介質
dispersoid 分散膠體

displacement chromatography 置換層析法；頂替色譜法
display cabinet 展示〔櫥〕櫃；陳列〔櫥〕櫃
display cooler 展示冷藏櫃
disposable dinner set 免洗餐具；一次性餐具
dissociation 解離；離解
dissociation constant 解離常數；離解常數
dissolution 溶解
dissolvent 溶劑
distant gilled milk cap 濕乳菇
distarch glycerol 二澱粉甘油酯
distarch phosphate 二澱粉磷酸酯；磷酸雙澱粉
distasteful 口味不佳的
distearin 二硬脂〔精〕；甘油二硬脂酸酯
distil(l) 蒸餾
distilland 被蒸餾物
distillate 餾出液；蒸餾物；精華
distillating still 蒸餾器；蒸餾釜
distillation 蒸餾；蒸餾物
distillation column 蒸餾塔
distillation residue 蒸餾殘渣
distillator 蒸餾器
distilled 蒸餾得來的
distilled beverage 蒸餾酒〔飲料〕
distilled liquor 蒸餾酒
distilled spirit 蒸餾酒
distilled vinegar 蒸餾醋
distilled water 蒸餾水
distiller 蒸餾器；蒸餾酒廠
distillers' grains 酒粕；酒糟
distillers' malt 釀酒麥芽
distillers' soluble 可溶酒槽
distillers' wort 酒醪；發酵醪；釀酒麥芽汁
distillers' yeast 釀酒酵母；酒麴 [= distillery yeast]
distillery 蒸餾室；釀酒廠
distillery residue 製酒殘渣；酒糟
distillery stillage 蒸餾廢液
distillery yeast 釀酒酵母；酒麴 [= distillers' yeast]
distributor 經銷商
disulfide 二硫化物 [= disulphide]
disulfide bond 雙硫鍵
disulphide 二硫化物 [= disulfide]
ditali 指套形通心粉
ditalini 小指套形通心粉
ditrema 海鯽魚
diuresis 利尿；多尿
diuretics 利尿劑
DNA (deoxyribonucleic acid) 脱氧核糖核酸；去氧核糖核酸
doch-an-dorris 離別酒
dock 尾心；尾基部；臀肉
docosatetraeneoic acid 二十二碳四烯酸
docosenoic acid 二十二〔碳〕烯酸；芥酸
γ-dodecalactone γ-十二內酯
δ-dodecalactone δ-十二內酯
dodecanoic acid 月桂酸；十二〔烷〕酸 [= lauric acid]
dodecenoic acid 十二〔碳〕烯酸
dodecyl gallate 沒食子酸十二酯；五倍子酸十二酯
doe 母鹿；母兔
dog rose 玫瑰果；大薔薇
dog salmon 駝背鮭魚；大麻哈魚 [= keta]
dog shark 白鯊貂鮫；星貂鮫；灰星鯊
dogberry 山茱萸
dogfish 角鯊魚
dogshark 灰星鯊
dolichol 多萜醇

Dolichos lablab 扁豆
dolphin 海豚
dolphin fish 鬼頭刀；藍普卡魚；劍魚
domestic animal 家畜
domestic bird 家禽
domestic cattle 家牛
domestic economy 家政
domestic fowl 家禽
domestic sausage 家庭製造香腸；家用香腸
domestic science 家政學
done 煮透的
donkey 驢
donkey pump 輔助泵
dope 黏稠物；摻雜物
doppio 特濃意大利咖啡 [= double espresso]
dormant 休眠的
dorsal fin 背鰭
dorsal muscle 背肌
Dortmund beer 多特蒙〔德〕啤酒
dory 海魴
dosage 劑量；用量；酒增味劑
dose 一劑；劑量；酒增味劑
dose meter 劑量計；照射量測定計 [= dosimeter]
dosimeter 劑量計；照射量測定計 [= dose meter]
dotted stalk bolete 點柄乳牛肝菌
dotted stem bolete 似黃褐牛肝菌
double beater 雙效打漿機
double bond 雙鍵；雙鏈
double cake 疊層蛋糕
double carbonatation process 二重碳酸飽充法
double cream 高脂鮮乳油；二次分離的稀奶油
double-drum dryer 雙轉筒乾燥機；雙滾筒式乾燥機
double effect evaporator 雙效蒸發器
double espresso 特濃意大利咖啡 [= dippo]
double kettle 雙層鍋
double pipe 套管的
double pipe condenser 套管冷凝器
double reduced tinplate 超薄馬口鐵
double refining 再精製
double scourer 雙層打麥機
double seam(ing) 二重捲封；二重捲邊
double seamed can 二重捲封罐
double seamer 二重捲封機；封罐機
double sulfitation 二次亞硫酸處理
double tube heat exchanger 套管式熱交換器
doubled dish 二重培養皿
doubling time 世代時間；細胞增代時間
dough 麵糰
dough ball 麵糰球
dough box 發麵槽
dough conditioner 麵糰調整劑；麵糰〔質量〕改進劑
dough divider 麵糰分切機
dough kneader 揉麵機
dough kneading 揉麵
dough maturing agent 麵糰熟成劑
dough mill 調麵機；和麵機
dough mixing 和麵
dough raising 麵糰發起
dough scraper 麵糰刮刀
dough sheeter 軋麵機；麵糰軋片機
dough spring 烘烤膨脹；爐內膨脹 [= oven kick; oven spring]
dough strength 麵糰筋力
dough time 麵糰發酵時間
dough trough 發麵槽；麵糰發酵槽
dough up 揉麵糰
doughboy 油炸麵糰

doughiness 麵糰狀；黏糊狀
doughing 麵糰揉和
doughing machine 調麵機；和麵機
doughnut 沙翁；甜甜圈；炸甜圈；炸麵圈；道納；糖納子
doughy 麵糰〔狀〕的
dove 鴿子
dover sole 多佛鰨
draff 餿水；渣滓；廢物
draft beer 生啤；生啤酒 [= beer on draught; draught beer]
draft cattle 役畜
drag screen 括板篩
dragee 糖衣錠；包有糖衣的糖果
dragee machine 糖衣機；滾糖衣機
dragon eye 龍眼；桂圓 [= longan; lungan]
drain 排放；排水；排水溝
drain valve 排水閥；排放閥
drainage system 排水系統
drained fruit 瀝乾的〔水果〕蜜餞
drained weight 滴乾重；瀝乾物重
draining 滴乾；瀝乾；排水
drake 公鴨
draught beer 生啤；生啤酒 [= beer on draught; draft beer]
draught cattle 役畜；挽畜
drawing 一撮茶；回火；去內臟
drawn butter 乳酪醬；融製奶油
drawn can 沖壓罐
drawn fish 去內臟魚
dredge corn 燕麥大麥混合飼料；脫下的玉米粒
dreg 渣滓；糟粕；沉澱物
dressed 去內臟；分割好的
dressed carcass 屠體
dressed chicken 光雞；雞屠體
dressed poultry 去內臟及毛的家禽
dressed weight 屠體重；分割後的重量
dressing 沙拉醬；調味醬；屠體切割；篩理；去內臟及毛
dressing table 屠體切割台
dried bean skim 腐竹
dried beancurd 乾〔製〕豆腐；豆腐乾
dried beef 牛肉脯；牛肉乾
dried bonito 柴魚；鰹魚乾
dried bread 麵包乾
dried cream 乳油粉；稀奶油粉
dried currant 無核葡萄乾
dried fish 魚乾；乾魚
dried food 乾燥食品
dried fruit 乾燥果實；果乾
dried lily flower 金針
dried longan 桂圓；龍眼乾
dried meat 肉乾；脫水肉
dried milk 乳粉；奶粉 [= milk powder]
dried mussel 淡菜；貽貝乾
dried oyster 蠔豉
dried peel 果皮乾
dried persimmon 柿乾
dried plum 梅乾；乾李子
dried product 乾製品
dried salted fish 鹹魚乾
dried scallop 干貝；乾扇貝
dried shark fin 魚翅
dried shred pork 肉絨；肉脯
dried shrimp 蝦〔米〕乾；蝦米
dried soup 湯粉
dried tangerine peel 陳皮；乾橘皮
dried vegetable 乾燥蔬菜；菜乾
dried whole egg 全蛋粉
dried yeast 乾酵母
drier 乾燥機；乾燥器；乾燥劑 [= dryer]
drink 喝；飲料；酒；一杯飲料
drink powder 飲料粉
drinkable 可飲用的
drinking bout 酒宴；縱飲

drinking fountain 飲水機
drinking water 飲用水
drip 滴液；滴汁；水滴
drip coffee 滴瀝咖啡
drip coffee maker 滴瀝咖啡機 [= drip maker]
drip maker 滴瀝咖啡機 [= drip coffee maker]
Dripolator 滴瀝咖啡壺；點滴式咖啡壺
dripping 滴液〔現象〕；油滴
driving spray 盤式噴霧；杯式噴霧
drool 流口水
drop 滴液；滴；落果
drop culture 點滴培養；懸浮培養；懸滴培養
droplet 小滴
droplet culture 小滴培養
dropped egg 水煮荷包蛋
dropper 滴管
dropsy 水腫；浮腫
drosophila (fly) 果蠅
drug 藥物；麻醉劑
druggist 藥〔材〕商；藥劑師
drugstore 雜貨西藥店；藥房
drum barker 轉筒剝皮機；鼓式剝皮機
drum dryer 轉筒乾燥機；轉鼓式乾燥機
drum drying 轉筒乾燥〔法〕；轉鼓式乾燥〔法〕
drum filter 筒式過濾器；鼓式過濾器；濾鼓
drum meat 胴肉；肉胴
drum mixer 轉筒式混合機
drum screen 轉筒篩
drummette 翅根
drumstick 雞腿；小雞腿；雞腿下段
drupaceous fruit 核果 [= drupe (fruit); stone fruit]
drupe (fruit) 核果 [= drupaceous fruit; stone fruit]
drupel(et) 小核果
dry 乾燥；乾的
dry air sterilization 乾熱滅菌
dry alum 燒明礬
dry and wet bulb thermometer 乾濕球濕度計
dry ashing 乾灰化；乾灰法
dry blanch 乾燥殺菁
dry bread 無乳酪麵包；無牛油麵包
dry bulb thermometer 乾球溫度計
dry caustic peeling 乾鹼去皮；乾式鹼法去皮
dry champagne 乾香檳酒；無甜度香檳酒
dry cook 乾蒸
dry cooling 乾〔式〕冷卻
dry cow 不產乳的牛
dry cured meat 乾醃肉
dry curing 乾醃〔法〕；鹽醃〔法〕 [= dry salting]
dry fruit 乾果；果乾
dry frying 無油油炸〔法〕
dry grinding 乾式研磨〔法〕
dry heat cooking 乾熱烹調法
dry heat sterilization 乾熱滅菌〔法〕
dry ice 乾冰
dry icing 碎冰冷藏
dry matter 乾物質
dry milk 乳粉；奶粉 [= milk powder]
dry milling 乾磨
dry mixing 乾混〔法〕
dry-naked shrimp 脱殼蝦米〔乾〕
dry pack 乾燥包裝；乾裝
dry packed meat 乾醃肉
dry rendering 乾法煉〔畜〕油
dry rice 旱稻
dry salt cure 乾鹽醃法

dry salted 乾醃的；鹽醃的
dry salting 乾醃〔法〕；鹽醃〔法〕 [= dry curing]
dry sausage 乾熟香腸；乾香腸
dry smoking 乾燻〔法〕
dry steam 乾蒸汽
dry storage 乾藏；乾貨貯藏
dry sulphitation 乾燻硫〔法〕；乾法亞硫酸處理 [= dry sulfitation]
dry sulfitation 乾燻硫〔法〕；乾法亞硫酸處理 [= dry sulphitation]
dry weight 乾重
dry whey 乳清粉
dry wine 乾葡萄酒；無甜度葡萄酒
dry yeast 乾酵母
dryad' s saddle 鱗多孔菌 (蕈類)
dryer 乾燥機；乾燥器；乾燥劑 [= drier]
drying 乾燥
drying agent 乾燥劑
drying aid 乾燥助劑
drying cabinet 乾燥箱
drying chamber 乾燥室
drying loss 乾燥損失；乾燥減重
drying oil 乾性油
drying oven 烘箱；乾燥爐；乾燥箱
drying shrinkage 乾燥失重；乾縮
dryness 乾燥度
Dublin bay prawn 都柏林灣斑節蝦
duck 鴨；鴨肉
duck egg 鴨蛋
duck gizzard 鴨肫
duck meat 鴨肉
duck wheat 苦蕎麥
duckbill wheat 赤小麥；圓錐小麥
ducker 養鴨者；潛水者
duckery 養鴨場
duckling 嫩鴨；小鴨〔子〕
duckweed 浮萍
duct 導管；管
ductility 展延性
duff 葡萄乾布丁；膺品
duffer 虎河豚
dug 乳房
dulcet 美味的；有香味的
dulcification 甜美化；加甜
dulcin(e) 甜精；甘素；對乙氧苯脲；衛矛醇 [= sucrol; valzin]
dulcitose 衛矛糖
dulcitol 甜醇；半乳糖醇
dulcite 甜醇；半乳糖醇
dulcose 甜醇；半乳糖醇
dulse 食用紫紅藻；掌狀紅皮藻；達爾斯海藻
dumbwaiter 迴轉式食品架；送飯菜升降機
dump packing 散裝〔填充〕
dump tank 傾卸槽；收奶槽
dumpling 餃子；合子；糰子
dung 糞〔肥〕；施糞肥
dunk 浸泡；蘸
dunst 細麩；中粗麵粉
duo-trio test 對比品評試驗
duodenum 十二指腸
duplex bag 雙層袋
duplex mill 雙面磨
durability 耐久性
Durham 都漢短角肉牛；達拉謨短角肉牛
Durham fermentation tube 杜蘭發酵管
durian 榴蓮
Duroc (Jersey) 杜洛賽豬；杜洛克豬；德洛克豬
durometer 硬度計
durra 高粱；杜拉高粱；非洲高粱；蜀黍 [= jowar]
durum (wheat) 杜蘭小麥；杜林小麥；硬質小麥

durum flour 杜蘭小麥粉；杜林小麥粉；硬質小麥粉
durum semolina 粗粒杜蘭小麥粉；粗粒杜林小麥粉；粗粒硬質小麥粉
dust 撒〔糖〕粉
dust arrester 吸塵器；捕塵器
dust bag 集塵袋
dust collector 集塵器
dust proof 防塵
dust salt 鹽粉
dust separator 除塵器
dust tea 茶末
dust trap 集塵器；補塵器
duster 除塵器；噴粉器；撒粉器；篩分器
dusting 篩去麩皮；撒粉
dusting powder 敷粉；供撒的粉
Dutch cheese 荷蘭芝士；荷蘭乾酪
Dutch process 可可料的鹼處理
dwarf 矮生果樹
dwarf banana 矮生香蕉；粉蕉
dwarf egg 短小蛋；侏儒蛋
dwarf eggplant 矮生茄子
dwarf round herring 尖圓腹鯡
dwarf yellow daylily 紅萱
dye 染料；染色
dyestuff 染料；著色劑
dynamic property 動態性〔質〕
dynamic viscosity 動黏度
dysentery 赤痢；痢疾
dysontogenesis 發育不良
dysosmia 嗅覺障礙
dyspepsia 消化不良
dystrophy 營養不良；營養失調

E

E. coli (*Escherichia coli*) 大腸桿菌；埃希氏大腸桿菌；大腸埃希氏桿菌
ear pick fungus 耳匙菌 (蕈類)
ear shell 鮑魚
earl grey 伯爵茶
early maturation 早熟
early ripening 早熟
early vegetable 早熟蔬菜；早收蔬菜
earth chestnut 花生
earth vegetable 根菜類〔蔬菜〕
earthen casserole 砂鍋
earthen pot 砂鍋
earthenware 陶器
earthnut 落花生；花生 [= goober; groundnut; monkey nut; peanut]
earthnut oil 花生油 [= arachis oil; groundnut oil; peanut oil]
earthworm 蚯蚓
earthy flavo(u)r 泥土味
earthy taste 泥土味；土腥味
East Indian geranium oil 東印度香葉油
Easter egg 復活節彩色蛋
easy open 易開的
easy open can 易開罐
easy opening tab 易拉環；易開罐拉環
eat out 上館子吃飯；外食
eatable 食品；食物；可食用的
eater 食客
eatery 餐館；食堂
eating between meal 點心；茶點
eating house 飲食店
eating maturity 食用成熟度
eau 水
ebony 黑檀樹；烏梅
ebriety 酒醉
ebullience 沸騰
ebulliometer 沸點〔測定〕計
ebullition 沸騰；冒泡
Eccles (cake) 葡萄乾餡〔酥〕餅 (英式)
eccrisis 排泄
echinacea root 紫錐花
echinococcosis 包蟲病
echinoderm 棘皮類動物
echinus 海膽
éclair 長條泡芙；手指形包糖霜衣酥餅
ecology 生態學
Edam cheese 荷蘭愛丹乾酪；荷蘭艾丹姆乾酪
eddo 芋根；野芋
edema 水腫
edestin 麻仁球蛋白
edge 罐身鉤丁；罐身摺邊；邊緣
edge mill 輪碾機
edge runner 輪碾機
edge sealing 邊封
edibility 可食性
edible 可食用的
edible amaranth 莧菜；食用莧
edible blended oil 食用混合油
edible casing 食用腸衣

edible crab 可食蟹
edible fat 食用脂肪
edible film 可食性薄膜；食用薄膜
edible fungus 食用真菌
edible greens 食用草本植物
edible oil 食用油
edible pigment 食用色素
edible salt 食鹽；食用鹽 [= sodium chloride; table salt]
EDTA (ethylenediamine tetraacetic acid) 乙二胺四乙酸；乙二胺四醋酸
edulcorant 甜味劑；加甜劑
edulcoration 甜化；加甜
EEA (essential amino acid) 必需胺基酸 [= indispensable amino acid]
eel 鱔；鰻；鰻鱺
eelgrass 苦草；鰻草
eelpout 江鱈；黏魚類
eelworm 蛔蟲；線蟲
effective dose 有效劑量
effervescence 起泡；冒泡；泡騰
effervescent 起泡的；冒泡的
effervescent beverage 發泡飲料；充氣飲料
efflorescence 開花；風化；粉化
effluent 流出的；流出液；廢水
effluvium 惡臭；臭氣
efflux 流出；排出；流出物
effusion 瀉流；溢流
E-Fu noodle 伊麵
egestion 排便；排泄
egg 蛋；卵；雞蛋
egg albumen 蛋白；卵白
egg albumin 卵白蛋白
egg beater 打蛋器
egg breaker 破蛋器；打蛋器
egg custard 蛋卡士達
egg-flip 熱雞蛋酒
egg fooyo(o)ng 芙蓉蛋
egg fruit 雞蛋果；仙桃
egg globulin 卵球蛋白；蛋球蛋白
egg noodle 蛋麵〔條〕；雞蛋麵
egg pasta 雞蛋麵食
egg powder 蛋粉
egg product 蛋製品
egg protein 卵蛋白質
egg roll 蛋捲
egg shell 蛋殼
egg slicer 蛋切片器
egg spoon 蛋匙；食蛋小匙
egg whisk 攪蛋器
egg white 蛋白；卵白
egg yolk 蛋黃；卵黃
egg yolk agar 卵黃洋菜培養基
eggcup 蛋杯
egghot 熱蛋啤酒；熱雞蛋酒
egging 加蛋〔調理〕；撿蛋
eggnog 冷〔雞〕蛋酒
eggplant 茄子 [= garden-egg; auber-gine; brinjal]
EH (equilibrium humidity) 平衡濕度
eicosadienoic acid 二十碳二烯酸
eicosanoic acid 花生酸；二十烷酸 [= arachic acid; arachidic acid]
eicosapentaenoic acid (EPA) 二十碳五烯酸
eicosatetraenoic acid 花生四烯酸；二十碳四烯酸 [= arachidonic acid]
Eijkman test 易克曼試驗；愛克曼試驗
ejector 噴射器；擠出器
ejector type agitator 噴射型攪拌器
elaeometer 驗油比重計；油重計 [= elaiometer]
elaidin 反油酸；甘油三反油酸酯
elaiometer 驗油比重計；油重計 [= elaeometer]
eland 非洲大羚羊
elastic 彈性的

elastic deformation 彈性變形
elastic fiber 彈性纖維
elasticity 彈性；彈力
elasticizer 彈性增進劑；增塑劑
elastin 彈性蛋白
elastogel 彈性凝膠
elastometer 彈性計；彈力計
elbow 肘；肘管；彎管
elderberry 接骨木〔醬果〕
elderberry oil 接骨木果油
elderberry tea 接骨木果茶
eldrin 芸香苷；蘆丁
electric boiler 電熱鍋爐
electric cat-fish 電鯰
electric eel 電鰻
electric furnace 電〔熔〕爐
electric heater 電熱器
electric hot plate 電熱板
electric oven 電烘爐；電烤爐
electric rice cooker 電〔飯〕鍋；電飯煲
electric still 電蒸餾器
electric stove 電爐
electric tinplate 電鍍馬口鐵皮；電鍍錫板
electric tongs 電擊器
electrical stunner 電擊昏器
electrical stunning 電殺法；電昏；電擊昏
electrochemistry 電化學
electrode 電極
electrolysis 電解
electrolyte 電解質；電解液
electrolytic dissociation 電離作用
electrolytic plate 電鍍馬口鐵皮；電鍍錫板
electrolytic solution 電解液
electrolytic tinning 電鍍錫
electrolytic tinplate 電鍍馬口鐵皮；電鍍錫板
electromagnetic plate 電磁盤；電磁爐
electromagnetic stirrer 電磁攪拌器
electron 電子
electron(ic) microscope 電子顯微鏡
electronic sterilization 電子滅菌〔法〕；高頻加熱滅菌〔法〕
electrophoresis 電泳
electropure process 牛奶電流殺菌法；牛奶低頻交流電導殺菌法
electrosol 電溶膠
element 元素
elephant foot 蒟蒻；魔芋 [=conjac; konjac; konnyaku]
elicoidali 螺旋紋通心粉
elixir 藥酒；配劑
elk 麋鹿；麋
eluant 溶析液；洗提液；洗脱液
eluent 溶析液；洗提液；洗脱液
elution 溶析；洗提；洗脱
elver 幼鰻
emaciation 瘦弱；憔悴
emasculation 閹割；去勢
embitterment 加苦味
emblic 醋栗；油柑；餘甘子
embossment 罐蓋打代號，浮雕
embowel 取出內臟
embrittlement 脆化
embryo 胚胎；胚芽
embryo sac 胚囊
embryonated egg 受精蛋
embryotrophy 胚胎營養
emergency food 應急食物；救急食品
emergency ration 應急口糧；救急口糧
emesis 嘔吐
emetic 催吐劑
Emment(h)al cheese 瑞士衣曼塔乾酪；瑞士埃曼塔爾硬乾酪
Emmentaler 瑞士硬芝士；瑞士〔硬〕乾酪 [= Swiss cheese]

emmer 二粒小麥
emollient 軟化劑；潤滑劑
empty body 空腹
empty can 空罐
empty herring 無魚卵的鯡魚
empyreuma 焦臭〔味〕
emulsifiability 乳化性
emulsification 乳化〔作用〕
emulsified mixture 乳化液；乳化物
emulsified oil 乳化油
emulsifier 乳化劑；乳化器
emulsify 使乳化
emulsifying ability 乳化性
emulsifying agent 乳化劑
emulsifying capacity 乳化力
emulsifying mixer 乳化混合器
emulsion 乳化液；乳濁液；乳膠；乳液
emulsion colloid 乳膠體
emulsion flavo(u)r 乳化香精
emulsoid 乳膠體
emulsor 乳化劑；乳化器
emzymic 酶的；酵素的
enamel 塗料(罐頭)；搪瓷；琺瑯
encapsulation 膠囊化；外包膠囊
enchylema 細胞漿
enchyma 營養液；組織液
end doming 底蓋的拱凸
end fermentation 終發酵
end hook 蓋鈎
end press 罐蓋沖壓機
end product 終產物；最後產品
end seamer 封罐機
endive 苦苣；菊苣
endive sprout 青筍
Endo's agar 遠藤〔氏〕培養基 [= Endo's medium]
Endo's medium 遠藤〔氏〕培養基 [= Endo's agar]
endocarp 內果皮
endocellular 胞內的
endocellular enzyme 胞內酶 [= intracellular enzyme]
endocrine 內分泌
endocrinology 內分泌學
endoenzyme 胞內酶
endogenous 內源的；內生的
endolysin 胞解素；內溶菌素
endolysis 胞〔漿溶〕解
endomycin 內黴素；內孢霉素
endomysium 內肌膜；肌內膜
endopeptidase 肽鏈內斷酶；內肽酶
endoskeleton 內骨骼
endosperm 胚乳
endospore 芽孢；內生孢子
endotherm 溫血動物
endotoxin 內毒素
end-over-end (processing) 底蓋翻轉殺菌；翻滾式殺菌
end-over-end agitation 罐頭的翻滾轉動；底蓋的翻滾式轉動
end-to-end sterilization 罐頭旋轉式滅菌；罐頭翻滾式殺菌
enema 灌腸劑
energy 能量
energy food 熱能食品
energy intake 能量攝取量
energy source 能源
energy value 能值；熱值
enfleurage 花香〔精〕油萃法(用油脂從鮮花提取香味法)
engineered food 工程食品
Engler degree 恩格勒黏度；恩氏黏度 [= Engler number]
Engler number 恩格勒黏度；恩氏黏度 [= Engler degree]
Engler viscosimeter 恩格勒黏度計；恩氏黏度計

Engler viscosity 恩格勒黏度；恩氏黏度
English chop 羊排
English hawthorn 西洋山楂
English herring 海鯡魚
English muffin 英式鬆餅
engorgement 貪食；暴飲暴食；充血
enhancer 增強劑
ennoblement 營養素強化；營養素增補
enocianina 葡萄果皮色素萃；葡萄花青素
enoki mushroom 金菇
enol 烯醇
enolase 烯醇酶
enolate 烯醇鹽
enolization 烯醇化〔作用〕
enologist 葡萄酒釀造學家；釀葡萄酒學家 [= oenologist]
enology 葡萄酒釀造學；釀葡萄酒學 [= oenology]
enriched barley 營養強化大麥
enriched bread 營養強化麵包
enriched culture 增殖培養；增菌培養
enriched food 營養強化食品；強化〔營養素〕食品
enriched medium 增殖培養基；強菌培養基
enriched rice 營養強化米；強化〔營養素〕米
enrichment 強化〔營養素〕
enrichment medium 殖培養基；強化培養基
enrobe 浸掛糖衣
enrober 糖衣機；浸掛糖衣機
enteral 腸的 [= enteric]
enteric 腸的 [= enteral]
enteric bacilli 腸內桿菌
enteric bacteria 腸內細菌；腸道細菌
enteritis 腸炎
enterobacteria 腸內細菌；腸道細菌
enterobiasis 蟯蟲病
enterococci 腸球菌 〔= enterococcus的複數〕
enterococcus 腸球菌
enterogastrone 胃腸激素；腸抑胃素
enterohepatic circulation 腸肝循環
enteropathogenic 致腸病的
enteropathogenic bacteria 腸內病原細菌
enterotoxin 腸毒素
enterovirus 腸病毒
enthalpy 焓；熱函
entire 黑啤酒(英國)
entire colony 整齊菌落
entire wheat bread 全麥麵包
entire wheat flour 全麥麵粉
Entner-Doudoroff pathway 醣解途徑；恩納杜道夫醣解途徑
entomology 昆蟲學
entozoa 體內寄生蟲
entrails 內臟
entrecote 肉排(牛腰部肉)
entrée 主菜；副菜；間菜
entremets 附加菜；餐後甜點心
entropy 熵
environmental protection 環境保護
Environmental Protection Agency (EPA) 環境保護署；環境保護局
enzymatic activity 酵素活性
enzymatic browning 酵素性褐變
enzymatic degradation 酵素性降解
enzymatic deterioration 酵素性劣變；酵素性敗壞
enzymatic hydrolysis 酵素水解
enzymatic oxidation 酵素性氧化
enzymatic oxidative browning 酵素性氧化褐變
enzymatic reaction 酵素反應
enzymatic spoilage 酵素性腐敗

enzyme 酵素；酶
enzyme action 酵素作用
enzyme activator 酶活化劑
enzyme activity 酵素活性
enzyme assay 酵素分析法
enzyme digestion 酶解〔作用〕
enzyme inhibitor 酵素抑制劑
enzyme-modified milkfat 酶改性乳脂〔肪〕
enzyme preparations 酶製劑
enzyme reaction 酵素反應
enzyme specificity 酵素特異性；酵素專一性
enzyme-substrate complex 酵素受質複合物
enzyme treated starch 酶處理澱粉；酵素處理澱粉
enzyme's product 酶製劑
enzymology 酵素學；酶學
enzymolysis 酵素分解〔作用〕；酶解〔作用〕
eosin(e) 伊紅；曙紅
EPA (eicosapentaenoic acid) 二十碳五烯酸
epazote 洋香藜；土荊芥
ephedrine 麻黃素
epicarp 外果皮
epicatechin 表兒茶素；表兒茶酸
epicure 美食家；老饕
epicurism 美食主義
epidemiology 流行病學
epidermal tissue 表皮組織
epidermis 表皮
epilation 拔毛；除毛；脱毛
epilator 脱毛劑
epinephrine 腎上腺素
epipelagic fish 上層魚類；淺海魚類
episperm 種皮
epithelial tissue 上皮〔細胞〕；上皮組織；皮膜組織 [= epithelium]
epithelium 上皮〔細胞〕；上皮組織；皮膜組織 [= epithelial tissue]
epoxide 環氧化物
Epsom salt 瀉鹽 [= bitter salt]
equicaloric diet 等熱量膳食；等熱量飲食；等卡〔熱〕飲食
equilibrium humidity (EH) 平衡濕度
equilibrium moisture 平衡分；平衡濕度
equilibrium relative humidity(ERH) 平衡相對濕度
equilibrium temperature 平衡溫度
equine 馬
equipment 設備；裝置；器材
equivalent weight 當量
eradicator 果皮刮肉機；菠蘿刮肉機
eradicator meat 果皮〔殘留的〕碎果肉；碎果肉
erepsin 腸肽酶；腸蛋白酶 [= ereptase]
ereptase 腸肽酶；腸蛋白酶 [= erepsin]
ergamine 組〔織〕胺
ergocalciferol 麥角鈣化〔固〕醇；鈣化醇；骨化醇；維生素D_2 [= calciferol; vitamin D_2]
ergosterin 麥角固醇；麥角甾醇 [= ergosterol]
ergosterol 麥角固醇；麥角甾醇 [= ergosterin]
ergot 麥角
ergotic acid 麥角酸
ergotism 麥角中毒〔症〕
ERH (equilibrium relative humidity) 平衡相對濕度
erose colony 鋸齒形菌落
erucic acid 芥子酸；芥酸
erucin 甘油三芥〔子〕酸酯；芥酸精；三芥子脂

erysipelotoxin 丹毒毒素
erythorbic acid 異抗壞血酸 [= isoascorbic acid]
erythrin 地衣紅質；紅血球素；赤蘚素
erythritol 赤蘚糖醇；赤藻糖醇
erythrocyte 紅血球 [= red blood cell]
erythrodextrin(e) 紅糊精
erythromycin 紅黴素
erythrophyll 葉紅素
erythrose 赤藻糖；赤蘚糖
erythrosine 赤蘚紅；新品酸性紅；赤藻紅；食品紅14號
erythrosine aluminium lake 食品紅色七號鋁色澱；赤蘚紅鋁色澱
erythrosine BS 食品紅色七號
escallop 海扇貝；扇貝
escalope 薄肉片；薄魚片
escargot 食用蝸牛；蝸牛
eschalot 油蔥；紅蔥頭；分蔥；大頭蔥；珠蔥 [= shallot]
Escherichia coli (*E. coli*) 大腸桿菌；埃希氏大腸桿菌；大腸埃希氏桿菌
Escherichia coli food poisoning 大腸桿菌中毒
escolar 玉梭魚（大西洋熱帶區）
esculent 可食用的；可食用物
espresso 意大利濃縮咖啡；意大利蒸餾咖啡；意式濃咖啡
espresso cup 意大利式咖啡杯；意式咖啡杯
espresso macchiato 馬奇雅圖咖啡 [= caffe macchiato; macchiato]
espresso machine 意式咖啡機
espresso ristretto 威士戚圖咖啡 [= caffe ristretto]
essence 香精
essential amino acid (EEA) 必需胺基酸 [= indispensable amino acid]
essential fatty acid 必需脂肪酸
essential oil 揮發油；精油；芳香油 [= ethereal oil]
estaminet 咖啡館；小餐館；小酒吧
ester 酯
ester gum 酯膠
ester number 酯〔化〕值 [= ester value]
ester value 酯〔化〕值 [= ester number]
ester wax 酯蠟
esterase 酯酶
esterification 酯化〔作用〕
estragole 草蒿腦；蒿腦
estragon 茵陳蒿；香艾菊；龍蒿 [= tarragon]
estragon oil 茵陳蒿油；香艾菊油；龍蒿油 [= tarragon oil]
esurient 貪食；貪吃者
ethanal 乙醛；醋醛 [= acetaldehyde]
ethane 乙烷
ethanoic acid 醋酸；乙酸 [= acetic acid]
ethanol 乙醇；酒精 [= alcohol; ethyl alcohol]
ethanolamine 單乙醇胺；乙醇胺
ethene 乙烯 [= ethylene]
ether 乙醚；二乙基醚 [= diethyl ether]
ethereal oil 揮發油；精油；芳香油 [= essential oil]
etherification 醚化〔作用〕
ethnic food 民族食品
ethoxy quin 乙氧基奎
ethoxylated mono- and diglycerides 乙氧基單雙甘油酯
ethyl acetate 乙酸乙酯；醋酸乙酯 [= acetic ether]
ethyl acetoacetate 乙酰乙酸乙酯
ethyl acrylate 丙烯酸乙酯
ethyl alcohol 乙醇；酒精 [= alcohol; ethanol]
ethyl *p*-anisate 茴香酸乙酯

ethyl anthranilate 鄰氨基苯甲酸乙酯
ethyl benzoate 苯甲酸乙酯
ethyl brassylate 甘藍酸乙烯
ethyl butyrate 丁酸乙酯；乙基丁酸鹽
ethyl cellulose 乙基纖維素
ethyl cinnamate 桂皮酸乙酯；肉桂酸乙酯
ethyl *cis*-4-octenate 順式-4-辛烯酸乙酯
ethyl decanoate 癸酸乙酯
ethyl ether 乙醚
2-ethyl fenchol 2-乙基小茴香醇；葑醇
ethyl formate 甲酸乙酯
ethyl heptanoate 庚酸乙酯
ethyl hexanoate 己酸乙酯
2-ethyl-1-hexanol 2-乙基-1-己醇
ethyl 3-hydroxybutyrate 3-羥基丁酸乙酯
ethyl-*p*-hydroxybenzoate 對羥基苯甲酸乙酯；尼泊金乙酯
ethyl hydroxyethyl cellulose 羥乙基纖維素乙基醚
ethyl isobutyrate 異丁酸乙酯
ethyl isovalerate 異戊酸乙酯
ethyl lactate 乳酸乙酯
ethyl laurate 月桂酸乙酯；十二烷酸乙酯
ethyl levulinate 乙酰丙酸乙酯
ethyl maltol 乙基麥芽酚
ethyl mercaptan 乙〔基〕硫醇
ethyl 3-methyl thiopropionate 3-甲硫基丙酸乙酯
ethyl 2-methylbutyrate 2-甲基丁酸乙酯
ethyl methylphenylglycidate 草莓醛；楊梅醛
2-ethyl-3-methylpyrazine 2-乙基-3-甲基吡嗪
ethyl myristate 肉荳蔻酸乙酯
ethyl nitrite 亞硝酸乙酯
ethyl nonanoate 壬酸乙酯
ethyl octanoate 辛酸乙酯
ethyl oleate 油酸乙酯
ethyl orange 乙基橙
ethyl 3-phenyl propionate 3-苯丙酸乙酯
ethyl phenylacetate 苯乙酸乙酯
ethyl 4-phenylbutyrate 4-苯丁酸乙酯
ethyl phenylglycidate 苯代環氧丙酸乙酯；苯基縮水甘油酸乙酯
ethyl propionate 丙酸乙酯
ethyl protocatechuate 原兒茶酸乙酯
ethyl salicylate 水楊酸酸乙酯；柳酸乙酯
ethyl tiglate 惕各酸乙酯
ethyl valerate 戊酸乙酯
ethyl vanillin 乙基香蘭素；乙基香草醛 [= bourbonal]
ethyl 10-undecenoate 10-十一烯酸乙酯
ethylamine 乙胺
ethylcellulose 乙基纖維素
ethylene 乙烯 [= ethene]
ethylene dichloride 二氯〔化〕乙烯；1,2-二氯乙烷
ethylene glycol 乙二醇
ethylene oxide 環氧乙烷；乙烯化氧
ethylenediamine tetraacetic acid (EDTA) 乙二胺四乙酸；乙二胺四醋酸
4-ethylguaiacol 4-乙基癒創木酚
ethylparaben 對羥基苯甲酸乙酯；尼泊金乙酯
p-ethylphenol 對乙基苯酚
etna 酒精燈煮水器
eubacteria 真細菌
eucalyptene 桉樹烯
eucalyptol 桉〔葉〕腦；桉樹腦；桉葉醇；桉油精；桉葉〔油〕素
eucalyptus 桉樹
eucalyptus gum 桉〔樹〕膠
eucalyptus oil 桉葉油；桉樹油；藍桉葉油

eucheuma 麒麟菜屬
eucheuma seaweed 瓊芝屬海藻
eugenol 丁香酚；丁子香酚
eugenyl acetate 乙酸丁香酚酯
eugenyl methyl ether 丁香酚甲醚；甲基丁香酚
eumycete 真菌
euosmia 嗅覺正常
euphausia 南極蝦；燐蝦目
euphausiid 燐蝦〔類〕；燐蝦目浮游甲殼動物
euphausiid shrimp 燐蝦
European bird cherry 稠李
European chestnut 歐洲栗子
European eel 歐洲鰻鱺
European filbert 歐洲榛子
European gooseberry 鵝梅
European pear 西洋梨
European perch 河鱸
European wildginger 歐洲細辛；歐洲野生薑
eutectic 共溶的；低共溶的；共晶
eutrophia 優良營養；營養佳良
eutrophy 優良營養；營養佳良
evaporated milk 無糖煉乳；蒸發乳；淡煉乳；淡奶
evaporated skim(med) milk 蒸發脫脂乳；脫脂〔淡〕煉乳
evaporating field 鹽場；蒸發場
evaporating pan 蒸發鍋
evaporation 蒸發〔作用〕
evaporation pond 鹽池；蒸發池
evaporator 蒸發器
even stem cabbage 球莖甘藍
evening primrose oil 月見草油
everfermentation 發酵過度
evergreen 常綠的；常綠植物
evergreen fruit tree 常綠果樹
evergreen pear 常綠梨
eviscerate 去內臟
eviscerated carcass 去內臟屠體
evisceration 取出內臟
eviscerator 去內臟機；取內臟器
ewe 母羊
ewe lamb 母羔羊；哺乳母羊
ewer 大口水壺
excitant food 刺激性食品；興奮性食品
excorticate 剝皮
excrement 排泄物；糞便
excreta 分泌物；排泄物
excrete 分泌；排泄
excretion 排泄物
exhaust fan 抽氣扇；排氣風扇；抽風機
exhaust steam 廢蒸氣
exhaust steam heating 廢蒸氣加熱
exhausted cossette 甜菜廢絲
exhausted liquid 廢液
exhausted molasses 廢糖蜜
exoamylase 外澱粉酶
exocarp 外果皮
exocellular enzyme 胞外酵素
exoenzyme 外酶；外酵素；胞外酶
exogenous 外源的；外生的
exopeptidase 外肽酶；肽鏈端解酶
exoskeleton 外骨骼
exosmosis 外滲
exospore 外生孢子
exotoxin 外毒素
expansimeter 膨脹〔度〕計
expeller 壓榨機；螺旋壓榨機
expeller cake 粕餅；油粕
expelling 榨油；壓榨
experiment 實驗；試驗
explosion proof 防爆
expressed juice 壓榨出的果汁
expressible drip 可榨出液滴
expressible oil 可榨出油

expressive drying 壓榨乾燥〔法〕
expressive machine 壓榨機
exsanguination 放血
exsector 切除器
exsiccant 乾燥劑
exsiccation 脱水〔作用〕；乾燥〔作用〕
exsiccator 乾燥器；保乾器；除濕劑
extender 增量劑；稀釋劑；增補劑
extensibility 伸展性；延伸性
extensimeter 伸展計；伸長計
extensin 伸展蛋白
extensograph 麵糰伸展儀；延伸性圖譜儀
extensometer 伸長計；伸展計
extra dry wine 特乾葡萄酒
extra heavy syrup 特濃糖漿
extra old (X.O.) 特陳年白蘭地酒
extra virgin olive oil 橄欖油
extracellular enzyme 胞外酵素；胞外酶
extract 萃取物；浸出物；抽提物；提取物
extractant 萃取劑；提取劑
extracting agent 萃取劑；提取劑
extraction 萃取；浸出；抽提；提取
extractive 萃取物；熬汁；提取物；浸出物
extractor 萃取器；浸出器；抽提器；提取器
extruded candy 擠製糖果
extruder 擠出機；擠壓機；擠製機
extruding granulation 擠壓造粒〔法〕
extruding machine 擠出機；擠壓機；擠製機
extrusion 擠出；擠壓；擠製
extrusion cooking 擠壓烹煮；擠壓熟化
extrusion food 擠壓食品
exudate 滲出物；流出物
exudation 滲出〔作用〕
eye hole 眼孔
eye round 外側後腿眼肉；鯉魚管

F

F value F值；滅菌值 [= sterilizing value]
fabricated food 組合食品
FACC (Food Additives and Contaminants Committee) 食品添加劑及污染物委員會
factory 工廠
factory end 罐底
facultative 兼性的
facultative anaerobe 兼性厭氣細菌；兼性厭氧細菌
facultative thermophile 兼性嗜熱細菌
facultative thermophilic bacteria 兼性嗜熱細菌
faecal 糞便的；排泄物的 [= fecal]
faeces 糞便；排泄物；渣滓 [= feces]
faggot 烤碎豬雜；豬肝肉丸
fagopyrin 蕎麥鹼
fagopyrism 蕎麥中毒
fagot 烹調用之香料紗布袋；烤碎豬雜；烤肝片
faham 花漢茶；野茶
fairy ring 蘑菇圈
fairy ring champignon 硬柄小皮傘(蕈類)
fairy ring mushroom 仙女環洋菇
fall crop 秋作
falling ball viscosimeter 落球黏度計
falling film evaporator 降膜式蒸發器
falling film still 降膜式蒸餾器
falling rate drying 減率乾燥；降速乾燥
fallout 落塵
false chanterelle 桔黃擬蠟傘(蕈類)
false fruit 假果
false-killer whale 偽虎鯨
false seam 不正常捲封；假封
false yeast 假酵母
falsification 偽造
family flour 多用途麵粉；家用麵粉
family-size package 家庭用〔量〕包裝
famine 饑荒；飢餓
fan shell 扇貝〔殼〕
fanchonette 奶油塔餅；用蛋白糖霜飾面的餡餅
fancy 花式的；特級的
fancy bread 花式麵包
fancy food 嗜好食品
fancy meat 雜碎
fancy selected 特選的
fannings 碎茶；茶〔葉〕末；茶屑
FAO (Food and Agriculture Organization) 糧農組織；聯合國糧農組織；聯合國糧食〔及〕農業組織 [= United Nations Food and Agriculture Organiza-tion; UNFAO]
FAO/WHO (Joint FAO/WHO Food Codex Alimentarius Commission) 聯合國糧農組織(FAO)世界衛生組織(WHO)聯合食品法規委員會

FAR (Food Additive Regulation) 食品添加劑規范
farci 有餡的
fare 食物；伙食
farfal 粒狀麵食製品；麵粒
farfalla 大蝴蝶粉；蝴蝶形通心粉 [= bow tie pasta]
farfalle 大蝴蝶粉；蝴蝶形通心粉 [= bow tie pasta]
farfel 粒狀麵食製品；麵粒
farina 粗殼粉；粗粒小麥粉；馬鈴薯澱粉；花粉
farina avena 燕麥片；燕麥粉
Farina dolce 栗子粉
farinha 木薯粉
farinogramme 糰物性圖；麵粉調粉性圖譜
farinograph 糰物性測定儀；麵粉調粉性圖譜儀；麵粉試驗儀
farinose 產〔澱〕粉的；含澱粉的；粉質的
farl(e) 燕麥薄餅；燕麥蛋糕
farm 農場；牧場；養殖場；耕種
farm crop 農作物
farm product 農產品
farm-style ham 農家式火腿
farmer sausage 農家式香腸
farnesol 金合歡醇；雙三醇
farthing bun 水果甜麵包
fascia 肌膜；筋膜
fascioliasis 片吸蟲病
fast 禁食；齋戒
fast cooking 速煮
fast day 禁食日；齋戒日
fast food 快餐；速食
fast green 堅牢綠
fast green FCF 食用綠色三號；堅牢綠
fast green FCF aluminum lake 食用綠色三號鋁色澱；堅牢綠鋁色澱
fast red 堅牢紅
fast red E 堅牢紅；食用大紅
fast red lake 堅牢紅色澱
fast yellow 堅牢黃
fasting 禁食〔的〕
fasting cure 禁食療法
fat 脂肪；肥的
fat and oil 油脂
fat body 脂肪體
fat cell 脂肪細胞 [= adipose cell]
fat crab 肥蟹
fat emulsion 脂肪乳濁液
fat extender 油脂稀釋劑；油脂補充劑
fat-free 無脂的；脱脂的；不含脂肪的
fat globule 脂肪球
fat hardening 油脂硬化
fat intake 油脂攝取〔量〕
fat rendering 脂肪熬煉
fat-soluble vitamin 脂溶性維生素；脂溶性維他命 [= lipid-soluble vitamin]
fat splitting 脂肪分解
fat tissue 脂肪組織
fatling 肥畜
fattener 育肥禽畜
fatty acid 脂肪酸
fatty acid composition 脂肪酸組成
fatty alcohol 脂肪〔族〕醇
fatty compound 脂肪〔族〕化合物
fatty fish 多脂肪魚
fatty food 脂肪性食品；高脂肪食品
fatty glyceride 脂肪酸甘油酯
fatty oil 脂油
fatty tissue 脂肪組織
faulty closure 封罐不良；罐頭捲邊不良
favantini 通心麵
fawn shield cap 灰光柄菇
FCC (Food Chemicals Codex) 食品添加物法規；美國食用化學法典

FD&C (Federal Food, Drug and Cosmetic Act) 聯邦食品藥品及化粧品法〔案〕(美國)
FDA (Food and Drug Adminstration) 食品〔及〕藥物管理局 (美國)
FDA regulation 食品及藥物管理規則
feast 宴會；餐宴；筵席；酒席
feather 羽毛
feather dressing 拔〔羽〕毛
feather game 野禽
feather picker 去〔羽〕毛機；拔〔羽〕毛機
featherator 去〔羽〕毛機；拔〔羽〕毛機
fecal 糞便的；排泄物的 [= faecal]
fecal bacteria 糞便細菌
fecal coliform 糞便大腸桿菌群
fecal streptococcus 糞便鏈球菌
feces 糞便；排泄物；渣滓 [= faeces]
fecula 澱粉製成的食品
feculence 污物；混濁；渣滓
fed-upness 過飽
fedeli 意大利圓細麵
fedelini 細麵條；麵線
Federal Food, Drug and Cosmetic Act (FD&C) 聯邦食品藥品及化粧品法〔案〕(美國)
feed 飼料；飼養
feed flavo(u)r 牧草味
feed stuff 飼料
feeder 進料器；飼槽；飼養者；肥育畜
feeding 進料；餵食；飼養；攝食
feeding bottle 奶瓶；哺乳瓶
Fehling's solution 斐林溶液；費林溶液
Fehling's test 斐林試驗；費林試驗
feijoa 鳳梨番石榴；費約果；費喬亞果
femur 股骨
fenchol 小茴香醇；葑醇
fenchyl alcohol 小茴香醇；葑醇
fenks 鯨魚油渣
fennel 茴香
fennel oil 茴香油；小茴香油
fennel seed 茴香籽；小茴香籽；甜小茴香籽
fenugreek 葫蘆巴
fenugreek seed 葫蘆巴籽
ferment 酵素；酶；發酵
ferment odor 發酵臭
fermentable 可發酵的
fermentation 發酵〔作用〕
fermentation adjuvant 發酵補佐物；發酵助劑
fermentation broth 發酵液
fermentation cask 發酵桶
fermentation cellar 發酵窖
fermentation medium 發酵培養基
fermentation product 發酵產物；發酵產品
fermentation room 發酵室
fermentation tank 發酵槽
fermentation vat 發酵〔大〕桶；發酵池
fermentative 具發酵性的；酵性的
fermentative carbohydrate 可發酵醣類
fermented cider 蘋果酒
fermented fish sauce 魚醬油；魚露
fermented food 發酵食品
fermented goods 發酵食品；發酵製品
fermented juice 發酵果汁
fermented mash 發酵醪
fermented milk 發酵乳
fermented milk drink 發酵乳飲料
fermented product 發酵產品
fermented sausage 發酵香腸
fermented soybean paste 味噌；日本豆醬；豆瓣醬
fermented tea 發酵茶

fermented tofu 豆腐乳；腐乳 [= bean curd cheese; soy cheese; soybean cheese; sufu]
fermented vinegar 發酵醋
fermenter 發酵槽；發酵罐 [= fermentor]
fermentology 發酵學
fermentor 發酵槽；發酵罐 [= fermenter]
ferox 大湖鱒；猛鮭
ferrate 鐵酸鹽
ferric ammonium citrate, brown 檸檬酸鐵銨，棕色
ferric ammonium citrate, green 檸檬酸鐵銨，綠色
ferric chloride 氯化鐵
ferric citrate 檸檬酸鐵
ferric orthophosphate 正磷酸鐵
ferric phosphate 磷酸鐵
ferric pyrophosphate 焦磷酸鐵
ferric sulfate 硫酸鐵
ferritin 鐵蛋白
ferrous fumarate 富馬酸亞鐵
ferrous gluconate 葡〔萄〕糖酸亞鐵
ferrous lactate 乳酸亞鐵
ferrous pyrophosphate 焦磷酸亞鐵
ferrous sulfate 硫酸亞鐵
fertility 繁殖力；肥沃度；豐產
fertilizer 肥料
festonati 費斯透那提麵；彩帶麵
fetor 惡臭
fettuccelle 闊麵條
fettuccine 寬條麵
fetuin 胎球蛋白
feverfew 小白菊
fiber 纖維 [= fibre]
fiber crop 纖維作物
fibre 纖維 [= fiber]
fibril 小纖維；原纖維；根毛
fibrillar protein 纖維狀蛋白質
fibrin 血纖維蛋白
fibrinogen 血纖維蛋白原
fibrinolysis 纖維蛋白溶解作用
fibroin 絲〔纖維〕蛋白；絲心蛋白
fibrous casing 纖維〔素人造〕腸衣
fibrous protein 纖維狀蛋白質
ficin 無花果蛋白酶
ficus awkeotsang 愛玉
fiddlehead fern 蕨菜
fiddler crab 沙蟹
field bean 蠶豆；菜豆；四季豆
field corn 飼料玉米
field crop 田間作物
field drying 野外乾燥
field eel 鱔魚
field mushroom 野洋菇；蘑菇
field pea 普通豌豆；紫花豌豆；田豆
field trash 甘蔗夾雜物；甘蔗田廢棄物
FIFO (first-in first-out) 先進先出；先入先出
fig 無花果
filament 纖絲；細絲
filamentous algae 絲〔狀〕藻類
filamentous fungus 絲狀真菌；絲狀黴菌
filamentous mould 絲狀真菌；絲狀黴菌
filamentous yeast 絲狀酵母
filbert 榛子
filefish 單棘魨；豚魚
filet 肉片；魚片
filet mignon 腰內肉片；牛里脊肉片
filled bun 有餡小麵包；夾心甜麵包
filled can 實罐
filled chewing gum 夾心口香糖
filler 填充機；填裝機；填料
filler linker 香腸灌紮機
fillet 肉片；魚片

fillet mignon 腰內肉片；牛里脊肉片
fillet steak 腓利牛排
filleter 切魚片機
filling 填料；餡料；填裝
filling machine 填充機；灌裝機
filling seal machine 填充封口機
fillmass 糖膏
film dryer 薄膜乾燥機
film drying 薄膜乾燥
film evaporator 薄膜蒸發器
film-forming yeast 產膜酵母 [= film yeast]
film packaging 薄膜包裝
film type condensation 薄膜式冷凝
film type deaerator 薄膜式脫氣機
film yeast 產膜酵母 [= film-forming yeast]
filter 過濾器；過濾
filter aids 助濾劑
filter bag 過濾袋；濾袋
filter bed 濾床；濾墊
filter cake 濾餅
filter cloth 濾布
filter liquor 濾液
filter mass 濾餅
filter medium 過濾介質
filter membrane 濾膜
filter pad 過濾襯墊
filter paper 濾紙
filter plate 濾板
filter press 壓濾機
filter residue 濾渣
filter screen 濾網；濾篩
filter sterilizer 過濾滅菌器
filterable 可過濾的
filteraid 助濾劑
filtercel 助濾硅藻土
filtering cartridge 濾筒
filtering sterilization 過濾滅菌〔法〕
filth 污物；污垢
filtrate 濾〔出〕液
filtration 過濾
filtros 濾石
fin 鰭；魚翅
fin whale 長鬆鯨
final molasses 廢糖蜜
final product 終產物；成品
final proof 後醒麵；最後一次發麵
finback 長鬚鯨
fine champagne 上等白蘭地酒
fine crusher 細壓碎機
fine cut 細切；細切煙草
fine egg noodle 龍鬚蛋麵
fine flour 精白麵粉
fine-grained sugar 細〔粒〕砂糖
fine granulated sugar 細〔粒〕砂糖
fine grinding 細磨；精碾
fine grit 細粉
fine meat 細碎肉；去骨肉
fine powder 細粉
fine salt 細鹽；精鹽
fine semolina 細粒小麥粉；細粒麵粉
fine sugar 精糖；細晶糖
fine wine 精製酒；高級酒
finfish 有鰭魚〔類〕
finger bowl 洗指碗
finger citron 佛手柑
finger food 手抓食品
finger milking 人工擠乳
fingerling 一指長的小魚
fining 澄清；淨化；澄清劑
fining agent 澄清劑；淨化劑
finished goods 加工成品，製成品
finished product 成品
finished rice 白米
finisher 篩濾機；果漿過濾機
finnan haddock 冷燻黑線鱈；燻鱈魚

finnan kiln 鱈魚燻窯；冷燻黑線鱈的窯
fir needle oil 冷杉〔葉〕油
fir seed oil 冷杉籽油
fire drying 直火乾燥；明火烘乾
fire heating 明火加熱
firefly squid 螢烏賊
fireless cooking 無火烹調；無續熱烹調
firewater 烈酒
firing 燒；烘烤
firkin 法肯(9英加侖)；弗金(40.91升)；木製小桶
firm butter 硬質牛油
firming agent 硬化劑；固化劑
firmness 堅實；堅硬度
first cut 初餾分；酒頭
first evaporator 第一效蒸發器
first-fruit 初上市水果
first-in first-out (FIFO) 先進先出；先入先出
first running 初餾物；楓糖初漿
fish 魚；魚肉
fish and shellfish poisoning 魚貝中毒
fish ball 魚丸
fish bar 魚條；魚棒
fish berry 魚卵；魚子
fish block 魚塊
fish cake 魚餅；魚糕
fish farm 魚塭；養魚場
fish farming 養魚〔業〕
fish fat 魚脂
fish fillet 魚排；魚柳；魚片
fish flesh 魚肉
fish fry 油炸魚；魚苗
fish gelatin 魚膠 [= fish glue; isinglass]
fish glue 魚膠 [= fish gelatin; isinglass]
fish gravy 魚露；魚汁
fish hold 魚艙
fish isinglass 魚膠；鱘鰾膠
fish liver oil 魚肝油
fish loin 魚背肉
fish maw 魚鰾；魚肚
fish meal 魚粉
fish odo(u)r 魚〔腥〕臭
fish oil 魚油
fish poison 魚毒
fish product 魚產品
fish protein concentrate 濃縮魚蛋白；魚蛋白濃縮物
fish protein isolate 分離魚蛋白
fish roe 魚卵
fish sauce 魚露；魚醬油；魚製調味料
fish sausage 魚〔肉〕香腸
fish scale 魚鱗
fish scaler 去〔魚〕鱗器
fish sealing 去〔魚〕鱗
fish skin 魚皮
fish soluble 魚溶漿；濃縮魚汁
fish soup 魚湯；魚羹
fish steak 魚排；魚片
fish stick 魚條
fish tallow 魚脂
fish waste 魚體廢棄物
fisheries animal 水產動物
fishery 水產業；漁業；漁場；養魚場
fishery industry 漁業；水產業
fishery product 水產品；漁產品
fishiness 魚腥味；腥臭
fishy 魚的；魚味的；魚腥味的
fishy flavo(u)r 魚腥味；魚風味
fishy odo(u)r 魚腥味
fishy smell 魚腥味
five spices powder 五香粉
fixative 固定劑；定香劑
fixed bed drying 固定床乾燥
fixed oil 非揮發性油；不揮發性油
fixing agent 固定劑

fizz 發泡及嘶嘶響的飲料；菲滋酒；發嘶嘶聲
flaccidity 肌肉組織鬆弛
flagellate 鞭毛蟲；鞭毛藻
flageolet bean 笛豆
flagon 酒壺；大肚酒壺
flake cutter 刨片機
flake ice 薄片冰；冰屑
flake pastry 酥餅；油酥
flake salt 片狀鹽
flake yard 曬魚場
flaker 刨片機
flaking machine 刨片機
flaking mill 壓〔薄〕片機；軋片機
flame ionization detector 火餤離子化檢定器；火餤游離偵測器
flame sterilization 火餤滅菌
flame sterilizer 火餤滅菌器
flame ware 耐熱器皿
flan 芙蘭派；果餡餅
flange 罐緣
flanger 罐頭翻邊機；除雪器
flank 腹脅〔肉〕；牛柳陰
flank and knuckle 腹脅肉
flank steak 腹脅肉排；元冧片；陰柳
flannel cake 薄麥餅；鬆軟烤餅
flapjack 薄煎餅；烙餅；煎餅；烤餅 [= griddlecake; slapjack]
flash cooling 急驟冷卻；閃蒸冷卻
flash deaeration 急驟脱氣
flash drying 急驟乾燥
flash evaporation 急驟蒸發；閃急蒸發
flash freezing 急驟冷凍；閃凍
flash heating 急驟加熱
flash pasteurization 巴氏瞬間殺菌；瞬間巴氏殺菌
flash vapourization 急驟汽化
flat can 扁罐
flat cooking 原味烹調；未調味烹調
flat flavo(u)r 平淡的風味
flat mushroom 扁平洋菇
flat peach 蟠桃
flat plate 平盤；平皿
flat silver 銀製食具；鍍銀食具
flat sour 平酸〔腐敗〕；罐頭不產氣酸敗
flat sour bacteria 平酸〔腐敗〕細菌
flat sours 平酸菌
flat surface heat exchanger 平面熱交換器
flat taste 淡味
flat tin 扁罐
flatbread 乾薄鬆餅；小麵包乾
flatfish 鰈〔魚〕；扁魚；比目魚
flathead flounder 紅擬庸鰈；太平洋擬庸鰈
flatspine prickly ash 花椒；山椒 [= Chinese prickly ash; Szechuan peppercorn]
flattened dried squid 魷魚乾
flatulence 腸胃氣脹；腹脹
flatware 淺皿；扁平的餐具
flavanol 黃烷醇
flavanone 黃烷酮
flavedo 外果皮（柑桔類）
flavin 黃素
flavo(u)r 風味；香味
flavo(u)r compound 風味化合物；香味化合物
flavo(u)r enhancer 風味增強劑；增味劑
flavo(u)r food 調味食品
flavo(u)r potentiator 風味增強劑；風味增效劑
flavo(u)r reversion 變〔風〕味
flavo(u)red drink 調味飲料
flavo(u)red milk 調味乳；增香乳
flavo(u)ring 調味料；食用香料
flavo(u)ring adjunct 香味輔佐料

flavo(u)ring agent 調味劑；調味料；增香劑
flavo(u)ring essence 食用香精
flavo(u)ring material 調味料；香料
flavo(u)ring substance 食用香料；調味料
flavo(u)rless 無〔香〕味的
flavo(u)rsome 香氣足
flavochrome 黃色素
flavone 黃酮
flavonoid 黃酮類化合物
flavonol 黃酮醇
flavoprotein 黃素蛋白
flavoxanthin 黃黃質
flax 亞麻
flaxseed 亞麻仁；亞麻籽
flaying knife 剝皮刀；去果皮刀
flesh 肉；食用鮮肉；果肉
flesher 屠夫
fleshiness 多肉；肥胖
fleshing 刮肉；去肉
fleshings 括下的肉；削掉的肉
fleshless 無肉的；瘦的
fleshpot 煮肉鍋
fleshpots 美食；豪華餐廳
fleshy 肉厚的；多肉的
fleshy fruit 肉質豐厚的水果；多肉果
flexible can 軟罐
flexible package 軟〔質〕包裝；柔韌性包裝
flexible pouch 軟包裝袋
flexing machine 罐身成圓機；罐身撓曲機
flight kitchen 空廚
flint corn 硬粒玉米
flint maize 硬粒玉米
flint wheat 硬〔質〕小麥
flip 弗利潑酒
flipper 輕度膨罐
flitch 醃豬肋肉；比目魚的肉片
floated oil 浮油
floc 絮凝物；絮狀物
flocculant 絮凝劑
flocculate 絮凝；絮凝物
flocculating agent 絮凝劑
flocculation 絮凝〔作用〕
flocculent yeast 絮狀酵母
floor malt 地板法的麥芽
flora 植物相；微生物群落
flora nectar 花蜜
floral essence 花香精
florence oil 純橄欖油
floridian starch 紅藻澱粉
florist violet 香堇
floss candy 棉花糖
flotation agent 浮選劑
flotation grader 漂浮分級機
flotation washer 浮動洗滌機；浮洗機
flounder 鰈〔魚〕；扁魚；比目魚
flour 麵粉；撒〔上麵〕粉
flour ag(e)ing 麵粉熟成；麵粉陳化
flour barm 發麵酵母；發麵起子
flour bleaching 麵粉漂白
flour bleaching agent 麵粉漂白劑
flour dresser 篩粉機
flour enhancing resistance agent II 麵粉強筋劑II號
flour enrichment 麵粉〔營養〕強化
flour improver 麵粉改良劑；小麥粉處理劑
flour middlings 中粗粒麵粉
flour mill 麵粉磨機；磨粉機
flour modifing agent 小麥粉改質劑
flour paste 麵粉糊
flour quesadilla 墨西哥薄餅；墨西哥油炸玉米粉餅
flour reducing resistance agent 麵粉減筋劑

flour strength 麵粉強度；麵粉筋力
flour treatment agent 小麥粉處理劑
floury 粉狀的；麵粉的；粉質的
flow chart 流程圖
flow diagram 流程圖
flowability 流動性
flower bud oil 花蕾油
flower flute porgy 細鱗紫茵鯛；細鹵紫魚
flower tea 花茶
flowery odo(u)r 花香；花嗅
flowy dough 塌麵糰；變稀的發酵麵糰
flue cured 煙燻的
flue dryer 煙道式乾燥器
fluffer 疏鬆機；攪打疏鬆機
fluid 流體
fluid bed dryer 流動層乾燥機；流化床乾燥機
fluid bed drying 流動層乾燥；流化床乾燥
fluid diet 流質膳食
fluid film 流體薄膜
fluid food 流體食物
fluid medium 液體培養基
fluid ounce (fl. oz.) 液唡；液盎司
fluid shortening 液態酥油
fluid wax 液蠟
fluidimeter 流度計
fluidity 流動性；流度
fluidization 流化〔作用〕；流體化
fluidized bed 流動層；流化床
fluidized bed dryer 流動層乾燥機；流化床乾燥機
fluidized bed drying 流動層乾燥；流化床乾燥
fluidized bed granulation 流動層造粒；流化床造粒
fluidized bed granulator 流動層造粒機；流化床造粒機
fluidized distillation 流體化蒸餾〔法〕
fluidness 流動性
flummery 麵粉糊；玉米粉布丁；麥片粥
fluorescein 熒光黃；熒光素
fluorescence 熒光
fluoridation 加氟
fluoride 氟化物
fluoridization 氟化處理
fluorination 氟化〔作用〕
fluorine 氟
fluorosis 氟中毒
flush 茶樹嫩芽；萌芽；沖洗
flusher 沖洗器
flutemouth 鱗烟管魚；康氏烟管魚 [= cornetfish]
fly 蒼蠅
fly ash 飛塵
fly net 防蠅網
flycatcher 捕蠅器；捕蟲草
flyflap 蠅拍
flying fish 飛魚
flypaper 捕蠅紙
flytrap 捕蠅器
foal 小馬；駒
foam 泡沫
foam breaker 消泡器
foam destroyer 消泡器；除沫器
foam drying 泡沫乾燥
foam drying process 泡沫乾燥法
foam fermentation 起泡發酵
foam-mat drying 泡沫層乾燥；真空泡沫乾燥
foam modifier 泡沫改良劑
foam separation 泡沫分離〔法〕
foam spray drying 泡沫噴霧乾燥
foam stabilizer 泡沫穩定劑
foam type cake 泡沫〔型〕蛋糕

foam vacuum drying 真空泡沫乾燥；泡沫層乾燥
foamer 起泡劑；發泡劑
foaming 發泡；起泡
foaming agent 發泡劑；起泡劑
foaming killer 消泡機；消泡器
foamy body 發泡體
fodder 飼料
fodder grain 飼料穀物
foie gras 肥鵝肝；鵝肝
foil 箔；薄片
foil laminated film 鋁箔積層膜；鋁箔複合薄膜
foil wrapper 鋁箔包裝材料
folacin 葉酸
folate 葉酸鹽
folic acid 葉酸；蝶酰麩胺酸；蝶酰谷胺酸；維生素Bc [=pteroylgluta-mic acid; vitamin Bc]
fondant 糖霜泥；軟糖料；微晶糖膏；方旦〔半軟〕糖
fondue 溶製乾酪；液乾酪羹
fontina 意大利果仁味羊奶〔硬〕乾酪
food 食物；食品；食糧
food additive 食品添加物；食品添加劑
Food Additive Regulation (FAR) 食品添加劑規範
Food Additives and Contaminants Committee (FACC) 食品添加劑及污染物委員會
food adulteration 食品摻假
food allergy 食物過敏
food analysis 食品分析
Food and Agriculture Organization (FAO) 糧農組織；聯合國糧農組織；聯合國糧食〔及〕農業組織 [= United Nations Food and Agriculture Organiza-tion; UNFAO]
Food and Drug Adminstration (FDA) 食品〔及〕藥物管理局（美國）
food balance 食物均衡
food-borne disease 食物疾病
food-borne infection 食物〔媒介〕感染
food-borne intoxication 食物毒素型中毒；食物傳染性中毒
food chain 食物鏈
Food Chemicals Codex (FCC) 食品添加物法規；美國食用化學法典
food chemist 食品化學家
food chemistry 食品化學
food chopper 食物切碎機
food colo(u)r 食用色素；食品色素；著色劑
food composition 食品成分
food contaminant 食品污染物
food crops 食用作物
food cycle 食物循環
food dye 食用色素
food engineering 食品工程〔學〕
food enrichment 食品〔營養〕強化
food equivalent 食物代用品
food flavo(u)r 食用香精
food flavo(u)ring 食用香料；調味料
food fortification 食品〔營養〕強化
food hygiene 食品衛生〔學〕
food industry 食品工業
food infection 感染型食物中毒；食物感染
food ingredient 食品〔配料〕成分
food inspection 食品檢驗
food intake 食物攝取量
food intoxication 毒素型食物中毒；食物中毒
food irradiation 食品照射
food labelling 食品標籤；食品標誌
food labels 食品標籤
food law 食品法規

food microbiology 食品微生物學
food packaging 食品包裝
food poisoning 食物中毒
food preparation 食品製備
food preservation 食品保藏
food preservative 食品防腐劑
food processing 食品加工
food processing aids 食品加工助劑
food product 食物產品；食物製品
food safety 食品安全
food sanitation 食品衛生
food science 食品科學
food scientist 食品科學家
food service 餐飲業
food shop 熟食店
food spoilage 食品腐敗
food stamp 食物卷
Food Standards Code 食品標準法規
food supplement 食品增補劑
food supply 食品供應
food technologist 食品技術學家；食品工藝師
food technology 食品工藝學；食品技術學
food toxicology 食品毒物學
food value 食品價；營養價值
food web 食物網
food yeast 食用酵母
foodstuff 食物；糧食
foot and mouth disease 口蹄疫
foot oil 渣滓油
foots 渣滓；皂腳；油腳
forage 草料
forage crop 飼料作物
forced air drying 鼓風乾燥
forced circulation evaporator 強制循環蒸發器
forcemeat 肉餡；五香碎肉〔餡〕；調味碎肉
fore feet 豬前蹄
foregrinding 預〔先磨〕碎；預粉碎
forehock 前腿肉
foreign flavo(u)r 異味
foreign substance 異物；夾雜物
foreleg 前腿
forelimb 前肢
foremilk 頭奶；初乳
foreribs 牛前肋骨肉；牛胴胸肋肉
foreshank 牛前脛
forewarm 預熱
fork 餐叉；叉子
form-fill-seal 成型充填密封法；成型灌裝封口法
formaggio cheese 福瑪喬山羊奶乾酪(意大利)
formal dinner 正式餐會
formal tea 正規茶點
formaldehyde 甲醛
formalin 甲醛水；福〔爾〕馬林
formic acid 甲酸；蟻酸
forming agent 賦形劑
forming machine 成形機
formogelatin 甲醛明膠
formose 甲醛聚糖
formula 配方；公式
formulated food 配製食品
fortification 營養強化
fortified bread 營養強化麵包；強化麵包
fortified food 營養強化食品；強化食品
fortified milk 營養強化牛乳；強化牛乳
fortified wine 烈性葡萄酒；加度葡萄酒
foul 腐爛；惡臭的；污穢的
fountain 飲水器；噴泉；泉水；汽水
fountain syrup 汽水用糖漿

fountaineer 汽水商
four-side seal pouch 四邊封袋
four-sided fin seal wrap 四邊密封包裝
fourres 富雷斯糖
fovantini 粗通心粉；細管狀通心粉
fowl 家禽；禽肉
fowl run 雞場
fox grape 美洲狐臭葡萄
foxberry 苔桃；灌木懸鉤子
foxshark 長尾鯊
foxtail millet 粱粟
FPC (fish protein concentrate) 濃縮魚蛋白；魚蛋白濃縮物
fractional column 分餾柱；分餾塔
fractional distillation 分餾
fractional pack 分裝；小包裝
fractional sterilization 分段滅菌；間歇滅菌 [= tyndallization; intermittent sterilization]
fractionating column 分餾柱；分餾塔
fractionating distillation 分餾
fractionation 分級；分餾；區分
fractionator 分餾器
fragmentation 碎斷
fragrance 芳香；香氣
frame filter (press) 板框〔式〕壓濾機
frame plate 框式壓濾板
franchise chain 加盟店
frangipane 杏仁蛋糕；杏仁塔餅
frank 法蘭克福香腸 [= frankfurt(er)]
frankfurt(er) 法蘭克福香腸 [= frank]
frappe 刨冰飲料；果汁刨冰；濃奶昔
free acid 自由酸；游離酸
free alkali 自由鹼；游離鹼
free amino acid 自由胺基酸；游離胺基酸
free drip 游離滴液
free fatty acid 自由脂肪酸；游離脂肪酸
free flowing 自由流動的
free moisture 自由水分；游離水分
free pack 非定量包裝；不定量包裝
free radical 自由基；游離基
free run juice 自流汁
free setting 自由沉降
free-stone peach 離核桃
free water 自由水；游離水
freestone water 無溶質水
freeze 冷凍；凍結
freeze concentration 冷凍濃縮
freeze dehydration 冷凍乾燥；凍乾；升華乾燥
freeze denaturation 冷凍變性
freeze-dried 冷凍乾燥的
freeze-dried meat 冷凍乾燥肉
freeze dryer 冷凍乾燥機 [= lyophilizer]
freeze drying 冷凍乾燥；升華乾燥；凍乾法 [= lyophilization]
freezer 冷凍庫；冷凍室；冷凍機
freezer burn 凍燒；凍灼；凍傷
freezer storage 冷藏
freezing 冷凍〔的〕；凍結〔的〕
freezing injury 凍傷；凍害
freezing mixture 冷凍混合劑
freezing point 冰點；凝固點
freezing resistance 耐寒性；抗凍性
freezing room 冷凍室；凍結室
freezing sterilization 冷凍滅菌〔法〕
freezing storage 凍藏；冷凍儲藏
freezing tunnel 冷凍隧道；凍結隧道
French bean 四季豆；法國菜豆
French bread 法式麵包
French fired potato 炸馬鈴薯條；炸薯條；法式炸馬鈴薯片 [= French fries; French fry]
French fries 馬鈴薯條；炸薯條；法式炸馬鈴薯片 [= French fired potato; French fry]

French fry 馬鈴薯條；炸薯條；法式炸馬鈴薯片 [= French fired potato; French fries]
French pastry 法式糕點；法式甜餡餅
French plum 法國黑李子
French roast 法式咖啡；法式炒咖啡法
French roll 法式麵包卷
French toast 法式土司；西多士
French vermouth 法式苦艾酒；法國味美斯酒
Frenched rack 法式羊肋排
fresh egg 鮮蛋
fresh fish 鮮魚
fresh fruit 鮮果
fresh meat 鮮肉
fresh milk 鮮乳
fresh sausage 生鮮香腸
fresh squeezed juice 現榨果汁
fresh vegetable 新鮮蔬菜
fresh weight 鮮重；活重
freshelize 使新鮮；復鮮
freshen 淡化；減少鹹味；使顯得新鮮；提神
freshener 提神飲料；清涼劑
freshness 新鮮度；鮮度
freshness agent 保鮮劑
freshwater 淡水
freshwater algae 淡水藻
freshwater fish 淡水魚
friability 易粉碎性；脆性
fricandeau 油燜小牛肉；炖小牛肉 [= fricando]
fricando 油燜小牛肉；炖小牛肉 [= fricandeau]
fricasse 烤肉
fricassee 油燜肉；濃汁白燴肉丁；炖濃汁肉丁
friction peeler 摩擦去皮機
friction peeling 摩擦去皮〔法〕
fridge 冰箱
fried 油炸的；炸的；炒的
fried bean curd (puff) 油豆腐；油炸豆腐；炸豆腐泡 [= fried bean curd; fried tofu]
fried cake 油炸糕點
fried chicken 炸雞
fried egg 煎蛋
fried fish stick 炸魚條
fried fishcake 炸魚糕；油煎魚餅
fried food 油炸食物
fried meat 烤肉
fried noodle 炒麵〔條〕
fried potato chip 炸薯片
fried rice 炒飯
fried tofu 油豆腐；油炸豆腐；炸豆腐泡 [= fried bean curd (puff)]
fried vegetable 炒青菜
frier 油炸鍋；煎鍋 [= fryer]
Friesian 荷蘭黑白花牛；好斯坦乳牛；弗里賽〔黑白花〕牛 [= Holstein (-Friesian)]
frigate mackerel 圓花鰹；扁舵鰹
frigorifico 冷凍肉類處理廠；冷凍肉類包裝廠
frigorimeter 低溫計
frijole 豇豆；菜豆
frill shark 皺鰓鯊
fringe shark 日本鬚鯊
fritted glass filter 燒結玻璃濾器；多孔玻璃濾器
fritter 油炸餡餅
friture 炸；炒；煎
frizzle 煎脆；酥炸；煎炸
frog 蛙
frogfish 琵琶魚
frost 霜；結霜；冰凍；糖霜
frost damage 凍傷；霜害
frosted glass 磨砂玻璃

frostfish 帶魚；大西洋叉尾帶魚；大西洋小鱈
frosting 糖霜；結霜
froth 泡沫；起泡〔沫〕
froth breaker 消泡器
frothed white 打起蛋白
frother 起泡劑；發泡劑
frothiness 起泡性
frothing agent 起泡劑；發泡劑
frothy 起泡的；發泡的
frozen bread 冷凍麵包
frozen cake 冷凍蛋糕
frozen carcass 冷凍屠體
frozen concentrated juice 冷凍濃縮果汁
frozen convenience food 冷凍簡便食品；冷凍方便食品
frozen dessert 冷凍甜點
frozen dough 冷凍麵糰
frozen food 冷凍食品
frozen fruit 冷凍水果
frozen meat 冷凍肉
frozen meat slicer 冷凍肉切片機
frozen pack(age) 冷凍包裝
frozen precooked food 冷凍熟食品
frozen prepared food 冷凍調理食品
frozen product 冷凍產品
frozen storage 冷凍貯藏；凍藏；冷凍儲藏
frozen sucker 冰棒；冰棍
frozen vegetable 冷凍蔬菜
fructan 聚果糖；果聚糖
fructification 結果實；果實
fructofuranose 聚果呋喃糖；呋喃果聚糖
fructooligosaccharide 寡果糖
fructopyranose 吡喃果糖
fructosan 聚果糖；果聚糖；聚左旋糖 [= levan]
fructose 果糖；左旋糖 [= fruit sugar; laevulose; levulose]
fructose phosphate 果糖磷酸
fructose syrup 果糖〔糖〕漿
fructosidase 果糖苷酶
fructoside 果糖苷
fructosuria 果糖尿〔症〕
fructuronic acid 果糖酮酸
fruit 水果；果實
fruit beverage 水果飲料
fruit butter 果膏
fruit cake 水果糕；果料蛋糕
fruit cocktail 什錦水果
fruit cordial 果汁雞尾酒
fruit dish 水果盤
fruit drink 果汁飲料
fruit drop(ping) 落果
fruit drops 水果硬糖
fruit essence 水果香精
fruit flesh 果肉
fruit fly 果蠅
fruit fool 奶油伴水果；乳油果泥甜點
fruit hardness tester 果實硬度計
fruit jam 果醬
fruit jelly 水果果凍
fruit juice 果汁
fruit juice beverage 果汁飲料
fruit juice cocktail 什錦果汁飲料；混合果汁
fruit juice concentrate 濃縮果汁
fruit juice drink 含果汁飲料
fruit juice gel 果汁凝膠；果汁凍
fruit knife 水果刀
fruit nectar 果泥飲料；帶果肉果汁飲料
fruit odo(u)r 水果氣味
fruit parer 水果削皮刀；水果削皮器
fruit paste 果糊；果泥
fruit pie 水果派；水果餡餅

fruit preserve 蜜餞
fruit press 水果榨汁機
fruit pudding 水果布丁
fruit pulp 果漿；果肉；果泥
fruit punch 果汁潘趣〔飲料〕；果汁賓治
fruit puree 果泥；果漿
fruit ranch 大果園
fruit salad 水果沙拉；水果色拉；水果沙律
fruit sauce 水果醬〔料〕；水果沙司
fruit seeder 水果去核器
fruit sherbet 果汁雪泥；果汁雪花冰
fruit skin 水果皮；果皮
fruit soda water 果汁汽水
fruit spread 水果〔塗〕醬
fruit squash 果汁飲料；帶果肉果汁
fruit stalk 果梗
fruit stone extractor 去果核機
fruit sugar 果糖 [= fructose; laevulose; levulose]
fruit syrup 水果糖漿；果汁糖漿
fruit tree 果樹
fruit vinegar 水果醋
fruit washer 洗果機
fruit wine 水果酒
fruit(ing) body 子實體
fruitage 結果實；果實〔產量〕
fruitarian 食果主義者；常吃水果的人
fruitbody 子實體
fruiter 果樹；運水果船
fruiterer 水果商
fruitery 果實；果園；水果店；水果〔貯〕窖
fruitlet 小水果；小果實
fruity 果味的
frumenty 麥乳粥；牛奶麥粥；果料牛奶小麥粥 [= furmety]
fry 油炸；油煎；油炸食品；魚苗；小魚
fryer 油炸鍋；煎鍋 [= frier]
frying 油炸；油煎；油炒
frying oil 油炸油；煎炸油
frying pan 煎鍋
fuchsin 品紅；洋紅 [= magenta]
fucosan 聚海藻糖；岩藻聚糖
fucose 海藻糖；岩藻糖
fudge 焦軟糖；富奇軟糖；法奇〔軟〕糖
fuel 燃料
fugal machine 離心機
fugitive flavo(u)r 易揮發的風味
fugu 河豚 [= puffer (fish); globefish]
fuguism 河豚中毒
fugutoxin 河豚毒素
ful medames 福美達美斯豆；埃及土豆
full aroma 濃郁香氣
full cream 全脂〔的〕
full cream milk 全脂牛乳
full diet 齊全膳食
full fish 成魚；性成熟的魚
full fluid diet 全流質膳食
full meal 盛餐
full ripe 完全成熟的
full roe 成熟魚卵
full straight 全麥統粉；統粉
full wrap 全封式外包裝
Fuller's earth 漂白土
fullmasse 糖膏 [= massecuite]
fully cooked 煮熟的；全熟
fumaric acid 延胡索酸；反丁烯二酸；富馬酸
fume 煙；煙燻
fume hood 通風櫥
fume smoking 煙燻
fumeol 煙燻液；精煉煙燻料
fumigant 燻蒸〔消毒〕劑；燻劑
fumigating agent 燻蒸〔消毒〕劑

fumigation 燻蒸〔消毒〕
fumigator 燻蒸〔消毒〕器
functional property 功能性；機能性
fundament 基本；臀部
fungal amylase 真菌澱粉酶
fungal enzyme 真菌酶
fungal proteinase 真菌蛋白酶
fungal spore 真菌孢子
fungi 真菌 〔= fungus的複數〕
fungicidal agent 殺真菌劑
fungicide 殺真菌劑
fungicidin 殺真菌素
fungistasis 抑真菌作用
fungistat(in) 制真菌劑
fungistatic agent 抑真菌劑
fungivorous 食真菌的
fungoid 類真菌
fungus 真菌
fungus resistance 耐真菌性；抗真菌性
fungusproof 抗真菌的；抗霉的
funk 刺鼻臭味
funnel 漏斗；煙窗
furan 呋喃
furanose 呋喃糖；五〔環〕糖
furcellaran 紅藻膠 [= Danish agar]
furfural 糠醛；呋喃甲醛
furfuryl mercaptan 糠基硫醇
furfuryl pyrrole 糠基吡咯
furfuryl thioacetate 硫代乙酸糠酯
furfuryl thiopropionate 硫代丙酸糠酯
furmety 麥乳粥；果料牛奶小麥粥 [= frumenty]
furnace 爐；熔爐
Fusarium 鐮刀菌屬；新月菌屬
fusel oil 雜醇油
fusidic acid 梭鏈孢酸
fusiform 梭狀的；流線形的；兩端漸細的
fusiform spore 梭狀孢子
fusilli 螺絲〔通心〕粉；螺旋〔通心〕粉 [= spiral pasta; spirali]
fusilli bucati 螺旋形布卡提通心粉
fustiness 霉臭〔味〕
fusty 霉臭的
futira 果料餡餅
fuzzy melon 節瓜；毛瓜

G

gadoid (fish) 鱈〔魚〕；鱈科的魚
gadoleic acid 鱈油酸；二十碳烯酸
gage 意大利李；來因克洛德李
gaidic acid 十六碳烯酸
gail 單發酵啤酒；發酵用麥芽汁 [= gyle]
galactan 聚半乳糖；半乳聚糖
galactaric acid 半乳糖二酸
galactase 半乳糖酶
galactitol 半乳糖醇；肌醇半乳糖苷
galactococcus 乳球菌
galactofuranose 半乳呋喃糖；呋喃半乳糖
galactogen 半乳多糖
galactolipid 半乳糖脂
galactometer 乳液比重計
galactonic acid 半乳糖酸
galactopathy 飲乳療法；乳療法
galactosan 聚半乳糖；半乳聚糖
galactose 半乳糖
galactosidase 半乳糖苷酶
galactoside 半乳糖苷
galactosis 乳汁分泌；乳液生成
galactotherapy 乳療法
galactotoxin 乳毒素
galactotrophy 乳營養法
galacturonic acid 半乳糖醛酸
galangal 南薑；良薑
galantine 加蘭丁肉凍；肉凍；扎肉
gale 美洲楊梅
galette 撒糖餅乾；素料餅乾；千層蔬菜派
gall 膽汁；沒食子；五倍子
gall bladder 膽囊
gallate 沒食子酸鹽；五倍子酸鹽
galley 廚房
gallia 高盧瓜(以色列)
gallic acid 沒食子酸；五倍子酸
gallon 加侖
gallotannin 鞣酸；單寧酸；沒食子單寧；倍單寧
gallstone 膽石
gam 鯨魚群；腿
gambrel 掛肉鈎架
game 獵物；野味
game bird 野禽；獵禽
game fowl 野禽；獵禽
game meat 野味肉
gaminess 野味的風味
gammon 煙燻火腿
gamy 野味的風味；味道強的
gander 雄鵝
Ganges amaranth 莧菜
gangfish 白鮭
gannet 塘鵝
Ganoderma 靈芝屬
Ganoderma lucidum 靈芝
ganoid 硬鱗魚；硬鱗的
ganoid scale 硬鱗
garbage 廢物；廢料；垃圾

garbage can 垃圾桶；餿水桶
garde-manger 冷盤小廚房；食物貯藏室
garden 花園；菜園
garden bean 菜豆；四季豆；青刀豆
garden beet 根甜菜；菜用甜菜；恭菜
garden cane 果蔗
garden chrysanthemum 茼蒿
garden cress 獨行菜
garden-egg 茄子 [= eggplant; aubergine; brinjal]
garden mint 留蘭香
garden pea 豌豆；青豆
garden rhubarb 食用大黃
garden sass 庭園蔬菜；蔬菜；青菜
garden stuff 蔬菜〔類〕；水果〔類〕
garden truck 市售蔬菜
garden violet 香堇
gardenia 梔子
gardenia blue 梔子藍色素
gardenia concrete 梔子花浸膏
gardenia green 梔子綠色素
gardenia red 梔子紅色素
gardenia yellow 梔子黃；黃梔子
garfish 圓尾頜針魚；頜針魚；鱵魚
garganelli 捲筒粉
garget 乳房炎（家畜）
gargle 漱口〔藥〕水；漱口
garland chrysanthemum 茼蒿
garlic 蒜頭；大蒜
garlic bread 蒜味麵包；蒜蓉〔麵〕包
garlic hop 蒜瓣
garlic oil 蒜油；大蒜油
garlic powder 蒜粉
garlic salt 蒜鹽
garlic spread 蒜泥
garlicin 大蒜精；大蒜素
garlicky 有大蒜氣味的
garner 倉庫；穀倉；糧倉
garnet berry 紅醋栗
garnish 菜餚的飾菜或飾物
garniture 菜餚的飾菜或飾物
garrupa 紅點石斑魚
garum 魚醬油（地中海沿岸）
gas 氣體；瓦斯；汽油；煤氣
gas bacillus 產氣桿菌
gas bladder 魚鰾
gas chromatography (GC) 氣相層析法；氣相色譜法
gas flush(ing) package 充氣包裝
gas flush(ing) pouch 充氣包裝袋
gas forming bacteria 產氣細菌
gas heater 瓦斯爐
gas hole 氣孔（乾酪或芝士）
gas-lift fermenter 氣升發酵槽；空氣升液發酵槽 [= air-lift fermenter]
gas-liquid chromatography (GLC) 氣液〔相〕層析法；氣液〔相〕色譜法
gas oven 烘箱；烤箱；煤氣烤爐
gas packing 充氣包裝〔法〕
gas permeability 透氣性
gas producing bacteria 產氣〔細〕菌
gas proof 不透氣的；氣密的
gas rice cooker 瓦斯炊飯鍋
gas singer 煤氣燒毛器；煤氣燎毛器
gas storage 控氣保藏；氣調貯藏；人工氣候保藏法；控制大氣貯藏 [= controlled atmosphere storage; CA storage]
gas stunning 氣體昏厥〔法〕
gas tight 氣密的；不透氣的
gaseous fluidization 氣態流化〔作用〕
gaseous sterilant 氣體滅菌劑
gaseous sterilization 氣體滅菌
gasification 氣化〔作用〕
gasify 氣化；充氣
gasket 墊圈；膠圈

gasogen(e) 飲料充氣機；充氣機；輕便飲料
gassing 產氣；充氣；薰蒸消毒
gassing property 產氣性
gassy can 氣罐；脹罐
gassy fermentation 產氣發酵
gastric 胃的
gastric acid 胃酸
gastric juice 胃液
gastritis 胃炎
gastroenteritis 胃腸炎；腸胃炎
gastroenteropathy 胃腸病
gastrointestinal 胃腸的；腸胃的
gastrointestinal lipase 胃腸道脂肪酶
gastrointestinal tract 胃腸道
gastrologer 美食〔學〕家；烹飪學家
gastrology 烹飪學；胃〔病〕學
gastronome(r) 美食家；講究飲食的人；品味員
gastronomic emporium 美食餐館
gastronomist 美食家
gastronomy 美食學；美食法
gastrosis 胃病
gateau 糕點
gauffre 薄脆鬆餅；威化餅乾
gaultheria 白珠樹
gaultheria forrestii oil 地檀香油
gaultheria oil 白珠木油
gauze strainer 網狀濾器
gazogene 飲料充氣機；汽水製選造機
gazpacho 西班牙涼菜；西班牙生蔬菜湯
GC (gas chromatography) 氣相層析法；氣相色譜法
gefillte fish 什錦魚糕或魚丸；蘇式魚糜製品冷盤
gel 形成凍膠；膠凝；凍膠；凝膠體
gel chromatography 凝膠層析法；凝膠色譜法
gel electrophoresis 凝膠電泳
gel filtration 凝膠過濾
gel forming property 膠凝性〔能〕
gel point 膠凝點；膠凝溫度
gel rigidity 凝膠剛度
gelatin jelly 明膠凍；明膠啫哩；明膠軟糖
gelatin(e) 明膠；動物膠；魚膠粉；吉利丁；凝膠〔體〕
gelatin(e) agar 明膠洋菜培養基
gelatin(e) culture 明膠培養基
gelatin(e) dessert 明膠甜點
gelatin(e) glaze 明膠包衣
gelatinase 明膠酶
gelatinate 明膠化；膠〔凝〕化；糊化；使成膠狀
gelatination 明膠化
gelatinization 膠化；膠凝作用；糊化〔作用〕
gelatinization degree 糊化程度
gelatinization starch 糊化澱粉
gelatinize 膠〔凝〕化；糊化；使成膠狀
gelatinized point 糊化點；糊化溫度
gelatinizer 膠化劑；膠凝劑；稠化劑 [= gelatinizing agent; gelling agent；jelling agent]
gelatinizing agent 膠化劑；膠凝劑；稠化劑 [= gelatinizer; gelling agent；jelling agent]
gelatinous 凝膠狀的；膠質的
gelatinous fiber 凝膠纖維
gelatinous substance 膠凝〔狀〕物質
gelation 膠化；膠凝〔作用〕；糊化〔作用〕
gelded ox 閹牛
gelded pig 閹豬
gelidium 石花菜
gellan gum 潔冷膠
gelled food(stuff) 凝膠食品；膠凝食品

gelling agent 膠化劑；膠凝劑；稠化劑 [= gelatinizing agent; gelatinizer; jelling agent]
gelling effect 膠凝效應；膠凝作用
gelling point 膠凝溫度；膠凝點
gelling power 膠凝力
gelling property 膠凝性〔能〕
gelling starch 糊化澱粉
gelling strength 膠凝強度
gelometer 膠凝〔強度〕計；凝膠堅度計
gelose 洋菜糖；瓊脂糖
gemelli 紐紋粉；螺絲粉
gemma 芽孢；胞芽
gemmation 發芽〔生殖〕
gene 基因
gene cloning 基因選殖
gene expression 基因表現
gene locus 基因位點
gene map 基因圖譜
gene mutation 基因突變
gene recombination 基因重組
gene therapy 基因療法；基因治療
generally recognized as safe (GRAS) 公認安全；公認無毒；公認為安全的
generally recognized as safe substance (GRASS) 公認安全物質
generation time 世代時間；細胞增代時間；一代時間
genetic code 遺傳密碼
genetic engineering 遺傳工程〔學〕
genetic factor 遺傳因子
genetic map 基因圖譜
genetically modified (GM) food 基因〔改造〕食物
genetics 遺傳學
genitan 龍膽花；龍膽根
genome 基因組
genotype 基因型
gentian root 龍膽根
genuine 真正的；純正的
genuine porgy 真鯛
genuine tea 純茶
genus 屬
geophagy 食土癖
geranial 香葉醛
geranic acid 香葉酸
geraniol 香葉醇
geranium (rose) oil 香葉油
geranium oil 香葉油
geranyl acetate 乙酸香葉酯
geranyl acetoacetate 乙酰乙酸香葉酯
geranyl benzoate 苯甲酸香葉酯
geranyl butyrate 丁酸香葉酯
geranyl formate 甲酸香葉酯
geranyl isobutyrate 異丁酸香葉酯
geranyl isovalerate 異戊酸香葉酯
geranyl phenylacetate 苯乙酸香葉酯
geranyl propionate 丙酸香葉酯
Gerber centrifuge 格氏乳脂〔計〕離心機
Gerber test 格氏〔乳脂〕測驗法；格伯〔氏〕乳脂測定法
Gerber test bottle 格氏乳脂計；格伯〔氏〕乳脂計
geriatric food 老人食品
germ 芽胞；胚芽；微生物；細菌
germ bread 胚芽粉麵包
germ cell 生殖細胞
germ-free 無菌的
germ-free milk 無菌牛奶；無菌乳
germ layer 胚層
germ meal 胚芽粉
germ oil 胚芽油
germ rice 胚芽米
germ separator 胚芽分離機；脱胚芽機
germ(i)static action 抑菌作用
German millet 粟；飼用粟
German wine 德國葡萄酒
germicidal 殺菌的

germicidal action 殺菌作用
germicidal activity 殺菌活性
germicidal agent 殺菌劑
germicidal lamp 殺菌燈
germicide 殺菌劑
germination 發芽
germinator 發芽器；發芽測定器
gerty 燕麥粉
geska 蓋斯卡乾酪
gestation 懷孕；妊娠
gestation period 懷孕期；妊娠期
ghatti gum 哥地膠；加貼膠 [= gum ghatti]
ghee 水牛酪油；印度奶油；酥油；牛酪油 [= ghi]
gherkin 醃漬小黃瓜；西印度小黃瓜
ghi 水牛酪油；印度奶油；酥油；牛酪油 [= ghee]
giant black marlin 立翅旗魚
giant colony 巨大菌落
giant crab 大腳毛蟹；大蟹
giant egg 巨〔大〕蛋
giant funnel cap 大白椿菇
giant garlic 胡蒜
giant grouper 花尾龍躉
giant pacific oyster 正牡蠣
giant puffball 巨型馬疕蘑菇；大禿馬勃(蕈類)
giant river prawn 羅氏沼蝦；泰國蝦
giant tiger prawn 斑節對蝦
giantarum 魔芋；蒟蒻 [= elephant foot; konjac; konnyaku]
gibberellic acid 赤黴酸；赤霉酸
gibberellin 赤黴素；激勃素
gibbing 去鰓臟；去魚鰓
gibbing machine 去魚鰓機；去鰓臟機
giblet 禽類內臟或雜碎
gigot 熟羊後腿〔肉〕
gill 鰓；垂肉；菌褶；蕈褶
gill cover 鰓蓋
gilled bolete 紅黃褶孔菌(蕈類)
gilt 小母豬
gimmer 母羔羊；小母羊
gin 琴酒；杜松子酒
gin palace 豪華酒館；豪華酒店
ginaca 鳳梨處理機；鳳梨去皮芯機
gingelly oil 芝麻油；香油 [= benne oil; jinjili oil; sesame oil]
ginger 薑
ginger ale 薑汁汽水；薑汁啤酒
ginger beer 薑汁汽水；薑汁啤酒
ginger nut 脆薑餅乾；薑汁餅乾；薑味小酥餅 [= gingersnap]
ginger oil 薑油；生薑油
ginger oily flavo(u)r 薑油香精
ginger oleoresin 薑油樹脂
ginger pop 薑汁汽水
ginger root 薑〔根〕
gingerbread 薑〔汁〕餅；薑味糕點
gingergrass oil 香薑草油
gingerin 薑油脂
gingersnap 脆薑餅乾；薑汁餅乾；薑味小酥餅 [= ginger nut]
gingko 銀杏；白果 [= ginkgo]
ginkgetin 銀杏黃酮；銀杏黃素
ginkgo 銀杏；白果 [= gingko]
ginnol 銀杏醇
ginseng 人參；高麗參 [= sang]
girasole 菊芋
gizzard 砂囊；肫；珍
glace 包糖衣；包冰衣；糖衣；糖漬的 [= glaze]
glacial acetic acid 冰醋酸
glair 蛋白；塗抹蛋白
glasgow magistrate 燻鹹鰊魚；冷燻鯡魚
glass 玻璃；玻璃杯
glassed food 玻璃瓶裝食品

glassful 一杯的容量
glassine (paper) 玻璃紙
glassware 玻璃器皿
glaze 包糖衣；包冰衣；糖衣；糖漬的 [= glace]
glaze fruit 糖衣蜜餞水果；糖漬水果
glazed rice 精碾米
glazing 精碾；包糖衣；包冰衣
glazing agent 包覆劑；發亮劑
glazing drum 軋光筒；拋光筒
glazing machine 軋光機；拋光機
GLC (gas-liquid chromatography) 氣液〔相〕層析法；氣液〔相〕色譜法
gliadin 穀膠蛋白；麥醇溶蛋白
globe artichoke 球狀朝鮮薊；朝鮮薊；雅芝竹；洋薊 [= artichoke]
globe mill 球磨機
globefish 河豚 [= puffer (fish); fugu]
globefish poisoning 河豚中毒
globin 血球蛋白；球蛋白
globular protein 球狀蛋白質
globulin 球蛋白
glossa 舌
glucagon 升血糖激素；胰增血糖素
glucan 聚葡萄糖；葡聚糖 [= glucosan]
glucanase 聚葡萄糖酶；葡聚糖酶
glucaric acid 葡〔萄〕糖二酸
glucemia 糖血症
glucitol 山梨〔糖〕醇；葡〔萄〕糖醇 [= sorbitol]
glucoamylase 葡〔萄〕糖澱粉酶；葡〔萄〕糖糖化酶
glucoascorbic acid 葡〔萄〕糖型抗壞血酸
glucocerebroside 葡〔萄〕糖腦苷脂
glucofuranose 葡萄呋喃糖；呋喃葡萄糖
glucogenesis 葡〔萄〕糖生成
glucokinase 葡〔萄〕糖激酶
glucomannan 聚葡甘露糖；葡甘露聚糖
gluconate 葡〔萄〕糖酸鹽
gluconic acid 葡〔萄〕糖酸
gluconic acid fermentation 葡〔萄〕糖酸發酵
glucono-delta-lactone 葡〔萄〕糖酸-δ-內酯
glucono-δ-lactone 葡〔萄〕糖酸-δ-內酯
gluconobacter 葡萄糖酸桿菌
gluconolactone 葡〔萄〕糖酸內酯
glucophore 生甜〔基〕團
glucoprotein 醣蛋白
glucopyranose 吡喃型葡〔萄〕糖；葡萄吡喃糖
glucosamine 葡萄胺糖；葡〔萄〕糖胺；胺基葡〔萄〕糖
glucosan 聚葡萄糖；葡聚糖 [= glucan]
glucose 葡萄糖；右旋糖 [= dextrose; grape sugar]
glucose agar 葡萄糖瓊脂〔培養基〕
glucose dehydrogenase 葡萄糖去氫酶
glucose isomerase 葡萄糖異構酶
glucose oxidase 葡萄糖氧化酶
glucose phosphate 葡萄糖磷酸
glucose syrup 葡萄糖漿 [= dextrose syrup]
glucosidase 葡〔萄〕糖苷酶
glucoside 葡〔萄〕糖苷；配糖體
glucosidic bond 葡萄糖苷鍵
glucosinolate 硫代葡萄糖苷
glucosone 葡萄醛酮糖；葡〔萄〕糖醛酮
glucosuria 葡萄糖尿症
glucosylation 葡萄糖苷化；葡〔萄〕糖基化
glucuronate 葡〔萄〕糖醛酸鹽
glucuronic acid 葡〔萄〕糖醛酸
glucuronidase 葡〔萄〕糖醛酸苷酶
glucuronide 葡〔萄〕糖醛酸苷
glucuronolactone 葡〔萄〕糖醛酸內酯

glue 膠〔水〕；黏合
gluside 糖精；沙卡林 [= saccharin]
glut 暴食
glutamate 麩胺酸鹽；麩胺酸酯；谷胺酸鹽；谷胺酸酯
glutamic acid 麩胺酸；谷氨酸；穀氨酸
glutamic acid hydrochloride 谷氨酸鹽酸鹽
glutaminase 麩醯胺酸酶
glutamine 麩醯胺；谷氨醯胺；穀氨醯胺
glutaral 戊二醛
glutaraldehyde 戊二醛
glutaric acid 戊二酸
glutathione 麩胱甘肽；谷胱甘肽
glutelin 穀蛋白
gluten 麵筋
gluten ball 麵筋球
gluten flour 高筋麵粉
gluten-free 無麵筋；不含麵筋
gluten-free food 無麵筋食品；不含麵筋食品
glutenin 小麥穀蛋白；麥穀蛋白
gluteus 臀肌
glutinosity 黏性
glutinous dumpling 湯丸
glutinous flour 高麵筋麵粉
glutinous rice 糯米 [= sticky rice; sweet rice; waxy rice]
glutinous rice flour 糯米粉
glutinousness 黏〔滯〕性
glutose 格魯糖；3-酮基己糖
glutton 貪食者；暴食者
gluttony 貪食；暴食
glycan 聚糖
glycemia 糖血症；血糖過高
glyceraldehyde 甘油醛
glycerate 甘油酸鹽；甘油酸酯
glyceric acid 甘油酸
glyceridase 甘油酯酶
glyceride 甘油酯
glycerin ester 甘油酯
glycerin fatty acid ester 甘油脂肪酸酯
glycerin monostearate 甘油單硬脂酸酯；單硬脂酸甘油酯 [= glycerol monostearate; glyceryl monostearate]
glycerin(e) 甘油；丙三醇 [= glycerol]
glycerol 甘油；丙三醇 [= glycerin(e)]
glycerol ester of partially dimerized rosin 部分二聚松香甘油酯
glycerol ester of polymerized rosin 多聚松香甘油酯
glycerol ester of tall oil rosin 妥爾油松香甘油酯
glycerol ester of wood rosin 木松香甘油酯；三松香酸甘油酯 [= glyceryl triabietate]
glycerol esters of fatty acids 脂肪酸甘油酯類
glycerol monostearate 甘油單硬脂酸酯；單硬脂酸甘油酯 [= glyceryl monostearate; glycerin monostearate]
glycerolipid 甘油脂
glycerophosphate 甘油磷酸；磷酸甘油 [= glycerophosphoric acid]
glycerophosphatide 甘油磷脂 [= glyceryl phosphatide]
glycerophosphoric acid 甘油磷酸；磷酸甘油 [= glycerophosphate]
glycerose 甘油糖
glyceryl behenate 扁油酸甘油酯
glyceryl caprylate-caprate 辛癸酸甘油酯
glyceryl diacetate 二乙酸甘油酯；二醋酸甘油酯
glyceryl monoacetate 一乙酸甘油酯；醋精；甘油醋酸酯

glyceryl monostearate 甘油單硬脂酸酯；單硬脂酸甘油酯 [= glycerol monostearate; glycerin monostearate]
glyceryl phosphatide 甘油磷脂 [= glycerophosphatide]
glyceryl triabietate 木松香甘油酯；三松香酸甘油酯 [=glycerol ester of wood rosin]
glyceryl triacetate 甘油三乙酸酯；三醋精；甘油三醋酸酯 [= triacetin]
glyceryl tributyrate 三丁酸甘油酯；三酪酯
glyceryl trilaurate 三月桂酸酯；甘油三月桂酸酯；三月桂精 [= trilaurin]
glyceryl trilinoleate 甘油三亞〔麻〕油酸酯；三亞麻油脂；亞油精 [= trilinolein; linolein; linolin]
glyceryl trilinolenate 甘油三次亞麻油酸酯；三次亞麻油脂；亞麻精 [= trilinolenin; linolenin]
glyceryl trioleate 甘油三油酸酯；三油酸甘油酯；三油脂 [= triolein; olein]
glyceryl tripalmitate 軟脂；三棕櫚精；棕櫚精；甘油〔三〕棕櫚酸酯 [= tripalmitin; palmitin]
glyceryl tripropanoate 三丙酸甘油酯
glyceryl tristearate 硬脂；三硬脂精；硬脂精；三硬脂酸甘油酯；甘油三硬脂酸酯 [= tristearin; stearin]
glycidol 縮水甘油
glycine 甘胺酸；胺基醋酸；胺基乙酸 [= aminoacetic acid; collagen sugar]
glycinin 大豆球蛋白
glycitol 糖醇
glycocoll 甘胺酸；胺基醋酸
glycogen 肝糖；糖原；動物澱粉 [= animal starch; liver starch]
glycogenase 肝糖酶；糖原酶
glycogenesis 肝糖生成；糖原生成
glycogenic 生糖原的；產糖原的
glycogenolysis 肝糖分解；糖原分解
glycol 乙二醇；甘醇
glycolic acid 乙醇酸；羥〔基〕醋酸；羥〔基〕乙酸 [= hydroxyacetic acid]
glycolipid 糖脂
glycolipin 糖脂
glycolysis 糖〔酵〕解
glycolytic enzyme 糖解酶
glyconate 糖酸鹽；葡糖酸鹽
glycopeptide 糖肽
glycophospholipin 糖磷脂
glycoprotein 糖蛋白〔質〕
glycosamine 胺醣；葡糖胺
glycose 單糖
glycosidase 糖苷酶
glycosidation 糖苷化；苷化
glycoside 糖苷；配糖物
glycosuria 糖尿〔症〕
glycuronic acid 醣醛酸
glycyrrhiza 甘草
glycyrrhizic acid 甘草素；甘草苷酸；甘草酸苷 [= glycyrrhizin]
glycyrrhizin 甘草素；甘草甜〔素〕；甘草苷酸；甘草酸苷 [= glycyrrhizic acid]
glycyrrhizinic acid 甘草酸
glyoxal 乙二醛
glyoxylic acid 乙醛酸
GMP (good manufacturing practice) 良好作業規範；良好生產操作規程
gnocchetti sardi 沙丁魚麵疙瘩
gnocchi 意大利雲吞；意大利餃子；麵疙瘩
goa 西藏羚羊
goa butter 羚羊奶油；羚羊乳酪
goat 山羊
goat acid 山羊酸

goat meat 山羊肉
goat milk 山羊乳
goat tallow 山羊脂
goatling 小山羊；山羊羔
gobbet 一塊；小片；一口
gobbler 雄火雞
goblet 高腳酒杯
goblin shark 劍嘴鯊
goboazami 菊牛蒡
goby 沙塘鱧；鰕虎魚
gold ripe(ness) 黃熟
golden banded caesio 金帶梅鯛
golden cap 金褐傘(蕈類)
golden carp 鯽〔魚〕
golden chanterelle 變黃雞油菌(蕈類)
golden currant 花醋栗
golden gram 綠豆
golden meal 細玉米粉
golden mushroom 金針菇
golden nugget 金瓜
golden shiner 金體美洲鯿魚
golden skinned porgy 魬鯛
golden thread 金線魚
golden tips 金尖茶
golden wax cap 硫磺濕傘(蕈類)
goober 落花生；花生 [= earthnut; groundnut; monkey nut; peanut]
Gooch crucible 古氏坩鍋
Gooch filter 古氏濾器
good manufacturing practice (GMP) 良好作業規範；良好生產操作規程
good trencherman 食量大的人；大胃王
goody 美食；糖果
goose 鵝；鵝肉
goose liver paste 鵝肝醬
gooseberry 醋栗
goosegog 醋栗
gooseneck round 外側後腿肉；上臀肉
gorcock 紅色雄松雞；雄赤松雞
gorge 貪食；暴食；胃；咽喉
gormandise 講究飲食；暴飲暴食；大吃大喝 [= gourmandise]
gormandize 講究飲食；暴飲暴食；大吃大喝 [= gourmandize]
gorp 高能量小吃
gosling 小鵝
gossypol 棉子酚
gossypose 棉子糖
Gouda cheese 哥打芝士；哥達乾酪；古烏達乾酪
goulash 格辣燉肉；菜燉牛肉；匈牙利燉肉
gourd 葫蘆；葫蘆屬
gourmand 嗜食家；食品鑑賞家
gourmandise 講究飲食；暴飲暴食；大吃大喝 [= gormandise]
gourmandism 美食主義
gourmandize 講究飲食；暴飲暴食；大吃大喝 [= gormandize]
gourmet 美食家
gracilaria 龍鬚菜
gracillis 內股肌肉
grader 分級機；分選機
grading machine 分級機；選別機
grading sieve 分級篩
grading sifter 分級〔細〕篩
graduated cylinder 量筒
graduated glass 量杯
grafted starch 接枝澱粉
graham (flour) 全穀粉；全麥麵粉
graham bread 全穀麵包
grahamism 素食主義
grain 穀物；穀粒；穀類；顆粒；晶粒
grain alcohol 乙醇；酒精；糧穀酒精
grain bin 穀倉；糧倉
grain blower 揚穀機
grain crop 穀類作物

grain distillery 釀酒廠；蒸餾室；酒精廠
grain elevator 穀倉；糧倉
grain feed 穀物飼料
grain hardening 糖晶粒固結；並晶
grain mill 碾穀機
grain of bread 麵包疏鬆度；麵包孔隙度
grain puffer 爆米花機
grain sorghum 高粱；穀蜀黍
grain thresher 穀物脱粒機
graininess 粒度；顆粒性
graining 粒化；結成粒狀
graining process 脂肪凝固過程
gram 克；鷹嘴豆
gram calorie 卡路里；卡 [= calorie; cal]
Gram negative bacteria 格蘭〔氏〕陰性細菌
Gram positive bacteria 格蘭〔氏〕陽性細菌
Gram variable bacteria 格蘭〔氏〕可變細菌；格蘭〔氏〕染色不定細菌
Gram's stain 格蘭〔氏〕染色
gramigna 搓草形通心粉
granadilla 百香果；雞蛋果；西番蓮果；熱情果 [= passionfruit; parchita]
granary 穀倉；糧倉
granatum 石榴皮
granula 顆粒
granular 粒狀的；顆粒狀的
granularity 粒性；粒度
granulated honey 帶晶粒蜂蜜
granulated sugar 晶粒砂糖；砂糖
granulating machine 造粒機
granulation 造粒；粒化；成粒狀
granulator 造粒機；製粒機
granule 顆粒；小粒；粒
granulose 粒質糖；澱粉粒質
grape 葡萄
grape flavo(u)r 葡萄香精
grape juice 葡萄汁
grape mass 破碎葡萄
grape oily flavo(u)r 葡萄油香精
grape pomace 葡萄〔榨〕渣；葡萄皮渣
grape press 葡萄榨汁機
grape seed 葡萄子
grape skin extract 葡萄皮萃；葡萄皮提取物；葡萄皮色素；葡萄皮紅
grape sugar 葡萄糖；右旋糖 [=dex-trose; glucose]
grape vinegar 葡萄醋
grapefruit 葡萄柚
grapefruit oil 葡萄柚油
grapery 葡萄園
grapestone 葡萄子
graphite 石墨
GRAS (generally recognized as safe) 公認安全；公認無毒；公認為安全的
GRASS (generally recognized as safe substance) 公認安全物質
grass carp 草魚；鯇
grass prawn 草〔對〕蝦
grass sorghum 草蜀黍；高粱
grassleaf daylily 黃花菜
grassy 草味的
grassy odo(u)r 草味；青草味
grate 搓碎；篦條篩
grated cheese 搓碎乾酪；粉碎乾酪
grater 擦菜板；擦細器；磨碎器
gravel filter 砂礫濾器
gravimetric analysis 重量分析〔法〕
gravity balance 比重天秤
gravity bottle 比重瓶
gravy 肉汁〔醬〕
gravy boat 肉汁〔醬〕器皿；盛肉汁船形器皿；醬油壺
gravy powder 肉汁醬粉；肉滷粉
gray cod 大頭鱈

grayling 河鱒；茴魚；百里香魚
grazer 草食動物
graziery 畜牧業；放牧業
grazing 放牧；牧草
grease 脂；潤滑脂；黃油
greaseproof 防油的；耐油的
greaseproof paper 防油紙
greaseproof wrapping 防油包裝〔材料〕；耐油包裝〔材料〕
greaser 抹油器
greasiness 油膩〔性〕
greasy 油膩的
greasy spot 油斑
great burdock 牛蒡
great calorie 大卡；仟卡 [= kiloca-lorie; kcal; Calorie; Cal]
greaves 食用油渣；油渣
green 綠的；新鮮的；青菜；蔬菜
green algae 綠藻
green asparagus 綠蘆筍
green bacon 未燻製培根〔醃肉〕
green bean 四季豆；菜豆
green bean sprout 綠豆芽；銀芽 [= silver sprout]
green bean vermicelli 粉絲；冬粉 [= bean noodle; bean thread (vermicelli); bean vermicelli]
green butter 植物脂
green capsicum 青〔辣〕椒
green cheese 未成熟乾酪；生乾酪
green coffee 生咖啡〔豆〕
green corn 嫩玉米
green crab 青蟹
green dough 未發酵麵糰
green duck 子鴨；小鴨
green ginger 生薑
green goods 新鮮蔬菜
green goose 子鵝；小鵝
green gram 綠豆 [= mung bean]
green kernel rice 青米
green lentil 綠扁豆
green malt 生麥芽
green mango 情人果；青芒果
green mature 青熟
green meat 綠變肉；鮮肉；未醃的肉
green mo(u)ld 綠霉；綠黴
green mussel 翡翠貽貝
green onion 小洋蔥；青蔥
green pea 豌豆〔仁〕；青豌豆
green pepper 青椒
green russule 變綠紅菇
green S 食用綠S
green shrimp 生蝦 (帶殼)；青蝦
green soybean 毛豆；青大豆
green stuff 青菜；蔬菜
green sturgeon 中吻鱘；庫頁島鱘
green tea 綠茶
green tuna 綠變鮪魚
green turnip 蕪菁；大頭菜；青蘿蔔
green vegetable 青菜；蔬菜
green weight 活重；鮮重
green zucchini 綠皮小胡瓜；綠皮西葫蘆
greenery 溫室
greengage 意大利李；青梅；西洋李
greengrocer 蔬菜〔水果〕販；菜販
greengrocery 蔬菜〔水果〕店；蔬菜；水果
greenhouse 溫室
greening 綠皮蘋果；青蘋果
greenling 大瀧六線魚
greenness 綠色；未熟；新鮮
greens 青菜；綠葉菜
grenadine 石榴糖漿；石榴汁
grenadine syrup 石榴糖漿
grey and yellow knight cap 灰口蘑 (蕈類)
grey knight cap 棕灰口蘑 (蕈類)

grey mullet 烏魚；灰鯔
grey trout 湖鮭；海鮭
griddle 大孔篩；平〔底〕鍋
griddlecake 薄煎餅；烙餅；煎餅；烤餅 [= flapjack; slapjack]
grig 小鰻〔魚〕
grigger 薄餅；烤餅
grill 烤架；烤肉；烤肉店
grill room 烤肉店
grilled meat 烤肉
grilling 燒烤；烤 [= broiling]
grilse 回游鮭；回游三文魚；逆河產卵的鮭
grind 研磨；磨〔碎〕；破碎
grinder 磨碎機；研磨器；破碎機
grinding 磨碎；破碎；研磨
grinding machine 磨床；研磨機；磨碎機
grinding mill 研磨機；破碎機；磨碎機
grinding stone 磨石
grindstone 磨石
griskin 豬里脊肉；豬腰部瘦肉；豬排
grist 製粉〔用〕穀物；粗穀粉；粗〔碎〕麥芽
gristle 軟骨
gristmill 磨坊；磨粉機；磨粉廠
grit 砂礫；粗粒
grits 粗磨穀粉；粗燕麥粉；粗玉米粉
groats 去殼穀粒；去殼小麥粒；去殼燕麥片
grocer 雜貨店
grocery 食品雜貨；食品雜貨店；雜貨店
grocery shop 食品雜貨店
grocery store 食品雜貨店；雜貨店
groceteria 自助食品雜貨店；食品雜貨自助售貨店
grog 烈酒；加水烈酒
groggery 酒吧；小酒館
grogshop 酒館；小酒店
gromwell 紫草屬植物
gromwell pigment 紫草紅
gross sample 總樣品
gross ton 長噸
gross weight 總重〔量〕；毛重
ground beef 絞細牛肉
ground cherry 酸漿莓
ground coffee 磨碎咖啡〔粉〕
ground fish 底棲魚類 [= bottom fish]
ground meat 碎肉；絞〔細〕肉；肉糜
ground pea 花生；落花生
ground water 地下水
groundling 底棲魚；水底魚
groundnut 花生；落花生 [= earthnut; peanut]
groundnut kernel 花生仁；花生米
groundnut oil 花生油 [= arachis oil; earth-nut oil; peanut oil]
grouper 石斑魚；紅鮨
grouse 松雞
grout 粗粉；帶殼穀粒；未過篩粉
grouts 穀粉粥；燕麥粥
grower 栽培者；飼養者；生長物
growing season 生長季節
growler 啤酒罐
growth curve 生長曲線
growth cycle 生長周期
growth factor 生長因素
growth hormone 生長激素
growth inhibitor 生長抑制劑
growth medium 培養基；微生物培養基
growth period 生長期
growth phase 生長期；增殖期
growth promoter 生長促進劑
growth rate 生長速率；增殖率
growth regulator 生長調節劑
growth stimulant 生長刺激劑

growth substance 生長物質
gruel 稀粥；薄糊
gruel diet 稀粥飲食
grunt 黑鰭髭鯛；橫帶髭鯛
gruntling 小豬
guaiac gum 癒創〔木〕樹脂 [= gum guaiac; guaiac resin; guaiacum (resin)]
guaiac resin 癒創〔木〕樹脂 [= gum guaiac; guaiac gum; guaiacum (resin)]
guaiac wood oil 癒創木油
guaiacol 癒創木酚；鄰甲氧基苯酚
guaiacum (resin) 癒創樹脂；癒創木樹脂 [= gum guaiac; guaiac gum; guaiac resin]
guanabana 酸番荔枝；刺番荔枝
guanidine 胍
guanine 鳥嘌呤
guanosine 鳥嘌呤核苷；鳥苷
guanosine monophosphate 鳥苷一磷酸
guanosine nucleotide 鳥〔核〕苷酸
guanosine-5'-monophosphoric acid (GMP) 5'-鳥苷酸
5'-guanylic acid 5'-鳥苷酸
guanylic acid 鳥嘌呤核苷酸；鳥苷酸
guar gum 關華豆膠；瓜耳豆膠；瓜爾豆膠
guarana 瓜拉那豆；瓜拉拿藤
guava 番石榴
Guernsey cattle 更塞牛；格恩西牛
guinea corn 幾內亞高粱
guinea fowl 珠雞；珠珍雞
Guinea pepper 幾內亞胡椒
guinea squash 茄子
gula kelapa 椰糖
gulf hake 海灣無鬚鱈
gullet 食道；咽喉
gulonic acid 古洛糖酸
gulose 古洛糖
gum 膠；齒齦；植物膠
gum acacia 阿拉伯樹膠 [= acacia gum; arabic gum; arabian gum; gum arabic]
gum arabic 阿拉伯樹膠 [= acacia gum; arabic gum; arabian gum, gum acacia]
gum chicle 糖膠樹膠；人心果樹膠 [= chicle (gum)]
gum dragon 黃蓍膠；特拉卡甘膠 [= gum tragacanth; tragacanth gum]
gum drops 軟糖；橡皮糖
gum elastic 軟性〔樹〕膠
gum ghatti 哥地膠；加貼膠 [= ghatti gum]
gum guaiac 癒創〔木〕樹脂 [= guaiac gum; guaiac resin; guaiacum (resin)]
gum karaya 刺梧桐膠；刺梧桐〔樹〕膠 [= karaya (gum)]
gum resin 膠樹脂；樹〔膠〕脂
gum sugar 膠糖；阿拉伯糖
gum tragacanth 黃蓍膠；特拉卡甘膠 [= gum dragon; tragacanth gum]
gumbo 秋葵；黃秋葵；毛茄 [= okra]
gummosity 膠黏性；黏性
gummy 膠黏的；黏性的
gummy shark 白斑星鯊
gun shooting 鎗殺法
gunny 麻布袋
gunpowder tea 珠〔綠〕茶；圓珠茶
gur 原糖；糖蜜
gurnard 角魚；魴魚
gurry oil 魚油
gustation 嘗味；味覺
gustatory 味覺的；嘗味的
gustometer 味覺計；味覺測定儀
gut 腸；內臟
gut stripper 腸衣去黏膜機
gutta-percha 馬來乳膠；膠木膠；馬來樹膠；杜仲樹膠；古塔〔波〕膠
gutted fish 去內臟魚
gutted weight 去臟淨重〔量〕

gutter 水槽；排水溝；下水溝
gutting 去內臟
guttler 大吃大喝者；貪吃者
guttur 喉；咽喉
guzzler 狂飲者；酒鬼
gyle 單發酵啤酒；發酵用麥芽汁 [= gail]
gynolactose 乳寡糖；人乳糖
gynura's deux couleurs 紅鳳菜；紅蓊菜；觀音莧
gypsum 硫酸鈣；石膏
gypsy 皺褶羅鱗傘 (蕈類)
gyratory breaker 旋轉軋碎機；迴轉破碎機
gyratory crusher 迴轉壓碎機
gyratory sifter 旋轉篩

H

haberdine 腌鱈魚
habitat 生活環境；棲息地
hacker 搗碎機；剁碎機
hacking machine 搗碎機；剁碎機
haddie 黑線鱈
haddock 黑線鱈
haem 血紅素 [=heme]
haemagglutinin 紅血球凝集素；血球凝集素 [= hemagglutinin]
haematein 蘇木紅 [= hematein]
haematin 氧化原血紅素；羥高鐵血紅素；鐵(III) 血紅素 [= hematin]
haematoxylin 蘇木精 [= hematoxylin]
haemocyte 血細胞；血球 [= hemocyte]
haemocytolysis 溶血〔作用〕 [= hemocytolysis]
haemoglobin 血紅素；血紅蛋白 [= hemoglobin]
haemolysis 溶血〔作用〕 [= hemolysis]
hagfish 盲鰻
haggerel 幼公豬(1—2歲)；幼母羊(1歲)
haggis 羊雜餡肚；羊雜碎布丁；肚包羊雜
hair 毛；頭髮
hair mould 麵包霉
hair remover 去毛器；去毛劑
hairtail 白帶魚；帶魚
hairy crab 毛蟹
hairy squash 節瓜
hake 無鬚鱈
halavah 哈爾瓦芝麻糖；芝麻酥糖；土耳其碎芝麻蜂蜜糖 [= halva(h)]
half-baked 半烤熟的；半生的
half carcass 二分屠體；半片屠體
half-cooked egg 半熟蛋
half cream milk 半脂〔牛〕乳
half-done 未熟透的；半生的
half-finished goods 半成品；半製品
half-finished product 半成品；半製品
half free morel 半離羊肚
half-frozen 半冷凍的；半凍結的
half life 半衰期
half mouthed sardine 日本鯷
half-salted 半醃的
half slice 半〔圓〕片
half sponge method 二次發酵法；分醪法
half-thawed 半解凍的
halfbeak 鱵
halibut 大比目魚；大扁魚；庸鰈
halide 鹵化物
haliotis 鮑魚
halisteresis 骨鈣流失或缺乏；軟骨症
halite 岩鹽
hallah 白麵包(猶太教的安息日用) [= challah]
halobacteria 嗜鹽細菌
halobiont 適鹽生物；嗜鹽生物
halobios 海洋生物
halogen 鹵素

halogen disinfectant 鹵素消毒劑
halogenide 鹵化物
halometer 鹽量計
halophile 嗜鹽〔微〕生物
halophilic bacteria 嗜鹽細菌；親鹽細菌
halophilous bacteria 嗜鹽細菌
halophobic 厭鹽的
halophyte 嗜鹽植物；鹽土植物
haloplankton 鹽水浮游生物；海洋浮游生物
halotolerant bacteria 耐鹽細菌
halotolerant mo(u)ld 耐鹽黴菌
halva(h) 哈爾瓦芝麻糖；芝麻酥糖；土耳其碎芝麻蜂蜜糖 [= halavah]
halve 平分；分成兩半；對開片
halving 切半
ham 火腿；後腿肉
ham butt 豬〔後〕臀肉
ham press 火腿〔模〕壓機
ham pumping 火腿泵醃〔法〕；火腿注射鹽水醃製〔法〕
ham retainer 火腿成形模
ham sandwich 火腿三明治；火腿三文治
ham skinning 豬後腿剝皮
ham slicer 火腿切片機
hamburg brandy 仿白蘭地
hamburg parsley 漢堡西洋芹
hamburg(er) 漢堡包；漢堡煎牛肉餅；絞碎牛肉
hamburger bun 漢堡麵包
hamburger steak 漢堡牛排
hammer grinder 錘式粉碎機；錘磨機
hammer mill 錘式粉碎機；錘碎機
hammy 火腿香味的；油脂的銅銹味的
hammy flavo(u)r 火腿風味
hamper 有蓋食物籃
Hampshire 漢普夏〔爾〕豬；罕布夏豬
hamster 田鼠；倉鼠
hand 手；一串〔香蕉〕
hand dough 手工揉製的麵糰
hand dressing 手工開膛去內臟
hand-kill 手工屠宰
hand milking 手工擠乳
hand pack filler 手工裝罐機
hand peeling 人工去殼；人工剝皮
hand picking 人工採摘；人工挑選
hand-rolled tea 人工揉捻的茶葉
hand screw press 手動〔螺旋式〕壓榨機
hand seamer 手動封罐機
hand whisk 手工打蛋器；手工攪拌器
handbarrow 手推車
hander-up 麵糰搓圓機
handling room 處理室
handmade 手工製作的
Hang-zhou chrysanthemum flower extract 杭白菊浸膏
hanging drop culture 懸滴培養〔物〕
hank 腸衣（羊）；一束；一卷
Hansenula yeast 漢遜〔氏〕酵母；漢森酵母
HAP (hydrolyzed animal protein) 水解動物蛋白；動物蛋白水解物
hard-baked 烤過頭變乾硬的
hard bean curd 豆乾
hard boiled 煮硬的；煮老的
hard boiled egg 煮硬蛋；白焓蛋；焓蛋；煮老蛋
hard candy 硬糖〔果〕 [= boiled sweet]
hard cheese 硬〔質〕芝士；硬〔質〕乾酪
hard clam 文蛤；硬殼蛤；簾蛤 [= hard-shelled clam]
hard crab 硬殼蟹 [= hard-shelled crab]
hard dry sausage 乾硬香腸
hard drying 乾硬；充分乾燥
hard-grained 硬粒的；粗粒的
hard-grains 粗晶糖

hard liquor 烈酒
hard nut 堅果
hard pack 硬盒裝
hard red spring (wheat) 硬紅春麥；硬質紅皮春小麥
hard red winter (wheat) 硬紅冬麥；硬質紅皮冬小麥
hard resin 硬樹脂
hard roll 硬小麵包
hard salting 重鹽醃
hard-shelled clam 文蛤；硬殼蛤；簾蛤 [= hard clam]
hard-shelled crab 硬殼蟹 [= hard crab]
hard smoked fish 燻魚乾；燻乾魚
hard sugar 硬糖；粗晶粒硬糖
hard swell (can) 硬膨罐
hard water 硬水
hard wheat 硬〔質〕小麥
hard wheat flour 硬〔質〕小麥粉；硬〔質〕麵粉
hardbake 杏仁甜點；杏仁硬糖
hardened oil 硬化油；氫化油
hardener 硬化劑；硬化機
hardening 硬化；氫化；固化
hardening agent 硬化劑
hardness 硬度；硬性
hardtack 硬餅乾；硬麵包
hardware 硬體；金屬器皿
hardy vegetable 耐寒蔬菜
hare 野兔
haricot 菜豆；雲豆；四季豆；扁豆；蔬菜燉羊肉
haricot bean 扁豆；菜豆；雲豆；四季豆
harpoon 魚叉
harsh taste 澀味
harsh wine 劣質酒
harshness 粗糙性；酸澀性
harslet 豬內臟；豬雜碎；雜碎餡
hart 雄鹿
hartshorn 鹿角；鹿茸
harvest 收穫；收成
hash house 經濟餐館
hasher 廚師；切碎機；切肉機
haslet 豬內臟；豬雜碎；雜碎餡
hasting fruit 早熟水果
hastings 早熟蔬果
hasty pudding 燕麥布丁；燕麥粥；麥粉布丁
hatcher 孵化器；孵卵器
hatchery 孵化場；孵卵所
haulm 稻草；麥桿；莖桿；莖葉
haunch 腰臀；後臀；腰部
hautboy 麝香草莓
haver 燕麥
haver cake 燕麥餅
haw 山楂 [= hawthorn; maythorn]
haw apple 山楂果
haw berry 山楂果
haw pit smoke flavo(u)ring 山楂核煙燻香味劑
hawthorn 山楂 [= haw; maythorn]
hawthorn fruit tincture 山楂酊
hay 乾草（飼料）
hay butter 冬季黃油；冬季乾酪
haybox cooking 無續熱烹調；無火蒸煮
hazard analysis critical control point (HACCP) 危害分析重點控制
hazel 榛樹；榛子
hazel grouse 松雞；榛雞；花尾榛雞
hazel hen 松雞；榛雞；花尾榛母雞
hazelnut 榛子；榛果
HDL (high density lipoprotein) 高密度脂蛋白
head cabbage 甘藍；結球甘藍；卷心菜 [= common head cabbage; headed cabbage; heading cabbage]

head cheese 豬頭雜凍；豬頭肉凍
head lettuce 結球萵苣；卷心萵苣 [= cabbage lettuce]
head product 初產品
head retention 泡沫穩定性
head rice 完整白米粒；整米
headed cabbage 甘藍；結球甘藍；卷心菜 [= common head cabbage; head cabbage; heading cabbage]
heading 去頭
heading cabbage 甘藍；結球甘藍；卷心菜 [= common head cabbage; head cabbage; headed cabbage]
heading cutter 劈頭機
heading machine 去頭機
headless 無頭的
headspace 頂部空間；頂隙
health 健康
health food 健康食品；保健食品
heap fermentation 堆積發酵
heart 心臟；中心〔部〕
hearth 爐；爐床
hearth bread 底火麵包；盤烤麵包
heat 熱
heat ag(e)ing 熱老化；熱熟成；熱陳化
heat-and-eat food product 加熱即食食品；快速食品
heat-and-serve food product 加熱即食食品；快速食品
heat balance 熱〔量〕平衡
heat capacity 熱容量
heat coagulation 熱凝結
heat conduction 熱傳導
heat consumption 耗熱量
heat content 熱含量
heat convection 熱對流
heat-cool-fill method 殺菌—冷卻—充填法
heat damage 熱損壞；熱損傷
heat denaturation 熱變性
heat energy 熱能
heat exchange 熱交換
heat exchanger 熱交換器；換熱器
heat-fast 耐熱的；熱穩定的
heat insulation 熱絕緣
heat insulator 絕熱體
heat liable 不耐熱
heat-moisture treatment 濕熱處理
heat of hydration 水合熱
heat of rigor 僵直熱
heat oxidation 熱氧化
heat pack 熱裝〔罐〕
heat penetration 熱穿透
heat polymerization 熱聚合
heat preservation 加熱保存
heat processing 熱加工；熱處理；熱殺菌
heat production 熱量產生；放熱
heat proof 耐熱的；抗熱的
heat resistance 耐熱〔度〕；抗熱〔性〕
heat resistant 耐熱的；抗熱的
heat resistant bacteria 耐熱細菌
heat seal(ing) 熱封
heat sealable bag 熱封口的袋子；熱熔接合袋
heat sealer 熱封機
heat sensitive 熱敏感的
heat shock 熱休克
heat-shrinkage film 熱收縮模
heat stable 耐熱的；抗熱的
heat sterilization 加熱滅菌〔法〕
heat sterilizer 加熱滅菌器
heat tolerant 耐熱的
heat transfer 熱傳
heat treatment 熱處理
heater 加熱器；發熱器
heater car 保溫貨車
heater plate 加熱板

heather 石南花
heating 加熱
heating calandria 加熱排管
heating chamber 加熱室
heating coil 加熱盤管
heating curve 加熱曲線
heating inactivation 熱生活；熱鈍化
heating jacket 加熱套
heating loss 加熱損失
heating medium 加熱介質
heavily boiling 煮濃
heavily salted 重鹽醃的；重鹽漬的
heavy body 濃稠體
heavy bread 未烤透麵包；膨發得不鬆的麵包
heavy concentration 高濃度
heavy cream 高脂鮮奶油；高脂稀奶油
heavy dough 未發透麵糰
heavy duty detergent 強力清潔劑
heavy gravy 濃稠肉汁〔醬〕；濃汁
heavy metal 重金屬
heavy paste 稠糊；稠漿
heavy salting 重〔鹽〕醃
heavy syrup 濃糖漿
heavy wine 烈性酒
heavy wort 濃麥芽汁
hedge wine 劣質〔葡萄〕酒
heifer 小母牛；女牛
heifer calf 仔女牛；母牛犢
helianthin(e) 甲基橙 [= methyl orange]
helical bacteria 螺旋狀細菌
heliotrope 向日葵；天芥菜
helium 氦
helmet 頭罩
helminth 蠕蟲；腸蟲
helminthic 殺蟲劑；驅〔腸〕蟲劑
helping 食物一份；一份飯菜
helvolic acid 煙麴黴酸
hemagglutinin 紅血球凝集素；血球凝集素 [= haemagglutinin]
hematein 蘇木紅 [= haematein]
hematin 氧化原血紅素；羥高鐵血紅素；鐵(III)血紅素 [= haematin]
hematoporphyrin 血紫質
hematoxylin 蘇木精 [= haematoxylin]
heme 血紅素 [=haem]
heme iron 血質鐵
hemiacetal 半縮醛
hemicellulase 半纖維素酶
hemicellulose 半纖維素
hemichrome 半血色素；高鐵血色原
hemicolloid 半膠體
hemocyte 血細胞；血球 [= haemocyte]
hemocytolysis 溶血〔作用〕 [= hae-mocytolysis]
hemoferrin 血鐵素
hemoglobin 血紅素；血紅蛋白 [= haemoglobin]
hemolysis 溶血〔作用〕 [= haemolysis]
hemophilic bacteria 嗜血細菌
hemopoiesis 造血〔作用〕
hemorrhagic colitis 出血性結腸炎
hemp 大麻 [= cannabis]
hemp seed 大麻子；火麻仁 [= cannabis seed]
hempseed oil 大麻籽油
hen 母雞；雌禽；雌魚
hen of the woods 多葉奇異菌(蕈類)
hennery 養雞場
hep 野薔薇果〔實〕
hepar 肝臟
hepatitis 肝炎
hepatoflavin 肝黃素；核黃素
heptadecanoic acid 十七〔烷〕酸 [= margaric acid]
γ-heptalactone γ-庚內酯；丙位庚內酯

heptanal 庚醛
heptane 庚烷
heptanoic acid 庚酸
heptanone 庚酮
heptose 庚糖；七碳糖
heptyl acetate 乙酸庚酯
heptyl alcohol 庚醇
heptyl isobutyrate 異丁酸庚酯
heptylparaben 對羥基苯甲酸庚酯；尼泊金庚酯
herb 香草；藥草；草藥；草本植物
herb tea 青草茶；草藥茶
herbaceous 草本的
herbal 草藥的；藥草的；香草的
herbicide 殺草劑；除草劑
herbivore 草食〔性〕動物
herbivorous 草食〔性〕的；食草的
herd 牧群；畜群
Hereford (cattle) 赫利福德牛；海佛牛；海福特牛；赫里福牛
hermit crab 寄居蟹
hermitage 法國郝密紅〔葡萄〕酒
herring 鯡魚
herring roe 鯡魚卵
hesperetin 橘皮苷素；橙皮素
hesperidene 橘皮烯；橙皮烯
hesperidin 橘皮苷；橙皮苷
hesperidinase 橙皮苷酶
hesperidium 柑橘；柑果
hesperitinic acid 橘皮酸；橙皮酸
heterocyclic 雜環的
heterocyclic amino acid 雜環胺基酸
heterofermentation 異型發酵
heterofermentative bacteria 異型發酵細菌
heterofermentative lactobacilli 異型發酵乳酸桿菌
heterogenous 不均勻的
heteroglycan 異元多糖；雜聚糖
heterohexosan 異元聚己醣；雜己聚糖
heteromorphic 多晶的
heteromorphism 多晶現象
heterophagous animal 雜食性動物
heteropolysaccharide 異元多糖；雜多糖
heteroside 雜糖苷；葡萄糖苷
heterotrophic 異營的；異養的
heterotrophic bacteria 異營細菌；異養細菌；有機營養細菌
heterotrophy 異營〔作用〕
hexacosadienoic acid 二十六碳二烯酸
hexacosanoic acid 二十六烷酸
hexadecadienoic acid 十六碳二烯酸
hexadecenoic acid 十六〔碳〕烯酸
hexadienoic acid 己二烯酸；山梨酸 [= sorbic acid]
γ-hexalactone γ-己內酯；丙位己內酯
δ-hexalactone δ-己內酯
hexametaphosphate 六偏磷酸鹽
hexamethylene tetramine 六亞甲基四胺
hexanal 己醛
hexane 己烷；正己烷
hexanedione 己二酮
hexanoic acid 己酸 [= caproic acid]
hexanol 己醇
hexenoic acid 己烯酸
hexenol 己烯醇
hexenone 己烯酮
hexitol 己糖醇
hexone 異己酮
hexonic acid 己糖酸
hexosan 聚己糖；己聚糖；聚六碳糖
hexose 己糖
hexose phosphate 己糖磷酸
hexuronic acid 己糖醛酸
hexyl 2-methylbutyrate 2-甲基丁酸己酯
hexyl acetate 乙酸己酯
hexyl alcohol 己醇
hexyl butyrate 丁酸己酯

α-hexyl cinnamaldehyde α-己基〔肉〕桂醛
hexyl formate 甲酸己酯
hexyl hexanoate 己酸己酯
hexyl isobutyrate 異丁酸己酯
hexyl isovalerate 異戊酸己酯
hexyl-2-butenoate 2-丁烯酸己酯；巴豆酸己酯
HFCS (high fructose corn syrup) 高果糖玉米糖漿
HFGS (high fructose glucose syrup) 高果糖葡萄糖漿
hibiscetin 玫瑰茄色素；玫瑰茄紅；木槿黃素
hickory 山核桃〔木〕；山胡桃〔木〕(煙燻材料)
hickory nut 山核桃；山胡桃
hide 獸皮；畜皮
hide cutter 切皮機；肉皮絞碎機
hide pulling 剝皮
hide stripper 剝皮機
high acid food 高酸性食品
high amylopectin (starch) 高支鏈澱粉
high bake 高溫烘烤
high boiled 高溫熬製；急火熬煮
high boiled sweet 硬糖
high calorie food 高熱量食品
high carbohydrate diet 高碳水化合物飲食；高醣飲食
high density lipoprotein (HDL) 高密度脂蛋白
high density lipoprotein cholesterol 高密度脂蛋白膽固醇
high ester pectin 高酯化果膠
high fat diet 高脂〔肪〕飲食
high fermentation 高層發酵
high fiber diet 高纖〔維〕飲食
high frequency drying 高頻乾燥
high frequency heating 高頻加熱
high frequency sterilization 高頻滅菌
high fructose corn syrup (HFCS) 高果糖玉米糖漿；果葡糖漿；高果糖漿
high fructose glucose syrup (HFGS) 高果糖葡萄糖漿
high fructose syrup 高果糖糖漿
high grown tea 高山茶
high lipid lipoprotein 高脂〔質〕脂蛋白
high maltose syrup 高麥芽糖〔糖〕漿
high methoxyl pectin 高甲氧基果膠
high milling 高效碾磨
high moisture food 高水分食品
high performance liquid chromatography (HPLC) 高效液相層析法；高效液相色譜法
high pressure steam peeling 高壓蒸汽去皮〔法〕
high pressure steam sterilization 高壓蒸汽滅菌〔法〕
high-proof beverage 高酒度飲料
high protein bread 高蛋白〔質〕麵包
high protein diet 高蛋白〔質〕膳食；高蛋白〔質〕飲食
high ratio cake 重蛋糕；高成分蛋糕；高糖高水分的蛋糕
high ratio fat 高甘油脂肪；高單甘油酯脂肪
high ratio flour 高糖用麵粉；高糖高水分蛋糕用的麵粉
high ratio shortening 高甘油〔起〕酥油；高單甘油酯〔起〕酥油
high salinity 高鹽度
high salt diet 高鹽飲食
high seasoned 高調味的；香辛味強烈的
high-short sterilization 高溫短時〔間〕滅菌
high sodium diet 高鈉飲食
high speed centrifuge 高速離心機

high strength flour 高筋麵粉
high-sugar 高含糖量的
high tea 午茶
high temperature pasteurization 高溫巴氏殺菌；高溫巴斯德殺菌
high-temperature short-time (HTST) 高溫短時
high-temperature short-time pasteurization (HTST pasteurization) 高溫短時巴氏殺菌；高溫短時巴斯德殺菌
high-temperature short-time process (HTST process) 高溫短時殺菌法
high-temperature short-time sterilization (HTST sterilization) 高溫短時滅菌
high test molasses 高級糖蜜；高糖量轉化糖蜜
high viscosity starch 高黏度澱粉
high wine 高酒精蒸餾酒
high yield 高產率
highball 高杯雞尾酒；摻冰〔蘇打〕水威士忌飲料
highbush blueberry 沼地藍莓；繖房花越橘
highly unsaturated fatty acid 高〔度〕不飽和脂肪酸
hils herring 鰣魚
hilsa 鰣；鯡鰣
hilum 種臍；澱粉臍；澱粉粒臍點
hind 後半部
hind end 臀部
hind quarters 後軀；牛半胴體的後半部
hind shank 後腿
hip 臀部；薔薇果
hip bone 髖骨；臀骨
hippophae rhamnoides yellow 沙棘黃
hippuric acid 馬尿酸；苯甲酰甘胺酸 [= benzoylglycine]
hiragonic acid 十六碳三烯酸
hircine smell 羊膻臭；山羊膻臭
histamine 組〔織〕胺
histidine 組胺酸；咪唑丙胺酸
histidine monohydrochloride 組氨酸鹽酸鹽
histiocyte 組織細胞
histohaematin 細胞色素 [=cytochrome]
histology 組織學
histone 組〔織〕蛋白
histy bread 未烤透的麵包
hive 蜂巢；蜂箱；蜂群
hive honey 蜂巢蜜
HLB (hydrophilic-lipophilic balance) 親水親脂均衡；親水親油平衡〔值〕
HMC (hydroxymethyl cellulose) 羥甲基纖維素
hock 萊茵〔河區〕白葡萄酒；霍克白葡萄酒；德國白葡萄酒；豬前腿肉；豬肘
hodgepodge 雜燴；大雜燴
hoecake 玉米餅
hog 肉豬；肥豬
hog bristle 豬鬃；剛毛
hog bung 豬直腸
hog carcass 豬屠體
hog casing 豬腸衣
hog ears 豬耳
hog face 豬頭皮
hog fat 豬脂〔肪〕
hog grease 工業用豬油
hog middle 豬結腸
hog skin 豬皮
hoggery 養豬場
hogget 幼畜；公豬（約兩歲）；羊（一歲以上）
hogshead 桶（容量單位）
hogwash 餿水；劣等飲食；豬食
hoki 鱈魚
holder pasteurization 低溫〔長時〕巴氏殺菌；低溫〔長時〕巴斯德殺菌

holding freezer 冷藏庫
holding pasteurization 低溫〔長時〕巴氏殺菌；低溫〔長時〕巴斯德殺菌 [= low-temperature long-time pasteurization]
holding sterilization 保持滅菌〔法〕
holding tank 保存槽；保溫槽
holding time 保存期限
holey bread 多孔洞的麵包
Holland gin 荷蘭杜松子酒；荷蘭〔琴〕酒
hollandaise sauce 荷蘭調味醬；荷蘭沙司
hollander 打漿機；大麥脫殼精碾機
Hollands 荷蘭杜松子酒
hollow kernel 空心穀粒；瘦〔穀〕粒
hollow ware 凹型器皿
holocellulose 全纖維素
holoenzyme 全酶
holophytic nutrition 全植物式營養
holoside 全糖苷；低聚糖
holozoic nutrition 全動物式營養
Holstein (-Friesian) 荷蘭黑白花牛；好斯坦乳牛；弗里賽〔黑白花〕牛 [= Friesian]
Holstein dairy cattle 荷蘭乳牛
home baking 家庭烘焙〔的〕
home brew 家釀酒
home freezer 家用凍箱；家用冰箱
home making 家庭式製造
homely meal 家常便飯
homemade 自製的；家庭自製的
homemade bread 自製麵包；家庭自製麵包
homemade noodle 自製麵條；家庭自製麵條
homemade sausage 家庭自製香腸
hominy 玉米粥
hominy grits 脫殼玉米粗粉
hominy meal 玉米粉
homoarginine 高精胺酸
homocitric acid 高檸檬酸
homocysteine 高半胱胺酸；同型半胱胺酸
homocystine 高胱胺酸；同型胱胺酸
homofermentative 同型發酵的
homofermentative lactic acid bacteria 同型發酵乳酸菌
homogenate 均質物；均勻混合物
homogeneity 均勻性；均一性；同質性
homogenization 均質化；均一化
homogenized milk 均質乳
homogenizer 均質機；均和機
homoiotherm 恒溫動物；溫血動物
homolactic acid bacteria 同型乳酸〔細〕菌
homolactic fermentation 同型乳酸發酵
homologous amino acid 同系胺基酸
homopolypeptide 同〔元〕多肽
homopolysaccharide 同〔元〕多醣
homoserine 高絲胺酸
homotherm 恒溫動物；溫血動物
honey 蜂蜜；蜜糖
honey bean 荷蘭豆；蜜糖豆
honey berry 蜜莓
honey date 蜜棗
honey extractor 搖蜜機；分蜜機
honey fungus 蜜環蕈
honey mushroom 蜂蜜菇
honey pea 甜豌豆
honey syrup 蜂蜜糖漿；人造蜂蜜
honeybee 蜜蜂
honeycomb 蜜蜂巢；蜂巢胃
honeycomb tripe 蜂巢胃
honeydew 甘露；蜜露；蜜瓜
honeydew honey 甘露蜜
honeydew melon 蜜〔香〕瓜；白蘭瓜
honeymead 蜂蜜酒
hooch 劣酒；私酒；私造劣酒

hooding machine 封蓋機；冠帽機
hoof 蹄
hoosh 濃湯
hop 蛇麻花；啤酒花；忽布
hop-back 啤酒花浸取槽
hop bitter substance 啤酒花苦味物質
hop extract 啤酒花萃；啤酒花浸膏；啤酒花提取物
hop oil 蛇麻花油；啤酒花油
hop resin 啤酒花樹脂
hopped 加過啤酒花的
hopped wort 加啤酒花的麥芽汁
hoppy 啤酒花的；啤酒花香的
hops 啤酒花
hops oil 酒花油
hordein 大麥醇溶蛋白；大麥蛋白
hordeum 大麥 [= barley]
horehound 苦薄荷；苦〔薄荷〕糖果
horizontal retort 臥式〔高壓〕殺菌釜
horizontal scourer 臥式打麥機
horizontal sterilizer 臥式滅菌鍋；臥式滅菌釜
hormone 激素；荷爾蒙
horn 角
horn of plenty 喇叭菌（蕈類）
hors-doeuvre 冷盤；拼盤；開胃菜
horse 馬
horse bean 蠶豆
horse hair crab 馬尾蟹
horse mackerel 真鰺；領圓鰺
horse mushroom 田野傘菌；馬菇；野蘑菇
horsemint 麝香薄荷；美洲薄荷
horseradish 辣根；山葵
horticulture 園藝學；園藝
hospital diet 醫院膳食；醫院飲食
hot 熱的；辣的；辛辣的
hot-air dryer 熱風乾燥機
hot-air drying 熱風乾燥〔法〕
hot-air oven 熱風烘箱；熱風烤箱；熱風〔烤〕爐
hot-air sterilization 熱風滅菌〔法〕；乾熱滅菌〔法〕
hot-air sterilizer 熱風滅菌器；乾熱滅菌器
hot cake 烙餅；班戟；薄煎餅；薄烤餅 [= pancake]
hot carcass 溫屠體；剛宰的胴體
hot curing 熱醃〔法〕
hot dog 熱狗；熱狗小腸
hot filling 熱〔充〕填；熱裝
hot marinating 熱醋醃
hot pack(ing) 熱裝；熱包裝
hot pepper 辣椒；番椒
hot plate 加熱板；輕便電爐
hot pot 什錦燉肉；罐燜燒肉及蔬菜
hot pressing 熱壓榨
hot red pepper 紅辣椒
hot saccharate 熱蔗糖鹽
hot salt curing 熱鹽醃〔法〕
hot sauce 辣醬；辣沙司 [= hot spice sauce]
hot seal 熱封
hot smoking 熱燻〔法〕；熱燻製〔法〕
hot spice sauce 辣醬；辣沙司 [= hot sauce]
hot swell can 熱脹罐
hot swelling 熱脹〔現象〕；熱脹罐
hot taste 辣味
hot weight 熱重；鮮重
hotchpotch 什錦大麥濃湯；雜燴
hotcross bun 肉桂重料麵包
hothouse 高溫溫室；乾燥室
house fly 家蠅
household 家庭的；家用的
hovis 胚芽麵粉烘製品；發芽小麥粉的烘製品；胚芽麵包
HPLC (high performance liquid chro-

matography) 高效液相層析法；高效液相色譜法

HTST (high-temperature short-time) 高溫短時

hubbard 筍瓜；冬南瓜

huckleberry 酸越橘；卵形越橘；黑果

hull 外殼；外皮；莢；糠

hull aspirator 皮殼分離器；穀殼吸風分離器

hulled 去殼的；去皮的；去莢的

hulled rice 糙米；脱殼稻米 [= cargo rice; brown rice; husked rice]

huller 脱殼機；去殼機；去皮機；去莢機；礱穀機 [= husker]

hulless 無殼的；無皮的

hulless barley 無殼大麥；裸大麥

hulless oat 無殼燕麥；裸燕麥

hulling 脱殼；去殼；去皮；去莢；礱穀 [= husking]

hulling mill 脱殼機；去殼機

human milk 人乳；母乳

humanized milk 母乳化牛乳

humble pie 鹿內臟餡餅；次級鹿肉餡餅

humbug 薄荷糖

humectant 保水劑；保濕劑；濕潤劑

humectation 保濕；潤濕；增濕

humic acid 腐植酸

humic substance 腐植〔物〕質

humid 濕的

humidification 增濕〔作用〕

humidifier 增濕器；濕潤器

humidistat 恆濕器；濕度調節器；測濕計

humidity 濕度

humidity drying 濕式乾燥

humidor 增濕器；保濕盒

hummel 除〔穀物的〕芒

hummeller 除芒器

hump 駝峰；峰肉

humpback salmon 駝背鮭魚；細鱗大麻哈魚

humpback whale 大鰭鯨；駝背鯨

humulene 蛇麻葫烯；葎草烯；草烯

humulone 蛇麻葫酮；葎草酮；草香苦酮

Humulus lupulus 啤酒花

humus 腐植質

hunch 厚片；大塊；峰肉

hundreds and thousands 七彩花飾糖果粒；糕餅的花飾糖果或蜜餞

hunger 飢餓

hunger cure 禁食療法；飢餓療法

hunting sausage 野餐〔小〕香腸

husk 外殼；莢；稻殼；糠；外皮；苞葉

husked 去殼的；去皮的；去莢的；去苞葉的

husked rice 糙米；脱殼稻米 [= cargo rice; brown rice; hulled rice]

husker 脱殼機；去殼機；去皮機；去莢機；去苞葉機；礱穀機 [= huller]

husking 脱殼；去殼；去皮；去莢；礱穀；去苞葉 [= hulling]

huso 鰉

huso sturgeon 鰉魚

HVP (hydrolyzed vegetable protein) 水解植物蛋白；植物蛋白水解物

hyacinth bean 肉豆；鵲豆；眉豆

hybrid corn 雜交玉米

hybrid variety 雜交品種

hybridization 雜交；配種

hydrargyrism 汞中毒；水銀中毒 [= mercurialism]

hydrase 水合酶；水化酶

hydratability 水合性；水化性

hydrate 水合物

hydration 水合〔作用〕

hydrazine 聯胺
hydro-extractor 脱水機
hydrocarbon 烴；碳氫化合物
hydrocellulose 水解纖維素
hydrochemistry 水〔質〕化學
hydrochloric acid 鹽酸；氫氯酸
hydrocholesterol 氫膽固醇
hydrocolloid 水膠體；親水〔性〕膠體；水狀膠體
hydrocooling 水冷；用水冷卻
hydrofined oil 加氫精製油
hydrogel 水凝膠
hydrogen 氫
hydrogen bond 氫鍵
hydrogen halide 氫鹵化物
hydrogen peroxidase 過氧化氫酶
hydrogen peroxide 過氧化氫
hydrogen sulfide 硫化氫
hydrogenase 氫化酶
hydrogenated ester gum 氫化松香甘油酯
hydrogenated fat 氫化脂；硬化油
hydrogenated glucose syrup 氫化葡萄糖醬
hydrogenated oil 氫化油；硬化油
hydrogenated shortening 氫化〔起〕酥油
hydrogenated starch hydrolysate (HSH) 氫化澱粉水解物
hydrogenation 氫化〔作用〕；加氫〔作用〕
hydrogenator 氫化器；氫化罐
hydroglucan 水解葡聚糖
hydrolase 水解酶
hydrolock retort 水封式〔連續〕殺菌釜
hydrolysate 水解〔產〕物 [= hydrolyzate]
hydrolysis 水解
hydrolyte 水解物；水解質
hydrolytic agent 水解劑
hydrolytic decomposition 水解分解
hydrolytic degradation 水解降解
hydrolytic enzyme 水解酵素；水解酶
hydrolytic rancidity 水解〔性〕酸敗
hydrolyzate 水解〔產〕物 [= hydrolysate]
hydrolyzed animal protein (HAP) 水解動物蛋白；動物蛋白水解物
hydrolyzed corn protein 水解玉米蛋白；水解粟米蛋白；玉米蛋白水解物 [= hydrolyzed vegetable protein; HVP]
hydrolyzed mash 水解麥芽汁
hydrolyzed plant protein (HPP) 水解植物蛋白；植物蛋白水解物 [= hydrolyzed vegetable protein; HVP]
hydrolyzed protein 水解蛋白〔質〕；蛋白〔質〕水解物
hydrolyzed vegetable protein (HVP) 水解植物蛋白；植物蛋白水解物 [= hydrolyzed plant protein]
hydrolyzer 水解器；水解罐
hydromel 蜂蜜酒
hydrometer 比重計；液體比重計
hydroperoxide 氫過氧化物；過氧化氫物
hydrophile 親水物；親水劑
hydrophile-lipophile balance 親水親脂均衡；親水親油平衡
hydrophilic 親水的
hydrophilic colloid 親水膠體
hydrophilic-lipophilic balance (HLB) 親水親脂均衡；親水親油平衡〔值〕
hydrophilic sol 親水溶膠
hydrophilic substance 親水物質
hydrophilicity 親水性
hydrophobe 疏水物
hydrophobic 疏水的
hydrophobic colloid 疏水膠體
hydrophobic sol 疏水溶膠
hydrophobic substance 疏水物質
hydrophobicity 疏水性
hydrophyte 水生植物

hydroponics 水耕法；水栽法
hydrosol 水溶膠
hydrosoluble 水溶性；水溶的
hydrosorbic acid 氫己二烯酸；己烯酸
hydrostatic cooker 水壓蒸煮器；靜水壓蒸煮器
hydrostatic retort 水壓殺菌釜；靜水壓殺菌釜
hydrostatic sterilization 水壓滅菌〔法〕；靜水壓滅菌〔法〕
hydrostatic sterilizer 水壓滅菌器；靜水壓滅菌器
hydrotropic agent 增溶劑
hydrous 含水的；水合的
hydroxide 氫氧化物
hydroxocobalamine hydrochloride 維生素B_{12}；羥鈷胺
hydroxy amino acid 羥〔基〕胺基酸
hydroxy ethyl starch 羥乙〔基〕澱粉
hydroxyacetic acid 乙醇酸；羥〔基〕醋酸；羥〔基〕乙酸 [= glycolic acid]
hydroxycholesterol 羥膽固醇 [= oxy-cholesterol]
hydroxycitronellal 羥基香茅醛
hydroxycitronellal dimethyl acetal 羥基香茅醛二甲縮醛
hydroxycitronellol 羥基香茅醇
hydroxycobalamin 羥鈷維生素
hydroxyglutamic acid 羥谷胺酸
hydroxyl group 羥基；氫氧基
hydroxylamine 羥胺
hydroxylase 羥〔化〕酶
hydroxylated lecithin 羥基化卵磷脂
hydroxylysine 羥〔基〕離胺酸；羥〔基〕賴胺酸
hydroxymethyl cellulose (HMC) 羥甲基纖維素
hydroxyproline 羥〔基〕脯胺酸 [= oxyproline]
hydroxypropyl cellulose (HPC) 羥丙基纖維素
hydroxypropyl distarch glycerol 羥丙基二澱粉甘油酯；羥丙基甘油雙澱粉
hydroxypropyl distarch phosphate 羥丙基二澱粉磷酸酯；羥丙基磷酸雙澱粉
hydroxypropyl methylcellulose 羥丙基甲基纖維素
hydroxypropyl starch 羥丙〔基〕澱粉
hydroxyvitamin D_3 羥維生素D_3
hygiene 衛生；衛生學
hygienic 衛生學的；衛生的
hygienics 衛生學
hygienist 衛生學家
hygrograph 濕度記錄儀
hygrometer 濕度計
hygromycin 濕黴素；潮霉素
hygroscope 濕度儀；濕度計
hygroscopic 吸濕的；吸潮的
hygroscopic agent 吸濕劑
hygroscopic moisture 吸濕水 [= hygroscopic water]
hygroscopic salt 吸濕鹽
hygroscopic water 吸濕水 [= hygroscopic moisture]
hygroscopicity 吸濕度；吸濕性
hygrostat 恆濕器；濕度調節器；測濕計
hyperalimentation 靜脈輸入高營養液
hyperbaric retorting 加氣高溫殺菌〔法〕
hyperchlorhydria 胃酸過多症
hyperchloric acid 過氯酸
hypercholesterolemia 高膽固醇血〔症〕；血膽固醇過多〔症〕
hyperfiltration 超微濾〔法〕；超濾
hyperglycemia 血糖過多〔症〕；高血糖〔症〕

hyperglyceridemia 高甘油酯血〔症〕；血甘油酯過多〔症〕
hyperlip(a)emia 高脂血〔症〕；血脂過多〔症〕
hypermaturity 過熟
hypernutrition 營養過度
hyperorexia 食慾過盛
hyperoxide 過氧化物
hyperphagia 嗜食症；攝食過量
hypertension 血壓高；高血壓
hypertonic 高滲〔壓〕的
hypertonic solution 高滲〔壓〕溶液
hypertriglyceridemia 高三酸甘油酯血〔症〕；血三酸甘油酯過多〔症〕
hypertrophy 營養過度
hyperuric(a)emia 高尿酸血〔症〕；血尿酸過多〔症〕
hypervitaminosis 維生素過多症
hypha 菌絲
hyphae 菌絲 〔= hypha的複數〕
hyphal 菌絲的
hyphal body 菌絲體
hyphomycete 絲黴菌；絲狀菌
hypobaric storage 低壓儲藏
hypochlorite 次氯酸鹽
hypochlorization 低鹽攝食法
hypochlorous acid 次氯酸
hypocholesterolemia 低膽固醇血〔症〕；血膽固醇過少〔症〕
hypoferric anemia 缺鐵性貧血
hypoglycemia 低糖血〔症〕；血糖過低〔症〕
hypoosmotic 低滲壓的
hyposite 低熱食品
hypotaurine 亞牛磺酸；氨乙基亞磺酸
hypotension 血壓〔過〕低；低血壓
hypotonic 低滲〔壓〕的
hypotonic solution 低滲〔壓〕溶液
hypotrophy 營養不足
hypovitaminosis 維生素缺乏症
hypoxanthine 次黃嘌呤；次黃質
hyson 熙春茶
hyssop 牛膝草；柳薄荷

I

I band I帶
ice 冰；冰凍
ice bar 冰棒
ice bar freezer 冰棒冷凍機 [= bar machine]
ice bath 冰浴
ice box 冰箱；冰桶
ice cellar 冰窖
ice chopper 碎冰機
ice coffee 冰咖啡
ice cooling 冰冷卻
ice cream 冰淇淋
ice cream bar 雪糕
ice cream brick 冰淇淋磚
ice cream cone 甜筒；圓錐形冰淇淋杯；蛋卷冰淇淋
ice cream custard 冰淇淋蛋糕
ice cream freezer 冰淇淋冷凍機；冰淇淋凍結機
ice cream mix 冰淇淋配料
ice cream overrun 冰淇淋膨脹率
ice cream sandwich 冰淇淋三明治；夾心冰淇淋
ice cream soda 冰淇淋蘇打水
ice crusher 碎冰機
ice crystal 冰〔結〕晶
ice drink 加冰飲料；刨冰飲料
ice flaker 刨冰〔片〕機
ice generator 製冰機
ice glaze 冰衣
ice glazing 包冰衣
ice lolly 糖衣雪糕；帶柄雪糕
ice machine 製冰機
ice maker 製冰機；製冰器
ice making machine 製冰機
ice milk 乳冰；冰牛奶
ice pick 碎冰錐；破冰鐵鑿
ice plant 製冰廠
ice point 冰點
ice sherbet 冰雪泥；冰凍果子露
ice slab 大冰塊
ice storage 冰藏
ice sucker 冰棒
ice tea 冰茶；冰紅茶
ice warehouse 冰庫
ice water 冰水
ichthulin 魚卵磷蛋白
ichthyism 魚中毒 [= ichthyotoxism]
ichthyocol(l) 魚膠；鰾膠；鱘鰾膠
ichthyocolla 魚膠；鰾膠；鱘鰾膠
ichthyology 魚類學
ichthyootoxin 魚卵毒素
ichthyosarcotoxin 魚肉毒素
ichthyosarcotoxism 肉毒魚類中毒；魚肉中毒
ichthyotoxin 魚毒〔素〕
ichthyotoxism 魚中毒 [= ichthyism]
iciness 冰屑感；冰屑結構
icing 包冰；掛糖衣；糖衣；加冰

icing machine 粉糖機；掛糖衣機；加冰機
icing storage 包冰儲藏
icing sugar 糖霜；粉糖；糖粉
iditol 艾杜糖醇
IDL (intermediate density lipoprotein) 中間密度脂蛋白
idla 蒸米餅
idonic acid 艾杜糖酸
idose 艾杜糖
iduronic acid 艾杜糖醛酸
IFJU (International Fruit Juice Union) 國際果汁聯盟；國際果汁聯合會
IFMA (International Food Manufacture Association) 國際食品製造協會
IFT (Institute of Food Technologists) 食品科技學會；美國食品科技學會；食品工藝學家學會
ignition point 燃點
IIR (International Institute of Refrigeration) 國際冷凍學會；國際製冷技術學會
ikan 魚
ikpan oil 西瓜子油
ilama 衣拉麻果；番荔枝
ilangilang 加拿楷樹；衣蘭香精；香水樹；衣蘭樹 [= ylangylang]
ileitis 迴腸炎
ileum 迴腸
imazalil 抑霉唑
imbibition 吸水；吸液
IMF (intermediate moisture food) 中濕〔性〕食品；半濕〔性〕食品；中等水分食品；半乾食品 [= semimoist food]
imidazole 咪唑
imine 亞胺
imino acid 亞胺基酸
imitated food 仿造食品；仿製食品
imitation 仿造；仿製品
imitation champagne 仿製香檳〔酒〕；仿香檳〔酒〕
imitation food 仿造食品；仿製食品
imitation meat 人造肉；仿真肉
imitation milk 人造乳；仿真牛奶
imitation wine 仿製〔葡萄〕酒
immature 未成熟的
immature dough 未醒麵糰；未發酵透的麵糰
immature spirit 未醇烈酒；未成熟的蒸餾酒
immature wine 未陳〔葡萄〕酒；未成熟的〔葡萄〕酒
immediate oxygen demand (IOD) 直接需氧量；即時耗氧量
immerser 浸漬器
immersion 浸入；浸漬
immersion cooling 浸沒冷卻〔法〕；沉浸冷卻〔法〕
immersion defrosting 浸沒解凍〔法〕；沉浸解凍〔法〕
immersion freezing 浸沒冷凍〔法〕；沉浸凍結〔法〕
immiscibility 不混溶性；不混合性
immiscible 不混溶的；不混合的
immobile phase 固定相
immobilized enzyme 固定化酶；固相酶
immobilized tannin 固化單寧
immune globulin 免疫球蛋白 [= immunoglobulin]
immunity 免疫性
immunization 免疫〔作用〕
immunoglobulin 免疫球蛋白 [= immune globulin]
immunology 免疫學
immunopolysaccharide 免疫多糖〔類〕
immunoprotein 免疫蛋白質

impact breaker 撞擊破碎機；衝擊式破碎機

impact crusher 撞擊粉碎機；衝擊式粉碎機

impact extruded can 衝擠型罐；撞擠罐

impact grinder 撞擊粉碎機；衝擊式粉碎機

impact mill 衝擊式研磨機

impact screen 衝擊篩；振動篩

impalpable powder 微粉末

impeller 葉輪

imperfect egg 不完全蛋

imperfect grain 癟〔穀〕粒；不飽滿穀粒

impermeability 不滲透性；不透性

impermeable 不滲透性的；不能滲透的

imporosity 無孔性

impregnated wrapping 防潮包裝

impregnation 浸滲；浸透

improver 改良劑；改進劑

impulse seal(ing) 瞬間封口；熱脈衝封口

impure 不純的

impurity 雜質

imputrescible 不腐敗的；不腐爛的

in-bottle pasteurization 瓶裝巴氏殺菌；裝瓶後巴氏殺菌

in-bottle sterilization 瓶裝滅菌；裝瓶後滅菌

in-line 生產線中的；作業線內的

in-package desiccant 包裝內乾燥劑

in-package desiccation 包裝內乾燥

in-place cleaning 就地清洗；原位清洗

in-plant 廠內的

in-plant chlorination 廠內用水加氯消毒；廠內水源加氯

in-sack 袋裝的

in vitro 試管內；體外；離體

in vivo 體內；活體中

inactivation 鈍化〔作用〕；去活〔作用〕；滅活〔作用〕

inactivator 鈍化劑；去活劑；滅活劑

inadequate diet 不當膳食；營養不足的膳食

inalimental 無營養的

inanition 餓虛；營養不足

inappetence 食慾不振

incense 燻香；燻香料

inchurn milking 機器擠奶(與奶桶相連)

incineration 燒灼滅菌〔法〕；灰化；焚化

incipient fermentation 初期發酵

incipient spoilage 初期腐敗

incitant 興奮劑

inclined tube evaporator 斜管〔式〕蒸發器

incoagulable 不凝固的；無法凝固的

incomplete protein 不完全蛋白質

inconnu 大北鮭(淡水魚)

incrustation 結殼；硬殼；鍋垢；水垢

incubation 保溫；孵化；培養；潛伏

incubation period 保溫〔培養〕期；潛伏期；孵化期

incubator 保溫箱；恆溫箱；孵卵器；培養箱

incubator house 保溫室；孵卵室

indehiscent fruit 閉果

index of nutritional quality (INQ) 營養品質指數；營養質量指標

Indian corn 玉米；玉蜀黍

Indian dwarf wheat 印度矮生小麥

Indian gooseberry 醋栗；印度醋栗

Indian gum 印度樹膠

Indian hemp 印度大麻

Indian jujube 印度棗

Indian laurel oil 印度月桂油

Indian maize 玉米；玉蜀黍

Indian meal 玉米粉；玉蜀黍粉

Indian millet 黑黍
Indian mustard 芥菜
Indian rice 茭白
Indian sago palm 孔雀椰子
Indian strawberry 蛇莓
indica type rice 秈稻；在萊米；印度秈
indican 靛苷
indicator 指示劑；指示器
indicator bacteria 指標細菌
indicator medium 指標培養基
indicator organism 指標微生物
indicator paper 試紙
indigenous flora 原宿群落；自然菌群
indigestible 不易消化的
indigestion 消化不良
indigo 靛；靛藍
indigo blue 靛藍
indigo carmine 食用藍色二號；酸性靛藍
indigotine 靛藍；靛青
indigotine alum lake 靛藍鋁色澱
indirect calorimetry 間接熱量測定法
indirect drying 間接乾燥
indirect fired dryer 間接火乾燥機
indirect food additive 間接食品添加物
indirect freezing 間接冷凍
indirect heating 間接加熱
indispensable amino acid 必需胺基酸 [= essential amino acid; EAA]
indispensable enzyme 必需酶
individual packaging 個別包裝；單件包裝
individual quick freezing 個別快速冷凍；個體速凍
individual storage 個別儲藏；單件儲藏
indole 吲哚
indophenol 靛酚
indophenol oxidase 靛酚氧化酶
indoxyl 吲哚酚
induced enzyme 誘導酶
inducible enzyme 可誘導酶
inducing agent 誘導劑；誘發劑
induction heating 感應加熱
induction heating sterilization 感應〔加〕熱滅菌
industrial alcohol 工業酒精
industrial crop 工業作物
industrial effluent 工業污水；工業廢水
industrial fermentation 工業〔規模〕發酵
industrial goods 工業產品
industrial microbiology 工業微生物學
industrial pollution 工業污染
industrial scale 工業規模
industrial waste 工廠廢料
industrial waste water 工廠廢水
industrial water 工廠用水
inedible 不適於食用的；不可食用的；非食用的
inedible oil 不可食用油；非食用油
inedible tallow 不可食用動物脂；非食用動物脂
inelasticity 無彈性；無伸縮性
inert food 無效食品
inert-gas packing 充惰性氣體包裝
infancy 嬰兒期
infant 嬰兒
infant food 嬰兒食品
infant formula 嬰兒乳；嬰兒配方食品
infect 感染；傳染
infected 受感染的；被感染的
infection 感染；傳染
infection type food poisoning 感染型食物中毒
infectious disease 感染性疾病；傳染病
infectious food poisoning 感染型食物中毒
inferior 低級的；下等的；劣質的
infertile egg 無精蛋

infiltration 滲入；透濾；滲透
inflammability 可燃性；易燃性
inflammation 燃燒；發炎
inflating agent 發泡劑；膨脹劑
inflation 充氣；膨脹
inflight meal 航空餐；飛機餐
informal dinner 便餐
infrared (ray) 紅外線
infrared dehydration 紅外線〔加熱〕乾燥
infrared desiccation 紅外線〔加熱〕乾燥
infrared dryer 紅外線〔加熱〕乾燥器
infrared drying 紅外線〔加熱〕乾燥
infrared heating 紅外線加熱
infrared moisture analyzer 紅外線水分分析儀
infrared moisture meter 紅外線水分計；紅外線水分測定儀
infrared peeling 紅外線去皮〔法〕
infrared radiation thawing 紅外〔線〕輻射解凍
infrared spectrophotometer 紅外線分光光度計
infuse 泡茶；浸出；泡製
infuser 浸出器
infusion 浸出液；浸劑；浸出
infusion broth 浸出肉汁；肉浸汁
infusion process 浸出法
infusion wort 浸出麥芽汁；浸製麥芽汁
infusorial earth 矽藻土；硅藻土
ingesta 飲食物；食物
ingestion 攝取；口服
ingredient 配料；成分
inhibited starch 穩化澱粉；交鏈澱粉；交聯澱粉；穩定澱粉
inhibition 抑制
inhibitor 抑制劑
inhibitory agent 抑制劑
inhibitory substance 抑制劑；抑制物
inhomogeneity 不均一〔性〕；多相〔性〕
initial handling 初處理；初加工
initiator 引發劑
injection 注射；注入
injection mo(u)lding 射出成型；注射成型
injection mo(u)lding machine 射出成型機；注射成型機
injector 注射器
ink 墨汁；墨水；烏賊墨汁；油墨
ink bag 墨囊（墨魚）
ink gland 墨腺（墨魚）
ink sac 墨囊（墨魚）
inkfish 墨魚；烏賊
inlet 入口
innards 內臟
inner membrane 內膜
inner packaging 內裝〔法〕；內包裝
inner thin albumen 內稀蛋白
inner veil 內菌幕
inner wall 內壁
innocuous 無害的；無毒的
innocuous substance 無害物質；無毒物質
innutrition 營養不良
inocula 接種體；菌種 〔= inoculum的複數〕
inoculate 接種；接菌
inoculating crystal 晶種；籽晶
inoculating hood 接種箱；接種廚
inoculating loop 接種環
inoculating needle 接種針
inoculation 接種
inoculator 接菌器；接種器
inoculum 接種體；菌種 〔= inocula的單數〕
inodorous 無氣味的；無臭的
inorganic compound 無機化合物
inose 肌醇；環己六醇

inosine 肉苷；肌苷；次黃〔嘌呤核〕苷
inosine monophosphate (IMP) 肉苷酸；肌苷酸；肌苷一磷酸；肉苷一磷酸；次黃〔嘌呤核〕苷酸 [= inosinic acid]
inosine-5'-monophosphoric acid 5'-肌苷酸
inosinic acid 肉苷酸；肌苷酸；肌苷一磷酸；肉苷一磷酸；次黃〔嘌呤核〕苷酸 [= inosine monophosphate]
5'-inosinic acid 5'-肌苷酸
inosite 肌醇；環己六醇 [= inositol]
inositol 肌醇；環己六醇 [= inosite]
insect 昆蟲
insect damage 蟲害
insect killer 殺蟲劑
insect powder 殺蟲粉
insect-proof packaging 防蟲包裝〔法〕
insect repellant 驅蟲劑
insecticide 殺蟲劑
insectifuge 驅蟲劑
insectofungicide 殺蟲滅〔真〕菌劑
inshore fish 近海魚〔類〕；沿海魚〔類〕
inshore fishery 近海漁業；沿海漁業
inside (round) 內側後腿肉
inside washing 體腔洗滌；胴體腹腔沖洗
insides 內臟
insipid 味淡的；乏味的；無味的
insipid food 味淡食物；乏味食物；無味食物
insobriety 酗酒
insolation 日曬；曬乾；日射病
insolubility 不溶性
insoluble 不溶的；不能溶解的
insoluble enzyme 不溶性酶；固相酶
insoluble solid 不溶性固形物
inspection 檢查；檢驗
inspector 檢查員；檢驗員
inspissate 濃縮；蒸濃
inspissated juice 濃縮果汁
inspissation 濃縮〔作用〕；蒸濃〔作用〕
inspissator 濃縮器；蒸濃器
instant coffee 即溶咖啡；速溶咖啡
instant dessert 速食甜點；即食甜點
instant food 速食食品；即食食品
instant milk powder 即溶奶粉；速溶奶粉
instant noodle 即食麵；速食麵；方便麵；泡麵
instant rice 速食飯
instant soup 速食湯；速溶湯
instant tea 即溶茶
instantizer 即溶製機；速溶化設備
instantizing 即溶化；速溶化；速食化
Institute of Food Technologists (IFT) 食品科技學會；美國食品科技學會；食品工藝學家學會
institutional food 團體食品
institutional pack 業務用包裝；餐館用包裝；公共飲食業用包裝
instrument 儀器
insulant 絕緣物
insulated cabinet 絕熱箱
insulated container 保溫容器
insulating foam 絕熱泡沫材料
insulating material 絕緣材料
insulating sleeve 隔熱杯套；絕熱杯套
insulation 絕緣
insulator 絕緣體
insulin 胰島素
intake 攝取；攝入
integument 外皮；覆蓋物
intensifier 增強劑；風味增強劑
interaction 交互作用
intercellular fluid 細胞間液
intercellular substance 細胞間質
intercooling 中間冷卻

intercostal 肋間的
intercostal meat 牛肋條；腩條
interesterification 交酯化〔作用〕；酯交換〔作用〕
interesterified lard 交酯化〔油〕脂；酯交換〔油〕脂
interfacial activity 界面活性
interfacial agent 界面活性劑
interfacial tension 界面張力
interfusion 混合；滲入
interheater 中間加熱器
interior packaging 內包裝
intermediate compound 中間化合物
intermediate culture 中間培養物
intermediate density lipoprotein (IDL) 中間密度脂蛋白
intermediate moisture food (IMF) 中濕〔性〕食品；半濕〔性〕食品；中等水分食品；半乾食品 [= semimoist food]
intermediate packaging 中層包裝
intermediate product 中間產物；中間產品
intermediate proof 中間醒麵；中間醒發
intermediate proofer 中間醒麵機；中間醒發機；中間檢驗台
intermediate sugar 粗製砂糖
intermingle 混合
intermittent dryer 間歇式乾燥機
intermittent drying 間歇式乾燥〔法〕
intermittent sterilization 分段滅菌；間歇滅菌 [= fractional sterilization; tyndallization]
internal corrosion 罐內腐蝕
internal organ 內臟
internal phase 內相
International Dairy Federation (IDF) 國際乳業聯盟；國際乳品聯合會
International Dietetic Association (IDA) 國際營養師協會
International Food Information Service 國際食品資訊服務中心
International Food Manufacture Association (IFMA) 國際食品製造協會
International Food standard 國際食品標準
International Fruit Juice Union (IFJU) 國際果汁聯盟；國際果汁聯合會
International Institute of Refrigeration (IIR) 國際冷凍學會；國際製冷技術學會
International Office of Consumers Union (IOCU) 國際消費者同盟
International Organization for Standardization (ISO) 國際標準〔化〕組織
International Pharmacopoeia (IP) 國際藥典
International Reference Standard 國際對照標準；國際仲裁標準
International Rice Research Institute (IRRI) 國際稻米研究所
International Society for Fat Research (ISFR) 國際油脂研究學會
International Standard 國際標準
international tolerance 國際容許〔殘留〕量
International Union of Biochemistry (IUB) 國際生物化學聯合會
International Union of Chemistry (IUC) 國際化學聯合會
International Union of Food Science and Technology (IUFoST) 國際食品科技聯盟；國際食品科學技術協會
international unit (IU) 國際單位
intestinal 腸內的；腸的
intestinal bacteria 腸〔內〕細菌
intestinal discharge 糞便

intestinal fat 網油；腸油；腸脂
intestinal flora 腸內菌叢
intestinal juice 腸液
intestinal wall 腸壁
intestine 腸
intolerance 不耐性
intoxicant 致醉物；毒物
intoxicating beverage 酒精飲料；致醉酒類
intoxication 中毒；酒醉
intracellular 胞內的
intracellular enzyme 胞內酶 [= endo-cellular enzyme]
intracellular fluid 胞內液
intracytoplasm 胞內漿
intramuscular fat 肌內脂肪；肌肉間脂肪
intravenous alimentation 靜脈營養補給
intrinsic energy 內能
intrinsic factor 內在因素；內因子
introfaction 加速浸泡
introfier 加速浸泡劑
inulase 菊糖酶；菊粉酶
inulic acid 阿蘭酸；土木香酸 [= alantic acid]
inulin 菊糖；菊粉 [= alantin]
inulinase 菊糖酶；菊粉酶
inunction 塗油
invalid diet 病人膳食
invalid food 病人食品
invert molasses 高糖量轉化糖蜜
invert sugar 轉化糖 [= inverted sugar; invertose]
invert syrup 轉化糖漿
invertase 蔗糖酶；轉化酶 [= sucrase; saccharase]
invertebrate 無脊椎動物
inverted sugar 轉化糖 [= invert sugar; invertose]
invertin 轉化酶
invertose 轉化糖 [= invert(ed) sugar]
invigoration 滋補；使強壯；使復壯
invigorator 補藥；強壯劑
involuntary muscle 不隨意肌
IOD (immediate oxygen demand) 直接需氧量；即時耗氧量
iodate 碘酸鹽
iodide 碘化物
iodimetry 碘〔量〕滴定法
iodinated casein 碘化酪蛋白
iodine 碘
iodine number 碘值；碘價 [= iodine value (IV)]
iodine test 碘試驗
iodine value (IV) 碘值；碘價 [= iodine number]
iodized salt 加碘食鹽；碘化食鹽
iodizing compound 加碘化合物；增碘劑；加碘劑
iodoform scented russule 特希紅菇
iodometric method 碘量滴定法
iodometry 碘〔量〕滴定法
iodophor 帶碘化合物；含碘化合物
iodophor sterilant 含碘滅菌劑
iodoprotein 碘蛋白
ion 離子
ion exchange 離子交換〔作用〕
ion exchange chromatography 離子交換色譜法
ion exchange column 離子交換〔管〕柱
ion exchange filtration 離子交換過濾
ion exchange group 離子交換基
ion exchange resin 離子交換樹脂
ion exchange water 離子交換水
ion exchanger 離子交換劑；離子交換器
ion filtration chromatography 離子過濾層析法

Iona 阿沃納葡萄；愛歐那葡萄
ionic bond 離子鍵
ionic detergent 離子性清潔劑
ionic strength 離子強度
ionization 電離〔作用〕；離子化
ionone 紫羅蘭酮；紫羅酮
Irish cobbler 愛爾蘭薯
Irish coffee 愛爾蘭咖啡
Irish mist 愛爾蘭〔混合〕甜酒
Irish moss 鹿角菜；角叉菜；鹿角菜苷 [= carrag(h)een]
Irish whisky 愛爾蘭威士忌〔酒〕
iron 鐵
iron balance 鐵均衡
iron caseinate 酪蛋白鐵
iron chloride 氯化鐵
iron citrate 檸檬酸鐵
iron deficiency anemia 缺鐵性貧血
iron enzyme 含鐵酶
iron oxide 氧化鐵
iron oxide black 氧化鐵黑
iron oxide red 氧化鐵紅
iron oxide yellow 氧化鐵黃
iron ration 救急口糧
iron sulfide blackening 硫化鐵黑變
iron sulfide discoloration 硫化鐵變色
iron taste 鐵〔腥〕味
iron turbidity 鐵混濁
iron, carbonyl 羰基鐵
iron, electrolytic 電解鐵
iron, reduced 還原鐵
irone 鳶尾酮
ironed can 沖劑延伸罐；壓薄拉深罐
irradiated food 照射食品；輻照食品
irradiated meat 照射肉
irradiation 照射；輻射
irradiation pasteurization 照射巴氏殺菌；照射巴斯德殺菌
irradiation sterilization 照射滅菌；輻照滅菌
irregular (shaped) can 異型罐
irreversible colloid 不可逆膠體
irreversible gel 不可逆凝膠
irreversible reaction 不可逆反應
irritant 刺激的；刺激物
isatin 靛紅
ischiatic 腰臀的
isinglass 魚膠 [= fish gelatin; fish glue]
islanditoxin 島黴毒素；島青霉毒素
ISO (International Organization for Standardization) 國際標準〔化〕組織
iso-osmotic 等滲〔壓〕的 [= isosmotic]
iso-osmotic pressure 等滲壓
iso-osmotic solution 等滲〔壓〕溶液
isoamyl acetate 乙酸異戊酯
isoamyl alcohol 異戊醇
isoamyl benzoate 苯甲酸異戊酯
isoamyl butyrate 丁酸異戊酯
isoamyl formate 甲酸異戊酯
isoamyl gallate 異沒食子酸戊酯；沒食子酸異戊酯；五倍子酸異戊酯
isoamyl hexanoate 己酸異戊酯
isoamyl isobutyrate 異丁酸異戊酯
isoamyl isovalerate 異戊酸異戊酯
isoamyl phenylacetate 苯乙酸異戊酯
isoamyl propionate 丙酸異戊酯
isoamyl salicylate 水楊酸異戊酯；柳酸異戊酯
isoamylase 異澱粉酶；澱粉剪支酶
isoascorbic acid 異抗壞血酸 [= erythorbic acid]
isobarometric 等壓的
isobarometric filler 等壓灌裝機；等壓裝瓶機
isobetanin 異甜菜苷
isoborneol 異龍腦；異冰片

isobornyl acetate 乙酸異龍腦酯；乙酸異冰片酯
isobutane 異丁烷
isobutanol 異丁醇
isobutyl acetate 乙酸異丁酯
isobutyl alcohol 異丁醇
isobutyl butyrate 丁酸異丁酯
isobutyl cinnamate 桂酸異丁酯；肉桂酸異丁酯
isobutyl furfuryl acetate 糠基乙酸異丁酯
isobutyl phenylacetate 苯乙酸異丁酯
isobutyl propionate 丙酸異丁酯；柳酸異丁酯
isobutyl salicylate 水楊酸異丁酯
isobutyl-2-butenoate 2-丁烯酸異丁酯
isobutylparaben 對羥基苯甲酸異丁酯；尼泊金異丁酯
isobutyraldehyde 異丁醛
isobutyric acid 異丁酸
isocatalysis 異構催化〔作用〕
isocitric acid 異檸檬酸
isodynamic food 等熱食品
isoelectric 等電的
isoelectric point (pI) 等電點
isoelectric protein 等電蛋白質
isoenzyme 同功酶 [= isozyme]
isoeugenol 異丁香酚；異丁子香酚
isoeugenyl acetate 乙酸異丁香酚酯
isoflavone 異黃酮
isoglutamine 異麩醯胺酸
isohumulone 異〔葎〕草酮；異蛇麻葫酮
isolate 分離物
isolated soybean protein 分離大豆蛋白〔質〕；大豆分離蛋白〔質〕
isolation 分離；隔離
isolation medium 分離培養基
isolation slaughter 隔離屠宰
isoleucine 異白胺酸；異亮胺酸
isomaltitol 異麥芽糖醇
isomaltose 異麥芽糖
isomaltulose 異麥芽酮糖
isomer 異構物；異構體
isomerase 異構酶
isomerism 異構現象；異構性
isomerization 異構化〔作用〕
isomerized lactose syrup 異構化乳糖液
isomerized syrup 異構化糖漿
isomerose 異構糖
isooleic acid 異油酸
isopentanol 異戊醇
isopentyl alcohol 異戊醇
isophagy 自溶；自體消化
isoprene 異戊二烯；橡膠基質
isopropanol 異丙醇 [= isopropyl alcohol]
isopropyl acetate 乙酸異丙酯
isopropyl alcohol 異丙醇 [= isopropanol]
isopropyl citrate mixture 檸檬酸異丙酯混合物
isopropyl isobutyrate 異丁酸異丙酯
isopropyl myristate 肉荳蔻酸異丙酯
isopropylparaben 對羥基苯甲酸異丙酯；尼泊金異丙酯
isopulegol 異蒲勒醇；異胡薄荷醇
isoquinoline 異哇啉
isorhamnose 異鼠李糖
isoriboflavin 異核黃素
isosaccharic acid 異糖二酸；異糖質酸
isosaccharinic acid 異糖林酸；異己糖酸
isosafrole 異黃樟油素
isoserine 異絲胺酸
iso- α-methyl ionone α-異甲基紫羅蘭酮
isosmotic 等滲〔壓〕的 [= iso-osmotic]
isosmoticity 等滲性；等滲壓
isostearic acid 異硬脂酸
isotherm 等溫線

isothermal 等溫的
isothiocyanate 異硫氰酸鹽
isothiocyanic acid 異硫氰酸
isothreonine 異羥丁胺酸；異蘇胺酸
isotonic 等滲〔壓〕的
isotonic diluent 等滲〔壓〕稀釋液
isotonic solution 等滲〔壓〕溶液
isotonicity 等滲性
isotope 同位素
isovaleraldehyde 異戊醛
isovaleric acid 異戊酸
isovitamin C 異維生素C
isozyme 同功酶 [= isoenzyme]
itaconic acid 分解烏頭酸；衣康酸；甲叉丁二酸
Italian black vinegar 意大利黑醋
Italian broccoli 青花菜
Italian fresh sausage 意大利生鮮香腸
Italian millet 粟；小米
Italian red onion 意大利紅洋葱
Italian rice 意大利米
Italian salami 意大利薩拉米香腸；色拉米腸；薩拉米香腸 [= salami (sausage)]
Italian vermouth 意式苦艾酒
item packaging 個別包裝；逐別包裝
IU (international unit) 國際單位
IUFoST (International Union of Food Science and Technology) 國際食品科技聯盟；國際食品科學技術協會
IV (iodine value) 碘值；碘價 [= iodine number]
ivory nut 象牙果；象牙棕櫚之果實
iwan tea 柳蘭茶；狹葉柳味茶

J

Jaborandi pepper 耶仆蘭胡椒
jaca 大樹波蘿；波蘿蜜；木波蘿；天波蘿 [= jackfruit; jak; jakfruit; nangka]
jack 鰺〔類〕
jack bean 洋刀豆；刀豆；闊刀豆
jack in oil 油浸鰺魚
jack mackerel 綠真鰺
jack salmon 銀大麻哈魚
jack screw press 螺旋壓榨機
jacketed evaporator 雙層蒸發器；夾層蒸發器
jacketed heat exchanger 夾套〔式〕熱交換器
jacketed kettle 雙層鍋；夾層鍋
jacketed steam kettle 雙層蒸氣鍋；夾層〔蒸〕氣鍋
jackfruit 大樹波蘿；波蘿蜜；木波蘿；天波蘿 [= jaca; jak; jakfruit; nangka]
jacky 杜松子酒；琴酒
jagdwurst 狩獵臘腸
jagger 麵糰鋸刀；花邊刀
jaggery 椰〔子〕糖
jak 大樹波蘿；波蘿蜜；木波蘿；天波蘿 [= jackfruit; jaca; jakfruit; nangka]
jake 牙買加薑萃；牙買加薑汁烈酒
jakfruit 大樹波蘿；波蘿蜜；木波蘿；天波蘿 [= jackfruit; jaca; jak; nangka]
jalap 紫茉莉
jam 果醬 [= confiture]
jam puff 果醬油酥餅
jam tart 果醬烘餅；果醬餡餅
Jamaica 牙買加甜酒
Jamaica cherry 牙買加櫻桃
Jamaica gooseberry 牙買加鵝莓；牙買加醋栗
Jamaica pepper 甜椒；甘椒；多香果；眾香子 [= allspice; pimento]
Jamaica sorrel 牙買加酢漿草；香酢漿草
jambalaya 詹巴什錦飯；什錦菜餚；肉菜雜燴
jambo 馬來蓮霧；蒲桃
jandagum 角豆膠
jannock 燕麥麵包；燕麥餅乾
Japanese Agricultural and Forestry Standard 日本農林規格；日本農林標準 [= Japanese Agricultural Standard; JAS]
Japanese Agricultural Standard (JAS) 日本農林規格；日本農林標準
Japanese anchovy 日本鯷魚
Japanese apricot 梅
Japanese bandfish 史氏赤刀魚；施氏赤刀魚
Japanese barracuda 日本金梭魚；大目金梭
Japanese blueberry 深紅越橘

Japanese butterfish 瓜子鯧；肉魚
Japanese catfish 鯰魚；日本鯰
Japanese crab 堪察加蟹
Japanese croaker 日本黃姑魚
Japanese eel 日本鰻鱺；白鰻
Japanese flowering quince 日本木瓜
Japanese herring 日本鯡魚
Japanese horse mackerel 真鰺；日本竹莢魚
Japanese horseradish 日本芥茉；芥茉；山葵；山萮菜 [= wasabi]
Japanese longtailed fowl 日本長尾雞
Japanese mackerel 馬加鰆
Japanese medlar 枇杷 [= loquat]
Japanese parrotfish 日本鸚鯉；日本纖鸚嘴魚
Japanese pear 白梨；日本梨
Japanese pepper 山椒
Japanese persimmon 柿子
Japanese radish 日本蘿蔔
Japanese rice wine 日本清酒 [= sake]
Japanese sandfish 日本叉齒魚；日本叉牙魚
Japanese sea bass 鱸魚
Japanese sea-perch 真鱸；七星鱸；海鮒
Japanese spearfish 尖吻四鰭旗魚
Japanese spider crab 大腳毛蟹
Japanese Standard of Food Additives (JSFA) 日本食品添加劑標準；日本食品添加物公定書
Japanese stauntonvine 日本野木瓜
Japanese stingray 赤魟
Japanese summer orange 日本夏橙
Japanese tamper 桂皮扁魚；桂皮斑鮃
Japanese velvet bean 藜豆
Japanese wine 清酒
japonica (type rice) 蓬萊米
jar 廣口瓶；罐
jar fermenter 小型發酵槽；小型發酵罐
jargonelle 早熟〔法國〕梨
jarovization 春化處理
JAS (Japanese Agricultural Standard) 日本農林規格；日本農林標準
jasmine 茉莉；素馨
jasmine absolute 素馨淨油；茉莉淨油
jasmine aldehyde 茉莉醛
jasmine concrete 茉莉浸膏
jasmine oil 茉莉〔精〕油
jasmine tea 茉莉〔花〕茶；香片；加味茶
jasmone 茉莉酮
jasmonyl 茉莉酯
Java coffee 爪哇咖啡
java ratio 蔗汁比例；蔗汁比率
Java tea 爪哇茶
javelle water 次氯酸鈉溶液；次氯酸鈉消毒液
javellization 次氯酸鈉消毒處理；漂白粉消毒
jaw 頜；顎
jaw breaker 顎式破碎機；頜式破碎機
jaw crusher 顎式壓碎機；頜式壓碎機
jaw oil 頜油；顎油
JECFA (Joint FAO/WHO Expert Committee on Food and Additives) 食品添加劑聯合專家委員會
jejune 營養缺乏的；缺乏營養的
jejune diet 營養缺乏膳食；缺乏營養膳食
jejunum 空腸
jell 凝膠；凍膠；膠凍
jellied meat 肉凍
jellification 膠凝〔作用〕
jelling agent 膠凝劑；膠化劑；稠化劑 [= gelatinizing agent; gelatinizer; gelling agent]
jello 果〔子〕凍；凝膠物；膠凍

jelly 果〔子〕凍；啫喱；膠凍；膠凍軟糖
jelly bean 糖豆；軟糖；軟心豆粒糖
jelly candy 軟糖
jelly drink 果凍粒飲料
jelly drops 軟糖〔粒〕
jelly-fig 愛玉子
jelly filling 果凍餡
jelly fungus 木耳；銀耳
jelly-like 膠〔凍〕狀的
jelly pocket 香腸內凝膠塊；香腸內膠凍塊
jelly powder 果凍粉；膠凍粉
jelly strength 膠強度；凝膠強度
jelly sweets 水晶軟糖；果凍軟糖
jelly tester 膠度計；膠凍強度測定儀
jellyfish 海蜇；水母
jellying point 膠凝點
jellying power 膠凝力
jelutong 節路頓膠
jerk 肉乾；乾肉條
jerked beef 風乾牛肉條；乾牛肉條；牛肉乾
jerked meat 風乾肉條；乾肉條
jerky 風乾肉條；乾〔牛〕肉條；牛肉乾
jeroboam 大酒瓶；大酒杯
jersey cow bolete 乳牛肝菌
Jersey dairy cattle 娟姍乳牛
Jersey tea 美洲茶
Jerusalem artichoke 菊芋；洋薑
jet agitator 噴射攪拌器
jet dryer 噴射式熱風乾燥機
jet mixer 噴射混合器
jew' s ear 黑木耳；木耳
jicama 沙葛；豆薯；涼薯 [= Chinese turnip]
jig water 酒
jigger 簸篩機；篩屑機；量酒杯；計量杯
jigging conveyer 振動式輸送機；跳動運輸帶
jigging machine 簸選機；跳汰機
jigging screen 簸動篩；跳動篩；跳汰篩
Jinhwa hog 金華豬
jinjili oil 芝麻油；香油 [= benne oil; gingelly oil; sesame oil]
jobstears 薏苡；薏仁；薏米 [= adlay]
Johannisberger 約翰內斯堡白葡萄酒
John Dory 海魴；魴
Johnny cake 玉米餅；約翰尼餅
joint cap 密封蓋
Joint FAO/WHO Expert Committee on Food and Additives (JECFA) 食品添加劑聯合專家委員會
Joint FAO/WHO Food Codex Alimentarius Commission (FAO/WHO) 聯合國糧農組織(FAO)世界衛生組織(WHO)聯合食品法規委員會
jointed wax gourd 毛豆；節瓜
jojoba 荷荷巴豆；黃楊栗；油栗
jojoba oil 荷荷巴油
Jonathan 紅玉蘋果
Jordan almond 約旦杏仁
Joule 焦耳
jowar 高粱；蜀黍；杜拉高粱；非洲高粱 [= durra]
jowl 垂肉；頰肉
joy water 酒
JSFA (Japanese Standard of Food Additive) 日本食品添加劑標準；日本食品添加物公定書
jug 壺；水壺
juglans 核桃；胡桃 [= walnut]
juice 果汁；汁；汁液
juice beverage 果汁飲料
juice clarification 果汁澄清
juice deliming 糖汁脫〔石〕灰

juice drink 果汁飲料
juice extraction 榨汁
juice extractor 榨汁機
juice liming 糖汁澄清；糖汁加石灰
juice mixer 果汁機
juice powder 果汁粉
juice sac 果實的汁囊
juice yield 出汁量
juiceless 無汁的
juicer 榨汁機；榨汁器
juiciness 多汁〔性〕；多液〔性〕；多汁程度
juicing 榨汁；壓汁；取汁
juicy 多汁的
jujube 棗 [= Chinese date]
jujube tincture 棗子酊
juke joint 小酒吧；小餐館
julep 茉莉〔薄荷〕甜酒；冰鎮薄荷酒
julienne 蔬菜絲；蔬菜絲肉湯
jumbal 輪狀薄糖果；甜環餅；環形小甜薄餅 [= jumble]
jumble 輪狀薄糖果；甜環餅；環形小甜薄餅 [= jumbal]
jumbo shrimp 大頭蝦
jumbuck 綿羊
jumped seam 跳封
jungle fowl 野雞；野禽
junior food 幼兒食品
juniper 杜松
juniper berries oil 杜松子油；刺柏油
juniper berry 杜松子
juniper berry oil 杜松子油
junk 鹹乾肉
junk bottle 深綠色厚壁玻璃瓶
junk food 解饞食物
junket 凝乳甜點；甜凝乳
jus 肉湯；肉汁
jute 黃麻
jute bag 麻袋
juvenile 幼體；青少年的；幼年的
juvenile prawn 幼蝦

K

kabob 羊肉；回式烤〔羊〕肉串；烤肉串 [= kebab]
kafir 高粱
kafir beer 高粱啤酒
kafir corn 高粱
kafirin 高粱醇溶蛋白
kafiroic acid 高粱酸
kaiser roll 凱撒卷
kaki 柿子 [= persimmon]
kakotrophy 營養不良
kalbwurst 牛肉香腸（德國）
kale 芥藍；羽衣甘藍
kaltrauch 冷燻
kamaboko 魚糕
kanamycin 康黴素；卡那黴素
kanemi oil 米糠油；米麩油 [= rice bran oil]
kangaroo 袋鼠
kangaroo meat 袋鼠肉
kanoy 肯諾依茶
kante 牛油表面的硬表皮
kaoliang 高粱 [= sorghum]
kaoliang colo(u)r 高粱色素
kaolin 高嶺土；瓷土；白陶土；硅酸鋁
kaolinite 高嶺石
kapok 木棉花；木棉
kapok oil 木棉籽油
kaporic 柳蘭茶；柳蘭
karaya gum 刺梧桐〔樹〕膠 [= gum karaya]
Karl Fischer method 卡耳費雪〔水分測定〕法；卡爾費希爾〔水分測定〕法
Karl Fischer reagent 卡耳費雪試劑；卡爾費希爾試劑
kasha 蕎麥片；蕎麥粥；麥糊
kasher 猶太教食品；猶太律法食品 [= kosher]
kasnudln 卡斯努得餃子（奧地利）
kat 卡他茶；阿拉伯茶
katathermometer 卡他溫度計；低溫溫度計；冷卻〔率〕溫度計
katenrauchwurst 鄉村燻腸
kava 卡瓦酒
kcal (kilocalorie) 大卡；仟卡 [= Calorie; Cal; great calorie]
kebab 羊肉；回式烤〔羊〕肉串；烤肉串 [= kabob]
kedgeree 印式雞肉蛋葱豆飯
keep frozen 注意冷凍；保持冷凍
keeping time 保存期；保藏期
keeve 發酵甕；大發酵桶
kefir 克菲爾酸牛乳酒；克弗〔酸牛乳〕酒；酸牛乳酒 [= kephir; kefyr]
kefir grain 克弗酒麴；酸牛乳酒麴
kefyr 克菲爾酸牛乳酒；克弗〔酸牛乳〕酒；酸牛乳酒 [= kefir; kephir]
keg 小桶（容量單位）
keg beer 桶裝啤酒
kek mill 臥式〔圓〕盤磨；單輪碾碎機

kelp 巨海帶；巨昆布；巨海藻；海藻粉；海藻灰
kelp bass 莫拉石斑魚；雲紋石斑魚
kelp cod 大鱈魚
Kempner diet 肯氏低鹽米食
kench 堆鹽漬；醃漬箱
kench curing 層醃；堆鹽漬；倉醃
kench salting 堆醃；層醃；倉醃
keng rice 粳稻；硬米
kenotoxin 疲倦毒素
kephalin 腦磷脂 [= cephalin]
kephir 克菲爾酸牛乳酒；克弗〔酸牛乳〕酒；酸牛乳酒 [= kefir; kefyr]
kerasin 角苷脂
keratein 還原角蛋白
keratin 角蛋白；角質素
keratinase 角蛋白酶
keratoprotein 角蛋白質；角質蛋白
kermes 胭脂〔紅〕
kernel 穀粒；仁；核
kernel oil 核油
kernel paste 核仁糊
kernel weight 穀粒重（千粒重）
kernelled 有核仁的；有仁的
Kerry 克雷牛（愛爾蘭黑色乳牛）
Kestner evaporator 克斯納蒸發器；凱斯特納蒸發器；長管式無循環蒸發器
kestose 蔗果三糖
keta 駝背鮭魚；大麻哈魚 [= dog salmon]
ketchup 番茄醬 [= catchup; catsup]
keto acid 酮酸
keto sugar 酮糖
ketobutyric acid 丁酮酸
ketogenic amino acid 產酮〔體〕胺基酸
ketogenic diet 產酮膳食；產酮食物
ketogluconic acid 酮葡〔萄〕糖酸
ketoglutaric acid 酮戊二酸；氧代戊二酮酸
ketolysis 酮解〔作用〕；解酮〔作用〕
ketone 酮
ketone body 酮體 [= acetone body]
ketonemia 酮血〔症〕
ketonic acid 酮酸
ketonuria 酮尿〔症〕 [= acetonuria]
ketose 酮糖
ketoside 酮〔糖〕苷
ketosis 酮症
kettle 壺；罐；鍋；釜
keturonic acid 糖酮酸
key opening can 捲開罐；鑰匙罐
key opening reclosure 復蓋捲開罐
keyless can 無鑰匙罐
khat 阿拉伯茶
kheer 印度煉乳
khorassan wheat 東方小麥
khosam 苦參子
kibble 粗磨；粗碎
kibbled grain 粗穀粉；粗碎穀粒
kibbler 粗磨機；粗碎機
kickshaw 精美菜餚
kid 羔羊；小山羊
kidney 腎〔臟〕；腰子
kidney bean 四季豆；敏豆；菜豆；蕓豆
kidney suet 腎區板油
kiese gel 硅膠
kieselguhr 硅藻土
kilfoam 消泡劑；除沫劑；防沫劑 [= antifoaming agent; defoaming agent]
killing bed 宰殺台；屠宰台
kiln 窯；爐
kiln dry(ing) 窯乾；窯式乾燥；烘爐乾燥
kiln dryer 窯式乾燥機

kilned malt 窯乾麥芽；烘乾麥芽
kilocalorie (kcal) 大卡；仟卡 [= Calorie; Cal; great calorie]
kilogram (kg) 公斤；千克
kilojoule 千焦〔耳〕
kimchi 韓國泡菜；朝鮮泡菜
kinase 激酶
kine 黃牛
kinetic energy 動能
king crab 王蟹；巨蟹；鱈場蟹
king mackerel 棘鰆；大西洋馬鮫
king nut 光皮山核桃
king salmon 大鱗大麻哈魚
king's paprika 匈牙利〔辣〕椒
kingfish 無齒鰺；高體若鰺；無鰾石首魚
kip 千磅；基普（453.59 kg或1000 lb）
kipper 開背冷燻魚；醃燻魚；基普斯燻鯡魚；雄鮭魚
kippered fish 醃燻魚；開背冷燻魚
kirsch 櫻桃酒；櫻桃白蘭地酒
kiskatom 山核桃
kiss 小糖果；蛋白甜酥餅
kitchen 廚房
kitchen garden 菜園
kitchen herb 草本調味香料；調味用草本香料
kitchen pepper 烹調用混合胡椒粉
kitchen stuff 烹調用料；廚房廢料或油垢
kitchen tongs 菜夾
kitchen utensil 廚房器具
kitchener 廚師
kitchenette 小廚房
kitchenware 廚具；廚房用具
kitefish 銀大眼鯧；銀鳶鯧
kitembilla 錫蘭醋栗
kitol 鯨〔肝〕醇
kiwi (fruit) 奇異果；獼猴桃；中國醋栗；中國鵝莓；藤梨 [= Chinese gooseberry]
kiwi flavo(u)r 奇異果香精；獼猴桃香精
Kjeldahl flask 克耳達燒瓶；凱氏燒瓶
Kjeldahl method 克耳達法；凱氏〔測定氮〕法
Kjeldahl nitrogen 克耳達氮；凱氏氮
Klebsiella 克萊桿菌屬；克雷伯氏菌屬；克氏桿菌屬
klipfish 鹹鱈魚乾；挪威醃鱈魚乾；歐洲鹽乾鱈
knackery 家畜屠宰場；宰馬場
knead(ing) 捏揉；捏製；捏和；揉和
kneader 捏揉機；捏和機；揉麵糰機
kneading arm 捏和桿；揉麵桿
kneading machine 捏揉機；捏和機；揉麵糰機
kneading mixer 捏和機
kneading trough 揉麵槽；捏和槽
knife 刀子
knife box 刀具箱
knife grinder 磨刀器
knife machine 磨刀機
knife pouch 刀套
knife rest 刀架
knife sharpener 磨刀器；磨刀床
knob 砂糖塊；方糖；旋鈕
knobbed wrack 念珠褐藻
knoblanch 蒜〔味〕香腸
knoblauchwurst 蒜味香腸
knock back 揉緊麵糰；發麵過度後的回縮
knockdown flange 罐身脱鈎；身鈎脱扣；裂緣
knocking 擊昏；敲
knotter 打結機
know-how 訣竅；技術訣竅

knuckle 後腿股肉；牛冧；圓冧牛肉；和尚頭

knuckle steak 後腿股肉排；圓冧牛排

knuring 罐身滾壓淺紋；滾壓線紋

Koestler number 凱斯特勒值

kohlrabi 球莖甘藍；結頭菜；大頭菜；莖藍

koji 麴；曲

koji agar 麴汁洋菜〔培養基〕；麴汁瓊脂〔培養基〕

koji amylase 麴澱粉酶；曲澱粉酶

koji bed 麴床；曲床

koji extract 麴汁；曲汁

koji flavo(u)r 麴味；曲味

koji-making 製麴；製曲

koji-making machine 製麴機；製曲機

koji mo(u)ld 麴黴；曲霉

koji tray 麴盤；曲盤

kojiboise 麴二糖；曲二糖

kojic acid 麴酸；曲酸

kola 可樂樹

kola nut 可樂果

komarov lilac 西蜀丁香

kombu 海帶；昆布

konjac 蒟蒻；魔芋 [= elephant foot; giantarum; konnyaku]

konjac flour 蒟蒻粉；魔芋粉；魔芋甘露聚糖

konjac mannan 蒟蒻聚甘露糖；蒟蒻甘露聚糖

konnyaku 蒟蒻；魔芋 [= elephant foot; giantarum; konjac]

koppe 梭麵包；兩端尖的麵包

Korean mackerel 高麗馬加鰆；朝鮮馬鮫

kosher 依猶太食品；猶太律法食品 [= kasher]

kosher method 斷頭法

koumiss 馬奶酒；馬乳酒 [= kumiss; kumys]

kraft (paper) 牛皮紙

krama 水酒（東正教會）

kraul salt 聚〔合〕偏磷酸鹽 [= polyme-taphosphate]

krausen 發酵麥芽汁；麥芽汁發酵時的泡沫

kraut 德式酸泡菜；泡菜；酸漬菜 [= sauerkraut; sourkrout; sourcrout]

Krebs cycle 克里勃司環；克雷布斯循環；三羧酸循環

Kreis test 克雷氏〔油脂氧化酸敗〕試驗；克雷薩〔油脂氧化酸敗〕試驗

kreotoxin 腐肉毒素；肉毒素

kreotoxism 腐肉中毒；食肉中毒；肉毒

krill 南極蝦；磷蝦

kruller 油煎餅；油炸麻花；油炸圓圈

kryptoxanthin 玉米黃質

kumiss 馬奶酒；馬乳酒 [= koumiss; kumys]

kummel 欽美酒；蒔蘿利口酒；香旱芹酒

kumquat 金橘；金桔；金柑；金棗 [= cumquat]

kumys 馬奶酒；馬乳酒 [= koumiss; kumiss]

kuo 焙茶〔平〕鍋

kuruma prawn 日本對蝦；斑節蝦

kuza 粉葛

kvass 苦哇啤酒；格瓦斯淡啤酒；克瓦斯淡啤酒

kwashiorkor 蛋白質營養缺乏症；加西卡病；夸希奧科病

kynuric acid 犬尿酸

L

50% lethal dose 半數致死劑量
[= LD50; medium lethal dose]
labdamum concrete 岩薔薇浸膏；賴白當浸膏
labdanum oil 岩薔薇油；賴百當油
label 標籤；商標
labelled can 貼上標籤的罐頭
labelling 標籤；貼商標；貼標籤
labile 不穩定的；易變的
labile protein 不穩定蛋白質
lablab bean 拉巴拉巴豆；扁豆
laboratory animal 實驗動物
laboratory scale 實驗室規模
labrusca 拉布魯斯卡〔釀酒用〕葡萄
lac 紫膠；蟲膠
lac dye 紫膠紅；蟲膠紅 [= lac red]
lac red 紫膠紅；蟲膠紅 [= lac dye]
laccaic acid 紫膠酸
lacceroic acid 紫膠蠟酸；蟲漆蠟酸
lace 佐料烈酒；烈性酒
lacepaper edging 蛋糕裝飾用花邊墊紙
lacks ham 拉克斯〔煙燻〕火腿
lacquered can 塗漆罐
lactacidase 乳酸菌酶
lactacidogen 半乳酸磷酸；乳酸精
lactalase 產乳酸酶
lactalbumin 乳白蛋白
lactam 乳胺；內酰胺
lactase 乳糖酶
lactase deficiency 乳糖酶缺乏症
lactate 乳酸鹽；泌乳
lactation 乳汁分泌；哺乳期
lacteal 乳糜管；乳狀的
lacteal gland 乳腺
lactein 煉乳
lactein bread 牛乳麵包
lactenin 乳〔煙〕素；乳抑菌素
lactescence 乳狀液；乳汁狀
lactic acid 乳酸
lactic acid and fatty acid esters of mono- and diglycerides 乳酸和脂肪酸甘油酯；乳酸化單雙甘油酯
lactic acid bacteria 乳酸菌
lactic acid bacteria drink 乳酸菌飲料
lactic acid casein 乳酸酪蛋白；乳酸乾酪素
lactic acid fermentation 乳酸發酵
lactic acid forming bacteria 產乳酸細菌
lactic bacillus 乳酸桿菌
lactic beverage 乳酸菌發酵飲料
lactic cheese 乳酸乾酪
lactic dehydrogenase 乳酸脫氫酶；乳酸去氫酶
lactic streptococci 乳酸鏈球菌
lactification 乳酸生成
lactitol 乳糖醇
lactoalbumin 乳清蛋白
lactobacillic acid 乳〔酸〕桿菌酸
lactobacillin 乳酸〔桿〕菌素
Lactobacillus 乳酸桿菌屬

Lactobacillus acidophilus 嗜酸乳〔酸〕桿菌
Lactobacillus arabinocus 阿拉伯乳〔酸〕桿菌
Lactobacillus bifidus 雙叉乳〔酸〕桿菌；雙岐乳〔酸〕桿菌
Lactobacillus brevis 短乳〔酸〕桿菌
Lactobacillus bulgaricus 保加利亞乳〔酸〕桿菌
Lactobacillus casei 酪蛋白乳〔酸〕桿菌；乾酪乳〔酸〕桿菌
Lactobacillus caucasicus 高加索乳〔酸〕桿菌
Lactobacillus delbruckii 戴白氏乳〔酸〕桿菌；德氏乳〔酸〕桿菌
Lactobacillus fermentum 發酵乳〔酸〕桿菌
Lactobacillus helveticus 瑞士乳〔酸〕桿菌
Lactobacillus lactis 乳酸乳〔酸〕桿菌
Lactobacillus pastorianus 酪酸乳〔酸〕桿菌
Lactobacillus plantarum 胚芽乳〔酸〕桿菌；植物乳〔酸〕桿菌
Lactobacillus thermophilus 嗜熱乳〔酸〕桿菌
lactobionic acid 乳糖酸
lactobiose 乳二糖
lactobutyrometer 乳脂計
lactochrome 核黃素；維生素B_2 [= lactoflavin; riboflavin(e)]
lactocidin 乳酸殺菌素（保存劑）
lactocrit 乳脂計；乳脂測定器
lactodensimeter 乳比重計 [= lactometer]
lactofarinaceous food 乳穀食品
lactoferrin 乳鐵蛋白
lactoflavin 核黃素；維生素B_2 [= lactochrome; riboflavin(e)]
lactogen 催乳激素
lactoglobulin 乳球蛋白
lactol 乳醇
lactolase 乳酸酶
lactometer 乳比重計 [= lactodensimeter]
lactomucin 乳黏蛋白
lactonase 內酯酶
lactone 內酯
lactonic acid 內酯酸
lactoperoxidase 乳過氧化物酶
lactoprotein 乳蛋白〔質〕
lactoscope 乳脂計
lactose 乳糖 [= milk sugar]
lactose fermenting yeast 乳糖發酵酵母
lactose-free diet 無乳糖膳食
lactose intolerance 乳糖不耐症
lactose molasses 乳糖〔廢〕蜜
lactoserum 乳清
lactoxin 乳毒素
lactulose 乳酮糖；乳果糖；二蔗酮糖
lactylic esters of fatty acids 脂肪酸乳酰酯
lacustrine 湖泊的
lacustrine fish 湖〔泊〕魚
ladanum oil 勞丹油；岩薔薇油
lady's finger 手指狀鬆脆餅；細小鬆脆餅；細長葡萄
laevoglucose 左旋葡萄糖 [= levog-lucose]
laevulose 果糖；左旋糖 [= fructose; fruit sugar; levulose]
lag phase 遲滯期；遲緩期；滯後期
lager (beer) 拉格啤酒；陳貯啤酒；釀造後再貯藏陳化的啤酒
lager fermentation 後發酵；啤酒後發酵
lager yeast 淡色啤酒酵母；陳貯啤酒酵母
lake herring 湖鯡
lake trout 湖鱒；湖紅點鮭

laker 湖魚
lamb 羔羊；小綿羊；羔羊肉
lamb meat 小羊肉；羔羊肉
lamb's lettuce 野苣
lambkin 小羊肉；羔羊肉
lamen 拉麵
laminaria 海帶；昆布
laminaribiose 海帶二糖；昆布二糖
laminarin 海帶多糖；昆布多糖
laminariose 海帶糖；昆布糖
laminaritol 海帶醇；昆布醇
laminated film 複合薄膜；積層薄膜
lampern 八目鰻；七鰓鰻
lamprey (eel) 八目鰻；七鰓鰻
lancet fish 帆蜥魚
land crab 陸地蟹
land cress 陸生芹菜
Landrace (swine) 藍瑞斯豬；蘭德瑞斯豬；丹麥長白豬；蘭德列斯豬
langouste 歐洲龍蝦；龍蝦
langsat 蘭撒漿果；榔色木漿果
lanolin 羊毛脂
lanosterol 羊毛固醇
lantern fish 燈魚；發光燈籠魚
lappa 牛蒡 [= burdock]
lapper milk 酸牛奶；凝乳
larch bolete 厚環乳牛肝菌
larch gum 松膠；阿拉伯半乳聚糖
lard 豬油；豬脂
lard dreg 豬油渣
lard filter press 豬油壓濾機
lard fruit 豬油果
lard hasher 肥肉絞碎機
lard hog 肥豬
lard oil 豬脂油
lard olein 液態豬油；豬油的液態部分
lard press 油渣壓榨機
lard stearin 固態豬油；豬油的固態部分
larder 食品室；肉貯藏處；食品櫥
lardine 人造固化脂；人造奶油；代豚脂
larding 肥肉餡充填；填肥肉餡
lardo(o)n 肥鹹肉片；夾心用鹹肥肉
lardy 豬油的；含豬油的
large fruited 大果實的；大果型的
large intestine 大腸
large red onion 大紅洋蔥
large scale production 大規模生產
large white bean 大白豆；白扁豆
large white swine 大白豬；大約克夏豬
large yellow croaker 大黃魚
largemouth bass 黑鱸
larva 幼蟲
larval fish 幼魚
larynx 喉
lasagna 闊麵；千層麵
lasagnette 小千層麵
Lasche value 拉氏麥芽糖化力值
lasting food 防腐食品；耐久存放的食品
late lactation milk 末期乳
late rice 晚稻
late variety 晚熟品種
latent heat 潛熱
latent infection 潛伏感染
latent period 潛伏期
lateral muscle 側肌
latex 乳汁；乳膠
lather 泡沫；發泡
lather booster 增泡器
lathering power 發泡力
latimeria 矛尾魚
latte 拿鐵咖啡 [= caffe latte]
launce 玉筋魚
lauraldehyde 月桂醛；十二醛
laurate 月桂酸鹽；月桂酸酯
laurel 月桂樹 [= sweet bay]

laurel butter 月桂脂
laurel leaf 月桂葉；香葉 [= bay leaf]
laurel leaf oil 月桂葉油；月桂油 [= laurel oil; bay leaf oil; bay oil]
laurel oil 月桂葉油；月桂油 [= laurel leaf oil; bay leaf oil; bay oil]
lauric acid 月桂酸；十二〔烷〕酸 [= dodecanoic acid]
lauroleic acid 月桂烯酸；十二碳烯酸
laurone 月桂酮
lauryl alcohol 月桂醇；十二〔烷〕醇
lauryl aldehyde 月桂醛；十二醛
lauryl gallate 沒食子酸月桂酯
lauter 過濾；澄清；過濾麥芽汁
lauter tank 過濾槽
lavandin oil 雜熏衣草油
lavender 熏衣草
lavender flower oil 熏衣草花油
lavender oil 熏衣草油
laver 紫菜；海苔；石蒓；紅藻 [= nori]
lawn sugar 粉糖
lawyer's wig 毛頭鬼傘(蕈類)
lax 鮭魚；腹瀉
laxative 輕瀉劑
layer 層；產蛋雞；產卵魚
layer cake 千層蛋糕；夾心蛋糕
layered tooth fungus 彎曲層齒菌(蕈類)
laying fowl 蛋禽；產卵禽
lazy susan 餐桌轉盤；轉臺
LD (lethal dose) 致死劑量
LD_{50} 半數致死劑量 [= 50% lethal dose; medium lethal dose]
LDL (low density lipoprotein) 低密度脂蛋白
leach 瀝濾；瀝濾器
leaching liquor 瀝濾液
lead 鉛
lead acetate 乙酸鉛；醋酸鉛
lead grey bovist 鉛色灰球菌(蕈類)
lead poisoning 鉛中毒
leaf 葉；薄片；頁
leaf alcohol 葉醇
leaf aldehyde 葉醛
leaf beet 牛皮菜；莙薘菜
leaf butyrate 丁酸葉醇酯
leaf caproate 己酸葉醇酯
leaf fat 板油
leaf filter 壓濾器；葉〔狀〕濾器
leaf formate 甲酸葉醇酯
leaf isinglass 魚膠片；片狀魚膠
leaf lard 板油
leaf lettuce 散葉萵苣
leaf mustard 雪裏紅；雪菜；芥菜
leaf protein 葉蛋白
leaf stalk 葉柄
leaf stewing 泡茶；沏茶
leaf vegetable 葉菜
leafage 葉子
leaflet 小葉；嫩葉
leaky butter 析水黃油；析水乳酪
lean 瘦肉；瘦的
lean cheese 低脂乾酪；脱脂乾酪
lean dough 未加糖麵糰
lean meat 瘦肉
lean muscle 瘦肌肉
leanness 瘦
leather carp 無鱗鯉魚
leather jacket 革魨
leathery cheese 堅韌質地的乾酪；皮質乾酪
leaven 使發酵膨鬆；膨鬆劑；膨發劑
leaven dough 膨鬆麵糰
leavening 膨發；膨鬆；發酵
leavening agent 膨鬆劑；膨發劑
leben 酸奶製品；力本發酵乳
leberkase 肝泥肉腸
lebkuchen 蜂蜜果脯酥餅

lecithalbumin 卵磷脂白蛋白；卵磷脂清蛋白
lecithin(e) 卵磷脂；磷脂酰膽鹼；膽鹼磷脂 [= phosphatidylcholine]
lecithinase 卵磷脂酶
lecithinated soy flour 卵磷脂強化大豆粉
lecithoprotein 卵磷〔脂〕蛋白
lectin 植物凝集素
leechee 荔枝 [= lychee; lichee; lichi; litchi]
leek 韭菜；韭葱；蒜苗
lees 酒糟；酒泥
leg chump off 羊腿肉
leg cuts 後腿肉
leg finisher 禽腿去毛機
Leghorn 來亨雞；來航雞
legume 豆；豆科植物
legumelin 豆白蛋白；豆清蛋白
legumin 豆球蛋白
Leguminosae 豆科
leguminous 豆科的；豆類的
leisure food 消閑食品
lembus rudderfish 短鰭舵魚；蘭勃舵魚
lemon 檸檬
lemon balm 檸檬香水薄荷
lemon beverage 檸檬飲料
lemon curd 檸檬酪醬；檸檬羹
lemon daylily 金針菜；黃花菜；黃花萱草
lemon drop 檸檬糖
lemon essence 檸檬香精
lemon flavo(u)r 檸檬香精
lemon grass 香茅〔草〕；檸檬草 [= citronella (grass)]
lemon grass oil 香茅油
lemon grating 檸檬皮碎粒
lemon juice 檸檬汁
lemon leaf oil 檸檬葉油
lemon lily 金針菜；黃花菜
lemon oil 檸檬油
lemon oily flavo(u)r 檸檬油香精
lemon peel 檸檬皮
lemon pie 檸檬派；檸檬餡餅
lemon pip oil 檸檬子油
lemon sole 檸檬鰨
lemon squash 檸檬汁飲料；檸檬汁汽水
lemon squeezer 擠檸檬器
lemon verbena 防臭木
lemonade 檸檬水；檸檬汁飲料
lemongrass oil 檸檬草油；香茅草油；楓茅油
lenten 齋戒的
lenthionine 香菇香精
lentil 小扁豆；金麥豌豆
lentinan 香菇多糖
Lentinula edodes 香菇 [= shiitake]
leopard grouper 七星斑
leopard palm 麻芋
lesser galangal 良薑
lesser grain borer 麵粉蠹；穀蠹
lethal 致死的；致命的
lethal dose (LD) 致死劑量
lethality 致死率；致命性
lettuce 生菜；萵苣；西生菜
leucine 白胺酸；亮胺酸
leuco 無色
leucoanthocyanin 無色花青素；無色花青苷
leucoflavin 無色核黃素 [= leucoribof-lavin]
leucomethylene blue 無色亞甲藍；無色美藍
Leuconostoc 白念珠球菌屬；明串珠菌屬
leucoriboflavin 無色核黃素 [= leuco-flavin]

leucosin 麥白蛋白；麥清蛋白
leukocyte 白血球 [= white blood cell]
leukodextrin 白色糊精
leukophyll 葉白素
levan 聚果糖；果聚糖；聚左旋糖 [= fructosan]
levansucrase 聚果糖蔗糖酶；果聚糖蔗糖酶
levoglucose 左旋葡萄糖 [= laevoglu-cose]
levulosan 聚果糖；果聚糖；聚左旋糖
levulose 果糖；左旋糖 [= fructose; fruit sugar; laevulose]
lichee 荔枝 [= lychee; leechee; lichi; litchi]
lichen 地衣；苔蘚；青苔
lichenase 地衣多糖酶
lichenin 地衣多糖
lichi 荔枝 [= lychee; leechee; lichee; litchi]
licopersicin 番茄素
licorice 甘草 [= liquorice]
licorice extract 甘草萃；甘草浸膏
licorice root 甘草
licorice root antioxidant 甘草抗氧物
licorice root extract 甘草流浸膏
lid 罐蓋；蓋子
lid curler 圓蓋機
lid embossing machine 罐蓋壓印機；罐蓋打印機
lid flipper 開瓶器；撬蓋器
lid maker 製蓋機
lid seamer 封罐機
lidding machine 糕餅表面裝飾機；蛋糕飾花機
life science 生命科學
lifter 封罐機的〔提升〕托板
ligament 韌帶
ligase 連接酶
ligating 結紮
light beer 淡啤酒；低酒度啤酒
light bread 白麵包
light cake 鬆蛋糕
light cappuccino 牛奶卡布奇諾咖啡 [= cappuccino chiaro]
light concentration 低濃度
light cream 低脂鮮乳油；低脂稀奶油
light flour 淡麵粉；白麵粉
light industry 輕工業
light liquor 淡烈酒
light malt 淡色麥芽
light meat 淡色肉；白色肉
light meat tuna 白色鮪魚
light off flavo(u)r 光變味
light packaging 輕包裝
light petroleum 石油醚
light-proof packaging 防光包裝
light-proof paper 防光紙
light roast 輕度焙烤
light salting 低鹽漬
light soup 清湯
light soy sauce 豉油；醬油；生抽 [= soy sauce; soya sauce; soybean sauce]
light syrup 淡糖漿；稀糖漿
light textured confection 酥糖
light wine 淡葡萄酒；低酒度葡萄酒
light wrapper 薄包裝材料
lightly salted 輕醃的；淡醃的
lights 食用的屠體肺臟；食用的牲畜肺臟
lignin 木質素
lignocellulose 木質纖維素
lilac 丁香；紫丁香；丁香花
lilac alcohol 丁香花醇
lillywhite cake 白鬆糕
lily bulb 百合
lily root 百合

lima bean 萊豆；皇帝豆；利馬豆；棉豆
limb 肢
Limburg(er) cheese 比利時寧堡乾酪；寧堡乾酪；比利時林堡乾酪；林堡乾酪
lime 酸橙；萊姆；萊姆酸橙；青檸
lime defecation 石灰澄清
lime juice 酸橙汁；萊姆果汁
lime milk 石灰乳
lime oil 萊姆油；萊姆酸橙油；白檸檬油 [= limette oil]
lime water 石灰水
limeade 酸橙飲料；萊姆飲料；萊姆汽水
limeberry 萊姆莓
limed juice 加石灰澄清的糖汁
limequat 萊姆金柑
limette 白檸檬
limette oil 萊姆油；萊姆酸橙油；白檸檬油 [= lime oil]
limettin 萊姆素；白檸檬素
liming 石灰法澄清；加石灰澄清
limit dextrin 極限糊精
limit dextrinase 極限糊精酶
limited storage 貯藏限期
limiting amino acid 限制胺基酸
limiting factor 限制因素
limiting nutrient 限制營養素
limnetic 淡水的
limnetic fish 淡水魚
limonene 檸檬烯；檸檬精油
limonin 檸檬苦素
limosis 極度飢餓；異常飢餓
limotherapy 飢餓療法
limpet 帽貝
limpid 清澈的
linaloe oil 沉香油；伽羅木油；里哪油
linaloe wood oil 芳樟油；沉香油
linalool 沉香醇；芳樟醇；里哪醇
linalyl acetate 醋酸芳樟酯；乙酸芳樟酯
linalyl benzoate 苯甲酸芳樟酯
linalyl butyrate 丁酸芳樟酯
linalyl formate 甲酸芳樟酯
linalyl isobutyrate 異丁酸芳樟酯
linalyl propionate 丙酸芳樟酯
linden 菩提樹
linear compound 直鏈化合物
lined bag 裏襯袋；襯裏袋
lined wrapping 裏襯包裝
liner 襯墊；襯裏
linerless cap 無襯墊瓶蓋
ling 長身鱈魚；予魚；海比目魚
linger 香腸結紮機
lingering bitterness 餘苦味；後苦味
lingua 舌
linguine 粗麵；扁麵條
lining paper 襯墊紙
link 聯結；香腸紮節；香腸段
linking 聯結；香腸紮節
linking machine 結紮機；紮腸機
linoleate 亞〔麻〕油酸鹽；亞〔麻〕油酸酯
linoleic acid 亞麻油酸；亞油酸
linolein 甘油三亞〔麻〕油酸酯；三亞麻油脂；亞油精 [= linolin; trilinolein; glyceryl trilinoleate]
linolenate 次亞麻油酸鹽；次亞麻油酸酯
linolenic acid 次亞麻油酸
linolenin 甘油三次亞麻油酸酯；三次亞麻油脂；亞麻精 [= trilinolenin; glyceryl trilinolenate]
linolic acid 亞〔麻〕油酸
linolin 甘油三亞〔麻〕油酸酯；三亞麻油脂；亞油精 [= linolein; trilinolein; glyceryl trilinoleate]
linoxanthine 亞麻黃素

linoxyn 氧化油膜；氧化亞麻仁油
linseed 亞麻仁；亞麻子
linseed oil 亞麻仁油
Lintner degree 林〔特納〕氏度
Lintner starch 林〔特納〕氏澱粉
Lintner value 林〔特納〕氏值
Lintnerization 林〔特納〕氏處理
lipase 脂酶；脂肪酶
lipid digestibility 脂質消化率
lipid-soluble vitamin 脂溶性維生素；脂溶性維他命 [= fat-soluble vitamin]
lipid(e) 脂質；脂類
lipidemia 脂血症
lipidic 脂質的；油脂的
lipin 脂質；脂類
lipoalbumin 脂白蛋白；脂清蛋白
lipoblast 脂母細胞
lipochrome 脂色素
lipoclasis 脂肪分解
lipocyte 脂肪細胞
lipodieresis 脂肪分解
lipodystrophia 脂質營養不良；脂肪代謝障礙
lipofuscin 脂褐素；脂褐質
lipoid 類脂質；類脂的
lipolysis 解脂〔作用〕；脂肪分解
lipolytic 解脂的
lipolytic bacteria 解脂細菌
lipolytic enzyme 解脂酶
lipolytic hydrolysis 脂肪水解
lipolytic rancidity 解脂性酸敗；脂肪分解酸敗
lipometabolism 脂肪代謝
liponic acid 脂酮酸
lipopeptide 脂肽
lipopexia 脂肪蓄積
lipophilic 親脂的
lipophobic 疏脂的
lipopolysaccharide 脂多糖
lipoprotein 脂蛋白
lipoprotein lipase 脂蛋白脂肪酶
liporhodin 脂紅質
lipositol 肌醇磷脂
liposoluble 脂溶〔性〕的
liposome 脂質體
lipotropy 趨脂性；親脂性
lipovitamin 脂溶性維生素
lipoxanthine 卵黃質；黃脂色素
lipoxgenase 脂肪加氧酶
lipoxidase 脂〔肪〕氧化酶
liquefacient 液化劑
liquefaction 液化〔作用〕
liquefied cheese 軟化乾酪
liquefied protein 液化蛋白〔質〕
liquefier 溶化鍋；液化器
liquefrozen 液氮冷凍
liquefying agent 液化劑
liquefying amylase 液化澱粉酶
liquefying enzyme 液化酶
liqueur 甜〔露〕酒；利口酒
liqueur center 酒〔糖〕心
liquid 液體；流體
liquid apple 帶肉蘋果汁
liquid-binding power 持水性；保水性
liquid chromatography 液體層析法；液相色譜法
liquid cooling 液體冷卻
liquid culture 液體培養
liquid diet 流質膳食；液態膳食
liquid egg 液〔態〕蛋
liquid extraction 液體萃取
liquid fat 液態脂肪
liquid filler 液體充填機；液體灌裝機
liquid film 液膜
liquid fish 濃縮魚湯
liquid food 流質食物；液態食品
liquid freezing 液體冷凍
liquid koji 液〔體〕麴

liquid-liquid extraction 液液提取；液液萃取
liquid medium 液態培養基
liquid nitrogen 液〔態〕氮
liquid nitrogen freezing 液〔態〕氮冷凍
liquid packing 充液包裝〔法〕
liquid phase 液相
liquid seal(ing) 液封
liquid shortening 液態〔起〕酥油
liquid smoke 燻液；液體燻製劑
liquid smoke solution 燻液
liquid smoking 液燻〔法〕
liquid solid extraction 液固提取；液固萃取
liquid spice 液態香辛料
liquid sugar 液〔態〕糖；糖漿
liquid-tight 不透液體的
liquidambar concrete 楓香浸膏；蕓香浸膏
liquidity 流動性
liquifier 稀釋劑；液化劑
liquifying enzyme 液化酶
liquor 液體；汁；溶液；烈酒
liquorice 甘草 [= licorice]
liquorice plum 甘草梅
litchi 荔枝 [= lychee; leechee; lichee; lichi]
liter 公升
lithographed can 彩印罐
lithol rubine BK 立索玉紅BK
litmus 石蕊
litmus milk 石蕊牛乳〔培養基〕
litmus paper 石蕊試紙
litsea cubeba oil 山蒼子油
little radish 田季蘿蔔
little yellow croaker 小黃魚
littlehip hawthorn 小果山楂
live-bearing fish 胎生魚
live farming 畜牧業
live weight 活重；活畜重量
liver 肝臟
liver cheese 肝泥捲；肝泥糕
liver fungus 肝色牛排菌
liver loaf 肝泥捲；肝泥糕
liver oil 肝油
liver paste 肝醬；肝糊
liver sausage 肝〔泥〕香腸；肝腸
liver starch 肝糖；動物澱粉 [= glycogen; animal starch]
liverwurst 肝腸
livestock 家畜；牲畜
livestock farm 牧場；畜牧場
livetin 卵黃球蛋白
living cell 活細胞
living insecticide 生物殺蟲劑
living yeast 活酵母
lixiviation 浸濾；浸提
Lo-Han-Kuo extract 羅漢果抽提液
loach 泥鰍
loading hopper 加料斗；進料斗
loaf 麵包方條；枕形麵包；羅夫；肉糕
loaf cheese 長方形乾酪
loaf filler 麵包餡
loaf form 麵包烤模
loaf making 麵糰成形；麵包製作
loaf sugar 方糖；塊糖
lobster 龍蝦
lobster tail 去頭龍蝦
local variety 當地品種；地方品種
locker paper 冷藏包裝紙
locksoy 米粉條；細米粉絲（中國）
locust bean 刺槐豆；角豆 [= carob bean]
locust bean germ 角豆莢胚芽色素
locust bean gum 刺槐豆膠；槐豆膠；角豆膠 [= carob bean gum]
loft drying 風乾

log phase 對數〔生長〕期 [= logarithmic phase]
loganberry 羅甘莓；大楊梅；洛根莓
logarithmic growth phase 對數生長期；對數增殖期
logarithmic phase 對數〔生長〕期 [= log phase]
logey 瘦魚
logger 自動記錄器
loin 腰〔部〕；里脊肉；背腰肉；腰肉；背脊肉
loin butt 豬里脊肉
loin ham 背腰肉火腿；里脊肉火腿
lolled-up ends can 拉舌〔蓋〕罐
lollipop 棒棒糖
long chain 長鏈
long chain fatty acid 長鏈脂肪酸
long clam 沙土貝
long dough 發酵過久的麵糰
long-fin(ned) albacore 長鰭金槍魚
long-fin(ned) flying fish 長鰭飛魚
long-fin(ned) herring 鰳魚；長鰳
long-fin(ned) tuna 長鰭鮪；長鰭金槍魚 [= albacore (tuna)]
long grain rice 長粒米；維長粒米
long-keeping 可長期儲藏的；耐儲藏的；保久的 [= long-life]
long-keeping milk 保久乳
long-life 可長期儲藏的；耐儲藏的；保久的 [= long-keeping]
long-life milk 保久乳
long pepper 長辣椒；長番椒
long rooted mushroom 金針菇
long-set curd 緩凝乳〔凝〕塊
long setting 緩慢凝結；緩凝
long spear 長〔蘆〕筍頭；長〔蘆〕筍尖
long sponge 長時間發酵的麵糰
long tail anchovy 長尾鯷魚；鳳尾魚
long thread moss 髮菜
long tube evaporator 長管蒸發器
long tube vertical evaporator 垂直長管蒸發器
long yellow daylily 金針〔菜〕；黃花菜
longan 龍眼；桂圓 [= dragon eye; lungan]
longitudinal muscle 縱肌
longnosed sucker 胭脂魚；長吻亞口魚
loofah 絲瓜
loosehead lettuce 散葉萵苣
looseners 李子
loquat 枇杷 [= Japanese medlar]
lotus 蓮
lotus leaf 荷葉
lotus nut 蓮子 [= lotus seed]
lotus nut paste 蓮蓉
lotus rhizome 蓮藕
lotus root 蓮藕；蓮根
lotus seed 蓮子 [= lotus nut]
lotus seed paste 蓮蓉
lotus tuber 蓮藕
lovage 圓葉當歸；獨活草
lovage oil 獨活草油；圓葉當歸油
love apple 番茄
Lovibond colo(u)r 羅偉邦色度；羅維邦色度
Lovibond tintometer 羅偉邦色調計；羅維邦色調計
low acid food 低酸性食品
low alcohol 低酒精度
low-baking 低溫烘焙的
low calorie 低熱量
low calorie food 低熱量食品
low carbohydrate diet 低醣膳食；低碳水化合物膳食
low cholesterol diet 低膽固醇膳食
low cost food 廉價食品
low density lipoprotein (LDL) 低密度脂蛋白

low density lipoprotein cholesterol 低密度脂蛋白膽固醇
low-diet 粗食
low energy diet 低熱量膳食；低熱能膳食
low ester pectin 低酯果膠
low fat 低脂〔肪〕的
low fat cocoa 低脂可可粉
low fat diet 低脂膳食
low fat milk 低脂乳
low fat soy flour 低脂大豆粉
low fermentation 低層發酵
low fire 文火；微火
low lactose milk 低乳糖乳
low methoxyl pectin (LMP) 低甲氧〔基〕果膠
low-moisture food 低水分食品
low-pressure storage 低壓儲藏
low-proof 酒精度低的
low protein diet 低蛋白膳食
low protein flour 低筋麵粉；低蛋白質麵粉
low ratio cake 輕蛋糕；低成分蛋糕
low salt diet 低鹽膳食
low sodium diet 低鈉膳食
low sodium food 低鈉食品
low sodium milk 低鈉乳
low sugar 低糖度的
low temperature cooking 低溫烹調
low temperature evaporation 低溫蒸發
low temperature heat method 低溫加熱法
low temperature holding pasteurization 低溫長時巴氏殺菌；低溫長時巴斯德殺菌
low-temperature long-time pasteurization (LTLT pasteurization) 低溫長時巴氏殺菌；低溫長時巴斯德殺菌 [= holding pasteurization]
low-temperature long-time process (LTLT process) 低溫長時殺菌法
low temperature ripening 低溫熟成
low temperature storage 低溫儲藏；低溫冷藏
low wine 初餾酒；低度葡萄酒
low yeast 低層發酵用的酵母
lower fatty acid 低級脂肪酸
Lowry method 勞立法
lox 燻鮭魚；燻大麻哈魚
LTLT pasteurization (low-temperature long-time pasteurization) 低溫長時巴氏殺菌；低溫長時巴斯德殺菌
LTLT process (low-temperature long-time process) 低溫長時殺菌法
lubricant 潤滑劑
lubricating grease 潤滑脂
lubricating oil 潤滑油
lubrication 潤滑
lubricator 潤滑器
lucern(e) 苜蓿；紫花苜蓿 [= alfalfa]
luffa 絲瓜
luffanine 絲瓜鹼
lug box 果菜箱；運果箱
lug cap 四旋〔瓶〕蓋；牙蓋
lukewarm water 溫水
lumache medie 蝸牛通心粉
lumiflavin 光黃素
luminous bacteria 發光細菌
luminous fungus 發光真菌
lump 塊；團；結塊
lump sugar 結塊的糖
lumpfish 圓鰭魚
lumpfish caviar 疙瘩魚卵
lumpiness 結塊性；團塊
lunch 午餐；便餐
luncheon 午餐
luncheon meat 午餐肉；便餐肉
luncheonette 小餐館

lunchroom 餐廳；學校餐廳
lung 肺〔臟〕
lungan 龍眼；桂圓 [= dragon eye; longan]
lungfish 肺魚
lungo vermicelli coupe 長段的細麵條〔通心粉〕；長段維米切利〔通心粉〕
luparol 盧杷醇
lupeose 水蘇四糖；羽扇豆糖 [= stachyose]
lupine 羽扇豆；魯冰豆
lupinic acid 羽扇豆酸
lupulin 蛇麻素；啤酒花苦味素
lupulone 蛇麻〔蘆〕酮
lupulus 蛇麻
luscious 甘美的；甜的
lush 味美的；烈性酒
lutefish 路特鱈魚乾
lutein 黃體素；葉黃素
luteoskyrin 黃變米〔霉〕毒素
lyase 解離酶；裂解酶；裂合酶
lychee 荔枝 [= leechee; lichee; lichi; litchi]
lychee flavo(u)r 荔枝香精
lychee oily flavo(u)r 荔枝油香精
lycine 甜菜鹼
lycium 枸杞
lycopene 番茄紅素
lycophyll 番茄紫素
lycoxanthin 番茄黃質；番茄黃色素
lye 鹼液
lye peeling 鹼液去皮〔法〕
lye washing 鹼洗
lymph 淋巴
lymphatic 淋巴管
lymphatic system 淋巴系統
lymphatic vessel 淋巴管
lyoenzyme 細胞外酶；胞外酶
lyogel 水凝膠；液凝膠
lyolipase 胞外脂酶；可溶性脂肪酶
lyophile 親液物；親液膠體
lyophilic 親液的
lyophilic colloid 親液膠體
lyophilization 冷凍乾燥；升華乾燥；凍乾法 [= freeze drying]
lyophilizer 冷凍乾燥機；升華乾燥機 [= freeze dryer]
lyophobe 疏液物；疏液膠體
lyophobic 疏液的
lyophobic colloid 疏液膠體
lyosol 水溶膠；水懸膠體
lysate 溶菌產物；溶解產物
lysergic acid 麥角酸
lysin 細胞溶素
lysine 離胺酸；賴胺酸
lysine monohydrochloride 賴氨酸鹽酸鹽
lysis 溶胞〔作用〕；溶菌〔作用〕
lysogenic 溶原性的；細胞溶素生成的
lysolecithin 溶血卵磷脂
lysosome 溶酶體
lysozyme 溶菌酶；溶解酶
lytic enzyme 分解酶
lyxoflavin 來蘇黃素
lyxonic acid 來蘇糖酸
lyxose 來蘇糖
lyxoside 來蘇糖苷

M

MA storage (modified atmosphere storage) 調氣保藏；氣調儲藏
mabinlang 馬檳榔
macadamia nut 澳洲胡桃；昆士蘭〔堅〕果；夏威夷果 [= Australian nut; Queensland nut]
macaroncelli 粗通心麵；粗通心粉；細管狀通心粉
macaroni 通心粉；通心麵
macaroni flour 通心粉用麵粉；高筋麵粉
macaroon 蛋白杏仁甜餅
macassar nutmeg 銀肉荳蔻
macchiato 馬奇雅圖咖啡 [= caffe macchiato; espresso macchiato]
MacConkey's agar 麥康凱洋菜〔培養基〕；麥氏瓊脂〔培養基〕
MacConkey's medium 麥康凱培養基；麥氏培養基
mace 肉荳蔻；肉荳蔻衣；肉荳蔻皮；荳蔻香料
mace butter 肉荳蔻脂
mace oil 肉荳蔻油；肉荳蔻衣油
macedoine 涼拌蔬果；蔬菜凍；什錦蔬果
macerating enzyme 離解酶；浸軟酶
maceration 泡軟；浸軟；浸漬
machete 彎刀；砍甘蔗刀
mackerel 鯖〔魚〕
mackerel in oil 油浸鯖魚
mackerel pike 秋刀魚
macon 羊臘肉；羊培根肉
macroanalysis 常量分析；宏觀分析
macrocystis 巨藻；大昆布
macrodispersoid 粗粒分散膠體
macroelement 巨量元素
macroglobulin 巨球蛋白
macrograin 巨粒；粗晶粒
macrography 肉眼檢查
macromolecule 巨分子；高分子；大分子
macronutrient 巨量營養素；常量營養素
macroscopic algae 大型藻類
macrospore 大孢子
mad apple 茄子
mad cow disease 瘋牛症
madefaction 潤濕；濕潤
maderia 馬德拉甜酒
maderia cake 馬德拉海棉蛋糕
madia oil 紫苑油；麻迪菊油
madid 濕潤的；潮濕的
magenta 品紅；洋紅 [= fuchsin]
maggot 蛆
magma 糖糊；糖膏；糖晶漿
magnesium 鎂
magnesium carbonate 碳酸鎂
magnesium chloride 氯化鎂
magnesium chlorophyll 葉綠素鎂；天然綠3號

magnesium diglutamate 二谷氨酸鎂
magnesium gluconate 葡糖酸鎂
magnesium hydrogen phosphate 磷酸氫鎂
magnesium hydroxide 氫氧化鎂
magnesium hydroxide carbonate 鹼式碳酸鎂
magnesium lactate 乳酸鎂
magnesium oxide 氧化鎂
magnesium phosphate, tribasic 磷酸鎂
magnesium silicate 硅酸鎂；三硅酸鎂
magnesium stearate 硬脂酸鎂
magnesium sulfate 硫酸鎂；瀉鹽
magnetic interceptor 磁力捕集機；磁力除鐵機
magnetic stirrer 電磁攪拌器；磁力攪拌器
magnum 大酒瓶
mahewu 發酵玉米粥（南非）
mahogany 琴酒飲料；甜杜松子酒；稀白蘭地酒
mahseer 馬西亞魚；印度淡水大魚
maid 女羊；幼母羊
maid of honor 檸檬杏仁餡〔酪〕餅
maidenhair tree 白果樹；銀杏樹
maidenhair tree fruit 白果；銀杏
maigre 素食的；無肉的
maigre food 素食
Maillard reaction 梅納反應；美拉德反應；羰胺反應；胺羰反應 [= carbonyl-amine reaction; aminocarbonyl reaction]
main course 主菜
main dish 主菜
main fermentation 主發酵
maintenance 保養；維護
maintenance food 維生食物；維持生命的食物
maintenance medium 維持培養基
maize 玉米；玉蜀黍；粟米 [= corn]
maize bread 玉米麵包
maize cob 玉米穗軸；玉米棒子
maize distillery 玉米釀酒廠
maize ear 玉米穗
maize flake 玉米雪片；玉米脆片
maize flour 玉米粉
maize gluten 玉米蛋白
maize gluten meal 玉米蛋白粗粉
maize grit 碎玉米
maize groats 玉米粒
maize gruel 玉米粥；玉米糜
maize husk 玉米殼
maize husker 玉米去殼機
maize kernel 玉米粒
maize meal 玉米〔粗〕粉
maize mill 玉米碾磨機
maize oil 玉米油
maize popper 爆玉米花機
maize puff 爆玉米花
maize rice 脱皮玉米粒
maize sheller 脱粒機；玉米剝粒機
maize starch 玉米澱粉
maize sugar 玉米糖
maize syrup 玉米糖漿；澱粉糖漿 [= corn syrup]
maize whisky 玉米威士忌酒
maize yellow 玉米黃
maizena 玉米澱粉
malabar chestnut 馬拉巴栗
malabsorption 吸收不良
malachite green 孔雀石綠
malacodermous 果皮軟的
Malaga 瑪拉加白葡萄酒（西班牙）
malaxation 揉捏；揉和
malay bean 馬來菜豆
Malay fowl 馬來雞
malay kumquat 長葉金橘
maldigestion 消化不良

malic acid 蘋果酸
malic enzyme 蘋果酸酶
mallard 野鴨〔肉〕
mallorizer 瞬間巴氏殺菌器；馬拉里氏殺菌器
mallorizing 瞬間巴氏殺菌；馬拉里氏殺菌
malnutrition 營養不良
malo-lactic fermentation 蘋果酸乳酸發酵
malo-succinic fermentation 蘋果酸琥珀酸發酵
malodo(u)r 惡臭；難聞氣味
malodorant 惡臭物
malodorous 惡臭的
malonic acid 丙二酸
malt 麥芽
malt amylase 麥芽澱粉酶
malt beverage 麥芽飲料
malt bread 麥芽麵包
malt brewing 麥芽釀酒
malt carbohydrase 麥芽糖酶
malt culm 麥芽小根
malt diastase 麥芽澱粉酶
malt dust 麥芽粉；麥芽渣
malt extract 麥芽精；麥精；麥芽萃；麥芽〔浸〕膏；大麥糖漿
malt extract bath 麥芽汁浸出槽
malt flake 麥芽片
malt flour 麥芽粉
malt house 麥芽廠；製麥芽室
malt liquor 麥芽酒
malt mash 麥芽糖化醪
malt meal 麥芽〔粗〕粉
malt mill 麥芽碾機；麥芽粉碎機
malt sugar 飴糖；麥芽糖飴；麥芽糖
malt syrup 麥芽糖漿；飴糖
malt vinegar 麥芽醋
malt whisky 麥芽威士忌酒
malt wort 麥芽汁
maltase 麥芽糖酶
malted barley 發芽大麥；麥芽
malted milk 麥芽乳；麥乳精
malted milk powder 麥芽乳粉
malting 發麥芽；糖化；製造麥芽
maltitol 麥芽糖醇
maltitol syrup 麥芽糖醇糖漿
maltobiose 麥芽糖
maltodextrin 麥芽糖糊精；麥芽糊精
maltoflavin 麥芽黃素
maltogenic amylase 產麥芽糖澱粉酶
maltol 馬爾托環醇；麥芽酚
maltose 麥芽糖
maltose syrup 麥芽糖漿
maltoside 麥芽糖苷
maltotriose 麥芽三糖
maltster 麥芽製造者；製造麥芽的工人
malty 麥芽的；麥芽味的；有麥芽的
malty flavo(u)r 麥芽〔風〕味
malvidin 錦葵色素；錦葵定；二甲花翠素
malvin 錦葵色素苷；二甲花翠苷
mamaliga 玉米麵糊；玉米粥
mamma 乳房
mammae 乳汁
mammal 哺乳動物
mammary gland 乳腺
mammee 曼蜜蘋果
mammey 曼蜜蘋果
mammoth grader 轉筒式分級機；有孔轉筒分級機
mammoth wildrye 巨野黑麥
manatee 海牛
mandarin oil 橘子油
mandarin orange 橘子
mandarin peel 桔皮；乾果皮
mandarin(e) 桔；柑橘；温州蜜柑
mandatory standard 法定標準

manducate 咀嚼
manganese 錳
manganese chloride 氯化錳；二氯化錳
manganese citrate 檸檬酸錳
manganese gluconate 葡糖酸錳
manganese glycerophosphate 甘油磷酸錳
manganese hypophosphite 次磷酸錳
manganese sulfate 硫酸錳
mangetout 嫩豌豆；脆甜豌豆
mangler 絞肉機
mango 芒果
mango melon 柑瓜
mangosteen 山竹〔果〕；鳳果；倒捻子；莽吉柿
manifold 反芻動物第三胃；毛肚
manihot 木薯；樹薯 [=manioc; cassava]
manioc 木薯；樹薯 [=cassava; manihot]
manmose 甘露糖
manna 甘露〔蜜〕；嗎哪；精神食糧
mannan 聚甘露糖；甘露聚糖
mannase 聚甘露糖酶；甘露聚糖酶
mannitan 甘露糖酯；無水甘露醇
mannitol 甘露醇；甘露糖醇
mannitol fermentation 甘露〔糖〕醇發酵
mannitol salt agar 甘露〔糖〕醇食鹽洋菜〔培養基〕
mannonic acid 甘露糖酸
mannosan 聚甘露糖；甘露聚糖
mannose 甘露糖 [= seminose]
mannoside 甘露糖苷
manometer 壓力計
mantis shrimp 富貴蝦；蝦蛄
manual 手動的；手控的；手冊；指南
manure 糞肥；廄肥；肥料
maple 楓樹；槭樹
maple fondant 楓糖霜泥；楓樹軟糖料
maple honey 楓〔樹〕糖蜜；槭〔樹〕糖蜜
maple sap 楓樹液；槭樹液
maple sugar 楓〔樹〕糖；槭〔樹〕糖
maple syrup 楓〔樹〕糖漿；槭〔樹〕糖漿
marang 麵包果
maranta 竹竽；葛〔澱〕粉
marasca (cherry) 馬拉斯加櫻桃；馬拉斯金櫻桃；歐洲酸櫻桃 [= maraschino cherry]
maraschino 馬拉斯加櫻桃酒；黑櫻桃酒
maraschino cherry 馬拉斯加櫻桃；馬拉斯金櫻桃；歐洲酸櫻桃 [= marasca (cherry)]
marasmus 消瘦；衰弱
marbled meat 大理石紋脂肪的肉；霜降肉
marbling 大理石紋〔脂肪〕
marc 果渣；水果榨渣
Marc brandy 果渣白蘭地酒；葡萄榨白蘭地酒
marching ration 行軍乾糧
marchpane 杏仁軟糖；蛋白杏仁糖果
mare 母馬；母驢
margaric acid 十七〔烷〕酸 [= heptade-canoic acid]
margarine 瑪珈琳；人造乳酪；人造黃油；人造奶油；人造白脫；植物黃油；植物牛油 [= butterine; oleomargarine]
maria 江鱈
mariculture 海鮮養殖
marigold 萬壽菊；金盞花
marinade 醃料；滷液；滷汁；醋漬液
marinate 醃漬；調味料浸漬
marinated sardine 醃漬沙丁
marine algae 海藻
marine bacteria 海洋細菌
marine biology 海洋生物學
marine carp 海鯽魚
marine crab 海蟹

marine eel 海鰻
marine fish 海魚；海洋魚類
marine fisheries 海洋漁業
marine food 海產食品；海味；海鮮
marine oil 水產油〔脂〕
marine organism 海洋生物
marine plant 海生植物
marine product 水產品；海產品
maripa 棕櫚
marjoram 甘牛至；馬郁蘭；牛膝草
marjoram oil 甘牛至油；馬郁蘭油
market fish 市售魚
market milk 鮮乳；市乳；消毒牛乳 [= city milk]
marketable life 儲藏期限；保存期限；櫥存期限；商品壽命
marketing fresh 鮮銷
markhor 大捲角山羊；捻角山羊
marlin 馬林魚；旗魚
marmalade 橘皮果醬；柑橘皮絲果凍
marmite 砂鍋；有蓋湯鍋；燉鍋；酵母調味品
marron glace 蜜栗；蜜餞栗子
marrow 骨髓；果肉；西葫蘆；歐南瓜
marrow squash 西葫蘆 [= vegetable marrow]
marrowbone 髓骨
marrowfat pea 大縐皮豌豆；皺粒豌豆；大粒青豌豆
marsala 馬沙拉甜點；馬沙拉白葡萄酒；馬爾薩拉葡萄酒
marshmallow 棉花〔軟〕糖；藥蜀葵〔蜜餞〕
marsoon 白鯨
martini 馬丁尼酒；馬提尼酒
marumi kumquat 圓金柑；金桔
marzipan 杏仁〔蛋白〕軟糖；杏仁酥糖
marzipan cake 杏仁蛋糕
mash 醪；糖化醪；麥芽醪
mash goods 糖化原料
mash tub 糖化鍋；醪桶
mash tun 糖化桶；醪桶
mashed 磨碎的；搗碎的
mashed fruit 果泥；果漿
mashed potato 馬鈴薯泥；薯泥；薯蓉；土豆泥
mashed vegetable 蔬果泥；菜泥
masher 搗碎機；搗爛機
mashing 製醪；糖化；搗碎
mask crab 紅蟹
maslin 混穀麵包；混合物
mass poisoning 集體〔性食物〕中毒
mass production 大量生產
massecuite 糖膏 [= fullmasse]
mastic 乳香；乳香酒
mastika 乳香酒
mastitis 乳腺炎；乳房炎
mastitis milk 乳腺炎乳
mate 馬黛茶；巴拉圭茶 [= yerba mate]
maternized milk 母乳化牛乳
matfull 性成熟的鯡魚
matie 性未成熟的肥鯡魚；懷卵幼鯡魚
matrimony vine 枸杞 [= boxthorn]
Matsusaka beef 松阪牛肉
matsutake 松茸；松口蘑；松蕈
matte finish 糙面精整
maturation 成熟
mature wine 陳年酒
matured dough 成熟麵糰；發酵完畢的麵糰
matured koji 熟麴
maturing agent 陳化劑；熟化劑；催熟劑；小麥粉處理劑
maturity 成熟〔度〕
maturometer 成熟〔度測定〕計；果蔬成熟計
matzo 無酵餅；無發酵麵包
maw 魚鰾；鰾；胃；真胃；肚；嗉囊

mawkish 沒風道的；風味不良的
maximum ration 最大日糧
maximum safety level 最大安全量
maximum tolerated dose 最大耐量
May hawthorn 夏花山楂
mayonnaise 蛋黃醬；蛋黃沙律醬
Maythorn 山楂 [= haw; hawthorn]
mazi morado colo(u)r 紫玉米色素
mazzard cherry 歐洲甜櫻桃
McIntosh 麥金塔蘋果
mea slicer 肉切片機；切肉片機
mead 蜜酒飲料；蜂蜜酒
meadow clover 紅苜蓿
meadow mushroom 洋蘑菇；野外鮮蘑菇
meadow puffball 革生瓦賽菌 (蕈類)
meagre 馬葛爾魚
meal 粗粉；膳食；一餐
meal planning 膳食計劃
meal ticket 飯票；餐券
mealie rice 脱皮玉米粒；碎米
mealie(s) 玉米；玉蜀黍
mealiness 粉狀；粉性
mealtime 進餐時間
mealy 粉狀的；含粉的；粉質的
measuring cup 量杯
measuring cylinder 量筒
measuring filler 計量裝填機；定量裝填機
measuring flask 量瓶
measuring rod 量桿
measuring spoon 量匙
meat 肉；肉類；果肉
meat analog 人造肉
meat and drink 飲食
meat ball 肉丸
meat bar 乾肉棒；塊狀肉乾
meat beater 搥肉器
meat by product 肉副產品
meat cattle 肉用牛
meat chopper 絞肉機
meat curing 肉醃製；醃肉
meat cut 肉塊；分割肉
meat cutter 絞肉機；切肉機
meat dicer 切肉丁機
meat diet 肉食；葷食
meat emulsion 煉合肉；肉糜
meat extender 肉增量劑
meat extract 肉萃；肉類抽提物
meat flavo(u)r 肉風味
meat grinder 絞肉機
meat hygiene 肉品衛生
meat infusion 肉浸液；肉浸汁
meat inspector 肉類檢驗員
meat juice 肉汁
meat loaf 肉糕；肉羅夫
meat mincing machine 碎肉機
meat mixer 混肉機；肉餡攪拌器
meat modifier 肉質改良劑；肉類改良劑
meat packer 肉類產業〔者〕
meat paste 肉餡；肉醬
meat pattie 肉餅
meat patty 肉餅
meat pie 肉派；肉餡餅
meat plasma protein 肉漿蛋白質
meat pocket 肉〔餡〕餃子
meat pounder 敲肉木槌
meat processing 肉品加工
meat product 肉〔類〕製品；肉品
meat safe 菜櫥；肉櫃；食品櫥
meat shredder 粗碎肉機；粗絞肉機
meat slice 肉片
meat stew 燉肉
meat strip 肉絲
meat sugar 肉糖；肌糖
meat tea 午茶；有肉食的茶點
meat tenderizer 嫩肉劑；肉嫩化劑

meat thermometer 肉溫度計
meat type 瘦肉型
meaty 肉〔類〕的；肉厚的
meaty food 似肉食品；人造肉食品
mechanical milking 擠奶機
mechanical peeling 機械去皮
mechanical press 機械壓榨機
mechanical sorting 機械分級；機械選別
mechanical stunning 機械擊昏
medic 苜蓿
medical inspection 檢疫；健康檢查
medicated diet 藥膳；食療
medicated food 加藥食品
medicated liquor 藥酒
medicinal crop 藥用作物
medicinal herb 藥草
medicinal liquor 藥酒
medicine 藥物
medium 培養基；介質；半熟的；中等的
medium acid food 中酸性食品
medium fat cocoa 中脂可可粉
medium flour 中筋麵粉
medium lethal dose 半數致死劑量 [= 50% lethal dose; LD_{50}]
medium salted 中度鹽漬的；中等鹽漬的
medium salted fish 中等鹹度的魚；中度鹹魚
medium strength flour 中筋麵粉；中力粉
medium wheat 中等硬度的硬粒小麥
medlar 波斯山渣；歐渣〔果〕；歐洲山渣
Medoc 美杜克紅葡萄酒；美多克紅葡萄酒
medulla 骨髓；脊髓
medusa 海蜇；水母
megaloblastic an(a)emia 幼巨紅細胞性貧血；巨紅血球貧血
megasse 蔗渣；甘蔗渣 [= bagasse]
melabar gourd 黑子南瓜；無花果葉瓜
melada 糖膏
melampyrin 衛矛〔己六〕醇；半乳糖醇
melangeur 混合機；攪和器
melanin(e) 黑色素
melanoid 類黑素
melanoidin 梅納汀；類黑精；梅納環丁
melanoidin malt 焦麥芽
melassigenesis 成〔糖〕蜜作用 [= molasses formation]
melba toast 脆薄吐司
melberry 桑椹
melezitose 松三糖
melibiose 蜜二糖
melicitose 松三糖
melissa 蜂花；蜜蜂花
melissic acid 蜂花酸；三十烷酸
melitose 蜜里糖
melitriose 蜜三糖；棉子糖 [= raffinose]
melizitose 松三糖
Melle process 米氏〔發酵〕法；梅勒〔發酵〕法
melliferous 產蜜的；含蜜的；甜的
mellorine 無乳油冰淇淋；植物油脂製的冰淇淋；低乳脂冰淇淋
mellow 軟化；醇化；成熟的；陳化的；甘美的
melon 甜瓜；香瓜
melon field 瓜田；瓜園
melon flavo(u)r 瓜香香精
melon seed 瓜子
melon tree 番木瓜
melonal 西瓜醛；甜瓜醛
melt-flake method 碎溶技術
melted 熔化的；融化的

melter 熔化鍋
melting point 熔點
meltwater 冰雪溶化成的水；溶雪水
membrane 膜
membrane filter 膜濾器
membrane filtration 膜〔過〕濾
membrane permeability 膜滲透性；膜通透性
membrane separation 薄膜分離
menadione 維生素K
menadione sodium bisulfite 維生素K_3；亞硫酸氫鈉甲萘醌
menhaden 鯡魚
Mentha 薄荷屬
Mentha arvensis 亞洲薄荷
Mentha arvensis oil 薄荷油；亞洲薄荷油
Mentha canadensis 北美薄荷
Mentha rotundifolia 圓葉薄荷
Mentha silvestris 銀薄荷
Mentha viridis 綠薄荷；荷蘭薄荷；留蘭香
menthol 薄荷醇；薄荷腦
menthone 薄荷酮
menthyl acetate 乙酸薄荷酯
menu 菜單；菜譜
mephitis 臭氣；惡臭
mercaptan 硫醇 [= thiol]
mercaptoamino acid 硫醇胺基酸；巰基胺基酸
3-mercapto-2-butanone 3-巰基-2-丁酮
2-mercaptopropionic acid 2-巰基丙酸
mercurial thermometer 水銀溫度計
mercurialism 汞中毒；水銀中毒 [= hydrargyrism]
mercury 汞；水銀
meringue 蛋白糖飾；蛋糖霜
meromyosin 裂肌球蛋白；酶解肌球蛋白
mescal 麥斯卡爾酒
mesenteric fat 腸脂
mesentery 腸繫膜
mesh 篩孔；篩眼
mesh filter 網狀濾器；篩網過濾器
mesh number 篩孔目數；篩號；篩目
mesh screen 網篩
mesh sieve 網篩
mesh size 篩孔〔大小〕
mesobacteria 中溫細菌
mesocarp 中果皮
mesomorph 中等身材；體育型體質
mesopelagic fish 中層魚類
mesophile 中溫生物；嗜溫生物
mesophilic 嗜溫的
mesophilic bacteria 中溫細菌；嗜溫細菌
mesophyll 葉肉
mesorectum 直腸繫膜
mesquite 螺絲豆；螺旋豆
mess 伙食；菜餚；會餐
mess gear 餐具
mess kit 餐具
mess rack 食品櫥
messhall 餐館；食堂
messhouse 餐館；食堂
messtin 軍用飲盒；飲盒
metabisulfite 偏酸式亞硫酸鹽
metabolic activity 代謝活性
metabolic pathway 代謝途徑
metabolic product 代謝產物
metabolic quotient 代謝商
metabolic rate 代謝〔速〕率
metabolism 新陳代謝；代謝〔作用〕
metabolite 代謝物
metafilter 層濾機
metal can 金屬罐
metal container 金屬容器

metal detector 金屬偵察器；金屬偵測器；金屬探測器
metal eliminator 除鐵機；去鐵機
metallic can 金屬罐
metallic flavo(u)r 金屬氣味；鐵腥味
metallic paper 鋁箔紙；金屬塗布紙
metallic taste 金屬味
metalloenzyme 金屬酶
metalloprotein 金屬蛋白質
metaphosphate 偏磷酸鹽
metaphosphoric acid 偏磷酸
metaprotein 變性蛋白質
metatartaric acid 偏酒石酸
metatrophic bacteria 腐生細菌
metatrophy 腐生營養；營養異常
methane 甲烷
methane fermentation 甲烷發酵
methanol 甲醇；木醇；木精 [= methyl alcohol; wood alcohol; wood spirit]
methanol poisoning 甲醇中毒
metheglin 蜂蜜酒
methemoglobin 鐵(III)血紅蛋白；高鐵血紅蛋白；氧化型血紅蛋白
methional 3-甲硫基丙醛
methionine 蛋胺酸；甲硫胺酸；甲硫基丁胺酸
methionol 3-甲硫基丙醇
methose 甲糖
p-methoxybenzaldehyde 大茴香醛；對甲氧基苯甲醛
methoxylase 甲氧基酶
2-methoxypyrazine 2-甲氧基吡嗪
methyl acetate 乙酸甲酯；醋酸甲酯
methyl acetyl ricinolate 乙酰蓖麻油酸甲酯
4-methyl acetophenone 對甲基苯乙酮
methyl alcohol 甲醇；木醇；木精 [= methanol; wood alcohol; wood spirit]
methyl amyl ketone 甲基戊基酮
p-methyl anisole 對甲酚甲醚；對甲基大茴香醚；對甲基苯甲醚
methyl anthranilate 鄰氨基苯甲酸甲酯
p-methyl benzaldehyde 對甲基苯甲醛
methyl benzoate 苯甲酸甲酯
methyl blue 甲基藍
methyl butyrate 丁酸甲酯
methyl cellulose (MC) 甲基纖維素
α-methyl cinnamaldehyde α-甲基〔肉〕桂醛
methyl cinnamate 桂〔皮〕酸甲酯；肉桂酸甲酯
methyl cobalamin 甲基鈷維生素
6-methyl coumarin 6-甲基香豆素
methyl cyclopentenolonne 甲基環戊烯醇酮
methyl dihydrojasmonate 二氫茉莉酮酸甲酯
methyl ester of rosin 松香甲酯
methyl ethyl cellulose (MEC) 纖維素甲乙醚
methyl eugenol 丁香酚甲醚
methyl formate 甲酸甲酯
2-methyl-3-furanthiol 2-甲基-3-巰基呋喃
methyl gallate 沒食子酸甲酯；五倍子酸甲酯
methyl green 甲基綠
methyl heptenone 甲基庚烯酮
methyl hesperidin 甲基橙皮苷；甲基橘皮苷；維生素P
methyl hexanoate 己酸甲酯
methyl 2-hexenoate 2-己烯酸甲酯
methyl histamine 甲基組〔織〕胺
α-methyl ionone α-甲基紫羅蘭酮
methyl isobutyrate 異丁酸甲酯
methyl isoeugenol 異丁香酚甲醚；甲基異丁香酚

methyl mercaptan 甲硫醇
methyl 2-methylbutyrate 2-甲基丁酸甲酯
methyl 3-methylthiopropionate 3-甲硫基丙酸甲酯
methyl myristate 肉荳蔻酸甲酯
methyl 2-nonenoate 2-壬烯酸甲酯
methyl 2-nonynoate 辛炔羧酸甲酯
methyl octanoate 辛酸甲酯
methyl 2-octynoate 2-辛炔羧酸甲酯；庚炔羧酸甲酯
methyl orange 甲基橙 [= helianthin(e)]
methyl paraben 對羥苯甲酸甲酯
2-methyl pentanal 2-甲基戊醛
methyl pentanoic acid 甲基戊酸
2-methyl-2-pentenoic acid 2-甲基-2-戊烯酸；草莓酸
3-methyl pentanoic acid 3-甲基戊酸；酐焓酸
methyl phenethyl ether 甲基苯乙醚
methyl phenylacetate 苯乙酸甲酯
methyl 4-phenylbutyrate 4-苯丁酸甲酯
methyl phenylcarbinyl acetate 乙酸蘇合香酯
methyl polysilicone 甲基聚矽氧烷；聚甲基硅氧
methyl red 甲基紅
methyl salicylate 柳酸甲酯；水楊酸甲酯；冬青油；冬綠油 [= wintergreen oil]
methyl silicone 甲基矽氧烷
methyl violet 甲基紫
methyl β-naphthyl ketone 甲基 β-萘酮
methylamine 甲胺
methylated spirit 含甲醇酒精；甲基化酒精
methylation 甲基化
2-methylbutyl isovalerate 異戊酸-2-甲基丁酯
2-methylbutyradehyde 2-甲基丁醛
2-methylbutyric acid 2-甲基丁酸
methylene blue 甲烯藍；亞甲〔基〕藍；美藍
methylene chloride 二氯甲烷
5-methylfurfural 5-甲基糠醛
methylhexanone 甲基己基酮
methylhistidine 甲基組胺酸
methylparaben 對羥基苯甲酸甲酯；尼泊金甲酯
2-methylpyrazine 2-甲基吡嗪
methyltheobromine 甲基可可鹼；咖啡因；咖啡鹼
3-methylthiobutanal 3-甲硫基丁醛
o-methylthiophenol 鄰甲硫基苯酚
2-methylundecanal 甲基壬乙醛
2-methylvaleric acid 2-甲基戊酸
metmyoglobin 鐵(III)肌紅蛋白；高鐵肌紅蛋白；正鐵肌紅蛋白；氧合肌紅蛋白
Mexican sage 墨西哥鼠尾草
mezzanelli 梅扎奈里〔花式〕通心粉
mezzani 梅扎尼〔花式〕通心粉
mezzani rigati 梅扎尼里格體〔波紋狀〕通心粉
MIC (minimum inhibitory concentration) 最低抑制濃度
micella(e) 微膠粒；膠粒；膠束；膠態離子；分子團；微團 [= micelle]
micelle 微膠粒；膠粒；膠束；膠態離子；分子團；微團 [= micella(e)]
micellisation 微膠粒化；膠化形成
michelia alba (flower) concrete 白蘭〔花〕浸膏
michelia alba flower oil 白蘭花油
michelia alba leaf oil 白蘭葉油
Mickey Finn 蒙汗酒；加催眠藥飲料
micro calorimeter 微量熱量計；微卡計
micro-constituent 微量成分

micro Kjeldahl method 微量克耳達法；微量凱氏〔測定氮〕法
microaerobe 微嗜氧細菌
microaerophile 微嗜氧細菌；微需氧性細菌
microaerophilic bacteria 微嗜氧細菌
microanalysis 微量分析
microbalance 微量天平
microbe 微生物
microbial 微生物的
microbial assay 微生物測定法
microbial contamination 微生物污染
microbial degradation 微生物降解
microbial flora 微生物群落；微生物區系；微生物菌叢 [= microflora]
microbial population 微生物族群；微生物數；微生物群體
microbial toxin 微生物毒素
microbicide 殺菌劑；殺微生物劑
microbiological 微生物學的
microbiological assay 微生物測定法
microbiological product 微生物產物；微生物產品；微生物製劑
microbiologist 微生物學家
microbiology 微生物學
microbion 微生物
microbiosis 微生物性感染
microbiotic 抗菌素
microcapsule 微膠囊
microchemical analysis 微量化學分析
Micrococcus 微球菌〔屬〕；小球菌〔屬〕
microcolony 微菌落；小菌落
microcrystal 微晶
microdispersoid 微粒分散膠體
microelement 微量元素
microencapsulated enzyme 微膠囊化酵素
microencapsulation 微膠囊化
microfibril 微纖維
microfilament 微絲
microfiltration 微〔量〕過濾
microflora 微生物群落；微生物區系；微生物菌叢 [= microbial flora]
microgerm 微生物
microgram (μg) 微克
microliter (μl) 微升
micrometer (μm) 微米 [= micron]
micromethod 微量法
micron 微米 [= micrometer]
micronutrient 微量營養素
microorganism 微生物
microphyte 微植物；細菌
microscope 顯微鏡
microscope slide 載玻片
microscopic count 顯微鏡數菌法
microscopic examination 顯微鏡檢查；鏡檢
microsome 微粒體
microspore 小孢子
microstructure 微結構；顯微結構
microtitration 微量滴定
microtubule 微管
microvilli 微絨毛
microwavable food 微波食品
microwave 微波
microwave cooking 微波烹調
microwave drying 微波乾燥
microwave heating 微波加熱
microwave irradiation 微波輻射
microwave oven 微波爐
microzyme 酵母菌
mid-joint 中翅；翅板 [= middle section; wing flat]
middle gut 中腸；結腸
middle section 中翅；翅板 [= mid-joint; wing flat]
middling flour 帶麩粗麵粉；粗麵粉；麥心粉

middlings 帶麩粗〔麵〕粉；粗〔麵〕粉；麥心
midgut 中腸
midriff 橫膈膜 [= diaphragm]
migratory fish 回游魚
milch 乳的；奶的；產乳的
milch cow 乳牛；奶牛 [= milk cow]
mild cure 輕醃
mild disinfectant 溫和消毒劑
mild fire 文火
mild flavo(u)r 柔和風味
mild milk cap 細質乳菇
mild smoke 輕燻
mildew 黴；霉
mildew proof 防黴；防霉
mildewed 發霉的；長霉的
military foodstuff 軍糧；軍用食品
milk 乳；牛奶；牛乳
milk-and-water 淡而無味的；無味的；水稀釋牛乳
milk beverage 乳〔製〕飲料
milk bottle 乳瓶；奶瓶
milk bottle filler 乳裝瓶機
milk cattle 乳牛；奶牛
milk chocolate 牛乳巧克力；牛奶朱古力
milk clarifier 乳淨化機；淨乳機
milk clotting enzyme 凝乳酶
milk coagulant 凝乳劑
milk coagulating enzyme 凝乳酶
milk cow 乳牛；奶牛 [= milch cow]
milk cream 乳油；奶油
milk curd 凝乳；乳凝塊
milk depot 擠乳站；收奶站
milk diet 乳餐；牛奶餐
milk drink 乳飲料；牛奶飲料
milk fat 乳脂〔肪〕
milk food 乳製食品
milk gauge 乳比重計
milk house 牛乳處理廠；牛奶場
milk mushroom 乳菇
milk plant 乳品工廠
milk powder 乳粉；奶粉
milk product 乳產品；牛奶產品
milk protein 乳蛋白質
milk separation 乳脂分離；牛奶脱脂
milk serum 乳清
milk shake 奶昔
milk solids 乳固形物；牛奶固形物；乳固體
milk stone 乳石
milk strainer 牛奶過濾器
milk sugar 乳糖 [= lactose]
milk tank 乳桶；牛奶儲槽
milk thistle 乳薊果
milker 擠乳器；擠乳者
milkfish 虱目魚；遮目魚；乳香魚
milking 擠乳；擠奶；榨奶
milking machine 擠乳機；擠奶機
milkmaid 擠奶的女工
milkman 售奶〔工〕人；送奶〔工〕人
milkotester 乳脂肪分析儀
milky fluid diet 牛奶流質膳食；乳〔狀〕流質膳食
mill 碾磨；磨碎；磨粉機；粉碎機；磨坊
mill aspirator 吸風分離器
mill bannock 麥製薄餅；燕麥粉製糕餅
mill surface 碾磨面
milled barley 碾白大麥
milled rice 白米；去殼米
miller 磨粉機；碾磨業〔者〕；製粉廠
miller wheat 普通小麥；軟質小麥
millerator 磨粉分離器
millet 粟；小米；稷
milligram (mg) 毫克
milliliter (ml) 毫升；公撮
millimeter (mm) 毫米

milling 碾磨；研磨；磨碎
milling machine 磨粉機
milling separator 磨粉篩；麵粉篩分機
millipore filter 微孔〔過〕濾器
Millon test 米隆〔酪胺酸〕試驗
millstone 磨石；磨盤
millstone mill 石磨；磨石碾磨機
milo 西非高粱；高粱；蜀黍
milo grit 碎高粱；粗高粱粉
mimosa 含羞草
mince 絞碎；切碎；絞碎物；肉糜
mince collops 碎肉；絞碎肉餡
mince pie 肉餡餅；肉餡派
minced beef 免治牛肉；碎牛肉；絞碎牛肉
minced chicken 雞肉糜；雞蓉
minced fish (meat) 魚漿；魚〔肉〕糜
minced ham 絞肉火腿
minced luncheon meat 絞肉午餐肉
minced meat 碎肉；肉糜；肉漿；肉餡 [= mincemeat]
minced meat ball 肉丸
minced pork ball 豬肉貢丸；貢丸
mincemeat 碎肉；肉糜；肉漿；肉餡 [= minced meat]
mincer 切碎機；搖潰機；碎肉機；絞肉機
mincing 切碎；搖潰；絞碎
mincing machine 切碎機；搖潰機；碎肉機；絞肉機
mine salt 岩鹽
mineral 礦物質
mineral oil 石蠟油；礦物油
mineral salts 無機鹽；礦物質
mineral supplementation 礦物質增補；增補礦物質
mineral water 礦泉水
minestrone 菜麵濃湯；蔬菜通心粉湯
mingler 混合機；混和機
miniature fermenter 小型發酵器
minimum daily requirement 最低日需〔要〕量；每日最低需要量
minimum inhibitory concentration (MIC) 最低抑制濃度
minimum medium 最低營養素培養基；基本培養基；貧乏培養基
minimum thermal process 最低熱殺菌條件
minnow 頜鬚鮈；鯉科小魚
minor element 微量原素；次要元素
mint 薄荷；薄荷糖
mint essence 薄荷香精
mint flavo(u)ring 薄荷調味料；薄荷香料
mint julep 薄荷茉莉酒
mint leaves 薄荷葉
mint oil 薄荷油
mint sauce 薄荷醬
minute steak 薄牛排；快熟薄肉排
mirabelle 米拉別〔里〕李子；布拉斯李
miracle berry 神奇果；神秘果
mirepoix 植物性調味香料
mirror carp 鏡鯉
misbleed 放血不淨
miscella 溶劑混合油；油與溶劑的混合物；混合油
miscella refining 混合油精煉；溶劑混合油精煉
miscible 互溶的；可混的
mishmash 混雜物；雜燴
miso 味噌；日本豆〔麵〕醬
mispelie 蜜思葡萄酒
mist spray 噴霧
mistorizer 霧化器；水霧化器
miticide 殺蟎劑
mitochondria 粒線體；線粒體
mitten crab 中華絨螯蟹；毛蟹
mix 混合；混料；混合物

mixed bed 混合床
mixed culture 混合菌種；混合培養〔物〕
mixed culture fermentation 多菌發酵；混合發酵
mixed curing 混合醃；混合鹽醃製
mixed diet 混合膳食
mixed drink 混合飲料
mixed fermentation 混合發酵
mixed fruit 什錦水果
mixed fruit jam 什錦果醬
mixed fruit jelly 什錦果凍；什〔錦水〕果啫喱
mixed glyceride 混酸甘油酯
mixed jam 什錦果醬
mixed juice 綜合果汁
mixed juice drink 綜合果汁飲料
mixed milk 混合乳
mixed packing 混合包裝；混合填充
mixed pickle 什錦醬菜；什錦醃菜；什錦泡菜
mixed salad 什錦沙拉；什錦色拉；什錦沙律
mixed syrup 混合糖漿
mixed triglyceride 混酸三甘油酯
mixed vegetable 什錦蔬菜
mixer 混合器；攪拌器
mixer-beater 攪打混合機；混合打漿機
mixer pasteurizer 攪拌式巴氏殺菌器
mixing 混合；攪拌
mixing agitator 混合攪拌機
mixing arm 攪拌機槳葉
mixing bowl 攪拌缽；混料罐
mixing chamber 混合室
mixing device 混合裝置；混合器
mixing granulation 混合造粒
mixing paddle 攪拌槳葉
mixing time 混合時間；攪拌時間
mixograph 糰性測定儀；麵粉調混性記錄儀；揉混儀
mixture 混合物；混合體
mobile catering 餐車業務；流動餐館膳食
mobile dough 軟麵糰；稀麵糰
mobile phase 流動相
mocha 摩卡咖啡；穆哈咖啡 [= caffe mocha]
mock cutlet 免治肉排；碎肉做的肉排
mock milk 配製牛乳
mock strawberry 蛇莓
modified atmosphere packaging 調氣包裝；氣調包裝
modified atmosphere storage (MA storage) 調氣保藏；氣調儲藏
modified cellulose 改性纖維素
modified diet 調整膳食
modified milk 調製乳
modified soybean phospholipid 改性大豆磷脂
modified starch 修飾澱粉；改性澱粉；變性澱粉
modifier 調節劑；改良劑；修飾劑
modifying agent 改良劑；修飾劑
moist 潮濕的；濕潤的
moist air 濕空氣
moist cabinet 濕氣箱
moist chamber 保濕室
moist curing 潮濕醃製
moist heat 濕熱
moist heat sterilization 濕熱滅菌〔法〕
moist steam 濕蒸汽；飽和水蒸汽
moistener 增濕器；增濕劑
moisture 水分；濕氣
moisture absorption 吸濕
moisture absorption isotherm 等溫吸濕曲線
moisture barrier 防潮層

moisture content 含水量；水分含量
moisture desorption isotherm 等溫脱濕曲線
moisture equilibrium 濕度平衡；水分平衡；均濕
moisture-free 乾的
moisture meter 水分計
moisture permeability 透濕性；透水汽性
moisture-proof 防潮；抗濕
moisture-proof packaging 防潮包裝；防濕包裝
moisture regain 回潮
moisture resistant 防潮；抗濕
moisture-retaining agent 保水劑；保濕劑；濕潤劑
moisture retention 保水性；保濕力
moisture sorption isotherm 等溫吸濕曲線
moisty 潮濕的
mojama 鹹金槍魚乾
Mojonnier deoiler 莫氏提香精油機；莫瓊尼爾提香精油機
Mojonnier fat test 莫氏乳脂試驗；莫瓊尼爾乳脂肪測定法
Mojonnier milk test 莫氏乳品試驗器；莫瓊尼爾乳品測定器
Mojonnier tube 莫氏管；莫瓊尼爾管
Mojonnier visco(si)meter 莫氏黏度計；莫瓊尼爾黏度計
molar concentration 莫耳濃度；摩爾濃度
molar solution 莫耳溶液；摩爾溶液
molarity 莫耳濃度；摩爾濃度
molasses 糖蜜；廢糖蜜
molasses formation 成〔糖〕蜜作用 [= melassigenesis]
molasses mash 糖蜜醪；廢糖蜜漿
molasses sugar 糖蜜塊
mold 黴菌；霉菌；模具 [= mould]
mold amylase 黴菌澱粉酶
mold bran 麩〔皮〕麴
mold culture 黴菌培養〔物〕
mold inhibitor 黴菌抑制劑；防霉劑
mold mycelia 黴菌菌絲體
mold polysaccharide 黴菌多糖類
mold proofing 防黴
mold protease 黴菌蛋白酶
mold rice 米麴
mold starter 黴菌元；種麴
mole 莫耳
molecular formula 分子式
molecular weight 分子量
molecule 分子
mollifier 軟化劑；軟化器
mollusc 軟體動物
mollusk 軟體動物
molybdenum 鉬
monascorubin 紅麴紅素
monascorubramine 紅麴紅胺；紅麴紫
monascus 紅麴菌；紅麴霉
Monascus anka 赤紅麴菌；赤紅麴霉
monascus colo(u)r 紅麴〔霉〕色素；紅麴米
Monascus purpureus 紫紅麴菌；紫紅麴霉
monellin 莫內林〔天然蛋白甜味劑〕；蒙那靈〔天然蛋白甜味劑〕
Mongolian cattle 蒙古牛
monkey fruit 麵包果
monkey head mushroom 猴頭菇
monkey nut 落花生；花生 [= earthnut; goober; groundnut; peanut]
monkfish 琵琶魚
mono-and diglycerides 單雙甘油酯
monoacid 一元酸
monoamino acid 單胺基酸

monoammonium glutamate 谷氨酸〔一〕銨
monoammonium glycyrrhizinate 甘草酸〔一〕銨
monoarachin 甘油〔一〕花生酸酯
monocalcium phosphate 磷酸〔一〕鈣
monocell culture 單細胞培養
monocular microscope 單筒顯微鏡
monoenoic acid 單雙鍵酸；單烯酸
monoenoic fatty acid 單雙鍵脂肪酸；單烯脂肪酸
monogastric animal 單胃動物
monoglyceride 甘油〔一〕酸酯；一酸甘油酯
monoglyceride citrate 檸檬酸單甘油酯
monoglyceryl ester 單酸甘油酯；一甘油酯
monoglycol ester 一甘醇酯
monohydrate 一水合物；一水化物
monohydroxy alcohol 一元醇
monolayer 單層
monomer 單體
monophagia 偏食；單食癖
monophenol oxidase 一元酚氧化酶；單酚氧化酶
monoplast 單細胞生物
monopotassium glutamate (MPG) 谷氨酸〔一〕鉀；麩胺酸〔一〕鉀
monosaccharide 單糖 [= monose; monosaccharose]
monosaccharose 單糖 [= monose; monosaccharide]
monose 單糖 [= monosaccharide; monosaccharose]
monoside 單苷
monosodium aspartate 天冬胺酸〔一〕鈉
monosodium fumarate 反丁烯二酸〔一〕鈉；富馬酸〔一〕鈉
monosodium glutamate (MSG) 味精；味素；麩氨酸鈉；麩胺酸〔一〕鈉；谷氨酸〔一〕鈉 [= aginomoto; sodium glutamate]
monosodium succinate 琥珀酸〔一〕鈉
monospore 單孢子
monostarch phosphate 單澱粉磷酸酯；磷酸單澱粉酯
monounsaturated fatty acid 單不飽和脂肪酸
monovitaminosis 單一維生素缺乏症；單種維生素缺乏症
moon cake 月餅
moonfish 眼鏡魚；翻車魚
moorcock 雄紅松雞；雄蘇格蘭雷鳥
moorfowl 紅松雞；蘇格蘭沙雞；蘇格蘭雷鳥
moorhen 雌紅松雞；雌蘇格蘭雷鳥
morat 桑葚蜂蜜汁；桑葚蜂蜜酒
Morchella 羊肚菌屬
morel 羊肚菌；龍葵；草笠竹
morel fungus 羊肚蕈；羊肚菌
morello (cherry) 莫櫻桃；莫利洛黑櫻桃
morning milk 晨乳；早晨擠的乳
moromi 醪
morphine 嗎啡
morpholine 嗎啡
morpholine fatty acid salt 嗎啡脂肪酸鹽
morpholine fatty acid salt fruit wax 嗎啡脂肪酸鹽果蠟
mortadella 摩泰臺拉香肚（生燻）
mortal rigor 死後僵直；死後強直；僵直；死僵
mortar 研缽
morus 桑
mosaic puffball 泡囊狀巾馬勃（蕈類）
Moselle 莫塞爾白葡萄酒；摩澤爾白葡萄酒

moss 苔；青苔；蘚；苔蘚食品
mossberry 摩西果醬；蔓越橘
most probable number (MPN) 最確數；最大可能數
moth 蛀蟲；蛾
moth ball 萘丸
mother culture 種菌；菌母；種麴
mother juice 初榨汁；頭榨汁
mother liquor 母液
mother milk 母乳
mother starter 菌元；種用酵母
mother vinegar 醋母；種醋
mothproofing agent 防蛀劑；防蛾劑
mottle 斑點；斑紋；雜色
mottled butter 不勻色乳酪；斑點奶油
Mou Tai (Chiew) 茅台酒
mould 黴菌；霉菌；模具 [= mold]
mould amylase 黴菌澱粉酶
mould bran 麩〔皮〕麴
mould inhibitor 黴菌抑制劑
mould mycelia 黴菌菌絲體
mould polysaccharide 黴菌多糖類
mould proofing 防黴
mould protease 黴菌蛋白酶
mould starter 黴菌元；種麴
mountain cowberry 越橘
mountain cranberry 越橘
mountain herring 白鮭
mountain rye 山黑麥
mountain spinach 法國菠菜
mousse 木斯；慕思(泡沫狀奶油甜點)
mouth 嘴；口
mouth feel(ing) 口感
mozzarella 意大利水牛芝士
MPN (most probable number) 最確數；最大可能數
MSG (monosodium glutamate) 味精；味素；麩胺酸一鈉；谷胺酸一鈉 [= aginomoto; sodium glutamate]
mucedin 穀膠蛋白；麵筋蛋白
mucic acid 黏酸；半乳糖二酸
mucification 黏液化
mucigen 黏蛋白原；黏蛋白元；黏液素元
mucilage 黏漿；黏液；黏膠
mucilaginous gum 植物黏膠；黏質樹膠
mucin 黏蛋白；黏液素
mucinase 黏蛋白酶；黏多糖酶
mucinogen 黏蛋白原；黏蛋白元；黏液素元
mucinoid 類黏蛋白；類黏液素
muck 糞肥；腐植質；垃圾
muckamuck 食物
mucoglobulin 黏球蛋白
mucoid 類黏蛋白；黏液狀的
mucoid colony 黏液狀菌落
mucoid substance 類黏蛋白物質
mucopeptide 黏肽
mucopolysaccharide 黏多糖
mucoprotein 黏蛋白
Mucor 毛黴菌〔屬〕；毛霉菌〔屬〕
Mucor mucedo 霉白黴菌
Mucor racemosus 葡萄狀白黴菌
Mucor rouxii 獸氏毛黴菌
mucosa 黏膜
mucosity 黏〔液〕性
mucotin 黏多糖；黏液素
mucous gland 黏液腺
mucous lining 黏膜；黏液外膜
mucous membrane 黏膜
mucus 黏液
mud carp 鯪魚；大魚；鯁魚
mud crab 紅蟳；鋸緣青蟹
mud dab 黃蓋鰈
mud eel 黃鱔〔魚〕
mud fish 弓鰭魚；鯰
mud press 濾泥壓乾機

mud skipper 彈塗魚；花跳
mudcake 濾〔泥〕餅
muddy flesh 肉泥
muffin 鬆餅；英式麥餅
muffineer 調味品瓶（蓋上有孔）
muffle furnace 灰化爐；回熱爐
mug 啤酒杯；杯子
mulberry 桑葚；桑樹；桑椹
mulberry flavo(u)r 桑椹香精
mulberry oily flavo(u)r 桑椹油香精
mule 騾
mull 香甜熱飲料；加糖加香料並加熱
mulled wine 香甜熱酒
muller 粉碎機；研磨機
mullet 鯔魚；烏魚；梭魚
mullet roe 烏魚子；鯔魚子
mulligan 燉什錦；蔬菜燴肉
mulligatawny paste 咖喱雞肉泥
mulligatawny soup 咖喱雞肉濃湯
mulser 乳化機
multieffect evaporator 多效〔應〕蒸發器
multiflora bean 紅花菜豆
multifoil 鋁箔與膠膜複合之包裝材料；多層箔 [= polyfoil]
multifunctional enzyme 多功能酶
multihead filling machine 多頭灌注機；多頭裝瓶機
multilayer 多層
multilayered food 多層化食品
multimixer 複式混合器；多用混合器
multipack 複式包裝；多層包裝；聯合包裝
multipackaging machine 複式包裝機；多用包裝機
multiplate chopper 多板絞肉機
multiplate freezer 多板冷凍機；多層平板式冷凍機
multiple effect evaporation 多效蒸發
multiple effect evaporator 多效蒸發器
multiple extraction 多次萃取
multiple fruit 複果；聚花果
multiple media filtration 多介質過濾
multiple-needle stitch pumping 多針頭肌肉注射醃漬法
multiple package 複式包裝；多層包裝
multiplicity 多樣性；多重性
multipurpose food 多用途食品
multiroll crusher 多輥壓碎機；多輥破碎機
multiscrew extruder 多軸擠壓機
multistage disk dryer 多段圓盤乾燥機
multistage extraction 多段萃取
multistage flash evaporation 多段急驟蒸發
multistage freezer 階梯式凍結裝置
multistation seamer 多頭封罐機；多頭封口機
multistory cold storage 多層冰箱；多層冷庫
multitube exchanger 多管熱交換器
multitubular condenser 多管冷凝器
multitubular heat exchanger 多管熱交換器
multivalve filler 多頭填充器；多頭裝填器
multivitamin 綜合維生素；多種維生素
multiwall bag 複層袋；多層袋
multiwall kraft 多層牛皮紙
multiwall paper bag 多層紙袋
multiwall sack 多層袋
mum 摩姆啤酒
mume 梅
mummy wheat 埃及小麥
munchen(er) beer 慕尼黑啤酒 [= Munich beer]
mung bean 綠豆 [= green gram]
mung bean sprout 綠豆芽

Munich beer 慕尼黑啤酒 [= munchen(er) beer]
Munich malt 慕尼黑麥芽；深色麥芽
muraenesox 海鰻
murlin 小海帶；小昆布
murphy 馬鈴薯；土豆
murraya paniculata concrete 九里香浸膏
muscadine 麝香葡萄酒
muscadine grape 圓葉葡萄；麝香葡萄
muscat 麝香葡萄酒；玫瑰香葡萄酒
muscat grape 麝香葡萄；圓葉葡萄；玫瑰香葡萄
muscle 肌；肌肉；瘦肉
muscle bundle 肌束
muscle cell 肌細胞
muscle fiber 肌纖維
muscle fibril 肌原纖維
muscle plasma 肌漿
muscle protein 肌蛋白質
muscle sugar 肌糖
muscott honey orange 茂谷桔橙
muscovado (sugar) 紅糖；黑砂糖；混糖
muscular fascia 肌膜
muscular layer 肌層
muscular tissue 肌〔肉〕組織
musculin 肌〔球〕蛋白
mush 玉米粥；玉米糊；糊狀食物
mushroom 菇；洋菇；蘑菇；蕈
mushroom cap 菇傘；蘑菇傘
mushroom mycelium 洋菇菌絲體；蘑菇菌絲體
mushroom paste 洋菇醬；蘑菇醬
mushroom sauce 洋菇醬油；蘑菇醬油
mushroom soy 洋菇醬油；蘑菇醬油
mushroom spawn 洋菇菌種；蘑菇菌柱；菌絲體
mushroom sugar 洋菇糖；蘑菇糖
mushy fruit 軟果
mushy potato(es) 馬鈴薯泥；薯蓉
musk 麝香
musk deer 麝香鹿
musk duck 麝香鴨
musk ketone 麝香酮
musk mallow 黃葵；麝香錦葵
musk onion 麝香洋蔥
musk root oil 麝香根油
musk xylene 二甲苯麝香
muskellunge 大梭魚；北美大梭魚；北美狗魚
muskmelon 甜瓜；香瓜
muskone 麝香酮
mussa 挪威幼鮴魚
mussel 青口；貽貝；淡菜；殼菜
must 麥芽汁；未發酵的葡萄汁或果汁；新〔葡萄〕酒；黴
mustang grape 白亮葡萄
mustard 芥菜；芥末
mustard dressing 芥末沙拉醬；芥末調味料
mustard flour 芥末〔粉〕
mustard green 芥菜
mustard oil 芥子油 [= mustard seed oil]
mustard seed 芥菜子；芥子
mustard seed oil 芥子油 [= mustard oil]
musty 霉臭的；發霉的
musty odor 霉味
musty smell 霉臭
mutagen 誘突變劑；誘變劑
mutagenesis 誘突變作用
mutagenic agent 誘突變劑；誘變劑
mutagenicity 誘突變性；突變性；誘變性
mutant 突變種；突變體；突變株
mutation 突變
mutlipass condenser 多程冷凝器

mutlipass heat exchanger 多程熱交換器
mutton 綿羊肉；羊肉
mutton chop 綿羊排；羊排
mutton fat 綿羊脂；羊脂
mutton odo(u)r 羊羶味
mutton tallow 羊脂
muttony 羊肉味的
mycelia 菌絲〔體〕 [= mycelium的複數]
mycelial 菌絲體的
mycelial filament 菌絲
mycelium 菌絲〔體〕
mycelium fungus 菌狀真菌
Mycobacterium 分枝桿菌屬
mycocide 殺真菌劑
mycoderm 菌膜
Mycoderma 生膜菌屬；醭酵母屬
mycolic acid 黴菌酸
mycolipenic acid 黴脂酸
mycology 真菌學；黴菌學
mycophenolic acid 黴酚酸
mycoprotein 真菌蛋白質；黴菌蛋白質；菌蛋白
mycorrhiza 菌根
mycose 繭蜜糖；海藻糖
mycostat 抑真菌劑；抑黴菌劑
mycostatic 抑真菌的；抑黴菌的
mycostatic agent 抑真菌劑；制黴菌劑
mycostatin 抑真菌素；抑黴菌素
mycotoxicosis 黴菌毒素中毒；真菌素病
mycotoxin 黴菌毒素；真菌毒素
myoalbumin 肌白蛋白
myocardium 心肌
myochrome 肌色素
myofibril 肌原纖維
myofibrillar protein 肌原纖維蛋白質
myofilament 肌纖絲；肌原絲；肌絲
myogen 肌漿蛋白 [= myosinogen]
myoglobin 肌紅蛋白；肌紅素
myoglobulin 肌〔肉〕球蛋白
myoinositol 肌醇；環己六醇
myokinase 肌激酶
myolemma 肌膜；肌纖維膜；肌鞘
myoprotein 肌蛋白質
myosin 肌凝蛋白；肌球蛋白
myosin filaments 肌球蛋白絲
myosinogen 肌漿蛋白 [= myogen]
myostromin 肌基質蛋白
myrcene 香葉烯；月桂烯
myrcenol 香葉烯醇；月桂烯醇
myrcia oil 香葉油；月桂油
myrica oil 楊梅油
myricinic acid 楊梅酸
myricoid 蜂花素
myricyl acid 蜂花酸
myristaldehyde 肉荳蔻醛
myristic acid 肉荳蔻酸；十四烷酸
myristic oil 肉荳蔻油
Myristica 肉荳蔻屬
myristica oil 肉荳蔻油
myristicene 肉荳蔻香精；肉荳蔻
myristyl alcohol 肉荳蔻醇
myrrh 沒藥
myrrh oil 沒藥油
myrtle 桃金娘
myrtle oil 桃金娘油
myrtol 桃金娘油；香桃木油
mysost 山羊乳清乾酪；挪威乳清乾酪；山羊乳清芝士
mytilite 貽貝糖醇；貽貝肌醇；貽貝酚
mytilitol 貽貝糖醇；貽貝肌醇；貽貝酚
mytilotoxin 貽貝毒素
Mytilus 貽貝屬
myxobacteria 黏液菌；黏絲菌

myxococcus 黏球菌
myxoglucosan 黏葡聚糖
myxovirus 黏病毒；黏液病毒
myxoxanthin 藍溪藻黃質；蘭溪藻黃質

N

nacarat 胭脂紅；洋紅
nacre 珍珠母；珠母貝
NAD (nicotinamide adenine dinucleotide) 菸鹼醯胺腺二核苷酸；二磷酸吡啶核苷酸；輔酶I
NADP (nicotinamide adenine dinucleotide phosphate) 菸鹼醯胺腺二核苷酸磷酸；三磷酸吡啶核苷酸；輔酶II
naiad 淡水貝
naked oat 裸燕麥
namaycush 湖紅點鮭
nangka 大樹波蘿；波蘿蜜；木波蘿；天波蘿 [= jackfruit; jaca; jak; jakfruit]
nanogram (ng) 奈克；毫微克
nanometer (nm) 奈米；毫微米；納米
napery 餐桌用巾；檯布
naphthalene
naphthol red 萘酚紅
naphthol yellow 萘酚黃
napkin 餐巾；餐巾紙
Napoleon 拿破崙〔白蘭地酒〕
Napoleon fish 蘇眉
nappy 酒味強烈的；酒精含量高的
narcissus concrete 水仙浸膏
narcotic 麻醉藥
nardgrass oil 亞香芳油
naringin 柚苷；柚皮
naringinase 柚皮酶；柚苷酶
nasal cavity 鼻腔
nasal rating 嗅覺評分
naseberry 納斯果；人心果
nasturtium 金蓮花；旱金蓮花
nasty smell 臭味
natamycin 海松素；游霉素
National Academy of Sciences 國家科學院
National Canners Association (NCA) 美國罐頭協會
National Institute of Health (NIH) 國立衛生研究所(美國)
National Standard 國家標準
native breed 土種；本地品種
native cellulose 天然纖維素
native compound 天然化合物
native product 土產
native protein 天然蛋白質
native species 本地種；土種
native variety 本地品種
native water 天然水
natron 天然碳酸鈉
natsudaidai 夏橙；葡萄柚
natto 納豆
natural 天然的；自然的
natural amaranthus red 天然莧菜紅
natural brewing 天然釀造
natural casing 天然腸衣
natural cheese 天然乾酪；未經調配的乾酪
natural colloid 天然膠體

natural color(ing) 天然色素；天然〔著〕色料
natural convection 自然對流
natural draft 自然通風
natural drying 天然乾燥
natural essence 天然香精
natural extractive flavo(u)ring substance 天然提取香料
natural feeding 天然飼養
natural fermentation 天然發酵
natural fiber 天然纖維
natural flavo(u)r 天然〔食用〕香料；天然風味
natural flavo(u)ring substance 天然香料
natural food 天然食品；天然食物
natural food additive 天然食品添加物；天然食品添加劑
natural food colo(u)r 天然食品〔著〕色料；天然食用色素
natural fruit juice 天然果汁
natural gas 天然氣；天然瓦斯
natural gum 天然〔樹〕膠
natural juice 天然果汁；原汁
natural medium 天然培養基
natural mineral water 天然礦泉水
natural pigment 天然色素
natural product 天然產物
natural resource 天然資源
natural ripening 自然成熟
natural rubber 天然橡膠
natural seasoning 天然調味料
natural spice 天然香辛料
natural stock 天然產物
natural sweetener 天然甜味料；天然甜味劑
natural thickener 天然增稠劑
natural thickening agent 天然增稠劑
natural toxicant 天然毒素
natural water 天然水
natural wine 自然發酵〔的葡萄〕酒
natural yoghurt 天然〔未經調味的〕酸乳酪
nausea 噁心；反胃
nauseating food 噁心食物
nauseous 噁心的；令人作嘔的
navaga 寬突鱈；遠東寬突鱈
navel 臍；肚臍
navel end 牛胸肉
navel orange 臍橙
navette 那維特蘿蔔
navy bean 四季豆；敏豆；莞豆；海軍豆
near beer 淡啤酒；低酒度啤酒
neat 牛；純的；未攙水的
neat brandy 純白蘭地〔酒〕
neat's foot 牛蹄
neat's tongue 牛舌
Nebbiolo 尼比羅〔釀酒紅〕葡萄；內比奧洛〔釀酒紅〕葡萄
nebulization 噴霧〔作用〕
nebulizer 噴霧器
neck 頸〔部〕；頸肉；脖子；瓶頸
neck end 牛頸肉
neck label 瓶頸標籤；瓶頸商標
necrobiosis 細胞壞死
necrophyte 腐肉植物
necropsy 宰後檢驗
necrosis 壞死
nectar 果泥飲料；花蜜；果蜜；甘露
nectar plant 蜜源植物
nectared 有花蜜的；甘美的
nectariferous plant 蜜源植物；分泌花蜜的植物
nectarine 油桃；桃駁李
nectary 花蜜腺；蜜腺
needle mackerel 中華九刺魚
needlefish 竹刀魚；突嘴頜針魚；頜針魚；鶴鱵魚

negative staining 陰性染色法
negus 尼格斯含酒飲料
nelma 大北鮭
nem 能母(1克母乳之營養值單位)
nematocide 殺線蟲劑
nematode 線蟲
neoagarobiose 新洋菜二糖；新瓊脂雙糖
neocarotene 新胡蘿蔔素
neochlorogenic acid 新漂木酸；新綠原酸
neoergosterol 新麥角固醇
neogala 初乳
neohesperidin 新橘皮苷；新桔皮苷；新橙皮苷
neohesperidin dihydrochalone 新橘皮苷二氫查酮；新桔皮苷二氫查耳酮(甜味劑)
neohesperidoside 新橘皮糖苷；新桔皮糖苷
neolycopene 新番茄紅素
neomycin 新黴素
neonate 新生嬰兒
neonicotine 新菸鹼；新煙鹼
neotype 新型
neovitamin 新維生素
neovitamin A 新維生素A
neoxanthin 新黃質；新葉黃素
nephelometer 濁度計
nephelometry 濁度測定法；比濁法
neptune 香螺
neral 橙花醛
neritic fish 淺海魚；近海魚
nerol 橙花醇
neroli bigarade oil 橙花油；苦橙油
neroli oil 橙花油
nerolidol 橙花叔醇
nerve fiber 神經纖維
nervonic acid 神經酸；二十四碳烯酸
neryl acetate 乙酸橙花酯
neryl isobutyrate 異丁酸橙花酯
Nessler reagent 奈氏試劑；奈斯勒氏試劑
nestotherapy 飢餓療法；斷食療法
net capacity 淨容量
net content 淨含量
net package 網包裝
net product 淨產物
net protein utilization (NPU) 淨蛋白質利用率
net sieve 網篩
net volume 淨容量
net weight 淨重
net yeast 純淨酵母
netted melon 網紋瓜；哈密瓜
netting 結網；金屬絲布
nettle 蕁麻
network structure 網狀結構
neurine 神經鹼
neurochitin 神經殼糖；神經幾丁質；神經甲殼質
neurokeratin 神經角蛋白
neurone 神經元；神經細胞
neuroprotein 神經蛋白〔質〕
neurospora 紅麵包黴；鏈孢黴
neurotoxin 神經毒素
neutral amino acid 中性胺基酸
neutral clay 中性白土
neutral detergent 中性清潔劑
neutral fat 中性脂肪
neutral glycolipid 中性醣脂
neutral lard 優良豬油；低溫煉製的豬油
neutral lipid 中性脂質
neutral spirit 精製酒精
neutrality 中性
neutralization 中和〔作用〕
neutralization value 中和值；中和價

neutralize 中和
neutralizer 中和劑
neutralizing agent 中和劑
neutrase 中性蛋白酶
new beer 新啤酒；未成熟啤酒
new coccine 新胭脂紅（食用紅色色素）
new-laid egg 剛生下的蛋
new red 新紅
new red aluminium lake 新紅鋁色澱
new wine 新釀〔葡萄〕酒
New Zealand spinach 番杏
newcastle kipper 背開冷燻鯡
newfish 澳洲鮭
Newtonian flow 牛頓〔型〕流動
Newtonian fluid 牛頓流體
Newtonian property 牛頓流體性
niacin 菸鹼酸；煙酸；尼克酸；維生素PP [= nicotinic acid]
niacinamide 菸鹼酰胺；尼克酰胺；煙酰胺 [= nicotinamide]
niacinamide ascorbate 抗壞血酸煙酰胺
niagara hawthorn 糙葉山楂
nib 咖啡豆；去胚芽可可豆；可可豆的豆肉
nibble 少量食物；一點點地咬；細咬；小口
nickle 鎳
Nicolas Appert 尼加拉斯阿伯
nicotiana 煙草；花煙草
nicotinamide 菸鹼酰胺；尼克酰胺；煙酰胺 [= niacinamide]
nicotinamide adenine dinucleotide (NAD) 菸鹼酰胺腺二核苷酸；二磷酸吡啶核苷酸；輔酶I
nicotinamide adenine dinucleotide phosphate (NADP) 菸鹼酰胺腺二核苷酸磷酸；三磷酸吡啶核苷酸；輔酶II
nicotinamide mononucleotide 菸鹼酰胺單核苷酸；烟酰胺單核苷酸
nicotinamide nucleotide 菸鹼酰胺核苷酸；烟酰胺核苷酸
nicotine 菸鹼；尼古丁
nicotinic acid 菸鹼酸；尼克酸；煙酸；維生素PP [= niacin]
nicotinic acid amide 菸鹼酰胺；煙酰胺
nidi 麵球
nidor 烹香；烹燒時之香味
niff 臭氣；惡臭
nigella 尼葛拉籽
nigerose 黑麴黴糖；黑糖
niggertoe 巴西栗；巴西果
night blindness 夜盲症 [= nyctalopia]
nightcap 睡前飲料
nigrescence 黑變；發黑
nigrometer 黑度計
nigrosine 尼格辛黑；苯胺黑
NIH (National Institute of Health) 國立衛生研究所（美國）
Nile tilapia 尼羅吳郭魚；尼羅羅非魚；尼羅非洲鯽魚；福壽魚
nilotic fish 尼羅魚
nine-eyes 鱔魚；八目鰻
ninhydrin 寧海準；茚三酮；寧希德林
ninhydrin reaction 寧海準反應；茚三酮反應；寧希德林反應
nisin 乳酸鏈球菌素；乳酸鏈球菌肽
nisinic acid 鯡〔油〕酸
niter 硝石；硝酸鉀 [= nitre]
nitrate (NO_3) 硝酸鹽；硝酸根
nitrate broth 硝酸鹽肉湯〔培養基〕
nitrate nitrogen 硝態氮
nitrate reducing bacteria 硝酸鹽還原細菌
nitrate reductase 硝酸鹽還原酶
nitration 硝化〔作用〕
nitre 硝石；硝酸鉀 [= niter]
nitric acid 硝酸
nitric oxide 氧化一氮

nitric oxide h(a)emoglobin 氧化氮血紅素；氧化氮血紅蛋白
nitrification 硝化〔作用〕
nitrifier 硝化細菌
nitrifying bacteria 硝化細菌
nitrite (NO_2) 亞硝酸鹽；亞硝酸根
nitrite bacteria 亞硝酸〔鹽〕細菌
nitrite burn 亞硝酸燒；亞硝酸鹽灼
nitrite reductase 亞硝酸鹽還原酶
nitro-compound 硝基化合物
Nitrobacter 硝化細菌屬
nitrobacteria 硝化細菌
nitrocellulose 硝酸纖維素；硝化纖維素
nitrodextrin 硝酸糊精；硝化糊精
nitrogen 氮
nitrogen conversion factor 氮轉換因數；氮轉換系數
nitrogen dioxide 二氧化氮
nitrogen fixation 固氮〔作用〕
nitrogen fixing bacteria 固氮細菌
nitrogen metabolism 氮〔素〕代謝
nitrogen oxide 氮氧化物
nitrogen-protein conversion factor 氮蛋白質轉換因數；氮蛋白質轉換系數
nitrogen solubility index 氮〔素〕溶出率
nitrogen storage 氮氣儲藏
nitrogenase 固氮酶
nitrogenization 氮化作用
nitrogenous compound 含氮化合物
nitrometer 氮素計；氮量計 [= azoto-meter]
2-nitropropane 2-硝基丙烷
nitroprotein 硝蛋白
nitrosaccharose 硝酸蔗糖；硝化蔗糖
nitrosamide 亞硝酰胺
nitrosamine 亞硝胺
nitrosation 亞硝化〔作用〕
nitroso (-NO) 亞硝基
nitroso compound 亞硝基化合物
nitrosoh(a)emoglobin 亞硝〔基〕血紅蛋白；亞硝血紅素
nitrosomyoglobin 亞硝〔基〕肌紅蛋白；亞硝肌紅素
nitrostarch 硝酸澱粉；硝化澱粉 [= starch nitrate]
nitrous acid 亞硝酸
nitrous oxide 一氧化二氮；笑氣；氧化亞氮
no-heat drying 不加熱乾燥
no-time bread 速製麵包
no-time dough 快速直接麵糰；速製麵糰
nobbed fish 去頭及內臟的魚
nobbing 去頭及內臟
nobhead 座頭鯨
nocake 炒玉米粉
nocuous 有毒的；有害的
node 結；節
nodule bacteria 根瘤〔細〕菌
nodulizer 成粒機；製粒機；球化劑
nog 諾格蛋黃〔乳〕酒；蛋乳酒；諾格濃啤酒
noggin 小杯；少量飲料；諾金（液量單位）
noisette 小塊〔精〕瘦肉；腰短骨肉小圓排
noisome 有毒的；有害的；有惡臭的
non-acid food 非酸性食品
non-acid fruit 非酸性水果
non-alcoholic beverage 無酒精飲料；非酒精飲料
non-amino nitrogen 非胺基態氮
non-aqueous 非水溶性
non-bacterial spoilage 非細菌性變質
non-caking 不結塊的
non-climacteric fruit 非更性水果
non-dairy creamer 非乳奶精
non-destructive analysis 非破壞性分析

non-destructive inspection 非破壞性檢驗
non-digestible 不易消化的
non-edible 不可食用；非食用
non-edible part 不可食用部份；非食用部份
non-effervescent 不起泡的
non-fecal bacteria 非糞便細菌
non-fermentable 非發酵性的；不發酵的
non-frozen water 非凍結水
non-glutinous rice 非糯米；籼米 [= non-sticky rice; non-sweet rice; non-waxy rice]
non-ionic detergent 非離子〔型〕清潔劑
non-ionic surfactant 非離子〔型〕界面活性劑；非離子〔型〕表面活性劑
non-lactic acidity 非乳酸酸度
non-leavened acidity 非發酵食品
non-lumping 不結塊的
non-metal 非金屬
non-Newtonian fluid 非牛頓流體
non-nitrogenous compound 無氮化合物
non-nitrogenous diet 無氮膳食
non-nutritive 無營養的
non-nutritive sweetener 無營養甜味劑
non-oily 低脂的
non-oily fish meal 低脂魚粉
non-pasteurized 未殺菌的
non-pathogenic bacteria 非病原菌；非致病菌
non-pungent pepper 甜椒
non-reducing sugar 非還原糖
non-returnable container 非回收容器；一次性容器
non-saccharide 不含糖的
non-saponifiable 不皂化的
non-saponifiable lipid 不皂化脂質
non-shelled egg 去殼蛋
non-soapy detergent 非皂性清潔劑
non-soluble 不溶解的
non-spore (forming) bacteria 無芽孢細菌
non-staple food 非主〔要〕食品；副食品
non-starchy 非澱粉的
non-sticking 無黏性；不黏的
non-sticky rice 非糯米；籼米 [= non-glutinous rice; non-sweet rice; non-waxy rice]
non-sucrose 非蔗糖
non-sugar 非糖物
non-sweet rice 非糯米；籼米 [= non-sticky rice; non-glutinous rice; non-waxy rice]
non toxic 無毒的
non toxic substance 無毒物質
non-volatile 不揮發性的；非揮發性的
non-volatile oil 不揮發油；非揮發性油
non-volatile substance 不揮發性物質
non-volatility 不揮發性；非揮發性
non-waxy rice 非糯米；籼米 [= non-sticky rice; non-sweet rice; non-glutinous rice]
non-wettable 不可濕的
non-wettable film 不濕膜；斥水性薄膜
non-yeasted dough 非酵母發酵之麵糰
nonadecane 十九烷
nonadecanoic acid 十九〔烷〕酸
δ-nonalactone δ-壬內酯
nonanal 壬醛
nonanoic acid 壬酸
nonanone 壬酮
noncarbonated beverage 非碳酸飲料
nonenzym(at)ic 非酵素的；非酶的
nonenzymatic browning 非酵素〔性〕褐變；非酶性褐變
nonessential amino acid 非必需胺基酸 [= dispensable amino acid]

nonessential enzyme 可缺酵素
nonessential fatty acid 非必需脂肪酸
nonessential trace element 非必需微量元素
nonesterified fatty acid 非酯化脂肪酸
nonfat 無脂肪的；非脂肪的；脱脂的
nonfat dry milk 脱脂乳粉；脱脂奶粉 [= skim milk powder]
nonfat milk 脱脂奶；脱脂乳 [= skim milk]
nonfat solids 無脂固形物；非脂固形物
nonfreezing 不凍的
nonfreezing solution 不凍溶液
nonheme iron 非〔原〕血紅素鐵
noni 蘿梨
nonose 壬糖；九碳糖
nonpolar amino acid 非極性胺基酸
nonpolar compound 非極性化合物
nonpolar solvent 非極性溶劑
nonprotein 非蛋白質；不含蛋白質的
nonprotein amino acid 非蛋白胺基酸
nonprotein nitrogen (NPN) 非蛋白態氮
nonyl acetate 乙酸壬酯
nonyl alcohol 壬醇
nonyl isovalerate 異戊酸壬酯
nonyl octanoate 辛酸壬酯
noodle 麵條；蛋麵條
noodle cutter 麵條切割機；切麵條機
noodle cutting machine 麵條切割機；切麵條機
noodle squash 魚翅瓜（一種西葫蘆）[= spaghetti squash; vegetable spaghetti]
nootkatone 圓柚酮
nordihydroguaiaretic acid 正二氫癒創酸
nori 紫菜；海苔；紅藻 [= laver]
norleucine 甘白胺酸；正白胺酸；正亮胺酸；己胺酸
normal diet 正常膳食；正常飲食
normal honey 花〔蒂〕蜜
normal milk 正常乳
normal saline 生理食鹽水
normal solution 當量溶液；規定溶液；規度溶液
normal temperature 常溫
normality (N) 當量濃度；規定濃度；規度
northern anchovy 加州鯷
northern bluefin 長腰鮪；青干金槍魚
northern dogfish 布氏角鯊
northern halibut 北方鰈
northern pike 白斑狗魚
norvaline 正纈胺酸；戊胺酸
Norway lobster 挪威海螯蝦
Norway pout 挪威鱈
notatin 葡萄糖氧化酶；點霉素；青霉氧化酶
notch 凹口；切口；缺口；切角
notching machine 罐身切角機
nougat 核果糖；牛軋糖；鳥結糖；果仁糖
nourishing 有營養的；滋養的
nourishing diet 滋補膳食
nourishment 滋補品；營養品；食物；滋養；營養狀況
noxious substance 有害物質
noyau 果仁酒
nozzle 噴嘴；噴頭
nubbin 短小穗；短小果
nubbing 去梗蒂；去梗未果頂皮
nuclease 核酸酶
nuclei 細胞核；核；晶核；原子核〔= nucleus的複數〕
nucleic acid 核酸
nucleic acid condiment 核酸調味料
nucleic acid seasoning 核酸調味料
nucleoalbumin 核〔白〕蛋白
nucleolus 核仁

nucleophosphatase 核〔酸〕磷酸酶
nucleoplasm 核漿；核原生質
nucleoprotein 核蛋白〔質〕
nucleosidase 核苷酶
nucleoside 核苷
nucleotidase 核苷酸酶
nucleotide 核苷酸
nucleus 細胞核；核；晶核；原子核
numbles 內臟；雜碎
nurse 護士；保姆；奶媽
nurser 奶瓶；奶樽；奶媽
nursery 育嬰室；托兒所；苗圃；育苗場
nurture 養育；營養品
nut 堅果；螺〔絲〕母
nut bush 歐洲榛
nut fruit 堅果
nut grass 香附子
nut meat 堅果仁
nut oil 堅果油
nut paste 堅果仁糊
nutcracker 堅果鉗；胡桃鉗
nutgall 五倍子；沒食子
nutlet 小堅果；果仁；果核
nutmeat 堅果核仁；碎堅果仁
nutmeg 肉荳蔻；玉果
nutmeg apple 肉荳蔻果〔實〕
nutmeg butter 肉荳蔻脂
nutmeg hickory 肉荳蔻山核桃
nutmeg mace 肉荳蔻乾皮
nutmeg melon 網紋甜瓜；醉瓜
nutmeg oil 肉荳蔻油
nutraceuticals 功能性食品
nutramin(e) 維生素；維他命
nutrient 營養物；營養素；養分；營養的
nutrient agar 營養洋菜〔培養基〕；營養瓊脂〔培養基〕
nutrient allowance 營養供給量
nutrient balance 營養平衡
nutrient broth 營養培養液
nutrient content claims 營養含量聲明
nutrient deficiency 營養不足；營養缺乏
nutrient-dense 高密度營養
nutrient-dense food 高密度營養食物
nutrient density 營養密度
nutrient yeast 飼料酵母
nutriment 營養物；滋養物；食物
nutriology 營養學
nutrition 營養；營養學
nutrition board 營養成分表；營養委員會
nutrition education 營養教育
nutrition enhancer 營養強化劑
nutrition enrichment 營養增富
nutrition facts 營養資料
nutrition fortification 營養強化
nutrition index 營養指數；營養指標
nutrition labeling 營養成分標籤；營養標籤；營養標示
nutrition restoration 營養增補
nutritional 營養的
nutritional anemia 營養性貧血
nutritional deficiency 營養不足；營養缺乏
nutritional disorder 營養失調
nutritional disturbance 營養障礙
nutritional edema 營養性水腫
nutritional enriching agent 營養強化劑
nutritional fortification substance 營養強化劑
nutritional guidance 營養指南
nutritional ingredient 營養成分
nutritional inhibitor 營養抑制劑
nutritional labeling 營養〔成分〕標簽
nutritional quality 營養品質；營養質量
nutritional requirement 營養需求〔量〕

nutritional science 營養科學
nutritional standard 營養標準
nutritional status 營養狀況
nutritional survey 營養調查
nutritional therapy 營養治療
[= therapeutic nutrition]
nutritional value 營養價值
nutritionist 營養學家
nutritious 有營養的；滋養的；營養價值高的
nutritive 營養的
nutritive factor 營養要素
nutritive index 營養指數；營養指標
nutritive ration 營養口糧
nutritive sweetener 營養性甜味劑
nutritive value 營養價值
nutriture 營養；滋養；營養狀況
nutrix 奶媽；乳母
nutrose 酪蛋白鈉；鈉酪蛋白
nutshell 堅果殼
nuttree 堅果樹；歐洲榛
nutty 產堅果的；有堅果味的；多堅果的
nyctalopia 夜盲症 [= night blindness]
nylon 尼龍
nystatin 制黴菌素

O

o/w emulsion (oil-in-water emulsion) 水中油型乳〔濁〕液；水包油型乳〔濁〕液
oak 橡樹；櫟樹
oak cask 橡木桶
oak moss concrete 橡苔浸膏
oak oval 橡木酒桶
oak wood 橡木桶
oarweed 大褐藻；昆布
oast 烘爐
oat 燕麥
oat flake 燕麥片
oat flour 燕麥粉
oat groats 去殼燕麥〔粒〕
oat meal 燕麥片；燕麥粥；燕麥粉
oat meal porridge 燕麥粥
oat middlings 燕麥渣
oatcake 燕麥餅乾
obese 肥胖的
obesity 肥胖症
obesity diet 肥胖症膳食
obesity drug 減肥藥
oblate 澱粉薄膜；糯米紙
obligate aerobe 絕對好氣菌；絕對嗜氧菌；專性需氧菌
obligate anaerobe 絕對嫌氣菌；絕對厭氧菌；專性厭氧菌
obligate halophile 嗜鹽菌
obligate psychrophile 絕對嗜冷菌；專性嗜冷菌
obligate saprophyte 絕對腐生菌；專性腐敗菌
obligate thermophilic bacteria 絕對嗜熱細菌；專性嗜熱細菌
oblong can 橢圓形罐；長方形罐
obnoxious species 有害微生物
obround can 長圓罐
obstipation 便秘；頑強便秘
occlusor 閉殼肌
ocean food 海洋食物
ocean perch 洋鱸
ocean quahaug 太平洋蛤蜊
ocean quahog 美人蛤
ocean sunfish 翻車魨
ocean whitefish 闊鱗方頭魚；海白魚
oceanic bonito 巴鰹；鰹
oceanic herring 大西洋鯡
oceanic skipjack 鰹
ocellated octopus 飯鮹
ocellated plesiope 黑鮗
ochratoxin 赭麴霉素；棕麴霉素
ochro 秋葵
octacosane 二十八〔碳〕烷
octadecadienoic acid 十八碳二烯酸；亞油酸
octadecanoic acid 十八烷酸；硬脂酸 [= stearic acid]
octadecenoic acid 十八碳烯酸
γ-octalactone γ-辛內酯
δ-octalactone δ-辛內酯

octanal 辛醛
octanoic acid 辛酸 [= caprylic acid]
octanol 辛醇 [= octyl alcohol]
October hawthorn 十月山楂
octopine 章魚肉鹼
octopus 章魚
octose 辛醣；八碳醣
octyl acetate 乙酸辛酯
octyl alcohol 辛醇 [= octanol]
octyl formate 甲酸辛酯
octyl gallate 沒食子酸辛酯
octyl isobutyrate 異丁酸辛酯
odo(u)r intensity 氣味強度；氣味濃度
odo(u)r threshold 嗅覺閾值；嗅覺低限
odo(u)rless 無臭的；無氣味的
odor 氣味 [= odour]
odorant 氣味劑；有氣味的物體
odoriferous herb 草本香料
odorimeter 氣味計
odorimetry 氣味測定法
odorometer 氣味計
odour 氣味 [= odor]
OEM 委託加工
oenanthic 酒香的
oenin 葡萄花青素；紅紫葡萄皮的花色素
oenocyanin 葡萄酒花色素
oenolin 葡萄酒色素
oenologist 葡萄酒釀造學家；釀葡萄酒學家 [= enologist]
oenology 葡萄酒釀造學；釀葡萄酒學 [= enology]
oenometer 酒度計；葡萄酒酒度計
oenometry 葡萄酒酒度測定法
oesophagus 食道
off colo(u)r 異色；變色
off flavo(u)r 異味；變味；風味變壞
off odo(u)r 臭氣；異味
off quality 不合標準的質量
off rigor 解僵
off scum 廢渣
off season 非生產季節；淡季
off standard 不合格的
off taste 變味
offal 內臟；廢棄物；碎屑
offal fat 內臟脂肪
offal fish 廉價魚
offensive odo(u)r 惡臭；令人討厭的氣味
offgrade product 低級產品；等外品；不合格產品
office catering 辦公室包飯
official method 法定方法
official method of analysis 法定分析法
official sample 法定試樣；正式樣品
offscouring 廢物；渣滓；垃圾
offspring 後代；產物；苗種
offtake 排水渠
ogen 歐根香瓜
Oikama minnow 寬鰭鯉
oil 油；油類；石油
oil and fat 油脂
oil bath 油浴
oil cake 油粕餅；油〔渣〕餅
oil cell 油細胞
oil content 含油量
oil crop 油料作物
oil degumming 油脂脱膠
oil droplet 油滴；油胞
oil extraction 油脂萃取；油脂提取
oil extractor 油脂萃取器；油脂提取器
oil foot 皂腳；油腳
oil gauge 油比重計；量油計
oil globule 脂肪球
oil grain 油籽
oil halibut 太平洋馬舌鰈
oil hardening 油硬化；油脂氫化
oil immersion 油浸

oil immersion lens 油鏡
oil-in-water emulsion (o/w emulsion) 水中油型乳〔濁〕液；水包油型乳〔濁〕液
oil meal 油粕粉
oil mill 油坊；油廠
oil nut 油果；油用堅果
oil off 油脂析出；出油；脱脂
oil orange 余甘子；油橙
oil palm 油棕〔櫚〕；油椰
oil palm fruit 油棕〔櫚〕果
oil palm kernel 油棕〔櫚〕仁
oil palm sterilization 油棕滅酶
oil plant 油料作物
oil polishing 油的濾清
oil preserves 油漬食品
oil press 榨油機
oil-proof 防油的；耐油的；不透油的；抗油的
oil-proof paper 防油紙
oil refining 油脂精製；油脂精煉
oil-repellent 拒油的；不吸油的
oil-resistant 耐油的
oil sac 油胞
oil sardine 長頭沙丁魚
oil separator 油分離機
oil soluble 油溶〔性〕的
oiled 表面塗油的；油漬的
oiled can 油漬罐頭
oilery 榨油廠
oiliness 油膩；油質
oilseed 油籽；含油種子
oily 油的；油質的；油膩的
oily flavo(u)r 油〔膩〕味
okara 豆渣
okra 秋葵；黃秋葵 [= gumbo]
old cheese 過熟乾酪
old dough 過熟麵欄；老麵欄；過度發酵麵欄
old flavo(u)r 陳舊風味
old ginger 老薑
old woman flounder 油鰈
oldwife 黑海鯛；油魣；灰姥鱸
oldworld arrowhead 慈菇
oleaginous 油質的；含油的
oleaginous food 含油食品
oleander 夾竹桃
olease 油酸酯酶
oleate 油酸鹽
olefin 烯烴；石蠟
oleic acid 油酸
olein 甘油三油酸酯；三油酸甘油酯；三油脂 [= glyceryl trioleate; triolein]
oleo 油；油脂的
oleo oil 黃油；人造奶油；液態牛油
oleo stock 食用牛脂；食用羊脂
oleomargarine 瑪珈琳；人造乳酪；人造黃油；人造奶油；人造白脱；植物黃油；植物牛油 [= butterine; margarine]
oleometer 油比重計
oleophilic 親油的
oleophobic 疏油的
oleophylic 親油的
oleorefractometer 油折射計
oleoresin 精油樹脂；含油樹脂
oleostearin(e) 牛油硬脂；壓製脂
oleovitamin 維生素油劑
olericulture 蔬菜栽培〔學〕；蔬菜園藝
oleuropein 橄欖的苦澀物質
olfaction 嗅覺
olfactometer 嗅覺計；氣味測定計
olfactory 嗅覺的
olfactory threshold 嗅覺閾值；嗅覺低限
olibanum 乳香
olibanum oil 乳香油
oligopepsia 消化力不足

oligopeptide 寡肽；低〔聚〕肽
oligosaccharidase 低聚糖酶；寡糖酶
oligosaccharide 寡糖；低聚糖；寡聚糖
oligose 寡聚糖；低聚糖
oligotrophy 營養不足
olio 什錦菜
olive 橄欖；油橄欖
olive kernel oil 橄欖仁油
olive oil 橄欖油；油橄欖油
olive oyster 晚生斑褶菇
omasum 重瓣胃；反芻動物第三胃
omelet(te) 煎蛋包；煎蛋〔餅〕；奄列
omental fat 網油
omentum 網膜
omnivore 雜食性動物；不偏食的人
omophagia 生吃食物；生食
oncotic pressure 膠體滲透壓；膨脹壓
one-piece container 沖壓罐；單片罐
one-spindle seamer 單頭封罐式封罐機
one-trip package 不回收包裝；一次性包裝
one-way bottle 不回收瓶
one-way container 不回收容器
one-way package 不回收包裝；一次性包裝
onion 洋葱
onion bulb 洋葱頭；洋葱鱗莖
onion oil 洋葱油
onion ring 炸洋葱圈
onion skin 洋葱皮
oolachan 太平洋細齒鮭
oolemma 卵膜；卵黃膜
oolong (tea) 烏龍茶
opacity 不透明度；混濁度
opah 月魚(溫帶海產)
open-air drying 風乾；露天乾燥
open fermentation 開放發酵；敞口發酵
open fermenter 敞口式發酵罐
open fire 明火
open kettle 開口鍋
open pan 開口鍋
open panel test 公開品評試驗
open stack 露天堆放
open steam 直接蒸汽
open top bread 圓頂麵包
open-top can 開頂式二重捲封機；衛生罐
open-top container 開頂貨櫃；開頂容器
open veil 菇類開傘；開傘
openness 鬆散；多孔
operational ration 軍糧
opium 鴉片；罌粟
optical activity 旋光性；旋光度
optical densitometer 光密度計
optical density 光密度
optical rotation 旋光性；旋光度
optimization 最適化；最佳化
optimum 最適〔宜〕的
optimum diet 最適膳食
optimum pH 最適酸鹼度；最適pH
oquassa trout 藍背紅點鮭
orach(e) 濱藜屬植物；法國菠菜
oral administration 口服
orange 橙；橘子；柳橙；柑；柑橘；柑桔
orange beverage 橙汁飲料
orange birch bolete 多皮疣柄牛肝菌
orange butter 橙膏
orange colo(u)r 橙色〔色素〕
orange drink 橙汁飲料
orange flavo(u)r 橘子香精
orange flesh sweet potato 黃肉甜〔甘〕薯
orange flower oil 橙花油
orange grisette 橘黃鵝膏(蕈類)
orange juice 橘子汁；柳橙汁；橙汁
orange lentil 橙扁豆
orange oil 橙〔皮〕油；甜橙油

orange oily flavo(u)r 橘子油香精
orange peel 橙皮
orange peel fungus 橙黃網孢盤菌(蕈類)
orange peel oil 橙皮油
orange peeler 柑橘去皮機；柑桔剝皮機
orange peeling machine 柑橘去皮機；柑桔剝皮機
orange pekoe 橘香紅茶；上等紅茶(產於印度或斯里蘭卡)；橙黃白毫紅茶
orange puree 橙泥
orange rock fish 海鱸魚
orange section 橘瓣；桔瓣
orange segmenting machine 柑橘分瓣機
orange soda 橘子汽水；橙汁汽水
orange squash 橘肉果汁；橘子水
orange wine 橘子酒；桔子酒
orangeade 橘子〔汽〕水；橙汁飲料
orangequats 橙金桔
orangery 柑橘溫室；柑桔溫室；橘園
orbfish 圓白鯧
orbitort 迴轉式殺菌釜
orcein 地衣紅；苔紅素
orchard 果園
orcharding 果樹栽培
orchil 地衣；地衣〔紅色〕染料；苔色素；苔蘚紅素 [= orselle]
ordering 加濕；增濕；潤濕；回潮
ordering room 增濕房；回潮房
ordinary pressure 常壓
ordinary temperature 常溫
orecchiette 耳形通心粉
oregano 牛至〔草〕；披薩草 [= origanum; wild marjoram]
Oregon sturgeon 歐鱘；俄勒岡鱘
orexis 食慾
orfe 圓腹雅羅魚
organ 器官
organic 有機的
organic acid 有機酸
organic acid seasoning 有機酸調味料
organic base 有機鹼
organic compound 有機化合物
organic farming 有機農業
organic food 有機食物；有機食品
organic nitrogen 有機氮
organic pesticide 有機農藥；有機殺蟲劑
organism 生物；有機體
organoleptic evaluation 官能品評；感官評價
organoleptic examination 官能檢驗；感官檢驗
organoleptic property 官能性質；感官性質
organoleptic test 官能試驗；感官試驗
organoleptic threshold 官能閾值；感官閾值
organophosphorus pesticide 有機磷農藥 [= phosphate pesticide]
organosol 有機溶膠
orgeat 杏仁糖漿
oriental bonito 東方狐鰹
oriental chimaera 黑線銀鮫
oriental pickling melon 越瓜
oriental sardine 遠東沙丁魚
orifice 孔口
orifice mixer 孔口混合器
origanum 牛至〔草〕；披薩草 [= oregano; wild marjoram]
origanum oil 牛至油
original breed 原種
ormer 鮑魚 [= abalone]
ornamental package 裝飾包裝
ornithine 鳥胺酸
orotic acid 乳清酸；維生素B_{13}

orris 昌蒲；鳶尾〔草〕
orris absolute 鳶尾淨油
orris oil 鳶尾凝脂
orrisroot 昌蒲根；鳶尾草根
orrisroot oil 昌蒲根油；鳶尾凝脂；鳶尾草根油
orselle 地衣；地衣〔紅色〕染料；苔色素；苔蘚紅素 [= orchil]
orymycin 稻霉素
Oryza 稻屬
oryzae farina 米粗粉
oryzanal 米糠素
oryzanin 硫胺；噻胺；維生素B_1
oryzanol 米糠醇
oryzenin 米穀蛋白
Osage orange 歐撒橘；桑橙
osan 聚糖
oscillating sieve 振動篩
oscillator 振盪器；振動器
osetr caviar 歐斯塔魚子醬
osmanthus 桂花；木犀屬
osmanthus concrete 桂花浸膏；木樨花浸膏
osmanthus oily flavo(u)r 桂花油香精
osmesthesia 嗅覺
osmics 嗅味學
osmology 嗅〔覺〕學；滲透學
osmometer 滲壓計；滲透壓力計
osmophile 嗜滲壓微生物；嗜高滲〔壓〕微生物；耐滲壓微生物
osmophilic bacteria 嗜滲壓細菌；嗜高滲細菌；耐滲壓細菌
osmophilic yeast 耐滲壓酵母；嗜滲壓酵母；嗜高滲酵母
osmosis 滲透〔作用〕
osomotic equilibrium 滲透平衡
osomotic membrane 滲透膜
osomotic permeability 滲透性
osomotic pressure 滲透壓
ossein 骨膠原
osteoporosis 骨質疏鬆症
ostracean 牡蠣；牡蠣的
ostreasterol 牡蠣固醇
ostreat 蠔菇
Ostwald viscosimeter 奧士瓦〔毛細管〕黏度計；奧斯瓦爾特〔毛細管〕黏度計
oswego flour 玉米粉；玉蜀黍粉
otto 玫瑰油
ounce (oz.) 安士；盎斯；盎司；啢
ouricury oil 小冠椰子油
outer container 外包裝容器
outer packaging 外包裝
outfall 排水口
outfit 用具；配備；裝備
outgassing 脱氣；去氣
outlet 出口
outside 外側後腿肉；上臀肉
outside flat 外側後腿板肉
outside steak 牛腿肉排；上臀肉排
ouzo 希臘茴香烈酒
ova 卵；卵細胞 〔= ovum的複數〕
oval can 橢圓〔形〕罐
oval colony 卵形菌落
oval kumquat 金橘；羅浮
oval shaped can 橢圓形罐
oval tin 橢圓罐
ovalbumin 卵白蛋白；卵清蛋白
ovary 卵巢；子房
oven 烘箱；烤箱；焗爐；烤爐
oven-bottom bread 底火麵包；直接火烤的麵包
oven dry weight 絕乾重量 [= absolute dry weight]
oven drying 烘箱乾燥；烘爐乾燥
oven flow 爐烤直徑增大量
oven grilling 烤爐燒烤
oven jump 爐烤持水增重；爐內增重

oven kick 烘烤膨脹；爐內膨脹
[= dough spring; oven spring]
oven spring 烘烤膨脹；爐內膨脹
[= dough spring; oven kick]
oven thermometer 爐溫計
oven toaster 小烘烤箱
over-aged dough 過熟麵糰；老麵糰；過度發酵麵糰
over fermentation 過度發酵
over grinding 過度粉碎；粉碎過度
over intake 過度攝取
over packaging 過多包裝
over production 生產過剩
overbake 過烘；烘焙過度
overbalance 失去平衡；過重
overbleach 漂白過度；過漂
overboiled 煮沸過度
overburn 過燒的；加熱過度
overburnt 過燒的
overcap placer 加蓋器
overcarbonated juice 過度充氣果汁；過飽充汁
overcook 煮過度；蒸煮過度
overcooked 煮過久的
overcooking 過度烹煮；蒸煮過度
overcooled 過度冷卻的
overcooling 過冷〔卻〕；過度冷卻
overcured 過度醃漬的；醃漬過久的
overcuring 過度醃漬；醃漬過久
overdone 煮過久的
overdose 劑量過度；超過劑量
overdried 過乾的
overdry 過度乾燥
overeat 飲食過量；吃得過多
overfeed 過量餵食；過量進料
overfill 裝填過量；過量裝填
overflow 溢流；溢出
overgassed juice 過度充氣果汁；過飽充汁
overgassing 過度充氣；充氣過量
overhead conveyor 高空輸送機；懸空輸送器
overheat 過熱；過度加熱
overheated steam 過熱蒸汽
overhydration 過度水合；水合過度
overladen 裝載過多的；超載的
overload 過載；超載
overmature 過熟的
overmilling 過度碾磨
overmix 過度拌合；過度混合
overnutrition 營養過度
overoxidation 氧化過度；過度氧化
overpasteurization 殺菌過度；巴氏殺菌過度
overprocessing 過度加工；過度加熱
overprocessing flavo(u)r 過度加工風味；加熱過度的焦味
overproof 酒精含量超過標準的
overproofed dough 過醒麵糰；醒麵過度的麵糰
overrefining 過度精製
overripe 過熟的
overripe dough 過度發酵的麵糰；成熟過度的麵糰
overrun 膨脹率；超出量
overrun tester 膨脹率測定器
overs 篩除物；篩上料
oversalted 鹽漬過久的；過鹹的
oversalting 加鹽過多
oversaturation 過飽和
oversize 過大
oversmoked 煙燻過度的
oversterilization 滅菌過度
overstock 進貨過多；存貨過多；庫存過剩
overtemperature 溫度過高；過熱；超溫
overweight 過重；超重

overwet 過濕的
ovoalbumin 卵白蛋白；卵清蛋白
ovoflavin 核黃素
ovoglobulin 卵球蛋白
ovomucin 卵黏蛋白
ovovitellin 卵黃磷蛋白 [= vitellin]
ovum 卵；卵細胞
ox 公牛；閹公牛
ox bile extract 牛膽抽提液
oxalate 草酸鹽
oxalic acid 草酸；乙二酸
oxalis 酢漿草
oxaloacetic acid 草乙酸
oxen 公牛 〔= ox的複數〕
oxidant 氧化劑 [= oxidizer; oxidizing agent]
oxidase 氧化酶
oxidation 氧化〔作用〕 [= oxidization]
oxidation accelerator 氧化促進劑
oxidation inhibitor 氧化抑制劑
oxidation number 氧化值；氧化數
oxidation preventive 抗氧化劑
oxidation-reduction enzyme 氧化還原酶
oxidation retarder 氧化抑制劑；氧化阻滯劑
oxidative 氧化的
oxidative decomposition 氧化分解
oxidative degradation 氧化降解
oxidative fermentation 氧化發酵
oxidative product 氧化產物
oxidative rancidity 氧化酸敗
oxidative stability 氧化安全性；抗氧化性
oxide 氧化物
oxidization 氧化〔作用〕 [= oxidation]
oxidized fat 氧化脂肪
oxidized flavo(u)r 氧化風味；氧化〔氣〕味
oxidized starch 氧化澱粉
oxidizer 氧化劑 [= oxidant; oxidizing agent]
oxidizing agent 氧化劑 [= oxidant; oxidizer]
oxidizing power 氧化力
oxidoreductase 氧化還原酶
oxybiosis 需氧生活
oxybiotic bacteria 需氧細菌
oxycalorimeter 氧卡計；氧量熱計
oxycellulose 氧化纖維素
oxycholesterol 羥膽固醇 [= hydroxycholesterol]
oxydase 氧化酶 [= oxidase]
oxyethylene higher aliphatic alcohol 環氧乙烷高級脂肪醇
oxygen 氧
oxygen absorber 脱氧劑；吸氧劑
oxygen bomb calorimeter 氧彈卡計；氧彈量熱計
oxygen scavenger 去氧劑；脱氧劑
oxygen scavenging agent 去氧劑；脱氧劑
oxygen tension 氧含量；氧壓
oxygenant 氧化劑
oxygenase 氧合酶；加氧酶
oxygenated myoglobin 氧合肌紅素；氧合肌紅蛋白 [= oxymyoglobin]
oxygenation 氧合〔作用〕；氧化〔作用〕
oxygenolysis 氧化分解
oxyh(a)emoglobin 氧合血紅素；氧合血紅蛋白
oxymyoglobin 氧合肌紅素；氧合肌紅蛋白 [= oxygenated myoglobin]
oxyneurine 甜菜鹼；甜菜素 [= betain(e)]
oxyntic cell 泌酸細胞
oxyproline 羥〔基〕脯胺酸 [= hydroxyproline]
oxystearin 氧化硬脂精

oyster 牡蠣；蠔
oyster bank 養蠔場；牡蠣養殖場
oyster bed 養蠔場；牡蠣養殖場
oyster blade 上肩胛脊底肌；板腱
oyster cap fungus 平菇
oyster dredge 開貝器
oyster drill 海蝸牛；荔枝螺；紅螺
oyster farm 蠔場；牡蠣〔養殖〕場
oyster fungus 蠔菇
oyster grass 巨海帶；巨昆布
oyster ground 養蠔場；牡蠣養殖場
oyster mushroom 蠔菇；月夜茸
oyster plant 婆羅門參；黑皮波羅門參 [= salsify; vegetable oyster]
oyster sauce 蠔油
oyster shell 牡蠣殼；蠔殼
oz (ounce) 安士；盎斯；盎司；啢
ozonation 臭氧化作用
ozone (O_3) 臭氧
ozonization 臭氧化〔作用〕；臭氧處理
ozonize 以臭氧處理
ozonized water 臭氧處理水
ozonizer 臭氧發生器
ozonometer 臭氧計

P

pabular 可食的；食品的
pabulum 食糧；食物；營養素
pachyman 茯苓聚糖
pachymic acid 茯苓酸
pachymose 茯苓糖
pacific black halibut 太平洋黑比目魚
pacific bonito 齒鰆；太平洋狐鰹
pacific capelin 太平洋毛鱗魚
pacific cod 鱈魚；太平洋鱈；大頭鱈；阿拉斯加鱈
pacific hake 太平洋無鬚鱈；太平洋狗鱈
pacific halibut 窄鱗庸鰈；太平洋庸鰈；狹鱗庸鰈
pacific herring 太平洋鯡
pacific lamprey 三齒七鰓鰻；太平洋鰻魚
pacific mackerel 日本〔花〕鯖
pacific marlin 箕作氏槍魚
pacific ocean perch 太平洋鱸魚
pacific oyster 太平洋〔大〕牡蠣
pacific pilchard 太平洋沙瑙魚；美洲沙瑙魚
pacific round herring 脂眼鯡
pacific sailfish 扁帆鰭魚
pacific salmon 大鱗大麻哈魚；春鮭
pacific sandlance 玉筋魚
pacific sardine 太平洋沙瑙魚；美洲沙瑙魚；沙瑙魚
pacific saury 秋刀魚
pacific tomcod 太平洋小鱈
pacific yellowfin tuna 黃鰭金槍魚；太平洋黃鰭鮪
pack 包裝；捆；包裹；填充；包裝容器
pack date 包裝日期；製造日期
package 包裝；打包；包裹；包裝容器
packaged goods 包裝貨品
packaged meat 包裝肉；定量分裝肉
packaging 包裝；包裹
packaging machine 包裝機
packaging material 包裝材料
packed 被包裝的
packed bed filter 填充床過濾器
packed column 填充柱
packed tower 填充塔
packed weight 毛重
packer 包裝工人；包裝機；打包機；裝填機；包裝食品生產廠商
packer's end 罐蓋
packet 小包裝；小包裹；小捆
packhouse 包裝廠；打包間
packing 包裝；打包；包裝填充料
packing board 包裝用紙板
packing canvas 麻布袋；帆布袋
packing house 肉類包裝廠；罐頭廠；食品加工廠
packing machine 包裝機；裝填機
packing material 包裝材料；密封材料；填充料

packing paper 包裝紙
packing plant 肉類包裝廠；罐頭廠；食品加工廠
packing ring 墊圈；密封圈
packing swelling 罐脹
paddle (type) agitator 槳式攪拌器
paddle mixer 槳式攪拌器；槳式混合器
paddle stirrer 槳式攪拌器
paddle tail 駝背笛鯛
paddlefish 匙吻鱘；白鱘
paddling 攪拌
paddy 稻；稻穀
paddy cleaner 清穀篩
paddy field 稻田；水田
paddy field rice 水稻
paddy machine 脱穀機；去稻穀機
paddy rice 未碾的稻穀；稻穀；糙米 [= rough rice]
paddy separator 脱穀機；去稻穀機
paddy straw mushroom 草菇
padi-straw mushroom 草菇
PAG (Protein Advisory Group) 聯合國蛋白質顧問組
pail 桶；提桶
pailful 一〔滿〕桶
painted bolete 斯普雷格乳牛肝菌
pakchoi 小白菜；白菜
palatability 可口性
palatability test 可口性品評；口味質量評定
palatable 美味的；可口的
palate 上顎；味覺
palate meat 上顎肉；顎肉
palatinose 巴拉金糖；異麥芽酮糖；帕拉金糖
pale ale (beer) 英國黃啤酒；黃啤〔酒〕；淡艾爾啤酒
pale beer 淺色啤酒
pale cure 淡醃；淡鹽醃食品
pale lager beer 淺色拉格啤酒；淺色貯藏啤酒
pale malt 淺色麥芽
pale sour cherry 黃色酸櫻桃
pale wine 中甜度酒；白葡萄酒
pale, soft, exudative muscle (PSE muscle) PSE肉；蒼白鬆軟和滲水肉；蒼白軟質汁液滲出肉；水樣肉
pall 走味；變得平淡無味
pallet 墊板；托盤
palm 棕櫚；棕櫚葉；手掌
palm butter 棕櫚油；棕櫚脂
palm cabbage 菜棕；椰菜
palm grease 棕櫚油
palm heart 棕櫚心
palm kernel 棕櫚仁
palm kernel oil 棕櫚仁油
palm nut 棕櫚核
palm oil 棕櫚油
palm sugar 棕櫚糖
palm wine 棕櫚酒
palmarosa 掌玫；馬丁香
palmarosa oil 掌玫油；玫瑰草油；老鸛草油
palmitic acid 棕櫚酸；軟脂酸；十六酸
palmitin 軟脂；三棕櫚精；棕櫚精；甘油〔三〕棕櫚酸酯 [= glyceryl tripalmitate; tripalmitin]
palmitoleic acid 棕櫚油酸；十六烯酸 [= zoomaric acid]
palmyra (palm) 扇椰子；扇葉樹頭櫚
palolo (worm) 磯沙蠶
pampano 鯧鯵；刺鯧
pan 平底鍋；平盤
pan arrest 盤托
pan bench 蒸發盤組
pan boiling 鍋煮〔沸〕
pan bread 土司麵包；模烤麵包
pan broiling 炕；平底鍋煎煮

pan dryer 盤式乾燥器；淺盤乾燥器
pan drying 盤式乾燥；淺盤乾燥
pan frying 煎
pan grease 烤盤油
pan greasing 烤盤塗油
pan loaf 土司麵包條；模烤長方形麵包
pan mill 盤磨機；碾碎機
pan milling 盤磨；碾碎
pan proof 烤盤醒麵
pan salt 鍋鹽；精鹽 [= stoved salt]
pan-searing 煎
pan-set cream 盤製乳油；靜〔置法〕製的奶油
panada 漿麵糊；麵糊；麵包糊
Panama orange 四季桔；金橘
panary fermentation 麵糰發酵
panbroil 平底鍋蒸煮；油煎
pancake 烙餅；班戟；薄煎餅；薄烤餅 [= hot cake]
pancake flour 薄〔煎〕餅用麵粉
pancake mix 薄煎餅混料粉；班戟粉
pancreas 胰臟；胰腺
pancreatic amylase 胰澱粉酶 [= amylopsin]
pancreatic gland 胰腺
pancreatic juice 胰液
pancreatic lipase 胰脂酶
pancreatic protease 胰蛋白酶
pancreatin 胰酶
panda 盤打酥油；熊貓
panda leaf 巴蘭葉
pandowdy 蘋果餅；蘋果布丁
panel 品評小組；評味員
panel test 小組品評；官感評定
panelist 官能檢查員；評味員
panettone 意大利式水果蛋糕
pangamic acid 平葡酸；潘胺酸
pangen 胚芽
panic grass 高粱；小米
pannikin 金屬小杯；小盤；小鍋
panning 掛糖衣；包糖衣
panocha 墨西哥粗糖
panoche 果仁奶糖
panose 潘〔諾〕糖；葡萄基麥芽糖
pansman 蒸煮工人
panthenol 泛醇；泛酰醇
pantoic acid 泛解酸
pantothenate 泛酸酯；泛酸鹽
pantothenic acid 泛酸；維生素B_3
pantoyltaurine 泛磺酸；泛解酰牛磺酸
pantry 食品室；餐具室；食物貯藏室；冷盤〔餐點〕配製室
pap 半流質食物；軟食
papain 木瓜酶；木瓜蛋白酶
papaver 罌粟
papaw 番木瓜
papaya 木瓜
papayotin 番木瓜汁
paper bag 紙袋
paper can 紙罐
paper chromatography 紙色譜法；濾紙層析法；紙色層分析法
paper cup 紙杯
paper electrophoresis 紙上電泳；濾紙電泳
papershell crab 軟殼蟹
papillote 用紙或鋁箔包裹後烹調〔之食品〕；烹調裹紙
paprika 甜辣椒〔粉〕；紅辣椒〔粉〕
paprika oleoresin 辣椒油樹脂
paprika orange 辣椒橙
paprika red 辣椒紅；辣椒色素
paraben 對羥〔基〕苯甲酸酯 [= parahydroxybenzoic acid ester]
paracasein 副酪蛋白；衍酪蛋白
parachymosin 副凝乳酶
paracolon bacilli 副大腸桿菌
paradise nut 樂園堅果；猴胡桃

paraffin 石蠟
paraffin oil 石蠟油
paraffin wax 固體石蠟；石蠟
paraglobulin 副球蛋白
Paraguay tea 巴拉圭茶
parahydroxybenzoic acid ester 對羥〔基〕苯甲酸酯 [= paraben]
paralactic acid 副乳酸
paralbumin 擬清蛋白；副白蛋白
parallel flow dryer 並流乾燥機；順流乾燥機
parallel flow evaporator 並流蒸發機；順流蒸發機
paralytic shellfish poisoning 麻痺性貝類中毒
paramyosin 副肌凝蛋白
paraprotein 副蛋白質
parasite 寄生蟲
parasitic bolete 寄生牛肝菌
parasitic infection 寄生蟲感染
parasitism 寄生；寄生蟲病
parasitosis 寄生蟲病
parasol mushroom 洋傘菇；高大環柄菇
paratha 烙餅；酥油餅；東印度薄餅
parathyroid gland 副甲狀腺；甲狀旁腺
paratrophy 營養不良；寄生性營養
parboil(ing) 預煮；預蒸煮；蒸成半熟
parboiled meat 蒸成半熟的肉
parch 烤乾；烘乾
parched barley flour 炒熟大麥粉
parched peas 乾豌豆
parchita 百香果；雞蛋果；時計果；西番蓮果；熱情果 [= granadilla; passionfruit]
parchment 咖啡豆的外殼；羊皮紙
parchment coffee 乾的帶殼咖啡豆
parchment skin 咖啡豆膜；咖啡豆角質皮
parcook(ing) 預煮；預蒸煮；蒸成半熟
pare 削皮；切去外皮
parer 削皮器；削皮刀；去皮刀
paring 削皮；切去外皮；削下的皮；皮屑
paring knife 削皮刀；去皮刀
paring machine 削皮機；去皮機
parliament cake 薄薑餅
Parmesan (cheese) 巴馬芝士；巴美乾酪；珀爾梅散乾酪；意大利式脱脂硬乾酪
parr 幼鮭
parsley 洋芫荽；洋芫茜；洋香菜；蕃荽；香芹；歐芹；荷蘭芹；西洋芹
parsley hawthorn 芹葉山楂
parsley herb oil 香芹莖油；洋芫荽莖油；歐芹莖油
parsley seed oil 香芹籽油；洋芫荽籽油；歐芹籽油
parsnip 蒲芹蘿蔔；防風〔草〕；歐洲防風草；美洲防風
partial freezing 部分冷凍
partial hydrolysis 部分水解
partial sterility 部分滅菌；部分不孕
partially incomplete protein 部分不完全蛋白質
particle 粒子；微粒
parting 分離
parting agent 脱膜劑
partition chromatography 分配層析法；分配色層法
partly dried 半乾的
partly skim milk cheese 半脱脂芝士；半脱脂乾酪
partly skimmed milk 半脱脂乳；半脱脂奶
partridge 山鶉；鷓鴣；鷓鴣；斑雞
parts per billion (ppb) 十億分點；十億分之一

parts per million (ppm) 百萬分點；百萬分之一
parvule 小粒；小丸
pasania fungus 香菇
passionfruit 百香果；時計果；雞蛋果；西番蓮果；熱情果 [= granadilla; parchita]
passivating agent 鈍化劑
passivation 鈍化
passivator 鈍化劑
pasta 麵食；麵
pasta product 麵食〔製品〕
paste 糊狀物；糊；泥；膏；漿
paste (like) food 糊狀食物
paste bag 擠花袋
paste mill 磨漿機
paste mixture 調漿機
Pasteur effect 巴斯德效應；巴氏效應
Pasteur filter 巴斯德濾菌器；巴氏濾菌器
Pasteur flask 巴斯德培養燒瓶
pasteurism 巴氏殺菌理論
pasteurization 巴斯德殺菌〔法〕；巴氏殺菌〔法〕；低溫殺菌〔法〕
pasteurize 巴氏殺菌
pasteurized beer 巴氏殺菌啤酒；低溫殺菌啤酒；熟啤酒
pasteurized lactic beverage 低溫殺菌乳酸飲料
pasteurized milk 巴氏殺菌牛乳；低溫殺菌牛乳
pasteurizer 巴氏殺菌器
pastille 錠劑；喉片
pastina 碎麵條；煮湯用小型通心粉
pasting agent 黏稠劑
pastometer 巴氏溫度計
pastrami 五香燻肉（羅馬尼亞）
pastry 餡餅皮；糕餅；麵粉製糕點
pastry flour 軟質〔小麥〕麵粉
pasty 餡餅；糊狀的；漿狀的
pasty mass 糊狀物
pat 小塊；小塊牛油
patchouli 天竺薄荷；廣藿香
patchouli oil 天竺薄荷油；廣藿香油
pate 肝醬；調味肉泥；醬狀食品；餡餅
patent 專利；專利權
patent blue V 專利藍V
patent flour 粉心〔麵〕粉
pathogen 病原體；病菌
pathogenic bacteria 病原〔細〕菌
pathogenic Escherichia coli 病原性大腸桿菌
pathogenic germ 病原菌
pathogenic microorganism 病原微生物；致病微生物
pathogenic organism 病原生物；致病生物
pathogenicity 病原性；致病性
pathology 病理學
patient 病人；患者
pattie 小餡餅；肉餅；小片糖；扁小的食物 [= patty]
pattie shell 小餡餅皮
patty 小餡餅；肉餅；小片糖；扁小的食物 [= pattie]
pattypan 烤餅鍋；小餡餅烤鍋
patum peperium 巴敦醬
pavement mushroom 大肥菇
paw 手爪；腳爪
pawpaw 木瓜；番木瓜
pea 碗豆；青豆
pea flour 豌豆粉
pea huller 豌豆脱粒機；豌豆去莢機；豌豆脱殼機
pea pod 豌豆莢
pea podder 豌豆去莢機
pea starch 豌豆澱粉

peach 桃；桃子；桃脯
peach aldehyde 桃醛；γ-十一碳烷內酯
peach brandy 桃子白蘭地酒
peach butter 桃〔子〕醬
peach flavo(u)r 桃子香精
peach jam 桃〔子〕果醬
peach kernel oil 桃仁油
peach pitter 桃〔子〕去核機
peanut 落花生；花生 [= earthnut; goober; groundnut; monkey nut]
peanut brittle 花生脆糖
peanut butter 花生醬
peanut cake 花生〔粕〕餅
peanut hull 花生殼
peanut kernel 花生米；花生仁
peanut meal 花生粕〔粉〕
peanut oil 花生油 [= arachis oil; earthnut oil; groundnut oil]
peanut sheller 花生脱殼機
pear 梨子；梨樹
pear butter 梨〔子〕醬
pear flavo(u)r 梨子香精
pear hawthorn 梨山楂
pear oily flavo(u)r 梨子油香精
pear sauce 梨醬
pear seed oil 梨籽油
pear shaped can 梨形罐
pearl 珍珠；珍珠母
pearl barley 珍珠麥；大麥仁；圓大麥米；珍珠粒大麥
pearl hen 珠雞
pearl onion 小洋葱；大頭蒜；珠葱
pearl oyster 大珍珠貝；馬氏珍珠貝
pearl sago 西米
pearl shell 珍珠貝
pearl white 魚鱗粉
pearling mill 碾穀機；大麥去殼機
pease 豌豆
pebble mill 卵石磨機
pecan 大胡桃；胡桃；美洲山核桃
pectase 果膠酶
pectate 果膠〔酸〕鹽；果膠酸酯
pectic acid 果膠酸
pectic enzyme 果膠酶
pectic substance 果膠質
pectin 果膠
pectin jelly 果膠果凍；果膠軟糖
pectin methoxylase 果膠甲氧基酶；果膠甲酯酶
pectin methylesterase 果膠甲酯酶
pectin polygalacturonase 果膠聚半乳糖醛酸酶
pectin sugar 阿拉伯糖
pectin(e) 果膠
pectinase 果膠分解酶；果膠酶
pectinate 果膠〔酯〕酸鹽
pectinesterase 果膠酯酶
pectinic acid 果膠酯酸
pectinose 果膠糖；阿拉伯糖
pectisation 果膠化〔作用〕；膠凝作用
pectization 果膠化〔作用〕；膠凝作用
pecto-cellulose 果膠纖維素
pectolase 果膠酶
pectolysis 果膠水解〔作用〕
pectolytic enzyme 果膠〔分解〕酶
pectosase 原果膠酶；果膠糖酶
pectose 原果膠；果膠糖
pectosinase 原果膠酶；果膠糖酶
pedicel 果梗；果柄；花梗
pedicle 果梗；果柄；花梗
pedigree cattle 譜系牛；純種牛
pediococcus 小球菌
peel 蔬菜皮；果皮；去皮；剝皮；去殼
peel oil 果皮油
peeled shrimp 蝦仁
peeler 去皮機；剝皮機；去殼機

peeling 剝皮；去皮；去殼
peeling knife 去皮刀；剝皮刀
peeling machine 去皮機；剝皮機；去殼機
peento 蟠桃
peg 含酒飲料；含醇飲料
pegnin 含酶牛奶
peinotherapy 飢餓療法
peka 白麴
Pekar test 畢卡〔麵粉粒度及色澤〕試驗；彼卡氏〔麵粉粒度及色澤〕試驗
Peking cabbage 大白菜；北京大白菜
Peking duck 北京鴨；北京烤鴨
Peking white pear 北京白梨；京白梨
pekoe (tea) 上等〔錫蘭〕紅茶；白毫；香紅茶
pekoe souchong 白毫疏宗茶；細毛尖香紅茶
pelagic fish 遠洋魚
pelagic fishery 遠洋魚業
pelargonic acid 壬酸；天竺葵酸
pelargonidin 天竺葵苷素；天竺葵定；天竺葵色素苷
pelargonin 花葵苷；天竺葵苷
pelargonium 天竺葵
pellagra 癩皮病
pellagramin 菸鹼酸；尼克酸
pellet 小片；小丸；小粒
pelleting press 壓片機
pelletization 造粒；壓片；製丸
pelletizer 造粒機；壓片機；製丸機
pelletizing 造粒；壓片；製丸
pelvis 骨盆
pemmican 乾肉瓶
Penaeus 對蝦屬
penetration 穿透；滲透
penetrometer 穿透度計；針入度計
penicill(in)ase 青黴素酶
penicillic acid 青黴酸
penicillin 青黴素；盤尼西林
Penicillium 青黴菌屬
penne 長通粉；筆管粉；筆尖形通心粉
penny bun 美味牛肝菌
pennyroyal oil 胡薄荷油；薄荷油；海地油
pentaerythritol ester of wood rosin 木松香季戊四醇酯
pentanal 戊醛；正戊醛
pentanoate 戊酸鹽；戊酸酯
pentanoic acid 戊酸 [= valeric acid]
pentanol 戊醇 [= pentyl alcohol; amyl alcohol]
pentanone 戊酮
pentenal 戊烯醛
pentosan 聚戊糖；戊聚糖；多縮戊糖
pentose 戊糖；五碳糖
pentoside 戊糖苷
pentyl alcohol 戊醇 [= pentanol; amyl alcohol]
pepo 瓜果；瓠瓜；南瓜子
pepper 胡椒；胡椒粉
pepper oil 胡椒油；辣椒油
pepper pot 胡椒〔粉〕瓶；辣味濃湯
pepper vine 樹蛇狀葡萄
pepperbox 胡椒〔粉〕瓶
peppercorn 胡椒子；胡椒粒
peppermint 薄荷油；胡椒薄荷；薄荷糖
peppermint camphor 薄荷醇
peppermint oil 胡椒薄荷油；薄荷油
pepperoni (sausage) 意大利辣肉腸；紅辣椒豬肉乾香腸
peppers 椒類蔬菜
peppery 胡椒〔味〕的；辣〔味〕的
peppery lactarius 乳蘑〔菇〕
peppery milk cap 辣乳菇
pepsin 胃蛋白酶
pepsinogen 胃蛋白酶原

pepsinum 胃蛋白酶
peptic 胃蛋白酶的；消化的；胃的
peptidase 肽酶
peptide 肽；縮胺酸
peptide bond 肽腱
peptide chain 肽鏈
peptidoglycan 肽聚醣
peptizate 膠溶體
peptization 膠溶〔作用〕；解膠〔作用〕；分散〔作用〕
peptizator 膠溶劑；解膠劑
peptone 腖；蛋白腖
peptone water 腖水；蛋白腖水
peptonization 腖化作用
peptonized milk 腖化乳；凝塊乳
PER (protein efficiency ratio) 蛋白質效率比〔值〕；蛋白質功效比〔值〕
per oral infection 經口腔感染
peracetic acid 過乙酸；過醋酸
peracidity 過酸性
peralkaline 過鹼性
perch 河鱸；真鱸
percolate 滲濾；滲濾液；濾出液
percolation 滲濾；滲漏；穿流
percolation filter 滲濾器
percolator 滲濾器
perdistillation 透析蒸餾
perforated 多孔的；穿孔的
perforated plate 多孔板
perforated plate column 多孔板塔；篩板塔；板板蒸餾塔
perforated sieve 多孔篩
perforation 穿孔
perfringens poisoning 產氣莢膜梭菌的食物中毒
perfume 香氣；香料；香水
perfumery 香料店；香料廠
perfumery crop 香料作物
perfusion 灌注；充滿
pergament 耐油紙；假羊皮紙
pericarp 果皮
perigord truffle 松露菌；黑孢塊菌(蕈類)
perilla 紫蘇
perilla aldehyde 紫蘇醛
perilla colo(u)r 紫蘇色素
perilla herb oil 紫蘇油
perilla oil 紫蘇油
perilla seed 紫蘇子
perillartine 紫蘇糖
perillatin 紫蘇精油
perimysium 肌束膜；肌周膜
periodic table 週期表
perishability 易腐性；變壞；變質
perishable 容易腐爛的
perishable food 易腐食品
perishables 易腐食品；易壞食品
perisperm 外胚乳
peristaltic pump 蠕動泵
peritoneal cavity 腹腔
peritoneum 腹膜
perlite 珍珠岩
permeability 滲透性；穿透性
permeable membrane 滲透膜；可透〔性〕膜
permeant 滲透物
permeaselectivity 滲透選擇性；選擇滲透性
permeate 滲透；透過液
permeation 滲透〔作用〕
permissible dose 容許劑量
permissible residue level 容許殘留量
permutite 人造沸石；濾〔水〕砂
permutite precipitator 砂濾沉澱器
pernicious anemia 惡性貧血
peroxidase 過氧化物酶
peroxidation 過氧化反應
peroxide 過氧化物

peroxide value (POV) 過氧化值
peroxisome 過氧化體；過氧化物酶體
peroxyacetic acid 過氧乙酸
perpetual raspberry 四季樹莓
perpetual strawberry 四季草莓
perry 梨酒
perry fruit 製梨酒用梨
Persian berry 波斯漿果
Persian lilac 花葉丁香
persimmon 柿子 [= kaki]
persimmon astringent 柿澀
persimmon cake 柿糕；柿餅
Peru balsam oil 秘魯香膠油
Peruvian anchovy 秘魯鯷
Peruvian ground cherry 秘魯酸漿
pest 害蟲
pest control 除蟲；殺蟲
pest control chemical 除蟲藥；殺蟲藥；農藥
pesticide 農藥；殺蟲劑；除蟲劑
pesticide residue 農藥殘留；殘留農藥
pestle 研杵；研棒；碾槌；用杵搗碎
pestle mill 研磨；搗錘磨
pestle shaped puffball 杯狀巾馬勃(蕈類)
petal 花瓣
petis 魚醬油(印尼)
petit four 法式小蛋糕；花飾小蛋糕
petitgrain dai-dai oil 玳玳葉油
petitgrain lemon oil 檸檬葉油
petitgrain oil 卑檸油；橙葉油
petits pois 小甜豆
petri dish 培養皿
petri plate 培養皿
petrolatum 凡士林；礦脂；石油凍
petroleum 石油
petroleum ether 石油醚
petroleum wax 石油石蠟
pettitoes 豬蹄；豬腳
petunidin 牽牛花素；甲花翠素
pewter 白蠟
pexin 凝乳酶
pexinogen 凝乳酶原
PG (propyl gallate) 沒食子酸丙酯；五倍子酸丙酯
pH 酸鹼度
pH adjusting agent pH調節劑
pH indicator pH指示劑；酸鹼度指示劑
pH meter 酸鹼度計；pH計；氫離子濃度測定計
pH paper pH試紙；酸鹼度試紙
pH scale 酸鹼值
pH value pH值；酸鹼值
phaeophorbide 去鎂葉綠酸；脫鎂葉綠酸
phaeophyta 褐藻〔門〕
phaeophytin 去鎂葉綠素，脫鎂葉綠素
phage 噬菌體 [= bacteriophage]
phagocyte 噬菌細胞；吞噬細胞
phagocytosis 吞噬細胞作用；噬菌作用
phagology 攝食學
phagomania 貪食癖
phagophobia 厭食症；恐食症
phallin 白蕈毒〔肽〕；溶血菌毒肽
pharmaceutical 藥品；製藥；藥用的
pharmaceutical botany 藥用植物學
pharmaceutical plant 藥用植物
pharmacology 藥理學
pharmacon 藥物
pharmacopoeia 藥典
pharmacotherapy 藥物療法
pharynx 咽〔喉〕
phase 相；階段
phase contrast microscope 位相差顯微鏡；相差顯微鏡
phase diagram 相〔轉變〕圖
phase equilibrium 相平衡
phase interface 相界面

phase inversion 相轉換；相變換
phaseolin 菜豆球蛋白
phaseoline 菜豆鹼
phaseolunatin 菜豆苷；菜豆亭
Phaseolus angularis 紅豆
Phaseolus calcaratus 赤小豆
Phaseolus vulgaris 菜豆
pheasant 雉；雉鳥；野雞
phellandrene 水芹烯
phenethyl acetate 乙酸苯乙酯
phenethyl alcohol 苯乙醇
phenethyl anthranilate 鄰氨基苯甲酸苯乙酯
phenethyl butyrate 丁酸苯乙酯
phenethyl cinnamate 桂酸苯乙酯；肉桂酸苯乙酯
phenethyl formate 甲酸苯乙酯
phenethyl isobutyrate 異丁酸苯乙酯
phenethyl isovalerate 異戊酸苯乙酯
2-phenethyl-2-methylbutyrate 2-甲基丁酸-2-苯乙酯
phenethyl octanoate 辛酸苯乙酯
phenethyl phenylacetate 苯乙酸苯乙酯
phenethyl propionate 丙酸苯乙酯
phenethyl salicylate 水楊酸苯乙酯
phenol 苯酚；石碳酸
phenol oxidase 酚氧化酶
phenol red 酚紅
phenol resin 酚樹脂
phenol-sulfuric acid method 酚—硫酸呈色法
phenol unit 酚單位(煙燻食品)
phenolase 酚酶
phenolate 酚鹽
phenolic compound 酚化合物
phenolic resin 酚醛樹脂
phenolphthalein 酚酞
phenomenal berry 現象莓
phenotype 表現型；顯型
2-phenoxyethyl isobutyrate 異丁酸-2-苯氧基乙酯
o-phenyl phenol (OPP) 鄰苯基苯酚；聯苯酚
phenylacetaldehyde 苯乙醛
phenylacetaldehyde dimethyl acetal 苯乙酸二甲縮醛
phenylacetic acid 苯乙酸
phenylalanine 苯〔基〕丙胺酸
phenylbenzene 聯苯
phenylhydrazone 苯棕
phenylketonuria (PKU) 苯酮尿症
3-phenylpropyl acetate 乙酸苯丙酯
phenylpropyl alcohol 苯丙酸
pheophorbide 去鎂葉綠酸；脱鎂葉綠酸
pheophytin 去鎂葉綠素；脱鎂葉綠素
phloem 韌皮部
phloretin 根皮素
phloridzin 根皮苷
phloxine 玫瑰紅；根皮紅；熒光桃紅
phoenix cap 扣帶式蓋
phoenix mushroom 鳳尾菇
phosphagen 磷肌酸；磷酸原
phosphatase 磷酸酶
phosphate 磷酸鹽；磷酸酯
phosphate pesticide 有機磷農藥 [= organophosphorus pesticide]
phosphated distarch phosphate 磷酸化二澱粉磷酸酯
phosphatidase 磷脂酶
phosphatide 磷脂
phosphatidic acid 磷脂酸
phosphatidyl glycerol 磷脂酰甘油
phosphatidyl sugar 磷脂酰糖
phosphatidylcholine 卵磷脂；磷脂酰膽鹼；膽鹼磷脂 [= lecithin(e)]
phosphatidylethanolamine 磷脂酰乙醇胺；腦磷脂

phosphatidylserine 磷脂酰絲胺酸；絲胺酸磷脂
phosphocholine 磷膽鹼
phosphocreatine 磷酸肌酸；肌酸磷酸 [= creatine phosphate]
phosphoesterase 磷酸酯酶
phosphoglobulin 磷〔酸〕球蛋白
phosphoglucoisomerase 磷葡萄糖異構酶；磷酸葡〔萄〕糖異構酶
phosphoglucomutase 磷葡萄糖變位酶；葡〔萄〕糖磷酸變位酶
phosphogluconic acid 磷葡萄糖酸；磷酸葡糖酸
phosphoglucoprotein 磷〔酸〕糖蛋白
phosphoglyceride 磷〔酸〕甘油酯
phosphoglycerol 磷〔酸〕甘油
phosphoinositide 磷肌醇甘油酯；磷酸肌醇
phospholipase 磷脂酶
phospholipid 磷脂
phospholipoprotein 磷脂蛋白
phosphoprotein 磷蛋白質
phosphoriboside 磷酸核糖苷
phosphoric acid 磷酸
phosphorism 磷中毒；慢性磷中毒
phosphorous acid 亞磷酸
phosphorus 磷
phosphorylase 磷酸解酶；磷酸化酶；加磷酸分解酶
phosphorylation 磷酸化〔作用〕
phosphoserine 磷絲胺酸
phosphosphingolipid 磷神經脂質；鞘磷脂
phosvitin 卵黃磷醣蛋白；卵黃高磷蛋白
photoautotroph 光合自營生物；光合自養生物
photocatalysis 光催化〔作用〕
photodecomposition 光〔分〕解作用；感光分解
photodegradation 光降解作用；感光降解
photodensitometer 光密度計
photoelectric colo(u)rimeter 光電比色計
photoheterotroph 光合異營生物；光合異養生物
photooxidation 光〔致〕氧化作用
photostability 耐光性
photosynthate 光合作用產物
photosynthesis 光合作用
photosynthetic pigment 光合的
phototroph 光合生物；光養生物
phreatic water 地下水；井水
phthalic acid 苯二酸；磷苯二甲酸
phycobilin 藻膽色素
phycochrome 藻色素
phycocyan 藻青蛋白
phycocyanobilin 藻青素；藻藍素
phycoerythrin 藻紅蛋白
phycoerythrobilin 藻紅素
phycology 藻類學
phycomycete 藻菌
phycophaein 藻褐素
phycophyte 藻類
phycoxanthin 藻黃質；藻黃素
phyllochlorin 葉綠〔素〕蛋白
phyllodulcin 葉甜精；葉甜素；甘茶素；葉甘素
phylloerythrin 葉紅素
phylloporphyrin 葉紫質
phylloquinone 葉〔綠〕醌；維生素K_1
phylloxanthin 葉黃質；葉黃素
physalin 酸漿果紅素
physical change 物理變化
physical hygrometer 乾濕球濕度計
physical refining 物理精煉
physicochemical 物理化學的

physiological 生理〔學〕的
physiological disorder 生理生調
physiological saline 生理〔食〕鹽水
physiology 生理學
phytagglutinin 植物凝集素
phytalbumin 植〔物〕白蛋白
phytase 植酸酶
phytate 植酸鹽；植酸酯
phytic acid 植酸；肌醇六磷酸
phytin 植酸鈣鎂
phytobenthos 水底植物；海底植物
phytochemical 植物化學的
phytochemicals 植物化合物
phytochemistry 植物化學
phytochrome 植物〔光敏〕色素；植〔物〕色蛋白
phytocide 除莠劑；除草劑
phytoglobulin 植物球蛋白
phytol 植醇；葉綠醇
phytolipopolysaccharide 植物脂多醣
phytology 植物學
phytomelane 植物黑素
phytonadione 維生素K_1；植物甲萘醌
phytone 植物蛋白腖
phytoplankton 浮游植物
phytosterin 植物固醇
phytosterol 植物固醇
phytotoxin 植物毒素
phytotrophy 植物自營；植物營養
phytoxanthin 葉黃質；葉黃素
pica 異食症；異食癖
pick(ing) 採摘；挑選；拔毛
picker 採收者；採摘者；拔毛機；採果器
pickerel 小梭魚；小狗魚
pickle 醃漬；醋漬；醃菜；酸菜；泡菜；醃漬食品；醋漬食品；醃漬液；鹽汁；滷汁
pickle curing 濕法鹽醃；液醃法
pickle injection method 醃液注射醃漬法
pickle milk cap 常見乳菇
pickle pumping needle 鹽水注射針
pickle salting 濕法鹽漬；鹽水醃漬
pickled cucumber 醃黃瓜；酸黃瓜
pickled food 醃漬食品；醋漬食品
pickled fruit 醃漬水果
pickled vegetable 醃菜；泡菜；酸漬菜；醋漬菜
pickling 醃漬；醋漬；浸酸
pickling tank 醃漬槽；醋漬槽
picnic 豬肩肉；野餐
picnic ham 肩肉火腿
picnic shoulder 豬肩肉；肩膀肉
picric acid 苦味酸
picul 擔（一百市斤）
pidan 皮蛋 [= alkalized (duck) egg; Chinese egg; thousand years' egg; pitan]
pie 餡餅；派
pie cherry 歐洲酸櫻桃
pie crust 餡餅皮；派皮
pie filling 派餡；餅餡
piece 塊；片；件；張
piepan 烤盤；餡餅烤盤
pieplant 食用大黃
piezometer 壓力計
pig 豬；豬肉
pigeon 白鴿；鴿子
pigeon grape 夏葡萄
pigeon pea 木豆；柳豆；樹豆；鴿豌豆
piggy 小豬
piglet 小豬；乳豬
pigling 小豬
pigment 色素；顏料
pigment cell 色素細胞
pigmentation 著色；色素形成

pignolia 松子；松果
pignut 苦山核桃；山胡桃果；光滑山核桃
pike 梭子魚；狗魚
pike eel 海鰻
pikelet 小煎薄餅；甜麵包
pilaf 西式炒飯；肉飯；辛香飯
pilau 皮羅〔肉〕飯；肉飯
pilchard 沙丁魚
pilchard oil 沙丁魚油 [= sardine oil]
pileus 菌傘；菌蓋
pilfer-proof cap (PP cap) 扭斷螺紋蓋；PP蓋
pill 丸劑；藥丸
pilose 多毛的；有毛的
pilot plant 中間工廠；領試工廠；先導工廠；試驗工場
pilot plant scale 中間工廠規模；領試工廠規模；先導工廠規模；試驗工場規模
pilot plant trial 中間工廠試產；領試工廠試產；先導工廠試產；試驗工場試產
Pilsner beer 皮爾遜啤酒；比爾森型啤酒
pimenta 多香果
pimenta oil 多香果油
pimento 眾香子；多香果；甘椒；甜椒 [= allspice; Jamaica pepper]
pimento oil 眾香子油；多香果油
pin 大頭針；別針；軋麵杖；通心粉的模心
pin bone 魚的細刺骨
pin hole can 穿孔罐；有微孔的罐
pin hole detector 針孔偵測器
pin point colony 針點狀菌落
pinang 檳榔
pinaster 松樹
pinaster seed oil 松果油
pincers 夾子；鉗子；螯（蝦或蟹等）
pine 松樹
pine mushroom 松蕈；松茸
pine needle oil 松葉油；松針油
pine nut 松果；松子
pine spike cap 淺紅色釘菇
pineapple 鳳梨；菠蘿 [= ananas]
pineapple bread 菠蘿包
pineapple flavo(u)r 菠蘿香精
pineapple guava 鳳榴；費約果
pineapple juice 鳳梨〔果〕汁；菠蘿汁
pineapple oil 鳳梨精油
pineapple oily flavo(u)r 菠蘿油香精
pineapple peeler-corer 鳳梨去皮通心機
pineapple slicer 鳳梨切片機；菠蘿切片機
pineapple spiral slice 鳳梨旋片；菠蘿旋片
pineapple tidbit 鳳梨扇片；菠蘿扇片
pineapple whole slice 鳳梨圓片；菠蘿圓片
pineapple wine 鳳梨〔甜〕酒
pinene 蒎烯；松萜
pinery 鳳梨溫室；菠蘿圓
pinfish 針頭魚；菱體兔牙鯛；重牙鯛
ping pong 蘋婆；鳳眼果
pink abalone 皺紋鮑〔魚〕
pink champagne 玫瑰香檳酒
pink fair head 肉色麗傘（蕈類）
pink laver 紅藻
pink salmon 細鱗大麻哈魚
pink tipped coral fungus 紅頂枝瑚菌（蕈類）
pink wine 粉紅葡萄酒；桃紅葡萄酒
pinking 紅變；變紅
pinning 軋薄麵糰
pinny 圍巾；圍嘴兒
pinole 炒玉米粉；甜炒玉米粉
pinon 矮松子；雪松果

pint 品脫
pinto bean 斑豆
pip 種子；果仁
pip fruit 仁果
pipe 管；煙斗；大酒桶
pipe cooler 管式冷卻器
pipe dryer 管導式乾燥機
pipe rigate 稜紋管通心粉
Piper nigrum 胡椒
piperic acid 胡椒酸
piperidine 胡椒啶
piperine 胡椒鹼
piperonal 胡椒醛；洋茉莉醛
piperonyl alcohol 胡椒醇
piperonyl butoxide 胡椒基丁醚
pipet(te) 吸管；移液管
piping sugar 糕點的裝飾糖衣；糕點的糖製花飾
pipkin 長柄鍋
pipless 無籽的；無果仁的
pippy 多種子的；多果仁的
piquancy 辛辣；開胃
piquant 辛辣的；開胃的
piquant sauce 辣醬；辣醬；開胃醬
piquillos iodosa 西班牙甜椒
pisatin 豌豆素
piscidic acid 番石榴酸
piscina 魚池；養魚池塘
piscine 魚的；養魚的
pistachio 阿月渾子樹；阿月渾子果實
pistachio nut 開心果
piston filler 活塞式裝填機
Pisum sativum 豌豆
pit 果核；去〔果〕核；貯藏窖；凹痕
pit cavity 核窩
pitaya 火龍果
pitcher 帶柄水罐；托壺
pitching yeast 種酵母；製酒用酵母
pitted 去核的
pitter 去核機
pitting 去核
pitting machine 去核機
pizza 薄餅；比薩；意大利薄餅
place mat 席位的餐具墊布或墊紙
plaice 鰈；河鰈；右鰈
plain bread 淡麵包
plain can 無塗漆罐
plain chocolate 素巧克力；純巧克力
plain condensed milk 無糖煉奶；濃縮乳
plain food 便餐
plain noodle 素麵條；普通麵條
plankton 浮游生物
plansifter 盤式篩選機；磨粉用套篩
plant 植物；工廠；車間
plant acid 植物酸
plant alkaloid 植物鹼
plant equipment 工廠〔生產〕設備
plant exudates 植物滲出物
plant food 植物性食品；素食；植物養料
plant hormone 植物激素
plant lectin 植物凝集素
plant nutrition 植物營養
plant product 植物性產品
plant protein 植物蛋白〔質〕
plant rennet 植物性凝乳酶
plant sanitation 工廠衛生
plant sap 植物汁液
plant tissue 植物組織
plant toxin 植物性毒素
plantain 大蕉；車前草
plantation 種植園
plantation sugar 耕地蔗糖
plantation white (sugar) 耕地白糖
plantose 菜子〔白〕蛋白
plasma 血漿；漿
plasma cholesterol 血漿膽固醇

plasma free amino acid 血漿游離胺基酸；血漿自由胺基酸
plasma free fatty acid 血漿游離脂肪酸；血漿自由脂肪酸
plasma lipid 血漿脂質
plasma protein 血漿蛋白質
plasmid 質體；質粒
plasmin 胞漿素；血纖維蛋白溶酶
plastic 塑膠；塑料製品；塑性的
plastic bottle 塑膠瓶
plastic container 塑膠容器
plastic cream 塑性奶油
plastic dough 可塑性麵糰
plastic fat 塑性脂肪；軟化脂
plastic film 塑性膜；塑料薄膜
plastic foam 泡沫塑料
plastic strip door 透明塑膠條門簾（冷庫用）
plastic tray 塑膠盤
plasticity 塑性；可塑性
plasticization 塑化；增塑
plasticizer 塑化劑；增塑劑；可塑劑
plasticizing agent 塑化劑；增塑劑
plastifier 塑化劑；增塑劑
plastigel 塑性凝膠
plastisol 塑料溶膠
plastograph 塑變記錄儀；稠度曲線記錄儀
plastometer 塑性計；塑度計
plate 板；盤；塔板；牛胸肉；平皿；平板培養皿
plate (type) heat exchanger 層板熱交換器；片式熱交換器；板式熱交換器
plate contact freezer 層板接觸冷凍機；平板式接觸冷凍機
plate cooler 層板冷卻器；平板式冷卻器
plate count agar 平板計數；平皿計數
plate count agar 總菌數洋菜培養基
plate culture 平皿培養〔物〕；平板培養〔物〕
plate dilution 平板稀釋；平皿稀釋
plate dryer 層板乾燥器；多層式乾燥器
plate end 牛前胸肉
plate evaporator 層板蒸發器；板式蒸發器
plate filter 薄板濾器
plate freezer 層板冷凍機；平板式冷凍機
plate freezing 層板冷凍〔法〕；平板式冷凍〔法〕；金屬板接觸冷凍〔法〕
plate heater 層板加熱器；片式加熱器
plate ice 片狀冰；板狀冰
plate method 平板培養法；平皿培養法
plate pasteurizer 層板〔巴氏〕殺菌器；板式〔巴氏〕殺菌器；片式〔巴氏〕殺菌器
plate press 板式壓榨機
plate rack 餐具架
plate sieve 板篩
platform inspection 收乳站檢查
platform test 收乳臺試驗：牛奶進廠時的平台試驗
plating 平板培養；平皿接種；鍍
plating medium 平板培養基
plating method 平板培養法；平皿培養法
platinum (wire) loop 白金絲耳；白金接種環
platinum needle 白金接種針
platinum wire 白金絲
Plato saccharometer 柏拉圖糖度計
Pleurotus 側耳屬（食用菌）
Pleurotus cystidiosus 鮑魚菇
Pleurotus obalonus 蠔菇
Pleurotus ostreatus 鮑魚菇；糙皮側耳
pleuston 水漂生物
pliability 柔韌性

plodder 螺旋擠壓機
pluck 內臟；雜碎；拔毛；採摘
pluck tea 採茶
plucker 拔毛機；採摘工人
plug cock 旋塞
plug ferment 塞子發酵罐(製啤酒用)
plug tobacco 口嚼煙〔草〕
plugging machine 塞瓶機
plum 梅；李子；洋李
plum cake 葡萄乾蛋糕
plum pudding 葡萄乾布丁
plumbism 鉛中毒
plumcot 李杏
plume 羽毛
plumelet 小羽毛
plumping agent 膨脹劑
plumpness 飽滿度；豐滿度
plumule 幼芽；胚芽
plural gel 複合凝膠
pluteus 光柄蘑菇；海膽類長腕幼蟲
ply 層板
plymouth rock 蘆花雞
plywood 夾板；三夾板
pneumatic classification 氣動選別；風力分級
pneumatic classifier 氣動選別機；風力分級機
pneumatic control 氣動控制
pneumatic conveying dryer 氣流乾燥機
pneumatic conveyor 氣動輸送機
pneumatic dryer 氣流乾燥機
pneumatic mill 風〔動〕磨
pneumatic rice pearler 噴風式碾米機
pneumatic separator 風選機；風力分選機
pneumatic stunner 氣動擊昏器
poached egg 水煮荷包蛋
poaching 灼；燙滾；水煮
pocket (type) can 袖珍罐；矮型罐
pod 莢；豆莢
podded 有莢的
pogy 鯡
poise 泊(黏度單位)
poison 毒藥；毒物；毒素
poisoning 中毒
poisonous 有毒的
poisonous fish 毒魚
poisonous fungus 有毒真菌
poisonous mushroom 毒菇
polar amino acid 極性胺基酸
polar cod 北極鱈
polar lipid 極性脂肪
polar solvent 極性溶劑
polariscope 旋光計；偏光鏡
polarizer 偏光鏡
polenta 玉米粥；粟粉粥；大麥〔片〕粥
Polish sausage 波蘭香腸
polished rice 精白米
polisher 拋光機
polishing 拋光；研光；磨光；擦光
polishing machine 拋光機；碾米機；磨光機
polishings 米糠
pollack 鱈魚；青鱈；鯖鱈 [=pollock]
pollan 白鮭(愛爾蘭)
pollard 碎麩皮；糠屑
polled 無角的
pollen 細磨麵粉；細麩皮；花粉
pollen grain 花粉粒
pollock 鱈魚；青鱈；鯖鱈 [=pollack]
pollutant 污染物〔質〕
polluted air 污染空氣
polluted water 污染水；污水
pollution 污染
pollution index 污染指數
polony sausage 半熟乾〔豬肉〕香腸；波羅泥〔豬肉〕香腸
poloxamer 331 波洛桑默331

poloxamer 407 波洛桑默407
polyacrylamide 聚丙烯酰胺
polyacrylamide gel 聚丙烯酰胺凝膠
polyamide 聚酰胺；尼龍 [= nylon]
polyamine 多胺；聚胺
polyavitaminosis 多種維生素缺乏性
polybag 聚乙烯〔薄膜〕袋 [= polythene bag]
polybasic acid 多元酸
polybasic alcohol 多元醇
polybasic ester 多元酸酯
polybutene 聚丁烯
polycarbonate 聚碳酸酯
polychlorinated biphenyl (PCB) 多氯聯苯
polycondensate 多縮合物；縮聚物
polycondensation 多縮合〔作用〕；縮聚〔作用〕
polydextrose 聚糊精；聚葡萄糖
polydimethylsiloxane 聚二甲基硅醚；聚二甲基硅氧烷
polyelectrolyte 高分子電解質；聚合〔高分子〕電解質
polyenoic fatty acid 多烯脂肪酸
polyester 聚酯
polyester film 聚酯薄膜
polyester shrink film 聚酯收縮膜
polyethylene (PE) 聚乙烯 [= polythene]
polyethylene film 聚乙烯薄膜
polyethylene glycol 聚乙二醇
polyethylene shrink film 聚乙烯收縮膜
polyethylene terephthalate (PET) 聚酯；聚對苯二甲酸乙二醇酯
polyfoil 鋁箔與膠膜複合之包裝材料；多層箔 [= multifoil]
polygalacturonase 聚半乳糖醛酸酶
polygalacturonic acid 聚半乳糖醛酸
polyglycerol ester 聚甘油酯
polyglycerol esters of fatty acid (PGFE) 脂肪酸聚甘油酯
polyglycerol esters of interesterified ricinoleic acid (PGPR) 交酯化蓖麻酸聚甘油酯
polyglycine 聚甘胺酯
polyhydric alcohol 多元醇 [= polyol]
polyisobutylene 聚異丁烯
polymer 聚合物；聚合體
polymerase 聚合酶
polymerism 聚合現象
polymerization 聚合作用
polymerized oil 聚合油
polymetaphosphate 聚〔合〕偏磷酸鹽 [= kraul salt]
polymorphism 同質多〔晶〕型現象
polymyxin 多黏〔桿〕菌素
polynucleotidase 多核苷酸酶
polynucleotide 多核苷酸
polyol 多元醇 [= polyhydric alcohol]
polyose 多糖；聚糖 [= polysaccharide; polysaccharose]
polyoxyethylene higher aliphatic alcohol 聚氧乙烷高級脂肪醇
polyoxyethylene sorbitan monooleate 聚氧乙烯脫水山梨醇單油酸酯；吐溫80；土溫80 [= Tween 80]
polyoxyethylene xylitan monostearate 聚氧乙烯木糖醇酐單硬脂酸酯
polyoxyethylene(40)stearate 聚氧乙烯(40)硬脂酸酯
polyoxyethylene(8)stearate 聚氧乙烯(8)硬脂酸酯
polyoxypropylene glycerol ether 聚氧丙烯甘油醚；消泡劑GP
polyoxypropylene oxyethylene glycerol ether 聚氧丙烯氧化乙烯甘油醚；消泡劑GPE
polypeptidase 多肽酶

polypeptide 多肽
polyphagia 雜食性；貪食；多食症
polyphagy 雜食性；貪食；多食症
polyphenol 多酚
polyphenolic compound 多酚化合物
polyphenolic oxidase 多酚氧化酶
polyphosphate 聚磷酸鹽；多磷酸鹽
polypore 多孔菌
polypropylene (PP) 聚丙烯
polypropylene glycol 聚丙烯乙二醇
polysaccharase 多糖酶 [= polysaccharidase]
polysaccharidase 多糖酶 [= polysaccharase]
polysaccharide 多糖；聚糖 [= polyose; polysaccharose]
polysaccharide phosphorylase 多糖磷酸解酶
polysaccharose 多糖；聚糖 [= polyose; polysaccharide]
polysarcia 肥胖
polysorbate 聚山梨〔糖〕醇酯；聚山梨酸酯(乳化劑)
polysorbate 20 聚山梨酸酯 20；吐溫20
polysorbate 40 聚山梨酸酯40；吐溫40
polysorbate 60 聚山梨酸酯60；吐溫60
polysorbate 65 聚山梨酸酯65；吐溫65
polysorbate 80 聚山梨酸酯80；吐溫80
polystyrene (PS) 聚苯乙烯
polystyrene shrink film 聚苯乙烯收縮膜
polythene 聚乙烯 [= polyethylene]
polythene bag 聚乙烯〔薄膜〕袋 [= polybag]
polythylene glycol 聚乙二醇
polytrophia 營養過度
polyunsaturated fat 多不飽和脂肪
polyunsaturated fatty acid 多不飽和脂肪酸
polyurethane 聚胺酯
polyurethane resin 聚胺酯樹脂
polyuronic acid 聚糖醛酸
polyvinyl acetate 聚乙酸乙烯酯；聚醋酸乙烯酯
polyvinyl chloride (PVC) 聚氯乙烯
polyvinyl chloride film 聚氯乙烯薄膜
polyvinyl chloride shrink film 聚氯乙烯收縮膜
polyvinyl resin 聚乙烯樹脂
polyvinylidene chloride (PVDC) 聚二氯亞乙烯；聚偏二氯乙烯
polyvinylidene chloride casing 聚二氯亞乙烯腸衣；聚偏二氯乙烯腸衣
polyvinylpolypyrrolidone (PVPP) 聚乙烯聚吡咯烷酮
polyvinylpyrrolidone (PVP) 聚乙烯吡咯烷酮
pomace 果漿；油粕；果渣
pomace wine 果渣酒
pomaceous 蘋果類的
pomaceous fruit 仁果類；梨果
pombe 非洲粟啤酒；非洲小米啤酒
pome 仁果；梨果
pome fruit 仁果；梨果
pomegranate 石榴
pomelin 蜜柑蛋白；柑橘球蛋白
pomelo 柚〔子〕；文旦 [= pummelo; pompelmous; shaddock]
pomewater 大甜蘋果
pomfret 鯧魚；銀鯧
pommard 波馬特〔紅葡萄〕酒
pomology 果樹〔栽培〕學
pompelmous 柚〔子〕；文旦 [= pomelo; pummelo; shaddock]
pompion oil 南瓜子油
ponceau 4R 胭脂紅；麗春紅4R; 食品紅7號

ponceau 4R aluminium lake 胭脂紅鋁色澱
pond 池塘
pond apple 圓滑番荔枝
pond fish 池魚
pond loach 土鰍
pone 玉米麵包；玉米餅
ponkan 椪柑；潮州柑
pony 小馬；圓球狀的小〔白蘭地酒〕杯；小杯
poonac 椰粕
popcorn 爆玉米花；爆穀；爆米花
poplar field cap 柱狀田頭菇
popover 脆薄空心鬆餅；淡烤酥餅
popper 爆玉米花機
poppy 罌粟
poppy seed 罌粟籽；芙蓉子
popsicle 冰棒；冰棍
porcelain 瓷器；瓷
porcelain clay 瓷土
porcelain crucible 瓷坩堝
porcelain enamel 搪瓷
porcelain filter 陶瓷濾器
porcelain fungus 霉狀小奧德菇
porcine 豬的
porcupine 箭豬；刺蝟；豪豬
pore 孔；氣孔
porgy 異黑鯛；真鯛
pork 豬肉
pork carcass 豬屠體
pork chop 豬排
pork fat 豬油
pork fillet 豬柳
pork floss 豬肉絲
pork leg schnitzels 炸豬腿排
pork leg steak 豬腿排
pork loin chop 帶骨豬腰肉排
pork loin medallion steak 圓狀豬腰肉排
pork loin steak 豬腰肉排
pork luncheon meat 午餐豬肉；午餐肉
pork pie 豬肉餡餅
pork pig 肉豬
pork rump steak 豬臀肉排
pork sausage 豬肉香腸
pork scrap 豬油粕
pork skin 豬皮
pork stomach 豬肚；豬胃
pork tapeworm 豬肉條蟲
pork tongue 豬舌
porker 肥小豬；食用豬
porkling 小豬；乳豬
porous body 素燒瓷體；多孔物體
porous sheet 素燒板
porphyra 紫菜屬；紅藻
porphyrin 紫質
porphyrine 紫菜鹼
porphyrization 研細；粉碎
porphyry mushroom 紫帶蘑菇
porpoise 白鯨；海豚；鼠海豚
porridge 粥；糜；麥片粥
porringer 粥碗；小湯碗
port 波特葡萄酒；砵酒；港口；艙口
port wine 波特紅〔葡萄〕酒；波爾特紅〔葡萄〕酒
portable 手提式的
portable stove 手提爐
porter 英國黑啤酒；黑啤酒
porterhouse 餐館；啤酒店；前腰脊肉；里脊牛排；一人份的上等牛排
porterhouse steak 前腰脊肉；里脊牛排；上等腰脊肉牛排
portion pack 一客份包裝；定量包裝
portion pack food 一客〔份〕包裝食品；定量包裝食品
portioning machine 定量分裝機
positive feed 強制餵食；強制進料
positive staining 陽性染色〔法〕
posset 牛奶甜酒；牛乳酒

post bleaching 後漂白
post defecated juice 後澄清汁
post defecation 後澄清
post fermentation 後發酵
post harvest handling 收穫後處理
post harvest ripening 收穫後熟成；後熟
post maturation 後熟
postmortem ag(e)ing 屠後熟成；屠後陳化；宰後嫩化；宰後排酸
postmortem change 屠後變化；宰後變化
postmortem inspection 屠後檢驗；宰後檢驗
postmortem rigidity 屠後僵直；宰後僵直；死後僵直；死僵
postprandial 餐後的；正餐後的
postprandial blood sugar 餐後血糖
postprandial hypoglycemia 餐後低血糖症
pot 罐；壺；鍋；盆；瓶
pot ale 醪糟；酒糟
pot barley 去殼大麥〔粒〕
pot cheese 酪農乾酪；酪農芝士；卡達乾酪；卡達芝士 [= cottage cheese]
pot mill 球〔形〕磨機
pot roast 燉肉；燜燒肉
potable 適於飲用的；可飲用的；飲料
potable spirit 飲料酒〔精〕
potable water 飲用水
potableness 可飲性
potage 濃湯；肉〔汁濃〕湯
potassium 鉀
potassium acetate 醋酸鉀；乙酸鉀
potassium acid tartrate 酒石酸氫鉀；塔塔粉；酒石 [= cream of tartar; potassium bitartrate; potassium hydrogen tartrate]
potassium alginate 褐藻酸鉀；海藻酸鉀
potassium alum 鉀〔明〕礬；硫酸鋁鉀；明礬鉀
potassium benzoate 苯甲酸鉀
potassium bicarbonate 碳酸氫鉀
potassium bisulfate 硫酸氫鉀
potassium bisulfite 亞硫酸氫鉀 [= potassium hydrogen sulfite]
potassium bitartrate 酒石酸氫鉀；塔塔粉；酒石 [= cream of tartar; potassium acid tartrate; potassium hydrogen tartrate]
potassium bromate 溴酸鉀
potassium carbonate 碳酸鉀
potassium chloride 氯化鉀
potassium citrate 檸檬酸鉀；檸檬酸三鉀
potassium copper chlorophyllin 葉綠素銅鉀鹽
potassium dihydrogen citrate 檸檬酸一鉀；檸檬酸二氫鉀
potassium dihydrogen phosphate 磷酸二氫鉀
potassium ferrocyanide 亞鐵氰化鉀
potassium gibberellate 赤霉酸鉀
potassium gluconate 葡糖酸鉀
potassium glycerophosphate 甘油磷酸鉀
potassium hydrogen sulfite 亞硫酸氫鉀 [= potassium bisulfite]
potassium hydrogen tartrate 酒石酸氫鉀；塔塔粉；酒石 [= cream of tartar; potassium bitartrate; potassium acid tartrate]
potassium hydroxide 氫氧化鉀
potassium iodate 碘酸鉀
potassium iodide 碘化鉀
potassium lactate 乳酸鉀

potassium malate 蘋果酸鉀
potassium metabisulfite 偏亞硫酸鉀；焦亞硫酸鉀
potassium metaphosphate 偏磷酸鉀
potassium nitrate 硝酸鉀
potassium nitrite 亞硝酸鉀
potassium oxalate 草酸鉀
potassium permanganate 高錳酸鉀；灰錳氧
potassium persulfate 過硫酸鉀
potassium phosphate 磷酸鉀
potassium polymetaphosphate 聚偏磷酸鉀；多偏磷酸鉀
potassium polyphosphate 聚磷酸鉀；多磷酸鉀
potassium propionate 丙酸鉀
potassium pyrophosphate 焦磷酸鉀
potassium saccharin 糖精鉀
potassium sorbate 山梨酸鉀；己二烯酸鉀
potassium stearate 硬脂酸鉀
potassium sulfate 硫酸鉀
potassium sulfite 亞硫酸鉀
potassium tartrate 酒石酸鉀
potassium tripolyphosphate 三聚磷酸鉀；三多磷酸鉀
potation 暢飲；飲酒；酒類飲料
potato 馬鈴薯；薯仔；洋芋；土豆
potato bean 利馬豆；菜豆
potato cake 炸馬鈴薯餅
potato chip 炸〔馬鈴〕薯片；炸土豆片；洋芋片；馬鈴薯片 [= potato crisp]
potato crisp 炸〔馬鈴〕薯片；炸土豆片；洋芋片；馬鈴薯片 [= potato chip]
potato dextrose agar 馬鈴薯葡萄糖洋菜〔培養基〕；馬鈴薯葡萄糖瓊脂〔培養基〕
potato flake 馬鈴薯雪片；土豆片
potato flour 馬鈴薯粉；薯粉
potato masher 馬鈴薯泥搗碎器
potato pancake 馬鈴薯餅
potato peeler 馬鈴薯去皮機
potato puree 馬鈴薯泥
potato starch 馬鈴薯澱粉
potato tuber 馬鈴薯塊莖
potato whisky 馬鈴薯威士忌〔酒〕
potaufeu 沙鍋燉菜
potcher 漂洗槽
potentiating agent 增效劑
potentiation 增效作用
potful 一鍋的量；一壺的量
potherb 野菜；熟食的葉用菜
potherb fameflower 土人參
potherb mallow 冬寒菜
potherb mustard 雪菜
potluck 自備聚餐；便飯
potomania 嗜酒狂
potpie 菜肉餡餅
potpourri 肉菜雜燴；百花香；混合香料
pottage 濃湯
potted ham 罐裝火腿
pottery 陶器
pottery clay 陶土
potting 裝填；裝瓶；裝罐
pottle 半加侖(舊的英國液量單位)
pouch 袋；囊
pouch food 袋裝食品
pouch packaging 袋包裝
poulard 閹母雞
poulard wheat 埃及小麥
poult 幼禽；雛
poultice 膏狀藥；糊劑
poultry 家禽
poultry yard 家禽場
poultryman 家禽飼養人
pound 磅；獸欄；魚塘；搗爛
pound cake 磅餅；磅蛋糕

pounder 破碎機；搗杵；搗錘
pounding 搥打；搗碎
pour plate 傾注平板〔培養基〕
pousse café 餐後混合甜酒；餐後的雞尾酒；餐後的含白蘭地酒咖啡
poussin 普桑雞；童子雞
pout 大頭魚類；條鱈
powder 粉；粉末；磨成粉
powder collector 集粉器
powder sugar 粉糖；糖粉
powdered 粉末狀的；磨成粉的
powdered cellulose 纖維素粉
powdered drink 粉末飲料
powdered flavo(u)r 粉末香料
powdered food 粉末食品
powdered green tea 綠茶粉
powdered meat 肉粉
powdered milk 乳粉；奶粉
powdered oil 粉末油脂
powdered skim milk 脱脂乳粉；脱脂奶粉
powdered soup 湯粉
powdered sugar 糖粉
powdered whey 乳清粉
powdering machine 粉碎機
powdery 粉狀的
powdery yeast 酵母粉
power actuated stunner 擊昏槍
PP (polypropylene) 聚丙烯
ppb (parts per billion) 十億分點；十億分之一
ppm (parts per million) 百萬分點；百萬分之一
ppt (precipitate) 沉澱物
prahoc 魚醬
praire fowl 野雞
praline 堅果糖；果仁糖
prawn 明蝦；中蝦；對蝦；斑節蝦
prawn crisp 蝦片
pre-enzyme 酶原
pre-evaporation 預蒸發
pre-evaporator 預蒸發器
pre-extraction 預萃取
prebreaker 預破碎機
prechill 預冷卻
prechill room 預冷卻室
precipitant 沉澱劑
precipitate (ppt) 沉澱物
precipitating agent 沉澱劑
precipitation 沉澱〔作用〕
precipitation tank 沉澱槽
precipitation vat 沉澱缸
precipitator 沉澱器
precipitin 沉澱素
precision balance 精密天平
precision weight 精細砝碼
precoat filter 預覆濾器；預先塗助濾劑的過濾器
precollagen 前膠原〔蛋白〕
preconcentrator 預濃縮器
precondenser 預冷凝器
preconditioning 預處理
precook(ing) 預煮
precooked food 預煮食品
precooked frozen food 預煮的冷凍食品
precooker 預煮器
precool(ing) 預冷〔卻〕
precooler 預冷器
precooling room 預冷室
precursor 先質；前驅物
precut 預切；按規格切割
predator 肉食性動物；掠食性動物
predefecation 預澄清；前澄清
predigestion 預消化
predistillation 預餾；初步蒸餾
predrying 預乾；預先乾燥
preference test 嗜好〔性〕試驗
prefermentation 前發酵

prefermenter 前發酵罐；前發酵槽
prefermentor 前發酵罐；前發酵槽
prefilter 預濾器；預過濾
prefiltration 預過濾
preform 雛型；預先塑造成型
prefractionation 預份化；初步分餾
prefractionator 預份化器；初步分餾塔
prefreeze 預冷凍；預凍
prefreezing 預冷凍；預凍結
prefrying 預炸
pregelatinization 預糊化；預膠凝
pregelatinized 預糊化的
pregelatinized food 預糊化食品
pregelatinized rice 預糊化米；速食飯
pregelatinized starch 預糊化澱粉
pregrinding 預磨
preground 預磨碎的；預粉碎的
preheat 預熱
preheater 預熱器
preheating 預熱
preheating chamber 預熱箱；預熱室
preheating tank 預熱槽
prehumidification 預回潮；預濕潤
prehydrolysis 預水解
preimpregnation 預浸漬
preincubation 預保溫；預培養
preliminary breaker 初碎機
preliminary crushing 初碎；初步壓碎
preliminary drying 初步乾燥
preliminary filter 初濾器；初步過器
preliminary filtration 初步過濾
preliminary heating 預加熱
preliminary test 初步試驗
preliminary washing 預洗
premash 預搗碎
prematernal nutrition 產前營養
premature 未成熟的；早熟的
prematurity 早熟
premix 預混合料；預混合
prepack 預包裝；定量包裝
prepack(age) 預〔先〕包裝
prepackaged food 分裝食品；定量包裝的食品
preparation 製備；預備；調理；預先加工；製品
prepared food 調理食品；預製食品
prepared frozen food 冷凍調理食品；預製冷凍食品
prepared lard 精製豬油
prepress 預壓〔榨〕
prerinse 預沖洗；預清洗
presalting 預醃〔製〕
preschooler 學前時期
preservable 可保藏的；可保存的
preservation 保藏；儲藏；防腐；保存
preservation limit 保藏期限；儲藏限度
preservation medium 保存培養基
preservative 防腐劑；保藏劑
[= conserving agent; preserving agent]
preservative solution 防腐液
preservative substance 防腐劑
preservative treatment 防腐處理
preserve 蜜餞；果脯；保藏；製成蜜餞
preserved apricot 杏脯
preserved bean curd 豆腐乳；腐乳
preserved date jujube 蜜棗
preserved duck 板鴨
preserved egg 鹹蛋；皮蛋
preserved food 蜜餞；保藏食品；罐頭食品
preserved fruit 蜜餞；果脯
preserving 保藏；保存；防腐
preserving agent 防腐劑；保藏劑
[= conserving agent; preservative]
presetting 預凝〔固〕
preslaughter care 屠〔宰〕前飼養；屠〔宰〕前護養
presoak(ing) 預浸〔泡〕

press 壓榨；壓榨機
press cake 壓濾餅；濾餅
press cylinder 壓榨筒
press filter 壓濾機；壓濾器
press juice 榨出汁；濾液
press liquor 榨出汁
press must 壓榨葡萄汁
press noodle 壓製麵條
press on cap 壓封式瓶蓋
press plate 壓榨板
press residue 榨渣；壓榨殘渣；壓濾殘渣
press-twist cap (PT cap) 壓旋蓋
pressed beet pulp 榨〔乾的〕甜菜渣
pressed cossette 榨〔乾的〕甜菜渣
pressed dripping 榨汁；壓榨出的液滴
pressed grape 葡萄渣
pressed ham 壓縮火腿；壓型火腿；模壓火腿
pressed oil 壓濾油
pressed sardine 壓型沙丁魚
pressed scrap 榨〔後〕油渣
pressed yeast 塊狀酵母；壓型酵母；壓榨酵母
presser 壓榨器；壓榨機
pressing 壓濾；壓榨
pressure 壓力
pressure atomization 加壓噴霧；壓力噴霧
pressure cooker 壓力鍋；加壓蒸煮器
pressure cooking 加壓蒸煮
pressure dialysis 壓力透析
pressure drying 加壓乾燥
pressure filter 壓濾機；壓濾器
pressure filtration 壓濾
pressure gauge 壓力計
pressure pan 壓力鍋
pressure regulating valve 壓力調節閥；調壓閥
pressure regulator 壓力調節器；調壓器
pressure resistant 耐壓性
pressure saucepan 壓力鍋
pressure screw extractor 螺旋榨汁機；螺旋壓榨機
pressure spray dryer 壓力式噴霧乾燥機
pressure steaming 加壓蒸煮
pressure sterilization 加壓滅菌
pressure sterilizer 加壓滅菌器
pressure vessel 耐壓容器
pressurization 加壓；增壓
pressurized food 氣溶膠罐裝食品
pressurized package 氣溶膠罐包裝
pressurized still 耐壓蒸餾器
presterilizarion 預滅菌
prethawing 預解凍
prethickening 預稠化；預增稠
pretreatment 預處理
pretzel 普澤餅；捲條脆餅；紐結狀椒鹽脆餅
pretzel stick 普澤餅棒；椒鹽脆餅棒
previtamin 維生素元；維生素原
prewash 預洗
prewasher 預洗機
prewet 預濕
prickly pear 仙人掌果；霸王梨；印度無花果
primal cut 粗切肉塊；初步分切
primary alcohol 伯醇；第一級醇
primary amine 伯胺；第一級胺
primary cane juice 初榨蔗汁；頭榨蔗汁
primary crusher 粗碎機；初步粉碎機
primary distillation 初級蒸餾
primary fermentation 初發酵；前發酵；主發酵
primary filter 初濾
primary filtrate 初濾液
primary juice 原汁

primary product 初產品；主要產品
primary sewage treatment 初級污水處理；一級污水處理
primary taste 原味
primary vitamin 維生素原
primary waste treatment 初級廢料處理
prime goods 高級商品
prime steam lard 蒸汽熬製的優質豬油
primrose 櫻草
primverose 冬綠糖；櫻草糖
principal fermentation 主發酵
print(ed) butter 定量包裝的印模牛油；定量包裝的印模乳酪
probe 探針
process butter 再製牛油；再製乳酪
process chart 流程圖
process tank 加工槽
process(ed) cheese 再製芝士；再製乳酪；融化乳酪
processed fat 加工油脂；再製油脂
processed food 加工食品
processing 加工；處理
processing aids 加工助劑
processing equipment 加工設備
processing line 加工生產線；作業線
processing machinery 加工機械
procoagulant 促凝劑
produce 生產；產品；農產品
producer 生產者
product 產品；產物
product inspection 產品檢驗
product water 淡化水
production 生產；產品
production line 生產線
productivity 生產力
proenzyme 酶元；酶原
proferment 酶原
profichi 春季無花果
progressive dryer 逐步乾燥器
progressive drying 逐步乾燥〔法〕
progressive freezing 逐步冷凍〔法〕
progressive hydrolysis 逐步水解
prolamin(e) 醇溶〔穀〕蛋白
proliferation 增殖；繁殖
proline 脯胺酸
promoter 促進劑；助催化劑
proof 證明；驗據；酒精度；醒麵；麵糰醒發成形
proof box 醒麵箱；麵糰後發酵箱
proof gallon 標準烈酒加侖；規定度加侖
proof spirit 標準烈度酒；規定濃度之烈度酒
proofer 醒麵箱；醒麵室
proofing 證明；麵糰後發酵；醒麵；麵糰醒發成形
proofing room 醒麵室；麵糰後發酵室；麵糰醒發成形室
prooxidant 氧化促進物質；氧化強化劑
propagate yeast 繁殖酵母
propagation tank 繁殖槽
propagative spore 繁殖孢子
propanal 丙醛
propane 丙烷
propanol 丙醇 [= propyl alcohol]
1-propanol 丙醇；正丙醇
propeller 螺旋槳；推進器
propeller agitator 螺旋槳攪拌器
propeller mixer 螺旋槳混合器
propeller stirrer 螺旋槳攪拌器
propenylguaethol 丙烯基乙基癒創木酚
propionaldehyde 丙醛
propionate 丙酸鹽；丙酸酯
propionic acid 丙酸
propionic acid bacteria 丙酸細菌
propionic acid fermentation 丙酸發酵
proportioner 配料器；定量給料器
p-propyl anisole 對丙基苯甲醚

propyl acetate 乙酸丙酯
propyl alcohol 丙醇 [= propanol]
propyl benzoate 苯甲酸丙酯
propyl gallate (PG) 沒食子酸丙酯；五倍子酸丙酯
propyl heptanoate 庚酸丙酯
propyl hexanoate 己酸丙酯
propyl-*p*-hydroxybenzoate 對羥基苯甲酸丙酯；尼泊金丙酯 [= propylparaben]
propylene 丙烯
propylene glycol 丙二醇；丙烯甘醇
propylene glycol alginate 藻酸丙二醇酯；海藻酸丙二醇酯
propylene glycol esters of fatty acids 脂肪酸丙二醇酯；丙二醇脂肪酸酯
propylene glycol lactostearate 乳酸脂肪酸丙二醇甘油混合酯
propylene glycol monostearate 單硬脂酸丙二醇酯
propylene glycol starch 丙二醇澱粉
propylmercaptan 丙硫醇
propylparaben 對羥基苯甲酸丙酯；尼泊金丙酯 [= propyl-p-hydroxybenzoate]
prorennin 凝乳酶原
proso millet 粟；小米；黍；稷
prosthetic group 輔基
protamin(e) 魚精蛋白
protaminase 魚精蛋白酶
protease 蛋白酶
protecting agent 保護劑
protective coating 保護塗層
protective colloid 保護膠體
protective wrapping 保護包裝〔材料〕
proteic 蛋白的
protein 蛋白質
Protein Advisory Group (PAG) 聯合國蛋白質顧問組
protein allowance 蛋白質容許量
protein biosynthesis 蛋白質生物合成
protein concentrate 濃縮蛋白質；蛋白質濃縮物
protein degrading enzyme 蛋白質分解酶
protein denaturation 蛋白質變性
protein digestibility 蛋白質消化率
protein dispersibility index 蛋白質分散指數
protein efficiency ratio (PER) 蛋白質效率比〔值〕；蛋白質功效比〔值〕
protein energy ratio 蛋白質能量比
protein equivalent 蛋白質當量
protein factor 蛋白質係數
protein food 蛋白質食品
protein-free 無蛋白質的；不含蛋白質的
protein-free diet 無蛋白質膳食
protein-free wheat starch 澄粉；無蛋白質小麥澱粉
protein gel 蛋白〔質〕凝膠
protein hydrolysate 蛋白質水解物；水解蛋白質
protein hydrolyzate 蛋白質水解物；水解蛋白質
protein isolate 蛋白質分離物；分離蛋白質
protein maintenance requirement 蛋白質維持需要量
protein milk 高蛋白乳
protein nitrogen 蛋白質〔態〕氮
protein quality index 蛋白質質量指數
protein ration 蛋白質日糧
protein requirement 蛋白質需要量
protein retention efficiency 蛋白質保留效率
protein score 蛋白質價；蛋白質評分
protein shift 蛋白質轉移；蛋白質轉換

protein structure 蛋白質構造
protein synthesis 蛋白質合成
protein therapy 蛋白質療法
protein titre 蛋白質滴定度
protein value 蛋白質價
proteinaceous 蛋白質的
proteinase 蛋白酶
proteinate 蛋白鹽
proteinochrome 蛋白色素
proteinuria 蛋白尿症
proteoglycan 蛋白聚糖；蛋白多糖
proteolipid 蛋白質脂；含蛋白質脂
proteolysin 蛋白水解素；蛋白溶素
proteolysis 蛋白質分解；蛋白質水解
proteolytic 蛋白質分解的；蛋白質水解的
proteolytic activity 蛋白〔質〕分解活性
proteolytic bacteria 蛋白〔質〕分解細菌
proteolytic enzyme 蛋白〔質〕分解酶
prothrombin 凝血酶原
protocatechuic acid 原兒茶酸
protochlorophyll 原葉綠素
protopectin 原果膠
protopectinase 原果膠酶
protoplast 原生質體
protoporphyrin 原紫質；蛋殼色素
prototrophic bacteria 原始營養細菌；原養型細菌
prototype 原型
protozoa 原生動物
provender 糧草；飼料
proventriculus 前胃
provisions 食品；糧食；供應
provitamin 原維生素；維生素原
proximate analysis 主成分分析
proximate composition 主成分組成
prune 梅乾；李乾；李脯；乾果李；西梅
prune juice 乾果李汁；西梅汁
prunello 小黃李乾
pry-off cap 撬開〔式〕蓋
PS (polystyrene) 聚苯乙烯
PSE muscle (pale, soft, exudative muscle) PSE肉；蒼白鬆軟和滲水肉；蒼白軟質汁液滲出肉；水樣肉
pseudoacid 假酸
pseudocarp 假果
pseudocolony 假菌落
pseudocrystal 假晶體
pseudofructose 假果糖
pseudoglobulin 假球蛋白
Pseudomonas 假單孢菌屬
Pseudomonas fluorescens 熒光假單孢菌
pseudomycelium 假菌絲體
pseudomycete 假黴菌
pseudoprotein 不完全蛋白
psicose 阿洛酮糖 [= allulose]
psoas 腰肌
psychrometer 乾濕球濕度計；乾濕計
psychrophile 嗜冷生物；嗜冷細菌
psychrophilic 嗜冷的
psychrophilic bacteria 嗜冷細菌
psychrotrophic bacteria 嗜冷細菌；適冷細菌
psyllium 車前子；車前草
psyllium seed 車前籽
pteroylglutamic acid 蝶酰麩胺酸；蝶酰谷胺酸；葉酸 [= folic acid]
ptyalase 唾液澱粉酶；涎澱粉酶 [= ptyalin]
ptyalin 唾液澱粉酶；涎澱粉酶 [= ptyalase]
ptyalose 涎糖
pub 酒館；酒吧
public house 酒吧
pudding 布丁；布甸
pudding rice 布丁米
puff 膨脹鬆化；泡芙

puff drying 膨化乾燥
puff pastry 酥皮；鬆餅；油酥點心
puffball 馬勃〔菌〕
puffed cereal 膨化穀類食品
puffed food 膨化食品
puffed grain 膨化穀物
puffed rice 爆米花
puffer 膨化器；胖罐；河豚
puffer (fish) 河豚 [= globefish; fugu]
puffer poison 河豚毒素
puffer poisoning 河豚中毒
puffing 膨化；膨脹
pull date 銷售期限
pull-off cap 易拉開蓋
pull tab 拉舌；拉環
pulled chicken 光雞；去毛雞
pulled fig 無花果片；無花果乾
pulled hard candy 絲光硬糖
puller 糖果拉製機；糖果絲光機
pullet 小母雞(一歲以下)
pulling machine 拉製機；糖料拉條機
pullman bread 三明治麵包；長形麵包
pullulan 聚三葡萄糖；茁霉多糖
pullulanase 支鏈澱粉酶
pullulane 出芽短梗孢糖；茁霉多糖
pulp 漿〔料〕；果肉；果漿；紙漿
pulp dryer 乾渣機
pulp filter 濾渣器
pulp press 壓渣器
pulp separator 渣分離機
pulper 打漿機；磨碎機
pulper finisher 打漿篩濾機
pulping machine 打漿機；磨碎機
pulpy 多果肉的；柔軟多汁的
pulpy fruit 軟果
pulque 龍舌蘭酒
pulsating screen 脈動篩
pulse 豆；豆類植物；脈衝
pulse flour 豆粉
pulverator 粉碎機
pulverization 粉碎；磨粉；研磨成粉末
pulverize 粉碎；磨粉；研磨成粉末；霧化
pulverized grain 穀物粉末
pulverized sugar 粉糖；糖粉
pulverizer 粉碎機；磨粉機；霧化機
pulverizing mill 粉碎機；磨粉機；細粉磨
pummelo 柚〔子〕；文旦 [= pomelo; pompelmous; shaddock]
pumpernickel 裸麥酸麵包；粗黑麥酸麵包
pumpkin 南瓜
punch 潘趣飲料；果汁混合飲料；賓治飲料；沖床；打孔器
puncture 穿孔；刺孔
pungency 辛辣；刺激性
pungent 辣的；辛辣的；刺激性的
pungent flavo(u)r 辛辣氣味；刺激性氣味
pungent sauce 辣醬
pungent taste 辛辣味；刺激味
punicic acid 石榴酸；天竺葵酸
punicin 石榴素
pure culture 純種培養；純種
pure food 無添加食品
pure natural juice 純天然果汁
pure salt 精鹽
pure water 純水；淨水
puree 食物泥；菜泥；果泥；濃湯
pureed bean 豆泥
pureed spinach 菠菜泥
pureness 純度；純淨；純潔
purgative 通便的；瀉劑
purification 純化；淨化
purified juice 淨化汁
purified oil 精煉油
purified starch 精製麵粉

purified water 淨化水
purifier 純化器；淨化器
purine 嘌呤
purity 純度；純淨；純潔
purple laver 紫菜
purple raspberry 紫莖樹莓
putament 果核的核殼；蛋殼角質膜；蛋殼內的卵膜
putrefaction 腐臭；腐敗
putrefactive 腐爛的；腐敗的
putrefactive bacteria 腐臭細菌；腐敗細菌
putrefactive fermentation 腐臭發酵；腐敗發酵
putrefactive spoilage 腐爛；腐臭腐敗
putrescence 腐爛；腐敗
putrescine 腐胺
putrid 腐爛的；腐敗的
putrid odo(u)r 腐臭
putrilage 腐爛物；腐臭物
PVC (polyvinyl chloride) 聚氯乙烯
PVDC (polyvinylidene chloride) 聚二氯亞乙烯；聚偏二氯乙烯
pycnometer 比重瓶
pyranofructose 吡喃果糖；六環果糖
pyranoglucose 吡喃葡糖；六環葡糖
pyranose 吡喃糖；六環糖
pyrene 小堅果
pyrenocarp 核果
pyrenoid 澱粉核
pyrethrin 除蟲菊酯
pyrethroid 除蟲菊素
pyrethrum 除蟲菊
pyridine 吡啶
pyridoxal 吡哆醛
pyridoxal phosphate 吡哆醛磷酸；磷酸吡哆醛
pyridoxine 吡哆素；吡哆醇；維生素B_6
pyridoxine hydrochloride 吡哆醇鹽酸鹽
pyrocatechol 鄰苯二酚；焦兒茶酚
pyrodextrin 焦糊精
pyrogallic acid 焦五倍子酸；連苯三酚
pyrogallol 焦五倍子酸；連苯三酚
pyroglutamic acid 焦麩胺酸；焦谷胺酸
pyroligneous acid 木醋酸；焦木酸
pyrolysis 熱解；高溫分解
pyrometer 高溫計
pyrophosphatase 焦磷酸酶
pyrophosphate 焦磷酸鹽；焦磷酸酯
pyrophosphoric acid 焦磷酸
pyrrole 吡咯
pyruvate 丙酮酸鹽；丙酮酸酯
pyruvic acid 丙酮酸
pyruvic dehydrogenase 丙酮酸去氫酶；丙酮酸脫氫酶
pyruvic oxidase 丙酮酸氧化酶
pythogenesis 腐敗；腐化

Q

QA (quality assurance) 品質保證；質量保證
QC (quality control) 品質管制；質量控制；品管
quaff 暢飲
quahaug 圓蛤類；簾蛤；硬殼蛤 [= quahog]
quahog 圓蛤類；簾蛤；硬殼 [= quahaug]
quail 鵪鶉
quail egg 鵪鶉蛋
qualitative analysis 定性分析
quality 品質；質量
quality assurance (QA) 品質保證；質量保證
quality booster 品質改進劑
quality control (QC) 品質管制；質量控制；品管
quality evaluation 品質評估
quality goods 優良產品
quality guarantee 品質保證
quality improvement agent 品質改良劑
quality inspection 品質檢查
quality specification 品質規格；品質標準
quality standard 品質標準
quantitative 定量的
quantitative analysis 定量分析
quantity 量；數量
quantity surveyor 估算師
quarantine 檢疫；隔離
quarantine inspector 檢疫員
quart 夸脱 (液體容積)
quarter 分成四份；四等分；四分之一
quartern 四分之一品脱；四磅重的麵包
quartz 石英
quassation 壓碎；搗碎
quassia 啤酒苦味劑；苦木〔科植物〕
quassia cup 苦味酒
quassis extract 苦木抽提物
quaternary structure 第四級結構
queen bee 蜂后；蜂王
queen crab 雪花蟹
Queensland nut 澳洲胡桃；昆士蘭〔堅〕果；夏威夷果 [= Australian nut; macadamia nut]
Quench(ing) 止渴；解渴；熄滅
quencher 止渴飲料；解渴飲料
quenelle 肉丸子
quercetin 槲皮素；槲皮酮；櫟皮酮；櫟精
quercitol 槲皮醇；櫟醇；櫟實糖；槲實糖 [= acorn sugar]
quercitrin 槲皮苷；櫟皮苷
quercitron 槲皮粉；櫟皮粉；美洲黑櫟
quercus 白櫟皮
quern 手石磨；手推磨
quey 小母牛
quick ag(e)ing 快速熟化；快速陳化
quick bar 快餐櫃台

quick bread 速發麵包
quick chilling 急速冷卻；快速冷卻
quick cooking food 快煮食品；速煮食品
quick cooling 快速冷卻
quick cure 快速醃製；速醃
quick curing 速醃〔法〕
quick defrosting 快速解凍；快速除霜
quick drying 快速乾燥；速乾
quick freezer 速凍機；快速凍結機
quick freezing 快速冷凍；速凍
quick freezing room 速凍室
quick frozen food 速凍食品
quick oven 快速烤箱
quick ripening 快速熟成
quick salting 快速鹽漬
quick serve meal 快餐食品
quick setting 快速凝固；速凝
quick smoking 快速煙燻
quick thawing 快速解凍
quick vinegar process 快速釀醋法
quicklunch 快餐
quill 翮羽
quillaia 皂樹皮
quillaia extract 皂樹皮抽提物
quillaja 皂樹〔皮〕(發泡劑)
quince 榲桲
quinic acid 奎尼酸；金雞納酸
quinine 奎寧；金雞納霜
quinine hydrochloride 鹽酸奎寧
quinine sulfate 硫酸奎寧
quinine water 奎寧水
quininic acid 奎寧酸
quinoa 鵝腳藜；奎藜籽
quinone 醌；苯醌
quintessence 精華；精萃

R

R-colony 粗糙〔型〕菌落 [= rough colony]
raad 電鯰；電鰻
rabbit 兔
rabbit meat 兔肉
rabbitfish 河豚；青籃子魚
rabbitry 養兔場；養兔業
race 種；族；薑根；特有風味
racemase 外消旋酶
racemic compound 外消旋化合物
raciness 獨特風味；保持原味
rack 架〔子〕；酒；燒酒；榨取；羊肋排〔肉〕
racking machine 灌裝機
racy 保持原味的
radappertization 輻射滅菌〔法〕；照射滅菌〔法〕
radiant energy 輻射能
radiant heat 輻射熱
radiant heater 輻射加熱器
radiate colony 放射狀菌落
radiation 輻射；放射；照射
radiation disinfestation 輻射滅蟲
radiation pasteurization 輻射巴氏殺菌〔法〕；照射巴氏殺菌〔法〕
radiation preservation 輻射保藏〔法〕；放射保藏〔法〕
radiation pyrometer 輻射高溫計
radiation sterilization 輻射滅菌〔法〕；照射滅菌〔法〕
radiation treatment 輻射處理；照射處理；放射處理
radiator 輻射器；散熱器
radical 自由基
radical reaction 自由基反應
radical scavenger 自由基捕捉劑
radicchio 紫萵苣
radicidation 輻射殺菌〔法〕
radioactive 放射性〔的〕
radioactive isotope 放射性同位素
radioactive source 放射源
radioactive substance 放射性物質
radioactivity 放射性；放射能
radiofrequency heating 射頻加熱
radiofrequency pasteurization 射頻巴氏殺菌
radioheating 射頻加熱
radiolysis 射解作用
radiopasteurization 輻射巴氏殺菌〔法〕；照射巴氏殺菌〔法〕
radiosterilization 輻射殺菌〔法〕；照射殺菌〔法〕
radiotoxicity 輻射毒性；放射毒性
radish 蘿蔔；菜頭
radish red 蘿蔔紅
radurization 輻射殺菌〔法〕；照射殺菌〔法〕
raffia 拉菲椰；酒椰 [= raphia]
raffinase 蜜三糖酶；棉子糖酶
raffinate 殘餘液

raffinose 蜜三糖；棉子糖 [= melitriose]
rag gourd 絲瓜
ragout 濃味蔬菜燉肉
rainbow herring 亞洲胡瓜魚
rainbow ice cream 多色冰淇淋
rainbow perch 虹鱸
rainbow trout 虹鱒；硬頭鱒；紅鱒
raised colony 隆起菌落
raisin 提子乾；葡萄乾
raisin bread 葡萄乾麵包
raisin grape 葡萄乾葡萄；無核葡萄
raisin seed oil 葡萄籽油
raisin wine 葡萄乾〔釀造〕酒
raising powder 發粉；發麵粉
rake conveyor 耙式輸送機
rake stirrer 耙式攪拌器
rakee 拉基〔葡萄〕酒；拉基梅酒 [= raki]
raki 拉基〔葡萄〕酒；拉基梅酒 [= rakee]
rakkyo 薤；蕎頭
Ralls 國光蘋果
ram 公羊；撞錘；填塞；衝壓機
ram lamb 小公羊
ramadan 齋月
rambutan 紅毛丹〔果〕
ramekin 小芝士蛋糕；小乾酪蛋糕；芝士餅；乾酪餅 [= ramequin]
ramen noodle 日本拉麵；即食麵；方便麵
ramequin 小芝士蛋糕；小乾酪蛋糕；芝士餅；乾酪餅 [= ramekin]
ranch 牧場；大農場
rancid 酸敗的；油耗的
rancid flavo(u)r 油耗味；酸敗味
rancid odo(u)r 油耗味
rancidification 酸敗
rancidify 酸敗
rancidity 酸敗
rancidness 酸敗；腐臭
randanite 矽藻土；硅藻土
random sampling 隨機採樣；任意採樣
rank 臭氣難聞的；惡臭的；等級；階級；排列
Rankine temperature 蘭金溫度
rape 油菜；野油菜
rape oil 菜籽油；菜油
rape(seed) cake 菜籽餅
rapeseed 菜籽；油菜籽 [= colza]
rapeseed meal 油菜籽粕；油菜籽粉
rapeseed oil 菜籽油 [= colza oil]
raphia 拉菲椰；酒椰 [= raffia]
rapic acid 菜籽酸
rapid ag(e)ing 快速熟成；快速陳化
rapid brine cure 快速〔濕〕鹽醃
rapid cooling 快速冷卻
rapid curing 速醃
rapid fermentation 快速發酵
rapid freezing 快速冷凍；速凍
rapid set 快速凝固；速凝
rare amino acid 稀有氨基酸
rarebit 威爾斯芝士麵包；威爾斯乾酪麵包；塗上熔化乾酪〔及啤酒〕的烤麵包；乾酪吐司 [= welsh rabbit; welsh rarebit]
rareripe 早熟的；早熟蔬果
rasher 鹹肉片；鹹火腿片
rasp 擦碎機；擦菜板；擦碎
raspberry 覆盆子；蔗莓；樹莓；懸鈎子
raspberry ketone 覆盆子酮；懸鈎子酮
rasper 磨碎機；擦菜板
rat 老鼠；鼠；大白鼠
ratafee 果酒；果汁興奮劑
ratafia 杏仁味甜餅乾；果仁酒；苦杏仁酒；果酒
rate egg 煮嫩蛋
ratfish 銀鮫；兔銀鮫

raticide 殺鼠藥
rating 等級；評價
ration 配給；配給量；定量配給；限量供應
rations 軍用口糧；日糧；糧食
rations bread 配給麵包；軍用麵包；粗硬麵包
raven 狼吞虎咽
ravioli 煮合子；餛飩；意式餃子；意大利餃
raw 生的；未加工的；粗的；未煮熟的
raw bread 生麵包
raw cream 生奶油；生乳酪
raw fat 粗〔製〕脂；原脂
raw juice 原汁
raw material 原料
raw meat 生肉；原料肉
raw milk 生乳；原料乳；未消毒乳
raw noodle 生麵〔條〕
raw oil 粗〔製〕油；原油
raw pack 生料包裝
raw product 半製成品；初級產品
raw sugar 粗糖；原糖
raw taste 原味；生腥味
raw water 未處理水；未淨化水；原水
rawboned 瘦的
ray fungus 放線菌
RDA (recommended daily allowance) 建議每日營養攝取量；推薦每日營養素供給量
RDNA (recommended daily nutrient allowance) 建議每日營養素攝取量；推薦營養素日供給量
reabsorption 再吸收
reactant 反應物
reaction 反應
reactivity 反應〔活〕性
reactor 反應器
ready-cooked 已煮熟的；預先煮好的
ready-cut 已切好的
ready made 現成的
ready-to-bake 備烤的；即可烘焙的
ready-to-cook 備煮的；即可烹煮的
ready-to-drink 即飲的；即可飲用的
ready-to-drink beverage 即飲飲料
ready-to-eat 即食的；即可食用的
ready-to-eat food 即食食品；熟食品；現成食品
ready-to-eat meal 即食〔便〕餐；快餐
ready-to-serve 即食的；即可食用的
ready-to-serve food 即食食品；熟食品；現成食品
ready-to-use 現成可用的；即可使用的
reagent 試劑
ream 榨汁；榨取
reamer 榨汁器(錐形)
reaming extractor 榨汁器(錐形)
reblending 再混合
reboil 再煮〔沸〕；翻煮
recarbonation 再碳酸化
receiving tank 接收槽；收集槽
rechauffe 回鍋菜〔餚〕；再翻熱的〔剩〕菜
recin 蓖麻毒素
recipe 食譜；配方；烹調法
reciprocating sieve 往復篩
recirculating type vinegar generator 環流式釀醋器
recirculating water 再循環水
recirculation 再循環；再流通
reclaim 回收；再生
reclamation 回收；再生；廢物利用
recoction 重煮；復煮；再次烹煮
recombined milk 重組乳；調製乳
recommended daily allowance (RDA) 建議每日營養攝取量；推薦每日營養素供給量

recommended daily nutrient allowance (RDNA) 建議每日營養素攝取量；推薦營養素日供給量
recommended dietary allowance 建議飲食攝取量；建議膳食供給量
recommended intake 建議攝取量
reconstituted food 復原食品；復水食品；還原食品
reconstituted juice 復原果汁
reconstituted milk 復原乳；還原乳；再製乳
reconstitution 復原；復水；還原
recooking 重蒸煮；二次蒸煮
recooling 再冷卻；回冷
recording thermometer 自記溫度計；記錄溫度計
recovery 回收；復原；回收率；廢料回收再用
recrement 廢物；渣滓
recrystallization 再結晶
rectal 直腸的
rectangular can 方形罐；長方〔形〕罐
rectangular tin 方形罐；長方〔形〕罐
recticulin fiber 網質纖維
rectification 精餾；淨化
rectified alcohol 精餾酒精
rectified spirit 精餾酒精
rectifier 精餾器；精餾塔
rectify 精餾
rectum 直腸
recuperation 回收；餘熱利用；回流換熱法；同流換熱法
recuperator 回流換熱器；同流換熱器；熱能回收器
recycle 再循環
recycling 再循環
red 2G, azogeramine 紅色2G，單偶氮
red algae 紅藻
red batfish 棘茄魚
red bean 紅豆；赤豆；小豆 [= adzuki bean; adsuki bean; azuki bean]
red beet 甜菜〔頭〕
red blood cell 紅血球 [= erythrocyte]
red cabbage 紅椰菜；紫色甘藍；紅色高麗菜
red cabbage pigment 紅球甘藍色素；捲心菜紅色素
red capped bolete 血紅牛肝菌
red caviar 紅魚子；醃鮭魚子醬
red cherry 紅櫻桃
red clover 紅三葉草
red cluster pepper 朝天辣椒
red colo(u)rs 紅色素
red cowpea 紅豇豆
red crab 紅蜘蛛蟹
red cracking bolete 牧草牛肝菌
red currant 紅醋栗
red date 紅棗
red date tincture 棗子酊
red delicious 紅〔五爪〕甜蘋果；紅帥蘋果
red durum wheat 硬質紅小麥；紅杜蘭小麥；紅杜林小麥
red eye 廉價威士忌酒；低劣酒
red grouper 黑邊石斑魚
red grouse 野雞
red hake 紅長鰭鱈；紅海鱈
red herring 燻鯡魚
red hot 肉桂味紅糖果
red huckleberry 越橘
red juice orange 血橙
red kernel 紅仁；紅核；紅穀粒
red kidney bean 紅菜豆；紅芸豆
red koji rice 紅麴米
red laver 紅藻
red man wheat 英國紅人小麥
red meat 紅肉；紅色肉；血合肉
red mulberry 紅桑果；紅桑葚

red mullet 秋姑魚；羊魚；金色洋魚；紅鯔
red muscle 紅肌肉
red oak bolete 櫟生疣柄牛肝菌
red oil 紅色油
red paprika 辣椒紅；辣椒色素
red pea 紅豌豆
red pepper 紅辣椒；蕃椒；辣椒〔粉〕
red perch 紅海鱸
red poll 紅色無角牛；無角紅毛牛
red porgy 嘉鱲；真鯛
red preserved beancurd 紅〔豆〕腐乳；南乳
red radish 紅衣蘿蔔
red raspberry 歐洲紅莓；覆盆子
red rice 紅米
red rice starter 紅麴
red salmon 紅鮭〔魚〕；紅大麻哈魚
red seaweed 紅藻
red short 細麩
red shrimp 紅蝦
red skate 赤魟
red snapper 勃氏笛鯛
red spring wheat 紅皮硬質小麥
red staining mushroom 木生蘑菇
red stemmed tough shank 紅柄金錢菇
red stingray 赤魟
red streak rice 紅線米
red sturgeon shrimp 北海蝦
red tail prawn 紅尾蝦
red tide 紅潮
red tilapia 尼羅紅魚；紅羅非魚
red vegetable 紅色根菜
red wheat 紅〔皮〕小麥
red wine 紅〔葡萄〕酒
red winter wheat 紅皮軟質小麥
red yeast 紅色酵母
redbellied dace 紅腹雅羅魚；紅腹鯉魚
redbelly 紅腹雅羅魚
redbelly tilapia 吉利無郭魚
redeye 紅眼魚
redfin 紅鰭淡水魚
redfin pickerel 美洲狗魚
redfish 雄鮭；紅鮭；紅大麻哈魚
redistillation 再蒸餾；復蒸餾
redistilled water 再蒸餾水
redox 氧化還原作用
redox reaction 氧化還原反應
redpotted trout 克氏紅點鮭
redry(ing) 再乾燥；重新曬乾；二次乾燥
reduced must 濃縮葡萄汁
reduced myoglobin 還原型肌紅蛋白
reducer 還原劑；稀釋劑
reducing agent 還原劑
reducing diet 減肥膳食
reducing power 還原力
reducing sugar 還原糖
reductant 還原劑
reductase 還原酶
reduction 還原；減少；粉碎
reductone 還原酮
reed 蘆葦
reefer 冷藏車；冰箱；冷藏船
reel 轉篩；圓篩；滾筒；轉筒
reextraction 再萃取
refection 小吃；點心；茶點；便餐；解除飢渴
refectory 食堂；餐廳
reference sample 參比樣品；參考樣品
reference standard 參比標準；參考標準
refill 再裝滿；再充填
refine 精煉；精製
refined lard 精製豬油；精煉豬油
refined oil 精製油；精煉油
refined salt 精鹽；精製鹽
refined starch 精製澱粉

refined sugar 精糖；精製糖
refinement 精煉；精製；精製〔產〕品
refiner 精製機；精煉機；精研磨機
refinery 精製廠；精煉廠
refinery molasses 精煉糖蜜；精製糖蜜
refinery syrup 精煉糖漿；精製糖漿
refining 精煉；精製
refining agent 精煉劑；精製劑
reflectometer 反射計
reflux 回流
reflux distillation 回流蒸餾
refractive index 折射率
refractometer 折射計
refractory 耐火的；耐熱的；耐熔的
refractory material 耐火材料；耐熱材料
refreezing 再冷凍；二次冷凍
refresher 點心；冷飲品；提神飲料
refreshment 點心；飲料；茶點；提神飲料
refreshment beverage 清涼飲料
refreshment drink 清涼飲料
refreshment room 小吃部；茶點室
refrigerant 冷凍劑；製冷劑；冷媒
refrigerate 冷卻；冷藏；冷凍
refrigerated container 冷藏容器；冷容器；冷凍容器
refrigerated dough 冷藏麵糰
refrigerated goods 冷凍貨物；冷藏貨物
refrigerated meat 冷凍肉；冷藏肉
refrigerated shelf 冷藏貨架
refrigerated storage 冷藏；凍藏
refrigerated storage life 冷藏期
refrigerated warehouse 冷藏庫；冷凍庫
refrigerating equipment 冷凍設備；製冷裝置
refrigeration 冷凍；冷藏；致冷
refrigeration equipment 冷凍設備；製冷裝置
refrigerator 冰箱；雪櫃；冷藏庫；冷藏室
refrigerator car 冷凍車；冷藏車
refrigerator compartment 冰箱冷藏室
refrigerator van 冷凍車；冷藏車
refrigeratory 冰箱；雪櫃；製冷的
regale 盛宴；佳餚；珍味；享受美食
regelation 再膠化；再膠凝；二次固化
regenerate 再生；新生
regeneration 再生；交流換熱〔法〕
regeneration heat exchanger 再生熱交換器；熱回收換熱器
regeneration pasteurizer 再生熱巴式殺菌器；交流換熱巴式殺菌器
regenerator 再生器；蓄熱器；交流換熱器
regimen 食物療法；養生法；攝生法
reglazing 再包糖衣；再包冰衣；二次掛糖衣
regrind mill 再磨機；細粉磨
regrinding 再磨研；二次粉碎
regular diet 普通膳食；正常膳食
regular ham 帶骨火腿
regular milk 正常乳；合格乳
regular twist-off cap 四旋蓋；爪旋蓋
regulation 規則；規章；法規；條例
regulator 調節器
regurgitation 反芻；反胃；回流
reheat(ing) 翻熱；再〔加〕熱；二次加熱
reheater 再熱器
rehydration 復水；再水合
rehydration aid 復水助劑
reicing 再包冰；再加冰；再加糖霜
reindeer 馴鹿
reinfection 再感染；二次感染
reinspection 複驗；複查；複檢
reject 廢品；廢料
rejected product 次品；廢品；不合格品

rejuvenation 恢復活力；復壯
relative density 相對密度
relative humidity (RH) 相對濕度
relaxation 鬆弛；放鬆
relaxation effect 鬆弛效應
release agent 脱膜劑
relish 風味；味道；津津有味；開胃小菜；調味品
remelt 復溶；再溶〔化〕
remoisten 再濕；潤濕；回潮
remoulade 調味蛋黃漿
remover 脱模劑；去除劑
remuage 香檳在生產期間沉澱物從瓶壁轉移到瓶塞
renaturation 恢復自然；復原性
render(ing) 熬油；煉油
rendered fat 煉好的油脂
rendering scraps 油渣
rennase 凝乳酶
rennet 凝乳酶
rennet casein 酶凝酪蛋白；酶凝乾酪素
rennet clotting 凝乳酶凝乳〔作用〕
rennet curd 酶凝乳塊
renneting 凝乳；酶凝乳作用
rennin 凝乳酶
renninogen 凝乳酶原
rennogen 凝乳酶原
renovated butter 再製牛油；再製乳酪
rentschlerizing 紫外線殺菌處理
reoxidation 再氧化
repack 改裝；重新包裝
repast 餐；飲食；餐宴；就餐；設宴；就餐時間
repasteurization 再〔巴氏〕殺菌；二次巴氏殺菌
repellent 拒斥劑；驅蟲劑
repercolation 再滲濾
replacement medium 置換培養基；更換培養基
replenish 補充；補足
repletion 過飽；飽餐
replica 複製〔法〕；複製品
reprecipitation 再沉澱
repressor 壓制劑；阻遏物
reproduction 生殖；繁殖
reproductive mycelium 生殖菌絲體；繁殖菌絲體
reprovision 補充糧食；補充食品
repurification 再淨化；再純化
resazurin 刃天青
rescreen 再篩分；再過篩
research 研究；調查
reseparation 再分離
reserve protein 儲藏〔性〕蛋白質
reserve starch 儲藏澱粉
reserve tank 儲藏槽
reservoir 儲存處；儲水槽；蓄水池；水庫
resid 殘油
residual 剩餘的；殘留的；殘餘的
residual fraction 殘餘餾分；剩餘部分
residual pesticide 殘留殺蟲劑；殘留農藥
residual product 副產品；殘餘產物
residue 殘餘〔物〕；渣滓；濾渣；殘留物；殘渣
residuum 殘留物；殘渣；殘油；沉澱
resiliency 彈性〔強度〕
resilker 殘鬚去除機
resin 樹脂
resin bed 樹脂床；離子交換樹脂層
resinous exchanger 樹脂〔質〕交換劑
resinous substance 樹脂物質
resoluble 可分解的；可溶解的
resolution 再溶解
resolution of rigor 解僵 [= rigor off]
resorption 再吸收
respiration 呼吸〔作用〕

respiration calorimeter 呼吸熱量計；呼吸卡計
respirator 口罩；防毒面具；呼吸器
respiratory calorimeter 呼吸熱量計；呼吸卡計
respiratory quotient 呼吸商
respiratory tract 呼吸道
restaurant 餐館；飯店；餐廳
restaurant contractor 包伙商；包伙食業者
restauranteur 餐館老闆；餐館〔營〕業者
resteeping 再浸漬
resting metabolism 靜止代謝
resting pen 繫留場
resting spore 休眠孢子
resting yeast 休眠酵母
restoration 復原；再生
restorative 營養食品；補品；滋補劑；恢復健康的
restored food 營養補償食品
restrainer 抑制劑；固定劑
restricted liquid diet 限制流質膳食
restructured meat 重組肉類
resublimation 再升華〔作用〕
resurgence 恢復活力
resurrectionlily 山柰；沙薑
retail 零售
retail cut meat 零售分割肉
retail dealer 零售商
retail market 零售市場
retail packaging 零售包裝
retail store 零售商店
retardant 阻滯劑；延緩劑
retarder 阻滯劑；緩凝劑
retarding agent 阻滯劑
retemper 再回火；再調節溫度；二次保溫
retention time 保持時間；滯留時間
retention volume 保持體積；滯留體積
reticular fiber 網狀纖維
reticular tissue 網狀組織
reticulin 網硬蛋白；網狀蛋白
reticulin fiber 網狀纖維
reticulorumen 反芻胃
reticulum 網狀結構；蜂巢胃；蜂窩胃
retinene 視網醛
retinoic acid 視網酸
retinol 松香油；視網醇；視黃醇；維生素A
retort 殺菌釜；高壓〔殺菌〕釜
retort basket 殺菌籃
retort blanching 加壓熱燙
retort crate 殺菌籃筐
retort food 殺菌食品
retort peeling 過熱蒸汽去皮
retort pouch 殺菌袋；蒸煮袋
retort pouch food 殺菌袋裝食品；蒸煮袋裝食品；軟罐頭食品
retort sterilization 殺菌釜滅菌
retort sterilizer 高壓滅菌釜
retreatment 再處理；再加工
retrogradation 澱粉回凝；澱粉老化；澱粉凝沉
retting 浸解
returnable bottle 回收瓶；可回用瓶
revaporizer 再蒸發器；二次蒸發器
revel 狂歡盛宴
revelry 狂歡宴會
reverse flow 逆流
reverse osmosis 逆滲透；反滲透
reverse tab 易開罐的反向拉舌；反向拉舌
reversibility 可逆性
reversible colloid 可逆膠體
reversible cycle 可逆循環
reversible process 可逆過程
reversible reaction 可逆反應

reversion 復原；回復現象；逆反應；逆轉現象
reversion flavo(u)r 油返味；回復味
revolving dryer 旋轉乾燥機；轉筒式乾燥機
revolving mill 旋轉研磨機
revolving pan 抛光機；包糖衣機；轉鍋
revolving rotary dryer 旋轉乾燥機
revolving screen 旋動篩；轉篩
revolving sieve 滾動篩；轉篩
revolving sifter 旋動篩；滾動篩
revolving silvering pan 滾糖衣機
rewasher 再洗機
reweighing 再稱量
rewetting 再增濕；再濕潤
Reynold's number 雷諾數
RH (relative humidity) 相對濕度
rhamnase 鼠李酶
rhamnitol 鼠李〔糖〕醇
rhamnose 鼠李糖
rhamnoside 鼠李糖苷
rheinic acid 大黃酸
Rhenish 萊茵〔河區〕白葡萄酒
rheology 流變學
rheometer 流變儀；流變計
rheopexy 觸變性；攪凝性；震凝現象
Rhine wine 萊茵〔河區〕白葡萄酒
Rhizobium 根瘤菌屬
rhizocorm 根狀球莖
rhizoid 假根
rhizome 地下莖；根莖；根狀莖
rhizomycelium 根狀菌絲體
Rhizopus 根黴屬
Rhizopus batatae 甘薯根黴
Rhizopus chinesis saito 中華根黴
Rhizopus japonicus 日本根黴
Rhizopus nigericans 黑根黴
Rhizopus oryzae 米根黴
rhodamine 玫紅；若丹明
rhodinol 玫紅醇；香茅醇；玫瑰醇
rhodinyl acetate 乙酸玫瑰酯
rhodinyl formate 甲酸玫瑰酯
rhodopsin 視紫質；視紫紅〔質〕
Rhodotorula 紅酵母屬
rhubarb 大黃
rib 排骨；肋骨
rib belly 豬胸肋肉
rib cartilage 軟肋骨
rib cut 肋條肉；排骨肉
rib eye 牛裏脊肉
rib eye roll 肋眼肉；柳脹；肉眼；沙朗；修清肉眼
rib fingers 牛肋條；腩條
rib set 肋脊部
rib steak 肋脊骨牛排；肋骨牛排
ribbon mixer 螺旋槳葉片攪拌器
ribbonfish 白帶魚
ribes 醋栗
ribitol 核糖醇
riboflavin(e) 核黃素；維生素B_2 [= lactochrome; lactoflavin]
riboflavin(e) tetrabutyrate 核黃素四丁酸脂
riboflavin-5'-phosphate sodium 核黃素-5'-磷酸鈉
ribonic acid 核糖酸
ribonuclease 核糖核酸酶
ribonucleic acid (RNA) 核糖核酸
ribonucleoprotein 核糖核蛋白〔質〕
ribonucleotidase 核糖核苷酸酶
ribonucleotide 核糖核苷酸
ribose 核糖
riboside 核糖〔核〕苷
ibosome 核糖體
ribosylzeatin 玉米素核〔糖〕苷；核糖苷玉米素
ribotide 核糖核苷酸

ribulose 核酮糖
riccini 呂契尼通心粉
rice 米；飯；稻；水稻
rice bean 飯豆；米豆
rice bin 米倉
rice bran 米糠；米麩
rice bran oil 米糠油；米麩油 [=kanemi oil]
rice bran wax 米糠蠟；米麩蠟
rice congee 粥；米粥 [=congee; conjee]
rice cooker 炊飯器；飯鍋；飯煲
rice cooking 炊飯；煮飯
rice corn 白高粱
rice crust 鍋巴；飯焦
rice field eel 黃鱔
rice flour 米粉
rice grain 稻粒；穀粒
rice hull 稻殼；粗糠；大糠 [= rice husk]
rice huller 稻穀脱殼機；碾穀機；礱穀機 [= rice husker]
rice hulling machine 稻穀脱殼機；碾穀機；礱穀機
rice husk 稻殼；粗糠；大糠 [= rice hull]
rice husker 稻穀脱殼機；碾穀機；礱穀機 [= rice huller]
rice koji 米麴
rice milk 米乳；加米的牛奶糊
rice mill 碾米廠
rice milling machine 碾米機
rice noodle 米粉〔絲〕
rice oil 米糠油
rice paper 米紙 (可食用)
rice pearling 碾米
rice pearling machine 精米機
rice polisher 碾米機；精米機
rice polishes 米糠；米麩
rice polishing machine 碾米機；精米機
rice polishings 米糠；米麩
rice porridge 米粥
rice scourer 稻米脱殼機
rice screenings 米糠；米麩
rice snack 米〔製〕點心
rice starch 米澱粉
rice sweep machine 碾米機；精米機
rice sweeper 碾米機；精米機
rice vermicelli 米粉；米線；米麵條
rice vinegar 米醋
rice water 米湯；米酒
rice whitening machine 精米機
rice wine 米酒
ricer 菜泥壓榨器；壓菜泥器
rich fruit 多汁的水果
rich milk 全脂乳
rich wine 濃酒；烈酒
ricin 蓖麻〔毒〕蛋白
ricing 米粒化
ricinic acid 蓖麻酸
ricinine 蓖麻鹼
ricinoleic acid 蓖麻酸
ricinus 蓖麻〔籽〕
ricinus oil 篦麻油
rickettsia 立克次體
rickettsiosis 立克次體病
rickey 利克酒；利克水
rickling onion 醃漬洋葱
rickstone funnel cap 白色杯傘 (蕈類)
Ricotta cheese 力可達乳清乾酪；力可達乳清芝士
rid eye steak 沙朗牛排
riddle 粗篩；篩選
riddling 篩餘料；篩下料
rigatoni 長通粉；波紋貝殼狀通心粉；線紋通心粉
rigid package 硬質包裝
rigor 僵直
rigor mortis 死後僵直；僵直；死僵

rigor off 解僵 [= resolution of rigor]
rigor phase 僵直期
rime 結霜；結晶
rind 外皮；果蔬皮；獸皮；外殼
rindless 去皮的；無皮的
rindless cheese 去皮乾酪；無皮乾酪；無皮芝士
ring doughnut 甜甜圈；炸麵圈
ring pull end 拉環蓋
ring roll crusher 環輥壓碎機
ring roll mill 環輥碾磨機
ring slicer 切圓片機
ringless honey fungus 發光蜜環菌（蕈類）
rinse 沖洗；漂洗；刷洗
rinsing 沖洗；漂洗
rinsing shaker 振動式洗菜機
rinsing tank 沖洗槽；刷洗槽
ripe dough 發酵完的麵糰；成熟麵糰
ripe green fruit 綠熟果
ripe wine 醇酒；陳酒
ripen cheese 成熟乾酪
ripen cream 發酵奶油；發酵乳酪
ripener 催熟劑；成熟槽
ripeness 成熟〔度〕
ripening 成熟；熟成；陳化；酸化
rising 膨發
rising film evaporator 升膜式蒸發器
risotto 意大利〔芝士洋葱雞肉〕調味飯；乾酪洋葱雞肉煨飯；里佐脫（意大利菜）
risotto rice 意大利米
rissole 煎肉餡餅；炸肉餡餅
rivelled fruit 萎縮果；皺皮果
river crab 河蟹
river eel 河鰻
river fish 河魚
river salmon 銀鮭；銀大麻哈魚
RNA (ribonucleic acid) 核糖核酸
roach 蟑螂
roast 烤；焙；炙
roast beef 烤牛肉
roast chicken 烤雞
roast coffee 焙炒咖啡
roast duck 烤鴨
roast meat 烤肉
roast meat fork 烤肉叉
roasted malt 炒麥芽；焦麥芽
roasted prawn 烤明蝦
roasted wheat flour 麵茶
roaster 烘烤爐；烘焙爐；烤箱；烤爐；嫩雞；乳豬
roasting 烘烤；烘焙
roasting pan 烤盤
roasting pig 烤用豬；烤乳豬
roborant 強壯劑
robust coffee 濃咖啡
robust soup 濃湯
Robusta 羅巴斯達咖啡；濃咖啡 [= coffee robusta]
Rochelle salt 酒石酸鉀鈉；羅雪鹽；羅謝爾鹽 [= sodium potassium tartrate]
rock 硬糖；冰糖
rock cake 表面粗硬的糕餅；岩皮餅
rock candy 硬糖；冰糖
rock crab 岩黃道蟹
rock maple 糖槭
rock melon 皺皮瓜；硬皮甜瓜
rock salt 岩鹽
rock scallop 曰月貝
rock sugar 冰糖
rocker screen 振盪篩
rocket 黃花南芥菜
rocking sieve 搖盪篩
rockling 岩予魚
Rockwell hardness 洛氏硬度
rod 棒；桿菌；桿狀體
rod chil(l)i 野山椒

rod mill 棒磨機
rod washer 隔柵式轉篩
rodent 囓齒動物
rodenticide 殺鼠機；滅鼠機
roe 魚卵；魚子
rokelax 挪威燻鮭魚
roll 小麵包；捲餡餅；捲狀物；輥筒；滾筒
roll cake 夾心蛋糕捲
roll crusher 輥碎機
roll drying machine 滾筒乾燥機
roll moulder 麵糰成圓機
roll on closure 旋轉瓶蓋；捲壓螺紋蓋
roll separator 滾筒分離機
roll tube culture 轉管培養〔物〕
roll tying machine 火腿捆紮機
rolled bacon 捲醃燻肉；醃燻肉捲
rolled food 捲狀食品
rolled oats 燕麥片
rolled rib 捲肋脊肉
rolled shoulder 捲肩腿肉
roller 滾筒；輥輪；轉鼓
roller breaker 輥式粉碎機
roller compaction 碾壓
roller conveyor 輥式輸送機；滾筒式輸送機
roller crusher 輥式軋碎機
roller dried milk 滾筒乾燥乳粉
roller dryer 滾筒乾燥機
roller dryer 滾筒乾燥機
roller drying 滾筒乾燥
roller extractor 滾筒萃取器
roller milk 滾筒乳粉
roller mill 滾筒碾磨機；輥軋機；軋粉機
roller press 輥壓機
roller scalder 滾筒式熱燙機
rolling kitchen 流動廚房
rolling machine 成圓機；空罐成〔圓〕型機
rolling mill 輥軋機
roly-poly 捲形布丁甜點
romaine (lettuce) 直立萵苣；長葉萵苣；羅馬萵苣
roman fennel oil 羅曼茴香油；甜茴香油
roman punch 羅曼酒；低酒度羅木酒
room temperature (RT) 室溫
rooster 公雞；雄禽
root 根；塊根
root beer 草根沙士；根啤酒
root crop 塊根農作物
root knot 根瘤
root mustard 諸葛菜；大頭菜
root nodule 根瘤
root peeler 去根機
root starch 塊根澱粉
root vegetable 根菜；塊根類蔬菜
rooting shank 長根小奧德菇
rootlet 細根；小根
rootstock 根莖；根砧木
rope 絲狀黏質；黏絲腐敗；繩；索
ropiness 黏性；黏絲狀
ropy fermentation 黏質發酵
ropy milk 稠變乳；黏乳
ropy wine 稠變酒；腐敗變黏的酒
Roquefort (cheese) 洛克福〔耳〕羊奶芝士；洛克福〔耳〕羊乳乾酪；羅奎福特羊奶芝士；羅奎福特羊乳乾酪
rosa 薔薇
rosa laevigata michx brown 金櫻子棕
rosaniline 玫瑰苯胺
rose 玫瑰
rose absolute 玫瑰淨油
rose apple 蒲桃
rose attar 玫瑰精油
rose bengal 孟加拉玫瑰紅

rose bengale 孟加拉玫瑰紅
rose crimson glory absolute 墨紅花淨油
rose crimson glory flower concrete 墨紅花浸膏
rose extract 玫瑰浸膏
rose flavo(u)r 玫瑰香精
rose hip 玫瑰果；薔薇果
rose hip syrup 玫瑰果糖漿
rose oil 玫瑰油；玫瑰花油
rose oily flavo(u)r 玫瑰油香精
rose otto 玫瑰油
rose oxide 玫瑰醚；氧化玫瑰
rose paprica 尖頭紅辣椒
rose tea 玫瑰茶
rose wine 洛西〔葡萄〕酒；玫瑰葡萄酒
rose wood oil 玫瑰木油
roselle 洛神葵；紅角葵；玫瑰茄；山茄
roselle drink 洛神茶
rosemary 迷迭香
rosemary oil 迷迭香油
rosin 松香；松脂；樹脂
rosy spike cap 玫瑰紅鉚釘菇
rot 腐爛；腐敗
rot proof 防腐；防霉
rot proofness 防腐性；防霉性
rotary autoclave 迴轉〔加壓〕殺菌釜
rotary bottle filler 迴轉式灌瓶機
rotary double-end flanger 旋轉式罐身雙頭翻邊機
rotary drum 轉鼓；滾筒
rotary drum fermenter 旋轉式發酵槽；轉鼓式發酵槽
rotary drum mixer 旋轉式混合槽；轉鼓式混合機
rotary dryer 迴旋式乾燥機；旋轉式乾燥機
rotary drying 迴旋式乾燥；旋轉式乾燥
rotary evaporator 旋轉蒸發器
rotary extractor 旋轉式萃取器
rotary filter 旋轉式過濾器；轉筒濾器
rotary flanging machine 旋轉式翻邊機
rotary kiln 旋轉窯；旋轉式乾燥爐
rotary refrigerator 旋轉式冷凍機
rotary screen 旋轉篩
rotary shaker 旋轉式搖床
rotary shelf dryer 旋轉板乾燥機
rotary sieve 滾筒篩
rotary sterilizer 迴轉式滅菌機
rotary table 轉盤；轉台
rotary washer 旋轉式洗滌機；轉鼓式洗滌機
rotating mixer 旋轉式混合器
rotating oven 轉爐
rotation 旋轉
rotational visco(si)meter 旋轉黏度計
rotgut 劣質〔威士忌〕酒；摻假的酒
roti 烤肉；炙肉
rotisserie 烤肉店；迴轉式電熱烤肉器
rotten 腐爛的；腐敗的；發臭的；發霉的
rotten flavo(u)r 腐敗味；臭味
rough colony 粗糙〔型〕菌落 [= R-colony]
rough filtration 粗濾
rough food 粗糧
rough picnic 帶肩胛骨的前腿肉
rough rice 稻穀；糙米；未碾的稻穀；未脱殼的米 [= paddy rice]
roughage 粗糙食物的纖維質；纖維性食物；粗飼料
roughing 粗加工；初步加工
round 牛股肉；牛後腿肉
round can 圓罐
round cardamon 圓荳蔻
round cure 全魚鹽醃；整魚鹽醃
round fish 全魚；整魚；圓體魚類
round herring 脂眼鯡；圓腹鯡

round kumquat 圓金柑；圓金橘
round salting 全魚鹽漬
round scad 赤尾鰺；藍圓鰺
round tin 圓罐
round top bread 圓頂麵包
rounder 麵糰成圓機；麵糰滾圓機
roundhead cabbage 圓頭甘藍
rounding machine 麵糰成圓機
roup 家禽流感；雞瘟〔疫〕
rouse 滾鹽；灑鹽；撒鹽
rouser 攪拌器
routine analysis 例行分析；常規檢驗
roux 摻油麵粉糊；奶油麵粉糊；油炒麵粉糊
rowan 歐洲花楸；美洲花楸
rowanberry 花楸〔漿〕果
royal feast 盛宴
royal jelly 蜂王漿；蜂乳
RT (room temperature) 室溫
rubber 橡皮；橡膠
rubber brush 盤菇
rubber latex 膠乳
rubbery texture 膠韌質地
rubbing board 擦菜板
rubbing machine 磨碎機；擦碎機
rubbing stone 磨石
rubbish 垃圾；廢物
rubble 碎石
rubixanthin 玉紅黃質
rubsen oil 菜〔子〕油
rudd 赤睛魚；紅眼魚
rudderfish 蘭勃舵魚；追船魚
rue oil 芸香油
ruffed grouse 松雞
ruffle fat 腸繫膜油；網油
rufous milk cap 噴紅乳菇
rum 蘭酒；糖蜜酒；朗姆酒；老姆酒
rum ether 老姆醚；朗姆醚
rumen 瘤胃
ruminant 反芻動物
rumination 反芻〔作用〕
rummer 大高腳酒杯；大酒杯
rump 臀部；後腰臀肉；臀肉；尾龍牛肉；冧肉；牛臀肉；禽尾〔椎〕部
rump steak 後腿牛排；牛臀肉排；臀肉牛排
run down pipe 溢流管
run off pipe 泄水管
rundlet 小琵琶桶；小桶
runlet 桶
runnability 流動性
runner 匍匐莖；蔓
runner bean 紅菜花豆
runnet 皺胃
running mallow 圓葉錦葵
running stock 經常庫存
runny dough 鬆軟麵糰
runt 矮小而發育不良的動物
ruoti 車輪狀通心粉
rusk 脆硬甜麵包；甜麵包乾；甜脆餅乾
rusk oven 麵包乾燥爐
russet tough shank 櫟金錢菇
Russian almond 矮扁桃
Russian olive 沙棗
Russian sturgeon 俄國鱘
russule 紅菇
rust 鐵
rust inhibitor 防銹劑
rust proof 防銹的
rusty dab 紅鰈
rusty stain 銹斑
rutabaga 瑞典蕪菁；蕪菁甘藍
rutin 芸香苷；芸香素；蘆丁
rutinose 芸香糖
rutinoside 芸香糖苷
rutoside 芸香苷；蘆丁
rye 黑麥；裸麥

rye bread 黑麵包；裸麥麵包
rye flour 黑麥粉；裸麥粉
rye malt 黑麥芽；裸麥芽
rye meal 黑麥粗粉；裸麥粗粉
rye whisky 黑麥威士忌〔酒〕；裸麥威士忌〔酒〕

S

S-colony 光滑〔型〕菌落 [=smooth colony]
sa-son seed gum 沙蒿籽膠
Saanen 撒能山羊；薩能奶山羊
sable 黑貂
sablefish 裸蓋魚
sac 袋；囊
saccharase 蔗糖酶；轉化酶 [= invertase; sucrase]
saccharate 葡萄糖二酸鹽；己醣二酸鹽；糖〔二〕酸鹽
saccharic acid 葡〔萄〕糖二酸；糖二酸；己醣二酸
saccharidase 糖酶
saccharide 糖；糖類
sacchariferous 含糖的；產生糖的
saccharification 糖化〔作用〕
saccharify 使糖化
saccharifying amylase 糖化澱粉酶
saccharifying enzyme 糖化酵素
saccharifying tank 糖化槽
saccharimeter 糖度計；糖量計
saccharin 糖精；沙卡林 [= gluside]
saccharin sodium 糖精鈉
saccharine 糖精；含糖的；糖的
saccharine sorghum 糖高粱；甜高粱 [= sugar sorghum]
saccharinic acid 糖精酸；己糖酸
saccharobiose 蔗糖 [= sucrose; saccharose]
saccharogenic amylase 澱粉〔糖化〕酶
saccharoid 粒狀物
saccharolytic bacteria 糖化〔細〕菌
saccharometer 糖度計；糖量計
Saccharomyces 酵母〔菌〕屬
Saccharomyces aceti 酸酒酵母〔菌〕
Saccharomyces acidifaciens 產酸酵母〔菌〕
Saccharomyces carlsbergensis 卡爾酵母〔菌〕；卡爾斯伯酵母〔菌〕
Saccharomyces casei 乳酪酵母〔菌〕；奶酪酵母〔菌〕
Saccharomyces cerevisiae 啤酒酵母〔菌〕
Saccharomyces diastaticus 糖化酵母〔菌〕
Saccharomyces ellipsoideus 橢圓酵母〔菌〕
Saccharomyces lactis 乳酸酵母〔菌〕
Saccharomyces mellis 蜂蜜酵母〔菌〕
Saccharomyces rouxii 魯酵母〔菌〕
Saccharomyces sake 清酒酵母〔菌〕
Saccharomyces soya 醬油酵母〔菌〕
saccharomycetolysis 酵母菌分解
saccharose 蔗糖 [= sucrose; saccharobiose]
saccharum 糖；蔗糖
Saccharum officinarum 甘蔗
sacculus 小囊；球囊
sachet 小袋；小包

sack 麻布袋；袋；沙克白葡萄酒（西班牙）
sack cleaner 淨袋機；清袋機
sack packer 裝袋機
sack storage 裝袋儲藏
sackcloth 麻袋布
sacker 裝袋機
sackful 袋（容量單位）；滿袋
sacking 麻袋布；打包；裝袋
sacrament 聖餐
sacramental wine 聖餐葡萄酒
sad bread 未烤透的麵包
saddening agent 深色劑；增色劑
saddle 羊腰部脊肉；牛腰部脊肉
safe 安全；安全櫃；食物櫥；冷藏櫃
safety factor 安全系數
safety glass 安全玻璃
safflower 紅花
safflower oil 紅花油
safflower seed oil 紅花籽油
safflower yellow 紅花黃
saffranal 番紅花醛
saffron 番紅花；藏紅花；藏花 [= crocus; safran]
saffron colo(u)r 番紅花色素；藏紅花色素
saffron oil 番紅花油
safran 番紅花；藏紅花；藏花 [= crocus; saffron]
safranine 番紅花紅；番紅精
safraninol 番紅花醇；藏紅醇
safranol 番紅定性素；藏花醇
safrene 黃樟烯
safrole 黃樟腦；黃樟素
saga oil 鼠尾草油 [= sage oil]
sage 鼠尾草；山艾
sage clary 鼠尾草
sage honey 鼠尾草蜜
sage oil 鼠尾草油 [= saga oil]
sago 西米；西谷米
sago flour 西谷粉；西米粉
sago palm 西谷椰子
sago starch 西谷澱粉
Saigon cassia 西貢肉桂；西貢桂皮
Saigon cinnamon 西貢肉桂；西貢桂皮
Saigon rice 西貢大米
saira 秋刀魚
saithe 綠鱈；青鱈
sake 日本清酒 [= Japanese rice wine]
sake koji 清酒麴
sake mash 清酒醪
salad 沙律；沙拉；色拉；涼拌菜
salad bar 沙律吧；沙拉吧
salad dish 沙律盤；沙拉盤；色拉盤
salad dressing 沙律醬；沙拉醬；色拉〔調味〕醬
salad mix 沙拉醬混料
salad oil 沙拉油；色拉油；食用植物油
salad plant 生食蔬菜類
salading 沙律用蔬菜；沙拉用蔬菜；色拉用蔬菜
salak 蛇皮果
salamander 烤盤；烤板
salami (sausage) 意大利色拉米香腸；色拉米腸；薩拉米香腸 [= Italian salami]
salep 食蘭粉（一種蘭花的塊莖澱粉）
saleratus 發粉；發酵粉；小蘇打
salicylaldehyde 水楊醛
salicylic acid 柳酸；水楊酸
salicylic ether 水楊酸酸乙酯；柳酸乙酯
salification 鹽化〔作用〕；加鹽
salify(ing) 鹽化；加鹽；成鹽的
salimeter 鹽度計；鹽水比重計
salina 鹽田；鹽場；鹽池；鹽湖
salination 鹽漬；鹽醃

saline 鹽水；生理〔食〕鹽水；鹽田；鹽場；鹽池；鹽湖
saline taste 鹹味
salineness 鹽度
salinimeter 鹽度計；鹽量計
salinity 鹽度
salinization 鹽醃；鹽漬〔化〕
salinometer 鹽度計；鹽量計
saliva 唾液
salivary amylase 唾液澱粉酶
salivary fluid 唾液
salmi 濃汁炖肉；紅燒鳥肉
salmon 鮭魚；三文魚；大麻哈魚
salmon oil 鮭魚油
salmon roe 鮭魚卵
salmon trout 真鱒
salmonberry 紅花覆盆子
Salmonella 沙門桿菌屬；沙門氏菌屬
Salmonella enterica 腸炎沙門桿菌；腸炎沙門氏菌
Salmonella food poisoning 沙門桿菌食物中毒
Salmonella paratyphi 副傷寒沙門桿菌；副傷寒沙門氏菌
Salmonella Shigella agar (SS agar) 沙門志賀桿菌〔洋菜〕培養基；沙門志賀氏菌〔瓊脂〕培養基
Salmonella typhi 傷寒沙門桿菌；傷寒沙門氏菌
Salmonella typhimurium 鼠傷寒沙門桿菌；鼠傷寒沙門氏菌
Salmonella typhosa 傷寒沙門桿菌；傷寒沙門氏菌
Salmonellosis 沙門桿菌病；沙門氏〔桿〕菌病；沙門桿菌食物中毒
salometer 鹽度計；鹽量計
saloon 沙龍；酒吧間；宴會廳；大會客室
salsify 黑皮波羅門參；婆羅門參 [= oyster plant; vegetable oyster]
salt 食鹽；鹽
salt bath 鹽浴
salt brine 鹽水；鹽液
salt burn 鹽灼
salt cellar 鹽碟；鹽瓶
salt cured 鹽醃的
salt curing 鹽醃
salt farm 鹽田；鹽場
salt field 鹽田
salt-free 無鹽〔的〕
salt-free diet 無鹽膳食
salt gauge 鹽度計；鹽水比重計
salt ice 鹽冰；冰鹽混合物
salt junk 鹹牛肉乾
salt loving mushroom 白鮮菇
salt meter 鹽度計；鹽水比重計
salt mine 鹽礦；岩鹽坑
salt pan 鹽田；鹽鍋
salt pickling 鹽醃
salt pond 鹽池
salt preserved product 鹽醃產品
salt-rising bread 傳統發酵的麵包；自然發酵的麵包
salt rust 鹽斑；鹽霜
salt shaker 撒鹽器
salt-soluble 鹽溶性；鹽溶的
salt-soluble protein 鹽溶性蛋白質
salt stock pickle 鹽醃物；鹽醃泡菜
salt substance 鹹味劑
salt substitute 代鹽品；代鹽劑
salt taste 鹹味
salt tolerance 耐鹽性
salt tolerant bacteria 耐鹽細菌
salt tolerant yeast 耐鹽酵母
salt water 鹽水；海水；鹹水
salt well 鹽井
salt works 製鹽廠；鹽場

salted 鹽醃的；鹽〔味〕的；鹹〔味〕的
salted and dried food 鹽乾食品
salted butter 加鹽白脱；加鹽乳酪
salted drying 鹽乾
salted duck egg 鹹鴨蛋；鹹蛋
salted fish 鹹魚
salted food 鹽漬食品
salted meat 鹹肉；醃肉
salted vegetable 鹹菜
saltern 鹽田；鹽場
saltery 醃製工廠
saltiness 鹹性；鹹度
salting 鹽漬；鹽醃
salting in 鹽溶
salting out 鹽析
salting stain 鹽斑
salting tank 醃製槽
saltish 微鹹的
saltless 無鹽的
saltpeter 硝石；硝酸鉀；鉀硝 [= saltpetre]
saltpetre 硝石；硝酸鉀；鉀硝 [= saltpeter]
salts of fatty acids 脂肪酸鹽
salty 鹹的；鹹味的；有鹽分的
salty flavo(u)r 鹹味
salty taste 鹹味
salver 盤；淺盤
salvia 鼠尾草；琴柱草
Salvia officinalis 鼠尾草
salvy butter 軟〔質〕奶油；軟〔質〕乳酪
samlet 幼鮭
samovar 俄式茶壺炊具
samp 玉米粥
samphire 海蒔蘿；歐洲海蓬子
sample 試樣；樣品；標本
sampler 取樣器
sampling 取樣
sampling device 取樣器
sand bath 砂浴
sand clam 沙蛤；沙土貝
sand crab 沙蟹
sand filter 砂濾器
sand filtering 砂濾
sand lance 玉筋魚
sand leach 砂濾
sand leek 胡蒜
sand pear 沙梨；白梨
sand plum 沙李
sand shrimp 沙蝦；褐蝦
sand sugar 砂糖
sand trap 除砂槽；捕砂器
sandal 檀香
sandalwood 檀香
sandalwood oil 檀香油
sandiness 砂粒感；砂質
sanding sugar 粗砂糖
sandwich 三明治；三文治；夾心麵包
sandwich biscuit 夾心餅乾
sandwich bread 三明治麵包；夾心麵包
sandwich cake 夾層蛋糕
sandy knight cap 鍍金口蘑（蕈類）
sang 人參；高麗參 [= ginseng]
sangaree 冷甜〔葡萄〕酒；桑加里酒
sanitary 衛生的
sanitary bacteriology 衛生細菌學
sanitary can 衛生罐；三片罐
sanitary condition 衛生條件；衛生環境
sanitary indicator bacteria 衛生指標細菌
sanitary inspection 衛生檢驗
sanitary inspector 衛生稽查員
sanitary science 衛生學
sanitation 衛生；衛生設備
sanitization 衛生處理；消毒
sanitize 消毒
sanitizer 消毒劑；殺菌劑
sanitizing compound 消毒劑

sans taste 無味
sansa 橄欖粕
sanshool 山椒素；山椒醇
Santa Rosa 山塔玫瑰李
santal 檀香〔木〕
santalol 檀香醇
santalum 檀香屬
santalyl acetate 乙酸檀香酯；乙酸白檀油酯
sap 樹汁；樹液
sap fruit 漿果
sapful 多汁的
sapid substance 滋味物質；調味料
sapidity 滋味
sapling 樹苗；幼苗
sapodilla 人心果
saponifiable 可皂化的
saponifiable lipid 可皂化脂質
saponification 皂化〔作用〕
saponification agent 皂化劑
saponification number 皂化值；皂化價
saponification value 皂化值；皂化價
saponifier 皂化劑
saponifying agent 皂化劑
saponin(e) 皂苷；皂素；皂角苷
sapor 風味；滋味；味；味覺
sapota 人心果；山欖果
sapour 風味；滋味；味；味覺
saprobe 腐生菌；腐生生物
saprogenic bacteria 腐生細菌
saprogenicity 腐敗性
saprophyte 腐物寄生植物；腐生植物
saprophytic bacteria 腐物寄生細菌；腐生細菌
sapucaya nut 猴胡桃
saratoga chip 油炸馬鈴薯片
saratoga potato 油炸馬鈴薯片
Sarcina 八疊球菌屬
sarcocarp 果肉；中果皮；肉果
sarcolactic acid 肌乳酸
sarcolemma 肌膜；肌纖維膜；肌鞘
sarcolysis 肌解；肌肉分解
sarcomere 肌節；肌小節
sarcoplasm 肌漿；肌質
sarcoplasmic protein 肌漿蛋白質
sarcoplasmic reticulum 肌漿網；肌質網
sarcosine 肉胺酸；肌胺酸
sarcosine oxidase 肉胺酸氧化酶；肌胺酸氧化酶
sarcous 肌肉的
sardelle 歐洲鯷
sardine 沙丁魚 [= pilchard]
sargasso 馬尾藻
sarsaparilla 土當歸
sashimi 生魚片；魚生
sass 醬料；調味醬；調味料 [= sauce]
sassafras oil 黃樟油；木察樹油
satay 沙嗲；串燒肉
satay sauce 沙嗲醬
satchel 紙袋；小背包
satiation 飽食
satiety 飽食；過飽
satsuma orange 溫州蜜柑
saturant 飽和劑
saturated fat 飽和脂肪
saturated fatty acid 飽和脂肪酸
saturated humidity 飽和濕度
saturated solution 飽和溶液
saturated steam 飽和蒸汽
saturated vapo(u)r 飽和蒸汽
saturation 飽和
saturation curve 飽和曲線
saturation isotherm 飽和等溫線
saturation juice 碳酸飽和汁；碳酸飽充糖汁
saturation level 飽和度
saturation point 飽和點
saturation value 飽和值

saturator 飽和器；飽充罐
saturnism 鉛中毒；慢性鉛中毒
sauce 醬料；調味醬；調味料 [= sass]
saucepan 單柄深鍋；平底鍋
saucepot 釜；雙柄鍋；蒸煮鍋
saucer 茶托；茶碟；淺盤；碟子
sauerkraut 德式酸泡菜；泡菜；酸漬菜 [= kraut; sourkrout; sourcrout]
saurine 秋刀魚毒素
saury 秋刀魚；竹刀魚
saury pike 秋刀魚；竹刀魚
sausage 香腸；臘腸
sausage casing 香腸腸衣；腸衣
sausage dough 香腸餡
sausage factor 香腸因數(瘦肉的比率)
sausage filler 灌腸機；灌香腸機
sausage goods 香腸製品
sausage link 小香腸(一節)
sausage meat 香腸肉餡
sausage mix(ture) 香腸餡料
sausage roll 香腸捲；香腸麵包
sausage skin 腸衣
sausage stuffer 灌腸機；灌香腸機
saute(ing) 煎；炒
sauted dish 炒菜
sauted meat 炒肉
sauted vegetable 炒蔬菜
saveloy 燻乾臘腸
savor 滋味；味道；風味；品嚐 [= savour]
savoriness 美味性；可口性 [= savouriness]
savorless 沒味道的；缺乏風味的 [= savourless]
savory 美味的；可口的；鹹的；不甜的；開胃菜餚；鹹味小菜；夏香薄荷；香薄荷；歐洲薄荷 [= savoury]
savory oil 歐洲薄荷油；香薄荷油 [= savoury oil]
savour 滋味；味道；風味；品嚐 [= savor]
savouriness 美味性；可口性 [= savoriness]
savourless 沒味道的；缺乏風味的 [= savorless]
savoury 美味的；可口的；鹹的；不甜的；開胃菜餚；鹹味小菜；夏香薄荷；香薄荷；歐洲薄荷 [= savory]
savoury oil 歐洲薄荷油；香薄荷油 [= savory oil]
savoy cabbage 皺葉包心菜；皺葉卷心菜；皺葉甘藍
saw shark 日本鋸鯊
saxitoxin 麻痺性貝毒素；貝類毒素
scald 燙〔洗〕；熱燙；褐斑點
scald picking 熱燙拔毛
scalded flour 燙麵；糊化麵粉
scalded milk 加熱煮過的牛乳
scalder 燙洗機；熱燙機；熱水燙毛缸；燙洗槽
scalder-peeler 熱燙去皮機
scalding 熱燙；燙洗
scalding tank 燙洗槽；燙毛缸；熱燙槽
scale 刻度；天秤；鱗片；魚鱗；水垢；鍋垢；剝落；去鱗
scale breaker 除鱗器；除鍋垢器
scale fish 有鱗魚
scale ice 片狀冰；小片冰
scale off 去鱗；去皮
scale pan 秤盤；天平盤
scale up 規模放大
scaled 去鱗片的
scaled fish 去鱗魚
scaler 除垢器；刮鱗器；去鱗機；秤量員
scaling 去垢；去鱗；秤量
scallion 葱；青葱；韮菜；蕎頭
scallion oil 葱油

scallop 海扇；扇貝；干貝
scallop muscle 帶子；扇貝肉
scalp(ing) 篩選；篩分
scalper 剝皮工；篩選機
scalpings 篩餘物；篩下料
scaly 鱗狀的；有鱗的
scaly fruit 蟲蛀水果；劣等水果
scaly stem 鱗莖
scaly stem vegetable 鱗莖菜
scampi 挪威龍蝦屬；食蟲蝦製的菜餚；挪威海螯蝦
scape 花梗；葉柄；花莖
scapula 肩胛骨
scar 傷痕；疤痕；結疤；種臍
scaritoxin 動物毒素
scarlet runner bean 紅花菜豆
scarlet wax cap 緋紅濕傘(蕈類)
scavenger 清除劑；清道夫
scavenging 清除；淨化
scent 氣味；香味；香料；嗅覺
scent bag 香袋；香囊；臭腺
scented 有香味的
scented tea 加味茶；花茶
scentless 無氣味的；無味
Schardinger dextrin 夏丁格糊精；沙丁格糊精；環糊精
Schardinger enzyme 夏丁格酶；黃質氧化酶；醛還原酶
schizocarp 離果；分裂果
schnap(p)s 烈酒；荷蘭杜松子酒
schnitzel 煎小牛肉片；炸肉片
school meal service 學校膳食服務
schooner 大啤酒杯；大玻璃啤酒杯
scilla 海葱；綿棗兒
scion 幼芽
sclerenchyma 厚壁組織
sclererythrin 麥角紅素；麥芽紅色素
sclerometer 硬度計
scleromucin 硬黏蛋白
scleroprotein 硬蛋白質
sclerosis 硬化
sclerotium 菌絲塊；硬菌絲團；菌核
scoff 食品；狼吞虎咽
scolic acid 二十碳三烯酸
scoliodonic acid 二十四碳五烯酸；斜齒鯊酸
Scomber 鯖魚屬
scombroid 鯖魚科
scone 英式的茶點小蛋糕；烙餅；烤餅；鬆軟圓烤餅
scoop 杓；勺；鏟
scooper 杓子；斗式升運機
scopometer 視測濁度計
scorch 焦化；燒焦；枯萎
scorched flavo(u)r 焦味
scorodose 大蒜糖
scorzonera 鴉葱；斯可左那拉參
Scotch and soda 威士忌蘇打
Scotch broth 麥菜肉濃湯(蘇格蘭)
Scotch gale 香楊梅
Scotch hake 蘇格蘭青鱈
Scotch hands 輥式榨油機；軋輥式榨油機
Scotch kale 蘇格蘭甘藍
Scotch on rock 冰塊威士忌
Scotch whisky 蘇格蘭威士忌〔酒〕
scour 洗滌；沖涮；清洗；去殼
scourer 洗滌機；碾麥機；去皮機；去殼機
scouring mill 碾麥機；去殼機
SCP (single cell protein) 單細胞蛋白〔質〕
scrag end 羊頸肉
scrambled egg 炒蛋
scran 食物；殘羹；食物屑
scrap 碎屑；廢料；殘羹
scrap fat 浮肉
scrap fish 下腳魚

scrap tallow 油渣
scrape 刮；擦
scraped surface heat exchange 刮面式熱交換器；刮板式熱交換器
scraped surface heater 刮面式加熱器；刮板式加熱器
scraper 剪毛機；刮刀；刮板；刮肉屑鏟
scrapple 玉米肉餅
screen 篩；過濾網
screen analysis 篩析
screen aperture 篩孔
screen cloth 篩絹
screen drum 篩筒
screen filter 篩濾器
screen mesh 篩目
screen opening 篩孔
screen size 篩號
screen-throughs 過篩部分
screening 篩選；篩分
screening machine 篩選機；篩分機
screening surface 篩面
screening test 篩選試驗
screening trommel 轉筒篩
screenings 篩渣；篩餘物
screw (type) extruder 螺旋擠壓機
screw (type) press 螺旋壓榨機
screw cap 螺旋蓋；螺帽
screw cap tightener 螺旋蓋擰蓋機
screw capper 螺旋蓋封蓋機
screw conveyor 螺運機；螺旋〔式〕輸送機
screw crusher 螺旋粉碎機
screw expeller press 螺旋擠出機
screw extractor 螺旋壓榨機；螺旋榨汁機
screw feeder 螺旋送料機
screw pump 螺旋泵
screw top 螺旋蓋；螺旋塞
screw top jar 螺旋口瓶
screwdriver 螺絲起子
scrod 幼鱈魚
scrubber 滌氣器；洗滌器；洗氣器
scruff 浮渣；頸背
scruple 微量
scullery 流理槽；洗碗槽；碗碟洗滌處
scum 浮渣；泡沫
scum off 撇浮渣
scum yeast 浮膜酵母；被膜酵母
scummer 撇浮渣杓；除渣杓
scumming 撇浮渣
scummy 滿浮渣的；浮渣般的
scuppernong 斯卡珀農葡萄；圓葉葡萄
scurvy 壞血病
scurvy weed 辣根菜類
scute 鱗甲
Scytonema 雙歧藻屬
SDS (sodium dodecyl sulfate) 十二基硫酸鈉
sea anenone 海葵
sea arrow 魷魚
sea bass 海鱸
sea beef 鯨魚肉；海豚肉
sea belt 海帶
sea biscuit 硬餅乾；麵包乾
sea bream 海鯛
sea brine 海〔鹽〕水
sea cabbage 濱菜；海白菜；海甘藍 [= sea kale]
sea calf 海豹
sea catfish 海鯰
sea chestnut 海膽
sea clam 海蛤
sea cook 船上廚師
sea crawfish 龍蝦
sea crayfish 龍蝦
sea cucumber 海參 [= trepang]
sea dog 海狗

sea fish 海水魚；鹹水魚
sea food 海產食物；海鮮；水產食品；海味
sea food extract 水產品抽提物
sea fowl 海鳥
sea fresh 海鮮
sea gherkin 海黃瓜
sea girdle 海帶
sea grape 馬尾藻；烏賊卵
sea grass 海草
sea hedgehog 海膽 [= sea urchin]
sea jelly 水母
sea kale 濱菜；海白菜；海甘藍 [= sea cabbage]
sea lily 海百合
sea lion 海獅
sea moss 海生苔蘚
sea mussel 貽貝
sea oak 墨角藻
sea oil 海豹油
sea perch 雙帶鱸魚；真鱸
sea pie 肉餡餅
sea salt 海鹽
sea sand 海砂
sea shell 海貝
sea slug 海參；海蛞蝓
sea snail 東鳳螺；海螺
sea snapper 紅笛鯛
sea stock 船糧
sea stores 船糧
sea turtle 海龜
sea urchin 海膽 [= sea hedgehog]
sea water 海水
sea whelk 海螺
sea wolf 狼魚
sea wrack 大葉藻
seafood platter 海鮮拼盤；海產食品拼盤
seakale 海芥蘭
seal 密封；蓋印；海豹
seal off 密封
sealability 密封性
sealant 密封劑
sealer 密封材料
sealing 封罐；密封；獵捕海豹業
sealing compound 封口膠
sealing machine 封口機；封罐機
sealing material 密封材料
seam 縫合；接縫；封罐
seam hook 捲封鈎；折邊
seamer 捲封機；封罐機
seaming 捲封；封罐
seaming chuck 封罐機軋頭；封罐上壓頭
seaming head 封罐機軋頭；封罐機捲邊封頭
seaming machine 封罐機
seaming roll 捲封滾輪；捲邊滾輪
seamless body 無縫罐身；沖壓罐身；沖拔罐身
seamless can 無縫罐；二片罐；沖壓罐；沖拔罐
seamless tin 無縫罐；二片罐；沖壓罐；沖拔罐
sear 烤焦；速烤
seared meat 烙肉；烤肉；速烤肉
season 季節；調味
seasonal food 季節性食品
seasoned 已調味的
seasoner 調味料；調味品
seasoning 調味；調味料
seasoning jar 調味品瓶
seatron 海帶製的蜜餞；褐藻類製的蜜餞
seaweed 海藻
sebaceous 脂肪的
sebacic acid 癸二酸
sebastan plum cordia 破布子

secalin 黑麥醇溶蛋白
secalose 黑麥糖
seckel 塞克爾梨
second carbonation 第二次碳酸飽和
second clear (flour) 二號二等〔麵〕粉
second evaporator 二次蒸發器；第二效蒸發器
second hand 二手貨；舊貨
second head 粗碎米；二級米
second juice 二榨汁
second molasses 淺色糖蜜；二號糖蜜
second patent (flour) 二號一等〔麵〕粉；二號粉心麵粉
second pickle 老滷
second press 二榨；二道壓榨
second product 二級產品；次級產品；副產品
second wort 二次麥芽汁
secondary alcohol 第二醇；仲醇
secondary amine 第二胺；仲胺
secondary butyl amine 仲丁胺
secondary crusher 二級壓碎機
secondary culture 次級培養〔物〕
secondary fermentation 後發酵〔作用〕
secondary filter 精過濾；二次過濾
secondary juice 二次榨汁
secondary package 二次包裝
secondary product 副產品；二次產品
secondary reaction 二次反應；次級反應
secondary spoilage 繼發性腐敗
secondary sterilization 二次滅菌
secondary yeast 後發酵用酵母
seconds 二級品；次貨；麩粉飼料
section 部份；切片；瓢瓣
section cutter 切片機
section honey 巢蜜
securo cap 螺紋保證蓋；安全蓋
sediment 沉積物；沉澱物
sediment beer 沉渣啤酒
sedimentation basin 沉澱池；沉澱槽
sedimentation tank 沉積槽
seed 種子；籽；卵；晶種
seed coat 種皮
seed corn 穀種；玉米種；玉蜀黍種
seed culture 種子培養物
seed drier 穀物乾燥機
seed dryer 穀物乾燥機
seed fat 油籽植物油
seed fermenter 種菌發酵槽
seed fish 產卵期魚
seed fruit 仁果
seed grain 晶種母粒；母粒
seed kernel oil 種〔子〕仁油；穀仁油
seed mash 種醪；酒母
seed oil 種子油
seed plant 種子植物
seed yeast 種酵母
seedcake 芳香植物種子的糕餅；油〔粕〕餅
seeder 去核器；去籽器；播種機；播種者
seeding yeast 種子酵母
seedless 無籽的；無核的
seedless raisin 無籽葡萄乾；無核葡萄乾
seedless variety 無籽品種
seedless watermelon 無籽西瓜；無核西瓜
seedling 秧苗；幼苗；播種
seedpod 莢果；種莢；豆莢
seep(ing) 滲出；漏出；滲漏
seepage 滲出現象；滲出量；滲漏；滲出
seethe 煮沸；沸騰
segment 果瓣；瓢瓣
segment separator 果瓣分離機；分瓣機
segregation 分離〔作用〕；分凝〔作用〕

Seitz filter 賽茨過濾器
selacholeic acid 鯊油酸；神經酸
selachyl alcohol 鯊油醇
selective and differential medium 選別性培養基
selective bacteriostasis 選擇性抑菌作用
selective culture 選擇性培養〔物〕
selective enrichment 選擇性增菌培養
selective medium 選擇性培養基
selective permeability 選擇滲透性
selective plating 選擇性平皿培養〔法〕
selectivity 選擇性
selenium 硒
self-catalyzed 自催化
self-digestion 自〔體〕消化
self-feeder 自動進料機；自給器
self-fermentation 自然發酵
self-filler 自動充填機；自動灌裝機
self-heating package 自〔然發〕熱包裝
self-raising flour 自發〔麵〕粉
self-sealing 自動密封；自封的
self-service 自助的
self-service store 自助〔商〕店
self-souring 自然酸化
selinene 芹子烯
seltzer water 德國賽爾脱兹礦泉水；德國色查礦泉水
semi-acid food 中酸性食品
semi-boneless 半去骨的
semi-condemned meat 高溫肉
semi-continuous distillation 半連續蒸餾
semi-continuous freeze drying 半連續凍乾
semi-defrost 半解凍
semi-dressed fish 半清理魚
semi-dried 半乾的
semidry 半乾〔的〕
semidry sausage 半乾香腸
semidry wine 中甜度葡萄酒；半乾葡萄酒
semi-drying oil 半乾性油
semi-fermented tea 半發酵茶
semi-finished product 半成品
semifluid 半流體；半流質
semifluid diet 半流質膳食
semi-frozen 半凍結的
semi-hard cheese 半硬〔質〕乾酪
semi-hardened oil 半硬化油
semiliquid 半流體；半流質
semiliquid diet 半流質膳食
semi-manufactures 半成品
semimoist food 中濕性食品；半濕性食品；中等水分食品；半乾食品 [=intermediate moisture food; IMF]
seminose 甘露糖 [= mannose]
semipermeable 半透性的；半滲透性的
semipermeable membrane 半透膜
semi-preservation 半保藏；短期保藏
semi-preserved food 半保藏食品；短期保藏食品
semi-product 半成品
semis 半成品
semi-skimmed milk 半脱脂乳
semisoft diet 半軟質膳食
semisolid 半固體
semi-soluble 半溶性
semistarvation 半飢餓
semi-sterile 半殺菌的
semi-sweet 低甜度的；低糖的
semi-thawing 半解凍
semi-voluntary muscle 半隨意肌
semolina 粗粒小麥粉；粗〔小〕麥粉；粗粒麵粉
senile 衰老的；老年的
senility 衰老；老年
senna 番瀉樹
senna leaf 番瀉〔樹〕葉

sense organ 感覺器官
sensible heat 顯熱
sensitivity 靈敏性；敏感性
sensor 感測器；傳感器
sensory evaluation 感官評價
sensory property 官感性；感官性質
sensory test 感官試驗；官能試驗
sepal 萼片
separate concentration 分離濃縮
separating screen 分選篩；選分篩
separating tank 分離槽；澄清槽
separation 分離
separation colume 分離柱
separator 分離器
sephadex 交聯葡聚糖
sepia 墨魚；烏賊；烏賊墨汁
sepia melanin 烏賊黑色素
sepium 烏賊骨
sepsis 膿毒病
septic tank 化糞池
sequester 螯合劑；鉗合劑
sequestering agent 鉗合劑；螯合劑；多價螯合劑 [= chelating agent; sequestrant]
sequestrant 鉗合劑；螯合劑；多價螯合劑 [= chelating agent; sequestering agent]
sequestration 鉗合〔作用〕；多價螯合〔作用〕
sera 血清；乳清；漿液 〔= serum的複數〕
seralbumin 血清白蛋白
serglobulin 血清球蛋白
serial dilution 連續稀釋；循序稀釋
serial effect 連續效應；順序效應
serine 絲胺酸
serolipase 血清脂酶
serolysin 血清溶素
seromucoid 血清黏蛋白
serpent gourd 蛇瓜
serpentine cucumber 刺黃瓜
serrated crab 蟳
serum 血清；乳清；漿液
serum albumin 血清白蛋白；乳清白蛋白
serum butter 乳清脂；乳清奶油
serum calcium 血清鈣
serum cholesterol 血清膽固醇
serum globulin 血清球蛋白；乳清球蛋白
serum iron 血清鐵
serum lipid 血清脂質
serum protein 血清蛋白〔質〕；乳清蛋白〔質〕
server 菜盤；托盤；端菜者；盛菜用具
service 服務；餐具
service berry 花楸果
service cabinet 自動販賣冷藏櫃
service plate 供餐首盤；引餐盤
service tree 花楸〔樹〕
serviette 餐巾
serving 一份〔食物〕；一客〔食物〕；服務；上菜
serving chopsticks 分菜筷；公筷
serving fork 分菜叉
serving knife 分菜刀
serving spoon 分菜匙；母匙
servings 供食份數
sesame 芝麻 [= benne]
sesame oil 麻油；芝麻油；香油 [= benne oil; gingelly oil; jinjili oil]
sesame paste 芝麻醬；芝麻糊
sesame sauce 芝麻醬
sesamin 芝麻明；芝麻素
sesamol 芝麻酚
set honey 結晶蜂蜜
set-up box 摺疊紙盒；組裝紙盒
seta 剛毛；刺毛

setting 靜置；沉澱；凝結；凝固
setting accelerator 促凝劑
setting agent 凝結劑；沉降劑
setting point 凝結點
settle 沉澱；沉降；澄清
settler 澄清器；沉降器；沉澱器
settling 沉澱的；沉降的；澄清的
settling pond 沉降池；澄清池
settling process 沉降法
settling tank 沉降槽；沉澱槽
settlings 沉澱物
seville orange 酸橙
sevruga 閃光鱘魚子醬
sevruga caviar 小粒魚子醬
sewage 污水；下水道
sewage bacteria 污水細菌
sewage disposal 污水處理
sewage farm 污水處理場
sewage treatment 污水處理
sewage water 污水；廢水
sewerage 污水；廢物；排水系統
sexual spore 有性孢子
shackle 羈絆；束縛物；腳鐐；吊腳鏈
shackle chain 扣腳鏈；吊腳鏈
shad 水滑魚；鰣魚；美洲西鯡
shaddock 柚子；文旦 [= pomelo; pummelo; pompelmous]
shaddock oil 柚皮油
shade drying 陰乾
shaggy parasol 粗鱗大環柄菇
shake 振動；搖動
shake and bake 搖烤〔調理〕
shake and bake food 搖烤食品
shake culture 振盪培養〔物〕
shaker 搖動器；振盪器
shaker screen 振動篩
shaking conveyor 振動〔式〕輸送帶
shaking screen 振動篩；振盪篩
shallot 油葱；紅葱頭；分葱；大頭葱；珠葱 [= eschalot]
shallow drawn can 淺沖壓罐；淺沖拔罐
shallow fry(ing) 煎；淺炸
shallow oval can 淺橢圓罐
shallow pan 淺盤；淺鍋
shambles 屠宰場；肉店
shamrock 酢漿草；三葉草
shandygaff 薑汁啤酒
shank 脛；小腿
shantung cabbage 山東大白菜
shao-hsing wine 紹興酒
shaper 成形器；模子
shaping die 成形模
shark 鯊魚
shark fin 魚翅；鯊魚鰭
shark liver oil 鯊魚肝油
sharp flavo(u)r 刺鼻風味；辛辣味
sharp freezer 低溫凍結器
sharp freezing 速凍；低溫凍結
sharp odo(u)r 刺鼻氣味；強烈氣味
sharp scaled parasol 粗糙環柄菇
sharp taste 烈味；辛辣味
sharp-toothed eel 尖齒鰻；海鰻
sharpener 削尖器；磨刀器
sharpening steel 磨刀鋼條
shaved ice 刨冰
shaver 刮刀
shaving machine 刮毛機
shea 非洲酪脂樹
shea butter 牛油樹油；牛油樹脂
shea butter tree 牛油樹
shea nut fat 牛油樹果脂
shear 剪力；剪切；切變
shear force 剪切力
shear strength 抗剪強度
shear viscosity 剪切黏度；切變黏度
shears 大剪刀；剪床

sheathed bacteria 有鞘細菌
shebeen 無照酒館；地下酒吧
shed 脱落；脱皮
sheep 綿羊；羊
sheep casing 綿羊腸衣；羊腸衣
sheep polypore 羊白孔菌 (蕈類)
sheep run 大牧羊場
sheet filter 片式過濾器
sheet tin 馬口鐵皮；鍍錫薄板
sheeter 壓片機；壓麵機
shelf 擱板；貨架；架子
shelf curing 架醃〔法〕
shelf dryer 盤架乾燥器
shelf life 儲藏期限；保存期限；櫥存期限
shelf-stable food 儲藏穩定食品
shelflife 保存期限；儲存期限
shell 貝殼；殼；介殼；甲殼；去殼
shell and tube heat exchanger 管殼式熱交換器；殼管式熱交換器
shell egg 帶殼蛋
shell membrane 蛋殼膜
shell pasta 貝殼粉；蜆殼粉 [= abassina rigata; conchiglie]
shellac 蟲膠；紫膠；蟲漆
shellac wax 蟲膠蠟
shelled corn 去鞘玉米穗；玉米粒
shelled cuticle 蛋殼角皮層；蛋殼護膜
shelled pea 去莢豌豆；去莢青〔豌〕豆
sheller 去殼機；去皮機
shellfish 貝類；甲殼類動物
shellfish poisoning 貝類中毒
shelling 去殼；去皮
shell-less 無硬殼的
shellolic acid 蟲膠酸
sherbet 果汁雪霜；雪果霜；冰糕；果子露
sherry 雪利〔葡萄〕酒；諧麗〔葡萄〕酒
shibuol 柿澀醇；柿澀單寧
shifter 篩子
Shigella 志賀桿菌屬；志賀氏菌屬
Shigella dysenteriae 痢疾志賀桿菌
shigellosis 志賀桿菌病；志賀氏菌性痢疾
shiitake 香菇；冬菇；椎茸
shikonin 紫草寧
shin 脛；脛骨
shin from foreshank 前小腿腱；前腲肉；金錢腲；前腱
shin from hindshank 後腲肉；腿腱子〔心〕；後腱；大腲
shinyleaf yellowhorn 文冠果
ship stuff 低級麵粉；麩皮
shiso 紫蘇
shisonin 紫蘇素
shiver 粉碎；碎片
shoal 淺灘；魚群
shoat 小豬 (不滿一歲)
shock 休克；震動；電擊
shocking pasteurization 瞬間巴氏殺菌；大溫差巴氏殺菌
shoemaker 鰹魚
shoestring fungus 蜜環菌
shoestring potato 炸馬鈴薯絲
shogaol 薑烯酚
shop fat 板油；肥膘肉
shopping bag 購物袋
shore crab 食草蟹
short-chain fatty acid 短鏈脂肪酸
short crust 油酥麵皮；酥皮
short cut 短切；捷徑
short damping 快速蒸燻
short fillet 腰椎里脊肉；小里肌
short gel 脆〔質〕凝膠；稠〔厚〕凝膠
short gluten 脆〔質〕麵筋；無韌性麵筋
short grain rice 短粒米
short loin 背脊肉
short meat 無黏性的肉；煮得鬆散的肉

short necked clam 日本綴錦蛤；菲律賓蛤仔
short nougat 脆核果糖；硬牛軋糖
short oil 短油；低聚合度的油；低乾性油
short order 快餐
short paste 脆質麵蝴；油酥麵糰
short pastry 油酥麵製糕餅
short plate 胸腹肉
short ribs 短肋肉；牛肋條肉；牛小排；牛仔骨
short-set(ting) 速凝；快速凝固
short texture 疏鬆的結構或質地（乾酪）
shortage 不足；缺乏
shortbread 油酥甜餅
shortbread cookie 厚酥餅；高油脂曲奇餅
shortcake 油酥糕餅；厚油酥甜餅
shortener 麵糰油酥料
shortening 縮短；酥油；起酥油；白油
shorthorn 短角牛（英國）
shortloin 前腰脊肉；腰眼骨肉
shortness 鬆脆性
shortometer 油酥性測定計；起酥計；酥鬆性測定計
shorts 麩粉；麩皮
shorts duster 刷麩機
shot 一口〔威士忌酒〕；一小杯〔威士忌酒〕
shoulder 肩胛肉；肩部
shoulder bacon 燻肩肉
shoulder blade 肩胛骨肉
shoulder butt (pork) 豬肩〔上〕端肉
shoulder fat 肩肥肉
shoulder ham 肩肉火腿；前腿肉製的火腿
shoulder meat 肩肉；肩胛肉
shoulder picnic 胛心肉
shoulder strap 牛前肩脊肉
showcase 展示櫃；陳列櫃
shower cooling 噴淋冷卻
shoyu 醬油
shoyu mash 醬〔油〕醪
shred 碎片；切絲
shred residue 碎屑
shredded cabbage 甘藍菜絲
shredded fish 魚鬆；魚絲
shredded meat 肉鬆；肉絲
shredder 切碎機；切絲機；撕碎機
shredding 切碎；撕碎；切絲
shredding machine 切碎機；切絲機；撕碎機
shrimp 小蝦；蝦子
shrimp bisque 蝦濃湯
shrimp butter 蝦白脱；蝦乳酪
shrimp meal 蝦粉
shrimp meat 蝦仁；蝦肉
shrimp paste 蝦醬；蝦膏 [= blacan; blachan]
shrimper 去捕殼機
shrink film 收縮薄膜
shrink package 收縮包裝
shrink wrap 收縮包裝
shrinkage film 收縮薄膜
shrinkproof 不縮的；防縮的
shrivel 皺縮；捲縮；起皺；枯萎
shrub 灌木；果汁甜雞尾酒
shuck 殼；莢；外皮；去殼；去皮
shucked 去殼的；去皮的
shucker 去殼機；去皮機
shuffle grader 擺動分級機
sialagogue 催涎劑；催涎的
sialic acid 唾液酸；涎酸
sialidase 唾液酸酶
siccation 乾燥
siccative 催乾劑；乾燥劑
sickle 鐮刀
sickle-cell anaemia 鐮狀細胞貧血症

side bacon 腹脇肉
side carcass 半邊屠體
side chain 側鏈；支鏈
side dish 配菜；小菜
side effect 副作用
side fermentation 副發酵〔作用〕
side group 側基
side plate 配菜
side pork 豬胸肉；豬肋肉
side seam 邊封；邊縫
side seam can 邊封罐
side seamer 邊封機
side strip 罐身〔開啟〕拉舌
sideboard 餐具櫃
sidebone 叉骨；肋骨
sidecar 塞卡雞尾酒
sideropenia 鐵質缺乏〔症〕
siege 爐底；爐床
sieva bean 雪豆；棉豆；金甲豆
sieve 篩；濾網；過濾
sieve analysis 篩析
sieve aperture 篩孔
sieve brush 清篩刷
sieve effect 篩孔效應
sieve grating 篩格
sieve mesh 篩目；篩網
sieve number 篩目號
sieve opening 篩孔
sieve plate 篩板
sieve plate column 篩板塔
sieve residue 篩渣；篩下料
sieve shaker 搖篩機；振動篩
sieve size 篩目孔徑；篩尺寸
sieve tray 篩盤
sieve tray column 篩盤塔
sieving 篩分；篩選
sieving machine 篩分機；篩選機
sift 篩分；篩選；過濾；詳細審查
sifter 篩；撒粉器
sifting machine 篩分機；篩選機
sifting screen 篩網
siftings 篩出物
sigma value 啤酒泡沫穩定度；啤酒泡沫持久度
significant difference 顯著差異
significant level 顯著水準
silage 青貯飼料
silent spirit 蒸餾酒精；高純度酒精
silex glass 石英玻璃
silica 二氧化矽；二氧化硅 [= silicone dioxide]
silica gel 矽膠；硅膠
silica wool 石英棉
silicate 矽酸鹽；硅酸鹽
silicon 矽；硅
silicon dioxide 二氧化硅
silicone 聚矽氧；硅酮；矽酮；硅樹脂
silicone dioxide 二氧化矽；二氧化硅 [= silica]
silicone oil 聚矽氧油；硅油；矽油
silicone resin 硅酮樹脂
silk 絲〔綢〕；玉米鬚
silk-cotton seed oil 木棉籽油
silk-cotton tree 木棉樹
silk sieve 綢篩
silker 除絲器；去〔玉米〕鬚機
silking machine 去〔玉米〕鬚機
silkworm 蠶；桑蠶
silky bird 烏骨雞
silky volvar 絲蓋小包腳菇
sillabub 乳潘趣飲料；乳酒凍 [= syllabub]
silo 圓穀倉；筒倉
silver 銀
silver belt fish 帶魚
silver bream 平鯛；黃錫鯛
silver carp 白鰱；竹葉鰱
silver cod 銀鱈

silver eel 銀鰻
silver liqueur 荷蘭〔銀箔〕甜酒
silver plate 銀盤；銀器皿
silver pomfret 白鯧
silver salmon 銀鮭；銀大麻哈魚
silver side 牛後腿外側肉
silver snapper 紅鮪
silver sprout 銀芽；綠豆芽 [= green bean sprout]
silverside 外側後腿肉；銀邊肉；牛腿端肉
silverskin onion 小白洋葱
silverware 銀器；銀製餐具
simmer(ing) 燜；燜煮
simmered meat 燉肉；文火燉肉
simple diet 粗茶淡飯
simple fruit 單果
simple lipid 單純脂質
simple microscope 單式顯微鏡
simple polysaccharide 單純多醣
simple protein 單純蛋白質；簡單蛋白質
simple sugar 單糖
simple triglyceride 單純三甘油酯
simulated food 仿製食品；模擬食品
simulated meat product 仿肉〔製〕品
sinapic acid 芥子酸
sinapine 芥子酸膽鹼〔酯〕
sinapyl alcohol 芥子醇
sinew 腱；筋肌
singe 燎毛；燒毛
singeing 燎毛；燒毛
singeing machine 燒毛機；燎毛爐
singeing torch 燒毛器；燎毛器
singer 燒毛器；燎毛器
singharanut 菱〔角〕
single baking powder 單效發粉
single bond 單鍵
single cell 單細胞
single cell protein (SCP) 單細胞蛋白〔質〕
single cream 稀〔鮮〕乳油；稀奶油
single cream cheese 稀奶油乾酪；稀乳油乾酪
single-drum dryer 單滾筒乾燥機
single-effect evaporator 單效蒸發器
single-layer 單層的
single-needle stitch pumping 單針頭肌肉注射醃漬法
single-pass dryer 單程乾燥機
single ply bag 單層袋
single-roll crusher 單輥壓碎機
single-screw extruder 單軸擠壓機；單螺杆擠壓機
single-service package 一次〔使用〕性包裝
single-stage evaporator 單效蒸發器；單級蒸發器
single-stage extraction 單級萃取
single-stage homogenizer 單級均質機
single strength 一倍濃度（未經稀釋或濃縮）
single strength juice 原汁
singleseed hawthorn 單籽山楂
sinigrin 黑芥酸鉀
sinistrin 海葱〔聚〕糖
sink 洗滌槽；水槽；沉下
sinkaline 膽鹼
sintered glass 燒結玻璃
sintered glass filter 燒結玻璃濾器
sip 啜飲；吸吮
siphon 虹吸；虹吸管 [= syphon]
sippet 小麵包片
Sippy diet 西蔗氏膳食；西比膳食
sirloin 牛腰脊肉；腰上部肉；後腰脊肉；沙朗
sirloin butt 牛厚里脊肉
sirloin steak 腰脊牛排

sirloin strip 里脊條〔肉〕
sirup 糖漿；糖汁；糖膏
sissy beer 淡啤酒
site test 現場試驗
sitology 膳食營養學；食品營養學；飲食營養學
sitomania 貪食症
sitophobia 畏食〔症〕
sitosterin 植固醇；穀固醇；穀甾醇
sitoterol 植固醇；穀固醇；穀甾醇
sitotherapy 飲食療法；營養療法
sitotoxin 食物毒素；穀物毒素
sitotoxism 食物中毒
size 大小；尺寸；膠料；漿料
size grader 大小分級機
sizer 篩選機；分級機；上膠器
sizing 篩分；分級
sizing screen 分級篩
sizings 小麥粗粒；粗粉；膠料
skate 鰩魚；魟魚
skeletal muscle 骨骼肌；隨意肌
skeletal muscle fiber 骨骼肌纖維
skeleton 骨骼；骨架
skewer 串肉簽；烤肉叉；烤肉的長鐵針；烤肉用的細長木簽
skillet 長柄平底鍋
skilly 麥片粥；羹湯
skim 撇去〔浮油〕；撇去〔浮沫〕
skim milk 脱脂奶；脱脂乳 [= nonfat milk]
skim milk cheese 低脂肪乾酪；脱脂乾酪
skim milk powder 脱脂乳粉；脱脂奶粉 [= nonfat dry milk]
skimmed cheese 脱脂芝士；脱脂乾酪
skimmer 撇渣器；去沫器
skimming 撇去〔浮沫〕；撇去〔油沫〕
skimmings 浮渣；泡沫層；浮沫
skin 皮膚；獸皮；皮；果皮
skin blemish 果皮斑點
skin drying 表面乾燥
skin formation 結膜；結皮；結衣
skin hide 獸皮
skin packaging 密著包裝；緊縮包裝
skin tight packaging 密著包裝；緊貼包裝
skinless 無皮的；去皮的
skinless ham 無皮腿肉；去皮腿肉
skinless sausage 無衣香腸
skinned 去皮的；已剝皮的
skinned carcass 去皮屠體
skinned fish 剝皮魚；去皮魚
skinned grain 去皮穀類；去殼穀類
skinner 去皮器；剝皮機；削皮刀；剝皮工人
skinning 去皮；剝皮；削皮
skinning machine 去皮機；剝皮機；削皮機
skipjack 鰹；飛魚
skirt 隔膜〔肉〕
skirt steak 牛腹肉
skiving machine 切片機
skull 頭骨
skuta 發酵綿羊乳清
skyr 冰島發酵凝乳；冰島酸牛奶
slab 厚塊；厚片；厚板
slab drying 厚層乾燥
slabber 切塊機；切片機
slack 鬆弛
slack baked 未烤熟的；未烤透的
slack fill 灌裝不足；裝罐不足
slack salted 輕醃的；淡醃的
slaked lime 熟石灰；消石灰；氫氧化鈣
slaker 石灰消和器
slant 斜面培養〔基〕
slant agar 斜面洋菜〔培養基〕；斜面瓊脂〔培養基〕

slant culture 斜面培養〔物〕
slapjack 薄煎餅；烙餅；煎餅；烤餅 [= flapjack; griddlecake]
slat 板條；薄細木條
slatted crate 板條箱
slaughter 屠宰
slaughter weight 屠體重
slaughterer 屠宰工
slaughterhouse 屠宰場 [= abattoir]
slaughtering facilities 屠宰設備
slaughtering plant 屠宰場
slaw 椰菜絲沙律；甘藍沙拉；卷心菜〔絲〕沙拉；涼拌卷心菜〔絲〕 [= coleslaw]
sledge mill 槌磨
sleeve fish 烏賊
slender diet 苗條膳食
slice 切片；薄片；片
sliced bacon 醃肉片；培根肉片
sliced bread 切片麵包；麵包片
sliced cheese 芝士片；乾酪片
sliced ham 火腿片
sliced meat 肉片
slicer 切片機
slicing 切片；切成薄片
slime 黏液；黏質物
slime bacteria 黏〔質〕細菌
slime fermentation 黏質發酵
slime-forming bacteria 黏質形成細菌
slime fungus 黏質真菌
slime layer 黏質層；黏液層
slime mol(u)d 黏質黴菌
slime starch 黏澱粉
slimer 去黏〔膜〕機；洗腸機
sliminess 黏性；發黏；滑溜
sliming 發黏；去黏質
slimming 減胖；減重
slimy 黏液的；發黏的
slimy casing 未洗腸衣
slimy meat 發黏肉
sling 斯林酒
slip-slop 淡而無味食物；乏味的酒
slippery jack 褐環乳牛肝菌
slitter 切板機；切條機
slitting machine 切板機；切條機
slivovitz 斯李白蘭地
slobber 流涎；流口水
sloe 黑刺李；野梅
sloe gin 黑刺李琴酒；野梅酒
slop 淡而無味的流質食物；食物殘渣；流質食物；溢出
slop basin 殘渣〔淺〕碟；殘茶盆
sloshed 喝醉的
slot machine 自動售貨機
slot mesh screen 長眼篩網
slotted plate 長眼篩網；長眼篩板
slow cooling 慢速冷卻
slow defrosting 慢速解凍；慢速除霜
slow fire 文火
slow freezing 慢速冷凍
slow set 緩凝；凝固緩慢
slow set pectin 緩凝果膠；慢凝果膠
slow starter 慢發酵劑；緩發酵元
sludge 污泥；淤渣；污物
sludgeless oil 無渣油
sluggish fermentation 延滯發酵；緩慢發酵
slurp 大聲地啜喝；啜食
slurry 泥漿；稀泥
slush 淤泥；爛泥
small abalone 雜色鮑；九孔
small ale 淡艾爾啤酒
small beer 淡啤酒；劣啤酒
small calorie 小卡；卡
small casing 小腸腸衣〔豬或牛〕
small fennel 小茴香
small fruit 小果；漿果
small grain 細粒禾穀；小粒穀類

small ice 小冰塊
small intestine 小腸
small macaroni 細通心粉
small yellow croaker 小黃魚
smare 棒鱸
smash 斯馬什雞尾酒；斯馬喜雞尾酒
smearing culture 塗抹培養〔物〕；塗片培養〔物〕
smell 嗅覺；氣味
smelt 香魚；胡瓜魚
smoke 煙；煙燻
smoke agent 發煙劑；煙燻劑
smoke curing 燻製
smoke dry 燻乾；煙燻脱水
smoke eliminator 消煙器
smoke flavo(u)ring 煙燻香料
smoke generator 發煙器
smoke house 煙燻室；燻房
smoke oven 煙燻爐
smoke solution 煙燻溶液；燻液
smoked bacon 燻培根
smoked cheese 燻芝士；燻乾酪
smoked fish 燻魚
smoked flavo(u)r 煙燻風味；煙燻香料
smoked food 煙燻食品；燻製食品
smoked ham 煙燻火腿
smoked meat 燻肉
smoked product 燻製品
smoked provision 燻製食品
smoked salmon 燻鮭
smoked sausage 燻香腸；燻製香腸
smokeless smoking 無煙燻製
smoker 煙燻室；燻製業者；吸煙者；吸菸者
smokery 煙燻室；燻房
smokiness 發煙性；煙燻程度
smoking 煙燻；燻製
smoking fluid 燻液
smoking room 煙燻室；燻房；吸煙室；吸菸室
smoking wood 煙燻材料；燻材
smoky 煙燻味的
smoky bean 燻豆
smoky flavo(u)r 煙燻風味；燻味
smolder 燻燒；悶燒
smolt 小鮭魚(二歲)
smooth chanterelle 磚紅雞油菌(蕈類)
smooth colony 光滑〔型〕菌落 [= S-colony]
smooth finish 光面精整
smooth flavo(u)r 爽口味
smooth flour 軟質麵粉
smooth muscle 平滑肌
smooth muscle fiber 平滑肌纖維
smooth pea 光皮豌豆
smorgasbord 北歐式自助餐；瑞典式自助餐
snack 小吃；點心；零食
snack bar 小吃部；快餐部；小吃店
snack food 小吃；點心；輕食食品；休閑零食
snack stall 小吃攤
snack stand 小吃攤
snail 蝸牛
snake 蛇
snake eel 蛇鰻
snake gourd 蛇瓜
snake head 烏鱧；鱧
snake melon 蛇甜瓜
snake soup 蛇羹
snake squash 蛇瓜
snap 薄脆餅
snap bean 四季豆；敏豆；菜豆
snap-chilling 急冷；快速冷卻；激冷
snap-on cap 搭鎖蓋；咬封蓋；按蓋
snapper 笛鯛
sniff analysis 鼻聞分析

snip 碎片；剪斷
snipe 鷸；沙錐鳥
snipefish 長嘴魚；長吻魚
snipper 切片機；剪切機
snoek 杖魚
snout 豬嘴；口鼻部
snowy wax cap 潔白濕傘（蕈類）
soak(ing) 浸泡；浸漬；浸透
soakage 浸漬；浸透
soaker 浸漬劑；浸漬槽；酒量大的人
soap 肥皂
soap fat 製皂用脂
soap stock 皂腳
soapbark 皂樹〔皮〕
soapbark extract 皂樹皮抽提物
soapy taste 皂味
sock 零食
sockeye (salmon) 紅鮭；紅大麻哈魚
soda 碳酸水；鹼；蘇打〔水〕
soda ash 無水碳酸鈉；鹼灰；蘇打灰
soda bread 蘇打麵包
soda cracker 蘇打餅乾
soda fountain 汽水〔供應〕器；蘇打水〔供應〕器；汽水櫃台；冷飲櫃
soda lime 鹼石灰
soda lye 氫氧化鈉溶液
soda pop 汽水；蘇打水；碳酸飲料
soda water 蘇打水；汽水
sodden 浸透的；浸濕的；未烤透的
sodium 鈉
sodium acetate 醋酸鈉
sodium acetate, anhydride 無水醋酸鈉
sodium acid pyrophosphate 焦磷酸鈉
sodium alginate 褐藻酸鈉；藻朊酸鈉；海藻酸鈉
sodium alum 硫酸鋁鈉；鈉明礬
sodium aluminium phosphate, acidic (SALP) 酸性磷酸鋁鈉
sodium aluminium phosphate, basic 鹼性磷酸鋁鈉
sodium aluminosilicate 硅鋁酸鈉
sodium ascorbate 抗壞血酸鈉
sodium benzoate 苯甲酸鈉；安息香酸鈉
sodium bicarbonate 碳酸氫鈉；小蘇打 [= sodium hydrogen carbonate; bicarbonate of soda; baking soda]
sodium bisulfate 硫酸氫鈉
sodium bisulfite 亞硫酸氫鈉
sodium borate 硼酸鈉
sodium carbonate 碳酸鈉
sodium carboxymethyl starch 羧甲〔基〕澱粉鈉
sodium carboxymethylcellulose 羧甲基纖維素鈉
sodium caseinate 酪蛋白酸鈉
sodium chloride 氯化鈉；食用鹽；食鹽 [= NaCl; edible salt; table salt]
sodium chlorite 亞氯酸鈉
sodium 4-chlorophenoxyacetate 4-氯苯氧乙酸鈉
sodium choleate 膽酸鈉
sodium chondroitin sulfate 硫酸軟骨素鈉
sodium cinnamate 桂皮酸鈉
sodium citrate 檸檬酸鈉
sodium copper chlorophyllin 葉綠素銅鈉鹽
sodium cyclamate 賽克拉美鈉；環己基氨基磺酸鈉；甜蜜素
sodium dehydroacetate 脱氫醋酸鈉
sodium diacetate 雙乙酸鈉
sodium dihydrogen citrate 檸檬酸一鈉
sodium dodecyl benzene sulfonate 十二烷基苯磺酸鈉
sodium dodecyl sulfate (SDS) 十二〔烷〕基硫酸鈉；月桂基硫酸鈉

sodium erythorbate 異抗壞血酸鈉
sodium ferric pyrophosphate 焦磷酸鐵鈉
sodium ferric succinic citrate 琥珀酸檸檬酸鐵鈉
sodium ferrocyanide 亞鐵氰化鈉
sodium ferrous citrate 檸檬酸亞鐵鈉
sodium fluoride 氟化鈉
sodium formate 甲酸鈉
sodium free 無鈉〔的〕
sodium free diet 無鈉膳食
sodium free salt 無鈉〔食〕鹽
sodium fumarate 富馬酸鈉
sodium gluconate 葡〔萄〕糖酸鈉
sodium glutamate 谷氨酸鈉；麩胺酸鈉；味精；味素 [= aginomoto; monosodium glutamate; MSG]
sodium glutamate 谷氨酸鈉；味精；麩氨酸鈉
sodium 5'-guanylate (5'-GMP-Na) 5'-鳥〔嘌呤核〕苷酸鈉
sodium hexametaphosphate 六偏磷酸鈉
sodium hydrogen carbonate 碳酸氫鈉；小蘇打 [= sodium bicarbonate; bicarbonate of soda; baking soda]
sodium (hydrogen) diacetate 二醋酸一鈉；雙乙酸鈉
sodium hydrogen malate 蘋果酸一鈉
sodium hydrogen sulfite 亞硫酸氫鈉
sodium hydrosulfite 低亞硫酸鈉；次硫酸鈉
sodium hydroxide 氫氧化鈉
sodium hypochlorite 次氯酸鈉
sodium hypophosphite 次磷酸鈉
sodium 5'-inosinate (5'-IMP-Na) 5'-肌苷酸鈉；5'-肉苷酸鈉
sodium iron chlorophyllin 葉綠酸鐵鈉鹽
sodium iron(Ⅲ) ethylenediaminetetraacetate 乙二胺四乙酸鐵鈉；EDTA鐵鈉鹽
sodium lactate 乳酸鈉
sodium L-ascorbate L-抗壞血酸鈉
sodium lauryl sulfate 十二〔烷〕基硫酸鈉；月桂基硫酸鈉
sodium L-tartrate L-酒石酸鈉
sodium magnesium aluminosilicate 硅鋁酸鈉鎂
sodium malate 蘋果酸鈉
sodium malonate 丙二酸鈉
sodium metabisulfite 偏亞硫酸鈉
sodium metaphosphate 偏磷酸鈉
sodium metasilicate 偏硅酸鈉；泡花鹼；水玻璃
sodium methylate 甲醇鈉
sodium nitrate 硝酸鈉
sodium nitrite 亞硝酸鈉
sodium oleate 油酸鈉
sodium *o*-phenylphenolate 鄰苯基苯酚鈉；聯苯酚鈉
sodium pantothenate 泛酸鈉
sodium phosphate 磷酸鈉
sodium phosphate dibasic 磷酸二鈉
sodium phosphate monobasic 磷酸一鈉
sodium phosphate starch 磷酸澱粉鈉；澱粉磷酸酯鈉
sodium phosphate tribasic 磷酸三鈉
sodium phosphate, chlorinated 氯化磷酸三鈉
sodium phosphate, monobasic 磷酸二氫鈉
sodium phosphate, tribasic 磷酸三鈉
sodium polyacrylate 聚丙烯酸鈉
sodium polyphosphate 聚磷酸鈉；多磷酸鈉
sodium potassium tartrate 酒石酸鉀鈉；羅雪鹽；羅謝爾鹽 [= Rochelle salt]

sodium propionate 丙酸鈉
sodium pyrophosphate 焦磷酸鈉
sodium pyrosulfite 焦亞硫酸鈉
sodium restricted diet 限鈉膳食
sodium saccharin 糖精鈉；可溶性糖精
sodium selenite 亞硒酸鈉
sodium sesquicarbonate 碳酸氫三鈉；倍半碳酸鈉
sodium silicate 矽酸鈉；硅酸鈉
sodium silicoaluminate 矽鋁酸鈉；硅鋁酸鈉
sodium sorbate 己二烯酸鈉；山梨酸鈉
sodium starch phosphate 磷酸澱粉鈉；澱粉磷酸鈉
sodium stearate 硬脂酸鈉
sodium stearoyl lactate (SSL) 硬脂酰乳酸鈉
sodium stearyl fumarate 硬脂酰富馬酸鈉
sodium succinate 琥珀酸鈉
sodium sulfate 硫酸鈉
sodium sulfite 亞硫酸鈉
sodium sulfocarbonate 維多酚
sodium tartrate 酒石酸鈉
sodium thiocyanate 硫氰酸鈉
sodium thioglycollate 巰〔基〕乙酸鈉
sodium thiosulfate 硫代硫酸鈉；大蘇打
sodium trimetaphosphate 三偏磷酸鈉
sodium tripolyphosphate 三聚磷酸鈉
soft biscuit 軟的；柔軟的；不含酒精的
soft boiled egg 輕煮蛋；嫩煮蛋；半熟蛋
soft bone 軟骨
soft can 軟罐頭
soft cheese 軟〔質〕乾酪
soft clot 軟凝塊；軟血餅
soft cook 輕煮
soft corn 軟質玉米
soft crab 軟殼蟹
soft curd milk 軟凝塊乳；輕度凝結的牛乳
soft diet 軟質膳食
soft dough 軟麵糰
soft drink 清涼飲料；不含酒精的飲料；軟飲料
soft dry sausage 軟乾香腸
soft fat 軟脂肪；低熔點脂肪
soft fire 文火
soft flesh 軟果肉
soft gluten 無筋力的麵筋
soft ice cream 軟冰淇淋；霜淇淋；軟雪糕
soft jelly 軟糖
soft red winter (wheat) 軟紅冬麥；軟紅〔冬〕小麥
soft resin 軟樹脂
soft roe 魚精；魚白
soft shelled crab 軟殼蟹
soft shelled egg 軟殼蛋
soft shelled turtle 甲魚；鱉
soft sugar 綿糖
soft sweets 軟糖
soft tallow 軟牛脂；低熔點動物脂
soft texture fondant 軟質糖霜泥；軟糖
soft water 軟水
soft wheat 軟質小麥
soft wheat flour 軟質小麥粉；軟質麵粉；筋力低的麵粉
softener 軟化劑；軟化器
softening 軟化〔處理〕
softness 柔軟性
soggy 未烤透的；生濕的
soggy ice cream 冰渣冰淇淋；帶冰結晶的冰淇淋
soilage 污水；污穢

soja (bean) 大豆；黃豆 [= soybean; soya (bean)]
sol 溶膠
solan 塘鵝
solanic acid 茄酸
solanine 茄鹼；茄靈
solar drying 太陽能乾燥；日光乾燥；曬乾
solar evaporation 日光蒸發
solar salt 曬製鹽
solation 膠溶〔作用〕；溶膠化〔作用〕
soldier crab 寄居蟹
sole 舌鰨魚；比目魚；底
sole bread 底烤麵包
solid 固體；固形物
solid content 固形物含量；乾物〔質〕含量
solid cream 固態鮮乳油；超脂稀奶油
solid fat 固體脂肪；固態脂；高熔點脂肪
solid foam 穩定泡沫；固體泡沫
solid food 固體食物
solid gum 固體膠
solid honey 結晶蜂蜜
solid-liquid extraction 固液萃取
solid matter 固形物；固體物質
solid medium 固態培養基
solid nonfat 非脂固體；非脂固形物
solid paraffin 固態石蠟
solid solution 固溶體
solid state 固態
solid waste 廢渣
solidification 凝固〔作用〕；固化〔作用〕
solidification reagent 凝固劑；固化劑
solidified alcohol 固態酒精
solidify 固化；凝固
solidness 硬度；硬性；堅硬性
solubility 溶解度；溶解性
solubilization 溶解〔化〕；增溶〔作用〕
solubilize 溶解；增溶溶解
solubilizer 助溶劑；溶解劑；增溶劑
soluble 可溶的；可溶物；溶解物
soluble cocoa 可溶可可粉
soluble dietary fiber 水溶性膳食纖維；可溶〔性〕膳食纖維
soluble gum 可溶〔性〕膠
soluble ingredient 可溶〔性〕配料；可溶〔性〕組分
soluble oil 水溶性油；溶性油
soluble packaging 可溶〔性〕包裝
soluble protein 可溶〔性〕蛋白質
soluble saccharin 可溶〔性〕糖精；糖精鈉
soluble solid 可溶〔性〕固形物
soluble starch 可溶〔性〕澱粉
solubleness 溶解性
solute 溶質；溶解物
solution 溶液
solvency 溶解〔能〕力
solvent 溶劑；溶媒
solvent extract 溶劑萃〔取物〕
solvent extraction 溶劑萃取
solvent fermentation 溶劑發酵
soma 身體；肉體；人體細胞
somatic cell 體細胞
somen 手拉麵
somerset brick 翻烤長麵包
sonication 聲波〔振動〕處理；音波〔振動〕處理
sonicator 音波〔振動〕器
sop 泡麵包片；浸漬
sophisticate 摻雜；摻假
sophorose 槐二糖
sorb 山梨〔樹〕
sorbate 山梨酸鹽；己二烯酸鹽；山梨酸酯；己二烯酸酯
sorbent 吸附劑
sorbet 果汁飲料；果汁凍；果汁雪泥

sorbic acid 山梨酸；己二烯酸 [= hexadienoic acid]
sorbin 山梨糖 [= sorbinose; sorbose]
sorbinose 山梨糖 [= sorbin; sorbose]
sorbit 山梨醇
sorbitan 去水山梨醇；山梨聚糖
sorbitan ester 山梨聚糖酯
sorbitan fatty acid ester (Span) 去水山梨醇脂肪酸酯；山梨糖醇酐脂肪酸酯；脂肪酸山梨醇酐酯
sorbitan monolaurate (Span 20) 單月桂酸山梨醇酐酯；斯潘20；司盤20
sorbitan monooleate (Span 80) 單油酸山梨醇酐酯
sorbitan monopalmitate (Span 40) 單棕櫚酸山梨醇酐酯
sorbitan monostearate 去水山梨醇單硬脂酸酯；硬脂酸山梨糖醇酯
sorbitan monostearate (Span 60) 單硬脂酸山梨醇酐酯
sorbitan tristearate (Span 65) 三硬脂酸山梨糖醇酐酯
sorbitol 山梨〔糖〕醇；葡〔萄〕糖醇 [= glucitol]
sorbol 山梨醇
sorbose 山梨糖 [= sorbinose; sorbin]
sorburonic acid 山梨糖酮酸
sorgho 甜高粱；蘆粟 [= sorgo; sweet sorghum]
sorghum 高粱 [= kaoliang]
sorghum molasses 高粱糖漿；高粱糖蜜
sorghum red 高粱紅
sorghum rice 高粱米
sorghum syrup 高粱糖漿
sorgo 甜高粱；蘆粟 [= sorgho; sweet sorghum]
sorgo syrup 高粱糖漿；蘆粟糖漿
sorosis 桑果；肉質聚合果
sorption 吸附作用
sorption agent 吸附劑
sorrel 酸模〔屬植物〕；含酸液的植物；葉帶酸味的植物
sort 等；級；種類
sorter 分級機；分選機
sorting 分級；分類；分選；揀選
sosera 肉菜香腸
sot 酒鬼；酗酒者
sotol 索托烈酒；百合烈酒
soubise sauce 洋葱奶油調味醬
souchong 小種〔紅〕茶；茶尖第四片葉
soufflé 舒夫里〔甜點〕；蛋奶酥
sound 鰾；聲音；健康的
sound coffee 優質咖啡
soup 湯
soup and fish 宴席菜
soup base 湯料
soup cup 湯杯
soup kitchen 流動廚房；施食處；施湯處
soup mix 湯粉
soup plate 湯盤
soup powder 湯粉
soup spoon 大湯匙
soup stock 高湯
soup tureen 有蓋湯碗；湯碗
soupcon 少許；微量
sour 酸的；酸味的；酸腐的；變酸；酸味飲料
sour cherry 酸櫻桃
sour cream 酸忌廉；酸奶油；酸乳油
sour cream butter 酸酪乳奶油；發酵乳酪
sour cucumber 酸黃瓜
sour dough 酸麵糰
sour mash 酸醪；酸性糖化醪
sour milk 酸乳
sour milk beverage 酸乳飲料

sour orange 酸橙 [= bigarade]
sour pickle 酸黃瓜；酸泡菜
sour rot 酸腐；酸敗
sour taste 酸味
sourcrout 德式酸泡菜；泡菜；酸漬菜 [= kraut; sauerkraut; sourkrout]
sourdough 酵母；發麵底子；酵頭
souring 酸化
sourish 微酸的
sourkrout 德式酸泡菜；泡菜；酸漬菜 [= kraut; sauerkraut; sourcrout]
sourness 酸性；酸度；酸味
soursop 刺番荔枝；酸番荔枝
souse 醃製；醃漬食品；醃漬用鹽水
sow 母豬；種母豬
Soxhlet extraction method 索氏萃取法
Soxhlet extractor 索氏萃取器
soy 黃豆；大豆；醬油
soy cake 黃豆粕；大豆粕
soy cheese 豆腐乳；腐乳 [= bean curd cheese; fermented tofu; soybean cheese; sufu]
soy flour 大豆粉；黃豆粉
soy koji 大豆麴
soy lecithin 大豆卵磷脂
soy meal 大豆粉；大豆粕
soy milk 豆漿；豆乳；豆奶
soy oil 黃豆油；大豆油 [= soya oil; soybean oil]
soy protein 大豆蛋白；黃豆蛋白
soy protein concentrate (SPC) 濃縮大豆蛋白；大豆蛋白濃縮物；大豆濃縮蛋白
soy protein food 大豆蛋白食品；黃豆蛋白食品
soy protein isolate (SPI) 分離大豆蛋白；大豆蛋白分離物
soy saponin 大豆皂苷；黃豆皂苷
soy sauce 豉油；醬油；生抽 [= soy (bean) sauce; light soy sauce]
soy sauce koji 醬油麴
soy sauce mash 醬〔油〕醪
soy skin 腐皮；豆腐皮；豆皮；腐竹 [= soya skin; soybean milk film; yuba]
soy sprout 大豆芽；黃豆芽
soya (bean) 大豆；黃豆 [= soybean; soja (bean)]
soya oil 黃豆油；大豆油 [= soy (bean) oil]
soya saponin 大豆皂苷；黃豆皂苷
soya sauce 豉油；醬油；生抽 [= soy (bean) sauce; light soy sauce]
soya skin 腐皮；豆腐皮；豆皮；腐竹 [= soy skin; soybean milk film; yuba]
soyasterol 大豆甾醇；大豆固醇
soybean 大豆；黃豆 [= soya (bean)；soja (bean)]
soybean cake 黃豆粕；大豆粕
soybean cheese 豆腐乳；腐乳 [= bean curd cheese; fermented tofu; soy cheese; sufu]
soybean curd 豆腐 [= tofu]
soybean flour 大豆粉；黃豆粉
soybean grit 大豆粗粉；碎大豆
soybean koji 大豆麴
soybean lecithin 大豆卵磷脂
soybean meal 大豆粉；大豆粕
soybean milk 豆漿；豆乳；豆奶
soybean milk film 腐皮；豆腐皮；豆皮；腐竹 [= soy (a) skin; yuba]
soybean miso 味噌；大豆味噌；豆醬
soybean oil 黃豆油；大豆油 [= soy (a) oil]
soybean paste 豆瓣醬
soybean protein 大豆蛋白；黃豆蛋白

soybean protein concentrate (SPC) 濃縮大豆蛋白；大豆蛋白濃縮物；大豆濃縮蛋白
soybean protein food 大豆蛋白食品；黃豆蛋白食品
soybean protein isolate (SPI) 分離大豆蛋白；大豆蛋白分離物
soybean sauce 豉油；醬油；生抽 [= soy (a) sauce; light soy sauce]
soybean sprout 大豆芽；黃豆芽
soybean trypsin inhibitor 大豆胰蛋白酶抑制劑
spa 礦泉；溫泉
spa water 礦泉水
space food 太空食品
spade 鏟子
spaghetti 意大利〔式〕直麵；意大利麵；意大利粉
spaghetti squash 意大利直麵南瓜；金線瓜；魚翅瓜 [= noodle squash; vegetable spaghetti]
Span 20 單月桂酸山梨醇酐酯；斯潘20；司盤20
Span 40 司邦40；山梨醇酐單棕櫚酸酯
Span 60 司邦60；山梨醇酐單硬脂酸酯
spanemia 貧血
Spanish garlic 胡蒜
Spanish ling 西班牙鱈魚
Spanish mackerel 鰆；西班牙鯖魚
Spanish onion 西班牙洋葱
Spanish pepper 西班牙辣椒
sparerib 肋〔條〕肉；排骨；帶肉豬肋骨
sparger 噴灑器
sparkling beverage 發泡飲料
sparkling water 汽水；蘇打水
sparkling wine 發泡〔葡萄〕酒；汽酒
sparling 美洲胡瓜魚
sparrow 麻雀
sparrowgrass 蘆筍；蘆筭
spartan 斯巴達〔蘋果〕
spat 貝卵；貝苗
spatchcock 現宰烤雞；宰後立即烤食的雞
spawn 菌絲；菌種；魚子；魚卵；產卵
spawner 產卵魚
spear 魚叉；幼芽；幼苗；筍尖
spearfish 箕作氏槍魚
spearmint 綠薄荷；矛型薄荷；荷蘭薄荷；薄荷；薄荷草；留蘭香
spearmint oil 薄荷油；薄荷草油；綠薄荷油；留蘭香油
speated 穿成串的
special diet 特殊膳食
special dietary product 特殊營養品
special food 特殊用途食品
special nutrition foodstuff 特殊營養食物
specialist 專家；行家
specialpurpose food 專用食品
specialty product 招牌食品；特級產品；特產
species 種；種類
specific density 比密度
specific gravity 比重
specific heat 比熱
specific humidity 比濕度
specific nutrient 特殊營養素
specific viscosity 比黏度
specific volume 比體積；比容
specific weight 比重
specimen 標本；樣本；樣品
speck 些微；斑點；肥〔豬〕肉；鯨脂
specked butter 斑點乳酪
specked trout 斑點鮭魚；紅鱒
spectrodensitometry 分光密度測定法；光譜密度測定法

spectrofluorometry 分光熒光測定法；光譜熒光測定法
spectrophotometer 分光光度計
spectrophotometry 分光光度測定法
spelt 斯卑爾脫小麥；斯卑爾脫小麥
spencer roll 去骨肋脊肉；沙朗
spense 伙食房；廚櫃；餐具室；食物貯藏室
spent brewer's grain 酒糟；穀渣；啤酒廢麥糟
spent fish 產卵後的〔廢〕魚
spent grain 酒糟；穀渣；啤酒廢麥糟
spent liquor 廢液
spent steam 廢汽
spent wash 酒餾殘液；酒糟水；底鍋水
spent water 廢水
sperling 小鯡魚
sperm 精子
sperm whale 抹香鯨
spermaceti 鯨蠟；鯨腦油
spermase 麥芽氧化酶；大麥氧化酶
spermol 鯨蠟醇
sphincter 括約肌
sphingolipid 神經鞘脂類；鞘脂類
sphingomyelin 神經鞘髓磷脂；鞘髓磷脂
sphingosine 神經鞘胺醇；鞘胺醇
spice 調味品；香料；香辛料；辛香料
spice curing 香辛醃漬
spice jar 香辛料罐
spice oleoresin 香辛料油樹脂
spice rack 香辛料架
spiced 香辛的；加香〔料調味〕的
spiced cheese 香辛乾酪；加香乾酪
spiced luncheon meat 調味午餐肉
spicery 香〔辛〕味；香辛料；香辛料店
spiciness 香辛性
spicy 加香料的；辛辣的
spider crab 蜘蛛蟹
spike 加入烈酒；穗；穗狀花
spike hawthorn 穗狀山楂
spike lavender oil 穗薰衣草油
spiked beer 酒精加強啤酒；加入酒精的啤酒
spikenard 甘松香；甘松油
spikenard oil 甘松油
spillage 溢出；溢出量
spinach 菠菜 [= spinage]
spinach beet 葉甜菜；莙薘菜
spinacin 角鯊素；菠菜素
spinage 菠菜 [= spinach]
spinal cord 脊髓
spinal fluid 脊髓液
spinal marrow 脊髓
spinasterol 菠菜固醇；菠菜甾醇
spindle stemmed bolete 附屬牛肝菌
spine 脊骨；尖刺
spinkganz 鵝肉乾 (德國)
spinning soy protein 紡絲大豆蛋白；噴絲大豆蛋白
spinning 旋轉；噴絲；紡紗
spiny 多刺的；刺狀的
spiny goby 刺鰕虎魚
spiny lobster 龍蝦
spiral 螺旋；螺旋形的；旋管的
spiral agitator 螺旋攪拌器
spiral conveyor 螺旋〔式〕輸送機
spiral feeder 螺旋進料器
spiral freezer 螺旋冷凍機
spiral heater 旋管加熱器
spiral pasta 螺絲粉；螺旋粉 [= fusilli; spirali]
spiral slice 螺旋片；鳳梨螺片；菠蘿旋片
spiral stirrer 螺旋攪拌器
spirale 扭紋形通心粉

spirali 螺絲粉；螺旋粉 [= spiral pasta; fusilli]
Spirillum 螺旋菌屬
spirit 烈酒；酒精；蒸餾酒
spirit of wine 酒精；乙醇；葡萄烈酒
spirit still 餾酒器；淡酒再餾器
spirit whisk(e)y 摻酒精威士忌；加酒精的威士忌
spirits 烈酒；酒精飲料
spirituous 酒精的；含酒精的
spirituous liquor 酒類；酒精飲料
spirituous vinegar 酒精醋
spirulina 螺旋藻
spirulina blue colo(u)r 藍藻色素
spit 烤肉叉；唾液；穿刺(用烤肉叉)
spitchcock 烤鰻；剖開而燒烤
spleen 脾臟
split 劈開；分裂；裂縫；剝皮；水果加冰淇淋的混合甜品；小瓶飲料
split carcass 半屠體
split champagne 小瓶香檳酒
split cure 剖醃
split fish 劈開兩半的魚
split pea 裂莢豌豆
splitter 劈開機；分割機；剝皮機
splitting 劈開；分離；分裂；裂縫；剝皮
splitting machine 劈開機；分割機；剝皮機
spoilage 損壞；腐敗；腐敗物
spoilage bacteria 腐敗〔細〕菌
spoilage organism 腐敗菌；腐敗生物
spoiled odo(u)r 腐敗氣味
sponge 海綿；海綿狀物；發麵糰
sponge cake 海綿蛋糕；鬆糕
sponge cucumber 絲瓜；菜瓜 [= dishcloth gourd; sponge gourd]
sponge dough 中種麵糰；發麵糰
sponge gourd 絲瓜；菜瓜 [= dishcloth gourd; sponge cucumber]
sponge meat 海綿狀肉
sponge mushroom 羊肚菌
sponginess 海綿狀
spongy 海綿狀；多微孔的；泡沫狀的
spongy cake 海綿蛋糕
spontaneous fermentation 自行發酵；自然發酵
spoon 調羹；匙
spoon bread 匙用麵包；軟質蛋奶麵包
spoon meal 流質食物；匙用食物
spoonful 一匙量；一滿匙
sporangiophore 孢囊梗；孢囊柄
sporangiospore 孢囊孢子
spore 芽孢；孢子
spore-bearing 產孢的
spore-bearing bacteria 產孢細菌
spore-former 產孢菌
spore-forming bacteria 產孢菌；產芽孢桿菌
spore membrane 孢子膜
spore sac 孢子囊
sporicidal activity 殺孢子〔能〕力
sporicide 殺孢子劑；殺芽孢劑
sporogenous yeast 產芽孢酵母；產孢子酵母
sport drink 運動飲料
sporulation 孢子形成
sporule 小孢子；小芽孢
spotted 有斑點的；有污點的
spotted dog 葡萄乾布丁
spouter 捕鯨船
sprat 小鯡魚；圓腹鯡；西鯡
spray 小枝；飛沫；噴霧；噴霧劑；噴霧器
spray atomizer 噴霧器
spray ball 噴〔水〕頭
spray chamber 噴霧室

spray cooler 噴淋冷卻器
spray cooling 噴淋冷卻
spray dehydration 噴霧乾燥
spray disc 噴霧盤
spray-dried 噴霧乾燥的
spray-dried milk 噴霧乾燥乳粉；噴乾乳粉
spray drier 噴霧乾燥機
spray dryer 噴霧乾燥機
spray drying 噴霧乾燥〔法〕
spray drying granulation 噴霧乾燥造粒
spray drying tower 噴霧乾燥塔
spray freezing 噴霧冷凍
spray glazing 噴霧包覆；噴霧包〔冰〕衣
spray granulation 噴霧造粒
spray gun 噴槍
spray head 噴嘴；噴頭
spray jet 噴霧器
spray nozzle 噴嘴
spray pond 噴淋池；噴水池
spray washer 噴淋式洗滌機；噴射洗滌機
spray washing 淋洗；噴洗
sprayer 噴霧器；噴射裝置
spraying nozzle 噴嘴
spread 膏；糊；醬；盛宴；宴會；塗抹食品；軟質食品；塗敷；塗抹；擺桌以備進膳
spreader 塗抹器；擴張器
spree 狂飲；狂歡
sprig 小枝；嫩枝
spring 春天；春季；彈簧；泉；水源
spring balance 彈簧秤
spring barley 春大麥
spring chicken 童子雞；雛雞
spring field cap 旱田頭菇
spring grain 春糧
spring green 春綠
spring onion 葱；青葱；大葱 [= cibol; ciboul; Welsh onion]
spring roll 春捲；炸春捲
spring-top jar 彈簧蓋廣口瓶
spring water 泉水
spring wheat 春小麥
springbok 南非鈴羊
springclean 大掃除(春季)
springer cow 臨產牛；懷孕母牛
springhouse 冷藏室
springiness 彈性
sprinkle 噴灑；灑；撒
sprinkler 灑水裝置
sprinkler head 噴水頭；蓮蓬頭
sprinkling machine 噴水器
sprout 芽；苗；發芽
sprouting broccoli 青花菜；嫩莖花椰菜；青花椰菜
spruce beer 雲杉啤酒；針樅啤酒
spruce cone toadstool 可食球果菌(蕈類)
spud 馬鈴薯；除草鏟
spumante wine 發泡葡萄酒
spume 浮沫；泡沫
spumescence 泡沫狀；泡沫性；起泡沫
spumoni 意大利式多色冰淇淋
spun protein 紡絲蛋白；纖維〔狀〕蛋白
spun sugar 棉花糖
squab 幼鴿；乳鴿
squalane 鯊烷；角鯊烷；三十〔碳〕烷
squalene 鯊烯；角鯊烯；三十碳六烯
squalid loaf 變形麵包；骯髒的麵包
squama 鱗片
squash 西葫蘆；南瓜；果泥；帶肉果汁加糖的不含酒精飲料；壓碎；擠壓；壓扁
squeezability 可壓縮性；可壓擠性
squeeze 擠壓；壓榨
squeeze cap 壓封蓋
squeeze juice 榨汁

squeezer 榨汁機；榨汁器
squeezing machine 壓榨機；擠壓機
squid 魷魚；烏賊；槍烏鰂；小卷
squid liver oil 魷魚肝油
squill 海葱
squilla 蝦蛄；螳螂蝦
squirrel 松鼠
squirrel dew 酒
Srilanka cinnamon 斯里蘭卡肉桂
SS agar (Salmonella Shigella agar) 沙門志賀桿菌〔洋菜〕培養基；沙門志賀氏菌〔瓊脂〕培養基
St. John's wort 貫葉連翹
St. George's mushroom 大柄基麗傘(蕈類)
stab culture 穿刺培養
stability 穩定性；安定性
stabilization 穩定作用；安定作用
stabilizer 穩定劑；安定劑
stabilizing agent 穩定劑；安定劑
stable 穩定的；廄；畜舍；畜棚
stachydrine 水蘇鹼
stachyose 水蘇四糖；羽扇豆糖 [= lupeose]
stag 雄鹿；公雞；剛長大的雄家禽；去勢〔成年〕公畜
stage drying 分級乾燥
stage extraction 分級萃取
staghorn 雄鹿角
stagnant water 滯留水
stain 沾污；污點；染料；著色劑
stainer 著色劑；染色劑
staining 染色；著色
staining reagent 著色劑；染色劑
stainless steel 不鏽鋼
stale 不新鮮的；發霉的；陳腐的；陳舊的；老化的；走味的
stale fish 不新鮮魚
stale flavo(u)r 陳腐味；霉味
stale flesh 不新鮮肉
stale odo(u)r 陳腐氣味；霉臭；不新鮮氣味
stale taste 陳霉味；霉味
stale wine 非發泡性葡萄酒
staleness 不新鮮；發霉；陳腐；陳舊；老化；走味
staling 老化；陳腐
stalk 柄；梗；莖
stalk machine 去梗機
stalky tea 帶梗茶葉
stall 廄；畜舍；畜棚；單頭牲畜的圍欄
stallion 公馬；種馬
stamen 雄蕊
stamp mortar 搗鎚研磨機
stamp pestle 搗杵
stamper 沖壓機；搗擊機，壓模
stamping mill 搗碎機；搗椿機
stamping press 打印機
standard 標準；水準；標準的
standard condition 標準狀況；標準條件
standard curve 標準曲線
standard deviation (SD) 標準偏差
standard diet 標準飲食
standard error (SE) 標準誤差
standard fare 標準菜單
standard pack 標準包裝
standard plate count 標準平板計數；生菌數
standard sample 標準樣品；標準試樣
standard screen 標準篩
standard screen scale 標準篩制；標準篩網尺度
standard solution 標準溶液
standard specification 標準規格
standardization 標準化
standfish 雷魚

standing pouch 自立袋
standing rib roast 前里脊肉
standing storage 長期儲藏；例行長期儲藏
standing tank 固定儲槽
stannous chloride 氯化亞錫
staphylococcal enterotoxin 葡萄球菌產的腸毒素
staphylococcal food poisoning 葡萄球菌食物中毒
staphylococcal poisoning 葡萄球菌中毒
staphylococcin 葡萄球菌毒素
Staphylococcus 葡萄球菌屬
Staphylococcus aureus 金黃色葡萄球菌
Staphylococcus epidermidis 表皮葡萄球菌
staphylolysin 葡萄球菌溶血素
staple 主食；主要產品；主要成分；廚房必備食品
staple crop 主要農作物
staple diet 主食
staple food 主食；主糧
star anise 大茴香；八角；八角茴香；茴芹 [= anise; Chinese aniseed]
star anise oil 大茴香油；八角〔茴香〕油
star apple 星蘋果
star fruit 楊桃 [= carambola]
starch 澱粉
α-starch α澱粉；阿伐澱粉 [= alpha-starch]
β-starch β澱粉；貝他澱粉 [= beta-starch]
starch acetate 醋酸澱粉〔酯〕
starch conversion syrup 澱粉轉化糖漿
starch converting enzyme 澱粉糖化酶
starch crop 澱粉作物
starch dextrin 澱粉糊精
starch dextrinization 澱粉糊精化
starch ester 澱粉酯
starch ether 澱粉醚
starch gel 澱粉凝膠
starch grain 澱粉粒
starch granule 澱粉顆粒
starch gum 澱粉膠；糊精 [= dextrin]
starch-iodine blue test 澱粉-碘試驗；澱粉和碘的藍色試驗
starch-iodine reaction 澱粉-碘反應
starch mill 澱粉廠
starch modifying agent 澱粉改性劑
starch nitrate 硝酸澱粉；硝化澱粉 [= nitrostarch]
starch noodle 粉絲
starch paste 澱粉糊；漿糊
starch pasting 澱粉糊化
starch product 澱粉製品
starch reduced bread 低澱粉麵包
starch retrogradation 澱粉回凝；澱粉老化；澱粉凝沉
starch settling 澱粉沉降
starch slurry 澱粉漿
starch sodium octenylsuccinate 辛烯基琥珀酸澱粉鈉
starch sugar 澱粉糖；葡萄糖
starch syrup 澱粉糖漿
starchiness 漿糊狀；澱粉質
starchy 含澱粉的；澱粉的
starchy material 澱粉質原料
starchy taste 澱粉味；生澱粉味
starchy vegetable 澱粉質蔬菜
starking 星王〔蘋果〕
starter 菌元；發酵劑；起子
starter culture 發酵劑培養物；起子培養物
starter distillate 發酵劑餾出液
starvation 飢餓；餓死
starvation therapy 飢餓療法
starveling 飢餓者；餓瘦的人；挨餓者
starver 絕食者

stassanization 加熱滅菌法；薄膜式巴氏殺菌法
stationary agitator 固定式攪拌器
stationary culture 靜止培養〔物〕
stationary phase 靜止相；穩定期；固定相
statistics 統計學
statospore 休眠孢子
stauntonvine 野木瓜
steak 肉排；牛排；魚排
steak frenching machine 牛排拍鬆機
steak hammer 損肉鎯頭
steak house 牛排館；牛排餐廳
steak knife 牛排餐刀
steak piece 肋骨肉；牛後腰肉
steam 蒸；蒸汽；水蒸汽
steam bath 汽浴；蒸汽浴
steam boiling 蒸煮；蒸汽蒸煮
steam box 蒸汽櫃；蒸汽箱
steam bubbling 蒸汽加熱攪拌
steam chamber 蒸汽室
steam chest 蒸汽櫃
steam condensate 蒸汽〔冷〕凝液
steam cooker 蒸煮鍋；蒸汽蒸煮器
steam cooking 蒸汽蒸煮；蒸煮 [= vapour cooking]
steam distillation 蒸汽蒸餾
steam dryer 蒸汽〔加熱〕乾燥機
steam drying 蒸汽〔加熱〕乾燥
steam exhausting 蒸汽脫氣
steam flow closure 蒸汽噴射排氣封罐；噴汽封罐
steam flow sealing 蒸汽噴射排氣封罐；噴汽封罐
steam gauge 汽壓表
steam generator 蒸汽發生器
steam heated dryer 蒸汽加熱乾燥機
steam heater 蒸汽加熱器
steam jacket 蒸汽套管；蒸汽夾套
steam jacket(ed) kettle 蒸汽夾層鍋；蒸汽夾套鍋
steam-jet atomizer 蒸汽噴霧器
steam-jet heater 蒸汽噴射加熱器
steam-jet refrigerating machine 汽噴冷卻機；蒸汽噴射製冷機
steam peeling 蒸汽去皮
steam rendering 蒸汽敖油；蒸汽煉油
steam sterilization 蒸汽滅菌
steam sterilizer 蒸汽滅菌器
steam still 蒸汽蒸餾器
steam superheater 蒸汽過熱器
steam table 蒸汽保溫檯；保溫檯
steam-tight 汽密的；不漏汽的
steam-vac method 蒸汽真空法
steamed 蒸的；清蒸的
steamed bread 饅頭
steamed fish 清蒸魚
steamed rice 蒸飯
steamer 蒸籠；蒸鍋
steaming 蒸煮
steapsin 胰脂酶
stearate 硬脂酸鹽；硬脂酸酯
stearic acid 十八烷酸；硬脂酸 [= octadecanoic acid]
stearin 硬脂；三硬脂精；硬脂精；甘油三硬脂酸酯 [= tristearin; glyceryl tristearate]
stearyl alcohol 十八烷醇
stearyl citrate 檸檬酸硬脂酰酯
stearyl monoglyceridyl citrate 檸檬酸硬脂酰單甘油酯
stearyl tartrate 硬脂酰酒石酸酯；酒石酸硬脂酰酯
steatolysis 脂肪分解
steel 鋼鐵；鋼
steelyard 提秤；杆秤
steenbok 小羚羊（非洲）
steep 浸；泡；浸透；浸漬；浸〔漬〕液

steeper 浸漬罐；浸漬槽
steeping 浸漬；浸泡
steer 閹公牛；去勢小公牛
stein 啤酒杯
stem 香焦串；莖；幹；梗；柄
stem crusher 去梗機；梗破碎機
stem vegetable 莖菜；莖用菜
stemmed grape 去梗葡萄
stemmer 去梗機；摘蒂機
stemming machine 去梗機；去柄機
stench 臭氣；惡臭
stenothermic bacteria 窄溫性細菌；狹溫性細菌
stercobilin 糞膽色素；糞後膽色素
sterculia 蘋婆
sterculia gum 蘋婆膠
sterculic acid 蘋婆酸
stercus 糞便
stereomicroscope 立體顯微鏡
stereoscope 立體鏡；體視鏡；實體鏡
stereoscopic microscope 立體顯微鏡
steriflame process 火燄滅菌法
steriflame sterilization 火燄滅菌法
sterilant 滅菌劑
sterile 無菌的；消毒〔過〕的
sterile culture 無菌培養；純菌培養
sterile filling 無菌填充；無菌灌裝
sterile filtration 無菌過濾；過濾滅菌
sterile food 高溫殺菌的食品；無菌食品
sterile milk 無菌乳；滅菌乳
sterile water 無菌水
sterility 不育；不孕；無菌；殺菌性能
sterility curve 滅菌率曲線
sterilization 滅菌；消毒；絕育
sterilize 殺菌；滅菌；消毒
sterilized food 已滅菌的食品
sterilized milk 滅菌〔牛〕乳
sterilizer 滅菌器
sterilizing 滅菌
sterilizing agent 滅菌劑
sterilizing chamber 滅菌箱；滅菌室
sterilizing effect 滅菌效果
sterilizing rate 滅菌率
sterilizing value 滅菌值
sterilvac 火燄滅菌；火燄殺菌
sternum 胸骨
stero-bile acid 固醇膽汁酸；甾族膽汁酸
steroid 類固醇
steroid hormone 類固醇激素
sterol 固醇；甾醇
sterone 固酮；甾酮
stevia 甜菊
stevioside 甜菊糖；甜菊糖苷；甜葉菊苷；卡哈苡苷
stew 燉〔煮〕；燜；燉肉；燉菜
stew pan 燉鍋；長柄燉鍋
stew pot 燉鍋；燉壺；雙把燉鍋
steward 膳食管理員；伙食採辦人
stewed beef 燉牛肉
stewed meat 燉肉；紅燒肉
stewing 燉〔煮〕；燜
stibium 銻
stick 棒；梗；棍；棒狀物；刺殺；黏貼；黏住
stick candy 棒糖
stick lac 樹枝蟲膠
stick packaging 棒狀包裝
sticker 刺殺工具；刺殺工人；標貼；黏着物
stickiness 黏性
sticking 黏貼；刺殺〔屠宰〕
sticking knife 刺殺刀；放血刀
stickjaw 黏牙的糖果
sticky dough 黏麵糰；發黏麵糰
sticky matter 黏性物質；黏稠物

sticky rice 糯米 [= glutinous rice; sweet rice; waxy rice]
sticky rice flour 糯米粉
stiff 僵硬的；僵直的；濃烈的
stiffener 硬化劑；增稠劑
stiffening agent 硬化劑；增稠劑
stigmastane 豆固烷；豆甾烷
stigmastanol 豆固烷醇；豆甾烷醇
stigmastenol 豆固烯醇；豆甾烯醇
stigmasterol 豆固醇；豆甾醇
still 蒸餾；蒸餾器；靜止；不含氣體的
still-air freezing 靜止空氣冷凍法
still beverage 不充氣飲料；非碳酸飲料
still culture 靜止培養〔物〕
still drink 不充氣飲料；不帶氣泡的飲料
still juice 不帶氣泡的果汁；不充氣果汁
still kettle 蒸餾釜
still retort 靜止〔式〕殺菌釜
still room 蒸餾室；食物貯藏室
still wine 不充氣葡萄酒；無氣葡萄酒
stillage 蒸餾廢醪；釜餾物
stimulant 刺激物；興奮劑
stimulation 刺激〔作用〕
stimulator 刺激器
stimulus 刺激；刺激物
stingo 烈性啤酒
stingray 黃貂魚
stink 發臭；惡臭
stinking 發臭的；有惡臭的
stiparogenic food 便秘性食物
stiparolytic food 通便性食物
stir 攪拌；攪動；攪和
stir-fried 炒的
stir-frying 炒
stirabout 玉米粥；麥片粥
stirred tank 攪拌槽
stirrer 攪拌器
stirrer arm 攪拌〔機〕槳葉
stirring 攪拌；攪動
stirring machine 攪拌機
stirring rod 攪〔拌〕棒
stirrup cup 餞別酒；上馬酒
stitch pumping 注射醃漬法；醃液注射法
stitcher 釘箱機
stock 清湯；高湯；原湯；湯料；儲備；存貨；家畜；紫羅蘭花
stock beer 庫存烈啤酒
stock breeder 畜牧業者
stock car 運載家畜之火車車廂
stock cube 湯料濃縮方塊
stock culture 保存菌種；儲備培養；原培養物
stock feed 牲畜飼料
stock pot 湯鍋；原湯罐
stock room 儲藏庫；陳列室；倉庫
stock solution 原液；儲備液
stock tank 貯槽；貯罐
stocker 育肥公犢；育肥小公牛
stocker cattle 育肥牛；食用牛
stocker heifer 育肥母牛
stockfish 魚乾；曬乾的魚類
stockyard 繫留場；畜欄
stodge 濃稠食物；濃膩食物；油膩乏味的食物
stodgy 油膩的；濃膩的；不易消化的
stollen 水果麵包
stomach 胃；食慾；胃口
stomachache 胃痛
stone 果核；去果核；石頭
stone finish 糙面精整
stone fruit 核果 [= drupe (fruit); drupaceous fruit]
stone grinder 石磨

stone leek 葱；大葱
stone mill 石磨
stone oil 核油
stone roll 石輥；石碾
stoneless 無核的；去核的
stoner 去〔果〕核機；除石機
stoneware 粗陶器；陶石器
stoning 去〔果〕核
stoning machine 去〔果〕核機；除石機
stool 糞便
stopcock 栓塞；活栓；龍頭
stopper 塞子；瓶塞
storability 保存性；保藏性；儲存性
storage 儲藏；貯存；倉庫；倉儲費
storage bin 儲倉
storage bunker 儲藏庫
storage cellar 儲藏地窖；地下儲藏室
storage condition 儲藏條件
storage container 儲藏容器
storage flavo(u)r 陳〔腐〕味；不新鮮味
storage life 儲藏期限
storage period 儲藏期；保藏期；儲存期
storage room 儲藏室；倉庫
storage tank 儲存槽
storage temperature 儲藏溫度；保藏溫度
storage yard 露天倉庫
storax 蘇合香
storax 蘇合香〔脂〕；安息香
storax oil 蘇合香油
store 儲藏；貯存；倉庫；商店
store room 儲藏室；儲存室
storehouse 倉庫；貨棧
stores 軍需品；儲藏室；倉庫；百貨店
stork 鸛
Stork process 斯脱克〔超高溫〕滅菌法；斯托克〔超高溫〕滅菌法
stoup 酒壺；酒杯；一杯之量
stout (beer) 黑濁啤酒；烈性黑啤酒
stout agaric 塊鱗灰鵝膏 (蕈類)
stove 爐；火爐；爐灶
stoved salt 鍋鹽；精鹽 [= pan salt]
stovepipe 煙囱
stoving 烘烤
straight 直的；不摻水的；純的
straight chain 直鏈
straight dough 直接〔發酵〕麵糰
straight dough method 直接〔發酵〕麵糰法；直接〔發酵〕麵糰製麵包法
straight firing 直火加熱
straight flour 統粉
straight juice 純果汁；原汁
straight salt 淨鹽；精鹽
straight whisk(e)y 純威士忌〔酒〕
strain 菌株；品種；品系；篩濾；粗濾
strained food 篩濾食品；粗濾食品
strained fruit 篩濾果泥；粗濾果泥
strained honey 篩濾蜂蜜；濾淨蜂蜜
strained tomato 篩濾番茄泥；番茄漿
strainer 濾器；濾網；濾篩；撇沫器
strainer-dipper 濾杓
straining cloth 濾布
strapper 包紮機；捆紮機
strapping machine 捆紮機
strata-cock process 成層殺菌法；分層殺菌法
strategic landfill 堆填區
stratification 成層〔現象〕；分層〔現象〕
stratiform cake 千層糕
stravidin 抗生蛋白菌素
straw 稻草；禾稈；麥稭；吸管
straw mushroom 草菇
strawberry 草莓；士多啤梨
strawberry flavo(u)r 草莓香精
strawberry guava 草莓番石榴
strawberry jam 草莓〔果〕醬
strawberry oily flavo(u)r 草莓油香精

strawberry tree 楊梅；樹梅
strawboard 硬紙板；馬糞紙
streak culture 畫線培養〔物〕
streaking 畫線接種；畫線分離〔培養〕
streaky 有條紋的
streaky bacon 五花醃肉；鹹五花肉
street market 街市
streptococcal infection 鏈球菌感染
streptococcal poisoning 鏈球菌中毒
streptococci 鏈球菌屬
〔= streptococcus的複數〕
streptococcus 鏈球菌屬
Streptococcus lactis 乳酸鏈球菌
Streptococcus faecalis 糞鏈球菌
Streptococcus thermophilus 嗜熱鏈球菌
streptolysin 鏈球菌溶血素
Streptomyces 鏈黴菌屬
streptomycin 鏈黴素
streptonigrin 鏈黑菌素
streptose 鏈黴糖
stretch film 伸縮薄膜；拉伸薄膜
stretch film package 伸縮薄膜包裝；拉伸薄膜包裝
stretch muscle 伸張肌
stretching machine 拉伸機；拉糖機
striated muscle 橫紋肌；骨骼肌
strict aerobe 絕對嗜氧菌；嚴格需氧菌
strict anaerobe 絕對厭氧菌；嚴格厭氧菌
strike pan 蒸發罐；濃縮罐
string bean 有筋青刀豆；紅菜花豆；菜豆；青豆
stringless bean 無筋菜豆
stringy curd 發黏凝塊
stringy meat 多筋肉
stringy milk 膠黏乳；牽絲牛乳
strip 窄條；脱去；剝去；拆除；採摘
strip cutter 切條機
striped marlin 紅肉旗魚；條紋槍魚
striped package 條帶綑包
striped perch 條紋鱸
striped tuna 鰹
striploin 前腰脊肉；西冷；紐約克
striploin steak 西冷牛排；紐約客牛排
strippant 洗滌劑
stripped loin 肋里脊肉
stripper 剝皮機；去皮機
stripping 剝皮；脱殼
stripping machine 剝皮機；脱殼機
stroma protein 基質蛋白質
strong acid 強酸
strong base 強鹼
strong brine 濃鹽水
strong drink 烈性飲料；高酒精飲料
strong flour 筋力強的麵粉；硬質麵粉
strong gluten 筋力強的麵筋
strong liquor 濃液；烈性酒精飲料
strong lye 強鹼水
strong scented blewit 彩虹香蘑(蕈類)
strong taste 刺激味；濃味
strontium 鍶
structural protein 結構蛋白質
strudel 薄水果餡甜餅；果餡甜酥餅；奧地利蘋果餡餅
strychnine 番木鱉鹼；馬錢子鹼
stubble field volvar 黏頭小包腳菇
stuff 材料；原料；物品；填塞；飽食
stuffed beef steak 碎牛肉〔製牛〕排
stuffed chicken 填餡雞
stuffed green pepper 釀青椒；填餡青椒
stuffer 填充機；灌〔香〕腸機；灌餡機
stuffer dispenser 分裝填充機
stuffing 填充料；填塞物；餡
stuffing machine 填充機；灌餡機
stum 未發酸葡萄汁
stump truffle 梨形馬勃(蕈類)
stun(ning) 擊昏；震昏

sturgeon 鱘魚
sturgeon roe 鱘魚子
sturin(e) 鱘精蛋白
styracin 桂皮酸桂皮酯；肉桂酸肉桂酯
styracitol(e) 蘇合香醇；安息香醇；苯乙醇
styracol 蘇合香腦
styralyl butyrate 丁酸蘇合香酯
styrax 蘇合香〔脂〕；安息香
styrene 苯乙烯
styrolene 苯乙烯
subacidity 微酸性；弱酸性
subalimentation 營養不足
subastringent 微澀的；輕收斂的
subatmospheric pressure 低壓；低於一個大氣壓的厭力
subcool 過冷
subcooler 過冷器；再冷却器
subculture 繼代培養〔物〕；次代培養〔物〕
subcutaneous fat 皮下脂肪
subcutaneous tissue 皮下組織
suber 軟木
suberate 辛二酸鹽
suberic acid 辛二酸
suberin 軟木脂；木栓質
sublimate 昇華物
sublimation 昇華〔作用〕
sublimation drying 昇華乾燥
submerged culture 浸沉培養〔物〕；沉沒培養〔物〕；深層培養〔物〕
submerged fermentation 浸沉發酵；深層發酵
submucosa 黏膜下層
subnutrition 營養不足
subsample 再取樣品；副樣；子樣品；二次抽樣樣品
subshrub 小灌木
subsidence 沉降；沉澱
subsider 沉降槽；沉澱池
subsider juice 澄清汁
subsidiary food 副食〔品〕
subsidy 補助金；津貼
subsistence 生存；餬口；勉強維生
subsistence crop 自種自食而不出售的作物
subsistence diet 維生〔最低〕飲食；生存餐食
subspecies 亞種
substance 物質
substandard 次〔級〕標準；標準以下的；不合格的
substitute 代替；代替物；代用品；替代品
substitution food 代食品；食物代用品
substitution product 取代〔產〕品
substrate 基質
subsurface fermentation 表面下發酵；深層發酵
subsurface storage 地下儲藏
subtilin 枯草菌素
subtle taste 難以捉摸的味道；微妙的味道
subtropical fruit 亞熱帶水果
subunit 小單位；次單元
subvitaminosis 維生素缺乏症
subzero storage 低溫冷藏庫
subzero temperature 零下溫度
succade 蜜餞
succinate 琥珀酸鹽；琥珀酸酯
succinic acid 琥珀酸；丁二酸
succinic anhydride 琥珀酸酐
succinic dehydrogenase 琥珀酸去氫酶；琥珀酸脱氫酶
succinylated monoglyceride 琥珀酰單甘油酯
succory 菊苣
succose 多汁的

succotash 沙可達玉米粥；玉米煮豆
succulence 多汁；多水份
succulent 多汁的；多肉汁的；肉質的；肥厚多肉的植物
succulometer 植物水份計；玉米成熟度測定計；濕度計
succus 汁；液
succus entericus 腸液
suck 吸吮；舐食；吸食
sucker 腋芽；側枝芽；根部的芽；哺乳的嬰兒；吸盤；棒棒糖
sucking pig 乳豬
suckling 哺乳的嬰兒；乳嬰；乳畜
sucralose 三氯蔗糖
sucrase 蔗糖酶；轉化酶 [= invertase; saccharase]
sucrate 蔗糖化合物
sucrochemistry 蔗糖化學
sucroclastic 解糖的；分解糖的
sucroclastic enzyme 解糖酶
sucroglyceride 蔗糖甘油酯
sucrol 甜精；甘素；對乙氧苯脲 [= dulcin(e); valzin]
sucrose 蔗糖 [= saccharose; saccharobiose]
sucrose acetate isobutyrate (SAIB) 乙酸異丁酸蔗糖酯；蔗糖乙酸異丁酸酯
sucrose distearate 蔗糖二硬脂酸酯
sucrose ester 蔗糖酯
sucrose esters of fatty acids (SE) 蔗糖脂肪酸酯；脂肪酸蔗糖酯
sucrose fatty acid ester 蔗糖脂肪酸酯；脂肪酸糖酯
sucrose intolerance 蔗糖不耐症
sucrose monostearate 蔗糖單硬脂酸酯
sucrose phosphorylase 蔗糖磷酸化酶
suction filter 吸濾器
suction flask 吸濾瓶
Sudan red 蘇丹紅
suds 肥皂泡沫；肥皂水
suet 板油
sufu 豆腐乳；腐乳 [= bean curd cheese; fermented tofu; soy(bean) cheese]
sugar 食糖；糖；蔗糖
sugar acid 糖酸
sugar alcohol 糖醇
sugar apple 番荔枝；釋迦；佛頭果 [= sweetsop]
sugar basin 糖罐；砂糖容器
sugar bean 拉馬豆
sugar beet 甜菜
sugar bloom 糖花〔白〕
sugar boiler 熬糖鍋
sugar bush 糖楓林；糖槭樹林
sugar camp 楓糖廠
sugar candy 糖果；冰糖；甜蜜的食物
sugar cane 甘蔗
sugar cane juice 蔗汁；甘蔗汁
sugar castor 糖瓶
sugar coated 裹上糖衣的
sugar coating 包糖衣；糖衣
sugar colo(u)ring 焦糖色；醬色
sugar content 含糖量
sugar corn 甜玉米
sugar crop 糖〔料〕作物
sugar crystal 糖結晶
sugar cube 方糖
sugar cured 糖漬的
sugar doctor 糖結晶抑制劑
sugar ester 糖酯
sugar juice 糖汁
sugar liquor 糖液；糖漿
sugar maple 糖楓；糖槭
sugar mass 糖膏
sugar mill 糖廠
sugar palm 糖棕；糖椰
sugar pan 熬糖鍋

sugar pea 甜豆；大粒豌豆；糖莢豌豆；荷蘭豆
sugar refinery 精煉糖廠
sugar refractometer 糖〔度〕折光儀
sugar sachet 糖包
sugar slab 片糖
sugar sorghum 糖高粱；甜高粱 [= saccharine sorghum]
sugar substitute 糖代用品；代糖品
sugar syrup 糖漿
sugar tolerance 耐糖性；糖耐量
sugar tongs 方糖鑷子；方糖鉗子
sugared 加糖的；糖漬的
sugared ginger 糖薑
sugarhouse 楓糖廠；製糖廠
sugariness 甜度；甜味
sugaring 加糖；糖漬；製糖
sugarless 無糖的；不含糖的
sugarloaf 塔糖；棒糖
sugarplum 小糖果
sugary 含糖的；甜的
sukiyaki 日式肉片火鍋；壽喜燒
sulfate 硫酸鹽；硫酸酯 [= sulphate]
sulfated butyl oleate 硫化油酸丁酯
sulfatide 硫酸腦苷酯
sulfating agent 硫酸化劑
sulfatocobalamin 硫酸鈷維生素；硫酸鈷胺素
sulfide 硫化物
sulfide blackening 硫化黑變
sulfide staining 硫化黑斑；硫化物黑斑污染
sulfinic acid 亞磺酸
sulfite 亞硫酸鹽；亞硫酸酯 [= sulphite]
sulfited juice 亞硫酸處理果汁
sulfolipid 硫脂質；腦硫脂；硫酸脂
sulfomucin 硫黏蛋白
sulfonation 磺化〔作用〕
sulfonic acid 磺酸
sulfoprotein 含硫蛋白質
sulfur 硫；硫磺 [= sulphur]
sulfur amino acid 含硫胺基酸 [= sulfur-containing amino acid]
sulfur blackening 硫〔化〕黑變
sulfur compound 含硫化合物
sulfur-containing amino acid 含硫胺基酸 [= sulfur amino acid]
sulfur dioxide 二氧化硫
sulfuration 硫化〔作用〕
sulfured fruit 硫燻果
sulfuric acid 硫酸 [= sulphuric acid]
sulfuring 燻硫；硫燻；亞硫酸處理；二氧化硫處理
sulfurous acid 亞硫酸 [= sulphurous acid]
sulphate 硫酸鹽；硫酸酯 [= sulfate]
sulphitation 亞硫酸〔化〕處理
sulphite 亞硫酸鹽；亞硫酸酯 [= sulfite]
sulphur 硫；硫磺 [= sulfur]
sulphur dioxide 二氧化硫
sulphuric acid 硫酸 [= sulfuric acid]
sulphuring 燻硫；硫燻；亞硫酸處理；二氧化硫處理
sulphurous acid 亞硫酸 [= sulfurous acid]
sultana 無子白葡萄；蘇丹娜葡萄乾
sultana bread 葡萄乾麵包
sulze 肉凍
summer bolete 網狀牛肝菌
summer crop 夏作；夏季作物
summer orange 夏橙
summer sausage 夏〔季〕香腸
summer savory 夏香薄荷；歐洲薄荷；香薄荷；夏香風輪菜
summer squash 矮生西葫蘆；夏季西葫蘆；夏季南瓜
summer truffle 夏塊菌（蕈類）

sump 沉澱槽；水坑
sun cured 曬製；曬乾
sun cured fruit 曬製水果
sun curing 曬製；曬乾
sun dehydration 曬乾
sun dried food 曬乾食品；乾貨
sun drying 曬乾
sunblister 曬焦
sunburn 曬焦；日灼；曬斑
sunburned fruit 曬斑水果；日灼斑果
sundae 新地；聖代冰淇淋
sundries 雜貨；雜物
sunfish 翻車魚；太陽魚
sunflower 葵花；向日葵
sunflower oil 葵花籽油；向日葵籽油
sunflower seed 葵花籽；向日葵籽
sunflower seed oil 葵花籽油；向日葵籽油
sunlight flavo(u)r 日曬味；日灼味
sunnyside up 單面煎的；單煎一面的
sunscald 曬斑；曬傷
sunset yellow 晚霞黃
sunset yellow FCF 日落黃；晚霞黃
sunset yellow FCF aluminium lake 日落黃鋁色澱
sunstruck flavo(u)r 日曬味；日灼味
sup 啜；啜飲
super grinder 細磨機；超細研磨機
superacid 酸過多的；大酸的
superacidulation 過酸化〔作用〕
superalimentation 灌食法；超量餵食法；超量營養法；強飼法
supercentrifugation 高速離心
supercentrifuge 高速離心機
superchilling 過冷卻
superconcentrated juice 超濃縮果汁
supercooling 過冷；過度冷卻
supercritical fluid extraction (SFE) 超臨界流體萃取
superficial drying 表面乾燥
superfine 精細的；特級的
superfine pulverizer 超細磨粉機
superglycerinated 超甘油化的
superglycerinated fat 超甘油化脂肪
superheat 過熱；過度受熱
superheated steam 超熱蒸汽；過熱蒸汽
superheated vapo(u)r 超熱蒸汽；過熱蒸汽
superheater 超熱器；過熱器
superheating 超溫加熱；過熱
supermarket 超級市場；超市
supernaculum 特級〔葡萄〕酒；最上等的酒
supernatant 上層液；上澄液
supernutrition 營養過度
superoxide 超氧化物；過氧化物
supersaturated 過飽和的
supersaturation 過飽和
supersonic disintegrator 超音波碎解器；超聲波粉碎機
supersonic homogenizer 超音波均質機；超聲波均質機
supervitaminosis 維生素過多症
supper 晚餐
supplement food 補充食品；營養補助食品
supplementary effect 互補效應
supply 供應；供給；給料
surface active agent 界面活性劑；表面活性劑 [= surfactant]
surface active substance 界面活性物質；表面活性物質
surface activity 表面活性
surface air cooler 表面空氣冷卻器
surface condenser 表面冷凝器
surface cooler 表面冷卻器

surface culture 表面培養〔物〕；液面培養〔物〕
surface drying 表面乾燥
surface fermentation 表面發酵；上面發酵
surface finish 表面精整
surface frozen 表面凍結的
surface layer 表層
surface smear 表面塗抹；表面抹片
surface tension 表面張力
surface viscosity 表面黏度
surfactant 界面活性劑；表面活性劑 [= surface active agent]
surfeit 食物過量；飲食過度
surfine 最高級的酒（法國）
surimi 魚漿
Surinam cherry 蘇利南櫻桃；扁櫻桃
surplus milk 剩餘乳；過剩乳
surrogate 代用品
survival curve 殘存曲線；存活曲線
susceptibility 感受性；敏感性
sushi 壽司
suspended centrifuge 懸掛〔式〕離心機；吊籃〔式〕離心機
suspended colloid 懸浮膠體
suspended particle 懸浮粒子
suspended phase 懸浮相
suspended pulp 懸浮果肉漿
suspended solid 懸浮固形物
suspending agent 懸浮劑；懸濁劑
suspension 懸浮；懸浮體；懸浮液
suspension colloid 懸浮膠體
suspension culture 懸浮培養〔物〕
suspensoid 懸浮膠體；懸膠體
sustenance 營養品；糧食；食物；維持體力的食品
sutler 軍中小販；隨軍小販
suttle (weight) 淨重
swab 藥棉；拖把；拭子；拭抹
swab method 擦抹取樣法；棉拭檢驗法
swallow 燕子；吞嚥；吞嚥量
swallow nest 燕窩
swallower 吞食者；暴食者
swan 天鵝
sweat box 回潮箱；發汗箱；濕潤箱
sweat damage 汗濕損壞；發汗損害
swede 瑞典蕪菁；蕪菁甘藍
Swedish landrace 瑞典藍瑞斯豬；瑞典蘭德瑞斯豬
sweepings 垃圾
sweet 甜的；芳香的；味道鮮美的；糖果；甜點
sweet almond 甜〔巴旦〕杏仁
sweet and sour 糖醋的；甜酸的
sweet and sour pork 咕咾肉；糖醋肉
sweet and sour sauce 糖醋醬；甜酸醬
sweet basil 丁香羅勒；羅勒；九層塔；洋紫蘇 [= basil]
sweet bay 月桂樹 [= laurel]
sweet bean paste 甜麵醬
sweet birch oil 樺木油
sweet biscuit 甜餅乾
sweet butter 無發酵乳酪；甜性奶油
sweet buttermilk 甜性酪乳
sweet cane 甘蔗
sweet cassava 甜樹薯；甜木薯
sweet cherry 歐洲甜櫻桃；西班牙櫻桃
sweet chocolate 甜巧克力
sweet cider 無發酵的甜蘋果汁
sweet corn 甜玉米；甜玉蜀黍
sweet course 宴席甜點；西餐甜點
sweet cream 鮮乳油；鮮奶油；甜性奶油
sweet cream butter 鮮乳油乳酪
sweet curd 鮮凝乳〔塊〕
sweet dessert wine 飯後甜〔葡萄〕酒
sweet dough 無發酵麵糰；鮮麵糰
sweet fennel oil 甜小茴香油

sweet flag 甜旗草；白菖；水菖蒲
sweet flavo(u)r 甜味
sweet gale 香楊梅
sweet goods 甜食；甜點
sweet grass 甜味草
sweet herb 芳香草料
sweet maize 甜玉米；甜玉蜀黍
sweet marjoram 馬郁蘭；甜馬郁蘭
sweet marjoram oil 甘牛至油；馬郁蘭油
sweet mash 未發酵麥芽漿；糖化醪
sweet melon 甜瓜
sweet mountain grape 北美秋葡萄
sweet oil 橄欖油；芳香的食〔用〕油
sweet olive 甜橄欖
sweet orange 甜橙；柳橙；柳丁
sweet orange oil 甜橙油
sweet pea 香碗豆
sweet pepper 青椒；甜椒；燈籠椒 [= bullnose pepper]
sweet persimmon 甜柿
sweet pickle 甜味醃漬〔品〕；甜酸泡菜
sweet potato 甘藷；甘薯；番薯；地瓜
sweet potato chip 甘薯〔脆〕片；番薯片
sweet potato starch 甘藷澱粉；甘薯澱粉
sweet rice 糯米 [= glutinous rice; sticky rice; waxy rice]
sweet rice flour 糯米粉
sweet roll 甜捲麵包
sweet root 甘草；甜旗草；白菖
sweet shop 糖果店
sweet sorghum 甜高粱；蘆粟 [= sorg(h)o]
sweet stuff 甜食；糖果；蜜餞
sweet taste 甜味
sweet tooth 嗜甜食者；愛吃甜食
sweet violet 香堇〔菜〕
sweet water 糖水
sweet wine 甜〔葡萄〕酒
sweet wort 麥芽浸出液；甜麥芽汁
sweet wrapping machine 糖果包裝機
sweetberry honeysuckle red 藍錠果紅
sweetbread 胰臟（牛或羊）；胸腺（牛或羊）
sweetened condensed milk 加糖煉乳；甜煉乳
sweetened juice 加糖果汁
sweetened milk powder 加糖乳粉
sweetener 甜味劑；甜味料；甜佐料
sweetening 甜味料；甜料
sweetening agent 甜味劑；甜味料
sweetfish 香魚 [= ayu (fish)]
sweetie 糖果
sweeting 甜蘋果
sweetish 甜膩的；帶甜味的
sweetmeat 甜食；糖果；蜜餞
sweetness 甜度；甜味
sweets 糖果；甜品；蜜餞；甜酒
sweetsop 番荔枝；釋迦；佛頭果 [= sugar apple]
sweetwood bark 苦香〔樹〕皮
sweety 糖果；蜜餞
swell 膨脹；膨罐
swelled can 膨罐；凸罐
swelling 膨脹；膨潤；腫脹
swelling agent 膨脹劑；膨潤劑
swelling capacity 膨脹度；膨潤度
swig 一大口；痛飲；牛飲；一口喝下
swill 沖洗；灌洗；痛飲；潲水；餿水；泔水；豬食
swim bladder 魚鰾
swine 豬
swineherd 養豬者；牧豬人
swinery 豬欄；豬群
swing sieve 搖動篩
swinging screen 搖動篩
swipes 劣啤酒；味淡的低劣啤酒

Swiss chard 厚皮菜；牛皮菜；莙薘菜；唐萵苣；瑞士恭菜；海甜菜
Swiss cheese 瑞士硬芝士；瑞士〔硬〕乾酪 [= Emmentaler]
Swiss roll 瑞士〔蛋糕〕捲；蛋糕捲
swizzle 碎冰甜雞尾酒
swizzle stick 攪酒棒
swollen granule 膨潤顆粒
sword bean 刀豆；闊刀豆；劍豆
swordfish 旗魚；箭魚；扁嘴旗魚
sycamore 無花果樹；歐洲〔大〕楓樹；美洲梧桐樹
syllabub 乳潘趣飲料；乳酒凍 [= sillabub]
symbiosis 共生；共棲
symptom 症狀；病徵
synanthrin 菊糖
synaptase 杏仁酶
syncarp 合心皮果；聚合果；聚花果
syndet 合成洗滌劑
syndrome 徵候群；綜合病症
syneresis 離水〔現象〕；脱水收縮〔作用〕
synergism 加乘作用；協合作用
synergist 增效劑；協合劑
synergistic effect 加乘效應
synthesis 合成〔法〕
synthesize 合成
synthetic 合成的；人造的
synthetic antioxidant 合成抗氧化劑
synthetic antiseptic 合成防腐劑
synthetic casing 合成腸衣；人造腸衣
synthetic dye 合成染料
synthetic essence 合成香精
synthetic fiber 合成纖維
synthetic flavo(u)r 人造〔食用〕香料；合成香料 [= artificial flavo(u)r]
synthetic food 合成食品
synthetic food flavo(u)ring 合成食用香料
synthetic medium 合成培養基 [= defined medium]
synthetic membrane 人工膜；合成膜
synthetic milk 人造乳；合成乳 [= artificial milk]
synthetic sake 合成清酒
synthetic sweetener 合成甜味料；合成甜味劑
synthetic vinegar 合成醋
syphon 虹吸；虹吸管 [= siphon]
syringa 丁香
syringa amurensis concrete 白丁香浸膏
syringaldehyde 丁香醛
syringe 注射器
syringic acid 丁香酸
syringone 丁香酮
syringyl alcohol 丁香醇
syrup 糖漿；糖汁
syrup blend 調合糖漿
syrup curing 糖漬
syruping 加糖漿
syrupy 含糖漿的；甜膩的；像糖漿的
system 系統；制度
Szechuan peppercorn 花椒；山椒 [= Chinese prickly ash; flatspine prickly ash]

T

T-bone steak 丁骨牛排；T-骨牛排
tab 標籤；拉環；易開罐拉環
tabasco 烏椒
tabasco pepper 指天椒
tabasco sauce 辣椒汁；塔巴斯哥酸辣醬
table 桌子；餐桌；台；表
table balance 托盤天平
table cream 餐用鮮乳油；餐用稀奶油
table fork 餐叉
table knife 餐刀
table linen 餐桌〔用〕布
table mat 餐桌用小墊子
table napkin 餐巾
table oil 餐桌用油
table plate 餐具
table ready 即可上餐桌食用的
table salt 食鹽；餐桌鹽；佐餐鹽；精製食鹽 [= edible salt; sodium chloride]
table set 餐具組；成套餐具
table setting 餐桌布置
table sugar 餐用砂糖；食用糖
table syrup 餐用糖漿
table vinegar 餐用食醋；佐餐醋
table wine 餐酒；佐餐葡萄酒；開胃酒
tablecloth 桌布；餐桌用台布；檯布
tablespoon 湯匙；大湯匙
tablespoonful (tbsp.) 一湯匙量；一大湯匙
tablet 片〔劑〕；錠〔劑〕
tablet candy 糖果片
tablet machine 打錠機；壓片機
tablet press 打錠機；壓片機
tablet sugar 糖錠
tabletting machine 打錠機；壓片機
tabletting press 打錠機；壓片機
tableware 餐具
tabloid 片劑；小片；藥片
tachometer 轉速計；流速計
tachyphagia 快速吞食〔癖〕；速食癖
tackifier 增黏劑；膠黏劑 [= tackifying agent]
tackifying agent 增黏劑；膠黏劑 [= tackifier]
tackiness 膠黏性；黏性
taco 墨西哥塔可包
taco sauce 墨西哥塔可醬
tael 両
taffy 太妃糖；拖妃糖；乳脂糖 [= toffee; toffy]
tafia 次級蘭酒；塔菲亞酒
tag 標誌；標籤
tagatose 塔格糖
tagetes 萬壽菊
tagetes oil 萬壽菊油
tagliarini 營養寬麵條；意大利細寬麵（加雞蛋）
tagliatelle 意大利寬麵（麵球）
tahini (paste) 芝麻醬
tail 尾

tail fin 尾鰭
tail meat 尾肉
tail water 廢水
tailings 廢料；殘渣；尾餾分
tailover 篩渣
taimen 哲羅魚
taint 腐敗；異味；污點；污染
taka-amylase 高峰〔氏〕澱粉酶 [= takadiastase]
takadiastase 高峰〔氏〕澱粉酶 [= taka -amylase]
takeaway food 外帶食物；外賣食物
takeout food 外帶食物
takeout packaging 外帶包裝
takeout service 外賣
talc 滑石〔粉〕 [= talcum]
talcum 滑石〔粉〕 [= talc]
talitol 塔羅糖醇
tall can 高罐；高身罐
tall russule 沼澤紅菇
tall tin 高罐；高身罐
tallow 牛脂；動物脂
tallow rendering 煉脂
tallow tree 油料樹；烏桕樹
tallowiness 脂肪氧化的臭味
tallowy 動物脂的；油膩的；油耗味的
talose 塔羅糖
tamale 墨西哥塔瑪玉米粉肉粽子
tamarillo 樹番茄；塔曼里羅果
tamarind 羅望子；羅望子果；酸豆
tamarind gum 羅望子膠；羅望子多糖膠
tamarind seed oil 羅望子油
tamaron 甲胺磷
tammy 濾布；布篩；絹篩
tamper 搗錘；杵；搗棒
tang flour 澄麵
tange navel 丹下臍橙
tangelo 橘柚
tangerine 紅橘；橘子
tangerine oil 紅橘油
tangerine peel 陳皮；果皮
tangic acid 昆布酸；海帶酸
tangle 昆布；海帶
tangleberry 越橘(美國東部)；藍越橘
tangor 焦柑；坦坡橙
tangy flavo(u)r 煙燻味
tania 芋類
tank 罐；槽；箱；池；桶
tankage 槽的容量；骨肉雜
tankan 桶柑
tankard 單柄高酒杯；大啤酒杯
tannase 單寧酸酶；鞣酸酶；丹寧酸酶
tannate 單寧酸鹽；鞣酸鹽
tannia 肉芋
tannic acid 單寧酸；鞣酸
tannin 單寧；鞣質；單寧酸；丹寧酸；鞣酸
tanoak brown 多穗柯棕
tansy 艾菊；艾菊的苦葉
tantalus 透明酒櫃；酒瓶架
tap 活栓；旋塞；水龍頭
tap water 自來水
tapered shape can 梯形罐；淺盆形罐；錐形罐
tapeworm 條蟲
taphouse 酒吧；酒館
tapioca 木薯；木薯粉
tapioca flour 木薯粉
tapioca starch 木薯澱粉
tapping 放液；排放
tapping conveyor 震動輸送帶
taproom 酒吧
tara gum 他拉膠
taraxacum 蒲公英；蒲公英乾根莖(藥用)
taraxanthin 蒲公英黃質
taraxol 蒲公英醇

tare 包裝重量；皮重；容器重；野豌豆；夾雜物
tare weight 皮重
taro 芋；芋頭 [= dasheen]
taro pudding 芋泥
tarragon 茵陳蒿；香艾菊；龍蒿 [= estragon]
tarragon oil 茵陳蒿油；香艾菊油；龍蒿油 [= estragon oil]
tart 水果小餡餅；果餡餅；酸的
tart fruit 酸果
tart rhubarb 原葉大黃；食用大黃
tartar 酒石；酒石酸氫鉀
tartar emetic 吐酒石
tartare sauce 韃靼調味醬；塔塔沙司
tartarian buckwheat 韃靼蕎麥；苦蕎麥
tartaric acid 酒石酸
tartary buckwheat 韃靼蕎麥；苦蕎麥
tartine 果醬麵包片；麵包片
tartlet 小水果餡餅
tartrate 酒石酸鹽；酒石酸酯
tartrazine 檸檬黃；酒石黃；食用黃色4號
tartrazine aluminum lake 檸檬黃鋁色澱；食用黃色4號鋁色澱
tasajo 乾肉條；牛肉乾
tassel 玉米雄花蕊；纓穗
taste 味；味覺；味道；品嚐
taste bud 味蕾
taste cell 味蕾細胞
taste compound 呈味化合物
taste modifier 味覺改變劑
taste organ 味覺器官
taste panel 品評小組；品味小組；品嚐小組
taste receptor 味覺受體
taste substance 呈味物質
taste test 品嚐試驗；嚐味試驗
taste threshold 味覺閾值
tasteful 美味的
tasteless 無味的
taster 品嚐者；試味者；評味員
tastiness 美味；味道；滋味
tasting 品嚐；評味；試味
tasty 可口的；味美的
taurine 牛磺酸；胺基乙磺酸
taurocholate 牛〔磺〕膽酸鹽；牛磺膽酸酯
taurocholic acid 牛〔磺〕膽酸
tavern 小酒館；小旅館；客棧
tawny funnel cap 柔弱香蘑 (蕈類)
tawny grisette 赤褐鵝膏 (蕈類)
tawny milk cap 多汁乳菇
TBA (thiobarbituric acid) 硫〔代〕巴比妥酸
TBA value (thiobarbituric acid value) 硫〔代〕巴比妥酸值；TBA值
TBHQ (tertiarybutylhydroquinone) 第三丁氫醌；叔丁基氫醌
tbsp. (tablespoonful) 一湯匙量；一大湯匙
TCA cycle (tricarboxylic acid cycle) 三羧酸循環；TCA循環；克雷布氏三羧酸循環
TDN (total digestible nutrients) 總可消化養分；可消化營養素總量
TDP (thermal death point) 熱滅點；受熱致死點；加熱死滅點；熱致死點
TDR (thermal death rate) 熱滅率；熱致死速度；加熱死滅率；熱致死率
TDT (thermal death time) 熱致死時間；加熱死滅時間；加熱致死時間
TDT curve (thermal death time curve) 加熱死滅時間曲線；熱致死時間曲線；加熱致死時間曲線
tea 茶；茶葉；茶樹
tea bag 茶包；茶袋
tea blend 混合茶；配製茶

tea board 茶盤
tea bread 茶點〔甜〕麵包
tea break 吃茶點和休息的時間
tea brick 茶磚；磚茶 [= brick tea; tile tea]
tea bush 茶樹叢
tea caddy 茶葉罐
tea cake 茶點的小甜點
tea catechin 茶葉兒茶素；茶葉兒茶酸
tea ceremony 茶道
tea chest 茶葉箱
tea cloth 茶具抹布；茶盤蓋布
tea cosy 茶壺〔保溫〕套
tea cozy 茶壺〔保溫〕套
tea cult 品茗會
tea cup 茶杯
tea equipage 茶具
tea fannings 茶〔葉〕末；茶屑；碎茶
tea fight 茶會；午後茶會
tea firing 焙茶
tea firing machine 焙茶機
tea garden 茶園
tea grower 茶葉種植業者
tea house 茶館；茶室
tea infuser 濾茶器
tea leaf 茶葉
tea leaf ball 珠茶
tea leaves 茶渣
tea liquor 茶汁
tea marker 沏茶器；泡茶器
tea panning 揉茶
tea party 茶會；午後茶會
tea plantation 茶園
tea plate 茶盤
tea polyphenols 茶多酚
tea rolling 茶葉揉捻
tea room 茶館；茶室；小吃店
tea seed oil 茶籽油
tea service 茶具組
tea set 茶具組
tea shoot 茶尖；毛尖；茶葉嫩芽
tea shop 茶館；茶室；小吃店
tea strainer 茶濾
tea table 茶几；茶桌
tea tannin 茶單寧
tea taster 品茗員；品茶員
tea tasting 品茗；品茶
tea tray 茶盤
tea trolley 茶點推車；有輪的茶具臺
teakettle 茶壺；燒水壺
teal (duck) 水鴨
teanine 茶胺酸
teapot 茶壺
teapoy 茶几；三腳茶几
tear-off lid 拉開蓋；撕開蓋
tear open can 拉開罐；撕開罐
tear tab end 拉舌罐蓋；易開罐蓋
teaspoon 茶匙
teaspoonful (tsp.) 一茶匙量
teat 奶嘴；乳頭；奶頭
teatime 茶點時間；下午茶時間；午後茶點時間
technology 工藝；技術；科技
teel oil 芝麻油
teetotal 滴酒不沾的；絕對戒酒的；不喝酒的
teetotal drink 無酒精飲料；不含酒精飲料
teetotalism 絕對禁酒主義
teewurst 茶腸
teff 畫眉草(埃塞俄比亞)
teflon 聚四氟乙烯；特氟隆；鐵氟龍
teg 二歲綿羊
tegument 外殼；外皮
teleost 硬骨魚
tempeh 天貝(印尼傳統食品)
temperance 戒酒；禁酒
temperance drink 無酒精飲料

temperature (temp.) 溫度
temperature scale 溫標
temperature tolerance 耐溫性
tempered wheat 調濕小麥；勻濕的小麥
temperer 調節器
tempering 調溫；回火；調濕；調和
tempering bin 調濕穀倉
tempering room 調溫室；調濕室
tempura 油炸裹麵糊的魚蝦；天婦羅
tenacity 黏滯性；堅韌；韌度
tender beef 嫩牛肉
tender meat 嫩肉
tenderization 嫩化
tenderizer 嫩化劑；嫩化器
tenderizing agent 嫩化劑
tenderloin 里脊肉；里肌肉；腰裏脊肉；裏肌肉；腰內肉；嫩里肌肉
tenderloin steak 牛柳；菲力；里脊牛排
tenderness 柔軟性；柔嫩度；嫩性
tenderometer 嫩度計
tenderside 牛臀肉
tendomucoid 腱黏蛋白
tendon 腱；筋
tengusa 製瓊脂的海藻；製洋菜的海藻
tensile 可伸張的；伸張性
tensile force 拉力；張力
tensile strength 抗拉強度；張力
tensility 伸張性；拉伸性
tensiometer 張力計；拉力計
tension 拉力；張力
tent 深紅甜葡萄酒
tepid 微溫的；微熱的
tepidity 微溫；微熱
tequila 龍舌蘭酒
teriyaki 紅燒的；照燒
termette 條麵
terminal amino acid 末端胺基酸
terminal amino group 末端胺基
terminal carboxyl group 末端羧基
terminal hydroxyl group 末端羥基
terminal market 農產品集散市場；集散市場
termination 終止
terpene 萜烯；萜〔烴〕
terpene resin 萜烯樹脂
terpeneless 無萜
terpeneless oil 無萜烯油；無萜油
terpinene 萜品烯
terpineol 萜品醇；松脂醇；松油醇
terpinolene 異松油烯；萜品油烯
terpinyl acetate 乙酸松油酯
terpinyl propionate 丙酸松油酯
terrapin 甲魚；鱉；海龜
terrestrial animal 陸生動物
terrine 有蓋湯碗；陶罐食品；沙鍋
tertiary butylhydroquinone 叔丁基對苯二酚
tertiary structure 第三級結構
tertiary waste treatment 第三級廢料處理
tertiarybutylhydroquinone (TBHQ) 第三丁氫醌；叔丁基氫醌
test 試驗；檢驗；介殼；甲殼
test diet 試驗膳食
test meal 試驗餐；試驗食物
test tube 試管
testa 外種皮；介殼；甲殼
testee 測試對象；受試驗者
tester 試驗器；試驗者；試驗裝置
testes 睪丸
testicles 睪丸
testosterone 睪固酮；睪丸激素
tetra-pack 無菌鋁箔包；四角包〔裝〕；利樂包 [= tetra pak]
tetra-pak 利樂包；無菌鋁箔包；四角包〔裝〕 [= tetra pack]

tetracoccus 四聯球菌
tetracosanoic acid 二十四〔烷〕酸
tetracosenoic acid 二十四碳烯酸
tetradecanoic acid 十四〔烷〕酸
tetradecenoic acid 十四碳烯酸
tetrahedral package 四角包裝
tetrahydrofurfuryl acetate 乙酸四氫糠酯
tetrahydrofurfuryl alcohol 四氫糠醇
tetrahydrogeraniol 四氫香葉醇
tetrahydrolinalool 四氫化芳樟醇
tetrahydrothiophen-3-one 四氫噻吩-3-酮
tetramethyl glucose 四甲基葡〔萄〕糖
tetramethyl glucoside 四甲基葡〔萄〕糖苷
tetraodon poisoning 河豚中毒；河魨中毒 [= tetrodon poisoning]
tetraodonic acid 河豚酸；河魨酸 [= tetrodonic acid]
tetraodotoxin 河豚毒素；河魨毒素 [= tetrodotoxin]
tetraphosphate 四磷酸鹽
tetrapotassium pyrophosphate 焦磷酸四鉀
tetrasaccharide 四醣
tetrasodium pyrophosphate 焦磷酸四鈉
tetrathionate broth 四硫磺酸鹽培養基；連四硫酸鹽培養基
tetritol 丁糖醇
tetrodon poisoning 河豚中毒；河魨中毒 [= tetraodon poisoning]
tetrodonic acid 河豚酸；河魨酸 [= tetraodonic acid]
tetrodotoxin 河豚毒素；河魨毒素 [= tetraodotoxin]
tetrose 丁糖；四〔碳〕糖
tetruronic acid 丁糖酮酸
teutlose 甜菜〔根〕糖
textured vegetable protein 組織化植物蛋白〔質〕 [= texturized vegetable protein]
texturized protein 組織化蛋白〔質〕
texturized soybean protein 組織化大豆蛋白〔質〕
texturized vegetable protein 組織化植物蛋白〔質〕 [= textured vegetable protein]
texturizer 調質劑
texturizing agent 調質劑；增稠劑
texturometer 稠度測定儀
thallophyte 藻菌植物；葉狀植物
thaumatin 索馬甜；非洲甜果素；非洲竹芋甜素；梭瑪丁
thaw dripping 解凍滴液〔現象〕
thaw exudates 解凍滲出液
thaw-in-bag food 連袋解凍食品；袋內解凍食品
thaw-in-pouch food 連袋解凍食品；袋內解凍食品
thaw point 解凍點；解凍溫度
thaw rigor 解僵；解凍僵直
thaw(ing) 解凍；融化
thawer 解凍裝置；解凍器
thawing drip 解凍滴液
thawing house 解凍房
thawing loss 解凍損失
thawing room 解凍室
thea saponin 茶皂苷
theaflavin 茶黃素
theamin 茶胺
theanine 茶胺酸
theine 茶鹼；咖啡鹼；咖啡因
theobroma oil 可可油；可可脂 [= cacao butter; cocoa butter]
theobromine 可可鹼；3,7-二甲基黃嘌呤
theol 茶葉香精油

theophylline 茶葉鹼
therapeutic nutrition 營養治療 [= nutritional therapy]
therapeutical diet 療效膳食；治療膳食
therapeutical food 療效性食品；治療食品
therapy 治療
theriac 解毒藥
thermal baffle 隔熱板
thermal capacity 熱容量
thermal coagulation 熱凝固
thermal conduction 熱傳導
thermal conductivity 傳熱性；導熱性；熱傳系數；導熱系數
thermal convection 熱對流
thermal cure 熱醃法 [= hot cure]
thermal death point (TDP) 熱滅點；受熱致死點；加熱死滅點；熱致死點
thermal death rate (TDR) 熱滅率；熱致死速度；加熱死滅率；熱致死率
thermal death time (TDT) 熱致死時間；加熱死滅時間；加熱致死時間
thermal death time curve (TDT curve) 加熱死滅時間曲線；熱致死時間曲線；加熱致死時間曲線
thermal decomposition 熱分解
thermal degradation 熱變形
thermal dehydration 加熱脱水
thermal denaturation 熱變性
thermal diffusion 熱擴散
thermal energy 熱能
thermal expansion 熱膨脹
thermal gel 熱凝膠
thermal insulation 隔熱；絕熱
thermal plasticity 熱塑性
thermal radiation 熱輻射
thermal reduction time 加熱殺菌時間
thermal resistance 熱阻
thermal shock 熱休克；熱震；熱沖擊
thermal stability 熱穩定性
thermal tolerance 耐熱性
thermal treatment 熱處理
thermally oxidized soya bean oil (TOS) 熱氧化大豆油
thermocoagulation 熱凝聚〔作用〕；熱凝固〔作用〕
thermocouple 熱電偶
thermocouple thermometer 熱電偶溫度計
thermocuring 熱醃
thermoduric 耐熱的
thermoduric bacteria 耐熱細菌
thermoduric culture 耐熱培養〔物〕
thermodynamics 熱力學
thermogenic bacteria 產熱細菌
thermolabile 不耐熱的；熱不穩定的
thermolability 不耐熱性；熱不穩定性
thermolysin 嗜熱菌蛋白酶
thermolysis 熱〔分〕解作用；散熱
thermometer 溫度計
thermopeeling 熱〔烤〕去皮
thermophile 嗜熱細菌；嗜熱生物
thermophilic bacteria 嗜熱細菌
thermophylactic 抗熱的
thermoplastic 熱塑性的
thermos (bottle) 保溫瓶；熱水瓶；暖〔水〕壺 [= thermos flask]
thermos flask 保溫瓶；熱水瓶；暖〔水〕壺 [= thermos (bottle)]
thermosensitive 熱敏的
thermosensitive food 熱敏性食物
thermosensitive substance 熱敏性物質
thermoset 熱〔凝〕固的
thermosetting resin 熱固性樹脂
thermostability 耐熱性；熱穩定性
thermostabilization 熱穩定化〔作用〕；加熱殺菌
thermostable 耐熱的；熱穩定的

thermostat 恆溫器
thermostatic oven 恆溫的
thermotolerant bacteria 耐熱細菌
thevetose 黃夾竹桃糖
thew 肌肉；腱
thiabendazole 噻苯咪唑；苯咪唑；涕必靈
thiaminase 噻胺酶；硫胺素酶
thiamine 噻胺；硫胺；硫胺素；抗神經炎素；維生素B_1 [= aneurin; vitamin B_1]
thiamine dicetylsulfate 硫胺素十六烷基硫酸鹽
thiamine dilaurylsulfate 硫胺素二月桂基硫酸鹽
thiamine hydrochloride 硫胺素鹽酸鹽；鹽酸硫胺〔素〕；氫氯化噻胺
thiamine (mono)nitrate 硫胺素硝酸鹽
thiamine naphthalene-1,5-disulfonate 硫胺素萘-1,5-二磺酸鹽
thiamine naphthalene-2,6-disulfonate 硫胺素萘-2,6-磺酸鹽
thiamine pyrophosphate (TPP) 噻胺焦磷酸鹽；硫胺素焦磷酸鹽
thiamine thiocyanate 硫胺素硫氰酸鹽
thiazole 噻唑
thick 厚的；濃的；稠的
thick boiling starch 稠糊澱粉；稠沸澱粉
thick cream 濃鮮乳油；高脂〔稀〕奶油
thick filaments 厚肌原絲
thick flank 厚腹脅肉；厚牛陰柳；臀腿
thick juice 濃汁；糖漿
thick mash 濃醪
thick runner 前肋肉
thick sauce 濃稠調味醬
thick slurry 稠漿
thick soup 濃湯
thick taste 濃味
thick yoghurt 濃固型酸乳酪
thickened must 濃葡萄汁
thickener 增稠劑
thickening 增稠；濃縮；稠化
thickening agent 增稠劑
thickness 稠度；厚度
thigh 大腿；上腿
thigh bone 股骨
thigh meat 腿肉；後腿肉
thimble cap 圓錐鐘菌（蕈類）
thimbleberry 果實樹莓；香莓；糙莓
thin 薄的；稀的；瘦的
thin boiling 輕沸
thin boiling starch 稀糊澱粉；輕沸澱粉；低黏度澱粉
thin filaments 薄肌原絲
thin film 薄膜
thin film evaporator 薄膜蒸發器
thin juice 稀汁；稀糖汁
thin layer chromatography (TLC) 薄層層析法；薄層色譜法
thin layer drying 薄層乾燥
thin paste 稀糊
thin runner 後肋肉
thin shirt 牛表層胸肉
thin skinned 薄皮的；皮薄的
thin taste 淡味
thin white 稀卵白；清蛋白
thinner 稀釋劑
thinness 薄度；瘦；細
thinning 稀化；稀釋
thinning agent 稀釋劑
Thiobacillus 硫桿菌屬
thiobacteria 硫細菌
thiobarbituric acid (TBA) 硫〔代〕巴比妥酸
thiobarbituric acid value (TBA value) 硫〔代〕巴比妥酸值；TBA值
thioctic acid 硫辛酸

thiocyanic acid 硫氰酸
thiodipropionic acid 硫代二丙酸
thioester 硫酯
thioether 硫醚
thioglucose 硫葡萄糖
thioglucosidase 葡糖硫苷酶；硫葡萄糖苷酶
thioglycerin 硫甘油
thiol 硫醇 [= mercaptan]
thiophilic bacteria 嗜硫細菌
thiosugar 硫糖
third press 三榨；三榨汁；三榨油
thirst 口渴；渴
thirstquenching 解渴的
thistle 薊
thixotropic fluid 觸變稀性流體
Thompson seedless 湯姆生無籽葡萄
thorax 胸腔；胸膛
thousand headed cabbage 孢子甘藍
thousand island dressing 千島沙拉醬；千島沙津醬
thousand years' egg 皮蛋 [= alkalized duck egg; alkalized egg; Chinese egg; pidan]
thrasher 打穀機；脱粒機；捶擊機
threadworm 蟯蟲
three grooved shrimp 三槽蝦
three-piece can 三片罐
threonine 羥丁胺酸；蘇胺酸
threose 異赤藻糖；蘇糖
threpsis 營養
threpsology 營養學
thresher 打穀機；脱粒機；捶擊機
threshing 打穀；脱粒；脱穀
threshold 閾限；閾值；低限
threshold limit value 閾限值；低限值；臨界值
threshold value 閾值
throat 咽喉；喉嚨
thrombin 凝血酶
thrombocyte 血小板
through circulation dryer 通風乾燥機
through flow dryer 通風乾燥機
throwaway bottle 用後即棄的瓶子
thuja oil 側柏葉油；金鐘柏油；雪松葉油 [= cedar leaf oil]
thuringer 圖林根香腸
thyme 百里香；麝香草；瑞香草
thyme oil 百里香油；瑞香草油；麝香草油
thymol 百里香酚；瑞香草酚；麝香草酚
thymus 百里香；胸腺
thyroid gland 甲狀腺
thyroxin(e) 甲狀腺立素
Tian Qing gum 田青膠
tibia 脛骨
tick bean 蚕豆
tidbit 小片的美味食物；扇〔形〕片 [= titbit]
tie 捆紮；結紮；捆包；綁
Tientsin cabbage 天津紹菜
tier 捆紮機；結紮機
tier dryer 多層式乾燥機
tierce 大桶；酒桶
tierce curing 裝桶醃漬
tiff 淡酒；一口酒
tiffin 中飯；午飯；午餐
tiger lucina 球形厚大蛤
tiger milk mushroom 虎奶菇
tiger prawn 草蝦；斑節對蝦
tiger puffer 東方魨
tiger saw dill 虎皮香菇
tigernut 非洲土杏紅
tight container 密閉容器；緊密容器
tightness 緊密性
tiglaldehyde 惕各醛
tikitiki 米糠；米糠的酒精提取液

til 胡麻
tilapia 黑邊吳郭魚；羅非魚
tile tea 茶磚；磚茶 [= brick tea; tea brick]
tilefish 馬頭魚；方頭魚
tiller 幼芽；嫩枝；農夫；耕種者
tilting kettle 可傾〔蒸煮〕鍋
timbale 模製餡餅
time table 時刻表；日程表
timer 定時器；計時器
tin 錫；馬口鐵皮；鍍錫鐵皮；罐；麵包烤模；鍍錫
tin bread 模製麵包
tin can 馬口鐵罐；鍍錫鐵罐；罐頭
tin coating 鍍錫層
tin foil 錫箔
tin free can 無錫〔鐵〕罐
tin labeller 罐頭貼標機
tin opener 開罐器；罐頭刀；開罐刀
tin pan 烤〔餅〕盤
tin sulfide 錫硫化物
tinct 色澤；色調
tincture 酊劑；藥酒；染色
tincture distillery 藥酒廠
tinker 幼鯖；炊具修補匠；修補匠
tinless can 無錫〔鐵〕罐
tinned butter 罐裝牛油；罐裝乳酪
tinned food 罐頭食品；罐裝食品；食品罐頭 [= canned food]
tinned goods 罐頭食品；罐裝食品；食品罐頭 [= canned goods]
tinned meat 罐裝肉；肉類罐頭
tinning 罐頭包裝法；鍍錫；製造罐頭食品
tinning factory 罐頭廠
tinny taste 錫味
tinplate 馬口鐵〔皮〕；鍍錫鐵皮
tintometer 色調計
tinware 錫器；鍍錫容器
tip 頂；尖端；末端；尖頭；小費
tip cap 芽冠；根冠
tipping tank 傾卸槽；傾卸桶
tipple 烈酒；飲烈酒
tippler 酒徒；酒量大的人
tippy leaf 頂芽；頂葉
tippy tea 毛尖茶
tipsiness 喝醉；微醉
tipsy 喝醉的；微醉的
tipsy cake 醉蛋糕；加葡萄酒的蛋糕
tirage 香檳酒發酵；香檳酒所用配料
tissue 組織；拭紙；薄紙
tissue culture 組織培養〔物〕
tissue fat 組織脂肪
tissue fluid 組織液
tissue protein 組織蛋白
titanium 鈦
titanium dioxide 二氧化鈦
titbit 小片的美味食物；扇〔形〕片 [= tidbit]
titration 滴定
titrimetric analysis 滴定分析〔法〕；滴定測量〔法〕
titrimetry 滴定分析〔法〕；滴定測量〔法〕
TLC (thin layer chromatography) 薄層層析法；薄層色譜法
toad 蟾蜍；癩蛤蟆
toad-in-the-hole 裹麵〔烤〕肉
toadfish 三齒躄魚；蟾魚
toadstool 毒菌；毒蕈；羊肚菌；傘菌
toast 烤麵包片；吐司；土司；多士；敬酒；烤；烘
toasted bread 土司；烤麵包片
toaster 烤麵包機；多士爐；舉杯祝賀者
tocopherol 生育酚；維生素E [= vitamin E]

α-tocopherol α生育酚；阿伐生育酚 [= alpha-tocopherol]
β-tocopherol β生育酚；貝他生育酚 [= beta-tocopherol]
tocopheronic acid 生育酸
tocopheryl acetate 醋酸生育酚
tocopheryl acid succinate 琥珀酸生育酚
tocopherylamine 生育胺
toddler 初學走路的孩童；剛學步行時期
toddy 棕櫚汁；棕櫚酒；烈性酒
toe 趾；腳趾
toffee 太妃糖；拖妃糖；乳脂糖 [= taffy; toffy]
toffy 太妃糖；拖妃糖；乳脂糖 [= toffee; taffy]
tofu 豆腐 [= soybean curd]
tofu puff 油泡豆腐
Tokay 匈牙利的托凱甜〔葡萄〕酒；匈牙利的多桂甜〔葡萄〕酒
toke 食物；乾麵包
tolerance 容忍；容許量；耐藥量
tolerance limit 容許限量；許可限度
tolu balsam 妥盧香脂；妥魯香液；妥路香油
toluene 甲苯
p-tolyl isobutyrate 異丁酸對甲苯酯
tom 雄性動物
Tom Collins 琴雞尾酒；杜松子果汁酒
tomatine 番茄素；番茄苷
tomato 番茄；西紅柿
tomato catchup 番茄醬；調味番茄醬；番茄調味醬 [= tomato catsup; tomato ketchup; tomato sauce]
tomato catsup 番茄醬；調味番茄醬；番茄調味醬 [= tomato catchup; tomato ketchup; tomato sauce]
tomato juice 番茄汁
tomato ketchup 番茄醬；調味番茄醬；番茄調味醬 [= tomato catsup; tomato catchup; tomato sauce]
tomato paste 番茄膏；番茄糊
tomato pomace 番茄粕；番茄渣
tomato pulp 番茄漿
tomato pulper 番茄打漿機
tomato puree 番茄泥
tomato sauce 番茄醬；調味番茄醬；番茄調味醬 [= tomato catsup; tomato catchup; tomato ketchup]
tommy 抵工資的食物；配給口糧；實物工資制
ton 噸
toned milk 調整乳；調脂〔水〕乳
toner 調色劑；增色劑；上色劑
tongs 夾子；鑷子；鉗子
tongue 舌〔頭〕
tongue sole 大鱗鞋底魚；短吻舌鰨
tonguefish 舌鰨
tonic 滋補的；補品；滋補劑
tonic food 滋補食品
tonic water 湯尼汽水；奎寧水
tonic wine 補酒
tonka bean 零陵香豆；香豆
tonsil 扁桃腺
tool 工具；用具
tooth 牙齒；胃口
toothed jelly 白色膠偽齒菌(蕈類)
toothful 一小口；微量
toothpick 牙籤
toothsome 可口的
top blade muscle 上肩胛脊底肌；板腱
top double seam 罐蓋二重捲封
top fermentation 上層發酵；頂層發酵；上面發酵
top icing 覆糖霜
top lid 頂蓋
top rib 上肋肉

top rump 上臀肉
top shell 海螺；蝾螺
top yeast 頂層酵母；上層酵母；上面酵母
tope 小鮫；翅鯊；星鯊
toper 酒徒；豪飲者
topping 頂飾；食物上加裝飾；去頂；頂部修整
topround 內牛腱；牛胴腱肉
topside (round) 牛後腿內側肉；內側後腿肉；上後腿肉；腰部上肉；砧排牛肉；頭刀
topside steak 內側後腿肉排；砧排
torgoch 紅腹鮭；紅點鮭
torigai 滑頂薄殼烏蛤
torpedo 日本單鰭電鰩
torrefaction 烘；烤；焙；炒
torrefy 烤；烘
torsion balance 扭力天平
torsion viscosimeter 扭力黏度計
torte 大蛋糕
tortellini 意大利雲吞；意大利餃子；餃形通心粉
tortilla 玉米薄餅；墨西哥玉米薄餅
tortoise 龜
tortoise shell 龜甲；龜板；玳瑁殼
torula (yeast) 圓酵母；食用酵母
torularhodin 圓酵母紅素
torulin 圓酵母素
tossed salad 拌好的生菜沙拉；拌好的生菜色拉；拌好的生菜沙律
tot 少量烈酒；小孩
total acidity 總酸度
total alkalinity 總鹼度
total analysis 全量分析
total carbohydrate 總醣
total count 總菌數
total digestible nutrients (TDN) 總可消化養分；可消化營養素總量
total dissolved solid (TDS) 總溶固形物；溶解總固體
total heat 總熱量
total lipid 總脂肪物質
total moisture 總水分
total nitrogen 總氮
total plate count 生菌數
total quality management (TQM) 全面質量管理
total solid 總固形物
total soluble solid 總可溶固形物
total solubles 總可溶物
total sugar 總糖〔量〕
total viable count 總活菌數
total weight 總重
total yield 總產量
toughness 堅韌；韌性；韌度
tournesol 向日葵；天芥菜
touslesmois 美人蕉澱粉
towel gourd 絲瓜 [= vegetable sponge]
tower dryer 塔式乾燥器
tower evaporator 塔式蒸發器
tower still 塔式蒸餾器
toxic 有毒的；毒的
toxic substance 有毒物質；毒物
toxicant 毒素；毒藥
toxication 中毒
toxicide 解毒劑
toxicity 毒性
toxicity test 毒性試驗
toxicogenic 產毒的
toxicological evaluation 毒物學評估；毒性評估
toxicology 毒物學；毒理學
toxigenic 產毒的
toxigenicity 產毒性
toxin 毒素
toxin-type food poisoning 毒素型食物中毒

toxis 中毒
toxisterol 毒固醇
toxoid 類毒素
toxone 弱化毒素；慢性毒素
toxoprotein 毒蛋白質
TQM (total quality management) 全面質量管理
trace 微量；痕量；蹤跡
trace element 微量元素；痕量元素
trace level 痕量級
trace metal 微量金屬；痕量金屬
trachea 氣管；導管
trademark 商標
traditional food 傳統食品
tragacanth 黃耆
tragacanth gum 黃芪膠；黃耆膠；特拉卡甘膠 [= gum dragon; gum tragacanth]
tragacanthin 黃耆溶膠；特拉卡甘溶膠；黃耆質；黃耆糖
train fish 鯨魚
train oil 海產動物油；鯨〔魚〕油
transacetylation 轉乙酰作用
transacylation 轉酰〔基〕作用
transamidation 轉酰胺〔基〕作用
transamination 轉胺作用
transcarboxylation 轉羧作用
transesterification 轉酯作用；酯基轉移
transfructosidase 轉果糖苷酶；果糖苷移轉酶
transglucosidase 轉葡〔萄〕糖苷酶；葡萄糖苷移轉酶
transglycosidase 轉糖苷酶
transglycosylation 轉糖苷作用；葡基轉移作用
transient temperature 瞬間溫度
transition temperature 過渡溫度
transitional period 過渡期
translucent 半透明的
transmethylation 轉甲基作用；甲基轉移作用
transmittance 透射係數；透射比；透明性
transparency 透明性；透明度
transparent 透明的
transparent package 透明包裝
transpeptidation 轉肽作用
transphosphorylation 轉磷酸作用
transpiration 蒸散作用；蒸騰作用
transplantation 移植
transport café 公路廉價餐廳；路邊小吃店
transudate 滲出液；滲出物
transudation 滲出；漏出；滲漏
trash 廢料；廢物；垃圾；下腳料
trash can 垃圾桶
trash fish 下腳魚
trash ice 含水碎冰
trassi 特拉西蝦醬
traumatic acid 癒創酸
travelling chain oven 鏈動連續烤爐；傳送鏈式烤爐
travelling dryer 傳送式乾燥機
travelling furnace 移動爐
travelling oven 連續式烤爐；傳送式烤爐
travelling plate oven 連續板式烤爐；傳送板式烤爐
travelling tray oven 連續盤式烤爐；傳送盤式烤爐
tray 盤；托盤；淺盤
tray agriculture 水耕法
tray conveyer 盤式輸送機
tray dryer 盤〔架〕式乾燥機
tray meal 盤菜
tray oven 盤式烤爐
tray washer 洗盤機
trayful 一盤量；滿盤

treacle 糖漿；糖蜜
treacly 甜蜜的；糖蜜的
tread 踐踏；腳踩
treat 處理；治療；款待；宴客
treated starch 改性澱粉
tree molasses 楓糖蜜
tree moss concrete 樹苔浸膏
tree sugar 楓糖
tree tomato 樹茄；樹番茄
treefish 梅花鱸
trehala 繭蜜；蟲繭蜜
trehalase 繭蜜糖酶；海藻糖酶
trehalose 繭蜜糖；海藻糖
tremella 白木耳
Tremella fuciformis 白木耳；銀耳
trencher 砧板；墊板；木盤；木碟
trencherman 食者
trepang 海參 [= sea cucumber]
triacetin 甘油三乙酸酯；三乙酸甘油酯；三醋精；甘油三醋酸酯 [= glyceryl triacetate]
triage 分級；篩餘料；碎咖啡豆
triage coffee 低級咖啡
trial 試驗；試用
trial and error 反覆試驗
trial manufacture 試製
trial produce 試製；試製品
trial run 試車；試驗運轉
triammonium citrate 檸檬酸〔三〕銨
triangle test 三角試驗
tribasic acid 三元酸
tribasic alcohol 三元醇
tributyrin 三丁酸甘油酯；三酪酯
tricalcium phosphate 磷酸三鈣
tricarboxylic acid cycle (TCA cycle) 三羧酸循環；TCA循環；克雷布氏三羧酸循環
trichina 旋毛蟲
Trichinella spiralis 旋毛蟲
trichinosis 旋毛蟲病
trichloroacetic acid (TCA) 三氯乙酸
trichloroethylene 三氯乙烯
trichloromethane 三氯甲烷；氯仿 [= chloroform]
1,1,2-trichlorotrifluoroethane 1,1,2-三氯三氟乙烷
trichuriasis 鞭蟲病；毛鞭蟲病
trichuris 鞭蟲；毛鞭蟲
tricin 麥黃酮
trickle 滴流；細流
trickling cooler 水淋式冷卻器；滴淋冷卻器
trickling filter 淋式濾床；滴濾池；滴濾器
2-tridecanone 2-十三酮
2-tridecenal 2-十三烯醛
trielaidin 甘油三反油酸酯；三反油酸甘油酯
triethanolamine 三乙醇胺
triethenoid fatty acid 三烯脂肪酸
triethyl citrate 檸檬酸三乙酯
triethylamine 三乙胺
trifle 海綿蛋糕；少量；雪利酒蛋糕
triglyceride 甘油三酸酯；三〔酸〕甘油酯
trilaurin 三月桂酸酯；甘油三月桂酸酯；三月桂精 [= glyceryl trilaurate]
trilinolein 甘油三亞〔麻〕油酸酯；三亞麻油脂；亞油精 [= glyceryl trilinoleate; linolein; linolin]
trilinolenin 甘油三次亞麻油酸酯；三次亞麻油脂；亞麻精 [= glyceryl trilinolenate; linolenin]
trim 修整；修飾；裝飾
trimethylamine 三甲胺
trimmings 配菜；裝飾品；切屑
triol 三元醇

triolein 甘油三油酸酯；三油酸甘油酯；三油脂 [= glyceryl trioleate; olein]
triose 丙醣；三碳醣
triose phosphate 丙醣磷酸；磷酸丙醣
tripalmitin 軟脂；三棕櫚精；棕櫚精；甘油〔三〕棕櫚酸酯；三棕櫚酸甘油酯 [= glyceryl tripalmitate; palmitin]
tripe 瘤胃；肚
tripe fungus 腸膜狀牛耳（蕈類）
tripeptidase 三肽酶
tripeptide 三肽
triple effect evaporator 三效蒸發器
triple helix 三螺旋
tripolini 小蝴蝶粉
tripolite 硅藻石；硅藻土；矽藻石；矽藻土
tripolyglycerol monostearate 三聚甘油單硬脂酸酯
tripolyphosphate 三聚磷酸鹽
tripotassium orthophosphate 磷酸鉀；正磷酸三鉀
tripotassium phosphate 磷酸三鉀
Tris buffer 托立斯緩衝液；Tris緩衝液
trisaccharidase 三糖酶
trisaccharide 三糖
trisodium glycyrrhizinate 甘草甜素三鈉；甘草酸三鈉
trisodium phosphate 磷酸三鈉
tristearin 硬脂；三硬脂精；硬脂精；甘油三硬脂酸酯；三硬脂酸甘油酯 [= glyceryl tristearate; stearin]
triticale 黑小麥
triticin 小麥糖
trituration 研碎；碾磨；磨碎
triturator 研磨器；搗碎機
troche 喉錠；片劑
trolley 手推車；餐車；吊運車
trolley conveyor 懸吊輸送機；吊軌式輸送機
trolley hoist 滾輪式提升機；電葫蘆
trollixanthin 金蓮花黃質
trommel 轉筒篩
tropaeolum 金蓮花；旱金蓮
trophedema 營養性水腫
trophic 營養的
trophic analysis 營養分析
trophic value 營養價值
trophism 營養代謝
trophology 營養學
trophon 神經營養素
trophonosis 營養失調症；營養性疾病
trophotherapy 營養療法；飲食療法
tropical crop 熱帶作物
tropical fruit 熱帶水果
tropocollagen 原膠原〔蛋白〕
tropoelastin 原彈性蛋白；彈性蛋白原
tropomyosin 原肌凝蛋白；原肌球蛋白
troponin 肌鈣蛋白
trotters 蹄
trout 鱒魚；鮭魚
troutlet 小鮭魚；小鱒魚 [= troutling]
troutling 小鮭魚；小鱒魚 [= troutlet]
truck crop 蔬菜作物
trucker 菜農
true acidity 真酸度
true bass 七斑石斑魚
true digestibility 真消化率
true fruit 真果
true mushroom 野蘑菇
true protein 純蛋白質
true solution 真溶液
true stomach 真胃；皺胃 [= abomasum]
true yeast 真酵母
truff 鱒魚；鮭魚
truffle 塊菌；麥蕈；松露
trug 淺底闊籃
trumpet chanterelle 管形雞油菌（蕈類）
trumpet fish 長嘴魚

trumpet oyster mushroom 白黃側耳(蕈類)
truss 捆綁
tryma 核果
trypsase 胰蛋白酶
trypsin 胰蛋白酶
trypsin inhibitor 胰蛋白酶抑制劑
trypsinogen 胰蛋白酶原
tryptamine 色胺
tryptic activity 胰蛋白酶活性
tryptic enzyme 胰蛋白酶
tryptophan 色胺酸
tryptophanase 色胺酸酶
tryptophol 色醇
tsp. (teaspoonful) 一茶匙量
tub 盆；桶
tube 管
tube cleaner 清管器
tube culture 試管培養〔物〕
tube dryer 管〔道〕式乾燥器
tuber 塊莖；球根
tuber crop 塊莖作物
tuber starch 塊莖澱粉
tuberin 馬鈴薯球蛋白
tuberose absolute 晚香玉淨油；月下香淨油；夜來香淨油
tuberose concrete 晚香玉浸膏；月下香浸膏；夜來香浸膏
tuberous polypore 塊莖形多孔菌(蕈類)
tubetti lunghi 長管形通心粉
tubful 一桶〔之量〕；一盆〔之量〕
tubular atomizer 管狀噴霧器
tubular boiler 管式鍋爐；火管鍋爐
tubular casing 管狀腸衣
tubular condenser 管式冷凝器
tubular evaporator 管式蒸發器
tubular heater 管式加熱器
tubule 小管；細管
tuck 甜食；學童的食品；大吃
tuckahoe 茯苓
tuckshop 糖果店
tufted tough shank 群生金錢菇
tullibee 美洲白鮭；白鮭
tumbler 平底玻璃杯；滾筒；轉筒；轉鼓
tumbler dryer 滾筒乾燥機
tumbler mixer 滾筒混合機
tumblerful 一杯的量；一滿杯
tumbling granulation 滾筒造粒
tun 酒桶；啤酒桶；啤酒發酵桶
tuna 鮪；金槍魚 [= tunny]
tuna fish 鮪魚罐頭；金槍魚罐頭
tuna sandwich 鮪魚三明治
tung oil 桐油
tunnel dryer 隧道式乾燥機
tunnel freezer 隧道式凍結機
tunnel oven 隧道式烤爐
tunny 鮪；金槍魚 [= tuna]
tup 公綿羊；公羊
turanose 松二糖
turban squash 扁形南瓜
turbid 混濁的
turbidimeter 濁度計
turbidity 混濁〔度〕；濁度
turbidness 濁度；混濁
turbine (type) agitator 渦輪型攪拌機
turbine type mixer 渦輪型攪拌機；渦輪型混合器
turbot 大比目魚；大菱鮃
turbulent flow 紊流
tureen 有蓋大湯碗；大湯盤；大湯碗
turkey 火雞；火雞肉
turkey cock 雄火雞
Turkish coffee 土耳其咖啡
Turkish delight 覆糖軟糖；土耳其軟糖
Turkish geranium oil 土耳其香葉油
Turkish paste 覆糖軟糖；拌糖粉軟糖
turmeric 薑黃；鬱金；郁金

turmeric acid 薑黃酸
turmeric oil 薑黃油
turmeric oleoresin 薑黃油樹脂
turmeric yellow 薑黃；郁金粉；薑黃素；酸性黃 [= curcumin]
turmerol 薑黃醇
turmerone 薑黃酮
turner 煎鏟
turnip 大頭菜；蕪菁
turnip cabbage 球莖甘藍
turnip radish 園蘿蔔
turnip-rooted cabbage 球莖甘藍
turnip-rooted chervil 香菜芹
turnip-rooted parsley 根用香芹
turnover 捲酥；餡角；半圓形酥餅；營業額
turntable 轉盤
turpentine (oil) 松節油；松脂
turtle 海龜；甲魚；玳瑁
turtle dove 斑鳩；鵰鳩
turtle meat 龜肉
turtle shell 龜甲；玳瑁殼；龜殼
tusk 獠牙；長牙
tutti-frutti 水果丁冰淇淋；什錦水果冰淇淋；什錦蜜餞
TV dinner 電視餐
Tween 吐溫〔乳化劑〕；土溫〔乳化劑〕
Tween 80 吐溫80；土溫80；聚氧乙烯脫水山梨醇單油酸酯 [= polyoxyethylene sorbitan monooleate]
twig 細枝；嫩枝
twin cylinder dryer 雙轉筒乾燥機
twin-drum dryer 雙轉筒乾燥機
twin screw extruder 雙軸擠壓機；雙螺桿擠壓機
twist 麵包捲；麻花；混合飲料
twist linker 自動充填結紮機；自動扭結機
twist linker machine 自動充填結紮機；自動扭結機
twist off cap 扭開蓋；爪旋蓋
twist wrapping machine 扭轉包覆機；扭結式包裝機
twister 麻花麵包
two-grained wheat 二粒小麥
two-piece can 兩片罐
two-rowed barley 二稜大麥
two-stage defrosting 二段解凍
two-stage drying 二段乾燥
two-step freezing 二段冷凍
two toned wood tuft 庫恩菇
tyee 王大馬哈魚
tylosin 泰黴素；泰樂菌素
tyndallization 分段滅菌；間歇滅菌 [= fractional sterilization; intermittent sterilization]
type culture 標準培養〔物〕
typhoid fever 傷寒熱
tyraminase 酪胺酶
tyramine 酪胺
tyrosinase 酪胺酸酶
tyrosine 酪胺酸
tyrosinosis 酪按酸代謝〔紊亂〕症
tyrosol 酪醇；對羥苯基乙醇
tyrothricin 短桿菌素
tyrotoxin 乾酪毒素
tyvelose 泰威糖

U

ubiquinone 輔酶Q；泛醌
udder 乳房
UDP (uridine diphosphate) 尿核苷二磷酸
UDPG (uridine diphosphoglucose) 尿核苷二磷酸葡萄糖
UF (ultrafiltration) 超濾〔法〕；超濾〔作用〕
ugli 柚橘；醜橘
UHT (ultrahigh temperature) 超高溫
UHT milk 超高溫滅菌乳
UHT sterilization (ultrahigh temperature sterilization) 超高溫滅菌；超高溫殺菌
ulcer 潰瘍
ulceration 潰瘍；潰爛
ultracentrifugation 超高速離心分離
ultracentrifuge 超高速離心機
ultrafilter 超濾器
ultrafiltrate 超濾液
ultrafiltration (UF) 超濾〔法〕；超濾〔作用〕
ultrafiltration membrane 超濾膜
ultrafine granulated sugar 微晶砂糖
ultrafine grinding mill 超細研磨機
ultrafine pulverizer 超細粉碎機
ultrahigh temperature (UHT) 超高溫
ultrahigh temperature sterilization (UHT sterilization) 超高溫滅菌；超高溫殺菌
ultramarine 群青；群青色素；海藍色素
ultramicro balance 超微量天平
ultramicroanalysis 超微量分析
ultramicrobe 超微生物
ultramicrocrystal 超微結晶
ultramicron 超微〔細〕粒；超細微粒
ultramicroorganism 超微生物
ultramicroscope 超顯微鏡
ultramicrosome 超微粒體
ultramicrotome 超薄切片機
ultrapure 超純的
ultrapurification 超純化；超提純
ultrasonic 超聲波；超音波的
ultrasonic cleaner 超音波清潔器；超聲波清潔器
ultrasonic homogenizer 超音波均質機；超聲波均質機
ultrasonic sterilization 超音波滅菌〔法〕；超聲波滅菌〔法〕
ultrastructure 超微組織；超微結構
ultrathin 超薄的
ultrathin film 超薄膜
ultraviolet (UV) 紫外線
ultraviolet detector 紫外線檢測器；紫外線偵測器 [= UV detector]
ultraviolet irradiation 紫外線照射
ultraviolet light 紫外光
ultraviolet spectrophotometer 紫外線分光光度計

ultraviolet spectrophotometry 紫外線分光光度測定〔法〕
ultraviolet sterilization 紫外線滅菌
ultraviolet water sterilizer 紫外線水消毒器滅菌
umami 鮮味
umbles 動物內臟；內臟
umbrella 海蜇傘膜；菌傘
umbrella polypore 傘形多孔菌 (蕈類)
ume 梅
unadulterated 無攙雜的；未摻雜的；純正的
unadulterated food 無攙雜食品；未摻雜食品；純正食品
unbaffled fermenter 無擋板發酵槽
unbaked 未熟的；未烘烤的
unbalanced diet 偏食；不均衡膳食
unbalanced growth 不均衡生長
unbleached 未漂白的；原色的
unbleached flour 未漂白麵粉
unblended wine 未調合酒
unbound water 未結合水
uncarbonated drink 非碳酸飲料；未飽充飲料
uncaser 開箱裝置
unclean flavo(u)r 不潔風味；風味不純；雜味
uncooked 未煮熟的；生的
uncooked food 未烹煮的食物；生的食物
uncooled 未冷卻的
uncork 去塞；拔去塞子；開蓋
unctuous 油質的；油膩的；含油脂的
unctuous food 油滑食品；油膩食品
unctuousness 油膩性
uncured meat 未醃肉
γ-undecalactone γ-十一碳烷內酯；桃醛
δ-undecalactone δ-十一內酯
2,4-undecadienal 2,4-十一碳二烯醛
undecanal 十一醛
undecanone 十一酮
undecenal 十一〔碳〕烯醛
undecenol 十一烯醇
undecyl alcohol 十一醇
under processing 加工不足
underback 里脊肉
underbaked 未烤熟的；未烤透的；烘烤不足的
underbaked bread 未烤熟麵包
underboiled 未煮透的
underbolted 未篩淨的
undercloth 墊布；襯布
undercooked 未煮透的；火候不足的
undercooked food 未煮熟食物
undercooled 冷卻不足的；過冷的
undercooling 過冷卻；過冷〔現象〕
undercuring 醃漬不足
undercut 里脊肉
underdone 煮得嫩的；半生不熟的
underdone meat 未煮熟的肉
underdose 劑量不足
underdried 未乾透的
underexhaustion 脱氣不足
underexpansion 膨脹不足
underfeed 餵養不足；營養不足
underfermented 發酵不足的
underfill 充填不足；裝填不足
underfilled can 未足量罐
underfired 底部加熱的；下部燃燒的
underfired cooking 底部加熱烹調
underfired food 欠熟食物；底部加熱食物
undergrade 質量低；不合規格
underground stem 地下根莖
underheating 加熱不足
underkneading 揉合不足；捏揉不足
undermilled rice 精碾不足的米

undermixing 混合不均；混和不足
undernourishment 營養不足
undernutrition 營養不良；營養不足
underpasteurization 巴氏殺菌不足；殺菌不足
underpickling 醃漬不足；酸洗不足
underpressure 壓力不足；真空壓力
underprocessed 加熱處理不足的
underproof 酒精濃度低於標準的
underproof spirit 不足度烈酒
underrefining 精煉不足
underripe 不夠成熟的；未全熟的
underripe fruit 未熟水果
underroast 未烤透的
undersaturated solution 未飽和溶液
undersaturation 未飽和；欠飽和
undersell 廉價出售；低於市價拋售
undersize 過篩物；尺寸過小
understable 不夠穩定的
understerilization 滅菌不足；殺菌不足
understock 庫存不足；存貨不足
undersupply 供應不足
underweight 重量不足
undigested 未經消化的
undiluted 未稀釋的；未攙水的
undissolved 不溶解的
undope 不摻雜
undressed meat 未處理肉
undrinkable 不能喝的
undulated colony 波形菌落
uneatable 不可食的
unedible 不可食的
unesterified fatty acid 未酯化脂肪酸
unevaporated 未蒸發的
UNFAO (United Nations Food and Agriculture Organization) 糧農組織；聯合國糧農組織；聯合國糧食〔及〕農業組織 [= Food and Agriculture Organization; FAO]
unfermentable sugar 非發酵〔性〕糖；不可發酵的糖
unfermented 未發酵的
unfermented beverage 未發酵飲料
unfermented tea 未發酵茶
unfilled 未充填的
unfiltered 未過濾的
unfinished product 未完成產品
unflanged can 無突緣罐
unfortified milk 未強化乳
unfreeze 解凍；融化
unfrozen food 未冷凍食品
unglazed fruit 未上蠟水果
unground 未磨的
unhairing 去毛；脫毛
unhairing machine 去毛機
unhealthy 不衛生的；有害健康的；不健康的
unheated 未加熱的
unhulled 未脫殼的
unhusked 去殼的
unhydrated 未水合的
unhydrogenated 未氫化的
unhydrolyzed 不水解的
unhydrolyzed protein 未水解蛋白質
unhydrous 無水的；不含水的
unicell 單細胞
uniflow 單流的
uniform cook 均勻烹調的
uniform flow 勻流
uniformity 均勻性；均一性
uniformization 均勻化；均一化
unilayer 單層
unintentional food additive 非故意添加的食品添加劑；無意食品添加劑
uniphase 單相
unirradiated food 未經照射的食物
unit 單位；組合件；部件
unit operation 單元操作

United Nations Food and Agriculture Organization (UNFAO) 糧農組織；聯合國糧農組織；聯合國糧食〔及〕農業組織 [= Food and Agriculture Organization; FAO]
United Stated Patent (USP) 美國專利
United States Department of Agriculture (USDA) 美國農業部
United States Pharmacopoeia (USP) 美國藥典
universal buffer 廣用緩衝液；通用緩衝液
universal product code (UPC) 通用產品記碼；商品萬用條碼
universal seamer 廣用捲封機；通用封罐機
unleachable 不可滲出的
unleavened dough 未膨發麵糰
unmalted 不含麥精的；未發麥芽的
unmingled 未混合的；純的
unmixed 未攙雜的
unmodified starch 未修釋澱粉；非改性澱粉
unofficial method 非法定方法；非正式方法
unpackaged 未包裝的
unpacked 未包裝的；打開的
unpalatable food 不可口食物
unpasteurized 未殺菌的；未經巴氏殺菌的
unpasteurized beer 未殺菌啤酒；生啤〔酒〕
unpathogenic 非病原性的
unpathogenic microorganism 非病原性微生物
unpigmented 未染色的
unpitted fruit 未去核的水果；帶核水果
unpolished rice 糙米
unprocessed 未經加工的
unpurified 未淨化的
unrefined 未精製的；未精煉的
unrefined oil 未精製油
unrefined sugar 粗糖；未精製糖
unrenewable 無法回收再用的
unripe fruit 未完熟水果；未熟透水果
unsalted 無鹽的；不加鹽的；新鮮的
unsalted dried fish 無鹽魚乾；淡乾魚；素乾魚
unsaponifiable 不皂化的
unsaponifiable matter 不皂化物
unsaturated 不飽和的
unsaturated fatty acid 不飽和脂肪酸
unsaturation 不飽和現象
unsavo(u)riness 無味；味道不佳；不可口
unsavo(u)ry 不可口的；不好吃的；味道不佳的
unscented 無香味的
unscreened 未過篩的；未篩選的
unsealed 未密封的；已開的
unsealing 開封；未密封
unseasoned 未調味的
unshelled 未去殼的
unsifted 未篩過的
unsophisticated food 未摻假食品
unstability 不穩定性
unstable 不穩定的
unsteady flow 非穩〔態〕流
unstrained 未過濾的
unstriated muscle 平滑肌
unsweetened 不加糖的；未加糖的
unsweetened condensed milk 不加糖煉奶
unsweetened juice 不加糖的果汁
unvoluntary muscle 不隨意肌
unwatered 不摻水的；去除水分的；乾燥的

unwatered alcohol 不摻水酒精；無水酒精
unwholesome 不衛生的；不健康的
unwholesome food 不衛生食品；不健康食品
unwholesomeness 不衛生；不健康
unwooling 脱毛；去毛
unwrap 打開
uperization 蒸汽噴射滅菌法；高壓蒸汽噴射滅菌法；汽噴滅菌法
uperizer 蒸汽噴射滅菌器；汽噴滅菌器
upland cress 獨行菜
upland rice 旱稻
upright freezer 立式冷凍機
upright retort 立式〔高壓〕殺菌釜
upright tomato 直立性番茄；無支架番茄
upstream heat exchanger 逆流式熱交換器
uptake 攝取；吸收
urate 尿酸鹽；尿酸酯
urchin 海膽
urd (bean) 黑綠豆；雞兒豆
urea 脲；尿素
urease 脲酶；尿素酶
urethane foam 胺甲酸乙酯泡沫
uric acid 尿酸
uridine diphosphate (UDP) 尿核苷二磷酸
uridine diphosphoglucose (UDPG) 尿核苷二磷酸葡萄糖
urinary bladder 膀胱
urine 尿
urn 茶壺；咖啡壺；甕
urochrome 尿色素
uronate 糖醛酸鹽；糖醛酸酯
uronic acid 糖醛酸
urosaccharometry 尿糖測定法
U.S. Canner 製罐級
U.S. Choice 特選級
U.S. Commercial 商用級
U.S. Cutter 切塊級
U.S. Prime 極佳級
U.S. Select 可選級
U.S. Standard 合格級
U.S. Utility 可用級
USDA (United States Department of Agriculture) 美國農業部
useful insect 益蟲
useful life 有效期；有效〔使用〕壽命
ustion 烙；燒灼
uszka 烏茲卡餃子(波蘭)
utensil 器皿；家庭用具；器具
UV detector 紫外線檢測器；紫外線偵測器 [= ultraviolet detector]

V

V-type mixer V型混合機
V.O. (very old) 老白蘭地酒(12—15年酒齡)
V.S.O. (very special old) 陳年白蘭地酒(15—20年酒齡)
V.S.O.P. (very special old pale) 陳年老白蘭地酒(25—30年酒齡)
vac-ice process 真空冷凍法
vaccenic acid 十八碳烯酸；異油酸
vaccumizer 抽真空機
vacreation 真空脱臭處理；真空殺菌脱臭
vacreator 真空脱臭器；真空殺菌器
vacu-therm instant sterilizer 快速真空加熱滅菌機；真空高溫瞬間滅菌機
vacu-therm pasteurizer 真空加熱〔巴氏〕殺菌器
vacuo 真空
vacuo heating 真空加熱；減壓加熱
vacuole 空泡；液泡
vacuum 真空
vacuum baking 真空烘焙〔法〕
vacuum bottle 保溫瓶；熱水瓶
vacuum can sealing machine 真空封罐機
vacuum capping machine 真空封蓋機
vacuum cleaner 真空清潔器；吸塵機
vacuum closing machine 真空封罐機；真空封瓶機
vacuum coffee maker 真空咖啡沖泡器
vacuum column 真空分餾塔；真空蒸餾塔
vacuum concentration 真空濃縮；減壓濃縮
vacuum concentrator 真空濃縮器
vacuum cooker 真空蒸煮鍋
vacuum cooling 真空冷卻
vacuum deaerator 真空脱氧器
vacuum degassing 真空脱氧
vacuum dehydration 真空脱水；真空乾燥
vacuum desiccator 真空乾燥器
vacuum distillation 真空蒸餾
vacuum drum dryer 真空滾筒乾燥機
vacuum dryer 真空乾燥器
vacuum drying 真空乾燥
vacuum drying oven 真空乾燥爐
vacuum evaporation 真空蒸發
vacuum evaporator 真空蒸發器
vacuum extractor 真空提取器
vacuum filler 真空裝填機；真空灌裝機
vacuum filling 真空充填；真空灌裝
vacuum filtration 真空過濾
vacuum flask 保溫瓶；熱水瓶
vacuum foam drying 真空泡沫乾燥
vacuum freeze drying 真空冷凍乾燥
vacuum fumigation 真空燻煙
vacuum gauge 真空計

vacuum kneader 真空捏揉機；真空揉混機
vacuum kneading machine 真空捏揉機；真空揉混機
vacuum meat mixer 真空混肉機
vacuum oven 真空乾燥爐
vacuum pack 真空包裝
vacuum package 真空包裝
vacuum packaging 真空包裝
vacuum packaging machine 真空包裝機
vacuum pan 真空鍋；真空罐
vacuum pasteurization 真空巴氏殺菌
vacuum pasteurizer 真空巴氏殺菌器
vacuum puff drying 真空泡沫乾燥
vacuum rotary dryer 真空旋轉乾燥機
vacuum seal(ing) 真空封密
vacuum seamer 真空封罐機
vacuum seaming 真空封罐
vacuum seaming machine 真空封罐機
vacuum still 真空蒸餾器
vacuum storage 真空儲藏
vacuum tight 真空密封的；真空密閉的
vacuum tightness 真空緊密度
vacuumize 抽真空；真空包裝
valencene 巴倫西亞橘烯；朱欒倍半萜
Valencia orange 晚崙西亞橙；伏令夏橙；華蘭西橙
valenta test 瓦蘭塔〔油脂〕試驗；伐蘭塔〔油脂〕檢驗
valentine chop 情人羊排
valeraldehyde 戊醛
valerate 戊酸鹽；戊酸酯
valerian 纈草
valerian oil 纈草油
valeric acid 戊酸；纈草酸 [= pentanoic acid]
γ-valerolactone γ-戊內酯
valine 纈胺酸
valinomycin 纈胺黴素
valley grape 山谷葡萄
value 值；價值；估價；效用
valzin 甜精；甘素；對乙氧苯脲 [= dulcin(e); sucrol]
vanaspati ghee 范氏植物牛油；混合氫化油
vanilla 香草；香子蘭
vanilla bean 香草豆；香子蘭豆；香莢蘭豆
vanilla bean concrete 香莢蘭豆浸膏
vanilla bean tincture 香莢蘭豆酊
vanilla extract 香草萃；香草浸液；香莢蘭豆浸膏
vanilla flavo(u)r 香草香精；香子蘭香精
vanilla ice-cream 香草冰淇淋
vanilla oily flavo(u)r 香草油香精
vanilla sugar 香草〔砂〕糖
vanillal 乙基香草醛；亞香草基
vanillic acid 香草酸
vanillin 香草醛；香蘭素
vapidity 無滋味；乏味
vapo(u)r 蒸汽；霧氣
vapo(u)r chamber 蒸汽室
vapo(u)r chromatography 氣相層析法；氣相色譜法
vapo(u)r cooking 蒸汽蒸煮；蒸煮 [= steam cooking]
vapo(u)r degreasing 蒸汽脱脂
vapo(u)r permeability 透汽性
vapo(u)r phase 汽相
vapo(u)r pressure 汽壓；蒸汽壓
vapo(u)r proof 汽密的；不透蒸汽的
vapo(u)r recovery 蒸汽回收
vapo(u)r seal 汽封；蒸汽密封
vapo(u)r superheater 蒸汽過熱器
vapo(u)r tension 蒸汽壓；蒸汽張力
vapo(u)r tight 汽密的

vapo(u)r vacuum sealing machine 蒸汽真空封口機
vapo(u)rimeter 蒸汽壓力計；揮發度計
vapo(u)rizability 汽化性
vapo(u)rizable 可汽化的
vapo(u)rization 汽化；蒸汽化；蒸發
vapo(u)rize 蒸發；汽化
vapo(u)rizer 汽化器；蒸發器；蒸餾器
varec(h) 海藻；海藻灰
variant 變種
variation 變化；變異
variegated bolete 彩斑狀乳牛肝菌
variety 種類；品種；變種
variety meat 雜碎肉；可食用的動物內臟
variety shop 雜貨店；雜貨舖
variety store 雜貨店；雜貨舖
vascular bundle 維管束
vascular system 血管系統；維管系統
vascular tissue 維管束組織
vaseline 凡士林
vasoconstriction 血管收縮
vasoconstrictor 血管收縮劑
vasodilatation 血管舒張
vasodilator 血管舒張劑
vat 大桶；大缸；甕
vat salting 甕漬；桶醃
vault 地窖；酒窖
veal 小牛肉
veal calf 肉用小牛
veal chop 小牛排
veal cutlet 小牛肉片
veal loaf 小牛肉捲；小牛肉糕
vector 病媒；媒介
vegan 純素食者；嚴格的素食主義者
veganism 純素食主義；素食主義
vegetable 蔬菜
vegetable butter 植物脂
vegetable casein 植物性乾酪素；植物性酪蛋白
vegetable cheese 植物性芝士；植物性乾酪；豆腐
vegetable colo(u)r 植物性色素
vegetable colo(u)ring 植物性色素著色；植物性著色劑
vegetable compartment 蔬果保鮮室
vegetable dehydrator 蔬菜脱水機
vegetable diet 素食
vegetable dish 蔬菜菜餚
vegetable dyed macaroni 蔬菜通心粉
vegetable egg 茄子
vegetable fat 植物脂
vegetable fat and oil 植物〔性〕油脂
vegetable fiber 植物纖維
vegetable food 素食；植物性食品；蔬菜性食品
vegetable garden 蔬菜園
vegetable glue 植物膠
vegetable gum 植物膠
vegetable juice 蔬菜汁
vegetable marrow 西葫蘆 [= marrow squash]
vegetable mix 什錦蔬菜
vegetable oil 植物油
vegetable oyster 婆羅門參；黑皮波羅門參 [= oyster plant; salsify]
vegetable pear 佛手瓜；梨瓜；香櫞瓜 [= chayote]
vegetable pigment 植物色素
vegetable plate 蔬菜菜餚
vegetable protein 植物〔性〕蛋白質
vegetable puree 菜泥
vegetable salad 蔬菜沙拉；蔬菜沙律
vegetable server 蔬菜盤
vegetable shortening 植物性〔起〕酥油
vegetable shredder 蔬菜切絲機
vegetable soup 蔬菜湯；菜湯

vegetable spaghetti 魚翅瓜(一種西葫蘆) [= noodle squash; spaghetti squash]
vegetable sponge 絲瓜 [= towel gourd]
vegetable tallow 植物性硬脂；硬化植物脂
vegetal 植物的；植物；蔬菜
vegetarian 素食〔主義〕者
vegetarian diet 素食；素餐
vegetarian food 素食
vegetarian restaurant 素食餐館
vegetarianism 素食主義
vegetation 植物；草本；增殖體
vegetative 植物〔性〕的；有生長力的；營養〔體〕的
vegetative cell 營養細胞
vegetative hypha(e) 營養菌絲
vegetative mycelium 營養菌絲體
vegol 濃縮維生素E
veil 幕；菌膜；蕈幕
veil fat 網膜脂；網油
vein 血管；葉脈；靜脈；紋理；蝦腸
veitchberry 維特契莓
vell 仔牛皺胃
velocimeter 速度計
velum 緣膜；菌膜
velvet bean 佛州黎豆
velvet shank 金針菇
velvet spirit 醇酒；溫和的酒
vendace 白鮭魚
vended food 自動販賣食物
vender 小販；流動攤販；自動販賣機；賣主 [= vendor]
vending machine 自動販賣機；自動售貨機
vendor 小販；流動攤販；自動販賣機；賣主 [= vender]
venenation 中毒
venerupin 貝毒素
venison 野味；鹿肉
venom 毒液
venomous 有毒的；分泌毒液的
vent 通風口；通風孔；出口；排氣；泄放
venter 腹部；胃
ventilated container 通風的
ventilation 通風
ventilator 通風器；通風設備
ventilator storage 通風儲藏
ventral fin 腹鰭
ventriculus 胃；沙囊；中腸
Venus ear 鮑〔魚〕
verbena 馬鞭草屬植物
verbena oil 馬鞭草油
verdigris russule 銅綠紅菇
verdoflavin 核黃素；膽綠黃素；維生素B_2
verdure 青菜；蔬菜；綠色植物；新鮮
verification 鑒定；確認；查證；檢定
verjuice 酸果汁
vermes 蠕蟲類
vermicelli 細麵條；細粉條；麵線；粉絲
vermicide 殺蠕蟲藥
vermifuge 驅蟲藥；殺蟲藥
vermin 害蟲；寄生蟲；害獸
verminosis 蠕蟲病
verminproof 防蟲的
verminproof package 防蟲包裝
vermis 蠕蟲
vermouth 苦艾酒；味美斯酒
vert haricot 青刀豆
vertebra(e) 脊椎骨
vertebrate 脊椎動物
vertical autoclave 立式殺菌釜
vertical boiler 立式鍋爐
vertical fin 豎鰭
vertical press 立式壓榨機

vertical retort 立式〔高壓〕殺菌釜
vertical screw press 立式螺旋壓濾機
vertical-tube (type) evaporator 豎管〔式〕蒸發器
vervain 馬鞭草
very high density lipoprotein (VHDL) 極高密度脂蛋白
very low density lipoprotein (VLDL) 極低密度脂蛋白
very old (V.O.) 老白蘭地酒（12—15年酒齡）
very special old (V.S.O.) 陳年白蘭地酒（15—20年酒齡）
very special old pale (V.S.O.P.) 陳年老白蘭地酒（25—30年酒齡）
vesicle 囊；泡
vessel 器皿；容器；導管
vetch 巢菜；野豌豆
veterinarian 獸醫
veterinary 獸醫；獸醫的
vetiver oil 香根油；岩蘭草油
VHDL (very high density lipoprotein) 極高密度脂蛋白
viable cell 活細胞
viable count 活菌數
vial 小瓶
viand 食物；一件食品；佳餚
vibrating conveyor 振動輸送機
vibrating feeder 振動進料機
vibrating screen 振動篩
vibrating sieve 振動篩
vibrating strainer 振動〔粗〕濾器
vibration 振動；搖動
vibration conveyor 振動輸送機
vibrator 振動器
vibratory screening 振動篩分；振動篩選
vibrio 弧菌
Vibrio cholera 霍亂〔弧〕菌
Vibrio parahaemolyticus 腸炎弧菌
vibriocin 弧菌毒素
vichyssoise 奶油濃湯
vicilin 碗豆球蛋白
vicine 蠶豆嘧啶葡萄糖苷
victual 供應儲備食物；食物；糧食
victual(l)er 食物供應者；餐館主；軍糧船；糧食補給船
victual(l)ing 糧食供應；糧食儲備
victual(l)ing house 飲食店；餐館
Vienna bread 維也納〔式〕麵包
Vienna roll 維也納〔式〕小麵包
Vienna sausage 維也納香腸 [= Wiener (sausage); wienerwurst]
Vigna 豇豆屬
vignin 豇豆球蛋白
villi 絨毛〔= villus 之複數〕
villus 絨毛
vimba 文鯿
vin 葡萄酒；酒
vin blanc 白葡萄酒
vin rouge 紅葡萄酒
vinaceous 葡萄酒的
vinage 加酒精的葡萄酒
vinaigrette 香料嗅瓶
vinaigrette sauce 酸辣醬油；調味醋醬；油醋汁
vinaigrous 酸味的
vinasse 酒槽
vine 藤；蔓；葡萄樹；葡萄葉
vine crop 藤本作物；蔓生作物
vine culture 葡萄栽培
vinegar 醋
vinegar bacteria 醋酸細菌
vinegar cure 醋漬
vinegar eel 醋蛆；醋線蟲
vinegar generator 釀醋器
vinegar mother 醋母
vinegar pickle 醋漬液；醋泡菜

vinegar pickled product 醋漬食品
vinegar pickling 醋漬
vinegary 有醋味的；醋的
vineland 葡萄園
viner 除蔓機；去蔓機
vinery 葡萄園；葡萄溫室；育葡萄暖房
vineyard 葡萄園
vinho 葡萄酒
vinic 葡萄酒的；酒的
viniculture 葡萄栽培 [= viticulture]
vinifera 釀酒用葡萄
vinification 葡萄酒釀製〔法〕
vinificator 酒精凝集裝置（釀酒用）
vino 葡萄酒；酒
vinometer 葡萄酒酒精比重計
vinosity 酒質；酒色；酒味
vinous 葡萄酒的
vinous fermentation 葡萄酒發酵
vintage 上等葡萄酒；佳釀〔葡萄酒〕；釀造年份；葡萄酒的釀造〔季節〕；葡萄收成量；葡萄收穫期
vintage champagne 葡萄香檳酒
vintage port 波特紅葡萄酒；博爾德葡萄酒
vintage year 葡萄酒的釀造年；佳釀酒的釀成年
vintner 葡萄酒商；酒商
vintry 酒窖
vinum 酒劑；藥酒
viny 葡萄樹的
vinyl alcohol 乙烯醇
vinyl chloride 氯乙烯
vinyl plastic 聚乙烯塑膠；乙烯基塑料
vinylidene chloride 氯化亞乙烯；偏二氯乙烯
violanin 堇菜苷
violaxanthin 堇菜黃質
violet 紫羅蘭
violet leaves concrete 紫羅蘭葉浸膏
violet red bile agar (VRBA) 紫紅膽汁洋菜培養基
violet web cap 堇紫絲膜菌（蕈類）
viomycin 紫霉素
viosterol 鈣化固醇；維生素D_2
viral infection 病毒感染
virgin comb 處女蜂巢
virgin honey 處女蜂巢取出的蜂蜜；自泌蜂蜜
virgin oil 初榨油；直餾油
virgin olive oil 輕榨優質橄欖油
Virginia ham 維吉尼亞火腿；弗吉尼亞火腿；維珍尼亞火腿
viricidal 殺病毒的
viricide 殺病毒劑 [= virucide]
virology 病毒學；濾過性微生物學
virosis 病毒性疾病
virucidal 殺病毒的
virucide 殺病毒劑 [= viricide]
virulence 毒性；毒害；毒力
virus 濾過性病毒；病毒
viscera 內臟；禽畜內臟
visceral cavity 腹腔；胴腔
visceral fat 內臟脂肪
visceral meal 內臟粉；內臟器官粉
visceral organ 內臟器官；臟器
viscerate 取出內臟；開膛〔去內臟〕
viscidity 黏性；黏質
viscoamylogram 連續黏度圖；黏度測定儀
viscoelasticity 黏彈性
viscogel 黏稠凝膠；黏性凝膠
viscogen 黏稠元；黏精
viscograph 黏度儀；黏度曲線儀
viscolization 均化作用；均質化
viscolizer 均化器
viscoloid 黏性膠體
viscometer 黏度計 [= viscosimeter]
viscometry 黏度測定法 [= viscosimetry]

viscose 黏膠〔液〕；鹼溶纖維素；纖維膠
viscose casing 黏膠腸衣；黏膠纖維腸衣；纖維素黏膠腸衣
viscosimeter 黏度計 [= viscometer]
viscosimetry 黏度測定法 [= viscometry]
viscosity 黏度
viscous 黏滯的；黏性的
viscous flow 黏滯流動；黏流
viscous liquid 黏滯液體
viscous product 黏性產品
viscousness 黏滯；黏性
viscus 內臟
vision package 透明包裝
visual estimation 肉眼估計；目視估測；目測
visual examination 肉眼檢驗；目視檢驗；目測檢驗
visual inspection 肉眼檢驗；目視檢驗
vitagen 維生〔素〕食物
vital staining 活體染色；活性染色
vitamer 同功維生素；擬維生素；類維生素
vitamin 維生素；維他命
vitamin A 維生素A；視黃醇
vitamin A alcohol 維生素A醇；視黃醇
vitamin A aldehyde 視黃醛
vitamin A_1 維生素A_1
vitamin A_2 維生素A_2；脱氫視黃醇
vitamin B complex 複合維生素B [= B complex]
vitamin B group 維生素B群
vitamin B_1 維生素B_1；噻胺；硫胺；硫胺素；抗神經炎素 [= aneurin; thiamine]
vitamin B_{12} 維生素B_{12}；氰鈷胺素；氰鈷維生素 [= cyanocobalamin]
vitamin B_{12a} 維生素B_{12a}；羥鈷胺
vitamin B_{13} 維生素B_{13}；乳清酸
vitamin B_{15} 維生素B_{15}
vitamin B_2 維生素B_2；核黃素 [= riboflavin(e)]
vitamin B_3 維生素B_3；泛酸 [= pantothenic acid]
vitamin B_6 維生素B_6；吡哆醇 [= pyridoxine]
vitamin Bc 葉酸；維生素Bc [= folic acid]
vitamin Bt 維生素Bt；肉〔毒〕鹼
vitamin C 抗壞血酸；維生素C [= ascorbic acid]
vitamin complex 複合維生素
vitamin D 維生素D
vitamin D_2 維生素D_2；麥角鈣化〔固〕醇；鈣化固醇；鈣化醇；骨化醇 [= calciferol; ergocalciferol; viosterol]
vitamin D_3 維生素D_3；膽骨化醇；膽鈣化固醇 [= cholecalciferol]
vitamin D_4 維生素D_4
vitamin D_5 維生素D_5
vitamin deficiency 維生素缺乏症
vitamin E 維生素E；生育酚
vitamin-enriched food 維生素強化食品
vitamin F 維生素F
vitamin fortification 維生素強化；維生素增補
vitamin fortified food 維生素強化食品
vitamin G 維生素G
vitamin H 維生素H；生物素
vitamin J 維生素J
vitamin K 維生素K；凝血維生素
vitamin K_1 維生素K_1；植物甲萘醌
vitamin K_2 維生素K_2
vitamin K_3 維生素K_3；亞硫酸氫鈉甲萘醌
vitamin K_4 維生素K_4
vitamin K_5 維生素K_5
vitamin K_6 維生素K_6

vitamin K_7 維生素K_7
vitamin L 維生素L
vitamin M 維生素M；葉酸
vitamin P 維生素P
vitamin PP 維生素PP；菸鹼酸；煙酸；尼克酸 [= niacin; nicotinic acid]
vitamin precursor 維生素原；維生素先質；維生素前驅物
vitamin R 維生素R
vitamin S 維生素S
vitamin supplementation 維生素增補
vitamin T 維生素T；圓酵母素
vitamin U 維生素U
vitamin V 維生素V
vitamin X 維生素X
vitamine 維他命；維生素
vitaminization 維生素強化；維生素添加；維生素化作用
vitaminized foodstuff 維生素強化食品；添加維生素食品
vitaminized juice 維生素強化果汁
vitaminized milk 維生素強化乳
vitaminology 維生素學
vitaminous 維生素的；維生素強化的
vitellase 卵黃〔凝固〕酶
vitellin 卵黃磷蛋白 [= ovovitellin]
vitellolutein 卵黃葉黃素
vitellomucoid 卵黃類黏蛋白
vitellus 卵黃
vitex cannabifolia leaf oil 牡荊葉油
viticulture 葡萄栽培 [= viniculture]
vitis 葡萄
vitreous malt 硬粒麥芽；硬脆的麥芽
vitriol 硫酸鹽；礬
vivers 食物；食品；糧食
VLDL (very low density lipoprotein) 極低密度脂蛋白
vodka 伏特加酒
volatile 揮發物；揮發〔性〕的
volatile acid 揮發酸
volatile amine 揮發胺
volatile basic nitrogen (VBN) 揮發性氨基態氮；揮發性鹼氮
volatile component 揮發〔性〕組分
volatile fatty acid 揮發〔性〕脂肪酸
volatile flavo(u)r 揮發性芳香物質
volatile oil 揮發油
volatile solvent 揮發性溶劑
volatility 揮發性；揮發度
volatilization 揮發〔作用〕；蒸發〔作用〕
volatilize 揮發
volauvent 法式肉蛋餡餅；魚肉香菇餡餅
Volgard titration 伏哈德滴定法；伏爾加德滴定法
volume 容量；體積；容積
volumeter 容量計；容積計
volumetric analysis 容量〔滴定〕分析；滴定分析
volumetric cylinder 量筒
volumetric flask 量瓶
volumetric pipet(te) 容量吸管
volumetrical 容量分析的
volumetry 容量測定法；容量分析法
voluntary muscle 隨意肌
volva 菌托；蕈托
vomiting 嘔吐
voracious 狼吞虎咽的；貪食的
voracity 貪食
vortex mixer 旋渦混合器
vortex stirrer 旋渦攪拌器
votator 刮面式熱交換器；螺旋式熱交換器
vulcanization 硫化；硬化
vulcanize 硫化；硬化

W

w/o emulsion (water-in-oil emulsion) 油中水型乳〔濁〕液；油包水型乳〔濁〕液
wad material 軟填料；填料；墊料
wafer 威化餅；薄鬆餅；薄脆餅；聖餅
wafer roll 薄餅捲
waferer 壓片機
waffle 夾心脆餅；華夫餅乾
waffle iron 脆餅烤模；華夫餅乾烤模
waffle roll 蛋捲
waiter 侍者；服務員；侍應
waiting lairage 待宰欄
wakame 裙帶菜
Waldhof fermenter 伍德夫發酵槽；瓦爾德霍夫發酵槽
walk-in freezer 大型冷凍庫
walnut 核桃；胡桃；合桃 [= juglans]
walnut cake 核桃蛋糕；胡桃蛋糕；合桃蛋糕
walnut cream 胡桃〔仁〕糊；核桃〔仁〕糊；合桃〔仁〕糊
walnut oil 胡桃油；核桃油；合桃油
wampee 黃皮
warden 冬梨
ware 器皿
warehouse 倉庫
wareroom 商品展銷室；商品儲藏室
Waring blender 瓦林混碎機
warm-blooded animal 溫血動物
warm carcass 溫〔屠〕體
warm filling 熱裝填
warm meat 溫〔體〕肉
warm salting 溫鹽漬
warmed-over 重新燒熱的；回熱的
warmer 加熱器；保溫器
warming plate 加熱板
warren 養兔場；魚塘
wasabi 芥茉；日本芥茉；山葵；山嵛菜 [= Japanese horseradish]
wash 洗；清洗；洗滌；餿水；淡而無味的湯
wash house 洗滌間；洗滌處 [= washery]
washer 洗滌劑；洗滌機；洗滌者；墊圈
washery 洗滌間；洗滌處 [= wash house]
washing 洗滌
washing agent 洗滌劑
washing bottle 洗滌瓶
washing liquid 洗滌液
washing machine 洗衣機；洗滌機
washing powder 洗滌粉
washing soda 晶鹼；洗滌鹼
washing tank 洗滌槽
washings 洗滌〔廢〕液
wasp 黃蜂；胡蜂；細腰蜂
wassail 飲宴；酒宴
wassail bowl 大酒杯
wassailer 祝酒者；豪飲者

wastage 浪費；損耗；損耗量
waste 浪費；損耗；廢料；廢物；殘料
waste disposal 廢物處理
waste heat 廢熱；餘熱
waste lagoon 廢水池；廢水氧化池
waste molasses 廢糖蜜
waste pipe 廢水管
waste product 廢產物；廢料；無用副產品
waste steam 廢汽
waste treatment 廢物處理
waste water 廢水；污水
waste water treatment 廢水處理；污水處理
water absorber 吸水劑
water absorption 吸水〔率〕
water activity (Aw) 水活性；水分活性
water bag 水袋
water balance 水分平衡
water bamboo 茭白；茭白筍；菰〔米〕
water bath 水浴
water binding capacity 保水力；持水力
water bird 水禽；水鳥
water biscuit 薄硬餅乾；無輔料餅乾
water-borne infection 飲水感染
water buffalo 水牛
water caltrop 菱〔角〕
water celery 水芹
water chestnut 荸薺；馬蹄；馬薯
water cistern 蓄水池；水塔
water condenser 水冷凝器
water content 水含量；含水量
water convolvulus 蕹菜；空心菜；通菜；通心菜 [= water spinach]
water cooling 水冷卻
water cress 水田芥；西洋菜；豆瓣菜
water culture 水耕
water curtain 水簾
water desorption isotherm 等溫脱濕曲線
water drinker 嗜飲水者；不喝酒者
water dropwort 水芹
water equivalent 水當量
water extract 水提取物；水萃〔取物〕
water extraction 水萃取；水提取
water feeder 供水器
water filter 濾水器
water fowl 水禽
water gauge 水表
water glaze 冰衣
water gruel 粥
water hardness 水硬度
water holding capacity 保水力；持水力
water ice 水冰；雪泥；果汁冰糕
water immersion 水浸〔泡〕
water-in-oil emulsion (w/o emulsion) 油中水型乳〔濁〕液；油包水型乳〔濁〕液
water insoluble 水不溶性
water jug 水壺；小口大壺
water lily 睡蓮；荷花
water metabolism 水〔分〕代謝
water meter 水表
water mill 水磨〔機〕
water phase 水相
water pipe 水管
water plant 水生植物
water pollution 水〔質〕污染
water purification 淨水處理
water purifier 水淨化器；濾水器
water quality 水質
water regain 復水
water resistance 防水性；抗水性；耐水性
water retting 水浸解
water rice 茭白；茭筍；菰
water seal(ing) 水封

water shield 蓴菜；莼菜；水葵
water soaking 水浸〔漬〕
water softener 軟水劑；水軟化劑；水軟化裝置
water solubility 水溶性
water-soluble 水溶〔性〕的
water-soluble solid 水溶性固形物
water-soluble vitamin 水溶性維生素；水溶性維他命
water sorption isotherm 等溫吸濕曲線；水分吸附等溫線
water source 水源
water spinach 蕹菜；空心菜；通菜；通心菜 [= water convolvulus]
water standard 水質標準
water strainer 濾水器
water supply 給水；供水；自來水
water tank 水槽；水箱
water tap 水龍頭
water vapo(u)r 水蒸氣；水汽
water works 供水系統；自來水廠；給水裝置
wateriness 水樣；多水狀；潮濕
watering 摻水；灌溉；灑水；濕潤
waterleaf 水葉；水迹葉
waterless 無水的
waterless cooker 無力密封鍋
watermelon 西瓜
watermelon nightshade 西瓜葉茄
waterproof 防水的；不透水的；耐水的
watertight 不透水的；不漏水的；水密的
watertightness 水密性；不透水性
watery 水份過多的；水汪汪的；含水的；似水的
watery pork 水樣豬肉
wattle 紅肉垂；魚觸鬚；銀栲皮樹；荊樹皮
wattle gum 銀栲皮樹膠
wax 蠟
wax apple 蓮霧
wax bean 四季豆；敏豆；黃刀豆
wax jambo 蓮霧
wax myrtle 楊梅 [= waxberry]
wax paper 蠟紙；油紙 [= waxed paper]
wax-sealed 蠟封
waxberry 楊梅 [= wax myrtle]
waxdip 上蠟；浸蠟
waxed paper 蠟紙；油紙 [= wax paper]
waxgourd 冬瓜 [= winter gourd; white gourd]
waxing 塗蠟；上蠟
waxy 蠟狀的；蠟質的
waxy corn 糯玉米
waxy oil 含蠟油
waxy rice 糯米 [= glutinous rice; sticky rice; sweet rice]
waxy rice flour 糯米粉
waxy starch 糯性〔穀物〕澱粉
wayaka yam bean 豆薯；涼薯
weak acid 弱酸
weak base 弱鹼
weak flour 筋力弱的麵粉
wean 斷奶；斷乳；離乳
weaner 斷奶幼畜；斷乳幼畜
weaner calf 斷奶小牛；離乳小牛
weanling 斷乳幼畜；剛斷乳嬰孩
weanling food 斷乳食品；離乳食品
wearproof 防磨損的；耐損
weasand 食道；食管；喉嚨
weathervane scallop 風信海扇〔貝〕
weatings 麩粉；粗麵粉；細麩
weaver 織布魚
weavers broom 鷹爪豆
webfat 網油
wedding cake 結婚蛋糕 [= bride cake]
wedding feast 喜筵
weed 雜草；野草；除草

weed control 除雜草；雜草防治
weed control chemicals 除草劑；殺草劑；除莠劑 [= weed killer]
weed faint 雜草臭
weed killer 除草劑；殺草劑；除莠劑 [= weed control chemicals]
weep 滲出；滲出液
weeping widow 絲絨狀垂齒菌（蕈類）
weever 鱸魚
weevil 象鼻蟲；穀蠹
weigh(ing) 稱重
weighing bottle 稱量瓶；稱重瓶
weighing paper 稱量紙
weight 重量；砝碼
weight bottle 稱量瓶；稱重瓶
weight gain 增重
weight loss 失重
weight loss diet 減肥餐
weisswurst 白香腸（德國）
welded seam can 電焊罐；熔接罐
well-done steak 全熟牛排
well-fed 育肥的；營養充足的
well-fed stock 育肥牲畜
well water 井水
Welsh onion 葱；青葱；大葱 [= spring onion; cibol; ciboul]
Welsh rabbit 威爾斯乾酪麵包；塗上溶化乾酪〔及啤酒〕的烤麵包；乾酪吐司 [= Welsh rarebit]
Welsh rarebit 威爾斯乾酪麵包；塗上溶化乾酪〔及啤酒〕的烤麵包；乾酪吐司 [= Welsh rabbit]
west Indian cherry 西印度櫻桃
west Indian pumpkin 西印度群島南瓜
Western dish 西餐；西式菜餚
wet 濕的；浸濕；酒
wet and dry bulb thermometer 乾濕球溫度計
wet ashing 濕灰化
wet-bulb thermometer 濕球溫度計
wet cleaning 濕式清潔；水洗
wet cooling 濕式冷卻〔法〕
wet crushing 濕碎〔法〕
wet curing 濕醃〔法〕
wet dehulling 濕法去皮
wet gluten 濕麵筋
wet goods 酒；液體物質；濕貨
wet grinding 濕式研磨〔法〕
wet heat 濕式加熱；濕熱
wet heat sterilization 濕熱滅菌〔法〕
wet milling 濕磨〔法〕
wet nurse 奶媽；乳母
wet pack 濕包裝
wet pasteurization 濕熱殺菌〔法〕
wet pressing 濕式壓榨
wet process(ing) 濕式處理；濕式加工
wet rendering 濕式熬油；濕式煉油
wet salt 濕鹽
wet salting 鹽水醃漬；濕醃
wet screening 濕式篩分〔法〕
wet steam 濕蒸汽
wet steep 水浸漬
wet vapo(u)r 濕蒸汽
wet weight 濕重；鮮重
wether 閹公羊；去勢公羊
wetproof 防濕的；防潮的
wetproof film 防濕膜；防潮膜
wettability 濕潤性；濕潤度
wetting agent 濕潤劑
whale 鯨魚
whale calf 小鯨；幼鯨
whale factory ship 捕鯨加工船
whale fishery 捕鯨〔業〕
whale liver oil 鯨肝油
whale shark 鯨鯊
whale tallow 鯨脂
whalemeat 鯨肉
whaler 捕鯨人；捕鯨船

whaling 捕鯨〔業〕
wheat 小麥
wheat ball stone 麥飯石
wheat belt 小麥產區帶
wheat berry 去麩小麥粒
wheat bran 小麥麩皮；麥麩
wheat bread 小麥麵包；白麵包
wheat cake 麵粉烙餅；麵粉煎餅
wheat flake 小麥片
wheat flour 麵粉；小麥麵粉
wheat germ 麥芽；小麥胚芽
wheat-germ bread 胚芽麵包；麥胚麵包
wheat-germ oil 小麥胚芽油；麥胚油
wheat gliadin 小麥醇溶蛋白
wheat gluten 麵筋；小麥麵筋
wheat grits 碎小麥粒
wheat groats 小麥片；小麥粒
wheat malt 小麥芽
wheat malt flour 小麥芽粉
wheat meal 去麩小麥粉；小麥粉
wheat starch 小麥澱粉
wheat straw 小麥稈
wheatear 小麥穗
wheaten 小麥〔製〕的
wheaten bread 小麥麵包
wheel mill 輪碾機
whelk 峨螺；海螺
whet 開胃物；促進食慾
whetstone 磨刀石；油石
whey 乳清
whey beverage 乳清飲料
whey butter 乳清奶油；乳清乳酪
whey cheese 乳清乾酪
whey concentrate 乳清濃縮物
whey drink 乳清飲料
whey powder 乳清粉
whey protein 乳清蛋白質
whey rennet 乳清凝乳酶
wheyey 乳清的；含乳清的
wheying off 乳清分離；分出乳清
whip 攪起泡沫；攪打器；攪泡器
whipped cream 泡沫乳油；攢奶油
whipper 攪打器；攪泡器
whipping 攪打
whipping ability 攪打起泡能力；起泡能力
whipping agent 發泡劑
whipping aid 攪打助泡劑；助泡劑
whipping machine 攪打器；打擦機
whipping power 攪打起泡力
whipping property 攪打起泡性；打擦度
whipping quality 起泡性；發泡性
whirlpool tank 旋渦槽；迴旋槽
whisk 攪打器；攪拌器
whiskey 威士忌〔酒〕 [= whisky]
whisky 威士忌〔酒〕 [= whiskey]
whisky soda 蘇打汁威士忌〔酒〕
whisky sour 檸檬汁威士忌〔酒〕
whisky straight 不加水威士忌〔酒〕；純威士忌〔酒〕
whisky tumbler 威士忌酒的玻璃杯
white agaric 落葉松蕈
white asparagus 白蘆筍
white bean 白豆
white belly rice 腹白米
white blood cell 白血球 [= leukocyte]
white bread 小麥麵包；白麵包
white cabbage 白球甘藍；高麗菜；荷蘭珍甘藍
white cake 白蛋糕
white chocolate 白巧克力
white core rice 白心米
white corn 白玉米
white cowpea 白豇豆
white croaker 白姑魚
white currant 白醋栗
white dextrin 白糊精

white Dutch runner bean 花豆；紅花〔菜〕豆
white egg bird's nest 白蛋巢菌(蕈類)
white flowered gourd 扁蒲
white gourd 冬瓜 [= winter gourd; waxgourd]
white grouse 白松雞
white kidney bean 白菜豆
white lady 白夫人雞尾酒
white maize 白玉米
white matsutake 具大環口蘑(蕈類)
white meat 白肉；白色肉
white molasses 白糖蜜
white mugwort 角菜
white mulberry 白桑果；白桑葚
white muscle 白色肌
white mustard 白芥菜；白芥
white oak 白橡木
white oil 白油(食品機械專用潤滑油)
white pepper 白胡椒
white perch 美洲狼鱸
white pomfret 白鯧；銀鯧；灰鯧
white potato 馬鈴薯；土豆
white potherb 野苣
white radish 白蘿蔔
white rice 白米
white rot 葡萄白腐病；白腐病
white roux 白色油炒麵粉〔糊〕
white sapota 白人心果
white sauce 白醬料；白汁
white sesame (seed) 白芝麻
white shellac 白蟲膠；漂白紫膠
white stonecrop 白景天
white sturgeon 美洲鱘
white sugar 白糖
white truffle 白松露；大塊菌(蕈類)
white turnip 白蘿蔔
white walnut 油胡桃；灰胡桃
white wheat 白麥；白小麥
white wine 白葡萄酒
white yolk 淺色卵黃；淺色蛋黃
whitebait 銀魚；幼鯡魚
whitefish 白魚
whiteheart cherry 甜櫻桃
whitener 漂白劑；白化劑
whitethorn 山楂
whiting 小鱈魚
whizzer 離心式乾燥機
WHO (World Health Organization) 世界衛生組織
whole 全部；整個；健康
whole bread 全麥麵包
whole carcass 全屠體；全胴體
whole chicken 全雞；整雞
whole egg 全蛋
whole egg powder 全蛋粉
whole grain 整穀粒；整粒
whole grain corn 整粒玉米
whole grain rice 整粒米
whole kernel (style) corn 整粒玉米
whole meal 全麥麵粉
whole milk 全乳；全脂乳
whole milk cheese 全乳乾酪；全脂乳乾酪
whole milk powder 全脂乳粉；全脂奶粉
whole turkey 整隻火雞
whole wheat 全麥的
whole wheat bread 全麥麵包
whole wheat flour 全麥麵粉
wholefoods 全營養食物
wholemeal 全麥的；未去麩的；粗麵粉製的
wholemeal bread 全麥麵包
wholemeal flour 全麥麵粉
wholesome 有益的；有益健康的；衛生的

wholesomeness 有益；衛生安全；有益健康
whortleberry 歐洲越橘；烏嘴莓
Wiener (sausage) 維也納香腸 [= Vienna sausage; wienerwurst]
wienerwurst 維也納香腸 [= Vienna sausage; Wiener (sausage)]
wig 香芹菜小麵包；香辛小麵包
wild boar 野公豬；野豬
wild cabbage 甘藍
wild celery 苦草
wild duck 野鴨
wild fermentation 自然發酵
wild fowl 野禽
wild goose 雁
wild grass 野草
wild leek 野葱
wild marjoram 牛至草；披薩草 [= oregano; origanum]
wild oat 野燕麥
wild radish 野蘿蔔
wild rice 野米；野生稻米；茭白；菰
wild succory 野生菊苣
wild type 野生型；原株
wild yeast 野生酵母；雜酵母
wildlife 野生動植物
willow herb 柳葉菜；千屈菜；柳蘭
wilt 枯萎；凋萎
wind cave storage 風穴儲藏
wind dried 風乾的
wind egg 未受精蛋
wind mill 風車磨坊
wind sifter 風力分選機
windfall 風落果
window bag 視窗袋；開窗袋
window carton 視窗紙盒；開窗孔紙盒
wine 酒；葡萄酒；果酒
wine and dine 飲宴；又吃又喝
wine bag 酒仙；豪飲者；皮酒囊
wine barrel 葡萄酒桶
wine body 酒體
wine bottle 酒瓶
wine bowl 大酒杯；酒碗；酒仙
wine cellar 酒窖
wine conner 驗酒員
wine cooler 冰酒器；香檳桶；淡酒飲料
wine cork 酒瓶軟木塞
wine glass 葡萄酒杯；酒杯
wine gum 含酒口香糖；酒膠糖
wine lees 酒渣；酒糟；酒泥
wine making 釀酒；葡萄酒釀造
wine marc 葡萄酒榨渣
wine plam 孔雀椰子；柔美棕
wine press 葡萄壓榨機；葡萄榨汁裝置
wine punch 葡萄酒飲料
wine raffiapalm 酒椰
wine skin 皮製酒袋
wine starter 葡萄酒菌元；葡萄酒酒母
wine stone 酒石
wine table 馬蹄形餐桌
wine taster 品酒師；品酒用的酒杯
wine tasting 品酒；評酒
wine vault 葡萄酒窖
wine vinegar 葡萄酒醋
wine yeast 葡萄酒酵母
wine yeast oil 葡萄渣油
wineberry 白樹莓；白里葉莓
winebibber 酒仙；酒徒；酗酒者
winebibbing 豪飲；酗酒；喝酒
winegrower 葡萄栽培兼釀酒者
winegrowing 葡萄栽培兼釀酒業
winery 釀酒廠；葡萄酒〔釀造〕廠
wing 翅膀
wing breast 帶翅胸肉
wing flat 中翅；翅板 [= middle section; mid-joint]
wing tip 翅尖

winged bean 翼豆；四棱豆
winkle 濱螺
winnow 吹掉穀物的糠；風選
winnower 風選機；揚穀器
winnowing machine 風選機
wino 酒鬼；葡萄酒愛好者
winter bamboo shoot 冬筍
winter banana 白香蕉
winter cress 山芥
winter crop 冬作；越冬作物；秋播作物
winter gourd 冬瓜 [= waxgourd; white gourd]
winter grape 北美秋葡萄
winter hardiness 耐冬寒性
winter melon 冬甜瓜
winter oil 冬化油；耐凍油
winter radish 冬季蘿蔔
winter savory 冬季風輪菜
winter squash 冬季西葫蘆；冬季南瓜；筍瓜
winter wheat 冬〔小〕麥
wintergreen 鹿蹄草
wintergreen oil 鹿蹄草油；冬青油；冬綠油 [= methyl salicylate]
winterization 冷濾；冬化〔處理〕
winterized oil 冬化處理的油
winy 有〔葡萄〕酒味的
wire hood 香檳酒瓶的金屬絲網罩
wire whipper 鋼絲攪打器；金屬絲攪打器
wish-wash 淡而無味的飲料；淡酒
wishy-washiness 淡而無味
wishy-washy 淡而無味的；稀的
wither 凋謝；枯萎
wok 炒鍋；鑊
wok-searing 煎
wolf down 狼吞虎咽地吃
wolf herring 寶刀魚
wolfberry (fruit) 杞子
wolffish 狼魚
wonton 雲吞；餛飩
wood alcohol 甲醇；木醇；木精 [= wood spirit; methanol; methyl alcohol]
wood blewit 紫丁香蘑 (蕈類)
wood ear 木耳
wood garlic mummy cap 蒜味小皮傘 (蕈類)
wood grouse 雄松雞
wood mushroom 白林地蘑菇
wood nut 大栗
wood oil 木油；桐油
wood sorrel 酢漿草
wood spirit 甲醇；木醇；木精 [= wood alcohol; methanol; methyl alcohol]
wood sugar 木糖 [= xylose]
wood vinegar 木醋液；焦木酸；人工醋
woodcock 山鷸；鳥鷸；丘鷸
wooden ear 木耳
wooden mushroom 冬菇
wooden ware 木器
woodruff 車葉草；香豬殃殃
woody 木本的；木質的
woody flavo(u)r 木味；木頭風味
wool 羊毛
woolly milk cap 疝疼乳菇
worcester(shire) sauce 喼汁；烏斯特〔調味〕醬；澳斯特沙司；烏斯達醬
worcesterberry 烏斯特醋栗
workhorse 役馬
working cattle 役牛
working table 工作臺
workshop 工場；車間；工作坊
World Health Organization (WHO) 世界衛生組織

worm 蚯蚓；蟲；蠕蟲；螺紋
worm conveyor 螺旋〔式〕輸送機
worm pipe 蝸形管；螺紋管
worm press 螺旋壓榨機
wormhole 蛀孔；蟲孔
wormwood 洋艾草
wormwood 苦艾；洋艾〔草〕
wormwood oil 洋艾草油；苦艾油
wormy 蟲蛀的；生蟲的；有蟲的
wormy fruit 蟲蛀果
wort 麥芽汁
wort sludge 麥芽汁濾渣；麥芽汁醪糟
wound infection 傷口感染
wrap 包裝；包裝材料
wraparound 裹包的；包裹的
wraparound case 裹包式紙箱；包裹式紙箱
wraparound casing 包裹式裝箱法
wrappage 包裝材料
wrapper 封套；包裝紙；包裝材料
wrapping 包裝材料；包裝紙
wrapping machine 包覆機
wrapping paper 包裝紙；外包裝紙
wreath 圈狀麵包；花圈
wring 絞；扭；榨；擠
wringer 擠壓機；榨汁機
wrinkled pea 皺皮豌豆；皺皮種青豌豆
wurld wheat 鹼脫皮麥粒
wurst 香腸 [= sausage]

X

X.O. (extra old) 特陳年白蘭地酒
xanthan gum 三仙膠；黃原膠
xanthate 黃原酸鹽；黃原酸酯
xanthematin 血黃素
xanthemia 黃色素血症；胡蘿蔔素血症
xanthine 黃嘌呤
xanthine oxidase 黃嘌呤氧化酶
xanthomycin 鏈黃黴素；黃黴素
xanthophane 視黃素；視網膜黃素
xanthophyll 葉黃素；胡蘿蔔醇
xanthoprotein 黃蛋白〔質〕
xanthopsin 視黃質
xanthopterin 黃喋呤
xanthosine 黃核苷；黃苷
xanthoxylum 秦椒；花椒
Xenical 賽尼可
xerogel 乾凝膠
xerophagia 吃乾燥食物；乾食〔法〕
xerophile 適旱生物；嗜旱生物；旱生植物
xerophilic 適旱的；嗜旱的
xerophilic bacteria 適旱細菌；嗜旱細菌
xerophilous plant 旱生植物
xerophyte 適旱生物；旱生植物
xerospore 乾孢子
xylan 聚木糖；木聚糖
xylanase 聚木糖酶；木聚糖酶
xylem 木質部
xylene 二甲苯
xylitan monostearate 單硬脂酸木糖醇酐酯；木糖醇酐單硬脂酸酯
xylitol 木糖醇
xyloascorbic acid 木糖〔型〕抗壞血酸
xylocarp 硬木質果
xyloketose 木酮糖 [= xylulose]
xylol 二甲苯
xylonic acid 木糖酸
xylophyta 木本植物
xylose 木糖；戊醛糖 [= wood sugar]
xylose isomerase 木糖異構酶
xyloside 木糖苷
xylosone 木糖醛酮
xylulokinase 木酮糖激酶
xylulose 木酮糖 [= xyloketose]

Y

yaito bonito 鮪
yak 犛牛；牦牛
yam 山藥；山芋；薯芋；甘薯；薯蕷
yam bean 豆薯；沙葛；涼薯
yangmei 楊梅
yardlong bean 豇豆；豆角；長豇豆；菜豆；裙帶豆
yarrow 西洋蓍草；千葉蓍
yautia 箭葉黃體芋
yearling 小牲畜（一至二歲大）；一歲牲畜
yeast 酵母〔菌〕
yeast autolysate 酵母自溶物
yeast cake 酵母餅；酵母〔壓〕塊；壓榨酵母〔餅〕
yeast cream 酵母乳；酵母醪
yeast culture 酵母培養〔物〕
yeast extract 酵母萃；酵母浸膏；酵母抽提物
yeast extract medium 酵母萃培養基
yeast fermentation 酵母發酵
yeast food 酵母活化劑；酵母養料
yeast hydrolyzate 酵母水解物
yeast mash 發酵醪
yeast paste 酵母泥
yeast powder 酵母粉
yeast press cake 壓榨酵母餅
yeast starter 酵母菌元；酵母菌母
yeast strain 酵母菌株
yeast wash 酵母醪；酵母發酵醪
yeastcidin 殺酵母素
yeastiness 發酵狀態；起泡
yeastrel 酵母浸膏
yeasts, riched selenium 富硒酵母
yeasty 酵母的；含酵母的；發酵的；起泡沫的
yeasty flavo(u)r 酵母〔風〕味
yellow bass 黃石鱸
yellow brain fungus 黃金銀耳
yellow chive 韭菜
yellow corn 黃玉米
yellow cracking bolete 似絨毛牛肝菌
yellow croaker 黃魚；黃花魚
yellow damaged rice 黃變米
yellow daylily 金針菜；金針；黃花菜；萱草；萱花
yellow dextrin 黃糊精
yellow enzyme 黃酶
yellow flowered gourd 珠瓜；觀葉南瓜
yellow green algae 黃綠藻；黃藻
yellow koji 黃麴
yellow lentil 黃扁豆
yellow maize 黃玉米
yellow mustard 黃芥菜；黃芥末
yellow pea 黃豌豆
yellow peach 黃桃
yellow perch 金鱸
yellow pored scaber stalk 裂紋疣柄牛肝菌

yellow staining knight cap 雕紋口蘑(蕈類)

yellow staining russule 美紅菇

yellow sugar 黃糖

yellow swamp russule 紫紅黃紅菇

yellow tomato 黃番茄

yellow vegetable 黃色根菜

yellow wax 黃蠟

yellow wine 黃酒

yellow yam 十二月薯

yellow zucchini 黃皮小胡瓜；黃皮筍瓜

yellowfin 黃鰭金槍魚；黃鰭魚

yellowfin albacore 黃鰭金槍魚

yellowfin menhaden 黃鰭油鯡

yellowfin puffer 條紋東方魨

yellowfin sea bream 黃鰭鯛；灰鰭鯛

yellowfin sole 刺黃蓋鰈

yellowfin tuna 黃鰭鮪；黃鰭金槍魚

yerba mate 巴拉圭茶；馬黛茶 [= mate]

yerbine 茶素；巴拉圭茶鹼

yield 生產量；產量；回收率

ylangylang 加拿楷樹；衣蘭香精；香水樹；衣蘭樹 [= ilangilang]

ylangylang oil 衣蘭油；衣蘭衣蘭油

yoghourt 酸凝酪；酸乳酪；優酪乳；優格；酸〔牛〕奶 [= yoghurt; yogurt]

yoghurt 酸凝酪；酸乳酪；優酪乳；優格；酸〔牛〕奶 [= yoghourt; yogurt]

yogurt 酸凝酪；酸乳酪；優酪乳；優格；酸〔牛〕奶 [= yoghurt; yoghourt]

yolk 卵黃；蛋黃

yolk colo(u)r 卵黃色；蛋黃色

yolk lecithin 卵黃卵磷脂

yolk membrane 卵黃膜

yolk powder 卵黃粉；蛋黃粉

yolk sac 卵黃囊

Yorkshire 約克夏豬

young 幼小的；新鮮的

yuba 腐皮；豆腐衣；豆腐皮；豆皮；腐竹 [= soy skin; soya skin; soybean milk film]

yucca 絲蘭；麟鳳蘭

yurotin 敗菌素

Z

Z line Z線
Z value Z值
zearin 地衣素；苔素
zeatin 玉米素
zeaxanthin 玉米黃質
zebu 單峯牛；瘤牛；駝牛
zedoary 莪蒁
zedoary root 莪蒁根；蓬莪蒁
zein 玉米〔醇溶〕蛋白
zeiosis 發泡作用
zeolite 沸石
zero point 零點
zeroing 歸零；零點調整
zerumbet ginger 球薑
zest 刺激性；滋味；橘子皮；檸檬皮
zinc 鋅
zinc gluconate 葡糖酸鋅
zinc methionine sulfate 硫酸蛋氨酸鋅
zinc oxide 氧化鋅
zinc resinate 樹脂酸鋅
zinc sulfate 硫酸鋅
Zinfandel 金飯〔葡萄〕酒
zingerone 薑油酮
zingiber 薑
ziti 短管通心粉
zitoni 光面中粗〔管狀〕通心粉（1/2英吋粗）
zitoni rigati 槽紋面中粗〔管狀〕通心粉（1/2英吋粗）
zomotherapy 肉食療法
zonal centrifugation 區帶離心〔法〕；帶域離心〔法〕
zone 帶；區
zone electrophoresis 區帶電泳〔法〕；帶域電泳〔法〕
zoobiotic 動物寄生菌
zooblast 動物細胞
zoochemistry 動物化學
zoogloea 菌膠團；細菌凝聚團
zoology 動物學
zoomaric acid 棕櫚油酸；十六烯酸 [= palmitoleic acid]
zoonosis 人畜共通傳染病
zoophagan 食肉動物
zoophagous 食肉的
zooplankton 浮游動物
zoospore 游走孢子
zoosterol 動物固醇；動物甾醇
zootoxin 動物毒素
zucchini 小胡瓜；筍瓜；小西葫蘆 [= courgette; summer squash]
zungenwurst 牛舌臘腸
zygomycete 接合菌
zygospore 接合孢子
zygote 接合子
zymase 釀酶；發酵酶；酒化酶
zyme 酶
zymogen 酶元；酶原
zymohexase 醛縮酶
zymohexose 可發酵己糖

zymohydrolysis 酶解作用
zymologist 發酵學家
zymology 發酵學
zymometer 發酵計；發酵能力測定器 [= zymosimeter; zymoscope]
zymophyte 發酵細菌
zymosan 酵母聚糖
zymoscope 發酵計；發酵能力測定器 [= zymometer; zymosimeter]
zymosimeter 發酵計；發酵能力測定器 [= zymometer; zymoscope]
zymosis 發酵
zymosterol 酵母固醇；酵母甾醇
zymotachegraph 酵母發麵能力測定計
zymotechnics 發酵技術；發酵工藝；釀造技術；釀造工藝
zymotechnique 發酵技術；發酵工藝；釀造技術；釀造工藝
zymurgy 釀造學

一畫

一〔滿〕桶 pailful
一乙酸甘油酯 acetin; glyceryl monoacetate
一人份的上等牛排 porterhouse
一人份量的肉排 cutlet
一口 gobbet
一口〔威士忌酒〕 shot
一口食物 bouchee
一口酒 tiff
一口喝下 swig
一口酥 bouchee
一大口 swig
一大湯匙 tablespoonful (tbsp.)
一小口 toothful
一小杯〔威士忌酒〕 shot
一元酚氧化酶 monophenol oxidase
一元酸 monoacid
一元醇 monohydroxy alcohol
一水化物 monohydrate
一水合物 monohydrate
一代時間 generation time
一甘油酯 monoglyceryl ester
一甘醇酯 monoglycol ester
一件食品 viand
一份〔食物〕 serving
一份飯菜 helping
一次性包裝 one-trip package; one-way package; single-service package
一次性容器 non-returnable container
一次性餐具 disposable dinner set
一串〔香蕉〕 hand
一批 batch
一束 hank
一卷 hank
一杯之量 cupful; stoup
一杯的容量 glassful
一杯的量 tumblerful
一杯飲料 drink
一客〔份〕包裝食品 portion pack food
一客〔食物〕 serving
一客份包裝 portion pack
一指長的小魚 fingerling
一盆〔之量〕 tubful
一致性 consistence; consistency
一倍濃度（未經稀釋或濃縮） single strength
一氧化二氮 nitrous oxide
一氧化碳 carbon monoxide; carbonic oxide
一級污水處理 primary sewage treatment
一茶匙量 teaspoonful (tsp.)
一匙量 spoonful
一桶〔之量〕 tubful
一袋量 bagful
一壺的量 potful
一湯匙量 tablespoonful (tbsp.)
一塊 gobbet
一歲牲畜 yearling
一道菜 course
一滿杯 tumblerful
一滿匙 spoonful

一酸甘油酯 monoglyceride
一撮茶 drawing
一盤量 trayful
一劑 dose
一餐 meal
一磷酸腺苷 adenosine monophosphate (AMP); adenylic acid
一鍋的量 potful
一點點地咬 nibble

二畫

乙二胺四乙酸 ethylenediamine tetraacetic acid (EDTA)
乙二胺四乙酸二鈉 disodium EDTA
乙二胺四乙酸二鈉鈣 calcium disodium EDTA
乙二胺四乙酸鐵鈉 sodium iron(III) ethylenediamine tetraacetate
乙二胺四醋酸 ethylenediamine tetraacetic acid (EDTA)
乙二酸 oxalic acid
乙二醇 ethylene glycol; glycol
乙二醇縮桂醛 cinnamaldehyde ethylene glycol acetal
乙二醛 glyoxal
乙炔 acetylene
乙氧基奎 ethoxy quin
乙氧基單雙甘油酯 ethoxylated mono- and diglycerides
乙胺 ethylamine
乙酐 acetic anhydride
乙偶姻 aceotin; acetylmethylcarbinol
乙基丁酸鹽 ethyl butyrate
乙基香草醛 ethyl vanillin; vanillal
乙基香蘭素 bourbonal; ethyl vanillin
乙基硫醇 ethyl mercaptan
乙基麥芽酚 ethyl maltol
乙基橙 ethyl orange
乙基纖維素 ethylcellulose
乙烯 ethene; ethylene
乙烯化氧 ethylene oxide
乙烯基塑料 vinyl plastic
乙烯醇 vinyl alcohol
乙烷 ethane
乙酸 acetic acid; ethanoic acid
乙酸乙酯 acetic ether; ethyl acetate
乙酸丁香酚酯 eugenyl acetate
乙酸丁酯 butyl acetate
乙酸二甲基苄基甲醇酯 dimethyl benzyl carbinyl acetate
乙酸二氫葛縷酯 dihydrocarvyl acetate
乙酸己酯 hexyl acetate
乙酸壬酯 nonyl acetate
乙酸丙酯 propyl acetate
乙酸四氫糠酯 tetrahydrofurfuryl acetate
乙酸甲酯 methyl acetate
乙酸白檀油酯 santalyl acetate
乙酸辛酯 octyl acetate
乙酸庚酯 heptyl acetate
乙酸松油酯 terpinyl acetate
乙酸玫瑰酯 rhodinyl acetate
乙酸的 acetic
乙酸芳樟酯 linalyl acetate
乙酸柏木酯 cedryl acetate
乙酸癸酯 decyl acetate
乙酸苯乙酯 phenethyl acetate
乙酸苯丙酯 3-phenylpropyl acetate
乙酸香芹酯 carvyl acetate
乙酸香茅酯 citronellyl acetate
乙酸香葉酯 geranyl acetate
乙酸桂酯 cinnamyl acetate

乙酸茴香酯 anisyl acetate
乙酸異丁香酚酯 isoeugenyl acetate
乙酸異丁酯 isobutyl acetate
乙酸異丁酸蔗糖酯 sucrose acetate isobutyrate (SAIB)
乙酸異丙酯 isopropyl acetate
乙酸異戊酯 isoamyl acetate
乙酸異冰片酯 isobornyl acetate
乙酸異龍腦酯 isobornyl acetate
乙酸鈣 calcium acetate
乙酸葛縷酯 carvyl acetate
乙酸解 acetolysis
乙酸鉀 potassium acetate
乙酸鉛 lead acetate
乙酸酯 acetate
乙酸對甲酚酯 *p*-cresyl acetate
乙酸銨 ammonium acetate
乙酸橙花酯 neryl acetate
乙酸龍腦酯 bornyl acetate
乙酸檀香酯 santalyl acetate
乙酸膽鹼 choline acetate
乙酸薄荷酯 menthyl acetate
乙酸蘇合香酯 methyl phenylcarbinyl acetate
乙酸纖維素 acetyl cellulose; cellulose acetate
乙酸鹽 acetate
乙酸苄酯 benzyl acetate
乙醇 alcohol; ethanol; ethyl alcohol; grain alcohol; spirit of wine
乙醇去氫酶 alcohol dehydrogenase
乙醇胺 ethanolamine
乙醇酸 glycolic acid; hydroxyacetic acid
乙縮醛 acetal
乙醣 biose; diose
乙醚 diethyl ether; ether; ethyl ether
乙醛 acetaldehyde; ethanal
乙醛二乙縮醛 acetaldehyde diethyl acetal
乙醛去氫酶 acetaldehyde dehydrogenase
乙醛脱氫酶 acetaldehyde dehydrogenase
乙醛酸 glyoxylic acid
乙酰乙酸 acetoacetic acid
乙酰乙酸去羧酶 acetoacetate decarboxylase
乙酰乙酸酯 acetoacetate
乙酰化〔作用〕 acetylation
乙酰化澱粉 acetylated starch
乙酰甲基甲醇 aceotin; acetylmethylcarbinol
乙酰值 acetyl number; acetyl value
乙酰胺酶 acetamidase
乙酰輔酶A acetyl CoA
乙酰價 acetyl number; acetyl value
乙酰磷酸酯 acetyl phosphate
乙酰膽鹼 acetylcholine
乙酰膽鹼酯酶 acetylcholinesterase
乙酰乙酸乙酯 ethyl acetoacetate
乙酰乙酸香葉酯 geranyl acetoacetate
乙酰化二澱粉己二酸酯 acetylated distarch adipate
乙酰化二澱粉甘油酯 acetylated distarch glycerol
乙酰化二澱粉磷酸酯 acetylated distarch phosphate
乙酰化己二酸雙澱粉 acetylated distarch adipate
乙酰化磷酸雙澱粉 acetylated distarch phosphate
乙酰丙酸乙酯 ethyl levulinate
乙酰苯 acetophenone; acetyl benzene
乙酰茴香醚 acetanisole
乙酰蓖麻油酸甲酯 methyl acetyl ricinolate
乙酰磺胺酸鉀 Acesulfame potassium (Acesulfame-K)
丁 cubelet
丁二酮 diacetyl

丁二酸 amber acid; succinic acid
丁二酸二乙酯 diethyl succinate
丁子香 *Caryophyllus aromaticus*
丁子香烯 caryophyllene
丁子香酚 eugenol
丁位癸內酯 δ-decalactone
丁苯橡膠 butadiene-styrene rubber
丁香 clove; lilac; syringa
丁香〔花蕾〕油 clove (bud) oil
丁香〔葡萄〕酒 clove wine
丁香油 clove oil; caryophyllene
丁香油素 caryophyllin
丁香花 lilac
丁香花醇 lilac alcohol
丁香莖油 clove stem oil
丁香酚 eugenol
丁香酚甲酯 methyl eugenol
丁香酚甲醚 eugenyl methyl ether
丁香葉油 clove leaf oil
丁香酮 syringone
丁香辣椒 allspice
丁香酸 syringic acid
丁香醇 syringyl alcohol
丁香樹 *Caryophyllus aromaticus*
丁香醛 syringaldehyde
丁香羅勒 basil; sweet basil
丁骨牛排 T-bone steak
丁基大茴香醚 butylated hydroxyanisole (BHA)
丁基羥基甲氧苯 butylated hydroxyanisole (BHA)
丁基羥基茴香醚 butylated hydroxyanisole (BHA)
丁基鄰苯二甲酰基乙醇酸丁酯 butyl phthalyl butyl glycol(l)ate
丁基橡膠 butyl rubber
丁烷 butane
丁酮酸 ketobutyric acid
丁酸 butyric acid; butanoic acid
丁酸-3-丁酮-2-酯 butan-3-one-2-yl butyrate
丁酸乙酯 ethyl butyrate
丁酸丁酯 butyl butyrate
丁酸二甲基苄基甲醇酯 dimethyl benzyl carbinyl butyrate
丁酸己酯 hexyl butyrate
丁酸戊酯 amyl butyrate
丁酸甲酯 methyl butyrate
丁酸芳樟酯 linalyl butyrate
丁酸苯乙酯 phenethyl butyrate
丁酸香茅酯 citronellyl butyrate
丁酸香葉酯 geranyl butyrate
丁酸桂酯 cinnamyl butyrate
丁酸異丁酯 isobutyl butyrate
丁酸異戊酯 isoamyl butyrate
丁酸細菌 butyric acid bacteria
丁酸發酵 butyric acid fermentation
丁酸葉醇酯 leaf butyrate
丁酸環己酯 cyclohexyl butyrate
丁酸蘇合香酯 styralyl butyrate
丁酸鹽 butyrate
丁酸苄酯 benzyl butyrate
丁醇 butanol; butyl alcohol
丁糖 tetrose
丁糖酮酸 tetruronic acid
丁糖醇 tetritol
丁醛 butyraldehyde
丁酰乳酸丁酯 butyl butyryl lactate
七星斑 leopard grouper
七星鱸 Japanese sea-perch
七彩花飾糖果粒 hundreds and thousands
七斑石斑魚 true bass
七碳糖 heptose
七鰓鰻 lampern; lamprey (eel)
九孔 baby abalone; small abalone
九斤〔黃〕雞 cochin
九成滅菌時間 D value; decimal reduction value

九里香浸膏 murraya paniculata concrete
九碳糖 nonose
九層塔 basil; sweet basil
二乙基乙酸 diethylacetic acid
二乙基乙醛 diethyl acetaldehyde
二乙基醚 diethyl ether; ether
二乙酸甘油酯 diacetin; glyceryl diacetate
二丁基硫醚 dibutyl sulfide
二丁基羥基甲苯 butylated hydroxytoluene (BHT)
二十二〔碳〕烯酸 docosenoic acid
二十二碳四烯酸 docosatetraeneoic acid
二十八〔碳〕烷 octacosane
二十六烷酸 hexacosanoic acid
二十六碳二烯酸 hexacosadienoic acid
二十四〔烷〕酸 tetracosanoic acid
二十四碳五烯酸 scoliodonic acid
二十四碳烯酸 nervonic acid; tetracosenoic acid
二十烷酸 arachic acid; arachidic acid; eicosanoic acid
二十碳二烯酸 eicosadienoic acid
二十碳三烯酸 scolic acid
二十碳五烯酸 eicosapentaenoic acid (EPA)
二十碳六烯酸 decosahexaenoic acid
二十碳四烯酸 arachidonic acid; eicosatetraenoic acid
二十碳烯酸 gadoleic acid
二元酸 dicarboxylic acid
二元醇 diol
二分屠體 half carcass
二手貨 second hand
二月桂精 dilaurin
二片罐 seamless can; seamless tin
二功能觸媒 bifunctional catalyst
二甘醇一乙醚 diethylene glycol monoethylether
二甲花翠素 malvidin
二甲花翠苷 malvin
二甲苯 xylene; xylol
二甲苯麝香 musk xylene
二甲胺 dimethylamine
二甲基二硫醚 dimethyl disulfide
二甲基三硫醚 dimethyl trisulfide
二甲基硫醚 dimethyl sulfide
二甲基酮 acetone; dimethyl ketone
二甲基對苯二酚 dimethoxybenzene
二甲基苄基原醇 dimethyl benzyl carbinol
二甲基吡咯 dimethylpyrrole
二甲基吡嗪 dimethylpyrazine
二甲硫 dimethyl sulfide
二年生植物 biennial
二年果 biennial fruit
二次分離的稀奶油 double cream
二次反應 secondary reaction
二次巴氏殺菌 repasteurization
二次加熱 reheat(ing)
二次包裝 secondary package
二次冷卻 after-cooling
二次冷凍 refreezing
二次亞硫酸處理 double sulfitation
二次固化 regelation
二次抽樣樣品 subsample
二次保溫 retemper
二次粉碎 regrinding
二次乾燥 redry(ing)
二次掛糖衣 reglazing
二次產品 secondary product
二次麥芽汁 after wort; second wort
二次發酵 after-fermentation
二次發酵法 half sponge method
二次感染 reinfection
二次滅菌 secondary sterilization
二次過濾 secondary filter
二次榨汁 after-pressed juice; secondary juice
二次蒸煮 recooking

二次蒸發器 revaporizer; second evaporator
二谷氨酸鎂 magnesium diglutamate
二亞油精 dilinolein
二油酯 diolein
二段冷凍 two-step freezing
二段乾燥 two-stage drying
二段解凍 two-stage defrosting
二苯甲酮 benzophenone
二苯胺 diphenylamine
二苯醚 diphenyl oxide
二苯酰硫胺素 dibenzoyl thiamine
二苯酰硫胺鹽酸鹽 dibenzoyl thiamine hydrochloride
二重培養皿 doubled dish
二重捲封 double seam(ing)
二重捲封機 double seamer
二重捲封罐 double seamed can
二重捲邊 double seam(ing)
二重碳酸飽充法 double carbonatation process
二倍體 diploid
二氧化物 dioxide
二氧化矽 silica; silicone dioxide
二氧化硫 sulfur dioxide; sulphur dioxide
二氧化硫處理 sulfuring; sulphuring
二氧化硅 silicon dioxide; silica; silicone dioxide
二氧化氮 nitrogen dioxide
二氧化氯 chlorine dioxide
二氧化鈦 titanium dioxide
二氧化碳 carbon dioxide
二氧化碳昏厥法 carbon dioxide stunning
二氧芑 dioxin
二級米 second head
二級品 seconds
二級產品 second product
二級壓碎機 secondary crusher
二胺基酸 diamino acid
二氫去水維生素 dihydroanhydrovitamin
二氫查耳酮 dihydrochalcone (DHC)
二氫茉莉酮酸甲酯 methyl dihydrojasmonate
二氫香豆素 dihydrocoumarin
二氫香芹酚 dihydrocarveol
二氫香芹酮 dihydrocarvone
二氫黃蒿萜酮 dihydrocarvone
二氫黃蒿萜醇 dihydrocarveol
二氫葉酸 dihydrofolic acid
二氫膽固醇 dihydrocholesterol
二烯酸 dienoic acid
二硫化物 disulfide; disulphide
二硫化苯甲酰噻胺 benzoyl thiamin(e) disulfide
二硫化苯酰硫胺 benzoyl thiamin(e) disulfide
二粒小麥 emmer; two-grained wheat
二酚 diphenol
二酚氧化酶 diphenol oxidase
二氯〔化〕乙烯 ethylene dichloride
二氯二氟甲烷 dichlorodifluoromethane
二氯化錳 manganese chloride
二氯甲烷 dichloromethane; methylene chloride
二氯異氰尿酸鈉 dichlorinated sodium isocyanurate
二硬脂〔精〕 distearin
二等小麥粉 consumer flour
二等麵粉 clear flour
二歲綿羊 teg
二稜大麥 two-rowed barley
二號一等〔麵〕粉 second patent (flour)
二號二等〔麵〕粉 second clear (flour)
二號粉心麵粉 second patent (flour)
二號糖蜜 second molasses
二道壓榨 second press
二羧酸 dicarboxylic acid
二態性 dimorphism
二態的 dimorphous

二榨 second press
二榨汁 second juice
二碳酸二甲酯 dimethyl dicarbonate (DMDC)
二碳醣 biose; diose
二聚〔合〕作用 dimerization
二聚物 dimer
二鉻酸鹽 dichromate
二蔗酮糖 lactulose
二醇 diol
二醋酸一鈉 sodium (hydrogen) diacetate
二醋酸甘油酯 diacetin; glyceryl diacetate
二噁甲 dioxin
二澱粉甘油酯 distarch glycerol
二澱粉磷酸酯 distarch phosphate
二糖 disaccharide; disaccharose
二糖酶 disaccharidase
二磷酸腺苷 adenosine diphosphate (ADP)
二磷酸吡啶核苷酸 nicotinamide adenine dinucleotide (NAD)
二糠基二硫醚 difurfuryl disulfide
二糠基硫醚 difurfuryl sulfide
二醣 biose; diose
二醛澱粉 dialdehyde starch
二苄醚 dibenzyl ether
二肽 dipeptide
二肽酶 dipeptidase
人工去殼 hand peeling
人工色素 artificial colo(u)r
人工挑選 hand picking
人工剝皮 hand peeling
人工氣候保藏 controlled atmosphere storage (CA storage)
人工氣候保藏法 gas storage
人工氣體〔環境〕 controlled atmosphere
人工採摘 hand picking
人工甜味料 artificial sweetener
人工陳化 artificial ag(e)ing
人工揉捻的茶葉 hand-rolled tea
人工催熟 artificial ripening
人工熟成 artificial ag(e)ing
人工膜 synthetic membrane
人工調味料 artificial seasoning
人工醋 wood vinegar
人工擠乳 finger milking
人心果 naseberry; sapota; sapodilla
人心果樹膠 chicle (gum); gum chicle
人乳 human milk
人乳糖 gynolactose
人畜共通傳染病 zoonosis
人參 ginseng; sang
人造〔食用〕香料 artificial flavo(u)r; synthetic flavo(u)r
人造牛油 artificial butter
人造奶油 artificial butter; butterine; lardine; margarine; oleo oil; oleomargarine
人造奶油製造器 complector
人造白脫 artificial butter; butterine; margarine; oleomargarine
人造肉 artificial meat; imitation meat; meat analog
人造肉食品 meaty food
人造乳 artificial milk; imitation milk; synthetic milk
人造乳酪 butterine; margarine; oleomargarine
人造固化脂 lardine
人造沸石 permutite
人造的 synthetic
人造香料 artificial flavo(u)ring substance
人造香精 artificial essence
人造脂肪 artificial fat
人造黃油 artificial butter; butterine; margarine; oleomargarine
人造腸衣 artificial casing; synthetic casing
人造蜂蜜 honey syrup
人蔘 asiatic ginseng

人體細胞 soma
入口 inlet
八目鰻 lampern; lamprey (eel); nine-eyes
八角 anise; anise seed; aniseed; Chinese aniseed; star anise
八角〔茴香〕油 star anise oil
八角茴香 star anise
八碳醣 octose
八疊球菌屬 *Sarcina*
刀叉餐具 cutlery
刀子 knife
刀豆 common bean; jack bean; sword bean
刀豆胺酸 canavanine
刀豆球蛋白 canavalin
刀豆屬 *Canavalia*
刀具 cutter
刀具箱 knife box
刀架 knife rest
刀套 knife pouch
力可達乳清芝士 Ricotta cheese
力可達乳清乾酪 Ricotta cheese
力本發酵乳 leben
十一〔碳〕烯醛 undecenal
十一烯醇 undecenol
十一酮 undecanone
十一醇 undecyl alcohol
十一醛 undecanal
十七烷酸 heptadecanoic acid; margaric acid
十九烷 nonadecane
十九烷酸 nonadecanoic acid
十二月薯 yellow yam
十二指腸 duodenum
十二烷基苯磺酸鈉 sodium dodecyl benzene sulfonate
十二烷基硫酸鈉 sodium dodecyl sulfate (SDS); sodium lauryl sulfate
十二烷酸 dodecanoic acid; lauric acid
十二烷酸乙酯 ethyl laurate
十二烷醇 lauryl alcohol
十二碳烯酸 dodecenoic acid; lauroleic acid
十二醛 lauraldehyde; lauryl aldehyde
十八烷酸 octadecanoic acid; stearic acid
十八烷醇 stearyl alcohol
十八碳二烯酸 octadecadienoic acid
十八碳烯酸 octadecenoic acid; vaccenic acid
十六烷醇 cetanol; cetyl alcohol
十六碳二烯酸 hexadecadienoic acid
十六碳三烯酸 hiragonic acid
十六碳烯酸 gaidic acid; hexadecenoic acid; palmitoleic acid; zoomaric acid
十六酸 palmitic acid
十分之一公升 deciliter
十月山楂 October hawthorn
十四烷酸 myristic acid; tetradecanoic acid
十四碳烯酸 tetradecenoic acid
十字花烤包 coburg
十億分之一 parts per billion (ppb)
十億分點 parts per billion (ppb)
又吃又喝 wine and dine

三畫

三乙胺 triethylamine
三乙酸甘油酯 triacetin
三乙醇胺 triethanolamine
三丁酸甘油酯 butyrin; glyceryl tributyrate; tributyrin
三丁酸甘油酯酶 butyrinase
三十〔碳〕烷 squalane
三十烷酸 melissic acid
三十碳六烯 squalene
三元酸 tribasic acid
三元醇 tribasic alcohol; triol
三反油酸甘油酯 trielaidin
三文治 sandwich
三文魚 salmon
三月桂精 glyceryl trilaurate; trilaurin
三月桂酸酯 glyceryl trilaurate; trilaurin
三片罐 built-up tin; sanitary can; three-piece can
三丙酸甘油酯 glyceryl tripropanoate
三仙膠 xanthan gum
三甲胺 trimethylamine
三多磷酸鉀 potassium tripolyphosphate
三次亞麻油脂 glyceryl trilinolenate; linolenin; trilinolenin
三夾板 plywood
三角形包裝 cornucopia
三角試驗 triangle test
三亞麻油脂 glyceryl trilinoleate; linolein; linolin; trilinolein
三明治 sandwich
三明治麵包 pullman bread; sandwich bread
三松香酸甘油酯 glycerol ester of wood rosin; glyceryl triabietate
三油脂 glyceryl trioleate; olein; triolein
三油酸甘油酯 glyceryl trioleate; olein; triolein
三芥子脂 erucin
三效蒸發器 triple effect evaporator
三偏磷酸鈉 sodium trimetaphosphate
三烯脂肪酸 triethenoid fatty acid
三硅酸鎂 magnesium silicate
三棕櫚精 glyceryl tripalmitate; palmitin; tripalmitin
三棕櫚酸甘油酯 tripalmitin
三氯乙烯 trichloroethylene
三氯乙酸 trichloroacetic acid (TCA)
三氯化氮漂白麵粉 agenized flour
三氯甲烷 chloroform; trichloromethane
三氯蔗糖 sucralose
三硬脂精 glyceryl tristearate; stearin; tristearin
三硬脂酸山梨糖醇酐酯 sorbitan tristearate (Span 65)
三硬脂酸甘油酯 glyceryl tristearate; tristearin
三腳茶几 teapoy
三葉草 clover; shamrock
三酪酯 glyceryl tributyrate; tributyrin

三羧酸循環 Krebs cycle; tricarboxylic acid cycle (TCA cycle)
三榨 third press
三榨汁 third press
三榨油 third press
三碳醣 triose
三聚甘油單硬脂酸酯 tripolyglycerol monostearate
三聚磷酸鈉 sodium tripolyphosphate
三聚磷酸鉀 potassium tripolyphosphate
三聚磷酸鹽 tripolyphosphate
三酸甘油酯 triglyceride
三槽蝦 three grooved shrimp
三醋精 glyceryl triacetate; triacetin
三齒七鰓鰻 pacific lamprey
三齒躄魚 toadfish
三糖 trisaccharide
三糖酶 trisaccharidase
三磷酸腺苷 adenosine triphosphate (ATP)
三磷酸吡啶核苷酸 nicotinamide adenine dinucleotide phosphate (NADP)
三螺旋 triple helix
三肽 tripeptide
三肽酶 tripeptidase
下午茶 afternoon tea
下午茶時間 teatime
下水溝 gutter
下水道 sewage
下肩胛肋眼心 chuck eye log
下肩胛眼肉捲 chuck eye roll
下部燃燒的 underfired
下等的 inferior
下腳料 trash
下腳魚 scrap fish; trash fish
下腹 abdomen
下雜魚 coarse fish
上皮〔細胞〕 epithelial tissue; epithelium
上皮組織 epithelial tissue; epithelium
上光精白米 coated mill rice
上肋肉 top rib
上色劑 toner
上肩肉 collar butt
上肩胛肉 blade; clod
上肩胛脊底肌 oyster blade; top blade muscle
上後腿肉 topside (round)
上面加冰淇淋的 alamode
上面發酵 surface fermentation; top fermentation
上面酵母 top yeast
上馬酒 stirrup cup
上等〔錫蘭〕紅茶 pekoe (tea)
上等白蘭地酒 fine champagne
上等紅茶（產於印度或斯里蘭卡） orange pekoe
上等腰脊肉牛排 porterhouse steak
上等葡萄酒 vintage
上菜 serving
上腿 thigh
上層液 supernatant
上層魚類 epipelagic fish
上層發酵 top fermentation
上層酵母 top yeast
上澄液 supernatant
上膠器 sizer
上館子吃飯 eat out
上臀肉 gooseneck round; outside; top rump
上臀肉排 outside steak
上顎 palate
上顎肉 palate meat
上蠟 waxdip; waxing
上癮 addictive
丸劑 pill
凡士林 petrolatum; vaseline
刃天青 resazurin
勺 scoop

千克 kilogram (kg)
千屈菜 willow herb
千島沙拉醬 thousand island dressing
千島沙津醬 thousand island dressing
千焦〔耳〕 kilojoule
千葉蓍 yarrow
千層蛋糕 layer cake
千層餅 abaisse
千層蔬菜派 galette
千層糕 stratiform cake
千層麵 lasagna
千磅 kip
叉子 fork
叉骨 sidebone
叉燒肉 basted meat; barbecued meat
口 mouth
口味不佳的 distasteful
口味質量評定 palatability test
口服 ingestion; oral administration
口香糖 cachou; chewing gum
口渴 thirst
口感 mouth feel(ing)
口罩 respirator
口鼻部 snout
口瘡 aphtha
口蹄疫 aphthous fever; foot and mouth disease
口嚼煙〔草〕 chewing tobacco; plug tobacco
土人參 potherb fameflower
土木香酸 alantic acid; inulic acid
土司 toast; toasted bread
土司麵包 pan bread
土司麵包條 pan loaf
土生種 autochthon
土耳其奶酒 busa
土耳其咖啡 Turkish coffee
土耳其香葉油 Turkish geranium oil
土耳其軟糖 Turkish delight
土耳其發酵乳 busa
土耳其碎芝麻蜂蜜糖 halavah; halva(h)
土杏紅 chufa nut
土豆 murphy; potato; white potato
土豆片 potato flake
土豆泥 mashed potato
土剎 catfish
土荊芥 epazote
土茴香 dill
土產 native product
土溫〔乳化劑〕 Tween
土溫80 polyoxyethylene sorbitan monooleate; Tween 80
土當歸 sarsaparilla
土腥味 earthy taste
土種 native breed; native species
土製香腸 country style sausage
土壤桿菌屬 *Agrobacterium*
土鰌 pond loach
士多啤梨 strawberry
夕奶 afternoon milk
大口水壺 ewer
大口鰈 big-mouthed flounder
大口鰨 big-mouthed sole
大小 size
大小分級機 size grader
大分子 macromolecule
大孔篩 griddle
大心芥菜 big stem mustard
大比目魚 halibut; turbot
大包裝 bulk pack
大包裝產品 bulk goods
大北鮭 nelma; inconnu
大卡 Calorie (Cal); great calorie; kilocalorie (kcal)
大平洋擬庸鰈 flathead flounder
大母松李 damson (plant)
大白豆 large white bean

大白菜 Chinese cabbage; celery cabbage; Peking cabbage
大白鼠 rat
大白樁菇 giant funnel cap
大白豬 large white swine
大目金梭 Japanese barracuda
大目鮪 big-eye tuna
大目鰱 bighead carp
大冰塊 ice slab
大吃 tuck
大吃大喝 gormandise; gormandize; gourmandise; gourmandize
大吃大喝者 guttler
大肉片 chump
大西洋叉尾帶魚 frostfish
大西洋小鱈 frostfish
大西洋沙丁魚 Atlantic sardine
大西洋金槍魚 Atlantic tuna
大西洋馬鮫 king mackerel
大西洋庸鰈 Atlantic halibut
大西洋鮪魚 Atlantic tuna
大西洋鮭 Atlantic salmon
大西洋鯖 Atlantic mackerel
大西洋鯡 Atlantic herring; oceanic herring
大西洋鱈 Atlantic cod
大禿馬勃（蕈類） giant puffball
大肚酒壺 flagon
大豆 soja (bean); soya (bean); soybean; soy
大豆分離蛋白〔質〕 isolated soybean protein
大豆卵磷脂 soy lecithin; soybean lecithin
大豆皂苷 soy saponin; soya saponin
大豆味噌 soybean miso
大豆固醇 soyasterol
大豆油 bean oil; soy oil; soya oil; soybean oil
大豆芽 soy sprout; soybean sprout
大豆甾醇 soyasterol
大豆粉 soy meal; soybean meal; soy flour; soybean flour
大豆胰蛋白酶抑制劑 soybean trypsin inhibitor
大豆球蛋白 glycinin
大豆粗粉 soybean grit
大豆粕 soy cake; soy meal; soybean cake; soybean meal
大豆蛋白 soy protein; soybean protein
大豆蛋白分離物 soy protein isolate; soybean protein isolate; SPI
大豆蛋白食品 soy protein food; soybean protein food
大豆蛋白濃縮物 soy protein concentrate; soybean protein concentrate; SPI
大豆腥味的 beany
大豆濃縮蛋白 soy protein concentrate; soybean protein concentrate; SPI
大豆麴 soy koji; soybean koji
大昆布 macrocystis
大果型的 large fruited
大果園 fruit ranch
大果實的 large fruited
大果蔓越橘 American cranberry
大牧羊場 sheep run
大肥菇 pavement mushroom
大孢子 macrospore
大型冷凍庫 walk-in freezer
大型藻類 macroscopic algae
大扁魚 halibut
大柄基麗傘（蕈類） St. George's mushroom
大玻璃啤酒杯 schooner
大珍珠貝 pearl oyster
大紅洋蔥 large red onion
大約克夏豬 large white swine
大缸 vat

大胃王 good trencherman
大胡桃 pecan
大栗 wood nut
大氣壓 atmospheric pressure
大氣壓力 barometric pressure
大氣濕度 atmospheric humidity
大海鰱 bastard mullet
大茴香 anise; anise seed; aniseed; Chinese aniseed; star anise
大茴香油 anise oil; star anise oil
大茴香椒粒 anise pepper
大茴香醛 *p*-methoxybenzaldehyde
大茴香黴菌 anisomycin
大酒杯 jeroboam; rummer; wine bowl; wassail bowl
大酒桶 butt; barricas; pipe
大酒瓶 jeroboam; magnum
大高腳酒杯 rummer
大剪刀 shears
大啤酒杯 schooner; tankard
大捲角山羊 markhor
大掃除(春季) springclean
大桶 tierce; vat
大梭魚 muskellunge
大理石紋〔脂肪〕 marbling
大理石紋脂肪的肉 marbled meat
大甜蘋果 pomewater
大眼桶 big-eyed flathead
大眼擬金槍魚 big-eye tuna
大眼鯖魚 big-eyed mackerel
大粒青豌豆 marrow fat pea
大粒魚子醬 beluga caviar
大粒豌豆 sugar pea
大蛋糕 abaisse; torte
大規模生產 large scale production
大部份蓋菌(蕈類) black staining polypore
大魚 mud carp
大麥 barley; hordeum
大麥〔片〕粥 polenta
大麥仁 pearl barley
大麥去殼機 pearling mill
大麥氧化酶 spermase
大麥浸出液 barley extract
大麥粉 barley flour; barley meal
大麥粒 barleycorn
大麥脫殼精碾機 hollander
大麥蛋白 hordein
大麥麥芽 barley malt
大麥煎汁 barley water
大麥醇溶蛋白 hordein
大麥糖 barley sugar
大麥糖漿 malt extract
大麥麴 barley koji
大麻 cannabis; hemp
大麻子 cannabis seed; hemp seed
大麻油 cannabic oil
大麻哈魚 calico salmon; dog salmon; keta; salmon
大麻籽油 hempseed oil
大麻酚 cannabinol
大麻醇 cannabinol
大喙篦鮫 cat shark
大晶粒砂糖 coarse granular sugar
大棒棒糖 all-day sucker
大湖鱒 ferox
大湯匙 soup spoon; tablespoon
大湯碗 tureen
大湯盤 tureen
大發酵桶 keeve
大菱鮃 turbot
大量生產 mass production
大黃 rhubarb
大黃花酢醬草 bermuda butter-cup
大黃魚 large yellow croaker
大黃酸 rheinic acid
大塊 hunch
大塊菌(蕈類) white truffle
大會客室 saloon

大楊梅 loganberry
大溫差巴氏殺菌 shocking pasteurization
大腸 large intestine
大腸〔桿〕菌群 coliform; coliform bacteria; coliform group; coliaerogenes organisms
大腸〔桿〕菌數 coliform count
大腸炎 colitis
大腸埃希氏桿菌 *Escherichia coli* (*E. coli*)
大腸桿菌 *Bacillus coli; Escherichia coli* (*E. coli*)
大腸桿菌中毒 *Escherichia coli* food poisoning
大腸桿菌毒素 colitoxin
大腸桿菌病 colibacillosis
大腸桿菌群 colon bacillus
大腸桿菌群乳房炎 coliform mastitis
大腸頭 bung cap
大腳毛蟹 giant crab; Japanese spider crab
大葉藻 sea wrack
大農場 ranch
大腿 thigh
大蒜 *Allium sativum*; garlic
大蒜油 garlic oil
大蒜素 garlicin
大蒜精 garlicin
大蒜糖 scorodose
大酸的 superacid
大蝴蝶粉 bow tie pasta; farfalla; farfalle
大褐藻 oarweed
大樹菠蘿 jaca; jackfruit; jak; jakfruit; nangka
大縐皮豌豆 marrow fat pea
大蕉 plantain
大頭針 pin
大頭魚類 pout
大頭菜 green turnip; kohlrabi; root mustard; turnip
大頭蒜 pearl onion
大頭蝦 jumbo shrimp
大頭鱈 bank cod; gray cod; pacific cod
大頭葱 eschalot; shallot
大糠 rice hull; rice husk
大聲地啜喝 slurp
大薔薇 dog rose
大鍋 caldron; cauldron
大雜燴 hodgepodge
大蟹 giant crab
大瀧六線魚 greenling
大蘇打 sodium thiosulfate
大鰭鯨 humpback whale
大鰭鱗鰓太陽魚 bluegill sunfish
大鰩魚 devilfish
大鱈魚 kelp cod
大鱗大麻哈魚 chinook salmon; king salmon; pacific salmon
大鱗鞋底魚 tongue sole
大鱗鮭魚 chinook salmon
大葱 cibo(u)l; stone leek; spring onion; Welsh onion
大腱 shin from hindshank
女牛 heifer
女羊 maid
子房 ovary
子葉 cotyledon
子實體 fruit(ing) body; fruitbody
子樣品 subsample
子鴨 green duck
子鵝 green goose
子囊菌〔類〕 ascomycete
小〔圓〕麵包 bun
小丸 pellet; parvule
小千層麵 lasagnette
小口 nibble
小口大壺 water jug
小山羊 goatling; kid
小公牛 bull calf; ram lamb

小公雞 cockerel
小方塊 cubelet
小水果 fruitlet
小水果餡餅 tartlet
小片 gobbet; pellet; tabloid
小片冰 scale ice
小片的美味食物 tidbit; titbit
小片糖 pattie; patty
小牛 calf
小牛肉 baby beef; veal
小牛肉片 veal cutlet
小牛肉捲 veal loaf
小牛肉糕 veal loaf
小牛排 veal chop
小包 sachet
小包裝 fractional pack; packet
小包裹 packet
小卡 small calorie
小母牛 heifer; quey
小母羊 gimmer
小母豬 gilt
小母雞（一歲以下） pullet
小白洋葱 silverskin onion
小白菊 feverfew
小白菜 pakchoi
小白麵包 cake bread
小冰塊 cube ice; small ice
小吃 refection; snack food; snack
小吃店 snack bar; tea room; tea shop
小吃部 refreshment room; snack bar
小吃攤 snack stall; snack stand
小米 Italian millet; millet; panic grass; proso millet
小羊肉 lamb meat; lambkin
小羽毛 plumelet
小肉片 cutlet; collop
小肉餡餅 bouchee
小肉雞 broiler
小西葫蘆 courgette; zucchini
小串烤肉 brochette
小牡蠣 blister
小豆 adsuki bean; adzuki bean; azuki bean; black gram; red bean
小荳蔻 cardamom; cardamon
小荳蔻油 cardamon oil
小里肌 short fillet
小卷 squid
小咖啡杯 demitasse
小昆布 murlin
小果 small fruit
小果山楂 littlehip hawthorn
小果實 fruitlet
小果籃 chip basket
小枝 spray; sprig
小杯 cannikin; canakin; noggin; pony
小杯清咖啡 demitasse
小杯黑咖啡 demitasse
小狗魚 pickerel
小芝士蛋糕 ramekin; ramequin
小芽孢 sporule
小孢子 microspore; sporule
小冠椰子油 ouricury oil
小型發酵槽 jar fermenter
小型發酵器 miniature fermenter
小型發酵罐 jar fermenter
小孩 tot
小扁豆 lentil
小指套形通心粉 ditalini
小洋葱 green onion; pearl cnion
小牲畜（一至二歲大） yearling
小玻璃瓶 ampoul
小背包 satchel
小胡瓜 courgette; zucchini
小香腸（一節） sausage link
小捆 packet
小旅館 tavern
小核果 drupel(et)
小根 rootlet

小海帶 murlin
小海魚 capelin
小烘烤箱 oven toaster
小茴香 cumin; cummin; small fennel
小茴香油 cumin oil; fennel oil
小茴香籽 cumin seed; cummin seed; fennel seed
小茴香酸 cuminic acid
小茴香醇 fenchol; fenchyl alcohol
小茴香醛 cumin aldehyde
小酒吧 estaminet; juke joint
小酒店 grogshop
小酒桶 barril
小酒館 groggery; tavern
小馬 foal; pony
小乾酪蛋糕 ramekin; ramequin
小堅果 nutlet; pyrene
小庸鰈 chicken halibut
小桶 cade; rundlet
小桶（容量單位） keg
小梭魚 pickerel
小球菌 pediococcus
小球菌〔屬〕 *Micrococcus*
小球藻 chlorella
小瓶 vial
小瓶香檳酒 split champagne
小瓶飲料 split
小甜豆 petits pois
小甜酥餅 butter cookie
小甜餅 cookie; cooky
小盒 caddy
小粒 granule; parvule; pellet
小粒魚子醬 sevruga caviar
小粒結晶砂糖 castor sugar
小粒穀類 small grain
小組品評 panel test
小羚羊（非洲） steenbok
小蛋糕 cup cake
小袋 sachet
小販 vender; vendor
小透明玻璃管 cuvette
小魚 baby fish; fry
小麥 wheat
小麥〔製〕的 wheaten
小麥片 wheat groats; wheat flake
小麥芽 wheat malt
小麥芽粉 wheat malt flour
小麥胚芽 wheat germ
小麥胚芽油 wheat-germ oil
小麥粉 wheat meal
小麥粉改質劑 flour modifing agent
小麥粉處理劑 flour improver; flour treatment agent; maturing agent
小麥產區帶 wheat belt
小麥粒 wheat groats
小麥粗粒 coarse sizings; sizings
小麥粗麵粉 alcomeal
小麥稈 wheat straw
小麥穀蛋白 glutenin
小麥醇溶蛋白 wheat gliadin
小麥麩皮 wheat bran
小麥澱粉 wheat starch
小麥糖 triticin
小麥穗 wheatear
小麥麵包 wheat bread; white bread; wheaten bread
小麥麵粉 wheat flour
小麥麵筋 wheat gluten
小單位 subunit
小湯碗 porringer
小琵琶桶 rundlet
小菌落 microcolony
小菜 side dish
小菜館 cookshop
小費 tip
小黃瓜 cuke
小黃李乾 prunello

小黃魚　little yellow croaker; small yellow croaker
小圓麵包　cob
小塊　pat
小塊〔精〕瘦肉　noisette
小塊牛油　pat
小煎薄餅　pikelet
小腸　small intestine
小腸腸衣（豬或牛）　small casing
小葉　leaflet
小滴　droplet
小滴培養　droplet culture
小種〔紅〕茶　souchong
小管　tubule
小綿羊　lamb
小腿　shank
小酸蘋果　crab apple
小廚房　kitchenette
小漿果　berry
小盤　pannikin
小蝴蝶粉　tripolini
小蝦　shrimp
小豬　gruntling
小豬　piglet; porkling; piggy; pigling; shoat
小賣部便餐　buffet lunch
小黎豆　Bengal gram
小糖果　bonbon; kiss; sugarplum
小餐車　dinette
小餐室　dinette
小餐館　cookshop; chophouse; estaminet; juke joint; luncheonette
小餡餅　pattie; patty
小餡餅皮　pattie shell
小餡餅烤鍋　pattypan
小鮑魚　baby abalone
小鴨　green duck; duckling
小龍蝦　crawfish; crayfish
小檗果　berberis
小檗屬植物　berberis
小燭樹蠟　candelilla wax
小鍋　pannikin
小鮫　tope
小鮭魚　troutlet; troutling; smolt
小點石斑　black spotted grouper
小雞　chick; chicken; chickling
小雞肉　baby chick
小雞腿　drumstick
小鵝　gosling; green goose
小鯨　whale calf
小鯡魚　brit(t); brisling; sprat; sperling
小蘋果　crabapple
小蘇打　baking soda; bicarbonate of soda; sodium bicarbonate; sodium hydrogen carbonate; saleratus
小麵包　bap; birlie; roll
小麵包片　sippet
小麵包乾　flatbread
小灌木　subshrub
小蠟燭樹蠟　candelilla wax
小囊　sacculus
小鰻〔魚〕　grig
小鱈魚　whiting
小纖維　fibril
小鱗莖　clove
小鱒魚　troutlet; troutling
小罐　cannikin; canakin; caddy; canister
山毛櫸木糖　beechwood sugar
山毛櫸堅果　beech mast; beech nut
山毛櫸樹　beech
山竹〔果〕　mangosteen
山羊　goat
山羊肉　goat meat
山羊乳　goat milk
山羊乳清芝士　mysost
山羊乳清乾酪　mysost
山羊羔　goatling
山羊脂　goat tallow

山羊酸 goat acid
山羊膻臭 hircine smell
山艾 sage
山芋 yam
山谷葡萄 valley grape
山東大白菜 shantung cabbage
山芥 winter cress
山胡桃 hickory nut
山胡桃〔木〕 hickory
山胡桃果 pignut
山茄 roselle
山核桃 bitternut; hickory nut; kiskatom
山核桃〔木〕 hickory
山核桃屬 *Carya*
山桑 bilberry
山桃 david peach
山茶油 camellia oil
山茱萸 dogberry
山梨〔樹〕 sorb
山梨〔糖〕醇 glucitol; sorbitol
山梨聚糖 sorbitan
山梨聚糖酯 sorbitan ester
山梨酸 hexadienoic acid; sorbic acid
山梨酸鈣 calcium sorbate
山梨酸鈉 sodium sorbate
山梨酸鉀 potassium sorbate
山梨酸酯 sorbate
山梨酸鹽 sorbate
山梨醇 sorbit; sorbol
山梨醇酐單棕櫚酸酯 Span 40
山梨醇酐單硬脂酸酯 Span 60
山梨糖 sorbin; sorbinose; sorbose
山梨糖酮酸 sorburonic acid
山梨糖醇酐脂肪酸酯 sorbitan fatty acid ester (Span)
山椒 Chinese prickly ash; flatspine prickly ash; Japanese pepper; Szechuan peppercorn
山椒素 sanshool
山椒醇 sanshool
山黑麥 mountain rye
山塔玫瑰李 Santa Rosa
山葵 horseradish; Japanese horseradish; wasabi
山楂 haw; hawthorn; Maythorn; whitethorn
山楂子 crab apple
山楂果 haw apple; haw berry
山楂酊 hawthorn fruit tincture
山楂核煙燻香味劑 haw pit smoke flavo(u)ring
山楸梅 chokeberry
山綠豆 beggarweed
山蒼子油 litsea cubeba oil
山藥 Chinese yam; cinnamon vine; yam
山螞蝗（西印度） beggarweed
山鶉 partridge
山蘿蔔 chervil
山鷸 woodcock
山欖果 sapota
山萮菜 Japanese horseradish; wasabi
山蒜 resurrectionlily
工作坊 workshop
工作臺 bench; working table
工具 tool
工場 workshop
工程食品 engineered food
工業〔規模〕發酵 industrial fermentation
工業用豬油 hog grease
工業污水 industrial effluent
工業污染 industrial pollution
工業作物 industrial crop
工業酒精 industrial alcohol
工業產品 industrial goods
工業規模 industrial scale
工業微生物學 industrial microbiology
工業葡萄糖 cerelose
工業廢水 industrial effluent

工廠 factory; plant
工廠〔生產〕設備 plant equipment
工廠用水 industrial water
工廠廢水 industrial waste water
工廠廢料 industrial waste
工廠衛生 plant sanitation
工藝 technology
己二烯酸 hexadienoic acid; sorbic acid
己二烯酸鈣 calcium sorbate
己二烯酸鈉 sodium sorbate
己二烯酸鉀 potassium sorbate
己二烯酸酯 sorbate
己二烯酸鹽 sorbate
己二酮 hexanedione
己二酸 adipic acid
己胺酸 norleucine
己烯酮 hexenone
己烯酸 hexenoic acid; hydrosorbic acid
己烯醇 hexenol
己烷 hexane
己聚糖 hexosan
己酸 caproic acid; hexanoic acid
己酸乙酯 ethyl hexanoate
己酸己酯 hexyl hexanoate
己酸丙酯 propyl hexanoate
己酸戊酯 amyl hexanoate
己酸甲酯 methyl hexanoate
己酸烯丙酯 allyl hexanoate
己酸異戊酯 isoamyl hexanoate
己酸葉醇酯 leaf caproate
己醇 hexanol; hexyl alcohol
己糖 hexose
己糖酸 hexonic acid; saccharinic acid
己糖醇 hexitol
己糖磷酸 hexose phosphate
己糖醛酸 hexuronic acid
己醣二酸 saccharic acid
己醣二酸鹽 saccharate
己醛 hexanal
已切好的 ready-cut
已剝皮的 skinned
已煮熟的 ready-cooked
已開的 unsealed
已滅菌的食品 sterilized food
已調味的 seasoned
乾貝 conpoy; dried scallop; scallop
干邑〔白蘭地〕酒 cognac
弓鰭魚 bowfin; mud fish

四畫

不匀色乳酪 mottled butter
不水解的 unhydrolyzed
不充氣果汁 still juice
不充氣飲料 still drink; still beverage
不充氣葡萄酒 still wine
不加水威士忌〔酒〕 whisky straight
不加熱乾燥 no-heat drying
不加糖的 unsweetened
不加糖的果汁 unsweetened juice
不加糖煉奶 unsweetened condensed milk
不加鹽的 unsalted
不可口 unsavo(u)riness
不可口的 unsavo(u)ry
不可口食物 unpalatable food
不可食用 non-edible
不可食用油 inedible oil
不可食用的 inedible
不可食用動物脂 inedible tallow
不可食用部分 non-edible part
不可食的 uneatable; unedible
不可逆反應 irreversible reaction
不可逆膠體 irreversible colloid
不可逆凝膠 irreversible gel
不可發酵的糖 unfermentable sugar
不可滲出的 unleachable
不可濕的 non-wettable
不孕 sterility
不正常風味 abnormal flavo(u)r
不正常捲封 false seam
不合格的 off standard; substandard
不合格品 rejected product
不合格產品 offgrade product
不合規格 undergrade
不合標準的質量 off quality
不回收包裝 one-trip package; one-way package
不回收容器 one-way container
不回收瓶 one-way bottle
不好吃的 unsavo(u)ry
不吸油的 oil-repellent
不含水的 unhydrous
不含氣體的 still
不含脂肪的 fat-free
不含酒精的 alcohol-free; soft biscuit
不含酒精飲料 soft drink; teetotal drink
不含蛋白質的 nonprotein; protein-free
不含麥精的 unmalted
不含糖巧克力〔原料〕 bitter chocolate
不含糖的 non-saccharide; sugarless
不含麵筋 gluten-free
不含麵筋食品 gluten-free food
不均一〔性〕 inhomogeneity
不均匀的 heterogenous
不均衡生長 unbalanced growth
不均衡膳食 unbalanced diet
不完全蛋 imperfect egg
不完全蛋白〔質〕 incomplete protein; pseudoprotein
不皂化物 unsaponifiable matter

不皂化的 non-saponifiable; unsaponifiable
不皂化脂質 non-saponifiable lipid
不育 sterility
不足 deficiency; shortage
不足度烈酒 underproof spirit
不定量包裝 free pack
不易消化的 indigestible; non-digestible; stodgy
不耐性 intolerance
不耐熱 heat liable
不耐熱性 thermolability
不耐熱的 thermolabile
不凍的 nonfreezing
不凍溶液 nonfreezing solution
不純的 impure
不能喝的 undrinkable
不能溶解的 insoluble
不能滲透的 impermeable
不起泡的 non-effervescent
不健康 unwholesomeness
不健康的 unhealthy; unwholesome
不健康食品 unwholesome food
不偏食的人 omnivore
不夠成熟的 underripe
不夠穩定的 understable
不帶氣泡的果汁 still juice
不帶氣泡的飲料 still drink
不混合性 immiscibility
不混合的 immiscible
不混溶性 immiscibility
不混溶的 immiscible
不甜的 savory; savoury
不產乳的牛 dry cow
不產芽孢細菌 asporulate bacteria
不產氣大腸桿菌 *Bacillus coli anaerogenes*
不符規格食品 adulterated food
不連續相 discontinuous phase
不連續攪動殺菌釜 discontinuous agitating retort
不透水性 watertightness
不透水的 watertight; waterproof
不透性 impermeability
不透明度 opacity
不透油的 oil-proof
不透氣的 airproof; airtight; gas proof; gas tight
不透液體的 liquid-tight
不透蒸汽的 vapo(u)r proof
不喝酒的 teetotal
不喝酒者 water drinker
不揮發性 non-volatility
不揮發性油 fixed oil
不揮發性物質 non-volatile substance
不揮發性的 non-volatile
不揮發油 non-volatile oil
不渴症 aposia
不發酵的 non-fermentable
不等滲溶液 anisotonic solution
不結塊的 non-caking; non-lumping
不新鮮 staleness
不新鮮肉 stale flesh
不新鮮味 storage flavo(u)r
不新鮮的 stale
不新鮮氣味 stale odo(u)r
不新鮮魚 stale fish
不溶性 insolubility
不溶性固形物 insoluble solid
不溶性酶 insoluble enzyme
不溶的 insoluble
不溶解的 undissolved; non-soluble
不當膳食 inadequate diet
不經酵母發麵的麵包 azyme
不飽和的 unsaturated
不飽和脂肪酸 unsaturated fatty acid
不飽和現象 unsaturation
不飽滿穀粒 imperfect grain
不摻水的 straight; unwatered
不摻水酒精 unwatered alcohol

不摻雜 undope
不漏水的 watertight
不漏汽的 steam-tight
不滲透性 impermeability
不滲透性的 impermeable
不腐敗的 imputrescible
不腐爛的 imputrescible
不潔風味 unclean flavo(u)r
不衛生 unwholesomeness
不衛生的 unhealthy; unwholesome
不衛生食品 unwholesome food
不適宜食用的肉類 condemned meat
不適宜食用的產品 condemned product
不適於食用的 inedible
不醇溶的 alcohol-insoluble
不凝固的 incoagulable
不隨意肌 involuntary muscle; unvoluntary muscle
不鏽鋼 stainless steel
不濕膜 non-wettable film
不縮的 shrinkproof
不黏的 non-sticking
不穩定性 unstability
不穩定的 labile; unstable
不穩定蛋白質 labile protein
中力粉 medium strength flour
中心〔部〕 heart
中凸蓋杯傘（蕈類） common funnel cap
中央屠宰 centralized slaughtering
中央屠宰設施 centralized slaughtering facility
中央排氣 central exhaust
中央廚房 central kitchen
中式美點 Chinese petit fours
中式食品 Chinese food
中式香腸 Chinese sausage
中式烹調 Chinese cooking
中式烹調法 Chinese cookery
中式甜點 Chinese petit fours
中式麵條 Chinese noodle
中吻鱘 green sturgeon
中和 neutralize
中和〔作用〕 neutralization
中和值 neutralization value
中和價 neutralization value
中和劑 neutralizer; neutralizing agent
中性 neutrality
中性白土 neutral clay
中性脂肪 neutral fat
中性脂質 neutral lipid
中性胺基酸 neutral amino acid
中性清潔劑 neutral detergent
中性蛋白酶 neutrase
中性醣脂 neutral glycolipid
中果皮 mesocarp; sarcocarp; albedo
中度鹹魚 medium salted fish
中度鹽漬的 medium salted
中毒 intoxication; poisoning; toxication; toxis; venenation
中翅 mid-joint; middle section; wing flat
中脂可可粉 medium fat cocoa
中國白酒 China white wine
中國肉桂 cassia; cassia bark; Chinese cinnamon; cinnamon
中國肉桂油 cassia oil; Chinese cinnamon oil; cinnamon oil
中國栗 Chinese chestnut
中國茶 Chinese tea
中國馬鮫 Chinese mackerel
中國農業標準 Chinese Agricultural Standard (CAS)
中國蜜餞 chowchow
中國豬 Chinese swine
中國醋栗 kiwi (fruit)
中國餐館症候群 Chinese restaurant syndrome
中國薊 Chinese artichoke

中國鵝莓 Chinese gooseberry; kiwi (fruit)
中國臘腸 Chinese sausage
中國鱘 Chinese sturgeon
中甜度酒 pale wine
中甜度葡萄酒 semidry wine
中粗粒麵粉 flour middlings
中粗麵粉 dunst
中等水分食品 intermediate moisture food (IMF); semimoist food
中等身材 mesomorph
中等的 medium
中等的葡萄酒 demifine
中等硬度的硬粒小麥 medium wheat
中等鹹度的魚 medium salted fish
中等鹽漬的 medium salted
中筋麵粉 medium flour; medium strength flour
中華九刺魚 needle mackerel
中華根黴 *Rhizopus chinesis saito*
中華絨螯蟹 mitten crab
中華鯖 Chinese mackerel
中間工廠 pilot plant
中間工廠規模 pilot plant scale
中間工廠試產 pilot plant trial
中間化合物 intermediate compound
中間加熱器 interheater
中間冷卻 intercooling
中間培養物 intermediate culture
中間密度脂蛋白 intermediate density lipoprotein (IDL)
中間產物 intermediate product
中間產品 intermediate product
中間醒發 intermediate proof
中間醒發機 intermediate proofer
中間醒麵 intermediate proof
中間醒麵機 intermediate proofer
中間檢驗台 intermediate proofer
中飯 tiffin
中溫生物 mesophile
中溫細菌 mesophilic bacteria; mesobacteria
中腸 middle gut; midgut; ventriculus
中種麵糰 sponge dough
中酸性食品 medium acid food; semi-acid food
中層包裝 intermediate packaging
中層魚類 mesopelagic fish
中蝦 prawn
中濕〔性〕食品 intermediate moisture food (IMF); semimoist food
丹下臍橙 tange navel
丹毒毒素 erysipelotoxin
丹麥長白豬 Danish landrace; Landrace (swine)
丹麥香腸 Danish sausage
丹麥酥餅 Danish bread
丹麥餡餅 Danish pastry
丹麥藍瑞斯豬 Danish landrace
丹麥蘭德瑞斯豬 Danish landrace
丹寧酸 tannin
丹寧酸酶 tannase
予魚 ling
井水 phreatic water; well water
互溶的 miscible
互補效應 supplementary effect
五〔環〕糖 furanose
五爪甜蘋果 Delicious
五花醃肉 streaky bacon
五香粉 five spices powder
五香碎肉〔餡〕 forcemeat
五香燻肉（羅馬尼亞） pastrami
五倍子 gall; nutgall
五倍子酸 gallic acid
五倍子酸十二酯 dodecyl gallate
五倍子酸丙酯 propyl gallate (PG)
五倍子酸甲酯 methyl gallate
五倍子酸異戊酯 isoamyl gallate

五倍子酸鹽 gallate
五絲馬鱍 bastard mullet
五碳糖 pentose
仁 kernel
仁果 pome; pome fruit; pip fruit; seed fruit
仁果類 pomaceous fruit
什〔錦水〕果啫喱 mixed fruit jelly
什果賓治 cocktail punch
什錦三明治 club sandwich
什錦大麥濃湯 hotchpotch
什錦水果 fruit cocktail; mixed fruit
什錦水果冰淇淋 tutti-frutti
什錦巧克力 assorted chocolate
什錦朱古力 assorted chocolate
什錦色拉 mixed salad
什錦沙拉 mixed salad
什錦沙律 mixed salad
什錦果汁 cocktail punch; cocktail juice
什錦果汁飲料 fruit juice cocktail
什錦果凍 mixed fruit jelly
什錦果醬 mixed fruit jam; mixed jam
什錦泡菜 assorted pickle; mixed pickle
什錦的 assorted
什錦魚湯 court bouillon
什錦魚糕或魚丸 gefillte fish
什錦菜 olio
什錦菜餚 jambalaya
什錦蜜餞 tutti-frutti
什錦酸菜 assorted pickle
什錦餅乾 assorted biscuit
什錦蔬果 macedoine
什錦蔬菜 assorted vegetable; mixed vegetable; vegetable mix
什錦醃菜 assorted pickle; mixed pickle
什錦燉肉 hot pot
什錦醬菜 mixed pickle
介殼 shell; testa; test
介電加熱 dielectric heating
介電乾燥 dielectric drying
介電常數 dielectric constant
介電解凍 dielectric defrosting; dielectric thawing
介壽果 *Anacardium occidentale*
介質 medium
元帥蘋果 Delicious
元素 element
元冧片 flank steak
內分泌 endocrine
內分泌學 endocrinology
內比奧洛〔釀酒紅〕葡萄 Nebbiolo
內牛腱 topround
內包裝 interior packaging; inner packaging
內生的 endogenous
內生孢子 endospore
內因子 intrinsic factor
內在因素 intrinsic factor
內收肌 adductor (muscle)
內肌膜 endomysium
內含物 content
內里脊肉 back fillet
內果皮 endocarp
內股肌肉 gracillis
內孢霉素 endomycin
內毒素 endotoxin
內相 internal phase
內能 intrinsic energy
內骨骼 endoskeleton
內側後腿肉 inside (round); topside (round)
內側後腿肉排 topside steak
內稀蛋白 inner thin albumen
內菌幕 inner veil
內溶菌素 endolysin
內源的 endogenous
內裝〔法〕 inner packaging
內酯 lactone
內酯酸 lactonic acid

內酯酶 lactonase
內聚力 cohesive force; cohesion
內聚性 cohesive
內聚現象 coherence; cohesion
內膜 inner membrane
內壁 inner wall
內酰胺 lactam
內臟 internal organ; entrails; gut; innards; insides; numbles; offal; pluck; umbles; viscera; viscus
內臟粉 visceral meal
內臟脂肪 offal fat; visceral fat
內臟器官 visceral organ
內臟器官粉 visceral meal
內黴素 endomycin
內肽酶 endopeptidase
六亞甲基四胺 hexamethylene tetramine
六偏磷酸鈉 sodium hexametaphosphate
六偏磷酸鹽 hexametaphosphate
六帶鰺 banded cavallal
六環果糖 pyranofructose
六環葡糖 pyranoglucose
六環糖 pyranose
公升 liter (L)
公斤 kilogram (kg)
公牛 bull; ox; oxen
公司三明治 club sandwich
公共飲食業用包裝 institutional pack
公合 deciliter
公式 formula
公羊 buck; ram; tup
公兔 buck
公泊 centipoise (cp)
公馬 stallion
公開品評試驗 open panel test
公筷 serving chopsticks
公路廉價餐廳 transport café
公綿羊 tup
公認安全 generally recognized as safe (GRAS)
公認安全物質 generally recognized as safe substance (GRASS)
公認為安全的 generally recognized as safe (GRAS)
公認無毒 generally recognized as safe (GRAS)
公撮 milliliter (ml)
公豬 boar; hogget
公鴨 drake
公雞 cock; rooster; stag
公犢 bull calf
分子 molecule
分子式 molecular formula
分子量 molecular weight
分子團 micella(e); micelle
分切肉 carved meat
分切圖 cuts chart
分出乳清 wheying off
分光光度計 spectrophotometer
分光光度測定法 spectrophotometry
分光密度測定法 spectrodensitometry
分光熒光測定法 spectrofluorometry
分成四份 quarter
分成兩半 halve
分批 batch
分批巴氏滅菌器 batch pasteurizer
分批送料機 batcher
分批殺菌器 batch sterilizer
分枝孢子菌屬 *Cladosporium*
分枝桿菌屬 *Mycobacterium*
分析化學 analytical chemistry
分析天平 analytical balance
分析柱 analytical column
分析員 analyser; analyzer
分析專家 assayer
分析試劑 analytical reagent
分析器 analyser; analyzer

分泌 excrete
分泌物 excreta
分泌花蜜的植物 nectariferous plant
分泌毒液的 venomous
分段滅菌 fractional sterilization; intermittent sterilization; tyndallization
分段解凍 differential thawing
分流 diffluence
分級 fractionation; sorting; sizing; triage
分級〔細〕篩 grading sifter
分級台 cull eliminator
分級乾燥 stage drying
分級萃取 stage extraction
分級圓筒篩 assorting cylinder
分級滾筒 assorting cylinder
分級器 classificator
分級機 classifier; grader; grading machine; sorter; sizer
分級篩 grading sieve; sizing screen
分配色層法 partition chromatography
分配層析法 partition chromatography
分配器 dispenser
分粒機 classificator
分割 cut up; cutting
分割好的 dressed
分割肉 carved meat; cut; meat cut
分割後的重量 dressed weight
分割機 splitter; splitting machine
分散〔作用〕 dispersion; peptization
分散介質 dispersion medium
分散相 dispersed phase
分散液 dispersion
分散媒介 dispersion medium
分散膠體 dispersoid
分散劑 dispersant; dispersing agent
分菜刀 serving knife
分菜叉 serving fork
分菜匙 serving spoon
分菜筷 serving chopsticks
分裂 split; splitting
分裂果 schizocarp
分裝 fractional pack
分裝食品 prepackaged food
分裝填充機 stuffer dispenser
分解〔作用〕 decomposition; disintegration
分解代謝 catabolism
分解烏頭酸 itaconic acid
分解產物 decomposition product
分解槽 decomposer
分解糖的 sucroclastic
分解體 decomposer
分解酶 lytic enzyme
分蜜機 honey extractor
分層〔現象〕 stratification
分層殺菌法 strata-cock process
分凝〔作用〕 dephlegmation; segregation
分選 sorting
分選機 cull eliminator; grader; sorter
分選篩 separating screen
分離 abstract; isolation; parting; separation; splitting
分離〔作用〕 segregation
分離大豆蛋白〔質〕 soy protein isolate; soybean protein isolate; SPI; isolated soybean protein
分離物 isolate
分離柱 separation colume
分離培養基 isolation medium
分離蛋白質 protein isolate
分離魚蛋白 fish protein isolate
分離槽 separating tank
分離劑 dispersing
分離器 separator
分離濃縮 separate concentration
分餾 dephlegmation; fractional distillation; fractionating distillation; fractionation

分餾柱 fractional column; fractionating column
分餾塔 dephlegmator; fractional column; fractionating column
分餾器 dephlegmator; fractionator
分醪法 half sponge method
分瓣機 segment separator
分類 sorting
分葱 eschalot; shallot
切丁機 cuber; cube cutter; dicer; dicing machine
切刀 cutter
切口 notch
切片 slice; slicing; section
切片機 section cutter; skiving machine; slicer; slabber; snipper
切片麵包 sliced bread
切半 halving
切去外皮 paring; pare
切皮機 hide cutter
切成方塊 diced
切成薄片 slicing
切肉 cutting
切肉丁機 meat dicer
切肉刀 butcher knife; carver
切肉片機 mea slicer
切肉板 cutting board
切肉機 hasher; meat cutter
切角 notch
切板機 slitter; slitting machine
切屑 trimmings
切除器 exsector
切條機 slitter; slitting machine; strip cutter
切粒 diced
切魚片機 filleter
切割 cutting
切割包裝機 cut-wrap machine
切割機 cutter
切斯特白豬 chester white
切絲 shred; shredding
切絲機 shredder; shredding machine
切圓片機 ring slicer
切塊級 U.S. Cutter
切塊機 block cutter; cube cutter; slabber
切碎 cut up; mincing; mince; shredding
切碎機 chopper; chopping machine; hasher; mincer; mincing machine; shredder; shredding machine
切達乾酪 cheddar cheese
切麵條機 noodle cutter; noodle cutting machine
切變 shear
切變黏度 shear viscosity
勻流 uniform flow
勻濕的小麥 tempered wheat
化合水 combined water; constitution water
化合物 compound
化能營養細菌 chemotrophic bacteria
化淡 desalination
化學 chemistry
化學分析 chemical analysis
化學去皮 chemical peeling
化學合成 chemosynthesis
化學成分 chemical composition
化學防腐劑 chemical preservative
化學性食物中毒 chemical food poisoning
化學的 chemical
化學消毒劑 chemical sanitizer
化學添加物 chemical additive
化學添加劑 chemical additive
化學發麵劑 chemical leavening
化學滅菌 chemical sterilization
化學滅菌劑 chemical sterilizer; chemosterilant
化學試劑 chemical reagent
化學調味料 chemical seasoning
化學穩定劑 chemical stabilizer

化學藥品 chemical
化糞池 septic tank
化驗 assay
化驗員 assayer
午後〔擠〕乳 afternoon milk
午後茶會 tea fight; tea party
午後茶點 afternoon tea
午後茶點時間 teatime
午茶 high tea; meat tea
午飯 tiffin
午餐 dejeuner; lunch; luncheon; tiffin
午餐肉 luncheon meat; pork luncheon meat
午餐豬肉 pork luncheon meat
升血糖激素 glucagon
升降門室烤爐 coverplate oven
升華乾燥 freeze drying; freeze dehydration; lyophilization
升華乾燥機 lyophilizer
升膜式蒸發器 climbing film evaporator; rising film evaporator
反丁烯二酸 fumaric acid
反丁烯二酸一鈉 monosodium fumarate
反向拉舌 reverse tab
反乳化 demulsify
反乳化〔作用〕 deemulsification; demulsification
反乳化劑 demulsifier; demulsifying agent; deemulsifying agent
反油酸 elaidin
反胃 nausea; regurgitation
反射計 reflectometer
反芻〔作用〕 regurgitation; rumination
反芻的食物 cud
反芻胃 reticulorumen
反芻動物 ruminant
反芻動物第三胃 manifold; omasum
反絮凝〔作用〕 deflocculation
反絮凝劑 deflocculating agent
反絮凝離心機 deflocculator
反滴定 back titration
反滲透 reverse osmosis
反應 reaction
反應〔活〕性 reactivity
反應物 reactant
反應器 reactor
反覆試驗 trial and error
壬酮 nonanone
壬酸 nonanoic acid; pelargonic acid
壬酸乙酯 ethyl nonanoate
壬醇 nonyl alcohol
壬糖 nonose
壬醛 nonanal
天〔門〕冬胺酸 aspartic acid
天〔門〕冬胺酸酯 aspartate
天〔門〕冬胺酸酶 aspartase
天〔門〕冬酰胺酶 asparaginase
天冬苯丙二肽酯 aspartame
天冬胺酸一鈉 monosodium aspartate
天冬甜素 aspartame
天冬酰胺〔酸〕 asparagine
天平 balance
天平盤 scale pan
天貝(印尼傳統食品) tempeh
天使蛋糕 angel cake
天使魚 angelfish
天使髮絲 angel hair; capellini
天波蘿 jaca; jackfruit; jak; jakfruit; nangka
天竺葵 pelargonium
天竺葵色素苷 pelargonidin
天竺葵定 pelargonidin
天竺葵酸 pelargonic acid; punicic acid
天竺葵苷 pelargonin
天竺葵苷素 pelargonidin
天竺薄荷 patchouli
天竺薄荷油 patchouli oil
天芥菜 heliotrope; tournesol

天門冬素 asparagine
天門冬酰苯丙氨酸甲酯 aspartame
天津紹菜 Tientsin cabbage
天秤 scale
天婦羅 tempura
天然〔未經調味的〕酸乳酪 natural yoghurt
天然化合物 native compound
天然水 native water; natural water
天然瓦斯 natural gas
天然色料 natural color(ing)
天然色素 natural color(ing); natural pigment
天然果汁 natural juice; natural fruit juice
天然的 natural
天然毒素 natural toxicant
天然風味 natural flavo(u)r
天然食用色素 natural food colo(u)r
天然食用香料 natural flavo(u)r
天然食物 natural food
天然食品 natural food
天然食品〔著〕色料 natural food colo(u)r
天然食品添加物 natural food additive
天然食品添加劑 natural food additive
天然香辛料 natural spice
天然香料 natural flavo(u)r; natural flavo(u)ring substance
天然香精 natural essence
天然氣 natural gas
天然乾酪 natural cheese
天然乾燥 natural drying
天然培養基 natural medium
天然康乃克油 cognac oil, green
天然甜味料 natural sweetener
天然甜味劑 natural sweetener
天然產物 natural product; natural stock
天然莧菜紅 natural amaranthus red
天然蛋白質 native protein
天然提取香料 natural extractive flavo(u)ring substance
天然發酵 natural fermentation
天然著色料 natural color(ing)
天然腸衣 natural casing
天然資源 natural resource
天然飼養 natural feeding
天然碳酸鈉 natron
天然綠3號 magnesium chlorophyll
天然增稠劑 natural thickener; natural thickening agent
天然膠體 natural colloid
天然調味料 natural seasoning
天然樹膠 natural gum
天然橡膠 natural rubber
天然礦泉水 natural mineral water
天然纖維 natural fiber
天然纖維素 native cellulose
天然釀造 natural brewing
天鵝 swan
太平洋〔大〕牡蠣 pacific oyster
太平洋小鱈 pacific tomcod
太平洋毛鱗魚 pacific capelin
太平洋沙瑙魚 pacific sardine; pacific pilchard
太平洋狗鱈 pacific hake
太平洋狐鰹 pacific bonito
太平洋馬舌鰈 oil halibut
太平洋庸鰈 pacific halibut
太平洋細齒鮭 oolachan
太平洋無鬚鱈 pacific hake
太平洋蛤蜊 ocean quahaug
太平洋黃鰭鮪 pacific yellowfin tuna
太平洋黑比目魚 pacific black halibut
太平洋鯡 pacific herring
太平洋鰻魚 pacific lamprey
太平洋鱈 pacific cod
太平洋鱸魚 pacific ocean perch
太妃糖 taffy; toffee; toffy
太空食品 space food
太陽能乾燥 solar drying

太陽魚 sunfish
孔 pore
孔口 orifice
孔口混合器 orifice mixer
孔雀石綠 malachite green
孔雀椰子 Indian sago palm; wine plam
少許 soupcon
少量 trifle
少量食物 nibble
少量烈酒 tot
少量飲料 noggin
尺寸 size
尺寸過小 undersize
巴比丘 barbecue; barbeque
巴比妥 barbital
巴比妥酸鹽 barbiturate
巴氏效應 Pasteur effect
巴氏殺菌 pasteurize
巴氏殺菌〔法〕 pasteurization
巴氏殺菌不足 underpasteurization
巴氏殺菌牛乳 pasteurized milk
巴氏殺菌啤酒 pasteurized beer
巴氏殺菌理論 pasteurism
巴氏殺菌過度 overpasteurization
巴氏殺菌器 pasteurizer
巴氏溫度計 pastometer
巴氏瞬間殺菌 flash pasteurization
巴氏濾菌器 Pasteur filter
巴布科克〔乳油〕瓶 Babcock bottle
巴布科克〔乳脂〕瓶 Babcock flask
巴布科克〔乳脂〕試驗 Babcock test
巴弗特試驗 Barfoed's test
巴伐利亞〔奶油〕甜點 Bavarian cream
巴伐利亞香腸 Bavarian sausage
巴伐利亞啤酒 Bavarian beer
巴式〔單醣〕試驗 Barfoed's test
巴西巴巴蘇棕櫚仁果 babassu kernel oil
巴西巴巴蘇棕櫚果 babassu nut
巴西肉乾 charqui
巴西果 brazil nut; cream nut; niggertoe
巴西栗 brazil nut; niggertoe
巴西棕櫚 cohune
巴西棕櫚果 babassu nut
巴西棕櫚果仁油 babassu kernel oil
巴西棕櫚酸 carnauba wax
巴西對蝦 brazilian prawn
巴克夏〔黑白〕豬 Berkshire
巴克啤酒 bock (beer)
巴豆酸己酯 hexyl-2-butenoate
巴拉他膠 balata
巴拉圭茶 mate; Paraguay tea; yerba mate
巴拉圭茶鹼 yerbine
巴拉金糖 palatinose
巴拉塔樹膠 balata
巴林比重度 Balling (scale)
巴林比重計 Balling hydrometer
巴林度 Balling (scale)
巴林計 Balling hydrometer
巴思甜麵包 Bath bun
巴思餅乾 Bath biscuit
巴美乾酪 Parmesan (cheese)
巴倫西亞橘烯 valencene
巴特納米 Carolina rice
巴馬芝士 Parmesan (cheese)
巴梨 Bartlett (pear)
巴袞蝦醬 bagoong
巴敦醬 patum peperium
巴斯馬蒂米 basmati rice
巴斯啤酒 bass
巴斯德效應 Pasteur effect
巴斯德培養燒瓶 Pasteur flask
巴斯德殺菌〔法〕 pasteurization
巴斯德濾菌器 Pasteur filter
巴羅洛葡萄酒 barolo
巴蘭葉 panda leaf
巴鰹 oceanic bonito
引發劑 initiator
引餐盤 service plate

心肌 cardiac muscle; myocardium
心肌纖維 cardiac muscle fiber
心磷脂 cardiolipin
心臟 heart
手 hand
手工打蛋器 hand whisk
手工屠宰 hand-kill
手工揉製的麵糰 hand dough
手工開膛去內臟 hand dressing
手工裝罐機 hand pack filler
手工製作的 handmade
手工擠乳 hand milking
手工攪拌器 hand whisk
手爪 paw
手冊 manual
手石磨 quern
手抓食品 finger food
手拉麵 somen
手指形包糖霜衣酥餅 éclair
手指狀鬆脆餅 lady's finger
手動〔螺旋式〕壓榨機 hand screw press
手動的 manual
手動封罐機 hand seamer
手控的 manual
手推車 barrow; handbarrow; trolley
手推磨 quern
手掌 palm
手提式的 portable
手提爐 portable stove
扎肉 galantine
支持者 backer
支鏈 side chain
支鏈反應 branched chain reaction
支鏈胺基酸 branched chain amino acid
支鏈澱粉 amylopectin
支鏈澱粉酶 pullulanase
文氏麴菌 *Aspergillus wentii*
文火 low fire; mild fire; slow fire; soft fire
文火燉肉 simmered meat
文火燉煮 coddle
文旦 pomelo; pompelmous; pummelo; shaddock
文冠果 shinyleaf yellowhorn
文蛤 hard clam; hard-shelled clam
文鯿 vimba
斗式升運機 scooper
斤 catty
方切後腿肉 butt square cut
方旦〔半軟〕糖 fondant
方形香腸 blockwurst
方形罐 rectangular can; rectangular tin
方便食品 convenience food
方便麵 instant noodle; ramen noodle
方塊 cube
方塊冰 cube ice
方糖 block sugar; brick sugar; cube sugar; cut sugar; knob; loaf sugar; sugar cube
方糖鉗子 sugar tongs
方糖鑷子 sugar tongs
方頭魚 tilefish
日本〔花〕鯖 pacific mackerel
日本叉牙魚 Japanese sandfish
日本叉齒魚 Japanese sandfish
日本方頭魚 blanquillo
日本木瓜 Japanese flowering quince
日本竹莢魚 Japanese horse mackerel
日本豆〔麵〕醬 miso
日本豆醬 fermented soybean paste
日本拉麵 ramen noodle
日本芥茉 Japanese horseradish; wasabi
日本金梭魚 Japanese barracuda
日本長尾雞 Japanese longtailed fowl
日本食品添加物公定書 Japanese Standard of Food Additives (JSFA)
日本食品添加劑標準 Japanese Standard of Food Additives (JSFA)
日本夏橙 Japanese summer orange

日本根黴 *Rhizopus japonicus*
日本馬頭魚 blanquillo
日本康吉鰻 beach conger
日本梨 Japanese pear
日本清酒 Japanese rice wine; sake
日本粗茶 bancha (tea)
日本野木瓜 Japanese stauntonvine
日本單鰭電鰩 torpedo
日本黃姑魚 Japanese croaker
日本農林規格 Japanese Agricultural and Forestry Standard; Japanese Agricultural Standard (JAS)
日本農林標準 Japanese Agricultural and Forestry Standard; Japanese Agricultural Standard (JAS)
日本對蝦 kuruma prawn
日本綴錦蛤 short necked clam
日本鳶魟 butterfly ray
日本鋸鯊 saw shark
日本鯖 chub mackerel
日本鯰 Japanese catfish
日本鯡魚 Japanese herring
日本鯷〔魚〕 half mouthed sardine; Japanese anchovy
日本鬚鯊 fringe shark
日本鰻鱺 Japanese eel
日本纖鸚嘴魚 Japanese parrotfish
日本蘿蔔 daikon (radish); Japanese radish
日本鸚鯉 Japanese parrotfish
日光乾燥 solar drying
日光蒸發 solar evaporation
日式肉片火鍋 sukiyaki
日灼 sunburn
日灼味 sunlight flavo(u)r; sunstruck flavo(u)r
日灼斑果 sunburned fruit
日射病 insolation
日常食品 daily food
日程表 time table
日落黃 sunset yellow FCF
日落黃鋁色澱 sunset yellow FCF aluminium lake
日糧 rations
日曬 insolation
日曬味 sunlight flavo(u)r; sunstruck flavo(u)r
曰月貝 rock scallop
月下香浸膏 tuberose concrete
月下香淨油 tuberose absolute
月見草油 evening primrose oil
月夜茸 oyster mushroom
月桂 bay
月桂油 bay leaf oil; bay oil; laurel leaf oil; laurel oil; myrcia oil
月桂脂 laurel butter
月桂基硫酸鈉 sodium dodecyl sulfate (SDS); sodium lauryl sulfate
月桂烯 myrcene
月桂烯酸 lauroleic acid
月桂烯醇 myrcenol
月桂葉 bay leaf; laurel leaf
月桂葉油 bay leaf oil; bay oil; laurel leaf oil; laurel oil
月桂酮 laurone
月桂酸 dodecanoic acid; lauric acid
月桂酸乙酯 ethyl laurate
月桂酸酯 laurate
月桂酸鹽 laurate
月桂醇 lauryl alcohol
月桂樹 laurel; sweet bay
月桂樹果實 bayberry
月桂醛 lauraldehyde; lauryl aldehyde
月魚 opah
月餅 moon cake
木水桶 cannikin
木本的 woody
木本植物 xylophyta
木瓜 Chinese quince; pawpaw; papaya
木瓜蛋白酶 papain
木瓜酶 papain

木生蘑菇 red staining mushroom
木耳 *Auricularia auricula*; jelly fungus; jew's ear; wood ear; wooden ear
木耳屬 *Auricularia*
木豆 pigeon pea
木味 woody flavo(u)r
木松香甘油酯 glycerol ester of wood rosin; glyceryl triabietate
木松香季戊四醇酯 pentaerythritol ester of wood rosin
木波蘿 jaca; jackfruit; jak; jakfruit; nangka
木油 wood oil
木炭 charcoal
木栓質 suberin
木桶 cask
木桶味 cask smell
木桶箍 cask hoop
木通 akebia
木斯(泡沫狀奶油甜點) mousse
木棉 kapok
木棉花 kapok
木棉籽油 kapok oil; silk-cotton seed oil
木棉樹 silk-cotton tree
木犀屬 osmanthus
木酮糖 xyloketose; xylulose
木酮糖激酶 xylulokinase
木察樹油 sassafras oil
木碟 trencher
木精 methanol; methyl alcohol; wood alcohol; wood spirit
木聚糖 xylan
木聚糖酶 xylanase
木製小桶 firkin
木盤 trencher
木質的 woody
木質素 lignin
木質部 xylem
木質纖維素 lignocellulose
木醇 methanol; methyl alcohol; wood alcohol; wood spirit
木醋液 wood vinegar
木醋酸 pyroligneous acid
木槿黃素 hibiscetin
木器 wooden ware
木糖 wood sugar; xylose
木糖〔型〕抗壞血酸 xyloascorbic acid
木糖異構酶 xylose isomerase
木糖酸 xylonic acid
木糖醇 xylitol
木糖醇酐單硬脂酸酯 xylitan monostearate
木糖醛酮 xylosone
木糖苷 xyloside
木頭風味 woody flavo(u)r
木樨花浸膏 osmanthus concrete
木薯 cassava; manihot; manioc; tapioca
木薯粉 cassava flour; cassava meal; farinha; tapioca flour; tapioca
木薯澱粉 cassava starch; tapioca starch
欠飽和 undersaturation
欠熟食物 underfired food
止咳糖 cough drop; cough sweets
止咳糖漿 cough syrup
止咳藥片 cough drop
止渴 quench(ing)
止渴飲料 quencher
比目魚 flatfish; flounder; sole
比目魚的肉片 flitch
比色分析〔法〕 colo(u)rimetric analysis
比色板 colo(u)r disc; colo(u)r disk
比色法 colo(u)rimetry
比色計 colo(u)rimeter; chromatometer
比色測定〔法〕 colo(u)rimetric determination
比色盤 colo(u)r disc; colo(u)r disk
比利時林堡乾酪 Limburg(er) cheese
比利時寧堡乾酪 Limburg(er) cheese

比思吉 biscuit
比重 specific gravity; specific weight
比重天秤 gravity balance
比重計 areometer; densimeter; densitometer; hydrometer
比重瓶 gravity bottle; pycnometer
比容 specific volume
比密度 specific density
比爾森型啤酒 Pilsner beer
比熱 specific heat
比濁法 nephelometry
比濕度 specific humidity
比黏度 specific viscosity
比薩 pizza
比體積 specific volume
毛 hair
毛瓜 fuzzy melon
毛地黃皂苷 digitonin
毛地黃毒苷 digitoxin
毛地黃苷 digitalin
毛尖 tea shoot
毛尖茶 tippy tea
毛肚 manifold
毛豆 green soybean; jointed wax gourd
毛油 crude oil
毛茄 gumbo
毛重 gross weight; packed weight
毛細管 capillary
毛細管現象 capillarity
毛細管黏度計 capillary viscosimeter
毛蚶 ark shell
毛霉菌〔屬〕 Mucor
毛頭鬼傘（蕈類） lawyer's wig
毛鞭蟲 trichuris
毛鞭蟲病 trichuriasis
毛蟹 hairy crab; mitten crab
毛鱗魚 capelin
毛黴菌〔屬〕 *Mucor*
水 aqua; eau
水不溶性 water insoluble
水中油型乳〔濁〕液 oil-in-water emulsion (o/w emulsion)
水分 moisture
水分代謝 water metabolism
水分平衡 moisture equilibrium; water balance
水分吸附等溫線 water sorption isotherm
水分含量 moisture content
水分活性 water activity (Aw)
水分計 moisture meter
水化性 hydratability
水化學 hydrochemistry
水化酶 hydrase
水牛 buffalo; water buffalo
水牛莓 buffalo berry
水牛酪油 ghee; ghi
水仙浸膏 narcissus concrete
水包油型乳〔濁〕液 oil-in-water emulsion (o/w emulsion)
水母 jellyfish; medusa; sea jelly
水生真菌類 aquatic fungi
水生動物 aquatic animal; aquatic plant; hydrophyte; water plant
水田 paddy field
水田芥 watercress
水份過多的 watery
水份調詳 conditioning
水冰 water ice
水合〔作用〕 hydration
水合性 hydratability
水合物 hydrate
水合的 hydrous
水合過度 overhydration
水合熱 heat of hydration
水合酶 hydrase
水污染 water pollution
水冷 hydrocooling
水冷卻 water cooling

水冷凝器 water condenser
水含量 water content
水坑 sump
水汪汪的 watery
水汽 water vapo(u)r
水芋屬 *Calla*
水底生物 benthon
水底魚 groundling
水底植物 benthophyte; phytobenthos
水果 fruit; greengrocery
水果〔奶油〕布丁 charlotte
水果〔貯〕窖 fruitery
水果〔塗〕醬 fruit spread
水果〔類〕 garden stuff
水果丁冰淇淋 tutti-frutti
水果刀 fruit knife
水果小餡餅 tart
水果加冰淇淋的混合甜品 split
水果去核器 fruit seeder
水果布丁 fruit pudding
水果皮 fruit skin
水果色拉 fruit salad
水果夾心奶油蛋糕 charlotte
水果夾心冰磚 auffait
水果沙司 fruit sauce
水果沙拉 fruit salad
水果沙律 fruit salad
水果店 fruitery
水果果凍 fruit jelly
水果削皮刀 fruit parer
水果削皮器 fruit parer
水果派 cobbler; fruit pie
水果香精 fruit essence
水果氣味 fruit odo(u)r
水果烤布丁 betty
水果酒 fruit wine
水果商 fruiterer
水果甜麵包 farthing bun
水果硬糖 fruit drops
水果飲料 fruit beverage
水果過熟腐爛 blet
水果榨汁機 fruit press
水果榨渣 marc
水果蜜餞 candied fruit
水果盤 fruit dish
水果醋 fruit vinegar
水果糕 fruit cake
水果糖漿 fruit syrup
水果餡餅 cobbler; fruit pie
水果醬〔料〕 fruit sauce
水果麵包 stollen
水果罐裝 canned fruit
水泡 blister
水狀膠體 hydrocolloid
水芹 cress; water celery; water dropwort
水芹烯 phellandrene
水表 water gauge; water meter
水垢 incrustation; scale
水封 water seal(ing)
水封式〔連續〕殺菌釜 hydrolock retort
水洗 degum; degumming; wet cleaning
水活性 water activity (Aw)
水玻璃 sodium metasilicate
水相 water phase
水庫 reservoir
水栗子 caltrop
水栽法 hydroponics
水浸〔泡〕 water immersion
水浸〔漬〕 water soaking
水浸解 water retting
水浸漬 wet steep
水浴 water bath
水浴〔燉〕鍋 bain-marie
水耕 water culture
水耕法 hydroponics; tray agriculture
水酒（東正教會） krama
水密性 watertightness
水密的 watertight

水梨花 Arabian jasmine
水淋式冷卻器 trickling cooler
水淨化器 water purifier
水產油〔脂〕 marine oil
水產物 aquatic product
水產品 aquatic product; fishery product; marine product
水產品抽提物 sea food extract
水產食品 sea food
水產動物 fisheries animal
水產業 aquaculture; fishery; fishery industry
水產養殖〔業〕 aquaculture
水袋 water bag
水軟化裝置 water softener
水軟化劑 water softener
水鳥 water bird
水壺 jug; water jug
水提取 water extraction
水提取物 water extract
水晶 crystal
水晶軟糖 jelly sweets
水晶龍頭魚 bombay duck
水煮 poaching
水煮食品 boiled food
水煮香腸 boiled sausage
水煮荷包蛋 dropped egg; poached egg
水煮餐 boiled dinner
水硬度 water hardness
水稀釋牛乳 milk-and-water
水萃〔取物〕 water extract
水萃取 water extraction
水越桔 bogberry
水菖蒲 sweet flag
水塔 cistern; water cistern
水楊酸 salicylic acid
水楊酸甲酯 methyl salicylate
水楊酸苯乙酯 phenethyl salicylate
水楊酸異丁酯 isobutyl salicylate
水楊酸異戊酯 isoamyl salicylate
水楊酸酸乙酯 ethyl salicylate; salicylic ether
水楊酸苄酯 benzyl salicylate
水楊醛 salicylaldehyde
水溶〔性〕的 hydrosoluble; water-soluble
水溶性 hydrosoluble; water solubility
水溶性固形物 water-soluble solid
水溶性油 soluble oil
水溶性維他命 water-soluble vitamin
水溶性維生素 water-soluble vitamin
水溶性膳食纖維 soluble dietary fiber
水溶液 aqueous solution
水溶膠 hydrosol; lyosol
水源 spring; water source
水滑魚 shad
水準 standard
水當量 water equivalent
水禽 water bird; water fowl
水腫 dropsy; edema
水葵 water shield
水葉 waterleaf
水解 hydrolysis
水解〔性〕酸敗 hydrolytic rancidity
水解分解 hydrolytic decomposition
水解玉米蛋白 hydrolyzed corn protein
水解物 hydrolyte
水解降解 hydrolytic degradation
水解動物蛋白 hydrolyzed animal protein (HAP)
水解產物 hydrolysate; hydrolyzate
水解蛋白〔質〕 hydrolyzed protein; protein hydrolysate; protein hydrolyzate
水解麥芽汁 hydrolyzed mash
水解植物蛋白 hydrolyzed plant protein (HPP); hydrolyzed vegetable protein (HVP)
水解粟米蛋白 hydrolyzed corn protein
水解葡聚糖 hydroglucan

水解酵素 hydrolytic enzyme
水解質 hydrolyte
水解劑 hydrolytic agent
水解器 hydrolyzer
水解纖維素 hydrocellulose
水解罐 hydrolyzer
水解酶 hydrolase; hydrolytic enzyme
水滴 drip
水漂生物 pleuston
水管 water pipe
水蒸汽 steam; water vapo(u)r
水銀 mercury
水銀中毒 hydrargyrism; mercurialism
水銀溫度計 mercurial thermometer
水樣 wateriness
水樣肉 pale, soft, exudative muscle (PSE muscle)
水樣豬肉 watery pork
水槽 gutter; sink; water tank
水稻 paddy field rice; rice
水箱 water tank
水膠體 hydrocolloid
水質 water quality
水質化學 hydrochemistry
水質污染 water pollution
水質標準 water standard
水凝膠 hydrogel; lyogel
水磨〔機〕 water mill
水鴨 teal (duck)
水龍頭 tap; water tap
水壓殺菌釜 hydrostatic retort
水壓滅菌〔法〕 hydrostatic sterilization
水壓滅菌器 hydrostatic sterilizer
水壓蒸煮器 hydrostatic cooker
水簾 water curtain
水霧化器 mistorizer
水懸膠體 lyosol
水蘇四糖 lupeose; stachyose
水蘇鹼 stachydrine
水迹葉 waterleaf
火候不足的 undercooked
火麻仁 cannabis seed; hemp seed
火管鍋爐 tubular boiler
火腿 ham
火腿〔模〕壓機 ham press
火腿三文治 ham sandwich
火腿三明治 ham sandwich
火腿切片機 ham slicer
火腿片 sliced ham
火腿成形模 ham retainer
火腿注射鹽水醃製〔法〕 ham pumping
火腿泵醃〔法〕 ham pumping
火腿風味 hammy flavo(u)r
火腿香味的 hammy
火腿捆紮機 roll tying machine
火腿罐頭 canned ham
火燄殺菌 sterilvac
火燄游離偵測器 flame ionization detector
火燄菜 chard
火燄滅菌 flame sterilization; sterilvac
火燄滅菌法 steriflame process; steriflame sterilization
火燄滅菌器 flame sterilizer
火燄離子化檢定器 flame ionization detector
火龍果 pitaya
火鍋 chafing dish
火雞 turkey
火雞肉 turkey
火爐 stove
爪 claw
爪哇咖啡 Java coffee
爪哇茶 Java tea
爪哇種矮雞 bantam
爪旋蓋 regular twist-off cap; twist off cap
爪篱 cullender
片 piece; slice
片〔劑〕 tablet

片式〔巴氏〕殺菌器 plate pasteurizer
片式加熱器 plate heater
片式過濾器 sheet filter
片式熱交換器 plate (type) heat exchanger
片吸蟲病 fascioliasis
片狀冰 plate ice; scale ice
片狀魚膠 leaf isinglass
片狀鹽 flake salt
片劑 tabloid; troche
片糖 brown slab sugar; brown sugar slab; sugar slab
牙買加甜酒 Jamaica
牙買加酢漿草 Jamaica sorrel
牙買加辣椒 allspice
牙買加醋栗 Jamaica gooseberry
牙買加薑汁烈酒 jake
牙買加薑萃 jake
牙買加鵝莓 Jamaica gooseberry
牙買加櫻桃 Jamaica cherry
牙膏管 collapsible tube
牙蓋 lug cap
牙齒 tooth
牙籤 toothpick
牛 cattle; neat
牛小排 short ribs
牛心梨 bullock's heart; custard apple
牛仔骨 short ribs
牛半胴體的後半部 hind quarters
牛奶 milk
牛奶卡布奇諾咖啡 cappuccino chiaro; light cappuccino
牛奶布丁 custard pudding
牛奶朱古力 milk chocolate
牛奶低頻交流電導殺菌法 electropure process
牛奶固形物 milk solids
牛奶流質膳食 milky fluid diet
牛奶凍 blancmange
牛奶甜酒 posset
牛奶產品 milk product
牛奶脫脂 milk separation
牛奶麥粥 frumenty
牛奶場 dairy; milk house
牛奶進廠時的平台試驗 platform test
牛奶飲料 milk drink
牛奶過濾器 milk strainer
牛奶電流殺菌法 electropure process
牛奶樹 cow tree
牛奶餐 milk diet
牛奶儲槽 milk tank
牛皮紙 kraft (paper)
牛皮菜 chard; leaf beet; Swiss chard
牛肉 beef
牛肉〔濃〕湯 beef broth
牛肉丸 beef ball
牛肉火腿 beef ham
牛肉香腸 beef sausage; kalbwurst
牛肉凍 beef royal
牛肉浸汁 beef extract
牛肉乾 dried beef; jerky; jerked beef; tasajo
牛肉培根 beef bacon
牛肉捲 beef roll
牛肉脯 dried beef
牛肉等級 beef grade
牛肉萃 beef extract
牛肉煙肉 beef bacon
牛肉漢堡 beefburger
牛肋條 intercostal meat; rib fingers
牛肋條肉 short ribs
牛至 oregano; origanum
牛至油 origanum oil
牛至草 oregano; origanum; wild marjoram
牛舌 neat's tongue
牛舌草 bugloss
牛舌臘腸 zungenwurst
牛血清白蛋白 bovine serum albumin
牛肝菌 bolete; boletus

牛肚 beef tripe
牛角麵包 croissant; crescent
牛豆 black eyed bean
牛里脊肉 rib eye
牛里脊肉片 filet mignon; fillet mignon
牛乳 cow milk; milk
牛乳巧克力 milk chocolate
牛乳酒 posset
牛乳處理廠 milk house
牛乳麵包 lactein bread
牛油 beef suet; butter
牛油曲奇餅 butter cookie
牛油豆 butter bean
牛油果 aguacate; alligator pear; avocado; butter-fruit
牛油表面的硬表皮 kante
牛油紙 butter paper
牛油硬脂 oleostearin(e)
牛油樹 shea butter tree
牛油樹果脂 shea nut fat
牛油樹油 shea butter
牛油樹脂 shea butter
牛的 bovine
牛盲腸 bung
牛股肉 round
牛肩肉 blade; clod
牛肩肉排 bolar blade steak
牛肩頸肉 chuck
牛初乳 beestings
牛表層胸肉 thin shirt
牛軋糖 nougat
牛前肋骨肉 foreribs
牛前肩脊肉 shoulder strap
牛前胸肉 plate end
牛前脛 foreshank
牛厚里脊肉 sirloin butt
牛後腰肉 steak piece
牛後腿內側肉 topside (round)
牛後腿外側肉 silver side
牛後腿肉 round
牛柳 tenderloin steak
牛柳陰 flank
牛脂 beef tallow; beef suet; tallow
牛胴胸肋肉 foreribs
牛胴腱肉 topround
牛胸肉 brisket; navel end; plate
牛排 beef steak; steak
牛排拍鬆機 steak frenching machine
牛排蕈 beefsteak fungus
牛排餐刀 steak knife
牛排餐廳 steak house
牛排館 steak house
牛眼豆 black eyed bean
牛蛙 bull frog
牛飲 swig
牛絛蟲 beef tapeworm
牛腰背肉 beef loin
牛腰脊肉 sirloin
牛腰部脊肉 saddle
牛腸衣 beef casing
牛腹肉 skirt steak
牛酪油 ghee; ghi
牛頓〔型〕流動 Newtonian flow
牛頓流體 Newtonian fluid
牛頓流體性 Newtonian property
牛腩 brisket point end, deckle-off
牛腿肉 beef ham
牛腿肉排 outside steak
牛腿端肉 silverside
牛蒡 burdock; great burdock; lappa
牛膝草 hyssop; marjoram
牛蹄 neat's foot
牛蹄凍 cowheel
牛頸肉 neck end
牛磺酸 taurine
牛磺膽酸 taurocholic acid
牛磺膽酸酯 taurocholate
牛磺膽酸鹽 taurocholate

牛臀肉 rump; tenderside
牛臀肉排 rump steak
牛臀肉塊 beef rump butt
牛膽抽提液 ox bile extract
牛膽酸 taurocholic acid
牛膽酸鹽 taurocholate
牛膻味 cowy flavo(u)r
牛犢 calf
牛犢肉小香腸 bockwurst sausage
牛脷 knuckle
犬尿酸 kynuric acid
王大馬哈魚 tyee
王蟹 king crab

五畫

丙二酸 malonic acid
丙二酸二乙酯 diethyl malonate
丙二酸鈉 sodium malonate
丙二醇 propylene glycol
丙二醇脂肪酸酯 propylene glycol esters of fatty acids
丙二醇澱粉 propylene glycol starch
丙三醇 glycerin(e); glycerol
丙位丁內酯 γ-butyrolactone
丙位己內酯 γ-hexalactone
丙位庚內酯 γ-heptalactone
丙位癸內酯 γ-decalactone
丙氨酸 alanine
丙烯 propylene
丙烯甘醇 propylene glycol
丙烯胺 allylamine
丙烯基乙基癒創木酚 propenylguaethol
丙烯酸 acrylic acid
丙烯酸乙酯 ethyl acrylate
丙烯醛 acrolein
丙硫醇 propylmercaptan
丙烷 propane
丙酮 acetone; dimethyl ketone
丙酮血症 aceton(a)emia
丙酮尿〔症〕 acetonuria
丙酮酸 pyruvic acid
丙酮酸去氫酶 pyruvic dehydrogenase
丙酮酸氧化酶 pyruvic oxidase
丙酮酸脫氫酶 pyruvic dehydrogenase
丙酮酸酯 pyruvate
丙酮酸鹽 pyruvate
丙酮醇 acetol
丙酮體 acetone body
丙酸 propionic acid
丙酸乙酯 ethyl propionate
丙酸松油酯 terpinyl propionate
丙酸芳樟酯 linalyl propionate
丙酸苯乙酯 phenethyl propionate
丙酸香茅酯 citronellyl propionate
丙酸香葉酯 geranyl propionate
丙酸桂酯 cinnamyl propionate
丙酸烯丙酯 allyl propionate
丙酸異丁酯 isobutyl propionate
丙酸異戊酯 isoamyl propionate
丙酸細菌 propionic acid bacteria
丙酸發酵 propionic acid fermentation
丙酸鈣 calcium propionate
丙酸鈉 sodium propionate
丙酸鉀 potassium propionate
丙酸酯 propionate
丙酸鹽 propionate
丙酸苄酯 benzyl propionate
丙醇 propanol; 1-propanol; propyl alcohol
丙醣 triose
丙醣磷酸 triose phosphate
丙醛 propanal; propionaldehyde
世代時間 doubling time; generation time
世界衛生組織 World Health Organization (WHO)

丘形牛肝菌 barrow's bolete
丘鷸 woodcock
主成分分析 proximate analysis
主成分組成 proximate composition
主要成分 staple
主要產品 primary product; staple
主要農作物 staple crop
主食 basic food; staple; staple food; staple diet
主教飲料 bishop
主發酵 main fermentation; principal fermentation; primary fermentation
主菜 entrée; main course; main dish
主廚 chef
主糧 staple food
乏味 vapidity
乏味的 insipid
乏味的酒 slip-slop
乏味食物 insipid food
以臭氧處理 ozonize
仔女鰻 heifer calf
仔牛皺胃 vell
他拉膠 tara gum
代可可脂 cocoa butter substitute
代用〔 surrogate; substitute
代用麵粉 alternative flour
代食品 substitution food
代豚脂 lardine
代替 substitute
代替物 substitute
代糖球 sugar substitute
代謝〔作用〕 metabolism
代謝餐具〕率 metabolic rate
代謝物 metabolite
代謝活性 metabolic activity
代謝商 metabolic quotient
代謝產物 metabolic product
代謝途徑 metabolic pathway
代鹽品 salt substitute
代鹽劑 salt substitute
令人作嘔的 nauseous
令人討厭的氣味 offensive odo(u)r
仙人掌 cactus
仙人掌〔芽孢〕桿菌 *Bacillus cereus*
仙人掌果 prickly pear
仙人掌桿菌食物中毒 *Bacillus cereus* food poisoning
仙女環洋菇 fairy ring mushroom
仙客來醛 cyclamen aldehyde
仙桃 egg fruit
充〔碳酸〕氣飲料 carbonated beverage; carbonated drink
充二氧化碳 carbonating
充二氧化碳水 aerated water
充二氧化碳機 aerated water machine
充分〔碳 hard drying
充血 engorgement
充氣 aerification; gassing; gasify; inflation
充氣〔葡萄〕酒 aerated wine
充氣水 aerated water
充氣包裝 gas flush(ing) package
充氣包裝〔法〕 gas packing
充氣包裝袋 gas flush(ing) pouch
充氣奶油 aerated cream
充氣疏鬆劑 aerating agent
充氣飲料 effervescent beverage
充氣過量 overgassing
充氣器 aerator
充氣機 gasogen(e)
充氣糖果 aerated candy
充氣麵包 aerated bread
充液包裝〔法〕 liquid packing
充惰性氣體包裝 inert-gas packing
充填不足 underfill
充滿 perfusion
冬〔小〕麥 winter wheat
冬化〔處理〕 winterization

冬化油 winter oil
冬化處理的油 winterized oil
冬瓜 waxgourd; white gourd; winter gourd
冬作 winter crop
冬季西葫蘆 winter squash
冬季南瓜 winter squash
冬季風輪菜 winter savory
冬季乾酪 hay butter
冬季黃油 hay butter
冬季櫻 alkekengi
冬季蘿蔔 winter radish
冬青油 methyl salicylate; wintergreen oil
冬青茶 cassina
冬南瓜 hubbard
冬粉 bean noodle; bean thread (vermicelli); bean vermicelli; green bean vermicelli
冬梨 warden
冬甜瓜 casaba; cassaba; winter melon
冬寒菜 potherb mallow
冬筍 bamboo shoot; winter bamboo shoot
冬菇 wooden mushroom
冬綠油 methyl salicylate; wintergreen oil
冬綠糖 primverose
冬蟲之一 *cordyceps sinensis*
凹口 notch
凹凸棒石黏土 attapulgite clay
凹型器皿 hollow ware
凹痕 pit
凹罐 collapsed can; dented can
出口 outlet; vent
出汁量 juice yield
出血 bleed(ing)
出血性結腸炎 hemorrhagic colitis
出油 oil off
出芽短梗孢糖 pullulane
凸罐 buckled can; bulged can; swelled can
加入咖哩〔粉〕的 curried
加入烈酒 spike
加入酒精的啤酒 spiked beer
加工 processing
加工不足 under processing
加工生產線 processing line
加工成品 finished goods
加工助劑 processing aids
加工油脂 processed fat
加工食品 processed food
加工設備 processing equipment
加工槽 process tank
加工機械 processing machinery
加水判定（用於牛乳試驗） added water judgement
加水烈酒 grog
加石灰澄清 liming
加石灰澄清的糖汁 limed juice
加冰 icing
加冰飲料 ice drink
加冰機 icing machine
加州鯷 California anchovy; northern anchovy
加州鰨魚 California sole
加成反應 addition reaction
加成劑 additive agent
加灰澄清 defecation
加米的牛奶糊 rice milk
加西卡病 kwashiorkor
加利福尼亞鯷 California anchovy
加利福尼亞鰨魚 California sole
加沙巴甜瓜 casaba; cassaba
加侖 gallon
加味茶 jasmine tea; scented tea
加度葡萄酒 fortified wine
加氟 fluoridation
加苦味 embitterment
加香〔料調味〕的 spiced
加香料的 spicy
加香乾酪 spiced cheese

加香甜酒 cordial
加香葡萄酒 aromatized wine
加乘作用 synergism
加乘效應 synergistic effect
加拿大白鮭 cisco
加拿楷樹 ilangilang; ylangylang
加拿薄點心 canape
加料斗 loading hopper
加氣高溫殺菌〔法〕 hyperbaric retorting
加氧酶 oxygenase
加氨處理 ammoniate
加酒精的威士忌 spirit whisk(e)y
加酒精的葡萄酒 vinage
加啤酒花的麥芽汁 hopped wort
加強環罐 beaded can
加氫〔作用〕 hydrogenation
加氫精製油 hydrofined oil
加甜 dulcification; edulcoration
加甜劑 edulcorant
加細冰屑 body icing
加蛇麻酮(啤酒花的苦味物質) adlupulone
加蛋〔調理〕 egging
加速浸泡 introfaction
加速浸泡劑 introfier
加速陳化 accelerated ag(e)ing
加速熟成 accelerated ag(e)ing
加速攪拌 accelerated churning
加貼膠 ghatti gum; gum ghatti
加催眠藥飲料 Mickey Finn
加盟店 franchise chain
加碘化合物 iodizing compound
加碘食鹽 iodized salt
加碘劑 iodizing compound
加葡萄酒的蛋糕 tipsy cake
加過啤酒花的 hopped
加葎草酮(啤酒花的苦味物質) adhumulone
加蓋器 overcap placer
加酸 acidulation
加酸皂化 acid saponification
加熱 heating
加熱不足 underheating
加熱介質 heating medium
加熱曲線 heating curve
加熱死滅時間 thermal death time (TDT)
加熱死滅時間曲線 thermal death time curve (TDT curve)
加熱死滅率 thermal death rate (TDR)
加熱死滅點 thermal death point (TDP)
加熱即食食品 heat-and-eat food product; heat-and-serve food product
加熱板 hot plate; heater plate; warming plate
加熱保存 heat preservation
加熱室 heating chamber
加熱致死時間 thermal death time (TDT)
加熱致死時間曲線 thermal death time curve (TDT curve)
加熱套 heating jacket
加熱排管 heating calandria
加熱殺菌 thermostabilization
加熱殺菌時間 thermal reduction time
加熱脱水 thermal dehydration
加熱處理不足的 underprocessed
加熱煮過的牛乳 scalded milk
加熱損失 heating loss
加熱滅菌法 heat sterilization; stassanization
加熱滅菌器 heat sterilizer
加熱過度 overburn
加熱過度的焦味 overprocessing flavo(u)r
加熱盤管 heating coil
加熱器 heater; warmer
加糖 sugaring
加糖加香料並加熱 mull
加糖乳粉 sweetened milk powder
加糖果汁 sweetened juice

加糖的 sugared
加糖的奶油漿 butter cream
加糖煉乳 condensed milk; sweetened condensed milk
加糖漿 syruping
加壓 pressurization
加壓乾燥 pressure drying
加壓滅菌 pressure sterilization
加壓滅菌器 pressure sterilizer
加壓蒸煮 pressure cooking; pressure steaming
加壓蒸煮器 pressure cooker
加壓噴霧 pressure atomization
加壓熱燙 retort blanching
加濕 ordering
加磷酸分解酶 phosphorylase
加藥食品 medicated food
加蘭丁肉凍 galantine
加鹽 salify(ing); salification
加鹽白脫 salted butter
加鹽乳酪 salted butter
加鹽過多 oversalting
功夫茶(中國紅茶) congou
功能性 functional property
功能性食品 nutraceuticals
包心菜 crambe cabbage; cabbage
包伙食業者 restaurant contractor
包伙商 restaurant contractor
包冰 icing
包冰衣 body icing; glace; glaze; glazing; ice glazing
包冰儲藏 icing storage
包有糖衣的糖果 dragee
包肉紙 butcher's paper
包肉麵皮捲 canelloni
包衣巧克力 covering chocolate
包封機 covering machine
包紮工具 binder
包紮機 binding machine; strapper
包裝 casing; pack; package; packing; packaging; wrap
包裝工人 packer
包裝內乾燥 in-package desiccation
包裝內乾燥劑 in-package desiccant
包裝日期 pack date
包裝用紙板 packing board
包裝肉 packaged meat
包裝材料 packing material; packaging material; wrap; wrapping; wrappage; wrapper
包裝重量 tare
包裝食品生產廠商 packer
包裝容器 pack; package
包裝紙 packing paper; wrapping paper; wrapper; wrapping
包裝貨品 packaged goods
包裝填充料 packing
包裝廠 packhouse
包裝機 packing machine; packaging machine; packer
包裹 pack; package; packaging
包裹式紙箱 wraparound case
包裹式裝箱法 wraparound casing
包裹的 wraparound
包糖 candying
包糖衣 glace; glaze; glazing; panning; sugar coating
包糖衣機 revolving pan
包蟲病 echinococcosis
包覆劑 glazing agent
包覆機 wrapping machine
北方鰈 northern halibut
北京大白菜 Peking cabbage
北京白梨 Peking white pear
北京芽菜 brussel sprout; baby cabbage
北京烤鴨 Peking duck
北京填鴨 crammed peking duck
北京鴨 Peking duck

北美大梭魚 muskellunge
北美牙鮃 California halibut
北美狗魚 muskellunge
北美洲大眼鯡魚 alewife
北美秋葡萄 sweet mountain grape; winter grape
北美薄荷 *Mentha canadensis*
北美鶉 bobwhite
北海蝦 red sturgeon shrimp
北極蝦 arctic shrimp
北極鱈 polar cod
北歐式自助餐 smorgasbord
仟卡 Calorie (Cal); great calorie; kilocalorie (kcal)
半〔圓〕片 half slice
半片屠體 half carcass
半加侖(舊的英國液量單位) pottle
半去骨的 semi-boneless
半生不熟的 underdone
半生的 half-baked; half-done
半成品 half-finished goods; half-finished product; semi-finished product; semimanufactures; semi-product; semis
半血色素 hemichrome
半冷凍的 half-frozen
半乳多糖 galactogen
半乳聚糖 galactan; galactosan
半乳酸磷酸 lactacidogen
半乳糖 galactose
半乳糖二酸 galactaric acid; mucic acid
半乳糖脂 galactolipid
半乳糖酸 galactonic acid
半乳糖醇 galactitol; melampyrin; dulcite; dulcitol; dulcose
半乳糖醛酸 galacturonic acid
半乳糖苷 galactoside
半乳糖苷酶 galactosidase
半乳糖酶 galactase
半乳呋喃糖 galactofuranose
半固體 semisolid
半保藏 semi-preservation
半保藏食品 semi-preserved food
半流質 semifluid; semiliquid
半流質食物 pap
半流質膳食 semifluid diet; semiliquid diet
半流體 semifluid; semiliquid
半凍結的 half-frozen; semi-frozen
半烤熟的 half-baked
半胱胺酸 cysteine (Cys)
半脂〔牛〕乳 half cream milk
半衰期 half life
半乾性油 semi-drying oil
半乾的 partly dried; semi-dried; semidry
半乾食品 intermediate moisture food (IMF); semimoist food
半乾香腸 semidry sausage
半乾葡萄酒 semidry wine
半屠體 split carcass
半殺菌的 semi-sterile
半清理魚 semi-dressed fish
半甜巧克力 bittersweet chocolate
半脫脂奶 partly skimmed milk
半脫脂乳 partly skimmed milk; semi-skimmed milk
半脫脂芝士 partly skim milk cheese
半脫脂乾酪 partly skim milk cheese
半軟質膳食 semisoft diet
半連續凍乾 semi-continuous freeze drying
半連續蒸餾 semi-continuous distillation
半透性的 semipermeable
半透明的 translucent
半透膜 semipermeable membrane
半發酵茶 semi-fermented tea
半硬〔質〕乾酪 semi-hard cheese
半硬化油 semi-hardened oil
半結晶體 crystalline granule

半圓形酥餅 turnover
半溶性 semi-soluble
半解凍 semi-defrost; semi-thawing
半解凍的 half-thawed
半滲透性的 semipermeable
半製成品 raw product
半製品 half-finished goods; half-finished product
半數致死劑量 50% lethal dose; LD_{50}; medium lethal dose
半熟的 medium
半熟乾〔豬肉〕香腸 polony sausage
半熟蛋 half-cooked egg; soft boiled egg
半膠體 hemicolloid
半醃的 half-salted
半隨意肌 semi-voluntary muscle
半濕〔性〕食品 intermediate moisture food (IMF); semimoist food
半縮醛 hemiacetal
半離羊肚 half free morel
半邊屠體 side carcass
半飢餓 semistarvation
半纖維素 hemicellulose
半纖維素酶 hemicellulase
卡 calorie (cal); gram calorie; small calorie
卡士達〔點心〕 custard
卡士達布丁 custard pudding
卡士達派 custard pie
卡士達粉 custard powder
卡士達糊 custard cream
卡士達醬 custard sauce
卡巴度斯蘋果白蘭地酒 apple jack; calvados
卡他茶 kat
卡他溫度計 katathermometer
卡布奇諾咖啡 cappuccino
卡布基諾咖啡 cappuccino
卡必力無花果 caprifig
卡瓦酒 kava
卡耳費雪〔水分測定〕法 Karl Fischer method
卡耳費雪試劑 Karl Fischer reagent
卡那黴素 kanamycin
卡拉蜜糖 caramel
卡拉膠 carrageenan; carrageenin; carragheenin
卡南加油 cananga oil
卡哈苡苷 stevioside
卡計 calorimeter
卡值 caloric value; calorie value; calorific value
卡特尼拉甜橙 cadenera orange
卡特蘭納雞 catalana
卡畢可拉豬肉香腸 cappicola sausage
卡斯努得餃子（奧地利） kasnudln
卡路里 calorie (cal); gram calorie
卡達芝士 coda cheese; cottage cheese; pot cheese
卡達乾酪 coda cheese; cottage cheese; pot cheese
卡爾斯伯酵母〔菌〕 *Saccharomyces carlsbergensis*
卡爾費希爾〔水分測定〕法 Karl Fischer method
卡爾費希爾試劑 Karl Fischer reagent
卡爾酵母〔菌〕 *Saccharomyces carlsbergensis*
卡薩巴甜瓜 casaba; cassaba
卡薩瑞齊亞麵 casareccia
卡羅萊納米 Carolina rice
卡藜油 cascarilla bark oil; cascarilla oil
去〔玉米〕鬚機 silking machine; silker
去〔羽〕毛 deplumation
去〔羽〕毛機 feather picker; featherator
去〔果〕核 pit; stoning
去〔果〕核機 stoner; stoning machine
去〔魚〕鱗 fish sealing
去〔魚〕鱗器 fish scaler
去〔鍋〕垢劑 descaler

去二氧化碳 decarbonation
去內臟 dressed; drawing; eviscerate; gutting
去內臟及毛 dressing
去內臟及毛的家禽 dressed poultry
去內臟屠體 eviscerated carcass
去內臟魚 drawn fish; gutted fish
去內臟機 eviscerator
去支酶 debranching enzyme
去毛 defeather; dehair; depilate; depilation; unhairing; unwooling
去毛膠料 depilating compound
去毛劑 depilatory; hair remover
去毛器 hair remover
去毛機 dehairer; defeatherer; defeathering machine; dehairing machine; depilator; unhairing machine
去毛雞 pulled chicken
去水 anhydration; dewater
去水山梨醇 sorbitan
去水山梨醇脂肪酸酯 sorbitan fatty acid ester (Span)
去水山梨醇單硬脂酸酯 sorbitan monostearate
去甲基〔作用〕 demethylation
去皮 dehusk; derind; dehull; husking; hulling; peeling; peel; skinning; scale off; shelling; shuck
去皮刀 peeling knife; paring knife; parer
去皮的 husked; hulled; rindless; skinned; shucked; skinless
去皮乾酪 rindless cheese
去皮屠體 skinned carcass
去皮魚 skinned fish
去皮腿肉 skinless ham
去皮穀類 skinned grain
去皮器 skinner
去皮機 dehuller; derinder; husker; huller; peeler; peeling machine; paring machine; sheller; shucker; skinning machine; stripper; scourer
去污染 decontamination
去污染劑 decontaminant; decontaminating agent
去灰 dedust
去肉 fleshing
去色素 depigmentation
去乳化 demulsify
去乳化〔作用〕 deemulsification; demulsification
去乳化劑 demulsifier; demulsifying agent
去乳糖乳 delactosed milk
去乳糖乳清 delactosed whey
去咖啡因 decaffeination
去咖啡因咖啡 decaffeinated coffee
去固態油〔作用〕 demargarination; destearination
去果心器 corer
去果皮刀 flaying knife
去果核 stone
去果核機 fruit stone extractor
去果膠 depectinization
去果膠果汁 depectinized juice
去沫器 skimmer
去油可可豆 broma
去油污劑 degreaser
去沫劑 defoamer
去芽機 disgerminator
去垢 scaling
去垢劑 detergent
去染劑 destainer
去柄機 stemming machine
去活〔作用〕 inactivation
去活化 deactivation
去活化劑 deactivator

去活劑 inactivator
去籽器 seeder
去胚芽 degerm
去胚芽可可豆 nib
去胚芽機 degerminator; degerminating machine
去胚芽磨 degerminating mill
去胚器 degerminator
去苦味 debittering; debitterize
去苦味酵母 debittering yeast
去苞葉 husking
去苞葉的 husked
去苞葉機 husker
去捕殼機 shrimper
去核 coring; pitting
去核刀 coring knife
去核的 pitted; stoneless
去核器 seeder
去核機 pitter; pitting machine
去根機 root peeler
去氣 outgassing
去氣〔作用〕 degasification
去氧〔作用〕 disoxidation; deoxidation; deoxidization
去氧核糖 deoxyribose; desoxyribose
去氧核糖核酸 deoxyribonucleic acid (DNA); desoxyribonucleic acid
去氧核糖核酸酶 deoxyribonuclease
去氧劑 deoxidant; deoxidizer; oxygen scavenger; oxygen scavenging agent
去氧糖 deoxysugar; desoxysugar
去氧酶 deoxygenase
去浮渣 despumation
去除水分的 unwatered
去除油污 degrease
去除泡沫 defoam
去除劑 remover
去除雜醇油 defuselation
去骨 boning; clear bone; debone
去骨工人 deboner
去骨火腿 boneless ham
去骨牛肩頸肉 chuck roll
去骨肉 boneless meat; fine meat
去骨肋脊肉 spencer roll
去骨的 boned; boneless
去骨的肉 boned meat
去骨豬肉 clear pork
去骨機 bone cleaner; deboner; deboning machine
去骨燻火腿 daisy ham
去骨雞塊 chicken nugget
去胺〔基〕作用 deamination; desamination
去胺〔基〕酶 deaminase; desaminase
去梗未果頂皮 nubbing
去梗蒂 nubbing
去梗葡萄 stemmed grape
去梗機 stemming machine; stem crusher; stemmer; stalk machine
去氫〔作用〕 dehydrogenation
去氫抗壞血酸 dehydroascorbic acid
去氫醋酸 dehydroacetic acid (DHA)
去氫膽固醇 dehydrocholesterol
去氫酶 dehydrogenase
去瓶蓋 decap
去瓶蓋器 decrowner
去硫〔作用〕 desulfurization
去莢 decapsulation; husking; hulling
去莢咖啡豆 cleaned coffee
去莢的 husked; hulled
去莢青〔豌〕豆 shelled pea
去莢豌豆 shelled pea
去莢機 husker; huller
去蛋白質 deproteinize
去蛋白質〔作用〕 deproteinization
去頂 topping
去魚鰓 gibbing
去魚鰓機 gibbing machine

去殼 dehull; decortication; husking; hulling; peel(ing); shell(ing); shuck; scour

去殼大麥粒 pot barley; barley grout

去殼小麥粒 groats

去殼米 milled rice

去殼的 husked; hulled; shucked; unhusked

去殼蛋 non-shelled egg

去殼穀粒 groats

去殼穀類 skinned grain

去殼機 dehuller; decorticator; husker; huller; hulling mill; peeler; peeling machine; sheller; shucker; scouring mill; scourer

去殼燕麥〔粒〕 oat groats

去殼燕麥片 groats

去氮〔作用〕 denitrification

去氮細菌 denitrifying bacteria

去氯〔作用〕 dechlorination

去硝〔作用〕 denitration

去硬脂〔作用〕 demargarination; destearination

去絮凝〔作用〕 deflocculation

去絮凝劑 deflocculating agent

去絮凝離心機 deflocculator

去鈣 decalcify

去鈣作用 decalcification

去勢 castrate; emasculation

去勢〔成年〕公畜 stag

去勢小公牛 steer

去勢公羊 wether

去塞 uncork

去溶劑 desolventization; desolventize

去腸〔對〕蝦 deveined shrimp

去腸〔線〕 devein

去飽和〔作用〕 desaturation

去羧基〔作用〕 decarboxylation

去羧酶 decarboxylase

去酯化〔作用〕 deesterification

去碳〔作用〕 decarbonization

去綠〔作用〕 degreening

去蓋器 cover remover

去酸〔作用〕 deacidification

去稻穀機 paddy machine; paddy separator

去膠〔作用〕 degelatinization

去蔓機 viner

去麩小麥粉 wheat meal

去麩小麥粒 wheat berry

去糖 desugar; desaccharify

去糖〔作用〕 desaccharification

去靜電劑 antistatic agent

去鞘玉米穗 shelled corn

去頭 decapitation; heading

去頭及內臟 nobbing

去頭及內臟的魚 nobbed fish

去頭機 beheading machine; heading machine

去頭龍蝦 lobster tail

去澀味 deastringent

去鎂葉綠素 phaeophytin; pheophytin

去鎂葉綠酸 phaeophorbide; pheophorbide

去鍋垢 descale

去黏〔膜〕機 slimer

去黏質 sliming

去離子〔化〕作用 deionization

去離子水 deionized water

去離子劑 deionizer

去離子器 deionizer

去雜味 deflavour

去穩磷酸鉀用〕 destabilization

去穩定劑 destabilizer

去礦物質〔作用〕 demineralization

去礦物質水 demineralized water

去礦物質裝置 demineralizer

去鰓臟 gibbing

去鰓臟機 gibbing machine

去鐵機 metal eliminator
去臟淨重〔量〕 gutted weight
去鱗 descale; scale off; scale; scaling
去鱗片的 scaled
去鱗魚 scaled fish
去鱗器 descaler
去鱗機 scaler
去鹽 desalting; desalination
去鹽乳清 demineralized whey
去鹽的 desalted
去醯胺〔基〕作用 deamidization
可〔蒸〕煮包裝 boilable pack
可口可樂 Coca Cola; Coke
可口性 palatability; savoriness; savouriness
可口道品評 palatability test
可口的 dainty; palatable; savory; savoury; tasty; toothsome
可口食品 dainty
可分散性 dispersibility
可分解性 decomposability
可分解的 resoluble
可可〔果〕殼 cacao shell; cocoa shell
可可豆 cacao; cocoa; cacao bean; cocoa bean
可可豆肉 cacao nib; cocoa nib
可可豆料素 cacao pigment
可可豆的豆肉 nib
可可豆粉 cacao nib; cocoa nib
可可豆殼皮 cacao husk; cocoa husk
可可豆碾磨機 cacao mill
可可油 cacao butter; cocoa butter; theobroma oil
可可油餅 cocoa cake
可可紅 cacao red
可可酊 cocoa tincture
可可料的鹼處理 Dutch process
可可粉 cacao; cocoa; cacao powder; cocoa powder
可可脂 cacao butter; cocoa butter; cacao oil; theobroma oil
可可脂肉用品 cocoa butter substitute
可可脂壓榨機 cacao press
可可蛋糕 cocoa cake
可可棕 cacao brown
可可殼 cacao husk; cocoa husk
可可飲料 cacao; cocoa
可可椰胛 coco plam
可可漿 bitter chocolate
可可樹 cacao; chocolate tree; cocoa
可可鹼 3,7-dimethylxanthine; theobromine
可尼西雞 Cornish
可用級 U.S. Utility
可同化氮 assimilable nitrogen
可吃的 comestible
可回用瓶 returnable bottle
可伸張的 tensile
可折疊紙盒 collapsible box
可汽化的 vapo(u)rizable
可皂化的 saponifiable
可皂化脂質 saponifiable lipid
可的松 cortisone
可長期儲藏的 long-keeping; long-life
可保存的 preservable
可保藏的 preservable
可食用物 esculent
可食用的 eatable; edible; esculent
可食用的動物內臟 variety meat
可食性 edibility
可食性薄膜 edible film
可食的 pabular
可食球果菌（蕈類） spruce cone toadstool
可食蟹 edible crab
可消化的 digestible
可消化營養素 digestible nutrient
可消化營養素總量 total digestible nutrients (TDN)
可缺酵素 nonessential enzyme

可能變酸的物質 acidoid
可逆反應 reversible reaction
可逆性 reversibility
可逆循環 reversible cycle
可逆過程 reversible process
可逆膠體 reversible colloid
可混的 miscible
可透〔性〕膜 permeable membrane
可斯達蘋果（英國） costard
可煮袋 boiling bag
可發酵己糖 zymohexose
可發酵的 fermentable
可發酵醣類 fermentative carbohydrate
可飲用的 drinkable; potable
可飲性 potableness
可傾〔蒸煮〕鍋 tilting kettle
可塑性 plasticity
可塑性麵糰 plastic dough
可塑劑 plasticizer
可溶〔性〕包裝 soluble packaging
可溶〔性〕固形物 soluble solid
可溶〔性〕配料 soluble ingredient
可溶〔性〕組分 soluble ingredient
可溶〔性〕蛋白質 soluble protein
可溶〔性〕膠 soluble gum
可溶〔性〕澱粉 soluble starch
可溶〔性〕糖精 soluble saccharin
可溶〔性〕膳食纖維 soluble dietary fiber
可溶可可粉 soluble cocoa
可溶性脂肪酶 lyolipase
可溶性澱粉 amylodextrin
可溶性糖精 sodium saccharin
可溶物 soluble
可溶的 soluble
可溶酒槽 distillers' soluble
可溶解的 resoluble
可蒂潘趣飲料 cordial punch
可蒂雞尾酒 cordial
可過蔬菜 filterable
可榨出油 expressible oil
可榨出液滴 expressible drip
可誘導酶 inducible enzyme
可樂 coke
可樂豆 cola; cola nut; cola seed
可樂果 cola; cola nut; cola seed; kola nut
可樂果抽提液 cola nut extract
可樂型香精 cola type flavo(u)r
可樂型飲料 cola (type) beverage
可樂樹 cola; kola
可樂類飲料 cola
可燃性 inflammability
可選級 U.S. Select
可壓擠性 squeezability
可壓縮性 squeezability
可擠軟管 collapsible tube
可體松 cortisone
古巴燒酒 aldabo
古巴糖 Cuban suagr
古氏坩鍋 Gooch crucible
古氏濾器 Gooch filter
古柯 coca
古柯葉抽提液 coca leaf extract
古洛糖 gulose
古洛糖酸 gulonic acid
古烏達乾酪 Gouda cheese
古塔〔波〕膠 gutta-percha
右旋〔作用〕 dextrorotation
右旋糖 dextrose; glucose; grape sugar
右鰈 plaice
司邦40 Span 40
司邦60 Span 60
司盤20 sorbitan monolaurate; Span 20
史氏赤刀魚 Japanese bandfish
台 table
台灣蜆 corbicula
四〔碳〕糖 tetrose
四分之一 quarter
四分之一品脫 quartern

四甲基葡〔萄〕糖 tetramethyl glucose
四甲基葡〔萄〕糖苷 tetramethyl glucoside
四角包〔裝〕 tetra-pack; tetra-pak
四角包裝 tetrahedral package
四季豆 French bean; field bean; green bean; garden bean; haricot; haricot bean; kidney bean; navy bean; snap bean; wax bean
四季桔 calamondin; Panama orange
四季草莓 perpetual strawberry
四季樹莓 perpetual raspberry
四菱豆 asparagus pea
四旋〔瓶〕蓋 lug cap
四旋蓋 regular twist-off cap
四氫化芳樟醇 tetrahydrolinalool
四氫香葉醇 tetrahydrogeraniol
四氫糠醇 tetrahydrofurfuryl alcohol
四氫噻吩-3-酮 tetrahydrothiophen-3-one
四硫磺酸鹽培養基 tetrathionate broth
四氯化碳 carbon tetrachloride
四等分 quarter
四棱豆 asparagus pea; winged bean
四磅重的麵包 quartern
四磷酸鹽 tetraphosphate
四聯球菌 tetracoccus
四醣 tetrasaccharide
四邊封袋 four-side seal pouch
四邊密封包裝 four-sided fin seal wrap
外包裝 outer packaging
外包裝容器 outer container
外包裝紙 wrapping paper
外包膠囊 encapsulation
外生的 exogenous
外生孢子 exospore
外皮 hull; husk; integument; rind; shuck; tegument
外果皮 epicarp; exocarp; flavedo
外毒素 exotoxin
外胚乳 perisperm
外食 eat out
外食者 diner-out
外消旋化合物 racemic compound
外消旋酶 racemase
外骨骼 exoskeleton
外側後腿肉 gooseneck round; outside; silverside
外側後腿板肉 outside flat
外側後腿眼肉 eye round
外帶包裝 takeout packaging
外帶食物 takeaway food; takeout food
外殼 casing; husk; hull; rind; tegument
外源的 exogenous
外滲 exosmosis
外種皮 testa
外酵素 exoenzyme
外賣 takeout service
外賣食物 takeaway food
外澱粉酶 exoamylase
外膳 boarding-out
外燴 catering
外肽酶 exopeptidase
外酶 exoenzyme
失去平衡 overbalance
失活 deactivation
失活劑 deactivator
失重 weight loss
奶牛 milch cow; milk cattle; milk cow
奶昔 milk shake
奶油 butter; cream; creme; milk cream
奶油〔酥皮〕角 butter horn
奶油上浮 creaming
奶油分離 creaming
奶油分離器 cream separator
奶油包裝紙 butter paper
奶油包裝機 butter printer
奶油色素 butter colo(u)r
奶油伴水果 fruit fool
奶油沙司 cream sauce

奶油乳清 butter serum; butter whey
奶油定型器 butter printer
奶油泡芙 cream puff
奶油芝士 cream cheese
奶油香精 butter flavo(u)r
奶油凍 custard
奶油蛋捲 brioche; cream roll
奶油蛋糕 butter cake
奶油蛋羹 custard
奶油發酵劑 butter starter
奶油硬度計 butter hardness tester
奶油硬糖 creamy rock
奶油硬糖（鹹味） butterscotch
奶油片 butter yellow
奶油圓球蛋糕 briocke
奶油塔餅 fanchonette
奶油洋小鱈 creamery
奶油餅乾 butterie
奶油層 cream layer
奶油濃湯 cream soup; vichyssoise
奶油糕點 brioche
奶油糖膏餡 cream filling
奶油醬 butter sauce
奶油蘇打餅 cream cracker
奶油麵包 butter bread; butterie
奶油麵粉糊 roux; cream roux
奶油洋鮭器 butter churn; cream beater; creamer
奶油攪拌機 churn
奶的 milch
奶粉 dried milk; dry milk; milk powder; powdered milk
奶瓶 feeding bottle; milk bottle; nurser
奶瓶洋鱈 bottle feeding
奶媽 nutrix; nurse; nurser; wet nurse
奶酪酵母〔菌〕 *Saccharomyces casei*
奶嘴 teat
奶樽 nurser
奶頭 teat

尼比羅〔釀酒紅〕葡萄 Nebbiolo
尼加拉斯阿伯 Nicolas Appert
尼古丁 nicotine
尼克酸 niacin; nicotinic acid; pellagramin; vitamin PP
尼克酰胺 niacinamide; nicotinamide
尼泊金乙酯 ethyl-*p*-hydroxybenzoate; ethylparaben
尼泊金丁酯 butyl *p*-hydroxybenzoate; butylparaben
尼泊金丙酯 propyl-*p*-hydroxybenzoate; propylparaben
尼泊金甲酯 methylparaben
尼泊金庚酯 heptylparaben
尼泊金異丁酯 isobutylparaben
尼泊金異丙酯 isopropylparaben
尼格辛黑 nigrosine
尼格斯含酒飲料 negus
尼葛拉籽 nigella
尼龍 nylon; polyamide
尼羅吳郭魚 Nile tilapia
尼羅非洲鯽魚 Nile tilapia
尼羅紅魚 red tilapia
尼羅魚 nilotic fish
尼羅羅非魚 Nile tilapia
巨〔大〕蛋 giant egg
巨大菌落 giant colony
巨分子 macromolecule
巨昆布 kelp; oyster grass
巨型馬疕蘑菇 giant puffball
巨紅血球貧血 megaloblastic an(a)emia
巨海帶 kelp; oyster grass
巨海藻 kelp
巨球蛋白 macroglobulin
巨粒 macrograin
巨野黑麥 mammoth wildrye
巨量元素 macroelement
巨量營養素 macronutrient
巨蟹 king crab

巨藻 macrocystis
巧克力 chocolate
巧克力奶 chocolate milk
巧克力乳漿 creamy chocolate
巧克力香精 chocolate flavo(u)r
巧克克夏 chocolate chip
巧克力捏磨機 conching machine
巧克力粉 chocolate powder
巧克力焦軟糖 chocolate fudge
巧克力酥糖 chocolate brittle
巧克力塊 block chocolate; chocolate bar
巧克力聖代 chocolate sundae
巧克力精煉機 conche; conching machine
巧克力精磨機 conche
巧克力製品 chocolate goods
巧克力漿 chocolate liquor
巧克力糖 chocolate candy
巧克力糖皮 chocolate coat(ing)
巧克力糖衣 couverture
巧克力糖漿 chocolate fudge
巧克力薄鬆餅 chocolate wafer
巧克力攪拌揉捏機 conche
左旋葡萄糖 laevoglucose; levoglucose
左旋糖 fructose; laevulose; levulose
市乳 city milk; market milk
市售魚 market fish
市售蔬菜 garden truck
市販肉 butcher's meat
布〔魯〕氏桿菌病 brucellosis
布丁 blancmange; pudding
布丁米 pudding rice
布氏角鯊 northern dogfish
布氏單位 Brabender unit
布氏黏度計 Brookfield viscometer
布甸 pudding
布里〔克氏〕糖度 Brix
布里比重計 Brix hydrometer; Brix scale hydrometer
布里糖度 Brix degree; Brix scale
布里糖度計 Brix hydrometer; Brix scale hydrometer; Brix spindle
布里糖酸比 Brix acid ratio
布拉班德爾單位 Brabender unit
布拉斯李 mirabelle
布拉德香腸 bratwurst (sausage)
布林森芝士 brinsen cheese
布林森乾酪 brinsen cheese
布枯葉油 buchu leaf oil
布格麥食 bulgur
布勞氏〔即溶奶粉〕製造 Blaw-knox process
布勞氏造粒機 Blaw-knox instantizer
布散拉斯葡萄酒 bucellas
布爾戈尼葡萄酒(法國) Burgundy
布爾戈尼酵母 Burgundy yeast
布魯氏桿菌 brucella
布魯克菲爾德迴轉式黏度計 Brookfield viscometer
布篩 tammy
布濾器 cloth filter
平〔底〕鍋 griddle
平分 halve
平皿 flat plate; plate
平皿計數 plate count agar
平皿培養〔物〕 plate culture
平皿培養法 plate method; plating method
平皿接種 plating
平皿稀釋 plate dilution
平底玻璃杯 tumbler
平底鍋 pan; saucepan
平底鍋煎煮 pan broiling
平底鍋蒸煮 panbroil
平板糖冷卻器 plate cooler
平板式冷凍〔法〕 plate freezing
平板式冷凍機 plate freezer
平板式接觸冷凍機 plate contact freezer
平板計數 plate count agar
平板培養 plating
平板培養〔物〕 plate culture

平板培養皿 plate
平板培養法 plate method; plating method
平板培養基 plating medium
平板酸釋 plate dilution
平花鰹 bulletmackerel
平面熱交換器 flat surface heat exchanger
平淡的風味 flat flavo(u)r
平菇 oyster cap fungus
平滑肌 smooth muscle; unstriated muscle
平滑肌纖維 smooth muscle fiber
平葡酸 pangamic acid
平酸〔腐敗〕 flat sour
平酸〔腐敗〕細菌 flat sour bacteria
平酸菌 flat sours
平盤 flat plate; pan
平衡 counter balance
平衡分 equilibrium moisture
平衡相對濕度 equilibrium relative humidity(ERH)
平衡溫度 equilibrium temperature
平衡濕度 equilibrium humidity (EH); equilibrium moisture
平鯛 silver bream
幼小的 young
幼公豬（1－2歲） haggerel
幼巨紅細胞性貧血 megaloblastic an(a)emia
幼母羊 haggerel; maid
幼年的 juvenile
幼兒食品 junior food
幼芽 plumule; scion; spear; tiller
幼苗 sapling; seedling; spear
幼畜 hogget
幼魚 baby fish; larval fish
幼禽 poult
幼嫩洋葱 button onion
幼蝦 juvenile prawn
幼鮭 parr; samlet; brandling
幼鴿 squab
幼蟲 larva
幼鯨 whale calf
幼鯖 tinker
幼鯡魚 brit(t); whitebait
幼鰻 elver
幼鱈 codling; scrod
幼體 juvenile
弗氏檸檬酸桿菌 *Citrobacter freundii*
弗吉尼亞火腿 Virginia ham
弗利潑酒 flip
弗里賽〔黑白花〕牛 Friesian; Holstein (-Friesian)
弗金（40.91升） firkin
必需脂肪酸 essential fatty acid
必需胺基酸 essential amino acid (EEA); indispensable amino acid
必需酶 indispensable enzyme
戊二麻 glutaric acid; glutaral; glutaraldehyde
戊胺酸 norvaline
戊基 amyl group
戊烯醛 pentenal
戊硫醇 amyl mercaptan
戊酮 pentanone
戊聚糖 pentosan
戊酸 pentanoic acid; valeric acid
戊酸乙酯 ethyl valerate
戊酸戊酯 amyl valerate
戊酸酯 pentanoate; valerate
戊酸鹽 pentanoate; valerate
戊醇 amyl alcohol; pentanol; pentyl alcohol
戊糖 pentose
戊糖苷 pentoside
戊醛 pentanal; valeraldehyde
戊醛糖 xylose
打孔器 punch
打毛 defeather

打毛機 dehairer; defeatherer; defeathering machine
打包 package; packing; sacking
打包工人 baler
打包間 packhouse
打包機 bailing press; baler; packer
打印機 stamping press
打起蛋白 frothed white
打蛋器 egg beater; egg breaker
打結機 knotter
打開 unwrap
打開果 unpacked
打開瓶蓋 abroach
打開罐頭 abroach
打葉 defoliation
打漿機 beater; hollander; pulper; pulping machine
打漿篩濾機 pulper finisher
打穀 threshing
打穀機 thrasher; thresher
打錠機 tablet machine; tablet press; tabletting machine; tabletting press
打擦糖 whipping property
打擦機 whipping machine
斥水性薄膜 non-wettable film
本氏試驗 Benedict's test
本尼地甜酒 Benedictine
本尼迪特試驗 Benedict's test
本生的動植物 autochthon
本地品種 native variety; native breed
本地種 native species
未上蠟水果 unglazed fruit
未水合的 unhydrated
未水解蛋白質 unhydrolyzed protein
未充填的 unfilled
未加工的 raw
未加熱的 unheated
未加糖的 unsweetened
未加糖麵糰 lean dough
未包裝的 unpacked; unpackaged
未去核的水果 unpitted fruit
未去殼的 unshelled
未去麩的 wholemeal
未全熟的 underripe
未成熟小蘋果 codling; codlin
未成熟水果 crude fruit
未成菜的 immature; premature
未成熟的〔葡萄〕酒 immature wine
未成麵包乾酪 curdy cheese
未成熟的蒸餾酒 immature spirit
未成熟乾酪 green cheese
未成熟啤酒 new beer
未冷卻的 uncooled
未冷凍食品 unfrozen food
未完成產品 unfinished product
未完熟水果 unripe fruit
未足量罐 underfilled can
未受精蛋 wind egg
未洗腸衣 slimy casing
未修釋澱粉 unmodified starch
未消毒乳 raw milk
未烘烤的 unbaked
未烤透的 soggy; slack baked; sodden; underroast; underbaked
未烤透的麵包 histy bread; sad bread; heavy bread
未烤熟的 slack baked; underbaked
未烤熟麵包 underbaked bread
未乾透的 underdried
未密封 unsealing
未密封的 unsealed
未強化乳 unfortified milk
未殺菌的 non-pasteurized
未氫化的 unhydrogenated
未混合的 unmingled
未淨化水 raw water
未淨化的 unpurified
未烹煮的食物 uncooked food

未脫殼的 unhulled
未脫殼的米 paddy rice; rough rice
未處理水 raw water
未陳〔葡萄〕酒 immature wine
末期乳 late lactation milk
未煮透的 undercooked; underboiled
未煮熟的 raw; uncooked
未煮熟的肉 underdone meat
未煮熟食物 undercooked food
未發透麵糰 heavy dough
未發麥芽的 unmalted
未發酵的 unfermented
未發酵的葡萄汁或果汁 must
未發酵茶 unfermented tea
未發酵透的麵糰 immature dough
未發酵麥芽漿 sweet mash
未發酵飲料 unfermented beverage
未發酵麵糰 green dough
未發酸葡萄汁 stum
未稀釋的 undiluted
未結合水 unbound water
未經巴氏殺菌的 unpasteurized
未經加工的 unprocessed
未經消化的 undigested
未經璃瓶的食物 unirradiated food
未經調配的乾酪 natural cheese
未過篩的 unscreened
未過篩粉 grout
未過濾的 unfiltered; unstrained
未飽充飲料 uncarbonated drink
未飽和 undersaturation
未飽和溶液 undersaturated solution
未酯化脂肪酸 unesterified fatty acid
未摻假食品 unsophisticated food
未摻雜的 unadulterated
未摻雜食品 unadulterated food
未漂白的 unbleached
未漂白麵粉 unbleached flour
未精煉的 unrefined
未精製油 unrefined oil
未精製的 unrefined
未精製糖 unrefined sugar
未蒸發的 unevaporated
未熟 greenness
未熟水果 underripe fruit
未熟的 unbaked
未熟透水果 unripe fruit
未熟透的 half-done
未碾的稻穀 paddy rice; rough rice
未調合酒 unblended wine
未調味的 unseasoned
未調味烹調 flat cooking
未醇烈酒 immature spirit
未醃的肉 green meat; uncured meat
未磨的 unground
未篩淨的 underbolted
未篩粗磨粉 coarse meal
未篩過的 unsifted
未篩選的 unscreened
未篩麵粉 atta; coarse flour
未膨發麵包 azyme
未膨發麵糰 unleavened dough
未醒麵糰 immature dough
未閹公牛 bull
未燻製培根〔醃肉〕 green bacon
未攙水的 neat; undiluted
未攙雜的 unmixed
未染色的 unpigmented
未殺菌的 unpasteurized
未殺菌啤酒 unpasteurized beer
末端 tip
末端胺基 terminal amino group
末端胺基酸 terminal amino acid
末端羥基 terminal hydroxyl group
末端羧基 terminal carboxyl group
正二氫癒創酸 nordihydroguaiaretic acid
正己烷 hexane
正丙醇 1-propanol

正戊醛 pentanal
正白胺酸 norleucine
正式樣品 official sample
正式粉會 formal dinner
正牡蠣 giant pacific oyster
正亮胺酸 norleucine
正常乳 normal milk; regular milk
正常飲食 normal diet
正常膳食 normal diet; regular diet
正規茶點 formal tea
正餐 dinner; dine
正餐醇溶 postprandial
正磷酸三鉀 tripotassium orthophosphate
正磷澱粉 ferric orthophosphate
正鐵肌紅蛋白 metmyoglobin
正纈胺酸 norvaline
母牛 cow
母牛犢 heifer calf
母羊 ewe
母羊肉 chilver
母乳 breast milk; human milk; mother milk
母乳化牛乳 humanized milk; maternized milk
母乳育嬰 breastfeeding
母兔 doe
母羔羊 ewe lamb; gimmer
母馬 mare
母匙 serving spoon
母液 mother liquor
母粒 seed grain
母鹿 doe
母豬 sow
母雞 hen
母驢 mare
民族食品 ethnic food
汁 juice; liquor; succus
汁液 juice
玄參紅酸 azafrin
玉米 corn; Indian corn; Indian maize; maize; mealie(s)
玉米〔粗〕粉 maize meal
玉米〔雪〕片 corn flake
玉米〔醇溶〕蛋白 zein
玉米片 cut corn; corn chip
玉米去殼工人 corn husker
玉米去殼機 corn husker; maize husker
玉米成熟度測定計 succulometer
玉米肉餅 scrapple
玉米油 corn oil; maize oil
玉米芯 cob; corncob
玉米威士忌酒 corn whisky; maize whisky
玉米剝粒機 maize sheller
玉米浸液 corn steep liquor
玉米粉 corn meal; corn flour; hominy meal; Indian meal; maize flour; oswego flour
玉米粉布丁 flummery
玉米素 zeatin
玉米果核〔糖〕苷 ribosylzeatin
玉米脆片 corn chip; maize flake
玉米酒精廠 corn distillery
玉米產區 corn belt
玉米粒 corn kernel; maize groats; maize kernel; shelled corn
玉米粗粉 corn meal
玉米脫粒機 corn-and-cob crusher; corn cutter; corn sheller
玉米蛋白 corn gluten; maize gluten
玉米蛋白水解物 hydrolyzed corn protein
玉米蛋白粗粉 corn gluten meal; maize gluten meal
玉米雪片 maize flake
玉米棒子 maize cob
玉米殼 corn husk; maize husk
玉米渣 corn grit
玉米煮豆 succotash

玉米筍 baby corn
玉米粥 corn gruel; hominy; mush; maize gruel; mamaliga; polenta; stirabout; samp
玉米雄花蕊 tassel
玉米黃 maize yellow
玉米黃質 cryptoxanthin; kryptoxanthin; zeaxanthin
玉米種 seed corn
玉米餅 hoecake; Johnny cake; pone
玉米碾磨機 corn mill; maize mill
玉米糊 cream style corn; mush
玉米麩質 corn gluten
玉米麩質粗粉 corn gluten meal
玉米澱粉 cornstarch; maize starch; maizena
玉米磨 corn mill
玉米糖 corn sugar; maize sugar
玉米糖漿 corn syrup; maize syrup
玉米穗 corn ear; maize ear
玉米穗絲 corn silk
玉米穗軸 cob; corncob; maize cob
玉米糜 corn gruel; maize gruel
玉米薄餅 tortilla
玉米醪 corn mash
玉米麵包 corn bread; maize bread; pone
玉米麵筋 corn gluten
玉米麵糊 mamaliga
玉米鬚 silk
玉米釀酒廠 corn distillery; maize distillery
玉果 nutmeg
玉紅黃質 rubixanthin
玉梭魚 escolar
玉蔥 *Allium cepa*
玉筋魚 launce; pacific sandlance; sand lance
玉蜀黍 corn; Indian corn; Indian maize; maize; mealie(s)
玉蜀黍粉 corn meal; Indian meal; oswego flour
玉蜀黍種 seed corn
瓜子 melon seed
瓜子鯧 Japanese butterfish
瓜田 melon field
瓜耳豆膠 guar gum
瓜形甜點心 bombe
瓜拉那豆 guarana
瓜拉拿藤 guarana
瓜果 pepo
瓜香香精 melon flavo(u)r
瓜園 melon field
瓜爾豆膠 guar gum
瓦林混碎機 Waring blender
瓦斯 gas
瓦斯炊飯鍋 gas rice cooker
瓦斯爐 gas heater
瓦爾德霍夫發酵槽 Waldhof fermenter
瓦蘭塔〔油脂〕試驗 valenta test
甘牛至 marjoram
甘牛至油 marjoram oil; sweet marjoram oil
甘白乳酸 norleucine
甘汞 calomel
甘松油 spikenard oil; spikenard; spikenard
甘油 glycerin(e); glycerol
甘油〔一〕花生酸酯 monoarachin
甘油〔三〕棕櫚酸酯 glyceryl tripalmitate; palmitin; tripalmitin
甘油一酸酯 monoglyceride
甘油二〔酸〕酯 diglyceride
甘油二月桂酸酯 dilaurin
甘油二亞麻油酸酯 dilinolein
甘油二油酸酯 diolein
甘油二硬脂酸酯 distearin
甘油三乙酸酯 glyceryl triacetate; triacetin
甘油三反油酸酯 elaidin; trielaidin

甘油三月桂酸酯 glyceryl trilaurate; trilaurin
甘油三次亞麻油酸酯 glyceryl trilinolenate; linolenin; trilinolenin
甘油三亞〔麻〕油酸酯 glyceryl trilinoleate; linolein; linolin; trilinolein
甘油三油酸酯 glyceryl trioleate; olein; triolein
甘油三芥〔子〕酸酯 erucin
甘油三癸酸酯 caprin
甘油三硬脂酸酯 glyceryl tristearate; stearin; tristearin
甘油三酸酯 triglyceride
甘油三醋酸酯 glyceryl triacetate; triacetin
甘油脂 glycerolipid
甘油脂肪酸酯 glycerin fatty acid ester
甘油單硬脂酸酯 glycerin monostearate; glycerol monostearate; glyceryl monostearate
甘油酯 glyceride; glycerin ester
甘油酯酶 glyceridase
甘油酸 glyceric acid
甘油酸酯 glycerate
甘油酸鹽 glycerate
甘油醋酸酯 acetin; glyceryl monoacetate
甘油糖 glycerose
甘油磷脂 glycerophosphatide; glyceryl phosphatide
甘油磷酸 glycerophosphate; glycerophosphoric acid
甘油磷酸鈣 calcium glycerophosphate
甘油磷酸鉀 potassium glycerophosphate
甘油磷酸錳 manganese glycerophosphate
甘油醛 glyceraldehyde
甘美的 luscious; mellow; nectared
甘素 dulcin(e); sucrol; valzin
甘草 glycyrrhiza; licorice; licorice root; liquorice; sweet root
甘草抗氧物 licorice root antioxidant
甘草流浸膏 licorice root extract
甘草浸膏 licorice extract
甘草素 glycyrrhizin; glycyrrhizic acid
甘草梅 liquorice plum
甘草甜〔素〕 glycyrrhizin
甘草甜素二鈉 disodium glycyrrhizinate
甘草甜素三鈉 trisodium glycyrrhizinate
甘草核 licorice extract
甘草酸 glycyrrhizinic acid
甘草酸一銨 monoammonium glycyrrhizinate
甘草酸二鈉 disodium glycyrrhizinate
甘草酸三鈉 trisodium glycyrrhizinate
甘草酸苷 glycyrrhizic acid; glycyrrhizin
甘草苷酸 glycyrrhizic acid; glycyrrhizin
甘茶素 phyllodulcin
甘胺酸 aminoacetic acid; collagen sugar; glycine; glycocoll
甘椒 allspice; Jamaica pepper; pimento
甘筍 carrot
甘菊 camomile
甘菊油 camomile oil
甘蔗 cane; *Saccharum officinarum*; sugar cane; sweet cane
甘蔗汁 cane juice; sugar cane juice
甘蔗田 cane field
甘蔗田廢棄物 field trash
甘蔗夾雜物 field trash
甘蔗板 celotex
甘蔗莖 cane stalk
甘蔗渣 bagasse; megasse
甘蔗園 cane plantation; cane field
甘蔗榨汁機 cane crusher
甘蔗廢蜜 cane molasses
甘蔗廢蜜伏特加 arak
甘蔗輥壓機 cane mill
甘蔗養植場 cane plantation
甘蔗糖蜜 cane molasses
甘蔗壓榨機 cane mill
甘醇 glycol

甘薯 batata; sweet potato; yam
甘薯〔脆〕片 sweet potato chip
甘薯根黴 *Rhizopus batatae*
甘薯澱粉 sweet potato starch
甘藍 cabbage; cole; colewort; common head cabbage; head cabbage; headed cabbage; heading cabbage; wild cabbage
甘藍沙拉 coleslaw; slaw
甘藍菜絲 shredded cabbage
甘藍酸乙烯 ethyl brassylate
甘藷 sweet potato
甘藷澱粉 sweet potato starch
甘露 honeydew; nectar
甘露〔蜜〕 manna
甘露〔糖〕醇食鹽洋菜〔培養基〕 mannitol salt agar
甘露〔糖〕醇發酵 mannitol fermentation
甘露聚糖 mannan; mannosan
甘露聚糖酶 mannase
甘露蜜 honeydew honey
甘露醇 mannitol
甘露糖 manmose; mannose; seminose
甘露糖酯 mannitan
甘露糖酸 mannonic acid
甘露用水 mannitol
甘露糖苷 mannoside
生化反應 biochemical reaction
生化需氧負荷量 biochemical oxygen demand loading
生化需氧量 biochemical oxygen demand
生奶油 raw cream
生吃食物 omophagia
生存 subsistence
生存餐食 subsistence diet
生肉 raw meat
生育胺 tocopherylamine
生育酚 tocopherol; vitamin E
生育酸 tocopheronic acid
生乳 raw milk
生乳酪 raw cream
生咖啡〔豆〕 green coffee
生命科學 life science
生抽 light soy sauce; soy sauce; soya sauce; soybean sauce
生物 organism
生物分析 biological analysis
生物分解〔作用〕 biolysis
生物化學 biochemistry
生物反應器 bioreactor
生物可降解的 biodegradable
生物危害 biohazard
生物合成 biosynthesis
生物色素 biochrome
生物技術 biotechnology
生物系統 biosystem
生物固氮〔作用〕 biological nitrogen fixation
生物毒素 biotoxin
生物活性 biological activity
生物活素 bios
生物科技 biotechnology
生物降解作用 biodegradation
生物降解性 biodegradability
生物素 antiavidin; biotin; vitamin H
生物脂 biolipid
生物能分解的 biodegradable
生物殺蟲劑 living insecticide
生物產品 biological product
生物量 biomass
生物催化作用 biocatalysis
生物催化劑 biocatalyst
生物試驗 biological assay; bioassay
生物碳酸鈣 biological calcium carbonate
生物需氧負荷量 biological oxygen demand loading
生物需氧量 biological oxygen demand
生物價 biological value
生物膠體 biocolloid

生物學 biology
生物類黃酮 bioflavonoid
生物鑑定 bioassay
生物鹼 alkaloid
生的 raw
生的 uncooked
生的食物 uncooked food
生長因素 growth factor
生長曲線 growth curve
生長抑制劑 growth inhibitor
生長刺激劑 growth stimulant
生長周期 growth cycle
生長季節 growing season
生長物 grower
生長物質 growth substance
生長味的劑 growth promoter
生長速率 growth rate
生長期 growth phase; growth period
生長調節劑 growth regulator
生長激素 growth hormone
生活環境 habitat
生食 omophagia
生食蔬菜類 salad plant
生料包裝 raw pack
生乾酪 green cheese
生啤〔酒〕 beer on draught; draft beer; draught beer; unpasteurized beer
生理〔食〕鹽水 physiological saline; saline
生理〔學〕的 physiological
生理生調 physiological disorder
生理食鹽水 normal saline
生理酸性食品 acid residue food
生理價值 biological value
生理學 physiology
生理鹼性食品 basic food; base residue food; base forming food
生甜〔基〕團 glucophore
生產 produce; production
生產力 productivity
生產者 producer
生產量 yield
生產過剩 over production
生產線 production line
生產線中的 in-line
生魚片 sashimi
生麥性 green malt
生殖 reproduction
生殖熱胞 germ cell
生殖菌絲體 reproductive mycelium
生菌數 standard plate count; total plate count
生菜 lettuce
生腥味 raw taste
生態學 ecology
生膜菌屬 *Mycoderma*
生蝦〔帶殼〕 green shrimp
生澱粉味 starchy taste
生糖原的 glycogenic
生濕的 soggy
生薑 green ginger
生薑油 ginger oil
生鮮香腸 fresh sausage
生蟲泡 wormy
生麵〔條〕 raw noodle
生麵包 raw bread
用水冷卻 hydrocooling
用具 outfit; tool
用咖哩〔粉〕調味的 curried
用咖哩粉調味 curry
用杵搗碎 pestle
用後即棄的瓶子 throwaway bottle
用埃淨漂白麵粉 agenize
用紙或鋁箔包裹後烹調〔之食品〕 papillote
用蛋白糖霜飾面的餡餅 fanchonette
用量 dosage
用糖熬煮 candying
月餐 dining

用餐溶 diner
田豆 field pea
田季蘿蔔 little radish
田芥菜 charlock
田青膠 Tian Qing gum
田野傘菌 horse mushroom
田間作物 field crop
田鼠 hamster
甲叉丁二酸 itaconic acid
甲位戊基桂醇 α-amylcinnamyl alcohol
甲位戊基桂醛 α-amylcinnamaldehyde
甲狀旁腺 parathyroid gland
甲狀腺 thyroid gland
甲狀腺立素 thyroxin(e)
甲花翠素 petunidin
甲苯 toluene
甲氧基酶 methoxylase
甲胺 methylamine
甲胺磷 tamaron
甲基β-萘酮 methyl β-naphthyl ketone
甲基液體甲的 acetoin
甲基丁香酚 eugenyl methyl ether
甲基己基酮 methylhexanone
甲基化 methylation
甲基化酒精 methylated spirit
甲基酒者醛 2-methylundecanal
甲基可可鹼 methyltheobromine
甲基戊基酮 methyl amyl ketone
甲基戊酸 methyl pentanoic acid
甲基庚烯酮 methyl heptenone
甲基矽氧烷 methyl silicone
甲基發 methyl red
甲基苯乙醚 methyl phenethyl ether
甲基異丁香酚 methyl isoeugenol
甲基組〔織〕胺 methyl histamine
甲基組胺酸 methylhistidine
甲基紫 methyl violet
甲基鈷維生素 methyl cobalamin
甲基綠 methyl green
甲基聚矽氧烷 methyl polysilicone
甲基橙 helianthin(e); methyl orange
甲基橙皮苷 methyl hesperidin
甲基環戊烯醇酮 methyl cyclopentenolonne
甲基藍 methyl blue
甲基轉移作用 transmethylation
甲基纖維素 methyl cellulose (MC)
甲烯藍 methylene blue
甲硫胺酸 methionine
甲硫基丁胺酸 methionine
甲硫醇 methyl mercaptan
甲硫醚 dimethyl sulfide
甲魚 soft shelled turtle; terrapin; turtle
甲烷 methane
甲烷發酵 methane fermentation
甲酚 cresol
甲殼 crusta; shell; testa; test
甲殼素 chitin
甲殼質 chitin
甲殼質酶 chitinase
甲殼類動物 crustacean; shellfish
甲酸 formic acid
甲酸乙酯 ethyl formate
甲酸丁酯 butyl formate
甲酸己酯 hexyl formate
甲酸和脂 amyl formate
甲酸甲酯 methyl formate
甲酸辛酯 octyl formate
甲酸玫瑰酯 rhodinyl formate
甲酸芳樟酯 linalyl formate
甲酸苯乙酯 phenethyl formate
甲酸香茅酯 citronellyl formate
甲酸水酒精 geranyl formate
甲酸桂酯 cinnamyl formate
甲酸茴香酯 anisy formate
甲酸異戊酯 isoamyl formate
甲酸鈣 calcium formate
甲酸鈉 sodium formate
甲酸葉醇酯 leaf formate

甲酸苄酯 benzyl formate
甲醇 methanol; methyl alcohol; wood alcohol; wood spirit
甲醇中毒 methanol poisoning
甲醇鈉 sodium methylate
甲糖 methose
甲醛 formaldehyde
甲醛水 formalin
甲醛明膠 formogelatin
甲醛聚糖 formose
白丁香浸膏 syringa amurensis concrete
白人心果 white sapota
白千層油 cajeput oil; cajuput oil
白土脱色糖 clayed sugar
白小麥 white wheat
白切肉 boiled meat
白切雞 boiled chicken
白化劑 whitener
白夫人雞尾酒 white lady
白心米 white core rice
白木耳 tremella; *Tremella fuciformis*
白木耳素 agaricin
白巧克力 white chocolate
白汁 cream sauce; white sauce
白玉〔葡萄酒〕 chauche
白玉米 white corn; white maize
白玉米粉 crream meal
白米 finished rice; milled rice; white rice
白肉 white meat
白色大鰉魚 beluga
白色肉 light meat; white meat
白色肌 white muscle
白色念珠菌 Candida albicans
白色杯傘（蕈類） rickstone funnel cap
白色油炒麵粉〔糊〕 white roux
白色糊精 leukodextrin
白色膠偽齒菌（蕈類） toothed jelly
白色鮪魚 light meat tuna
白血球 leukocyte; white blood cell
白利比重計 Brix hydrometer; Brix scale hydrometer
白利糖度 Brix; Brix degree; Brix scale
白利糖度計 Brix hydrometer; Brix scale hydrometer; Brix spindle
白豆 white bean
白豆蔻屬 *Amomum*
白里葉莓 wineberry
白姑魚 white croaker
白屈菜 celandine
白念珠球菌屬 *Leuconostoc*
白果 gingko; ginkgo; maidenhair tree fruit
白果樹 maidenhair tree
白林地蘑菇 wood mushroom
白松雞 white grouse
白松露 white truffle
白油 shortening
白油（食品機械專用潤滑油） white oil
白芝麻 white sesame (seed)
白花色素 anthocyanogen
白芥〔菜〕 white mustard
白芷 angelica
白金接種針 platinum needle
白金接種環 platinum (wire) loop
白金絲 platinum wire
白金絲耳 platinum (wire) loop
白亮葡萄 mustang grape
白扁豆 large white bean
白胡椒 white pepper
白香腸（德國） weisswurst
白香蕉 winter banana
白桑果 white mulberry
白桑葚 white mulberry
白珠木油 gaultheria oil
白珠樹 gaultheria
白高粱 rice corn
白鬼筆（蕈類） common stinkhorn
白焓蛋 hard boiled egg
白胺酸 leucine

白豇豆 white cowpea
白乾 China white wine
白假絲酵母 *candida albicans*
白帶魚 hairtail; ribbonfish
白斬雞 boiled chicken
白梨 Chinese pear; Japanese pear; sand pear
白毫 pekoe (tea)
白毫茶 baikhovi tea
白毫疏宗茶 pekoe souchong
白球甘藍 white cabbage
白脫 butter
白脫刀 butter knife
白脫用羊皮紙 butter paper
白脫用鮮奶油 butter cream
白脫色素 butter colo(u)r
白脫乳 buttermilk
白脫乳清 butter whey; butter serum
白脫定型器 butter printer
白脫油 butter oil
白脫脂 butterfat
白脫硬糖 butterscotch
白脫農業 butter starter
白脫黃 butter yellow
白脫餅 butter cake
白脫醬 butter sauce
白脫麵包 butter bread
白蛋白 albumin
白蛋白的分解〔作用〕 albuminolysis
白蛋巢菌(蕈類) white egg bird's nest
白蛋糕 white cake
白軟皮 albedo
白陶土 kaolin
白雪莉酒 amontillado
白魚 barbel; whitefish
白麥 white wheat
白斑狗魚 northern pike
白斑星鯊 gummy shark
白景天 white stonecrop
白棉布 calico
白菜 pakchoi
白菜豆 white kidney bean
白黃側耳(蕈類) trumpet oyster mushroom
白菖 sweet flag; sweet root
白菖油 calamus oil
白落葵 basella
白葡萄酒 pale wine; vin blanc; white wine
白腐病 white rot
白糊精 white dextrin
白調味醬 bechamel (sauce)
白醋栗 white currant
白橙蓋鵝膏(蕈類) Caesar's mushroom
白橡木 white oak
白糖 white sugar
白糖蜜 white molasses
白蕈毒〔肽〕 phallin
白薯 batata
白鮮菇 salt loving mushroom
白鮭 chub; gangfish; mountain herring; tullibee; pollan; vendace
白鴿 pigeon
白檸檬 limette
白檸檬油 lime oil; limette oil
白檸檬素 limettin
白蟲膠 white shellac
白醬料 white sauce
白鬆糕 lillywhite cake
白鯊貂鮫 dog shark
白鯨 beluga; marsoon; porpoise
白鯧 silver pomfret; white pomfret
白麴 peka
白櫟皮 quercus
白蘆筍 white asparagus
白麵包 light bread; wheat bread; white bread

白麵包(猶太教的安息日用) challah; hallah
白麵粉 light flour
白蘭〔花〕浸膏 michelia alba (flower) concrete
白蘭瓜 honeydew melon
白蘭地〔烈性酒〕 akvavit; aquavit
白蘭地酒 brandy
白蘭地酒小脆餅 brandy snap
白蘭地酒杯 brandy glass
白蘭地蛋酒 brandy eggnog
白蘭地糖果 brandyball
白蘭花油 michelia alba flower oil
白蘭葉油 michelia alba leaf oil
白鰱 silver carp
白鰻 Japanese eel
白蘿生 white radish; white turnip
白鑞 pewter
白鱘 paddlefish
白鯵 big-eye scad
皮 skin
皮下脂肪 subcutaneous fat
皮下脂肪層 adipose layer
皮下組織 subcutaneous tissue
皮重 tare weight; tare
皮屑 paring
皮酒囊 wine bag
皮蛋 alkalized duck egg; alkalized egg; Chinese egg; pidan; preserved egg; thousand years'egg
皮殼分離器 hull aspirator
皮爾遜啤酒 Pilsner beer
皮製臍橙 wine skin
皮膜組織 epithelial tissue; epithelium
皮膚 skin
皮質長白 cortisol
皮質乾酪 leathery cheese
皮質酮 corticosterone; cortisone
皮薄的 thin skinned
皮羅〔肉〕飯 pilau
目測 visual estimation
目測檢驗 visual examination
目視估測 visual estimation
目視檢驗 visual examination; visual inspection
矛尾魚 latimeria
矛型薄荷 spearmint
矢車菊 bluebottle; centaury
石刁柏 asparagus
石灰 calcium oxide
石灰化 calcification
石灰水 lime water
石灰乳 lime milk
石灰法澄清 liming
石灰消和器 slaker
石灰碳酸澄清法 defeco-carbonation process
石灰酸 carbolic acid
石灰澄清 lime defecation
石竹素 caryophyllin
石竹烯 caryophyllene
石油 oil; petroleum
石油石蠟 petroleum wax
石油凍 petrolatum
石油醚 light petroleum; petroleum ether
石花子 agar weed; gelidium
石南花 heather
石英 quartz
石英玻璃 silex glass
石英棉 silica wool
石首魚 croaker
石斑糖 grouper
石榴 carthaginian apple; common pomegranate; pomegranate
石榴汁 grenadine
石榴皮 granatum
石榴素 punicin
石榴酸 punicic acid

石榴糖漿 grenadine; grenadine syrup
石碳水 phenol
石膏 calcium sulfate; gypsum
石磙 stone roll
石輥 stone roll
石墨 graphite
石磨 buhr mill; millstone mill; stone grinder; stone mill
石蕊 litmus
石蕊牛乳〔培養基〕 litmus milk
石蕊試紙 litmus paper
石蕈 cep
石頭 stone
石蠟 olefin; paraffin; paraffin wax
石蠟油 mineral oil; paraffin oil
石蓴 laver
禾稈 straw
禾穀類 cereal
穴鰻 conger
立方體 cube
立生萵苣 cos lettuce
立式〔高壓〕殺菌釜 upright retort; vertical retort
立式冷凍機 upright freezer
立式殺菌釜 vertical autoclave
立式壓榨機 vertical press
立式螺旋壓濾機 vertical screw press
立式鍋爐 vertical boiler
立克次體 rickettsia
立克次體病 rickettsiosis
立索玉紅BK lithol rubine BK
立翅旗魚 giant black marlin
立體鏡 stereoscope
立體顯微鏡 stereomicroscope; stereoscopic microscope

六畫

交叉污染 cross contamination
交互作用 interaction
交流換熱〔法〕 regeneration
交流換熱巴式殺菌器 regeneration pasteurizer
交流換熱器 regenerator
交酯化〔作用〕 interesterification
交酯化〔油〕脂 interesterified lard
交酯化蓖麻酸聚甘油酯 polyglycerol esters of interesterified ricinoleic acid (PGPR)
交聯葡聚糖 sephadex
交聯凝膠 cross linking gel
交聯澱粉 cross bonded starch; cross linked starch; inhibited starch
交鏈凝膠 cross linking gel
交鏈澱粉 cross bonded starch; cross linked starch; inhibited starch
仿白蘭地 hamburg brandy
仿肉〔製〕品 simulated meat product
仿冒食品 copied food
仿香檳〔酒〕 imitation champagne
仿真牛奶 imitation milk
仿真肉 imitation meat
仿造 imitation
仿造食品 imitated food; imitation food
仿製〔葡萄〕酒 imitation wine
仿製品 imitation
仿製食品 imitated food; imitation food; simulated food
仿製香檳〔酒〕 imitation champagne
仿製蛋白 analog protein
伙食 board; fare; mess
伙食房 spense
伙食採辦人 steward
伊紅 eosin(e)
伊麵 E-Fu noodle
伍德夫發酵槽 Waldhof fermenter
伐蘭塔〔油脂〕檢驗 valenta test
休克 shock
休眠的 dormant
休眠孢子 cyst; resting spore; statospore
休眠酵母 resting yeast
休閑零食 snack food
休閑餐飲 casual dining
伏牛花 barberry
伏牛花果 barberry
伏令夏橙 Valencia orange
伏哈德滴定法 Volgard titration
伏特加酒 vodka
伏爾加德滴定法 Volgard titration
仲丁胺 secondary butyl amine
仲丁醇 2-butanol
仲胺 secondary amine
仲醇 secondary alcohol
件 piece
任意採樣 random sampling
光〔分〕解作用 photodecomposition
光〔致〕氧化作用 photooxidation
光化器 actinator

光皮山核桃 big shellbark hickory; king nut
光皮豌豆 smooth pea
光合生物 phototroph
光合自養生物 photoautotroph
光合自營生物 photoautotroph
光合作用 photosynthesis
光合作用產物 photosynthate
光合的 photosynthetic pigment
光合異養生物 photoheterotroph
光合異營生物 photoheterotroph
光析管 cuvette
光柄蘑菇 pluteus
光降解作用 photodegradation
光面中粗〔管狀〕通心粉（1/2英吋粗） zitoni
光面精整 smooth finish
光密度 optical density
光密度計 densitometer; optical densitometer; photodensitometer
光黃素 lumiflavin
光催化〔作用〕 photocatalysis
光滑〔型〕菌落 smooth colony; S-colony
光滑山核桃 pignut
光電霉素計 photoelectric colo(u)rimeter
光潔麵糰 clear dough
光養生物 phototroph
光雞 dressed chicken; pulled chicken
光譜密度測定法 spectrodensitometry
光譜熒光測定法 spectrofluorometry
光變味 light off flavo(u)r
先入先幕 first-in first-out (FIFO)
先天萎縮 aplasia
先進先出 first-in first-out (FIFO)
先質 precursor
先導工廠 pilot plant
先導工廠規模 pilot plant scale
先導工廠試產 pilot plant trial
全用途麵粉 all purpose flour
全乳 whole milk
全乳乾酪 whole milk cheese
全封式外包裝 full wrap
全流質膳食 full fluid diet
全面質量管理 total quality management (TQM)
全脂〔的〕 full cream
全脂牛乳 full cream milk
全脂奶粉 whole milk powder
全脂乳 rich milk; whole milk
全脂乳粉 whole milk powder
全脂乳乾酪 whole milk cheese
全胴體 whole carcass
全動物式營養 holozoic nutrition
全屠體 whole carcass
全蛋 whole egg
全蛋粉 dried whole egg; whole egg powder
全部 whole
全魚 round fish
全魚鹽漬 round salting
全魚鹽醃 round cure
全麥的 wholemeal; whole wheat
全麥粗粉 crushed wheat
全麥統粉 full straight
全麥麵包 entire wheat bread; whole bread; whole wheat bread; wholemeal bread
全麥麵粉 entire wheat flour; graham (flour); whole meal; whole wheat flour; wholemeal flour
全植物式營養 holophytic nutrition
全發酵茶 completely fermented tea
全量分析 total analysis
全熟 fully cooked
全熟牛排 well-done steak
全穀粉 graham (flour)
全穀麵包 graham bread
全糖苷 holoside
全營養食物 wholefoods

全雞 whole chicken
全纖維素 holocellulose
全酶 holoenzyme
共〔同〕沉澱 coprecipitation
共生 symbiosis
共沸冷凍乾燥 azeotropic freeze-drying
共沸乾燥 azeotropic drying
共沸蒸餾 azeotropic distillation
共軛 conjugation
共軛酸 conjugate acid
共軛鹼 conjugate base
共晶 eutectic
共棲 symbiosis
共溶的 eutectic
共聚物 copolymer
再〔巴氏〕殺菌 repasteurization
再〔加〕熱 reheat(ing)
再分離 reseparation
再升華〔作用〕 resublimation
再水合 rehydration
再充填 refill
再加工 retreatment
再加冰 reicing
再加糖霜 reicing
再包冰 reicing
再包冰衣 reglazing
再包糖衣 reglazing
再生 reclaim; reclamation; regenerate; regeneration; restoration; actification
再生熱巴式殺菌器 regeneration pasteurizer
再生熱交換器 regeneration heat exchanger
再生器 regenerator
再回火 retemper
再次烹煮 recoction
再冷卻 recooling
再冷凍 refreezing
再冷卻器 subcooler
再吸收 reabsorption; resorption
再沉澱 reprecipitation
再取樣品 subsample
再流通 recirculation
再洗機 rewasher
再氧化 reoxidation
再浸漬 resteeping
再純化 repurification
再乾燥 redry(ing)
再混合 reblending
再淨化 repurification
再處理 after-treatment; retreatment
再循環 recirculation; recycle; recycling
再循環水 recirculating water
再煮〔沸〕 reboil
再結晶 recrystallization
再萃專 reextraction
再感染 reinfection
再溶〔化〕 remelt
再溶解 resolution
再裝滿 refill
再過篩 rescreen
再滲濾 repercolation
再碳酸化 recarbonation
再稱量 reweighing
再精製 double refining
再蒸發器 revaporizer
再蒸餾 redistillation
再蒸餾水 redistilled water
再製毒液 process butter; renovated butter
再製乳 reconstituted milk
再製乳酪 process butter; process(ed) cheese; renovated butter
再製油脂 processed fat
再製芝士 process(ed) cheese
再增濕 rewetting
再熱器 reheater
再膠化 regelation
再膠凝 regelation
再調節溫度 retemper
再磨研 regrinding

再磨機 regrind mill
再篩分 rescreen
再濕 remoisten
再濕潤 rewetting
再翻熱的〔剩〕菜 rechauffe
冰 ice
冰水 ice water
冰片 borneol; bornyl alcohol
冰牛奶 ice milk
冰衣 ice glaze; water glaze
冰冷卻 ice cooling
冰咖啡 ice coffee
冰紅茶 ice tea
冰凍 frost; ice
冰凍果子露 ice sherbet
冰屑 flake ice
冰屑結構 iciness
冰屑感 iciness
冰島發酵凝乳 skyr
冰島酸牛奶 skyr
冰庫 ice warehouse
冰浴 ice bath
冰茶 ice tea
冰酒器 wine cooler
冰桶 ice box
冰淇淋 cream ice; ice cream
冰晶 ice crystal
冰結晶 ice crystal
冰淇淋三明治 ice cream sandwich
冰淇淋冷凍機 ice cream freezer
冰淇淋凍結機 ice cream freezer
冰淇淋配料 ice cream mix
冰淇淋蛋糕 ice cream custard
冰淇淋磚 ice cream brick
冰淇淋膨脹率 ice cream overrun
冰淇淋蘇打〔汽水〕 cream soda
冰淇淋蘇打水 ice cream soda
冰雪泥 ice sherbet
冰雪溶化成的水 meltwater
冰棒 frozen sucker; ice bar; ice sucker; popsicle
冰棒冷凍機 bar machine; ice bar freezer
冰棍 frozen sucker; popsicle
冰渣冰淇淋 soggy ice cream
冰窖 ice cellar
冰塊 block ice
冰塊威士忌 Scotch on rock
冰箱 cooler; cold store; fridge; ice box; refrigerator; refrigeratory; reefer
冰箱冷藏室 refrigerator compartment
冰醋酸 glacial acetic acid
冰糕 sherbet
冰糖 crystal sugar; candy; rock sugar; rock; rock candy; sugar candy
冰點 freezing point; ice point
冰藏 ice storage
冰鎮薄荷酒 julep
冰鹽 cryohydrate; cryosel
冰鹽混合物 salt ice
列管式蒸發器 calandria evaporator
劣味 cacogeusia
劣酒 hooch
劣啤酒 swipes; small beer
劣等水果 scaly fruit
劣等飲食 hogwash
劣質〔威士忌〕酒 rotgut
劣質〔葡萄〕酒 hedge wine
劣質的 inferior
劣質酒 harsh wine
劣質餾酒 aguardiente
匈牙利〔辣〕椒 king's paprika
匈牙利的多桂甜〔葡萄〕酒 Tokay
匈牙利的托凱甜〔葡萄〕酒 Tokay
匈牙利燉肉 goulash
印式雞肉蛋葱豆飯 kedgeree
印度篩麻 Indian hemp
印度月桂油 Indian laurel oil
印度奶油 ghee; ghi

印度大 bastimic rice
印度的酸辣調味果醬 chutnee; chutney
印度的醃漬果蔬 chutnee; chutney
印度秈 indica type rice
印度南瓜 autumn squash
印度枸杞 bael
印度淡水大魚 mahseer
印度棗 Indian jujube
印度無花果 prickly pear
印度煉乳 kheer
印度矮生小麥 Indian dwarf wheat
印度醋栗 Indian gooseberry
印度樹膠 Indian gum
印度鐮齒魚 bombay duck
危害分析重點控制 hazard analysis critical control point (HACCP)
危險區域 danger zone
吉士白豬(美國) chester white
吉士粉 custard powder
吉利丁 gelatin(e)
吉利無郭魚 redbelly tilapia
同〔元〕多醣 homopolysaccharide
同〔元〕多肽 homopolypeptide
同化〔作用〕 assimilation
同化反應 anabolic reaction
同化類固醇 anabolic steroid
同功維生素 vitamer
同功酶 isoenzyme; isozyme
同位素 isotope
同系胺基酸 homologous amino acid
同型半胱胺酸 homocysteine
同型乳酸〔細〕菌 homolactic acid bacteria
同型乳酸發酵 homolactic fermentation
同型胱胺酸 homocystine
同型發酵乳酸菌 homofermentative lactic acid bacteria
同型發酵的 homofermentative
同流換熱法 recuperation
同流換熱器 recuperator
同源胺基酸 cognate amino acid
同質多〔晶〕型現象 polymorphism
同質性 homogeneity
同質異晶〔現象〕 allomorphism
吊軌式輸送機 trolley conveyor
吊腳鏈 shackle chain; shackle
吊運車 trolley
吊籃〔式〕離心機 suspended centrifuge
吐司 toast
吐酒石 tartar emetic
吐溫〔乳化劑〕 Tween
吐溫20 polysorbate 20
吐溫40 polysorbate 40
吐溫60 polysorbate 60
吐溫65 polysorbate 65
吐溫80 polyoxyethylene sorbitan monooleate; polysorbate 80; Tween 80
向日葵 heliotrope; sunflower; tournesol
向日葵籽 sunflower seed
向日葵籽油 sunflower oil; sunflower seed oil
合子 dumpling
合心皮果 syncarp
合成 synthesize
合成〔法〕 synthesis
合成代謝 anabolism
合成代謝反應 anabolic reaction
合成代謝產物 anabolite; anastate
合成色素 artificial colo(u)r
合成抗氧化劑 synthetic antioxidant
合成機腐劑 synthetic antiseptic
合成乳 artificial milk; synthetic milk
合成果糖 acrose
合成的 synthetic
合成染料 synthetic dye
合成洗滌劑 syndet
合成苦杏仁油 benzaldehyde; benzoic-aldehyde

合成食用香料 synthetic food flavo(u)ring
合成食品 synthetic food
合成香料 artificial flavo(u)r; synthetic flavo(u)r
合成香精 artificial essence; synthetic essence
合成培養基 defined medium; synthetic medium
合成清酒 synthetic sake
合成甜味料 synthetic sweetener
合成甜味劑 synthetic sweetener
合成腸衣 synthetic casing
合成膜 synthetic membrane
合成醋 artificial vinegar; synthetic vinegar
合成纖維 synthetic fiber
合格牛乳 accredited milk
合格的 regular milk
合格級 U.S. Standard
合格水量水平 acceptable quality level (AQL)
合桃 walnut
合桃〔仁〕糊 walnut cream
合桃油 walnut oil
合桃蛋糕 walnut cake
合蛇麻酮 colupulone
合葎草酮 cohurmulone
吃茶點和休息的時間 tea break
吃乾燥食物 xerophagia
吃得合成 overeat
吃飯 dining
回火 drawing; tempering
回式烤〔羊〕肉串 kabob; kebab
回收 reclaim; recovery; reclamation; recuperation
回收率 recovery; yield
回收添 returnable bottle
回冷 recooling
回流 back flow; reflux; regurgitation
回流換熱法 recuperation
回流換熱器 recuperator
回流蒸餾 reflux distillation
回洗 afterwash
回旋式乾燥 rotary drying
回旋藥品燥機 rotary dryer
回旋槽 whirlpool tank
回復味 reversion flavo(u)r
回復現象 reversion
回游三文魚 grilse
回游魚 migratory fish
回游鮭 grilse
回潮 conditioning; moisture regain; ordering; remoisten
回潮房 ordering room
回潮箱 sweat box
回熱的 warmed-over
回熱爐 muffle furnace
回鍋菜〔餚〕 rechauffe
回轉〔加壓〕殺菌釜 rotary autoclave
回轉式殺菌釜 orbitort
回轉式殺菌器 agitating sterilizer
回轉式殺菌機 agitating cooker
回轉式殺菌鍋 agitating retort
回轉式間歇殺菌釜 discontinuous agitating retort
回轉式滅菌機 rotary sterilizer
回轉式電熱烤肉器 rotisserie
回轉式灌瓶機 rotary bottle filler
回轉破碎機 gyratory breaker
回轉壓碎機 gyratory crusher
地下水 ground water; phreatic water
地下根莖 underground stem
地下的食 shebeen
地下球莖 corm
地下莖 rhizome
地下儲藏 subsurface storage
地下儲藏室 storage cellar
地方品種 local variety

地瓜　sweet potato
地衣　lichen; orchil; orselle
地衣〔紅色〕染料　orchil; orselle
地衣多糖　lichenin
地衣多糖酶　lichenase
地衣紅　orcein
地衣紅質　erythrin
地衣素　zearin
地板法的麥芽　floor malt
地窖　cellar; vault
地窖儲藏　cellar storage
地榆　burnet
地檀香油　gaultheria forrestii oil
在萊米　indica type rice
多士　toast
多士爐　toaster
多不飽和脂肪　polyunsaturated fat
多不飽和脂肪酸　polyunsaturated fatty acid
多介質過濾　multiple media filtration
多元酸　polybasic acid
多元酸酯　polybasic ester
多元醇　polybasic alcohol; polyhydric alcohol; polyol
多孔　openness
多孔板　perforated plate
多孔板塔　perforated plate column
多孔物體　porous body
多孔的　perforated
多孔洞的麵包　holey bread
多孔玻璃濾器　fritted glass filter
多孔菌　polypore
多孔篩　perforated sieve
多毛的　pilose
多水分　succulence
多水狀　wateriness
多功葵酶　multifunctional enzyme
多汁　succulence
多汁〔性〕　juiciness
多汁乳菇　tawny milk cap
多汁的　juicy; succulent; sapful; succose
多汁的水果　rich fruit
多汁程度　juiciness
多用包裝機　multipackaging machine
多用混合器　multimixer
多用途食品　multipurpose food
多用途麵粉　family flour
多皮疣柄牛肝菌　orange birch bolete
多次萃取　multiple extraction
多肉　fleshiness
多肉汁的　succulent
多肉果　fleshy fruit
多肉的　fleshy
多色冰淇淋　rainbow ice cream
多佛鰨　dover sole
多尿　diuresis
多尿症　diabetes
多刺的　spiny
多刺魚　bony fish
多果仁的　pippy
多果肉的　pulpy
多板冷凍機　multiplate freezer
多板絞肉機　multiplate chopper
多泡的　bubbly
多段急驟蒸發　multistage flash evaporation
多段萃取　multistage extraction
多段圓盤乾燥機　multistage disk dryer
多相〔性〕　inhomogeneity
多籽的　acinaceous
多重性　multiplicity
多食症　polyphagia; polyphagy
多香果　allspice; Jamaica pepper; pimenta; pimento
多香果油　allspice oil; pimenta oil; pimento oil
多香果葉油　allspice leaf oil
多效蒸發　multiple effect evaporation

多效蒸發器 multiple effect evaporator; multieffect evaporator
多核細胞 apocyte
多核苷酸 polynucleotide
多核苷酸酶 polynucleotidase
多特蒙〔德〕啤酒 Dortmund beer
多脂的 adipic
多脂肪魚 fatty fish
多針頭肌肉注射醃漬法 multiple-needle stitch pumping
多骨膠 bony
多骨魚 bony fish
多胺 polyamine
多偏磷酸鉀 potassium polymetaphosphate
多堅果的 nutty
多液〔性〕 juiciness
多烯脂肪酸 polyenoic fatty acid
多酚 polyphenol
多酚化合物 polyphenolic compound
多酚氧化酶 polyphenolic oxidase
多晶的 heteromorphic
多晶現象 heteromorphism
多氯聯苯 polychlorinated biphenyl (PCB)
多程冷凝器 mutlipass condenser
多程熱交換器 mutlipass heat exchanger
多筋肉 stringy meat
多菌發酵 mixed culture fermentation
多軸擠壓機 multiscrew extruder
多微孔的 spongy
多碎屑的麵包 crumbly bread
多葉奇異菌（蕈類） hen of the woods
多種子的 pippy
多種洋狐鰹 multivitamin
多種維生素缺乏性 polyavitaminosis
多管冷凝器 multitubular condenser
多管熱交換器 multitube exchanger; multitubular heat exchanger
多聚松香甘油酯 glycerol ester of polymerized rosin
多價螯合〔作用〕 sequestration
多價螯合劑 chelating agent; sequestering agent; sequestrant
多層 multilayer
多層化食品 multilayered food
多層牛皮紙 multiwall kraft
多層包裝 multipack; multiple package
多層平板式冷凍機 multiplate freezer
多層冰箱 multistory cold storage
多層式乾燥器 plate dryer
多層式乾燥機 tier dryer
多層冷庫 multistory cold storage
多層洋鰻 multiwall paper bag
多層乾燥機 chest drying machine
多層洋 multiwall sack; multiwall bag
多層箔 multifoil; polyfoil
多樣性 multiplicity
多線魚 Atka fish; Atka mackerel
多輥破碎機 multiroll crusher
多輥壓碎機 multiroll crusher
多糖 polyose; polysaccharide; polysaccharose
多糖磷酸解酶 polysaccharide phosphorylase
多糖酶 polysaccharase; polysaccharidase
多頭封口機 multistation seamer
多頭封罐機 multistation seamer
多頭填充器 multivalve filler
多頭裝瓶機 multihead filling machine
多頭裝填器 multivalve filler
多頭灌注機 multihead filling machine
多磷酸鈉 sodium polyphosphate
多磷酸鉀 potassium polyphosphate
多磷酸鹽 polyphosphate
多穗柯棕 tanoak brown
多縮戊糖 pentosan
多縮合〔作用〕 polycondensation
多縮合物 polycondensate
多黏〔芽孢〕桿菌 *Bacillus polymyxa*

多黏〔桿〕菌素 polymyxin
多黏菌素 aerosporin
多綫魚 boregata
多肽 polypeptide
多肽酶 polypeptidase
多萜醇 dolichol
夸希奥科病 kwashiorkor
夸脱(液體容積) quart
好氣的 aerobic
好氣細菌 aerobic bacteria; aerophilic bacteria
好氧生活 aerobiosis
好氧培養 aerobic culture
好氧發酵 aerobic fermentation
好氧菌 aerobe; aerobium; aeromicrobe
好氧菌種 aerobic culture
好氧微生物 aerobic microorganism
好望角醋栗 cape gooseberry
好斯坦乳牛 Friesian; Holstein (-Friesian)
字母〔意大利〕麵 alfabeto
存活曲線 survival curve
存貨 stock
存貨不足 understock
存貨過多 overstock
安士 ounce (oz.)
安卡蓋 anchor cap
安古〔斯圖臘〕樹皮 angostura
安全 safe
安全系數 safety factor
安全玻璃 safety glass
安全蓋 securo cap
安全利 safe
安定作用 stabilization
安定性 stability
安定劑 stabilizer; stabilizing agent
安哥拉山羊 angola goat
安息香 storax; styrax
安息香酸 benzoic acid
安息香酸鈉 sodium benzoate
安息香酸鹽 benzoate
安息香膠 benzoin
安息香醇 styracitol(e)
安息香樹膠 benzoin gum
安息香醛 benzaldehyde; benzoic aldehyde
安咯斯牛 Angus
安琪〔利〕卡白葡萄酒 angelica
安琪〔兒〕蛋糕 angel cake
安瓿 ampoule
安瑪多立重排〔反應〕 Amadori rearrangement
尖吻四鰭旗魚 Japanese spearfish
尖刺 spine
尖圓腹鯡 dwarf round herring
尖端 tip
尖齒鰻 sharp-toothed eel
尖頭 tip
尖頭長蘋果 codlin
尖頭紅辣椒 rose paprica
帆布袋 packing canvas
帆蜥魚 lancet fish
并晶 grain hardening
成〔糖〕蜜作用 melassigenesis; molasses formation
成分 constituent; ingredient
成形模 shaping die
成形器 shaper
成形機 forming machine
成品 finished product; final product
成型計填密封法 form-fill-seal
成型灌裝封口法 form-fill-seal
成套的餐具 dinner service; dinner set
成套餐具 table set
成粒狀 granulation
成粒機 nodulizer
成魚 full fish
成圓機 rolling machine

成塊急速冷凍法 block quick freezing (BQF)
成層〔現象〕 stratification
成層殺菌法 strata-cock process
成熟 maturation; ripening
成熟〔度〕 maturity; ripeness
成熟〔葡萄酒〕計 maturometer
成熟的 mellow
成熟乾酪 ripen cheese
成熟魚卵 full roe
成熟過度的麵糰 overripe dough
成熟槽 ripener
成熟麵糰 matured dough; ripe dough
成鹽的 salify(ing)
扣帶式蓋 phoenix cap
扣腳鏈 shackle chain
托立斯緩衝液 Tris buffer
托兒所 nursery
托瓶盤 bottle pedestal
托壺 pitcher
托盤 pallet; server; tray
托盤天平 counter balance; table balance
收奶站 milk depot
收奶槽 dump tank
收成 harvest
收乳站檢查 platform inspection
收乳臺試驗 platform test
收集槽 receiving tank
收斂〔作用〕 astriction
收斂性 astringency
收斂劑 astringent
收縮包裝 shrink package; shrink wrap
收縮性蛋白質 contractile protein
收縮薄膜 shrink film; shrinkage film
收穫 harvest
收穫後處理 post harvest handling
收穫後熟成 post harvest ripening
早午餐 brunch
早田頭菇 spring field cap
早收蔬菜 early vegetable
早晨擠的乳 morning milk
早熟 early maturation; early ripening; prematurity
早熟〔法國〕梨 jargonelle
早熟水果 hasting fruit
早熟的 premature; rareripe
早熟蔬果 hastings; rareripe
早熟蔬菜 early vegetable
早餐 breakfast; dejeuner
早餐軟小麵包 bap
早餐稀粥 breakfast gruel
早餐粥品 breakfast gruel
早餐穀類食品 breakfast cereal
早餐點心 coffee cake
早點 breakfast
旭蟹 crimson crab
曲 koji
曲二糖 kojiboise
曲汁 koji extract
曲床 koji bed
曲味 koji flavo(u)r
曲奇餅 cookie
曲酸 kojic acid
曲盤 koji tray
曲霉 koji mo(u)ld
曲澱粉酶 koji amylase
曲臂式攪拌器 arm mixer
有〔葡萄〕酒味的 winy
有大蒜氣味的 garlicky
有仁的 kernelled
有孔轉筒分級機 mammoth grader
有毛的 pilose
有生長力的 vegetative
有污方頭 spotted
有肉食的茶點 meat tea
有芒小麥 awned wheat; bearded wheat
有性孢子 sexual spore
有花蜜的 nectared

有咬勁的 chewy
有毒物質 toxic substance
有毒的 deleterious; noisome; nocuous; poisonous; toxic; venomous
有毒真菌 poisonous fungus
有香味的 dulcet; scented
有害物質 noxious substance
有害的 deleterious; noisome; nocuous
有害健康的 unhealthy
有害微生物 obnoxious species
有效〔使用〕壽命 useful life
有效成分 active component
有效組分 active constituent
有效期 useful life
有效劑量 effective dose
有效營養素 available nutrient
有核仁的 kernelled
有氣味的物體 odorant
有氧呼吸 aerobic respiration
有氧培養箱 air incubator
有氧運動 aerobics
有益 wholesomeness
有益的 wholesome
有益健康 wholesomeness
有益健康的 wholesome
有骨的 bone-in
有堅果味的 nutty
有條紋的 streaky
有莢的 podded
有魚肉的烤麵包 canape
有麥芽的 malty
有惡臭的 noisome; stinking
有斑點的 spotted
有硬殼的 crusty
有筋青刀豆 string bean
有微孔的罐 pin hole can
有蓋大湯碗 tureen
有蓋食物籃 hamper
有蓋湯碗 soup tureen; terrine
有蓋湯鍋 marmite
有蓋罐 canister
有輪的茶具臺 tea trolley
有醋黄的 vinegary
有麩皮的 branny
有機化合物 organic compound
有機的 organic
有機食物 organic food
有機食品 organic food
有機殺蟲劑 organic pesticide
有機氮 organic nitrogen
有機溶膠 organosol
有機農業 organic farming
有機農藥 organic pesticide
有機酸 organic acid
有機酸調味料 organic acid seasoning
有機營養細菌 heterotrophic bacteria
有機磷農藥 organophosphorus pesticide; phosphate pesticide
有機體 organism
有機鹼 organic base
有鞘細菌 sheathed bacteria
有餡小麵包 filled bun
有餡的 farci
有營養的 alimental; nutritious; nourishing
有營養的食物 alible food
有蟲的 wormy
有鰭魚〔類〕 finfish
有鱗的 scaly
有鱗魚 scale fish
有鹽分的 salty
有鹽味的 brackish
朱古力 chocolate
朱古力奶 chocolate milk
朱古力威化餅 chocolate wafer
朱古力香精 chocolate flavo(u)r
朱古力塊 chocolate bar
朱古力新地 chocolate sundae
朱古力製品 chocolate goods

朱古力糖 chocolate candy
朱古力糖衣 couverture
朱草 alkanet
朱欒倍半萜 valencene
次〔級〕標準 substandard
次代培養〔物〕 subculture
次亞麻油酸 linolenic acid
次亞麻油酸酯 linolenate
次亞麻油酸鹽 linolenate
次品 rejected product
次要元素 minor element
次氧化醋菌 *Acetobacter suboxydans*
次級反應 secondary reaction
次級培養〔物〕 secondary culture
次級產品 second product
次級鹿肉餡餅 humble pie
次級蘭酒 tafia
次硫酸鈉 sodium hydrosulfite
次貨 seconds
次單元 subunit
次氯酸 hypochlorous acid
次氯酸鈣 calcium hypochlorite
次氯酸鈉 sodium hypochlorite
次氯酸鈉消毒液 javelle water
次氯酸鈉消毒處理 javellization
次氯酸鈉溶液 javelle water
次氯酸鹽 hypochlorite
次黃〔嘌呤核〕苷 inosine
次黃〔嘌呤核〕苷酸 inosine monophosphate (IMP); inosinic acid
次黃嘌呤 hypoxanthine
次黃質 hypoxanthine
次磷酸鈉 sodium hypophosphite
次磷酸錳 manganese hypophosphite
死亡期 death phase
死水 dead water
死後強直 mortal rigor
死後僵直 death stiffening; mortal rigor; postmortem rigidity; rigor mortis
死僵 mortal rigor; postmortem rigidity; rigor mortis
汗濕損壞 sweat damage
江鱈 burbot; eelpout; maria
池 tank
池魚 pond fish
池塘 pond
污水 polluted water; sewerage; sewage; soilage; sewage water; waste water
污水池 cesspool
污水細菌 sewage bacteria
污水處理 sewage disposal; sewage treatment; waste water treatment
污水處理場 sewage farm
污泥 sludge
污物 dirt; feculence; filth; sludge
污垢 dirt; filth
污染 contamination; pollution; taint
污染水 polluted water
污染物 contamination
污染物〔質〕 contaminant; pollutant
污染空氣 polluted air
污染指數 pollution index
污點 stain; taint
污穢 soilage
污穢的 foul
灰口蘑（蕈類） grey and yellow knight cap
灰分 ash
灰化 ashing; incineration
灰化用坩堝 ashing crucible
灰化爐 muffle furnace
灰水 buck
灰光柄菇 fawn shield cap
灰姥鱸 oldwife
灰星鯊 dogshark
灰胡桃 butternut; white walnut
灰胡桃南瓜 butternut squash
灰康吉鰻 black conger eel

灰焙玉米餅 ashcake
灰蚶 blood cockle
灰錳氧 potassium permanganate
灰燼烤餅 ashcake
灰鯨 devilfish
灰鯧 white pomfret
灰鯖糖 bonito shark
灰鯔 grey mullet
灰鰭鯛 yellowfin sea bream
百刁柏 asparagus
百子蓮 African lily
百合 lily bulb; lily root
百合烈酒 sotol
百里的 thyme; thymus
百里香油 thyme oil
百里香魚 grayling
百里香酚 thymol
百花香 potpourri
百香果 granadilla; parchita; passionfruit
百貨店 stores
百萬分之一 parts per million (ppm)
百萬分點 parts per million (ppm)
竹 bamboo
竹刀魚 needlefish; saury; saury pike
竹串 bamboo skewer
竹芋 arrowroot
竹芋澱粉 arrowroot starch
竹豆 border bean
竹竽 maranta
竹筍 bamboo shoot
竹葉 bamboo leaf
竹葉鰱 silver carp
竹籤 bamboo skewer
米 rice
米氏〔發酵〕法 Melle process
米豆 rice bean
米乳 rice milk
米拉別〔里〕李子 mirabelle
米倉 rice bin
米根黴 *Rhizopus oryzae*
米粉 rice vermicelli; rice flour
米粉〔絲〕 rice noodle
米粉條 locksoy
米紙(可食用) rice paper
米酒 rice wine; rice water
米粒化 ricing
米粗粉 oryzae farina
米湯 rice water
米粥 congee; conjee; rice porridge; rice congee
米隆〔酪胺酸〕試驗 Millon test
米飯 cooked rice
米製點心 rice snack
米穀蛋白 oryzenin
米線 rice vermicelli
米醋 rice vinegar
米麩 rice bran; rice polishes; rice polishings; rice screenings
米麩油 kanemi oil; rice bran oil
米麩蠟 rice bran wax
米澱粉 rice starch
米糠 bran; polishings; rice bran; rice polishes; rice polishings; rice screenings; tikitiki
米糠魚 kanemi oil; rice bran oil; rice oil
米糠的酒精提取液 tikitiki
米糠素 oryzanal
米糠醇 oryzanol
米糠蠟 rice bran wax
米麴 mold rice; rice koji
米麴菌 *Aspergillus oryzae*
米麵條 rice vermicelli
羊 sheep; hogget
羊毛 wool
羊毛固醇 lanosterol
羊毛脂 lanolin
羊白孔菌(蕈類) sheep polypore
羊皮紙 parchment

羊肉 kabob; kebab; mutton
羊肉味的 muttony
羊肋排〔肉〕 rack
羊肚菌 common morel; morel; morel fungus; sponge mushroom; toadstool
羊肚菌屬 *Morchella*
羊肚蕈 morel fungus
羊後腰脊肉 chump
羊背肌 backstrap
羊脂 mutton tallow; mutton fat
羊脂酸 capric acid; decanoic acid
羊胸肉 breast and flap
羊培根肉 macon
羊排 English chop; mutton chop
羊魚 red mullet
羊腰部脊肉 saddle
羊腸衣 sheep casing
羊腿肉 leg chump off
羊頸肉 scrag end
羊臊臭 hircine smell
羊雜碎布丁 haggis
羊雜餡肚 haggis
羊羶味 mutton odo(u)r
羊臘肉 macon
羊蠟 caprin
羽毛 feather; plume
羽衣甘藍 borecole; collard; kale
羽狀葉棕櫚 cohune palm
羽狀葉棕櫚堅果 cohune nut
羽狀葉棕櫚樹 cohune
羽扇豆 lupine
羽扇豆酸 lupinic acid
羽扇豆糖 lupeose; stachyose
老人食品 geriatric food
老化 ageing; aging; staling; staleness
老化的 stale
老母綿羊 crone
老白蘭地酒(12－15年酒齡) very old (V.O.)
老年 senility
老年的 senile
老姆酒 rum
老姆醚 rum ether
老抽 dark soy (sauce)
老鼠 rat
老鼠斑 barramundi cod
老滷 second pickle
老薑 old ginger
老麵糰 old dough; over-aged dough
老饕 epicure
老鸛草油 palmarosa oil
考力代羊 Corriedale
考利台爾羊 Corriedale
耳形通心粉 orecchiette
耳匙菌(蕈類) ear pick fungus
肉 flesh; meat
肉〔汁濃〕湯 potage
肉〔毒〕鹼 carnitine; vitamin Bt
肉〔餡〕餃子 meat pocket
肉〔類〕的 meaty
肉〔類〕製品 meat product
肉丁 cubed meat; diced meat
肉丸 meat ball; minced meat ball
肉丸子 ballotines; quenelle
肉切片機 mea slicer
肉片 cutlet; filet; fillet; meat slice; sliced meat
肉汁 broth; jus; meat juice
肉汁〔醬〕 gravy
肉汁〔醬〕器皿 gravy boat
肉汁洋菜〔培養基〕 broth agar; bouillon agar
肉汁培養 broth culture
肉汁培養基 broth medium; bouillon medium
肉汁清湯 bouillon
肉汁醬粉 gravy powder
肉用小牛 veal calf

肉用公牛 butcher bull; beef bull
肉用牛 butcher cattle; beef cattle; meat cattle
肉用牛種 beef breed
肉用母牛 butcher cow
肉用牲畜 butcher cattle
肉用〔塗 butcher's breast
肉用閹牛 beef steer
肉皮絞碎機 hide cutter
肉色暗淡 dead colo(u)r
肉色蒼白 dead colo(u)r
肉色麗傘(蕈類) pink fair head
肉芋 tannia
肉豆 hyacinth bean
肉豆蔻 mace; nutmeg
肉豆蔻山核桃 nutmeg hickory
肉豆蔻皮 mace
肉豆蔻衣 mace
肉豆蔻果〔實〕 nutmeg apple
肉豆蔻油 mace oil; myristic oil; nutmeg oil
肉豆蔻香精 myristicene
肉豆蔻脂 mace butter; nutmeg butter
肉豆派乾皮 nutmeg mace
肉豆蔻酸 myristic acid
肉豆蔻屬 *Myristica*
肉豆蔻萜 myristicene
肉店 shambles
肉果 sarcocarp
肉泥 muddy flesh
肉厚的 fleshy; meaty
肉品 meat product
肉品加工 meat processing
肉品衛生 meat hygiene
肉毒 kreotoxism
肉毒〔芽孢〕桿菌 *Bacillus botulinus*
肉毒〔梭狀芽孢〕桿菌 *Clostridium botulinum*
肉毒〔梭孢〕桿菌 *Clostridium botulinum*
肉毒中毒 botulism
肉毒盤 kreotoxin
肉毒桿菌 botulinum; botulinus
肉毒桿菌中毒症 botulism
肉毒桿菌毒素 botulin; botulinal toxin; botulinum toxin; botulinus toxin
肉毒桿菌毒素菌株 botulinogenic strain
肉毒梭菌 botulinum; botulinus
肉毒梭菌中毒症 botulism
肉毒梭菌毒素 botulin; botulinal toxin; botulinum toxin; botulinus toxin
肉毒梭菌毒素菌株 botulinogenic strain
肉毒魚毒素 ciguatera toxin; ciguatoxin
肉毒魚類中毒 ciguatera poisoning; ichthyosarcotoxism
肉派 meat pie
肉風味 meat flavo(u)r
肉食 meat diet
肉食動物 carnivore; predator
肉食療法 zomotherapy
肉凍 aspic jelly; aspic; galantine; jellied meat; sulze
肉桂 cassia; cassia bark; Chinese cinnamon; cinnamon; cinnamon bark
肉桂味紅糖果 red hot
肉桂油 cassia oil; Chinese cinnamon oil; cinnamon oil; cinnamon bark oil; cinnamic oil
肉桂重料麵包 hotcross bun
肉桂葉油 cinnamon leaf oil
肉桂酸 cinnamic acid
肉桂酸乙酯 ethyl cinnamate
肉桂酸甲酯 methyl cinnamate
肉桂酸肉桂酯 styracin
肉桂酸苯乙酯 phenethyl cinnamate
肉桂酸異丁酯 isobutyl cinnamate
肉桂醇 cinnamic alcohol
肉桂醛 cinnamaldehyde; cinnamic aldehyde

肉浸汁 infusion broth; meat infusion
肉浸液 meat infusion
肉粉 powdered meat
肉胴 drum meat
肉胺酸 sarcosine
肉胺酸氧化酶 sarcosine oxidase
肉乾 dried meat; dehydrated meat; jerk
肉副產品 meat by product
肉捲 collar
肉排 chop; collop; steak
肉排（牛腰部肉） entrecote
肉眼 rib eye roll
肉眼估計 visual estimation
肉眼檢查 macrography
肉眼檢驗 visual examination; visual inspection
肉脯 dried shred pork
肉販 butchery; butcher
肉魚 Japanese butterfish
肉鹵粉 gravy powder
肉荳蔻衣油 mace oil
肉荳蔻油 myristica oil
肉荳蔻酸乙酯 ethyl myristate
肉荳蔻酸甲酯 methyl myristate
肉荳蔻酸異丙酯 isopropyl myristate
肉荳蔻醇 myristyl alcohol
肉荳蔻醛 myristaldehyde
肉湯 broth; brewis; bouillon; jus
肉湯粉 bouillon powder
肉湯培養 broth culture
肉湯培養基 broth medium
肉湯瓊脂〔培養基〕 broth agar
肉絨 dried shred pork
肉絲 meat strip; shredded meat
肉萃 meat extract
肉菜香腸 sosera
肉菜雜燴 jambalaya; potpourri
肉貯品處 larder
肉飯 pilaf; pilau
肉塊 cut; meat cut
肉溫度計 meat thermometer
肉嫩化劑 meat tenderizer
肉餅 bakemeat; meat pattie; meat patty; pattie; patty
肉增量劑 meat extender
肉漿 minced meat; mincemeat
肉漿蛋白質 meat plasma protein
肉豬 hog; pork pig
肉質改良劑 meat modifier
肉質的 succulent
肉質聚合果 sorosis
肉質豐厚的水果 fleshy fruit
肉醃製 meat curing
肉糕 loaf; meat loaf
肉糖 meat sugar
肉餡 chopped meat; comminuted meat; forcemeat; meat paste; minced meat; mincemeat
肉餡取 mince pie
肉餡烤餅 bakemeat
肉餡餅 mince pie; meat pie; sea pie
肉餡攪拌器 meat mixer
肉糜 ground meat; mince; minced meat; mincemeat; meat emulsion
肉櫃 meat safe
肉醬 meat paste
肉雞 broiler
肉鬆 shredded meat
肉羅夫 meat loaf
肉類 meat
肉類分切圖 cuts chart
肉類包裝廠 packing house; packing plant
肉類改良劑 meat modifier
肉類抽提物 meat extract
肉類產業〔者〕 meat packer
肉類檢驗員 meat inspector
肉類罐頭 canned meat; tinned meat
肉鐵質 carniferrin

肉體 soma
肉苷 inosine
肉苷一磷酸 inosine monophosphate (IMP); inosinic acid
肉苷酸 inosine monophosphate (IMP); inosinic acid
肉苷酸二鈉 disodium inosinate
肋〔條〕肉 sparerib
肋里脊肉 stripped loin
肋狀皺盤菌（蕈類） cup like morel
肋脊骨牛排 rib steak
肋脊部 rib set
肋骨 costa rib; rib; sidebone
肋骨牛排 blade steak; rib steak
肋骨肉 steak piece
肋條肉 rib cut
肋眼肉 rib eye roll
肋軟骨 costal cartilage
肋間的 intercostal
肌 muscle
肌小節 sarcomere
肌內脂肪 intramuscular fat
肌內膜 endomysium
肌白蛋白 myoalbumin
肌肉 muscle; thew
肌肉分解 sarcolysis
肌肉的 sarcous
肌肉球蛋白 myoglobulin
肌肉組織 muscular tissue
肌肉組織鬆弛 flaccidity
肌肉間脂肪 intramuscular fat
肌色素 myochrome
肌束 muscle bundle
肌束膜 perimysium
肌乳酸 sarcolactic acid
肌周膜 perimysium
肌紅素 myoglobin
肌紅蛋白 myoglobin
肌原絲 myofilament
肌原纖維 muscle fibril; myofibril
肌原纖維蛋白質 myofibrillar protein
肌胺酸 creatine; sarcosine
肌胺酸氧化酶 sarcosine oxidase
肌酐 creatinine
肌動球蛋白 actomyosin
肌動蛋白 actin
肌動蛋白絲 actin filament
肌動蛋白纖維 actin filament
肌動凝蛋白 actomyosin
肌基質蛋白 myostromin
肌球蛋白 myoglobulin; musculin; myosin
肌球蛋白絲 myosin filaments
肌細胞 muscle cell
肌組織 muscular tissue
肌蛋白〔質〕 musculin; muscle protein; myoprotein
肌絲 myofilament
肌鈣蛋白 troponin
肌節 sarcomere
肌解 sarcolysis
肌酸 creatine
肌酸酐 creatinine
肌酸磷酸 creatine phosphate; phosphocreatine
肌層 muscular layer
肌漿 muscle plasma; sarcoplasm
肌漿蛋白 myogen; myosinogen
肌漿蛋白質 sarcoplasmic protein
肌漿網 sarcoplasmic reticulum
肌膜 fascia; myolemma; muscular fascia; sarcolemma
肌質 sarcoplasm
肌質網 sarcoplasmic reticulum
肌醇 inose; inosite; inositol; myoinositol
肌醇六磷酸 phytic acid
肌醇半乳糖苷 galactitol
肌醇磷脂 lipositol
肌凝蛋白 myosin

肌激酶 myokinase
肌糖 muscle sugar; meat sugar
肌鞘 myolemma; sarcolemma
肌纖絲 myofilament
肌纖維 muscle fiber
肌纖生物 myolemma; sarcolemma
肌苷 inosine
肌苷一磷酸 inosine monophosphate (IMP); inosinic acid
肌苷酸 inosine monophosphate (IMP); inosinic acid
肌苷酸二鈉 disodium inosinate
自〔身〕催化反應 autocatalytic reaction
自〔動〕氧化 auto-oxidation
自〔然〕發酵〔作用〕 autofermentation
自〔然發〕熱包裝 self-heating package
自〔體〕消化 autodigestion; self-digestion
自由水 free water
自由水分 free moisture
自由沉降 free setting
自由流動的 free flowing
自由脂肪酸 free fatty acid
自由胺基酸 free amino acid
自由基 free radical; radical
自由基反應 radical reaction
自由基捕捉劑 radical scavenger
自由酸 free acid
自由標 free alkali
自立袋 standing pouch
自行發酵 spontaneous fermentation
自助〔商〕店 self-service store
自助午餐 buffet lunch
自助的 self-service
自助食品雜貨店 groceteria
自助晚餐 buffet supper
自助餐 buffet
自助餐館 cafeteria
自身催化〔作用〕 autocatalysis
自來水 tap water; water supply
自來水廠 water works
自泌蜂蜜 virgin honey
自花交配作物 autogamous crop
自花受粉作物 autogamous crop
自花受精果 autocarp
自封的 self-sealing
自流汁 free run juice
自家消化 autolysis
自記溫度計 recording thermometer
自動〔水果〕去皮去芯機 automatic peeler and corer
自動〔水果〕去皮機 automatic peeling machine
自動分批配料秤 automatic batch scale
自動天平 automatic balance
自動充填結紮機 twist linker; twist linker machine
自動充填機 self-filler
自動包裝機 auto-wrapper
自動去骨機 automatic deboning machine
自動扭結機 twist linker; twist linker machine
自動取樣器 automatic sampler
自動恆溫器 automatic thermostat
自動食品櫃 automat
自動秤 automatic balance
自動記錄器 logger
自動迴轉式炸鍋 automatic round type fryer
自動骨肉分離機 automatic bone meat separator
自動售貨機 automat; slot machine; vending machine
自動密封 self-sealing
自動販賣冷藏櫃 service cabinet
自動販賣食物 vended food
自動販賣機 dispenser; vending machine; vender; vendor
自動進料器 automatic feeder

自動進料機 autofeeder; self-feeder
自動催化〔作用〕 autocatalysis
自動填充機 automatic filler
自動裝袋稱量秤 automatic bagging scale
自動裝填機 automatic filler; automatic filling machine
自動滴定器 autotitrator
自動監控程序 automonitor
自動監測器 automonitor; automatic monitor
自動餵料機 autofeeder
自動翻邊機 automatic flanger
自動灌裝機 automatic filling machine; self-filler
自動罐蓋打印機 automatic marker
自備聚餐 potluck
自然成熟 natural ripening
自然的 natural
自然敗壞 autolytic deterioration
自然通風 natural draft
自然發酵 self-fermentation; spontaneous fermentation; wild fermentation
自然發酵〔的葡萄〕酒 natural wine
自然發酵的麵包 salt-rising bread
自然菌群 indigenous flora
自然對流 natural convection
自然酸化 self-souring
自發〔麵〕粉 self-raising flour
自給器 self-feeder
自催化 self-catalyzed
自溶 autolysis; isophagy
自溶素 autolysin
自溶產物 autolysate
自溶變質 autolytic deterioration
自溶酶 autolytic enzyme; autoenzyme
自解酶 autolytic enzyme
自種自食而不出售的作物 subsistence crop
自製的 homemade
自製麵包 homemade bread
自製麵條 homemade noodle
自養生物 autotroph
自養細菌 autotrophic bacteria
自營生物 autotroph
自營細菌 autotrophic bacteria
自觸媒反應 autocatalytic reaction
自體消化 autolysis; isophagy
自體消化酶 autolytic enzyme
臼齒形玉米 dent corn
舌 glossa; lingua
舌〔頭〕 tongue
舌鰨〔魚〕 sole; tonguefish
色拉 salad
色拉〔調味〕醬 salad dressing
色拉用蔬菜 salading
色拉米腸 Italian salami; salami (sausage)
色拉油 salad oil
色拉盤 salad dish
色板 colo(u)r disc; colo(u)r disk
色度 colo(u)r intensity
色度計 chromatometer
色差計 colo(u)r difference meter
色素 colo(u)r; pigment
色素生成 chromogenesis; pigmentation
色素細胞 pigment cell
色胺 tryptamine
色胺凍 tryptophan
色胺酸酶 tryptophanase
色密度 colo(u)r density
色密度計 colo(u)r densitometer
色紳質硬乾燥肉 dark, firm, dry muscle (DFD muscle)
色溫度劑 colo(u)r thermometer
色調 tinct
色調計 tintometer
色醇 tryptophol
色澤 tinct
色覺計 chromatometer

艾伯內西餅乾 abernethy (biscuit)
艾杜糖 idose
艾杜糖酸 idonic acid
艾杜糖醇 iditol
艾杜糖醛酸 iduronic acid
艾美立加糖改進發酵法 amelioration
艾美利法 amerio method
艾美利接觸冷凍法 amerio method
艾菊 tansy
艾菊的苦葉 tansy
艾爾啤酒 ale
艾屬 *Artemisia*
血三酸甘油酯過多〔症〕 hypertriglyceridemia
血小板 thrombocyte
血平板瓊脂 blood agar
血甘油酯過多〔症〕 hyperglyceridemia
血白蛋白 blood albumin
血合肉 dark muscle; dark meat; red meat
血色乳 bloody milk
血尿酸過多〔症〕 hyperuric(a)emia
血乳 bloody milk
血亞碲酸鹽培養基 blood tellurite medium
血便 bloody stool
血紅牛肝菌 red capped bolete
血紅素 haem; haemoglobin; heme; hemoglobin
血紅蛋白 haemoglobin; hemoglobin
血香腸 black pudding; black sausage; bluewurst sausage
血粉 blood meal; blood powder
血脂過多〔症〕 hyperlip(a)emia
血液洋菜培養基 blood agar
血液培養基 blood medium
血清 blood serum; sera; serum
血清白蛋白 serum albumin; seralbumin
血清粉 blood plasma powder
血清脂質 serum lipid
血清脂酶 serolipase
血清球蛋白 serum globulin; serglobulin
血清蛋白〔質〕 blood albumin; serum protein
血清鈣 serum calcium
血清溶素 serolysin
血清膽固醇 serum cholesterol
血清黏蛋白 seromucoid
血清鐵 serum iron
血球 haemocyte; hemocyte
血球蛋白 globin
血球凝集素 haemagglutinin; hemagglutinin
血細胞 haemocyte; hemocyte
血蚶 blood cockle
血斑點 blood spot
血紫質 hematoporphyrin
血黃素 xanthematin
血腸 blood sausage
血腥糖莉〔雞尾酒〕 Bloody Mary
血管 blood vessel; vein
血管收縮 vasoconstriction
血管收縮劑 vasoconstrictor
血管系統 vascular system
血管舒張 vasodilatation
血管舒張劑 vasodilator
血管營養不良 angiodystrophia
血漿 plasma
血漿自由脂肪酸 plasma free fatty acid
血漿自由胺基酸 plasma free amino acid
血漿脂質 plasma lipid
血漿蛋白質 plasma protein
血漿游離脂肪酸 plasma free fatty acid
血漿游離胺基酸 plasma free amino acid
血漿膽固醇 plasma cholesterol
血質鐵 heme iron
血橙 blood orange; red juice orange
血糖 blood sugar
血糖過多〔症〕 hyperglycemia
血糖過低〔症〕 hypoglycemia
血糖過高 glycemia
血壓〔過〕低 hypotension

血壓高 hypertension
血膽固醇 blood cholesterol
血膽固醇過少〔症〕 hypocholesterolemia
血膽固醇過多〔症〕 hypercholesterolemia
血瓊脂 blood agar
血鯛 crimson sea bream
血鐵素 hemoferrin
血纖維蛋白 fibrin
血纖維蛋白原 fibrinogen
血纖維蛋白溶酶 plasmin
行軍乾糧 marching ration
行家 specialist
衣拉麻果 ilama
衣康酸 itaconic acid
衣蘭衣蘭油 ylangylang oil
衣蘭油 ylangylang oil
衣蘭香精 ilangilang; ylangylang
衣蘭膽 ilangilang; ylangylang
西比膽酸 Sippy diet
西瓜 watermelon
西瓜子油 ikpan oil
西瓜葉茄 watermelon nightshade
西瓜醛 melonal
西瓜屬 *Citrullus*
西生菜 lettuce
西印度小黃瓜 gherkin
西印度苦香樹 cascarilla
西印度群島南瓜 west Indian pumpkin
西印度櫻桃 acerola; Barbados cherry; west Indian cherry
西多士 French toast
西式炒飯 pilaf
西式菜餚 Western dish
西米 pearl sago; sago
西米粉 sago flour
西冷 striploin
西冷牛排 striploin steak
西谷米 sago
西谷粉 sago flour
西谷椰子 sago palm
西谷澱粉 sago starch
西芹 celery
西非高粱 milo
西洋山楂 English hawthorn
西洋白花菜 caper
西洋白花菜蕾蒡 caper
西洋李〔子〕 common plum; damson (plant); greengage
西洋芹 parsley
西洋梨 European pear
西洋菜 water cress
西洋蓍草 yarrow
西洋蔘 American ginseng
西紅柿 tomato
西班牙口利左豬肉香腸 chorizo (sausage)
西班牙生蔬菜湯 gazpacho
西班牙洋蔥 Spanish onion
西班牙苦橙 bitter orange
西班牙涼菜 gazpacho
西班牙甜椒 piquillos iodosa
西班牙辣椒 Spanish pepper
西班牙鯖魚 Spanish mackerel
西班牙櫻桃 sweet cherry
西班牙鱈魚 Spanish ling
西貢大米 Saigon rice
西貢肉桂 Saigon cassia; Saigon cinnamon
西貢桂皮 Saigon cassia; Saigon cinnamon
西梅 prune
西梅汁 prune juice
西陪特克餃子（匈牙利） csipetke
西粥氏膳食 Sippy diet
西番蓮果 granadilla; parchita; passionfruit
西葫蘆 marrow squash; marrow; squash; vegetable marrow
西蜀丁香 komarov lilac

西餐 Western dish
西餐甜點 sweet course
西藏羚羊 goa
西鯡 sprat
西蘭花 broccoli
両 tael
吖啶黃〔素〕acriflavine
吖啶橙 acridine orange

七畫

串 cluster
串肉簽 skewer
串燒肉 satay
位相差顯微鏡 phase contrast microscope
位處理肉 undressed meat
伴白蛋白 conalbumin
伴清蛋白 conalbumin
佛手瓜 chayote; vegetable pear
佛手柑 bergamot; finger citron
佛手柑油 bergamot oil
佛州黎豆 velvet bean
佛頭果 sugar apple; sweetsop
估算師 quantity surveyor
估價 value
佐料 accessory food; auxiliary material
佐料烈酒 lace
佐餐葡萄酒 table wine
佐餐醋 table vinegar
佐餐鹽 table salt
佐藥 adjuvant
伽羅木油 linaloe oil
伸長計 extensometer; extensimeter
伸展性 extensibility
伸展計 extensimeter; extensometer
伸展蛋白 extensin
伸張肌 stretch muscle
伸張性 tensile; tensility
伸縮薄膜 stretch film
伸縮薄膜包裝 stretch film package
似大腸〔埃希氏〕桿菌 coliform
似水的 watery
似肉食品 meaty food
似絨毛牛肝菌 yellow cracking bolete
似黃褐牛肝菌 dotted stem bolete
似麩皮的 branny
似糠的 branny
作物 crop
作業線 processing line
作業線內的 in-line
伯克夏〔黑白〕豬 Berkshire
伯杰氏細菌分類手冊 Bergey's manual
伯胺 primary amine
伯森莓 boysenberry
伯爾尼軟芝士 bellayay
伯爾尼軟乾酪 bellayay
伯樂納腸 bologna (sausage)
伯醇 primary alcohol
伯爵茶 earl grey
低〔聚〕肽 oligopeptide
低水分食品 low-moisture food
低甲氧〔基〕果膠 low methoxyl pectin (LMP)
低共溶的 eutectic
低劣酒 red eye
低成分蛋糕 low ratio cake
低血壓 hypotension
低乳脂冰淇淋 mellorine
低乳糖乳 low lactose milk
低亞硫酸鈉 sodium hydrosulfite

低於一個大氣壓的壓力 subatmospheric pressure
低於市價抛售 undersell
低芥酸菜籽油 canola oil
低度葡萄酒 low wine
低限 threshold
低限值 threshold limit value
低級咖啡 triage coffee
低級的 inferior
低級脂肪酸 lower fatty acid
低級產品 offgrade product
低級麵粉 ship stuff
低脂〔肪〕的 low fat
低脂大豆粉 low fat soy flour
低脂可可粉 low fat cocoa
低脂乳 low fat milk
低脂的 non-oily
低脂肪乾酪 skim milk cheese
低脂乾酪 lean cheese
低脂魚粉 non-oily fish meal
低脂稀奶油 light cream
低脂膳食 low fat diet
低脂鮮乳油 light cream
低酒度啤酒 light beer; near beer
低酒度葡萄酒 light wine
低酒度羅木酒 roman punch
低酒精度 low alcohol
低乾性油 short oil
低密度脂蛋白 low density lipoprotein (LDL)
低密度脂蛋白膽固醇 low density lipo-protein cholesterol
低甜度的 semi-sweet
低蛋白質麵粉 low protein flour
低蛋白膳食 low protein diet
低筋麵粉 low protein flour
低鈉乳 low sodium milk
低鈉食品 low sodium food
低鈉膳食 low sodium diet
低溫〔長時〕巴氏殺菌 holder pasteurization; holding pasteurization
低溫〔長時〕巴斯德殺菌 holder pasteurization; holding pasteurization
低溫化學 cryochemistry
低溫巴氏殺菌法 cold pasteurization
低溫巴斯德殺菌法 cold pasteurization
低溫加熱法 low temperature heat method
低溫冷藏 low temperature storage
低溫冷藏庫 subzero storage
低溫長時巴氏殺菌 low temperature holding pasteurization; low-temperature long-time pasteurization (LTLT pasteurization)
低溫長時巴斯德殺菌 low temperature holding pasteurization; low-temperature long-time pasteurization (LTLT pasteurization)
低溫長時殺菌法 low-temperature long-time process (LTLT process)
低溫保藏 cryopreservation
低溫計 cryometer; frigorimeter
低溫凍結 cryogenic freezing; sharp freezing
低溫凍結器 sharp freezer
低溫烘焙的 low-baking
低溫乾燥 cold drying
低溫殺菌〔法〕 pasteurization
低溫殺菌牛乳 pasteurized milk
低溫殺菌乳酸飲料 pasteurized lactic beverage
低溫殺菌啤酒 pasteurized beer
低溫烹調 low temperature cooking
低溫發酵 cold fermentation
低溫溫度計 katathermometer
低溫煉製的豬油 neutral lard
低溫試驗 cold test
低溫實驗法 cryogenics
低溫蒸發 low temperature evaporation
低溫蒸餾 cold distillation

低溫熟成 low temperature ripening
低溫熬油 cold rendering
低溫學 cryogenics
低溫學的 cryogenic
低溫儲藏 low temperature storage
低溫儲藏庫 cold stocker
低酯果膠 low ester pectin
低滲〔壓〕的 hypotonic
低滲〔壓〕溶液 hypotonic solution
低滲壓的 hypoosmotic
低熔點脂肪 soft fat
低熔點動物脂 soft tallow
低碳水化合物膳食 low carbohydrate diet
低聚合度的油 short oil
低聚糖 holoside; oligosaccharide; oligose
低聚糖酶 oligosaccharidase
低酸性食品 low acid food
低層發酵 low fermentation
低層發酵用的酵母 low yeast
低熱食品 hyposite
低熱能膳食 low energy diet
低熱量 low calorie
低熱量食品 low calorie food
低熱量膳食 low energy diet
低醇黍啤酒 bantu beer
低澱粉麵包 starch reduced bread
低濃度 light concentration
低糖血〔症〕 hypoglycemia
低糖的 semi-sweet
低糖的酒 brut
低糖度的 low sugar
低壓 subatmospheric pressure
低壓儲藏 hypobaric storage; low-pressure storage
低檔魚類 coarse fish
低膽固醇血〔症〕 hypocholesterolemia
低膽固醇膳食 low cholesterol diet
低醣膳食 low carbohydrate diet
低黏度澱粉 thin boiling starch
低鹽漬 light salting
低鹽膳食 low salt diet
低鹽攝食法 hypochlorization
余甘子 oil orange
克 gram
克氏紅點鮭 redpotted trout
克氏桿菌屬 *Klebsiella*
克弗〔酸牛乳〕酒 kefir; kefyr; kephir
克弗酒麴 kefir grain
克瓦斯淡啤酒 kvass
克安尼娜牛(意大利) chianina
克耳達法 Kjeldahl method
克耳達氮 Kjeldahl nitrogen
克耳達燒瓶 Kjeldahl flask
克利沙酒 curacao
克里勃司環 Krebs cycle
克門提柑 clementine
克斯納蒸發器 Kestner evaporator
克萊門氏小柑橘 clementine
克萊桿菌屬 *Klebsiella*
克菲爾酸牛乳酒 kefir; kefyr; kephir
克雷氏〔油脂氧化酸敗〕試驗 Kreis test
克雷牛(愛爾蘭黑色乳牛) Kerry
克雷布氏三羧酸循環 tricarboxylic acid cycle (TCA cycle)
克雷布斯循環 Krebs cycle
克雷伯氏菌屬 *Klebsiella*
克雷薩〔油脂氧化酸敗〕試驗 Kreis test
免治牛肉 minced beef
免治肉排 mock cutlet
免洗餐具 disposable dinner set
免疫〔作用〕 immunization
免疫多糖〔類〕 immunopolysaccharide
免疫性 immunity
免疫球蛋白 immune globulin; immunoglobulin
免疫蛋白質 immunoprotein
免疫學 immunology

冷〔雞〕蛋酒 eggnog
冷子番荔枝 cherimoya
冷充填 cold pack(ing)
冷打漿 cold break
冷汁加灰 cold liming
冷式滅菌 cold sterilization
冷收縮 cold shortening
冷灰〔澄清〕法 cold liming
冷肉 cold meat
冷杉〔葉〕油 fir needle oil
冷杉籽油 fir seed oil
冷卻 chill; chilling; cooling; refrigerate
冷卻〔率〕溫度計 katathermometer
冷卻不足的 undercooled
冷卻介質 cooling medium
冷卻水 cooling water
冷卻台 cold plate
冷卻池 cooling pond
冷卻肉 chilled meat; cooled meat
冷卻果汁 chilled juice
冷卻的 chilling; cooling
冷卻室 chilled room; chiller; chilling room; cold room; cooler; cooling chamber; cooling room
冷卻研磨機 cooling grinder
冷卻食品 chilled food
冷卻浴 cooling bath
冷卻混濁 chill haze
冷卻程度 coolness
冷卻裝置 cooling unit
冷卻槽 cooling tank
冷卻劑 coolant; cooler; cooling agent; cryogen
冷卻器 chiller; cooler
冷卻檢驗 chillproofing
冷洗 cool wash
冷風乾燥 cold drying
冷風乾燥機 cold air dryer
冷凍 freeze; refrigerate; refrigeration
冷凍〔的〕 freezing
冷凍方便食品 frozen convenience food
冷凍水果 frozen fruit
冷凍包裝 frozen pack(age)
冷凍肉 frozen meat; refrigerated meat
冷凍肉切片機 frozen meat slicer
冷凍肉類包裝廠 frigorifico
冷凍肉類處理廠 frigorifico
冷凍車 refrigerator car; refrigerator van
冷凍室 freezer; freezing room
冷凍食品 frozen food
冷凍倉庫 cold storage warehouse
冷凍容器 refrigerated container
冷凍庫 freezer; refrigerated warehouse
冷凍粉碎 cryo-milling
冷凍乾燥 freeze dehydration; freeze drying; lyophilization
冷凍乾燥肉 freeze-dried meat
冷凍乾燥的 freeze-dried
冷凍乾燥機 freeze dryer; lyophilizer
冷凍屠體 frozen carcass
冷凍混合劑 freezing mixture
冷凍甜點 frozen dessert
冷凍產品 frozen product
冷凍脫水〔法〕 dehydrofreezing
冷凍脫水的 dehydrofrozen
冷凍蛋糕 frozen cake
冷凍設備 refrigerating equipment; refrigeration equipment
冷凍貨物 refrigerated goods
冷凍貯藏 frozen storage
冷凍滅菌〔法〕 freezing sterilization
冷凍熟食品 frozen precooked food
冷凍蔬菜 frozen vegetable
冷凍調理食品 frozen prepared food; prepared frozen food
冷凍劑 refrigerant
冷凍機 freezer
冷凍濃縮 freeze concentration

冷凍濃縮果汁 frozen concentrated juice
冷凍隧道 freezing tunnel
冷凍儲藏 freezing storage; frozen storage
冷凍簡便食品 frozen convenience food
冷凍麵包 frozen bread
冷凍麵糰 frozen dough
冷凍變性 freeze denaturation
冷容器 refrigerated container
冷甜〔葡萄〕酒 sangaree
冷處理 cold treatment
冷媒 refrigerant
冷發酵 cool fermentation
冷飲 cold beverage; cold drink; cooler
冷飲品 refresher
冷飲料 cold beverage
冷飲櫃 soda fountain
冷傷 chilling injury; cold injury
冷傷害 cold damage
冷煤 coolant
冷碎〔法〕 cold break; cold pulping
冷裝〔法〕 cold pack(ing)
冷榨 cold press
冷榨油 cold drawn oil; cold pressed oil; coldpressed oil
冷僵直 cold rigor
冷盤 cold dish; hors-doeuvre
冷盤〔餐點〕配製室 pantry
冷盤小廚房 garde-manger
冷盤肉 cold meat
冷盤室 cold food pantry
冷盤菜〔餚〕 cold dish
冷凝 condensation
冷凝物 condensate
冷凝液 condensate
冷凝器 condenser
冷壓〔榨〕 cold press
冷壓〔榨〕機 cold press
冷壓油 cold pressed oil
冷縮 chilling shrink
冷濾 winterization
冷燻 kaltrauch
冷燻〔法〕 cold smoking
冷燻黑線鱈 finnan haddock
冷燻黑線鱈的窯 finnan kiln
冷燻鯡魚 glasgow magistrate
冷藏 chilled storage; cold preservation; cold storage; cool storage; freezer storage; refrigerated storage; refrigerate; refrigeration
冷藏包裝紙 locker paper
冷藏肉 refrigerated meat
冷藏車 reefer; refrigerator car; refrigerator van
冷藏房 cold room
冷藏室 chilling room; cold closet; cold storage room; refrigerator; springhouse
冷藏倉庫 cold storage warehouse
冷藏容器 refrigerated container
冷藏庫 cold store; holding freezer; refrigerated warehouse; refrigerator
冷藏船 reefer
冷藏貨物 refrigerated goods
冷藏貨架 refrigerated shelf
冷藏期 refrigerated storage life
冷藏箱 cold closet
冷藏櫃 safe
冷藏麵糰 refrigerated dough
冷變性 cold denaturation
冷鹽 cryosel
別尼迪克還原糖試驗 Benedict's test
別乳糖 allolactose
別針 pin
別異白胺酸 alloisoleucine
別葉綠素 allochlorophyll
別黴素 allomycin
利口酒 liqueur
利克水 rickey
利克酒 rickey

利尿 diuresis
利尿劑 diuretics
利馬豆 butter bean; lima bean; potato bean
利樂包 tetra-pack; tetra-pak
刨片機 flake cutter; flaker; flaking machine
刨冰 shaved ice
刨冰〔片〕機 ice flaker
刨冰飲料 frappe; ice drink
助泡劑 whipping aid
助消化的 digestive
助消化劑 digestant; digestive
助催化劑 promoter
助溶劑 solubilizer
助濾硅藻土 filtercel
助濾劑 filter aids; filteraid
即可上餐桌食用的 table ready
即可使用的 ready-to-use
即可食用的 ready-to-eat; ready-to-serve
即可烘焙的 ready-to-bake
即可烹煮的 ready-to-cook
即可飲用的 ready-to-drink
即食〔便〕餐 ready-to-eat meal
即食的 ready-to-eat; ready-to-serve
即食食品 instant food; ready-to-eat food; ready-to-serve food
即食甜點 instant dessert
即食麵 instant noodle; ramen noodle
即時耗氧量 immediate oxygen demand (IOD)
即飲的 ready-to-drink
即飲飲料 ready-to-drink beverage
即溶化 instantizing
即溶奶粉 instant milk powder
即溶咖啡 instant coffee
即溶茶 instant tea
即溶製機 instantizer
卵 egg; ova; ovum; seed
卵白 albumen; egg albumen; egg white
卵白蛋白 egg albumin; ovalbumin; ovoalbumin
卵白攪泡器 albumen whisk
卵石磨機 pebble mill
卵形菌落 oval colony
卵形越橘 huckleberry
卵巢 ovary
卵清蛋白 egg globulin; ovalbumin; ovoalbumin; ovoglobulin
卵細胞 ova; ovum
卵蛋白質 egg protein
卵黃 egg yolk; vitellus; yolk
卵黃〔凝固〕酶 vitellase
卵黃色 yolk colo(u)r
卵黃卵磷脂 yolk lecithin
卵黃洋菜培養基 egg yolk agar
卵黃粉 yolk powder
卵黃高磷蛋白 phosvitin
卵黃球蛋白 livetin
卵黃葉黃素 vitellolutein
卵黃膜 oolemma; yolk membrane
卵黃質 lipoxanthine
卵黃磷蛋白 ovovitellin; vitellin
卵黃磷醣蛋白 phosvitin
卵黃類黏蛋白 vitellomucoid
卵黃囊 yolk sac
卵膜 oolemma
卵磷〔脂〕蛋白 lecithoprotein
卵磷脂 lecithin(e); phosphatidylcholine
卵磷脂白蛋白 lecithalbumin
卵磷脂強化大豆粉 lecithinated soy flour
卵磷脂清蛋白 lecithalbumin
卵磷脂酶 lecithinase
卵黏蛋白 ovomucin
吞食者 swallower
吞食徐緩〔症〕 bradyphagia
吞噬細胞 phagocyte
吞噬細胞作用 phagocytosis

吞嚥 deglutition; swallow
吞嚥量 swallow
呈色反應 colo(u)r reaction
呈色試驗 colo(u)r test
呈味化合物 taste compound
呈味物質 taste substance
呂契尼通心粉 riccini
君遷子 dateplum
吹波糖 bubble (chewing) gum
吹氣簾 air curtain
吹掉穀物的糠 winnow
吸水 imbibition
吸水〔率〕 water absorption
吸水性 bibulosity
吸水劑 water absorber
吸光系數 absorbance index
吸光度 absorbance; absorbancy
吸光度指數 absorbance index
吸收 uptake
吸收〔作用〕 absorption; assimilation
吸收不良 defective absorption; malabsorption
吸收分光光度計 absorption spectrophotometer
吸收比 absorbance; absorbancy
吸收光度計 absorptiometer
吸收式冷凍機 absorption refrigerating machine
吸收式冷凍櫃 absorption refrigerator
吸收式製冷機 absorption refrigerating machine
吸收系數 absorptivity
吸收性 absorptivity
吸收指數 absorbance index
吸收計 absorptiometer
吸收能力 absorbance; absorbancy
吸收率 absorbance; absorbancy; absorptivity
吸收障礙 defective absorption
吸收劑 absorbent; absorbing agent
吸吮 suck; sip
吸附〔作用〕 adsorption
吸附水 adsorption water
吸附色譜法 adsorption chromatography
吸附作用 sorption
吸附脱色 adsorption bleaching
吸附過濾 adsorption filtration
吸附層析法 adsorption chromatography
吸附劑 adsorbent; sorbent; sorption agent
吸風分離器 mill aspirator
吸食 suck
吸氧劑 oxygen absorber
吸液 imbibition
吸菸者 smoker
吸菸室 smoking room
吸煙者 smoker
吸煙室 smoking room
吸塵器 dust arrester
吸塵機 vacuum cleaner
吸管 pipet(te); straw
吸潮的 hygroscopic
吸盤 sucker
吸濕 moisture absorption
吸濕水 hygroscopic moisture; hygroscopic water
吸濕性 hygroscopicity
吸濕的 hygroscopic
吸濕度 hygroscopicity
吸濕劑 hygroscopic agent
吸濕鹽 hygroscopic salt
吸濾瓶 suction flask
吸濾器 suction filter
含水的 hydrous; watery
含水酒精 aqueous alcohol
含水量 moisture content; water content
含水碎冰 trash ice
含奶油的 creamy
含甲醇酒精 methylated spirit

含杏仁的糖果 almond hardbake
含乳脂的 creamy
含乳清的 wheyey
含果汁飲料 fruit juice drink
含果肉的果汁 crushed juice
含果粒果汁 comminuted juice
含油的 oleaginous
含油食品 oleaginous food
含油脂的 unctuous
含油量 oil content
含油種子 oilseed
含油樹脂 oleoresin
含粉的 mealy
含酒口香糖 wine gum
含酒飲料 peg
含酒精的 alcoholic; spirituous
含酒精的食物 British compounds
含酒精的興奮性飲料 alcoholic cordial
含酒精飲料 alcoholic beverage; alcoholic drink; ardent drink
含硫化合物 sulfur compound
含硫胺基酸 sulfur amino acid; sulfur-containing amino acid
含硫蛋白質 sulfoprotein
含羞草 mimosa
含蛋白質脂 proteolipid
含氮化合物 nitrogenous compound
含量 content
含碘化合物 iodophor
含碘滅菌劑 iodophor sterilant
含蜜的 melliferous
含酵母的 yeasty
含酸液的植物 sorrel
含豬油的 lardy
含醇飲料 peg
含澱粉的 amyliferous; amyloid; farinose; starchy
含糖的 sugary; sacchariferous; saccharine
含糖量 sugar content
含糖漿的 syrupy
含蠟油 waxy oil
含鐵酶 iron enzyme
含鹼的 alkalinous
含酶牛奶 pegnin
均一化 homogenization; uniformization
均一性 homogeneity; uniformity
均勻化 uniformization
均勻性 homogeneity; uniformity
均勻混合物 homogenate
均勻烹調的 uniform cook
均化作用 viscolization
均化器 viscolizer
均和機 homogenizer
均質化 homogenization; viscolization
均質乳 homogenized milk
均質物 homogenate
均質機 homogenizer
均衡 balance
均衡生長 balanced growth
均衡飲食 balanced diet
均衡膳食 balanced diet
均濕 moisture equilibrium
坎台里蠟 candelilla wax
坎尼里諾豆 cannellino bean
夾子 pincers; tongs
夾心口香糖 filled chewing gum
夾心用鹹肥肉 lardo(o)n
夾心冰淇淋 ice cream sandwich
夾心脆餅 waffle
夾心甜麵包 filled bun
夾心蛋糕 layer cake
夾心蛋糕捲 roll cake
夾心餅乾 sandwich biscuit
夾心糖 bonbon
夾心糖果 centre sweets
夾心糖餡用的糖料 cream fondant
夾心麵包 sandwich bread; sandwich
夾竹桃 oleander

夾板 plywood
夾套〔式〕熱交換器 jacketed heat exchanger
夾瓶器 bottle gripper
夾袋器 bag holder
夾層〔蒸〕氣鍋 jacketed steam kettle
夾層蛋糕 sandwich cake
夾層蒸發器 jacketed evaporator
夾層鍋 jacketed kettle
夾雜物 foreign substance; tare
妊娠 gestation
妊娠期 gestation period
妥路香油 tolu balsam
妥爾油松香甘油酯 glycerol ester of tall oil rosin
妥魯香液 tolu balsam
妥盧香脂 tolu balsam
孜然油 cumin oil
孜然芹 cumin; cummin
孜然芹籽 cumin; cummin
孜然籽 cumin seed
完全成熟的 full ripe
完全食品 complete food
完全培養基 complete medium
完全蛋白質 complete protein
完全發酵 complete fermentation
完全滅菌 complete sterilization
完整白米粒 head rice
宏觀分析 macroanalysis
尿 urine
尿色素 urochrome
尿核苷二磷酸 uridine diphosphate (UDP)
尿核苷二磷酸葡萄糖 uridine diphosphoglucose (UDPG)
尿素 carbamide
尿素 urea
尿素酶 urease
尿酸 uric acid
尿酸酯 urate
尿酸鹽 urate
尿糖測定法 urosaccharometry
尿囊素 allantoin
尿囊酸 allantoic acid
尾 tail
尾心 dock
尾肉 tail meat
尾基部 dock
尾龍牛肉 rump
尾龍肉 D-Rump
尾餾分 tailings
尾鰭 caudal fin; tail fin
希臘茴香烈酒 ouzo
弄碎 crumble
形成乳狀物 creaming
形成凍膠 gel
形成鹼基的食品 base forming food
役牛 working cattle
役畜 draft cattle; draught cattle
役馬 workhorse
忌廉 cream
忌廉批 cream pie
忌廉梳打〔汽水〕 cream soda
忌廉梳打餅 cream cracker
忌廉湯 cream soup
志賀氏菌性痢疾 shigellosis
志賀氏菌屬 *Shigella*
志賀桿菌病 shigellosis
志賀桿菌屬 *Shigella*
快速〔濕〕鹽醃 rapid brine cure
快速冷卻 quick chilling; quick cooling; rapid cooling; snap-chilling
快速冷凍 quick freezing; rapid freezing
快速吞食〔癖〕 tachyphagia
快速直接麵糰 no-time dough
快速食品 heat-and-eat food product; heat-and-serve food product
快速凍結機 quick freezer
快速烤箱 quick oven

快速真空加熱滅菌機 vacu-therm instant sterilizer
快速除霜 quick defrosting
快速乾燥 quick drying
快速陳化 quick ag(e)ing; rapid ag(e)ing
快速發酵 accelerated fermentation; rapid fermentation
快速煙燻 quick smoking
快速解凍 quick defrosting; quick thawing
快速蒸燻 short damping
快速熟化 quick ag(e)ing
快速熟成 quick ripening; rapid ag(e)ing
快速醃製 quick cure
快速凝固 quick setting; rapid set; short-set(ting)
快速釀醋法 quick vinegar process
快速鹽漬 quick salting
快煮食品 quick cooking food
快熟薄肉排 minute steak
快餐 fast food; quicklunch; ready-to-eat meal; short order
快餐食品 quick serve meal
快餐部 snack bar
快餐菜 bubble and squeak
快餐櫃台 quick bar
戒酒 abstinence; temperance
抗水性 water resistance
抗生 antibiosis
抗生物素蛋白 avidin
抗生素 antibiotic
抗生蛋白菌素 stravidin
抗冷性 cold resistance
抗拉強度 tensile strength
抗油的 oil-proof
抗毒素 antitoxin
抗凍性 freezing resistance
抗原 antigen
抗氧化性 antioxidative activity; oxidative stability
抗氧化劑 antioxidant; oxidation preventive
抗病毒素 antiviral agent
抗真菌性 fungus resistance
抗真菌的 fungusproof
抗真菌活性 antifungal activity
抗真菌劑 antifungal agent; antimycotic agent
抗神經炎素 aneurin; thiamine; vitamin B1
抗神經炎素酶 aneurinase
抗胰蛋白酶 antitrypsin
抗剪強度 shear strength
抗發酵劑 antiferment
抗結塊劑 anticaking agent; caking inhibitor
抗結劑 anticaking agent
抗菌的 antibacterial; antimicrobial
抗菌活性 antimicrobial activity
抗菌素 bacteriophage; microbiotic
抗菌劑 antibacterial agent; antimicrobial; antimicrobial agent
抗微生物活性 antimicrobial activity
抗微生物劑 antimicrobial agent
抗維生素 antivitamin
抗酵母的 anti-yeast
抗酵素 antienzyme
抗酸性 acid resistance
抗酸的 acid resistant
抗酸染色 acid-fast stain
抗酸細菌 acid-fast bacteria
抗酸劑 antacid; antiacid
抗熱〔性〕 heat resistance
抗熱的 heat proof; heat resistant; heat stable; thermophylactic
抗褐變劑 antibrowning agent
抗戰的 fungusproof
抗凝劑 anticoagulant
抗靜電劑 antistatic agent

抗濕 moisture-proof; moisture resistant
抗壞血酸 ascorbic acid; cevitamic acid; vitamin C
抗壞血酸〔基〕棕櫚酸酯 ascorbyl palmitate
抗壞血酸〔基〕硬脂酸酯 ascorbyl stearate
抗壞血酸-6-棕櫚酸鹽 ascorbyl palmitate
抗壞血酸氧化酶 ascorbic acid oxidase
抗壞血酸鈣 calcium ascorbate
抗壞血酸鈉 sodium ascorbate
抗壞血酸煙酰胺 niacinamide ascorbate
抗體 antibody (Ab)
抗黴的 antimildew
抗黴素 antimycin
抗酶 antienzyme; antiferment
技術 technology
技術訣竅 know-how
扭 wring
扭力天平 torsion balance
扭力黏度計 torsion viscosimeter
扭紋形通心粉 spirale
扭結式包裝機 twist wrapping machine
扭開蓋 twist off cap
扭斷螺紋蓋 pilfer-proof cap (PP cap)
扭轉包覆機 twist wrapping machine
折流板 baffle (plate)
折射計 refractometer
折射率 refractive index
折邊 seam hook
抑〔細〕菌劑 bacteriostatic agent
抑制 inhibition
抑制物 inhibitory substance
抑制劑 depressant; inhibitory substance; inhibitor; inhibitory agent; restrainer
抑真菌作用 fungistasis
抑真菌的 mycostatic
抑真菌素 mycostatin
抑真菌劑 fungistatic agent; mycostat; mycostatic agent

抑菌作用 bacteriostasis; germ(i)static action
抑菌劑 bacteriostat
抑酸劑 antacid; antiacid
抑霉唑 imazalil
抑黴菌的 mycostatic
抑黴菌素 mycostatin
抑黴菌劑 mycostat
改良米 converted rice
改良劑 improver; modifying agent; modifier
改性大豆磷脂 modified soybean phospholipid
改性澱粉 modified starch; treated starch
改性纖維素 modified cellulose
改進劑 improver
改裝 repack
旱生植物 xerophilous plant; xerophyte; xerophile
旱芹 *Apium grareolens*
旱芹屬 *Apium*
旱金蓮 tropaeolum
旱金蓮花 nasturtium
旱稻 dry rice; upland rice
更換培養基 replacement medium
更塞牛 Guernsey cattle
束縛物 shackle
李子 looseners; plum
李杏 plumcot
李乾 prune
李脯 prune
杏 apricot
杏子酒 abricotine; apricotine
杏子醬 apricot jam
杏子罐頭 apricot conserve
杏仁 almond; amygdala; apricot kernel
杏仁〔蛋白〕軟糖 marzipan
杏仁乳 almond milk
杏仁味甜餅乾 ratafia
杏仁油 almond oil; apricot oil

杏仁油香精 almond oily flavo(u)r
杏仁香精 almond flavo(u)r
杏仁核 chufa nut
杏仁粉 almond meal
杏仁脆餅 almond hardbake
杏仁甜點 hardbake
杏仁粕 almond cake
杏仁脱殼機 almond decorticator
杏仁蛋糕 almond cake; frangipane; marzipan cake
杏仁軟糖 marchpane
杏仁硬糖 almond rock; hardbake
杏仁酥糖 marzipan
杏仁塔餅 frangipane
杏仁酪 almond cream
杏仁餅 almond cake
杏仁燙漂去皮機 almond blanching machine
杏仁磨粉機 almond mill
杏仁糖 almond dragees
杏仁糖漿 almond syrup; orgeat
杏仁霜 almond cream
杏仁露 almond milk
杏仁酶 synaptase
杏汁酒 abricotine
杏李 apricot plum; cotplum
杏果醬 apricot jam
杏脯 preserved apricot
杏菌紅素 canthaxanthin
杏菇 chanterelle
杏樹 almond
材料 stuff
杜仲樹膠 gutta-percha
杜拉高粱 durra; owar
杜林小麥 durum (wheat)
杜林小麥粉 durum flour
杜松 cade; juniper
杜松子 juniper berry
杜松子果汁酒 Tom Collins
杜松子油 juniper berries oil; juniper berry oil
杜松子酒 gin; jacky
杜洛克豬 Duroc (Jersey)
杜洛賽豬 Duroc (Jersey)
杜蘭小麥 durum (wheat)
杜蘭小麥粉 durum flour
杜蘭發酵管 Durham fermentation tube
杖魚 snoek
杞子 wolfberry (fruit)
杆秤 steelyard
杓 scoop
杓子 scooper
每日容許攝入量 acceptable daily intake (ADI)
每日容許攝取量 acceptable daily intake (ADI)
每日最低需要量 minimum daily requirement
汞 mercury
汞中毒 hydrargyrism; mercurialism
沙丁格糊精 Schardinger dextrin
沙丁魚 pilchard; sardine
沙丁魚油 pilchard oil
沙丁魚麵疙瘩 gnocchetti sardi
沙土貝 long clam; sand clam
沙卡林 gluside; saccharin
沙可達玉米粥 succotash
沙克白葡萄酒（西班牙） sack
沙李 sand plum
沙拉 salad
沙拉用蔬菜 salading
沙拉吧 salad bar
沙拉油 salad oil
沙拉盤 salad dish
沙拉醬 dressing; salad dressing
沙拉醬混料 salad mix
沙門氏〔桿〕菌病 Salmonellosis
沙門氏菌屬 *Salmonella*

沙門志賀氏菌〔瓊脂〕培養基 Salmonella Shigella agar (SS agar)
沙門志賀桿菌〔洋菜〕培養基 Salmonella Shigella agar (SS agar)
沙門桿菌食物中毒 Salmonella food poisoning; Salmonellosis
沙門桿菌病 Salmonellosis
沙門桿菌屬 *Salmonella*
沙律 salad
沙律用蔬菜 salading
沙律吧 salad bar
沙律盤 salad dish
沙律醬 salad dressing
沙朗 rib eye roll; spencer roll; sirloin
沙朗心 chuck eye log
沙朗牛排 cube roll steak; rid eye steak
沙翁 doughnut
沙梨 Chinese pear; sand pear
沙棘黃 hippophae rhamnoides yellow
沙棗 Russian olive
沙蛤 sand clam
沙塘鱧 goby
沙瑙魚 pacific sardine
沙葛 Chinese turnip; jicama; yam bean
沙嗲 satay
沙嗲醬 satay sauce
沙蒿籽膠 sa-son seed gum
沙蝦 sand shrimp
沙龍 saloon
沙薑 resurrectionlily
沙鍋 terrine
沙鍋燉菜 potaufeu
沙藍泰斯瓜 charentais
沙蟹 fiddler crab; sand crab
沙雕鳥 snipe
沙囊 ventriculus
沉下 sink
沉沒培養〔物〕 submerged culture
沉降 subsidence; settle
沉降池 settling pond
沉降法 settling process
沉降的 settling
沉降槽 subsider; settling tank
沉降劑 setting agent
沉降器 settler
沉香油 linaloe oil; linaloe wood oil
沉香醇 linalool
沉浸冷卻〔法〕 immersion cooling
沉浸凍結〔法〕 immersion freezing
沉浸解凍〔法〕 immersion defrosting
沉渣啤酒 sediment beer
沉澱 residuum; setting; settle; subsidence
沉澱〔作用〕 precipitation
沉澱池 sedimentation basin; subsider
沉澱物 dreg; precipitate (ppt); sediment; settlings
沉澱的 settling
沉澱缸 precipitation vat
沉澱素 precipitin
沉澱槽 clarifying tank; precipitation tank; sedimentation basin; settling tank; sump
沉澱劑 precipitant; precipitating agent
沉澱器 precipitator; settler
沉積作用 deposition
沉積物 deposition; sediment
沉積脂肪 deposited fat; depot fat
沉積槽 sedimentation tank
沖床 punch
沖拔罐 seamless can; seamless tin
沖拔罐身 seamless body
沖洗 flush; rinse; rinsing; swill
沖洗槽 rinsing tank
沖洗器 flusher
沖淡 deliquate; dilution
沖淡〔的〕 dilute
沖淡劑 diluent
沖涮 scour

沖煮〔咖啡〕 brewing
沖煮用具 brewer
沖煮咖啡 brew
沖蓋機 can end press
沖劑延伸罐 ironed can
沖壓機 stamper
沖壓罐 drawn can; one-piece container; seamless can; seamless tin
沖壓罐身 seamless body
沖擊式研磨機 impact mill
沖擊式破碎機 impact breaker
沖擊式粉碎機 impact crusher; impact grinder
沖擊篩 impact screen
沖擊攪動器 dash churn
沖擠型罐 impact extruded can
沖霜 defrost; defrostation
沒味道的 savorless; savourless
沒風道的 mawkish
沒食子 gall; nutgall
沒食子單寧 gallotannin
沒食子酸 gallic acid
沒食子酸十二酯 dodecyl gallate
沒食子酸月桂酯 lauryl gallate
沒食子酸丙酯 propyl gallate (PG)
沒食子酸甲酯 methyl gallate
沒食子酸辛酯 octyl gallate
沒食子酸異戊酯 isoamyl gallate
沒食子酸鹽 gallate
沒藥 myrrh
沒藥油 myrrh oil
汽化 vapo(u)rization; vapo(u)rize
汽化性 vapo(u)rizability
汽化器 vapo(u)rizer
汽水 brause; club soda; carbonated water; fountain; soda pop; soda water; sparkling water
汽水〔供應〕器 soda fountain
汽水用糖漿 fountain syrup
汽水商 fountaineer
汽水製選造機 gazogene
汽水櫃台 soda fountain
汽油 gas
汽封 vapo(u)r seal
汽相 vapo(u)r phase
汽浴 steam bath
汽酒 alcoholic carbonated drink; carbonated wine; sparkling wine
汽密的 steam-tight; vapo(u)r proof; vapo(u)r tight
汽蒸〔增濕〕桶 damping vat
汽噴冷卻機 steam-jet refrigerating machine
汽噴滅菌法 uperization
汽噴滅菌器 uperizer
汽壓 vapo(u)r pressure
汽壓表 steam gauge
灼 poaching
牡荊葉油 vitex cannabifolia leaf oil
牡蠣 ostracean; oyster
牡蠣固醇 ostreasterol
牡蠣的 ostracean
牡蠣殼 oyster shell
牡蠣養殖場 oyster bank; oyster bed; oyster farm; oyster ground
狂飲 spree
狂飲者 guzzler
狂歡 spree
狂歡宴會 revelry
狂歡盛宴 revel
皂化〔作用〕 saponification
皂化值 saponification number; saponification value
皂化價 saponification number; saponification value
皂化劑 saponification agent; saponifier; saponifying agent
皂角苷 saponin(e)
皂味 soapy taste

皂素 saponin(e)
皂腳 foots; oil foot; soap stock
皂樹〔皮〕 soapbark; quillaja; quillaia
皂樹皮抽提物 quillaia extract; soapbark extract
皂苷 saponin(e)
私酒 hooch
私造劣酒 hooch
系統 system
罕布夏豬 Hampshire
肝〔泥〕香腸 liver sausage
肝色牛排菌 liver fungus
肝泥肉腸 leberkase
肝泥捲 liver cheese; liver loaf
肝泥糕 liver cheese; liver loaf
肝油 liver oil
肝炎 hepatitis
肝黃素 hepatoflavin
肝腸 liverwurst; liver sausage
肝糊 liver paste
肝糖 animal starch; glycogen; liver starch
肝糖分解 glycogenolysis
肝糖生成 glycogenesis
肝糖酶 glycogenase
肝醬 liver paste; pate
肝臟 hepar; liver
肘 elbow
肘管 elbow
肛門 anus
肚 maw; tripe
肚包羊雜 haggis
肚臍 navel
育肥小公牛 stocker
育肥公犢 stocker
育肥牛 butcher cattle; stocker cattle
育肥母牛 stocker heifer
育肥的 well-fed
育肥牲畜 well-fed stock
育肥禽畜 fattener
育苗場 nursery
育葡萄暖房 vinery
育種 breed
育嬰室 nursery
良好生產操作規程 good manufacturing practice (GMP)
良好作業規範 good manufacturing practice (GMP)
良薑 galangal; lesser galangal
芒 awn
芒果 mango
芋 dasheen; taro
芋泥 taro pudding
芋根 eddo
芋頭 dasheen; taro
芋類 tania
角 horn
角叉〔菜〕膠 carrageenan; carrageenin; carragheenin
角叉菜 carrageen; carragheen; *Chondrus crispus*; Irish moss
角叉藻二糖 carrabiose
角豆 carob bean; locust bean
角豆莢胚芽色素 locust bean germ
角豆膠 carob bean gum; jandagum; locust bean gum
角孢粉褶蕈 aborted pink gill
角蛋白 keratin
角蛋白質 keratoprotein
角蛋白酶 keratinase
角魚 gurnard
角菜 white mugwort
角質素 keratin
角質蛋白 keratoprotein
角鯊素 spinacin
角鯊烯 squalene
角鯊魚 dogfish
角鯊烷 squalane
角苷脂 kerasin

谷氨酸 glutamic acid
谷氨酸〔一〕鉀 monopotassium glutamate (MPG)
谷氨酸〔一〕銨 monoammonium glutamate
谷氨酸鈣 calcium glutamate
谷氨酸鈉 sodium glutamate
谷氨酸鹽酸鹽 glutamic acid hydrochloride
谷氨酰胺 glutamine
谷胱甘肽 glutathione
谷胺酸〔一〕鈉 aginomoto
谷胺酸一鈉 monosodium glutamate (MSG)
谷胺酸酯 glutamate
谷胺酸鹽 glutamate
豆 bean; dhal; legume; pulse
豆奶 soy milk; soybean milk
豆白蛋白 legumelin
豆皮 soy skin; soya skin; soybean milk film; yuba
豆沙 bean puree
豆角 asparagus bean; yardlong bean
豆乳 soy milk; soybean milk
豆固烯醇 stigmastenol
豆固烷 stigmastane
豆固烷醇 stigmastanol
豆固醇 stigmasterol
豆泥 bean puree; pureed bean
豆芽 bean sprout
豆甾烯醇 stigmastenol
豆甾烷 stigmastane
豆甾烷醇 stigmastanol
豆甾醇 stigmasterol
豆科 Leguminosae
豆科的 leguminous
豆科植物 legume
豆風味 beany flavo(u)r
豆粉 bean flour; pulse flour
豆臭 beany flavo(u)r
豆乾 hard bean curd
豆清蛋白 legumelin
豆球蛋白 legumin
豆莢 bean pod; pod; seedpod
豆渣 okara
豆腐 bean curd; soybean curd; tofu; vegetable cheese
豆腐皮 soy skin; soya skin; soybean milk film; yuba
豆腐衣 yuba
豆腐乳 bean curd cheese; Chinese cheese; fermented tofu; preserved bean curd; soy cheese; soybean cheese; sufu
豆腐乾 dried beancurd
豆餅 bean cake
豆漿 soy milk; soybean milk
豆蔻香料 mace
豆糟 bean cake
豆薯 jicama; wayaka yam bean; yam bean
豆醬 bean paste; soybean miso
豆瓣菜 water cress
豆瓣醬 bean paste; fermented soybean paste; soybean paste
豆類分級機 bean grader
豆類去皮機 bean huller
豆類去莢機 bean huller
豆類的 leguminous
豆類植物 pulse
豆類篩選機 bean grader
貝氏〔乳脂〕瓶 Babcock bottle; Babcock flask
貝氏〔乳脂〕試驗 Babcock test
貝他生育酚 beta-tocopherol; β-tocopherol
貝他胡蘿蔔素 beta-carotene; β-carotene
貝他澱粉 beta-starch; β-starch
貝他澱粉酶 beta-amylase; β-amylase
貝耳定律 Beer-Lambert law; Beer's law
貝卵 spat
貝毒素 venerupin
貝苗 spat

貝殼 conch; shell
貝殼粉 abassina rigata; conchiglie; shell pasta
貝菲咖啡 breve
貝蒂布丁 betty
貝類 shellfish
貝類中毒 shellfish poisoning
貝類毒素 saxitoxin
赤〔砂〕糖 brown sugar
赤小豆 *Phaseolus calcaratus*
赤小麥 duckbill wheat
赤尾鰺 round scad
赤豆 adsuki bean; adzuki bean; azuki bean; red bean
赤貝 ark shell
赤紅麴菌 *Monascus anka*
赤紅麴霉 *Monascus anka*
赤痢 dysentery
赤睛魚 rudd
赤魟 Japanese stingray; red skate; red stingray
赤褐鵝膏（蕈類） tawny grisette
赤霉酸 gibberellic acid
赤霉酸鉀 potassium gibberellate
赤糖糊 blackstrap molasses
赤龍蘋果 Baldwin
赤藻紅 erythrosine
赤藻糖 erythrose
赤藻糖醇 erythritol
赤蘚紅 erythrosine
赤蘚紅鋁色澱 erythrosine aluminium lake
赤蘚素 erythrin
赤蘚糖 erythrose
赤蘚糖醇 erythritol
赤黴素 gibberellin
赤黴酸 gibberellic acid
走味 pall; staleness
走味的 stale
足夠的飲食 adequate diet
身體 soma
身鈎脫扣 knockdown flange
車打芝士 cheddar cheese
車前子 psyllium
車前籽 psyllium seed
車前草 plantain; psyllium
車厘子 cherry
車間 plant; workshop
車葉草 woodruff
車輪狀通心粉 ruoti
辛二酸 suberic acid
辛二酸鹽 suberate
辛炔羧酸甲酯 methyl 2-nonynoate
辛癸酸甘油酯 glyceryl caprylate-caprate
辛香料 spice
辛香飯 pilaf
辛烯基琥珀酸澱粉鈉 starch sodium octenylsuccinate
辛辣 acridity; pungency; piquancy
辛辣味 biting taste; pungent taste; sharp flavo(u)r; sharp taste
辛辣性 acrimony
辛辣的 hot; piquant; pungent; spicy
辛辣氣味 pungent flavo(u)r
辛辣菜餚 devil
辛酸 caprylic acid; octanoic acid
辛酸乙酯 ethyl octanoate
辛酸壬酯 nonyl octanoate
辛酸戊酯 amyl octanoate
辛酸甲酯 methyl octanoate
辛酸苯乙酯 phenethyl octanoate
辛醇 octanol; octyl alcohol
辛醣 octose
辛醛 octanal
邦比甜品 bombe
那維特蘿蔔 navette
里肌肉 tenderloin
里佐脫（意大利菜） risotto

里哪油 linaloe oil
里哪醇 linalool
里脊牛排 porterhouse steak; porterhouse; tenderloin steak
里脊肉 loin; tenderloin; underback; undercut
里脊肉火腿 loin ham
里脊條〔肉〕 sirloin strip
防水性 water resistance
防水的 waterproof
防光包裝 light-proof packaging
防光紙 light-proof paper
防冰器 deicer
防沫 antifoam
防沫劑 antifoaming agent; defoaming agent; kilfoam
防油包裝〔材料〕 greaseproof wrapping
防油的 greaseproof; oil-proof
防油紙 greaseproof paper; oil-proof paper
防毒面具 respirator
防風〔草〕 parsnip
防凍液 deicing fluid
防凍劑 antifreeze; antifreezing agent; deicer
防氣包裝 barrier package
防臭木 lemon verbena
防蛀劑 mothproofing agent
防蛾劑 mothproofing agent
防塵 dust proof
防滲血紙 butcher's paper
防腐 antisepsis; conservation; preservation; preserving; rot proof
防腐性 rot proofness
防腐的 antiseptic
防腐食品 lasting food
防腐液 preservative solution
防腐處理 asepsis; preservative treatment
防腐劑 antiseptic; conserving agent; conservative; preservative; preserving agent; preservative substance
防潮 moisture-proof; moisture resistant
防潮包裝 barrier package; impregnated wrapping; moisture-proof packaging
防潮的 dampproof; wetproof
防潮層 moisture barrier
防潮箱 desiccator
防潮膜 wetproof film
防霉 mildew proof; rot proof
防霉性 rot proofness
防霉劑 mold inhibitor
防磨損的 wearproof
防銹的 rust proof
防銹劑 rust inhibitor
防濕包裝 moisture-proof packaging
防濕的 dampproof; wetproof
防濕膜 wetproof film
防縮的 shrinkproof
防黏 antisticking
防黏劑 antisticking agent; antistickings
防蟲包裝 verminproof package; insect-proof packaging
防蟲的 verminproof
防爆 explosion proof
防蠅網 fly net
防黴 mildew proof; mold proofing; mould proofing
防黴劑 antimold
沏 brewing
沏茶 brew; leaf stewing
沏茶器 tea marker
吡咯 pyrrole
吡哆素 pyridoxine
吡哆醇 pyridoxine; vitamin B_6
吡哆醇鹽酸鹽 pyridoxine hydrochloride
吡哆醛 pyridoxal
吡哆醛磷酸 pyridoxal phosphate
吡啶 pyridine
吡喃果糖 fructopyranose; pyranofructose
吡喃型葡〔萄〕糖 glucopyranose

吡喃葡糖 pyranoglucose
吡喃糖 pyranose
吲哚 indole
吲哚 indoxyl
呋喃 furan
呋喃半乳糖 galactofuranose
呋喃甲醛 furfural
呋喃果聚糖 fructofuranose
呋喃葡萄糖 glucofuranose
呋喃糖 furanose

八畫

並流式乾燥機 cocurrent dryer
並流式噴霧乾燥機 cocurrent spray dryer
並流乾燥機 parallel flow dryer
並流蒸發機 parallel flow evaporator
乳 milk
乳〔烴〕素 lactenin
乳二糖 lactobiose
乳化〔作用〕 emulsification
乳化力 emulsifying capacity
乳化性 emulsifiability; emulsifying ability
乳化油 emulsified oil
乳化物 emulsified mixture
乳化香精 emulsion flavo(u)r
乳化液 emulsion; emulsified mixture
乳化混合器 emulsifying mixer
乳化劑 emulsifier; emulsor; emulsifying agent
乳化器 emulsifier; emulsor
乳化機 mulser
乳比重計 lactodensimeter; lactometer; milk gauge
乳牛 cow; dairy cattle; milch cow; milk cattle; milk cow
乳牛肝菌 jersey cow bolete
乳牛臭 cowy flavo(u)r
乳牛場 dairy farm
乳牛場場主 dairyman
乳母 nutrix; wet nurse
乳汁 latex; mammae
乳汁分泌 galactosis; lactation
乳汁狀 lactescence
乳用牛種 dairy breed
乳白色 creamy white
乳白蛋白 lactalbumin
乳石 milk stone
乳冰 ice milk
乳抑菌素 lactenin
乳固形物 milk solids
乳固體 milk solids
乳房 dug; mamma; udder
乳房炎 mastitis; garget
乳果糖 lactulose
乳油 cream; créme; milk cream
乳油〔乾酪〕製造廠 creamery
乳油上浮 creaming
乳泊小麵包 cream puff; cream bun
乳泊分層性 creaming property
乳泊分層劑 creaming agent
乳泊分離 creaming
乳油分離器 cream separator; creamer
乳油化 creaming
乳油用乳脂計 cream butyrometer
乳油白醬料 cream sauce
乳油夾心 cream filling
乳油果泥甜點 fruit fool
乳油炒麵粉 cream roux
乳油狀巧克力 creamy chocolate
乳油粉 cream powder; dried cream
乳油乾酪 cream cheese

乳油蛋捲 cream roll
乳油硬糖 creamy rock
乳油層 cream layer
乳油攪拌器 cream beater
乳狀的 lacteal
乳狀流質膳食 milky fluid diet
乳狀液 lactescence
乳的 milch
乳品〔加工〕業 dairy industry
乳品工廠 milk plant
乳品商店 creamery
乳品業 dairy
乳毒素 galactotoxin; lactoxin
乳洋葱 button onion
乳香 boswellia; mastic; olibanum
乳香油 olibanum oil
乳香酒 mastika; mastic
乳香魚 milkfish
乳畜 suckling
乳粉 dried milk; dry milk; milk powder; powdered milk
乳脂〔肪〕 milk fat
乳脂上浮促進劑 creaming agent
乳脂分離 milk separation
乳脂狀的 creamy
乳脂肪 butter oil; butterfat
乳脂肪分析儀 milkotester
乳脂計 butyrometer; butter test bottle; lactocrit; lactobutyrometer; lactoscope
乳脂測定器 lactocrit
乳脂凝聚析出現象 creaming property
乳脂糖 taffy; toffee; toffy
乳脂酶 butyrinase
乳酒凍 sillabub; syllabub
乳胺 lactam
乳桶 milk tank
乳液 emulsion
乳液分泌缺乏 agalactosis
乳液比重計 galactometer
乳液生成 galactosis
乳清 lactoserum; milk serum; sera; serum; whey
乳清分離 wheying off
乳清奶油 serum butter; whey butter
乳清白蛋白 serum albumin
乳清乳酪 whey butter
乳清的 wheyey
乳清粉 dry whey; powdered whey; whey powder
乳清脂 serum butter
乳清乾酪 whey cheese
乳清球蛋白 serum globulin
乳清蛋白〔質〕 lactoalbumin; serum protein; whey protein
乳清飲料 whey beverage; whey drink
乳清酸 orotic acid; vitamin B_{13}
乳清凝乳酶 whey rennet
乳清濃縮物 whey concentrate
乳淨化機 milk clarifier
乳球蛋白 lactoglobulin
乳球菌 galactococcus
乳瓶 milk bottle
乳產品 milk product
乳蛋白〔質〕 lactoprotein; milk protein
乳飲料 milk beverage; milk drink
乳菇 milk mushroom
乳塊切碎機 curd mill
乳腺 lacteal gland; mammary gland
乳腺炎 mastitis
乳腺炎乳 mastitis milk
乳裝瓶機 milk bottle filler
乳過氧化物酶 lactoperoxidase
乳酪 butter; créme; cream
乳酪刀 butter knife
乳酪分離器 creamer
乳酪用鮮奶油 butter cream
乳酪色素 butter colo(u)r
乳酪乳清 butter whey; butter serum

乳酪狀金錢菇 buttery tough shank
乳酪脂 butterfat
乳酪蛋捲 brioche
乳酪硬糖 butterscotch
乳酪菌元 butter starter
乳酪黃 butter yellow
乳酪酵母〔菌〕 *Saccharomyces casei*
乳酪餅 butter cake
乳酪醬 butter sauce; drawn butter
乳酪麵包 butter bread
乳酪攪製器 butter churn
乳酮糖 lactulose
乳寡糖 gynolactose
乳製品 dairy food; dairy product
乳製品加工工人 dairyman
乳製食品 milk food
乳製飲料 milk beverage
乳酸 lactic acid
乳酸〔桿〕菌素 lactobacillin
乳酸乙酯 ethyl lactate
乳酸丁酯 butyl lactate
乳酸化單雙甘油酯 lactic acid and fatty acid esters of mono- and diglycerides
乳酸去氫酶 lactic dehydrogenase
乳酸生成 lactification
乳酸乳〔酸〕桿菌 *Lactobacillus lactis*
乳酸亞鐵 ferrous lactate
乳酸和脂肪酸甘油酯 lactic acid and fatty acid esters of mono- and diglycerides
乳酸脂肪酸丙二醇甘油混合酯 propylene glycol lactostearate
乳酸乾酪 lactic cheese
乳酸乾酪素 lactic acid casein
乳酸桿菌 lactic bacillus
乳酸桿菌酸 lactobacillic acid
乳酸桿菌屬 *Lactobacillus*
乳酸殺菌素(保存劑) lactocidin
乳酸脱氫酶 lactic dehydrogenase
乳酸發酵 lactic acid fermentation
乳酸發酵短桿菌 *Brevibacterium lactofermentus*
乳酸菌 lactic acid bacteria
乳酸菌發酵飲料 lactic beverage
乳酸菌飲料 lactic acid bacteria drink
乳酸菌酶 lactacidase
乳酸鈣 calcium lactate
乳酸鈉 sodium lactate
乳酸酪蛋白 lactic acid casein
乳酸鉀 potassium lactate
乳酸精 lactacidogen
乳酸酵母〔菌〕 *Saccharomyces lactis*
乳酸鎂 magnesium lactate
乳酸鏈球菌 lactic streptococci; *Streptococcus lactis*
乳酸鏈球菌素 nisin
乳酸鏈球菌肽 nisin
乳酸鹽 lactate
乳酸酶 lactolase
乳潘趣飲料 sillabub; syllabub
乳穀食品 lactofarinaceous food
乳膠 emulsion; latex
乳膠體 emulsion colloid; emulsoid
乳豬 piglet; porkling; roaster; sucking pig
乳醇 lactol
乳齒 deciduous tooth
乳凝塊 milk curd
乳樹 cow tree
乳濁液 emulsion
乳糖 lactose; milk sugar
乳糖〔廢〕蜜 lactose molasses
乳糖不耐症 lactose intolerance
乳糖發酵酵母 lactose fermenting yeast
乳糖酸 lactobionic acid
乳糖醇 lactitol
乳糖醛酸鈣 calcium lactobionate
乳糖酶 lactase; β-galactosidase

乳糖酶缺乏症 lactase deficiency
乳頭 teat
乳餐 milk diet
乳嬰 suckling
乳營養法 galactotrophy
乳療法 galactotherapy; galactopathy
乳糜 chyle
乳糜微粒 chylomicron
乳糜管 lacteal
乳薊果 milk thistle
乳鴿 squab
乳黏蛋白 lactomucin
乳蘑〔菇〕 peppery lactarius
乳鐵蛋白 lactoferrin
些微 speck
亞〔麻〕油酸 linolic acid
亞〔麻〕油酸酯 linoleate
亞〔麻〕油酸鹽 linoleate
亞力酒 arrack
亞牛磺酸 hypotaurine
亞甲〔基〕藍 methylene blue
亞油精 glyceryl trilinoleate; linolein; linolin; trilinolein
亞油酸 linoleic acid; octadecadienoic acid
亞洲胡瓜魚 rainbow herring
亞洲箭齒鰈 arrow-toothed halibut
亞洲薄荷 *Mentha arvensis*
亞洲薄荷油 *Mentha arvensis* oil
亞香芳油 nardgrass oil
亞香草基 vanillal
亞胺 imine
亞胺基酸 imino acid
亞硫酸 sulfurous acid; sulphurous acid
亞硫酸〔化〕處理 sulphitation
亞硫酸氫鈣 calcium bisulfite; calcium hydrogen sulfite
亞硫酸氫鈉 sodium bisulfite; sodium hydrogen sulfite
亞硫酸氫鈉甲萘醌 menadione sodium bisulfite; vitamin K_3
亞硫酸氫鉀 potassium bisulfite; potassium hydrogen sulfite
亞硫酸處理 sulfuring; sulphuring
亞硫酸處理果汁 sulfited juice
亞硫酸鈉 sodium sulfite
亞硫酸鉀 potassium sulfite
亞硫酸酯 sulfite; sulphite
亞硫酸銨 ammonium sulfite
亞硫酸鹽 sulfite; sulphite
亞麻 flax
亞麻子 linseed
亞麻仁 flaxseed; linseed
亞麻仁油 linseed oil
亞麻油酸 linoleic acid
亞麻籽 flaxseed
亞麻黃素 linoxanthine
亞麻精 glyceryl trilinolenate; linolenin; trilinolenin
亞硒酸鈉 sodium selenite
亞氯酸鈉 sodium chlorite
亞硝〔基〕肌紅蛋白 nitrosomyoglobin
亞硝〔基〕血紅蛋白 nitrosoh(a)emoglobin
亞硝化〔作用〕 nitrosation
亞硝肌紅素 nitrosomyoglobin
亞硝血紅素 nitrosoh(a)emoglobin
亞硝胺 nitrosamine
亞硝基 nitroso (-NO)
亞硝基化合物 nitroso compound
亞硝酸 nitrous acid
亞硝酸〔鹽〕細菌 nitrite bacteria
亞硝酸乙酯 ethyl nitrite
亞硝酸根 nitrite (NO_2)
亞硝酸鈉 sodium nitrite
亞硝酸鉀 potassium nitrite
亞硝酸燒 nitrite burn
亞硝酸鹽 nitrite (NO_2)
亞硝酸鹽灼 nitrite burn
亞硝酸鹽還原酶 nitrite reductase

亞硝酰胺 nitrosamide
亞種 subspecies
亞熱帶水果 subtropical fruit
亞歷山大香腸 Alessandria sausage
亞歷山大椰子 Alexandra kingpalm
亞磷酸 phosphorous acid
亞磺酸 sulfinic acid
亞鐵氰化鈉 sodium ferrocyanide
亞鐵氰化鉀 potassium ferrocyanide
享受美食 regale
京白梨 Peking white pear
依猶太食品 kosher
依靠氧氣的 aerobic
依蘭油 cananga oil
侍者 waiter
侍應 waiter
佳味酚 chavicol
佳餚 ambrosia; regale; viand
佳釀〔葡萄酒〕 vintage
佳釀酒的釀成年 vintage year
使成膠狀 gelatinize; gelatinate
使乳化 emulsify
使強壯 invigoration
使復壯 invigoration
使發酵膨鬆 leaven
使新鮮 freshelize
使糖化 saccharify
使顯得新鮮 freshen
供水 water supply
供水系統 water works
供水器 water feeder
供食份數 servings
供給 supply
供撒的粉 dusting powder
供膳宿舍 boardinghouse
供餐首盤 service plate
供應 provisions; supply
供應不足 undersupply
供應餐食 catering
供應儲備食物 victual
例行分析 routine analysis
例行長期儲藏 standing storage
來因克洛德李 gage
來亨雞 Leghorn
來航雞 Leghorn
來蘇黃素 lyxoflavin
來蘇糖 lyxose
來蘇糖酸 lyxonic acid
來蘇糖苷 lyxoside
侏儒蛋 dwarf egg
兔 rabbit
兔耳草醛 cyclamen aldehyde
兔肉 rabbit meat
兔銀鮫 ratfish
兒茶 catechu; cutch
兒茶素 catechin
兒茶酚 catechol
兒茶酚胺 catecholamine
兒茶酚酶 catecholase
兒茶黑色素 catechu
兒茶酸 catechin
兩片罐 two-piece can
兩性乳化劑 amphoteric emulsifier
兩性的 amphoteric
兩性表面活性劑 amphoteric surfactant; ampholytic surfactant; ampholytic surfactant
兩性溶劑 amphiprotic solvent
兩性澱粉 amphoteric starch
兩性離子交換樹脂 amphoteric ion-exchange resin
兩棲類 amphibian
兩端尖的麵包 koppe
兩端漸細的 fusiform
具大環口蘑（蕈類） white matsutake
具發酵性的 fermentative
刻度 scale
刷洗 rinse

刷洗槽 rinsing tank
刷麩機 shorts duster
刺孔 puncture
刺毛 seta
刺狀的 spiny
刺柏油 juniper berries oil
刺梧桐〔樹〕膠 gum karaya; karaya gum; gum karaya
刺殺 stick
刺殺〔屠宰〕 sticking
刺殺刀 sticking knife
刺殺工人 sticker
刺殺工具 sticker
刺棘薊 cardoon
刺番荔枝 soursop; guanabana
刺菜薊 cardoon
刺黃瓜 serpentine cucumber
刺黃蓋鰈 yellowfin sole
刺槐豆 carob bean; locust bean
刺槐豆膠 carob bean gum; locust bean gum
刺鼻風味 sharp flavo(u)r
刺鼻氣味 sharp odo(u)r
刺鼻臭味 funk
刺蝶魚 angelfish
刺蝟 porcupine
刺激 stimulus
刺激〔作用〕 stimulation
刺激味 pungent taste; strong taste
刺激性 pungency; zest
刺激性的 pungent
刺激性食品 excitant food
刺激性氣味 pungent flavo(u)r
刺激物 irritant; stimulant; stimulus
刺激的 irritant
刺激器 stimulator
刺鯧 pampano
刺鰕虎魚 spiny goby
刮 scrape
刮刀 shaver; scraper
刮毛 dehair
刮毛機 shaving machine
刮肉 fleshing
刮肉屑鏟 scraper
刮板 scraper
刮板式加熱器 scraped surface heater
刮板式熱交換器 scraped surface heat exchange
刮面式加熱器 scraped surface heater
刮面式熱交換器 scraped surface heat exchange; votator
刮鱗器 scaler
制念珠菌素 candidin
制度 system
制真菌劑 fungistat(in)
制假絲菌素 candidin
制黴菌素 nystatin
制黴菌劑 mycostatic agent
剁肉板 chopping board
剁碎機 hacker; hacking machine
協合作用 synergism
協合劑 synergist
卑檸油 petitgrain oil
卷心菜 cabbage; common head cabbage; head cabbage; headed cabbage; heading cabbage
卷心菜〔絲〕沙拉 slaw; coleslaw
卷心萵苣 cabbage lettuce; crisphead lettuce; head lettuce
卷緣齒菌〔蕈類〕 common hedgehog fungus
卷壓螺紋蓋 roll on closure
取內臟器 eviscerator
取代〔產〕品 substitution product
取出內臟 embowel; evisceration; viscerate
取汁 juicing
取樣 sampling
取樣器 sampler; sampling device
叔丁基氫醌 tertiarybutylhydroquinone (TBHQ)

叔丁基對苯二酚 tertiary butylhydroquinone
叔丁基對羥基茴香醚 butylated hydroxyanisole (BHA)
受感染的 infected
受試驗者 testee
受精蛋 embryonated egg
受熱致死點 thermal death point (TDP)
受體 acceptor
味 sapor; sapour; taste
味之素 aginomoto
味美的 lush; tasty
味美斯酒 vermouth
味素 monosodium glutamate (MSG); sodium glutamate
味淡的 insipid
味淡的低劣啤酒 swipes
味淡食物 insipid food
味道 relish; savor; savour; taste; tastiness
味道不佳 unsavo(u)riness
味道不佳的 unsavo(u)ry
味道強的 gamy
味道鮮美的 sweet
味精 aginomoto; monosodium glutamate (MSG); sodium glutamate
味噌 fermented soybean paste; miso; soybean miso
味蕾 taste bud
味蕾細胞 taste cell
味覺 gustation; palate; sapor; sapour; taste
味覺改變劑 taste modifier
味覺受體 taste receptor
味覺的 gustatory
味覺計 gustometer
味覺喪失 ageusia
味覺測定儀 gustometer
味覺器官 taste organ
味覺閾值 taste threshold
咖哩 curry
咖哩牛肉 curried beef; curry beef
咖哩味的 curried
咖哩食品 curry
咖哩粉 curry; curry powder
咖哩飯 curry rice
咖哩雞肉 curried chicken; curry chicken
咖哩雞肉泥 mulligatawny paste
咖哩雞肉濃湯 mulligatawny soup
咖啡 coffee
咖啡〔香〕油 caffeol
咖啡奶精 coffee whitener
咖啡用白油 coffee mate; coffee whitener
咖啡用非乳製白油 coffee creamer
咖啡用鮮奶油 coffee cream
咖啡因 caffeine; methyltheobromine; theine
咖啡伴侶 coffee mate
咖啡吧 coffee bar
咖啡沖濾壺 coffee dripper
咖啡豆 coffee bean; nib
咖啡豆去殼機 coffee bean decorticator
咖啡豆角質皮 parchment skin
咖啡豆的外殼 parchment
咖啡豆膜 parchment skin
咖啡果 coffee berry; coffee cherry
咖啡杯 coffee cup
咖啡泡濾壺 coffee siphon
咖啡直流杯 direct flowing coffee cup
咖啡室 coffee room
咖啡研磨機 coffee grinder
咖啡酊 coffee tincture
咖啡時間 coffee break
咖啡軟糖 coffee fondant
咖啡啶 caffeidine
咖啡壺 coffee pot; coffee pot; urn
咖啡渣 coffee grounds
咖啡焦油 caffeol; caffeone
咖啡稀奶油 coffee cream
咖啡飲料 coffee drink

咖啡過濾紙 coffee filter bag; coffee filter paper
咖啡過濾袋 coffee filter bag
咖啡過濾漏斗 coffee dripper
咖啡滲濾壺 coffee percolator
咖啡酸 caffeic acid
咖啡醇 caffeol; caffeone
咖啡樹 coffee tree
咖啡磨 coffee mill; coffee grinder
咖啡館 café; coffee bar; coffee palace; estaminet
咖啡鹼 caffeine; methyltheobromine; theine
咖啡鹼中毒 caffeinism
咖啡廳 coffee palace
咕咾肉 sweet and sour pork
咀嚼 chew; manducate
咀嚼性 chewiness
咀嚼度 chewiness
呼吸〔作用〕 respiration
呼吸卡計 respiration calorimeter; respiratory calorimeter
呼吸商 respiratory quotient
呼吸道 respiratory tract
呼吸熱量計 respiration calorimeter; respiratory calorimeter
呼吸器 respirator
和尚頭 knuckle
和麵 dough mixing
和麵機 developer; dough mill; doughing machine
周圍空氣 ambient air
周圍條件 ambient condition
周圍溫度 ambient temperature
固化 hardening; solidify
固化〔作用〕 solidification
固化單寧 immobilized tannin
固化劑 coagulator; firming agent; solidification reagent
固色劑 colo(u)r retention agent
固形物 solid matter; solid
固形物含量 solid content
固定化酶 immobilized enzyme
固定式攪拌器 stationary agitator
固定床乾燥 fixed bed drying
固定相 immobile phase; stationary phase
固定劑 fixative; fixing agent; restrainer
固定儲槽 standing tank
固相酶 immobilized enzyme; insoluble enzyme
固液萃取 solid-liquid extraction
固氮〔作用〕 nitrogen fixation
固氮〔細〕菌 azobacteria; azotobacter(ia); nitrogen fixing bacteria
固氮生物 diazotroph
固氮黃素 azotoflavin
固氮酶 nitrogenase
固著促進劑 anchoring agent
固著劑 anchoring agent
固溶體 solid solution
固酮 sterone
固態 solid state
固態石蠟 solid paraffin
固態脂 solid fat
固態酒精 solidified alcohol
固態培養基 solid medium
固態豬油 lard stearin
固態鮮乳油 solid cream
固醇 sterol
固醇膽汁酸 stero-bile acid
固體 solid
固體石蠟 paraffin wax
固體泡沫 solid foam
固體物質 solid matter
固體食物 broma; solid food
固體脂肪 solid fat
固體膠 solid gum

垃圾 garbage; muck; rubbish; sweepings; trash
垃圾桶 garbage can; trash can
坩堝 crucible
坦坡橙 tangor
夜來香浸膏 tuberose concrete
夜來香淨油 tuberose absolute
夜盲症 night blindness; nyctalopia
奇異果 Chinese gooseberry; kiwi (fruit)
奇異果香精 kiwi flavo(u)r
奈氏試劑 Nessler reagent
奈米 nanometer (nm)
奈克 nanogram (ng)
奈斯勒氏試劑 Nessler reagent
奄列 omelet(te)
委託加工 OEM
孟加拉玫瑰紅 rose bengal; rose bengale
季節 season
季節性食品 seasonal food
定位洗滌〔法〕 cleaning in place (CIP)
定位洗滌系統 cleaning in place system (CIP system)
定性分析 qualitative analysis
定香劑 fixative
定時器 timer
定氮儀 azotometer
定量分析 quantitative analysis
定量分裝肉 packaged meat
定量分裝機 portioning machine
定量包裝 portion pack; prepack
定量包裝〔的〕食品 prepackaged food; portion pack food
定量包裝的印模牛油 print(ed) butter
定量包裝的印模乳酪 print(ed) butter
定量的 quantitative
定量配給 ration
定量給料器 proportioner
定量裝填機 measuring filler
官能性質 organoleptic property
官能品評 organoleptic evaluation
官能試驗 organoleptic test; sensory test
官能閾值 organoleptic threshold
官能檢查員 panelist
官能檢驗 organoleptic examination
官感性 sensory property
官感評定 panel test
岩予魚 rockling
岩皮餅 rock cake
岩黃道蟹 rock crab
岩薔薇油 labdanum oil; ladanum oil
岩薔薇浸膏 labdamum concrete
岩藻聚糖 fucosan
岩藻糖 fucose
岩蘭草油 vetiver oil
岩鹽 halite; mine salt; rock salt
岩鹽坑 salt mine
帕拉金糖 palatinose
庚炔羧酸甲酯 methyl 2-octynoate
庚烷 heptane
庚酮 heptanone
庚酸 heptanoic acid
庚酸乙酯 ethyl heptanoate
庚酸丁酯 butyl heptanoate
庚酸丙酯 propyl heptanoate
庚酸戊酯 amyl heptanoate
庚酸烯丙酯 allyl heptanoate
庚醇 heptyl alcohol
庚糖 heptose
庚醛 heptanal
底 sole
底火麵包 hearth bread; oven-bottom bread
底烤麵包 sole bread
底部加熱的 underfired
底部加熱食物 underfired food
底部加熱烹調 underfired cooking
底部供液蒸發器 bottom-fed evaporator
底部發酵 bottom fermentation

底部酵母 bottom yeast
底棲生物 benthon
底棲魚 benthic fish; groundling
底棲魚類 bottom fish; demersal fish; ground fish
底蓋的拱凸 end doming
底蓋的翻滾式轉動 end-over-end agitation
底蓋翻轉殺菌 end-over-end (processing)
底層魚類 demersal fish
底層發酵酵母 bottom yeast
底鍋水 spent wash
延伸性 extensibility
延伸性圖譜儀 extensograph
延胡索酸 fumaric acid
延滯發酵 sluggish fermentation
延緩劑 retardant
弧菌 vibrio
弧菌毒素 vibriocin
往復篩 reciprocating sieve
彼卡氏〔麵粉粒度及色澤〕試驗 Pekar test
忽布 hop
念珠菌屬 *Candida*
念珠褐藻 knobbed wrack
性未成熟的肥鯡魚 matie
性成熟的魚 full fish
性成熟的鯡魚 matfull
拉力 tensile force; tension
拉力計 tensiometer
拉巴拉巴豆 lablab bean
拉氏麥芽糖化力值 Lasche value
拉布魯斯卡〔釀酒用〕葡萄 labrusca
拉舌 pull tab
拉舌〔蓋〕罐 lolled-up ends can
拉舌罐蓋 tear tab end
拉伸性 tensility
拉伸機 stretching machine
拉伸薄膜 stretch film
拉伸薄膜包裝 stretch film package
拉克斯〔煙燻〕火腿 lacks ham
拉格啤酒 lager (beer)
拉馬豆 sugar bean
拉基〔葡萄〕酒 rakee; raki
拉基梅酒 rakee; raki
拉菲椰 raffia; raphia
拉開蓋 tear-off lid
拉開罐 tear open can
拉製機 pulling machine
拉糖機 stretching machine
拉環 pull tab; tab
拉環蓋 ring pull end
拉麵 lamen
拌好的生菜色拉 tossed salad
拌好的生菜沙拉 tossed salad
拌好的生菜沙律 tossed salad
拌有砂糖的蜜餞 crytal fruit
拌糖粉軟糖 Turkish paste
抹布 dishcloth
抹油器 greaser
抹香鯨 cachalot; sperm whale
拒斥劑 repellent
拒油的 oil-repellent
拒食症 cibophobia
招牌食品 specialty product
披薩草 oregano; origanum; wild marjoram
拔〔羽〕毛 deplumation; feather dressing
拔〔羽〕毛機 feather picker; featherator
拔毛 defeather; epilation
拔毛 pluck; pick(ing)
拔毛機 plucker; picker
拔去塞子 uncork
拔瓶塞 corkage
拔爾散冷杉油 balsam fir oil
拋光 polishing
拋光筒 glazing drum
拋光機 glazing machine; polishing machine; polisher; revolving pan
抽風機 exhaust fan
抽氣扇 exhaust fan

抽真空 vacuumize
抽真空機 vaccumizer
抽提 extraction
抽提物 extract
抽提器 extractor
抽濕機 dehumidifier
抵工資的食物 tommy
抱子甘藍 baby cabbage; brussel sprout
拖妃糖 taffy; toffee; toffy
拖把 swab
拆除 strip
放出 bleed
放血 bleed; bleeding; exsanguination
放血刀 sticking knife
放血工 bleeder
放血不淨 misbleed
放血用的傳送帶 bleeding conveyor
放牧 grazing
放牧業 graziery
放射 radiation
放射性 radioactivity
放射性〔的〕 radioactive
放射性同位素 radioactive isotope
放射性物質 radioactive substance
放射狀菌落 radiate colony
放射保藏〔法〕 radiation preservation
放射毒性 radiotoxicity
放射能 radioactivity
放射處理 radiation treatment
放射源 radioactive source
放氣 deflation; degassing
放氣閥 bleeder
放液 tapping
放熱 heat production
放線菌 ray fungus
放線菌病 actinomycosis
放線菌屬 *Actinomyces*
放鬆 relaxation
易克曼試驗 Eijkman test
易拉開蓋 pull-off cap
易拉環 easy opening tab
易粉碎性 friability
易揮發的風味 fugitive flavo(u)r
易開的 easy open
易開罐 easy open can
易開罐拉環 easy opening tab; tab
易開罐的反向拉舌 reverse tab
易開罐蓋 tear tab end
易溶食鹽 dendritic salt
易碎的 crumbly
易腐性 perishability
易腐食品 perishables; perishable food
易餓病 bulimia
易燃性 inflammability
易壞食品 perishables
易變的 labile
昌蒲 orris
昌蒲根 orrisroot
昌蒲根油 orrisroot oil
昆士蘭〔堅〕果 Australian nut; macadamia nut; Queensland nut
昆布 kombu; laminaria; oarweed; tangle
昆布二糖 laminaribiose
昆布多糖 laminarin
昆布酸 tangic acid
昆布醇 laminaritol
昆布糖 laminariose
昆蟲 insect
昆蟲學 entomology
明太鱈魚 Alaska pollack
明火 open fire
明火加熱 fire heating
明火烘乾 fire drying
明串珠菌屬 *Leuconostoc*
明泰魚 Alaska pollack
明膠 gelatin(e)
明膠化 gelatinate; gelatination
明膠包衣 gelatin(e) glaze

明膠洋菜培養基 gelatin(e) agar
明膠凍 gelatin jelly
明膠培養基 gelatin(e) culture
明膠甜點 gelatin(e) dessert
明膠軟糖 gelatin jelly
明膠啫喱 gelatin jelly
明膠酶 gelatinase
明蝦 prawn
明礬 alum
明礬鉀 potassium alum
昇華〔作用〕 sublimation
昇華物 sublimate
昇華乾燥 sublimation drying
服務 service; serving
服務員 waiter
杭白菊浸膏 Hang-zhou chrysanthemum flower extract
枕形麵包 loaf
東方小麥 khorassan wheat
東方狐鰹 belted bonito; oriental bonito
東方魨 tiger puffer
東方魨屬 blowfish
東印度香葉油 East Indian geranium oil
東印度薄餅 paratha
東鳳螺 sea snail
果〔子〕凍 jello; jelly
果子露 sherbet
果仁 nutlet; pip
果仁奶糖 panoche
果仁酒 noyau; ratafia
果仁餅乾冰淇淋 bisque ice-cream
果仁糖 nougat; praline
果心 core
果汁 fruit juice; juice
果汁冰糕 water ice
果汁刨冰 frappe
果汁汽水 fruit soda water
果汁凍 fruit juice gel; sorbet
果汁粉 juice powder
果汁混合飲料 punch
果汁甜雞尾酒 shrub
果汁雪泥 fruit sherbet; sorbet
果汁雪花冰 fruit sherbet
果汁雪霜 sherbet
果汁飲料 ade; fruit squash; fruit drink; fruit juice beverage; juice beverage; juice drink; sorbet
果汁賓治 fruit punch
果汁澄清 juice clarification
果汁潘趣〔飲料〕 fruit punch
果汁凝膠 fruit juice gel
果汁機 juice mixer
果汁糖漿 fruit syrup
果汁興奮劑 ratafee
果汁雞尾酒 fruit cordial
果汁罐裝 canned fruit juice
果皮 fruit skin; pericarp; peel; skin; tangerine peel
果皮〔殘留的〕碎果肉 eradicator meat
果皮刮肉機 eradicator
果皮油 peel oil
果皮乾 dried peel
果皮軟的 malacodermous
果皮斑點 skin blemish
果皮蜜餞 candied peel; candy peel
果肉 fruit flesh; fruit pulp; flesh; marrow; meat; pulp; sarcocarp
果肉緊黏於果核的 clingstone
果味的 fruity
果泥 fruit puree; fruit paste; fruit pulp; mashed fruit; puree; squash
果泥飲料 fruit nectar; nectar
果孢子 carpospore
果柄 carpopodium; pedicel; pedicle
果凍粉 jelly powder
果凍粒飲料 jelly drink
果凍軟糖 jelly sweets
果凍裝飾配菜 aspic

果凍餡 jelly filling
果料牛奶小麥粥 furmety; frumenty
果料蛋糕 fruit cake
果料餡餅 futira
果核 core; nutlet; pit; stone
果核的核殼 putament
果酒 ratafee; ratafia; wine
果乾 dried fruit; dry fruit
果梗 fruit stalk; pedicel; pedicle
果脯 candied fruit; preserve; preserved fruit
果渣 marc; pomace
果渣白蘭地酒 Marc brandy
果渣酒 pomace wine
果菜箱 lug box
果園 fruitery; orchard
果腹食品 belly-full food
果葡糖漿 high fructose corn syrup (HFCS)
果實 fruitery; fructification; fruit
果實〔產量〕 fruitage
果實的汁囊 juice sac
果實硬度計 fruit hardness tester
果實樹莓 thimbleberry
果聚糖 fructosan; fructan; levan; levulosan
果聚糖蔗糖酶 levansucrase
果膏 fruit butter
果蜜 nectar
果漿 fruit pulp; fruit puree; mashed fruit; pomace; pulp
果漿過濾機 finisher
果盤 compote
果糊 fruit paste
果膠 pectin; pectin(e)
果膠分解酶 pectinase; pectolytic enzyme
果膠化〔作用〕 pectisation; pectization
果膠水解〔作用〕 pectolysis
果膠甲氧基酶 pectin methoxylase
果膠甲酯酶 pectin methylesterase; pectin methoxylase
果膠果凍 pectin jelly
果膠軟糖 pectin jelly
果膠酯酸 pectinic acid
果膠酯酸鈣 calcium pectinate
果膠酯酸鹽 pectinate
果膠酯酶 pectinesterase
果膠聚半乳糖醛酸酶 pectin polygalacturonase
果膠酸 pectic acid
果膠酸鈣 calcium pectinate
果膠酸酯 pectate
果膠酸鹽 pectate
果膠酸鹽 pectinate
果膠質 pectic substance
果膠糖 pectinose; pectose
果膠糖酶 pectosase; pectosinase
果膠纖維素 pecto-cellulose
果膠酶 pectase; pectic enzyme; pectolase; pectinase; pectolytic enzyme
果蔗 garden cane
果蔬皮 rind
果蔬成熟計 maturometer
果樹 fruiter; fruit tree
果樹〔栽培〕學 pomology
果樹栽培 orcharding
果糖 fructose; fruit sugar; laevulose; levulose
果糖〔糖〕漿 fructose syrup
果糖尿〔症〕 fructosuria
果糖酮酸 fructuronic acid
果糖磷酸 fructose phosphate
果糖苷 fructoside
果糖苷移轉酶 transfructosidase
果糖苷酶 fructosidase
果餡甜酥餅 strudel
果餡餅 flan; tart
果穗軸 cob
果醬 confiture; fruit jam; jam
果醬油酥餅 jam puff

果醬烘餅 jam tart
果醬餡餅 jam tart
果醬麵包片 tartine
果瓣 segment
果瓣分離機 segment separator
果蠅 drosophila (fly); fruit fly
枇杷 Japanese medlar; loquat
枝晶食鹽 dendritic salt
林〔特納〕氏度 Lintner degree
林〔特納〕氏值 Lintner value
林〔特納〕氏處理 Lintnerization
林〔特納〕氏澱粉 Lintner starch
林堡乾酪 Limburg(er) cheese
杯 chalice
杯子 cup; mug
杯子洋菇 cup fungus; cup mushroom
杯式噴霧 driving spray
杯狀巾馬勃(蕈類) pestle shaped puffball
杯裝速食飯 cup rice
杯裝速食麵 cup noodle
杯麵 cup noodle
板 plate
板式〔巴氏〕殺菌器 plate pasteurizer
板式蒸發器 plate evaporator
板式熱交換器 plate (type) heat exchanger
板式壓榨機 plate press
板板蒸餾塔 perforated plate column
板油 leaf fat; leaf lard; shop fat; suet
板狀冰 plate ice
板框〔式〕壓濾機 frame filter (press)
板栗 Chinese chestnut; chestnut
板條 slat
板條箱 crate; slatted crate
板腱 oyster blade
板篩 plate sieve
板鴨 preserved duck
松二糖 turanose
松三糖 melezitose; melicitose; melizitose
松口蘑 matsutake
松子 cedar nut; pignolia; pine nut
松阪牛肉 Matsusaka beef
松乳菇 delicious milk cap
松果 pine nut; pignolia
松果油 pinaster seed oil
松油醇 terpineol
松柏抗氧化酚 conidendrin
松香 colophony; rosin
松香甲酯 methyl ester of rosin
松香油 retinol
松香酸 abietic acid
松脂 rosin; turpentine (oil)
松脂酸 abietic acid
松脂醇 terpineol
松茸 matsutake; pine mushroom
松針油 pine needle oil
松節油 turpentine (oil)
松葉油 pine needle oil
松鼠 squirrel
松膠 arbinogalactan; larch gum
松樹 pinaster; pine
松蕈 matsutake; pine mushroom
松蕈酸 agaric(ic) acid; agaricinic acid
松蕈屬 *Armillaria*
松雞 capercaillie; capercailzie; grouse; hazel hen; hazel grouse; ruffed grouse
松露 truffle
松露菌 perigord truffle
松萜 pinene
析水乳酪 leaky butter
析水黃油 leaky butter
杵 tamper
武夷茶 bohea
肴入 injection
注射 injection
注射成型 injection mo(u)lding
注射成型機 injection mo(u)lding machine
注射醃漬法 artery pumping; stitch pumping

注射器 injector; syringe
注意冷凍 keep frozen
泌乳 lactate
泌酸細胞 oxyntic cell
泥 paste
泥土味 earthy taste; earthy flavo(u)r
泥漿 slurry
泥鰍 beardie; loach
河豚 fugu; globefish; puffer; puffer (fish); rabbitfish
河豚中毒 fuguism; globefish poisoning; puffer poisoning; tetraodon poisoning; tetrodon poisoning
河豚毒素 fugutoxin; puffer poison; tetraodotoxin; tetrodotoxin
河豚酸 tetraodonic acid; tetrodonic acid
河魚 river fish
河魨中毒 tetraodon poisoning; tetrodon poisoning
河魨毒素 tetraodotoxin; tetrodotoxin
河魨酸 tetraodonic acid; tetrodonic acid
河蟹 river crab
河鯡 allis
河鰈 butt; plaice
河鰻 river eel
河鱒 grayling
河鱸 common perch; European perch; perch
沾 dip
沾上砂糖 crystallizing
沾上麵包屑的 crumbly
沾污 contamination; stain
沼地藍莓 highbush blueberry
沼澤紅菇 tall russule
波士頓豆 Boston bean
波士頓黑麵包 Boston brown bread
波士頓褐饅頭 Boston brown bread
波本威士忌 Bourbon (whisky)
波本茶 Bourbon tea
波克夏〔黑白〕豬 Berkshire
波希米亞香腸 Bohemian sausage
波希米亞黑麥 Bohemian rye
波形切片 corrugated slice
波形菌落 undulated colony
波洛尼亞香腸 bologna (sausage)
波洛桑默331 poloxamer 331
波洛桑默407 poloxamer 407
波美 Baumé (Bé); Beaume
波美比重 Baumé gravity
波美比重度 Baumé degree
波美比重計 Baumé hydrometer
波美度 Baumé degree; Baumé scale; Baumé (Bé); Beaume
波旁威士忌 Bourbon (whisky)
波旁茶 Bourbon tea
波旁醛 bourbonal
波特紅〔葡萄〕酒 port wine; vintage port
波特葡萄酒 port
波紋貝殼狀通心粉 rigatoni
波馬特〔紅葡萄〕酒 pommard
波斯山渣 medlar
波斯漿果 Persian berry
波森莓 boysenberry
波爾多葡萄酒 Bordeaux
波爾特紅〔葡萄〕酒 port wine
波羅泥〔豬肉〕香腸 polony sausage
波羅納香腸 bologna (sausage)
波蘭香腸 Polish sausage
波蘿蜜 jaca; jackfruit; jak; jakfruit; nangka
法令 act
法式土司 French toast
法式小蛋糕 petit four
法式羊肋排 Frenched rack
法式肉蛋餡餅 volauvent
法式咖啡 French roast
法式炒咖啡法 French roast
法式的棒形脆麵包 bread stick

法式炸馬鈴薯片 French fired potato; French fries; French fry
法式苦艾酒 French vermouth
法式甜餡餅 French pastry
法式黑咖啡 café noir
法式糕點 French pastry
法式麵包 French bread
法式麵包卷 French roll
法奇〔軟〕糖 fudge
法定分析法 official method of analysis
法定方法 official method
法定食用色素 certified colo(u)r
法定試樣 official sample
法定標準 mandatory standard
法肯（9英加侖） firkin
法案 act
法國〔長〕麵包 baguet(te)
法國牛奶咖啡 café au lait
法國布里芝士 brie cheese
法國布里乾酪 brie cheese
法國布爾戈尼葡萄酒 Bergundy
法國白蘭地酒 cognac
法國味美斯酒 French vermouth
法國阿爾薩斯白〔葡萄〕酒 alsatian wine
法國保根地葡萄酒 Bergundy
法國柏甘地葡萄酒 Bergundy
法國郝密紅〔葡萄〕酒 hermitage
法國菠菜 mountain spinach; orach(e)
法國菜豆 French bean
法國黑李子 French plum
法國蘋果白蘭地酒 apple jack; calvados
法規 act; regulation
法蘭克福香腸 frank; frankfurt(er)
沸水 boiling water
沸石 zeolite
沸點 boiling point (b.p.)
沸點〔測定〕計 ebulliometer
沸騰 boil; ebullition; ebullience; ebullience; seethe
沸騰狀發酵 boiling fermentation
泄水管 run off pipe
泄放 vent
泄氣的麵包 bleeding bread
油 oil; oleo
油中水型乳〔濁〕液 water-in-oil emulsion (w/o emulsion)
油分離機 oil separator
油比重計 oil gauge; oleometer
油包水型乳〔濁〕液 water-in-oil emulsion (w/o emulsion)
油瓜 calabazilla
油用堅果 oil nut
油石 whetstone
油坊 oil mill
油折射計 oleorefractometer
油豆腐 fried bean curd (puff); fried tofu
油果 oil nut
油泡豆腐 tofu puff
油炒 frying
油炒麵粉糊 roux
油的 oily
油的濾清 oil polishing
油返味 reversion flavo(u)r
油柑 emblic
油炸 deep fat frying; deep frying; fry; frying
油炸豆腐 bean curd crouton; fried bean curd (puff); fried tofu
油炸油 frying oil
油炸的 fried
油炸食物 fried food
油炸食品 deep fried product; fry
油炸香蕉 banana fritter
油炸馬鈴薯片 saratoga chip; saratoga potato
油炸魚 fish fry
油炸麻花 cruller; kruller
油炸菜肉丸 croquette

油炸圓圈 cruller; kruller
油炸圓餅 berlinois
油炸碎麵包片 crouton
油炸裹麵糊的魚蝦 tempura
油炸糕點 fried cake
油炸餡餅 fritter
油炸鍋 deep fryer; deep frying pan; frier; fryer
油炸麵糰 doughboy
油籽 oilseed; oil grain
油籽植物油 seed fat
油胡桃 butternut; white walnut
油胞 oil sac; oil droplet
油重計 elaeometer; elaiometer
油料作物 oil crop; oil plant
油料樹 tallow tree
油栗 jojoba
油桃 nectarine
油浸 oil immersion
油浸鯖魚 mackerel in oil
油浸鰺魚 jack in oil
油浴 oil bath
油紙 wax paper; waxed paper
油耗味 rancid flavo(u)r; rancid odo(u)r
油耗味的 tallowy
油耗的 rancid
油脂 axunge; fat and oil; oil and fat
油脂析出 oil off
油脂的 lipidic; oleo
油脂的銅鏽味的 hammy
油脂氫化 oil hardening
油脂脫膠 oil degumming
油脂提取 oil extraction
油脂提取器 oil extractor
油脂硬化 fat hardening
油脂稀釋劑 fat extender
油脂萃取 oil extraction
油脂萃取器 oil extractor
油脂補充劑 fat extender
油脂精煉 oil refining
油脂精製 oil refining
油脂凝固點試驗 cold test
油脂攝取〔量〕 fat intake
油脂鹼精煉 alkali refining; caustic refining
油梨 aguacate; alligator pear; avocado; butter-fruit
油粕 expeller cake; pomace
油粕粉 oil meal
油粕餅 oil cake; seedcake
油細胞 oil cell
油莎草 chufa
油斑 greasy spot
油棕〔櫚〕 oil palm
油棕〔櫚〕仁 oil palm kernel
油棕〔櫚〕果 oil palm fruit
油棕滅酶 oil palm sterilization
油渣 greaves; rendering scraps; scrap tallow
油渣餅 oil cake
油渣壓榨機 lard press
油硬化 oil hardening
油菜 cole; rape
油菜子 coleseed
油菜籽 colza; rapeseed
油菜籽固醇 campesterol
油菜籽粉 rapeseed meal
油菜籽粕 rapeseed meal
油酥 flake pastry
油酥性測定計 shortometer
油酥甜餅 shortbread
油酥糕餅 shortcake
油酥點心 puff pastry
油酥麵皮 short crust
油酥麵製糕餅 short pastry
油酥麵糰 short paste
油椰 oil palm
油溶〔性〕的 oil soluble
油滑食品 unctuous food

油煎 fry; frying; panbroil
油煎魚餅 fried fishcake
油煎餅 cruller; kruller
油腳 foots; oil foot
油滴 dripping; oil droplet
油漬的 oiled
油漬食品 oil preserves
油漬罐頭 oiled can
油與溶劑的混合物 miscella
油酸 oleic acid
油酸乙酯 ethyl oleate
油酸鈉 sodium oleate
油酸酯酶 olease
油酸鹽 oleate
油廠 oil mill
油質 oiliness
油質的 oleaginous; oily; unctuous
油醋汁 vinaigrette sauce
油墨 ink
油橙 bergamot; oil orange
油橙油 bergamot oil
油橄欖 olive
油橄欖油 olive oil
油燜小牛肉 fricandeau; fricando
油燜肉 fricassee
油膩 oiliness
油膩〔性〕 greasiness
油膩乏味的食物 stodge
油膩味 oily flavo(u)r
油膩性 unctuousness
油膩的 greasy; oily; stodgy; tallowy; unctuous
油膩食品 unctuous food
油鏡 oil immersion lens
油類 oil
油鯡 oldwife
油鰈 arrow-toothed halibut; old woman flounder
油葱 eschalot; shallot
沿海魚〔類〕 inshore fish
沿海漁業 inshore fishery
治療 therapy; treat
治療食品 therapeutical food
治療膳食 therapeutical diet
泡 steep; vesicle
泡沫 foam; froth; lather; scum; spume
泡沫層 skimmings
泡沫〔型〕蛋糕 foam type cake
泡沫分離〔法〕 foam separation
泡沫改良劑 foam modifier
泡沫乳油 whipped cream
泡沫性 spumescence
泡沫狀 spumescence
泡沫狀的 spongy
泡沫乾燥 foam drying
泡沫乾燥法 foam drying process
泡沫蛋白鬆糕 chiffon cake
泡沫塑料 plastic foam
泡沫噴霧乾燥 foam spray drying
泡沫層乾燥 foam-mat drying; foam vacuum drying
泡沫穩定性 head retention
泡沫穩定劑 foam stabilizer
泡泡口香糖 bubble (chewing) gum
泡泡酒 bubbly
泡泡糖 bubble (chewing) gum
泡芙 puff
泡花鹼 sodium metasilicate
泡茶 infuse; leaf stewing
泡茶器 tea marker
泡盛麴菌 *Aspergillus awamori*
泡軟 maceration
泡湯的麵包 brewis
泡菜 kraut; pickle; pickled vegetable; sauerkraut; sourcrout; sourkrout
泡葉藻 bladder wrack
泡鼓包裝 blister pack
泡製 infuse

泡騰 effervescence
泡麵 instant noodle
泡麵包片 sop
泡囊狀巾馬勃(蕈類) mosaic puffball
泛解酸 pantoic acid
泛解酰牛磺酸 pantoyltaurine
泛酸 pantothenic acid; vitamin B_3
泛酸鈣 calcium pantothenate
泛酸鈉 sodium pantothenate
泛酸酯 pantothenate
泛酸鹽 pantothenate
泛醇 panthenol
泛磺酸 pantoyltaurine
泛醌 ubiquinone
泛酰醇 panthenol
泊(黏度單位) poise
炕 pan broiling
炒 friture; saute(ing); stir-frying; torrefaction
炒玉米粉 nocake; pinole
炒肉 sauted meat
炒的 fried; stir-fried
炒青菜 fried vegetable
炒蛋 scrambled egg
炒麥芽 coloured malt; roasted malt
炒焦麥芽 caramel malt
炒菜 sauted dish
炒菜肉 bubble and squeak
炒飯 fried rice
炒熟大麥粉 parched barley flour
炒蔬菜 sauted vegetable
炒鍋 wok
炒雜碎 chop suey
炒雜燴菜 chop suey
炒麵 chow-mein
炒麵〔條〕 fried noodle
炊具 cooking utensil; cooker
炊具修補匠 tinker
炊飯 rice cooking
炊飯器 rice cooker
炙 roast
炙肉 roti
牧牛 cattle raising
牧草 grazing
牧草牛肝菌 red cracking bolete
牧草味 feed flavo(u)r
牧場 animal farm; cattle range; dairy farm; farm; livestock farm; ranch
牧群 herd
牧豬人 swineherd
物品 stuff
物理化學的 physicochemical
物理精煉 physical refining
物理變化 physical change
物質 substance
狗母魚 brush-toothed lizard fish
狗魚 pike
狗鱈 cape hake
狐鰹 bloch
玫紅 rhodamine
玫紅醇 rhodinol
玫瑰 rose
玫瑰木油 rose wood oil
玫瑰果 dog rose; rose hip
玫瑰果糖漿 rose hip syrup
玫瑰油 attar; otto; rose oil; rose otto
玫瑰油香精 rose oily flavo(u)r
玫瑰花油 rose oil
玫瑰毒鮋 devilfish
玫瑰紅 phloxine
玫瑰紅鉚釘菇 rosy spike cap
玫瑰茄 roselle
玫瑰茄色素 hibiscetin
玫瑰茄紅 hibiscetin
玫瑰苯胺 rosaniline
玫瑰香葡萄 muscat grape
玫瑰香葡萄酒 muscat
玫瑰香精 rose flavo(u)r

玫瑰香檳酒 pink champagne
玫瑰浸膏 rose extract
玫瑰草油 palmarosa oil
玫瑰茶 rose tea
玫瑰淨油 rose absolute
玫瑰葡萄酒 rose wine
玫瑰精油 rose attar
玫瑰醇 rhodinol
玫瑰醚 rose oxide
疝疼乳菇 woolly milk cap
疙瘩魚卵 lumpfish caviar
盲腸 caecum; cecum
盲檢試驗 blind test
盲鰻 hagfish
直火加熱 straight firing
直火乾燥 fire drying
直立性番茄 upright tomato
直立萵苣 cos lettuce; romaine (lettuce)
直的 straight
真空密封的 vacuum tight
真空密閉的 vacuum tight
真空緊密度 vacuum tightness
直接〔發酵〕麵糰 straight dough
直接〔發酵〕麵糰法 straight dough method
直接〔發酵〕麵糰製麵包法 straight dough method
直接火 direct fire
直接火烤的麵包 oven-bottom bread
直接火燄滅菌 direct flame sterilization
直接加熱 direct heating
直接冷凍 direct freezing
直接浸漬式速凍 direct immersion freezing
直接浸漬冷凍 direct immersion freezing
直接發酵〔法〕 direct fermentation
直接蒸汽 open steam
直接蒸汽 direct steam
直接蒸汽噴射加熱 direct steam injection heating
直接蒸汽瞬間巴斯德殺菌 direct flash pasteurization
直接需氧量 immediate oxygen demand (IOD)
直接鏡檢法 direct microscopy
直接鏡檢數 direct microscopic count
直腸 rectum
直腸的 rectal
直腸頭 bung cap
直腸繫膜 mesorectum
直餾油 virgin oil
直鏈 straight chain
直鏈化合物 linear compound
直鏈澱粉 amylose
直鏈澱粉—碘的絡合物 amylose iodine complex
矽 silicon
矽油 silicone oil
矽酮 silicone
矽酸鈣 calcium silicate
矽酸鈉 sodium silicate
矽酸鹽 silicate
矽膠 silica gel
矽鋁酸鈉 sodium silicoaluminate
矽藻 diatom
矽藻土 diatom earth; diatomaceous earth; diatomite; infusorial earth; randanite; tripolite
矽藻石 tripolite
秈稻 indica type rice
空心菜 water convolvulus; water spinach
空心穀粒 hollow kernel
空白 blank
空白試驗 blank test
空泡 vacuole
空氣〔中〕解凍 air defrosting
空氣中熟成 air cure
空氣分離器 air separator
空氣升液發酵槽 air-lift fermenter; gas-lift fermenter

空氣升液攪拌 air-lift agitation
空氣升液攪拌器 air-lift agitator
空氣污染 air pollution; airborne contamination
空氣冷卻 air cooling
空氣冷卻的 air cooled
空氣冷卻器 air cooler
空氣冷凝器 air condenser
空氣凍結 air freezing
空氣氧化 air oxidation
空氣浴 air bath
空氣乾燥器 air dryer
空氣淨化裝置 air purifier
空氣焚化爐 air incinerator
空氣傳染 airborne infection; aerial infection
空氣傳播細菌 airborne bacteria
空氣感染 aerial infection
空氣解凍 air thawing
空氣過濾網 air screen
空氣過濾器 air filter
空氣鼓泡器 air-lift agitator
空氣對流乾燥 air convection drying
空氣對流乾燥機 air convection dryer
空氣噴射攪拌 air-jet agitation
空氣調詳 air conditioning
空氣調節機 air conditioner
空氣濾淨器 air purifier
空浮細菌 airborne bacteria
空腸 jejunum; empty body
空廚 flight kitchen
空調 air conditioning
空調機 air conditioner
空罐 empty can
空罐成〔圓〕型機 rolling machine
空罐乾燥機 can dryer
空罐焊錫機 can soldering machine
空罐整形機 can straightener
羌活 angelica
肺〔臟〕 lung
肺魚 lungfish
肥〔豬〕肉 speck
肥小豬 porker
肥肉絞碎機 lard hasher
肥肉餡充填 larding
肥肉薄片 bard
肥沃度 fertility
肥皂 soap
肥皂水 suds
肥皂泡沫 suds
肥皂臭 alkaline soapy odor
肥育畜 feeder
肥的 fat
肥厚多肉的植物 succulent
肥胖 adiposity; fleshiness; polysarcia
肥胖的 adipose; obese
肥胖病 adiposis
肥胖症 adiposity; obesity
肥胖症膳食 obesity diet
肥料 fertilizer; manure
肥畜 fatling
肥膘肉 shop fat
肥豬 butcher bull; hog; lard hog
肥鵝肝 foie gras
肥蟹 fat crab
肥鹹肉片 lardo(o)n
肢 limb
股骨 femur; thigh bone
肫 gizzard
肩肉 shoulder meat
肩肉火腿 picnic ham; shoulder ham
肩肥肉 shoulder fat
肩背肉 butt
肩胛小排肉 chuck rib meat
肩胛肉 chuck and blade; shoulder; shoulder meat
肩胛里肌 chuck tender
肩胛骨 bladebone; scapula

肩胛骨肉 shoulder blade
肩部 shoulder
肩部肉 blade meat
肩膀肉 picnic shoulder
肯氏低鹽米食 Kempner diet
肯諾依茶 kanoy
臥式〔高壓〕殺菌釜 horizontal retort
臥式〔圓〕盤磨 kek mill
臥式打麥機 horizontal scourer
臥式滅菌釜 horizontal sterilizer
臥式滅菌鍋 horizontal sterilizer
芳〔香〕族化合物 aromatics
芳〔香〕族胺基酸 aromatic amino acid
芳香 aroma; fragrance
芳香性 aromaticity
芳香油 essential oil; ethereal oil
芳香物質 aromatic substance
芳香的 aromatic; sweet
芳香的食〔用〕油 sweet oil
芳香度 aromaticity
芳香料作物 aromatic crop
芳香草 aromatic herb
芳香草料 sweet herb
芳香族化合物 aromatic compound
芳香植物 aromatic plant
芳香植物種籽的糕餅 seedcake
芳香調味料 aromatic condiment; aromatic seasoning
芳香劑 aromatics
芳樟油 linaloe wood oil
芳樟醇 linalool
芝士 cheese
芝士孔 cheese eyeing
芝士片 sliced cheese
芝士皮 cheese rind
芝士肉羅夫 cheese and meat loaf
芝士刨 cheese grinder
芝士取樣器 cheese trier
芝士風味 cheese flavo(u)r; cheesy flavo(u)r
芝士食品 cheese food
芝士粉 cheese powder
芝士桶 chessel; chesset; chisset
芝士蛋糕 cheese cake
芝士餅 ramekin; ramequin
芝士熟成 cheese curing
芝士凝塊 cheese coagulum; cheese curd
芝麻 benne; sesame
芝麻明 sesamin
芝麻油 benne oil; gingelly oil; jinjili oil; sesame oil; teel oil
芝麻素 sesamin
芝麻酚 sesamol
芝麻酥糖 halavah; halva(h)
芝麻糊 sesame paste
芝麻醬 sesame paste; sesame sauce; tahini (paste)
芙蓉子 poppy seed
芙蓉蛋 egg fooyo(o)ng
芙蘭派 flan
芭芭可果 babaco
芽 bud; sprout
芽甘藍 baby cabbage; brussel sprout
芽生孢子 blastospore
芽生菌 *Blastomyces*
芽生菌病 blastomycosis
芽枝黴屬 *Cladosporium*
芽孢 endospore; gemma; spore
芽孢桿菌 bacilli
芽孢桿菌溶素 bacilysin
芽孢菌黴素 bacillomycin
芽冠 tip cap
芽胞 germ
芹子烯 selinene
芹菜 *Apium grareolens*; celery
芹菜籽 celery seed
芹菜籽油 celery oil; celery seed oil
芹菜腦 apiole
芹菜糖 apiose

芹菜屬 *Apium*
芹菜苷 apiin
芹葉山楂 parsley hawthorn
花〔蒂〕蜜 normal honey
花生 earth chestnut; earthnut; ground pea; groundnut; goober; monkey nut; peanut
花生〔粕〕餅 peanut cake
花生仁 groundnut kernel; peanut kernel
花生仁蛋白酶 arachain
花生四烯酸 arachidonic acid; eicosatetraenoic acid
花生米 groundnut kernel; peanut kernel
花生油 arachis oil; earthnut oil; groundnut oil; peanut oil
花生脆糖 peanut brittle
花生球蛋白 arachin
花生粕〔粉〕 peanut meal
花生脱殼機 peanut sheller
花生殼 peanut hull
花生酸 arachic acid; arachidic acid; eicosanoic acid
花生醬 peanut butter
花生屬 *Arachis*
花式肉凍冷盤 aspic
花式的 fancy
花式甜麵包 coffee cake
花式麵包 fancy bread
花色素酶 anthocyanase
花色苷 anthocyanin
花尾榛母雞 hazel hen
花尾榛雞 hazel grouse
花尾龍躉 giant grouper
花束 bouquet
花豆 white Dutch runner bean
花枝 cuttle; cuttlefish
花油 attar; caul fat; caul web fat
花青素 anthocyanin
花青苷 anthocyanin
花香 flowery odo(u)r
花香〔精〕油萃法（用油脂從鮮花提取香味法）enfleurage
花香精 floral essence
花粉 farina; pollen
花粉粒 pollen grain
花茶 flower tea; scented tea
花圈 wreath
花梗 pedicel; pedicle; scape
花莖 scape
花莖甘藍 asparagus broccoli; calabrese
花椒 Chinese prickly ash; flatspine prickly ash; Szechuan peppercorn; xanthoxylum
花菜 cauliflower
花黃質 anthoxanthin
花嗅 flowery odo(u)r
花園 garden
花椰菜 broccoli; cauliflower
花葵苷 pelargonin
花葉丁香 Persian lilac
花跳 mud skipper
花飾小蛋糕 petit four
花飾蛋糕 decorated cake
花楸〔漿〕果 rowanberry
花楸〔樹〕 service tree
花楸果 chokeberry; service berry
花煙草 nicotiana
花旗蔘 American ginseng
花漢茶 faham
花蜜 flora nectar; nectar
花蜜腺 nectary
花醋栗 golden currant
花蕾油 flower bud oil
花點糯鰻 conger eel
花瓣 petal
花邊刀 jagger
芥子 mustard seed
芥子油 mustard oil; mustard seed oil
芥子酸 erucic acid; sinapic acid

芥子酸膽鹼〔酯〕 sinapine
芥子醇 sinapyl alcohol
芥末 mustard
芥末〔粉〕 mustard flour
芥末沙拉醬 mustard dressing
芥末調味料 mustard dressing
芥茉 Japanese horseradish; wasabi
芥菜 Indian mustard; leaf mustard; mustard; mustard green
芥菜籽 mustard seed
芥菜類 *Brassica*
芥酸 docosenoic acid; erucic acid
芥酸精 erucin
芥藍〔菜〕 Chinese kale; cabbage mustard; kale
芥蘭菜 Chinese kale
蕓薹 cole; colewort
芸香油 amyris oil
芸香素 amyrin
芸香苷 eldrin
虎奶菇 tiger milk mushroom
虎皮香菇 tiger saw dill
虎河豚 duffer
虱目魚 milkfish
初上市水果 first-fruit
初加工 initial handling
初步分切 primal cut
初步分餾 prefractionation
初步分餾塔 prefractionator
初步加工 roughing
初步粉碎機 primary crusher
初步乾燥 preliminary drying
初步試驗 preliminary test
初步過濾器 preliminary filter
初步過濾 preliminary filtration
初步蒸餾 predistillation
初步壓碎 preliminary crushing
初乳 colostrum; foremilk; neogala
初級污水處理 primary sewage treatment
初級產品 raw product
初級蒸餾 primary distillation
初級廢料處理 primary waste treatment
初產品 head product; primary product
初處理 initial handling
初期發酵 incipient fermentation
初期腐敗 incipient spoilage
初發酵 primary fermentation
初碎 preliminary crushing
初碎機 preliminary breaker
初榨汁 mother juice
初榨油 virgin oil
初榨蔗汁 primary cane juice
初學走路的孩童 toddler
初濾 primary filter
初濾液 primary filtrate
初濾器 preliminary filter
初餾分 first cut
初餾物 first running
初餾酒 low wine
表 table
表皮 epidermis
表皮組織 epidermal tissue
表皮葡萄球菌 *Staphylococcus epidermidis*
表兒茶素 epicatechin
表兒茶酸 epicatechin
表面下發酵 subsurface fermentation
表面冷卻器 surface cooler
表面冷凝器 surface condenser
表面抹片 surface smear
表面空氣冷卻器 surface air cooler
表面活性 surface activity
表面活性物質 surface active substance
表面活性劑 surface active agent; surfactant
表面凍結的 surface frozen
表面乾燥 skin drying; superficial drying; surface drying
表面培養〔物〕 surface culture

表面張力 surface tension
表面粗硬的糕餅 rock cake
表面發酵 surface fermentation
表面塗抹 surface smear
表面塗油的 oiled
表面精整 surface finish
表面黏度 surface viscosity
表現型 phenotype
表層 surface layer
表觀消化率 apparent digestibility
表觀黏度 apparent viscosity
軋片機 flaking mill
軋光筒 glazing drum
軋光機 glazing machine
軋粉機 roller mill
軋輥式榨油機 Scotch hands
軋薄麵糰 pinning
軋麵杖 pin
軋麵機 dough sheeter
近海魚〔類〕 neritic fish; inshore fish
近海漁業 inshore fishery
金瓜 golden nugget
金甲豆 sieva bean
金合花浸膏 cassie concrete
金合歡 blackthorn
金合歡浸膏 cassie concrete
金合歡醇 farnesol
金尖茶 golden tips
金色洋魚 red mullet
金柑 kumquat
金桔 cumquat; kumquat; marumi kumquat
金酒 aurum
金針 citron day-lily; daylily; dried lily flower; yellow daylily
金針菜 common yellow daylily; lemon daylily; lemon lily; long yellow daylily; yellow daylily
金針菇 golden mushroom; long rooted mushroom; velvet shank
金帶梅鯛 banana fish; gclden banded caesio
金梭魚 barracuda
金眼鯛 alfoncino
金麥豌豆 lentil
金棗 kumquat
金華豬 Jinhwa hog
金飯〔葡萄〕酒 Zinfandel
金黃色沙司 aurore sauce
金黃色葡萄球菌 *Staphylococcus aureus*
金黃烈酒 aurum
金黃素 chrysin
金黃調味醬 aurore sauce
金黃質 auroxanthin
金黃醬 aurore sauce
金菇 enoki mushroom
金盞花 marigold
金槍魚 bluefin tuna; tuna; tunny
金槍魚罐頭 tuna fish
金線瓜 spaghetti squash
金線魚 besugo; golden thread
金蓮花 nasturtium; tropaeolum
金蓮花黃質 trollixanthin
金褐傘（蕈類） golden cap
金橘 cumquat; calamondin; kumquat; oval kumquat; Panama orange
金錢腿 shin from foreshank
金雞納酸 quinic acid
金雞納霜 quinine
金雞魚 black striped snapper
金雞菊 coreopsis
金鐘柏油 cedar leaf oil; thuja oil
金屬小杯 pannikin
金屬味 metallic taste
金屬底蓋的紙罐 composite can; composite paper can
金屬板接觸冷凍〔法〕 plate freezing
金屬容器 metal container
金屬氣味 metallic flavo(u)r

金屬偵測器 metal detector
金屬偵察器 metal detector
金屬探測器 metal detector
金屬蛋白質 metalloprotein
金屬絲布 netting
金屬絲攪打器 wire whipper
金屬塗布紙 metallic paper
金屬器皿 hardware
金屬罐 metal can; metallic can
金屬酶 metalloenzyme
金櫻子 Cherokee rose
金櫻子棕 rosa laevigata michx brown
金體美洲鯿魚 golden shiner
金鱸 yellow perch
長方形乾酪 loaf cheese
長方形罐 oblong can; rectangular can; rectangular tin
長牙 tusk
長白菜豆 cannellini
長吻亞口魚 longnosed sucker
長吻魚 snipefish
長尾海鱸魚 band-tailed sea perch
長尾鯊 foxshark
長尾鯷魚 long tail anchovy
長形麵包 birlie; pullman bread
長角豆 algaroba; carob
長身鱈魚 ling
長柄平底鍋 skillet
長柄杓 dipper
長柄燉鍋 stew pan
長柄鍋 pipkin
長段的細麵條〔通心粉〕 lungo vermicelli coupe
長段維米切利〔通心粉〕 lungo vermicelli coupe
長時間發酵的麵糰 long sponge
長根小奧德菇 rooting shank
長豇豆 asparagus bean; yardlong bean
長條〔大麥〕麵包 bar bread
長條泡芙 éclair
長眼篩板 slotted plate
長眼篩網 slotted plate; slot mesh screen
長粒米 long grain rice
長通粉 penne; rigatoni
長期儲藏 standing storage
長番椒 long pepper
長筍尖 long spear
長筍頭 long spear
長圓罐 obround can
長腰鮪 northern bluefin
長葉金橘 malay kumquat
長葉萵苣 romaine (lettuce)
長管式無循環蒸發器 Kestner evaporator
長管形通心粉 tubetti lunghi
長管蒸發器 long tube evaporator
長辣椒 long pepper
長嘴魚 snipefish; trumpet fish
長霉的 mildewed
長噸 gross ton
長頸南瓜 crookneck (pumpkin); crookneck squash
長頸葫蘆 crookneck squash
長頭沙丁魚 oil sardine
長鏈 long chain
長鏈脂肪酸 long chain fatty acid
長蘆筍尖 long spear
長蘆筍頭 long spear
長鰭金槍魚 albacore (tuna); long-fin(ned) albacore; long-fin(ned) tuna
長鰭飛魚 long-fin(ned) flying fish
長鰭鮪魚 long-fin(ned) tuna; albacore (tuna)
長鬚鯨 fin whale
長鬚鯨 finback
長鰳 long-fin(ned) herring
阿力甜 alitame
阿丹奇維茲反應 Adamkiewiez reaction
阿月渾子果實 pistachio

阿月渾子樹 pistachio
阿比可糖 abequose
阿加鯖魚 Atka fish; Atka mackerel
阿卡拉莓 akala
阿伐生育酚 alpha-tocopherol; α-tocopherol
阿伐葡聚糖 alpha-glucan
阿伐澱粉 alpha-starch; α-starch
阿伐澱粉酶 alpha-amylase; α-amylase
阿伐螺旋 alpha-helix
阿伐纖維素 alpha-cellulose
阿米諾法 amylo process
阿米露法 amylo process
阿沃納葡萄 Iona
阿貝折光儀 abbe refractometer
阿貝折射計 abbe refractometer
阿佩爾滅菌法 appertization; commercial sterilization
阿卓糖 altrose
阿奇樹 akee
阿拉比卡咖啡 arabica (coffee); coffee arabica
阿拉伯〔樹〕膠 arabic gum
阿拉伯半乳聚糖 arbinogalactan; larch gum
阿拉伯乳〔酸〕桿菌 *Lactobacillus arabinocus*
阿拉伯咖啡 arabica (coffee); coffee arabica
阿拉伯咖啡樹 arabian coffee
阿拉伯茶 Arabian tea; khat; kat
阿拉伯聚糖 araban; arabinan
阿拉伯辣椒 akola
阿拉伯酸 arabic acid
阿拉伯樹膠 acacia gum; arabian gum; arabian gum; gum acacia; gum arabic
阿拉伯糖 arabinose; gum sugar; pectin sugar; pectinose
阿拉伯糖酸 arabonic acid
阿拉伯糖醇 arabitol
阿拉克燒酒 arak
阿拉希酒 allasch
阿拉斯加青鱈 Alaska pollack
阿拉斯加黑魚 Alaska blackfish
阿拉斯加鱈 pacific cod
阿東〔戊〕糖 adonose
阿法〔白脱連續〕製法 Alfa process
阿柯糖 acrose
阿洛拉紅 allura red AC
阿洛拉紅鋁色澱 allura red AC aluminium lake
阿洛酮糖 allulose; psicose
阿洛糖 allose
阿根廷巴里爾小麥 Baril wheat
阿索爾蜜酒 Athole brose
阿馬尼亞克酒 armagnac
阿散酸 arsanilic acid
阿斯匹靈 aspirin
阿斯巴甜 aspartame
阿鳩彎 ajowan
阿瑪白蘭地 armagnac
阿樸脂蛋白 apolipoprotein
阿諾克索默 anoxomer
阿諾滅菌器 Arnold sterilizer
阿龍尼亞苦味果 chokeberry
阿薩姆茶 assam tea
阿魏膠 asafoetida
阿蘭酸 alantic acid; inulic acid
阻遏物 repressor
阻隔包裝 barrier package
阻滯劑 retardant; retarder; retarding agent
附加物 additament; adjunct
附加菜 entremets
附加價值 added value
附生植物 aerophyte
附耳湯碗 cream-soup bowl
附果 accessory fruit
附蛇麻葫酮 adhumulone
附蛇麻蘆酮 adlupulone
附著 adhesion

附碟杯 cup and saucer
附蓋烤爐 coverplate oven
附屬牛肝菌 spindle stemmed bolete
青〔辣〕椒 green capsicum
青刀豆 garden bean; vert haricot
青口 mussel
青大豆 green soybean
青干金槍魚 northern bluefin
青少年的 juvenile
青瓜 cucumber
青米 green kernel rice
青杆小麥 bluestem wheat
青芒果 green mango
青豆 garden pea; pea; string bean
青花菜 asparagus broccoli; broccoli; calabrese; Italian broccoli; sprouting broccoli
青花椰菜 sprouting broccoli
青春期 adolescence
青苔 lichen; moss
青紋乾酪 bleu cheese
青草味 grassy odo(u)r
青草茶 herb tea
青梅 greengage
青魚 black carp
青椒 bullnose pepper; green pepper; sweet pepper
青筍 endive sprout
青紫色食用傘菌 blewits
青菜 green; greens; green stuff; green vegetable; garden sass; verdure
青貯飼料 silage
青熟 green mature
青蝦 green shrimp
青豌豆 green pea
青霉氧化酶 notatin
青縢 blue mishima pufferfish
青檸 lime
青蠅 bluebottle
青蟹 blue crab; green crab
青籃子魚 rabbitfish
青蘋果 greening
青鱈 pollack; pollock; saithe
青蘿蔔 green turnip
青黴 blue mo(u)ld
青黴素 penicillin
青黴素酶 penicill(in)ase
青黴菌屬 *Penicillium*
青黴酸 penicillic acid
青葱 cibo(u)l; green onion; scallion; spring onion; Welsh onion
非〔原〕血紅素鐵 nonheme iron
非水溶性 non-aqueous
非牛頓流體 non-Newtonian fluid
非主〔要〕食品 non-staple food
非必需脂肪酸 nonessential fatty acid
非必需胺基酸 dispensable amino acid; nonessential amino acid
非必需微量元素 nonessential trace element
非正式方法 unofficial method
非生產季節 off season
非回收容器 non-returnable container
非改性澱粉 unmodified starch
非更性水果 non-climacteric fruit
非皂性清潔劑 non-soapy detergent
非乳奶精 non-dairy creamer
非乳酸酸度 non-lactic acidity
非定量包裝 free pack
非法定方法 unofficial method
非金屬 non-metal
非故意添加的食品添加劑 unintentional food additive
非洲土杏紅 tigernut
非洲大羚羊 eland
非洲小米啤酒 pombe
非洲白花菜 African spiderherb
非洲竹芋甜素 thaumatin

非洲高粱 durra; jowar
非洲甜果素 thaumatin
非洲粟啤酒 pombe
非洲菠菜 Ceylon spinach
非洲蜂蜜酒 asali
非洲酪脂樹 shea
非致病菌 non-pathogenic bacteria
非食用 non-edible
非食用油 inedible oil
非食用的 inedible
非食用動物脂 inedible tallow
非食用部分 non-edible part
非凍結水 non-frozen water
非病原性的 unpathogenic
非病原性微生物 unpathogenic micro-organism
非病原菌 non-pathogenic bacteria
非破壞性分析 non-destructive analysis
非破壞性檢驗 non-destructive inspection
非脂固形物 nonfat solids; solid nonfat
非脂固體 solid nonfat
非脂肪的 nonfat
非酒精飲料 non-alcoholic beverage
非胺基態氮 non-amino nitrogen
非細菌性變質 non-bacterial spoilage
非蛋白胺基酸 nonprotein amino acid
非蛋白態氮 nonprotein nitrogen (NPN)
非蛋白質 nonprotein
非揮發性 non-volatility
非揮發性油 fixed oil; non-volatile oil
非揮發性的 non-volatile
非晶形 amorphism
非晶形的 amorphous
非晶形糖 amorphous sugar
非發泡性葡萄酒 stale wine
非發酵〔性〕糖 unfermentable sugar
非發酵性的 non-fermentable
非發酵食品 non-leavened acidity
非微生物的 amicrobic
非極性化合物 nonpolar compound
非極性胺基酸 nonpolar amino acid
非極性溶劑 nonpolar solvent
非酯化脂肪酸 nonesterified fatty acid
非碳酸飲料 noncarbonated beverage; still beverage; uncarbonated drink
非精製鹽 crude salt
非酵母發酵之麵糰 non-yeasted dough
非酵素〔性〕褐變 nonenzymatic browning
非酵素的 nonenzym(at)ic
非酸性水果 non-acid fruit
非酸性食品 non-acid food
非蔗糖 non-sucrose
非澱粉的 non-starchy
非糖物 non-sugar
非糞便細菌 non-fecal bacteria
非還原糖 non-reducing sugar
非離子〔型〕表面活性劑 non-ionic surfactant
非離子〔型〕界面活性劑 non-ionic surfactant
非離子〔型〕清潔劑 non-icnic detergent
非穩〔態〕流 unsteady flow
非糯米 non-glutinous rice; non-sticky rice; non-sweet rice; non-waxy rice
非酶性褐變 nonenzymatic browning
非酶的 nonenzym(at)ic
孢子 spore
孢子甘藍 thousand headed cabbage
孢子形成 sporulation
孢子膜 spore membrane
孢子囊 spore sac
孢桿桿菌 bacillus
孢囊 cyst
孢囊孢子 sporangiospore
孢囊柄 sporangiophore
孢囊梗 sporangiophore
泔水 swill

炔 alkine; alkyne
炖 braise; braising; braize
炖小牛肉 fricandeau; fricando
炖熟的 braised
炖濃汁肉丁 fricassee
甾族膽汁酸 stero-bile acid
甾酮 sterone
甾醇 sterol
芫茜 coriander
芫荽 coriander
芫荽籽 coriander seed
芫荽籽油 coriander oil
苄丁醚 benzyl butyl ether
苄基異丁香醚 benzyl isoeugenyl ether
苄基異丁基甲醇 benzyl isobutyl carbinol
苄醇 benzyl alcohol
肽 peptide
肽腱 peptide bond
肽聚醣 peptidoglycan
肽鏈 peptide chain
肽鏈內斷酶 endopeptidase
肽鏈端解酶 exopeptidase
肽酶 peptidase
牦牛 yak

九畫

亮胺酸 leucine
亮茜素藍 brilliant alizarin blue
亮黑(食用色素) brilliant black
亮黑BN(食用色素) brilliant black BN
亮滑米 coated mill rice
亮綠乳糖膽汁培養液 brilliant green lactose bile broth
亮藍FCF brilliant blue FCF
亮藍鋁色澱 brilliant blue FCF aluminium lake
便秘 astriction; constipation; obstipation
便秘性食物 stiparogenic food
便飯 potluck; bread and cheese; informal dinner; lunch; plain food; refection
便餐肉 luncheon meat
保久乳 long-keeping milk; long-life milk
保久的 long-keeping; long-life
保水力 water binding capacity; water holding capacity
保水性 liquid-binding power; moisture retention
保水劑 humectant; moisture-retaining agent
保加利亞〔發酵〕酸乳 Bulgarian (sour) milk
保加利亞乳〔酸〕桿菌 *Lactobacillus bulgaricus*
保存 conservation; preserving; preservation
保存性 storability
保存培養基 preservation medium
保存期 keeping time
保存期限 holding time; marketable life; shelflife; shelf life
保存菌種 stock culture
保存槽 holding tank
保色劑 colo(u)r retention agent; colo(u)r fixative
保姆 nurse
保持冷凍 keep frozen
保持原味 raciness
保持原味的 racy
保持時間 retention time
保持滅菌〔法〕 holding sterilization
保持體積 retention volume
保乾器 desiccator; exsiccator
保健食品 health food
保溫 incubation
保溫〔培養〕期 incubation period
保溫室 incubator house
保溫容器 insulated container
保溫瓶 thermos (bottle); thermos flask; vacuum bottle; vacuum flask
保溫貨車 heater car
保溫蒸盤 bain-marie
保溫槽 holding tank
保溫箱 incubator
保溫器 warmer
保溫檯 steam table
保養 maintenance

保濕 humectation
保濕力 moisture retention
保濕室 moist chamber
保濕盒 humidor
保濕劑 humectant; moisture-retaining agent
保鮮劑 freshness agent
保藏 conservation; cure; curing; preservation; preserve; preserving
保藏性 storability
保藏食品 preserved food
保藏期 keeping time; storage period
保藏期限 preservation limit
保藏溫度 storage temperature
保藏劑 conserving agent; preservative; preserving agent
保護包裝〔材料〕 protective wrapping
保護塗層 protective coating
保護膠體 protective colloid
保護劑 protecting agent
促進食慾 whet
促進劑 accelerator; promoter
促凝劑 procoagulant; setting accelerator
俄式茶壺炊具 samovar
俄勒岡鱘 Oregon sturgeon
俄國鱘 Russian sturgeon
冒泡 effervescence; ebullition
冒泡的 effervescent
冠帽機 hooding machine
削下的皮 paring
削皮 pare; paring; skinning
削皮刀 parer; paring knife; skinner
削皮器 parer
削皮機 paring machine; skinning machine
削尖器 sharpener
削掉的肉 fleshings
削減 cut back
前小腿腱 shin from foreshank
前肋肉 thick runner
前里脊肉 standing rib roast
前肢 forelimb
前肩肉 blade meat
前胃 proventriculus
前胸肉 brisket; brisket point end, deckle-off
前發酵 prefermentation; primary fermentation
前發酵槽 prefermenter; prefermentor
前發酵罐 prefermenter; prefermentor
前腱 shin from foreshank
前腰牛排 club steak
前腰脊肉 porterhouse steak; porterhouse; striploin; shortloin
前維生素A carotene; carotin
前腿 foreleg
前腿心 chuck eye roll
前腿肉 forehock
前腿肉製的火腿 shoulder ham
前澄清 predefecation
前膠原〔蛋白〕 precollagen
前驅物 precursor
前腒肉 shin from foreshank
刺槐樹膠 bassora gum
勉強維生 subsistence
勃氏笛鯛 red snapper
匍匐莖 runner
南瓜 calabazilla; cushaw; cucurbit; pumpkin; squash
南瓜子 pepo
南瓜子油 pompion oil
南瓜子醇 cucurbitol
南乳 red preserved beancurd
南非鈴羊 springbok
南美的發酵玉米粉食品 cassiri
南美番荔枝 cherimoya
南美蘭酒 aguardiente
南極蝦 euphausia; krill
南極磷蝦 antarctic krill

南歐朝鮮薊 cardoon
南薑 galangal
厚片 hunch; slab
厚牛陰柳 thick flank
厚皮菜 Swiss chard
厚肌原絲 thick filaments
厚板 slab
厚油酥甜餅 shortcake
厚的 thick
厚度 thickness
厚桂皮 cassia lignea
厚酥餅 shortbread cookie
厚塊 slab
厚腹脇肉 thick flank
厚層乾燥 slab drying
厚層瓊脂〔培養〕法 deep agar method
厚壁組織 sclerenchyma
厚環乳牛肝菌 larch bolete
咬封蓋 anchor cap; snap-on cap
咽〔喉〕 pharynx
咽喉 gullet; guttur; gorge; throat
咪唑 imidazole
咪唑丙胺酸 histidine
品系 strain
品味小組 taste panel
品味員 gastronome(r)
品紅 fuchsin; magenta
品茶 tea tasting
品茶員 tea taster
品茗 tea tasting
品茗員 tea taster
品茗會 tea cult
品酒 wine tasting
品酒用的酒杯 wine taster
品酒員 cooper
品酒師 wine taster
品脫 pint
品評小組 panel; taste panel
品嚐 degustation; tasting
品嚐小組 taste panel
品嚐者 taster
品嚐室 cupping room
品嚐試驗 cup test; taste test
品種 breed; strain; variety
品管 quality control (QC)
品質 quality
品質改良劑 quality improvement agent; quality booster
品質保證 quality assurance (QA); quality guarantee
品質規格 quality specification
品質評估 quality evaluation
品質管制 quality control (QC)
品質標準 quality specification; quality standard
品質檢查 quality inspection
品嚐 savor; savour; taste
哈密瓜 cantaloup(e); netted melon
哈爾瓦芝麻糖 halavah; halva(h)
垂肉 gill
垂肉 jowl
垂肉（喉部） dewlap
垂直長管蒸發器 long tube vertical evaporator
垂直攪拌鍋 crutcher
契安尼娜牛 chianina
契普拉塔豬肉香腸（西班牙） chipolate sausage
奎尼酸 quinic acid
奎寧 quinine
奎寧水 quinine water; tonic water
奎寧酸 quininic acid
奎藜籽 quinoa
威士忌〔酒〕 barley-bree; barley-broo; barley broth; whiskey; whisky
威士忌酒的玻璃杯 whisky tumbler
威士忌蜜飲料 Athole brose
威士忌蘇打 Scotch and soda

威士戚圖咖啡 caffe ristretto; espresso ristretto
威化餅 wafer
威化餅乾 gauffre
威爾斯芝士麵包 rarebit
威爾斯乾酪麵包 rarebit; Welsh rabbit; Welsh rarebit
室溫 room temperature (RT)
客棧 tavern
封口膠 compound; sealing compound
封口機 capper; closing machine; sealing machine
封套 wrapper
封瓶 capping
封瓶塞機 bottle corking machine; bottle capping machine
封袋機 bag sealer; bag sealing machine
封閉式濾乳機 close milk filter
封塞機 corking machine; corker
封蓋 can closing; capping
封蓋機 capper; capping machine; closing machine; hooding machine
封灌機 closing machine
封罐 can closing; sealing; seam; seaming
封罐上壓頭 seaming chuck
封罐不良 faulty closure
封罐滾輪 can seaming wheel
封罐機 can sealer; can sealing machine; can seamer; can seaming machine; double seamer; end seamer; lid seamer; sealing machine; seaming machine; seamer
封罐機的〔提升〕托板 lifter
封罐機軋頭 seaming chuck; seaming head
封罐機捲邊封頭 seaming head
屋外廚房 cookhouse
建議每日營養素攝取量 recommended daily nutrient allowance (RDNA)
建議每日營養攝取量 recommended daily allowance (RDA)
建議飲食攝取量 recommended dietary allowance
建議膳食供給量 recommended dietary allowance
建議攝取量 recommended intake
待宰家畜 butcher' s breast
待宰欄 waiting lairage
後〔期〕烘焙 after-bake
後代 offspring
後半部 hind
後肋肉 thin runner
後冷卻 after-cooling
後味〔覺〕 after-taste
後研磨 after grinding
後苦味 lingering bitterness
後乾燥 after-drying
後產品 afterproduct
後處理 after-treatment
後期（細胞分裂） anaphase
後發酵 after-fermentation; lager fermentation; post fermentation
後發酵〔作用〕 secondary fermentation
後發酵用酵母 secondary yeast
後腱 shin from hindshank
後腰脊肉 sirloin
後腰臀肉 rump
後漂白 post bleaching
後腿 hind shank
後腿牛排 rump steak
後腿肉 butt square cut; ham; leg cuts; thigh meat
後腿股肉 knuckle
後腿股肉排 knuckle steak
後腿腱子〔心〕 shin from hindshank
後澄清 post defecation
後澄清汁 post defecated juice

後熟 after-maturation; after-ripening; post maturation; post harvest ripening
後糖化〔作用〕 after-saccharification
後醒麵 final proof
後擠乳 after-milking
後臀 haunch
後軀 hind quarters
後腿肉 shin from hindshank
急火熬煮 high boiled
急冷 snap-chilling
急性毒性 acute toxicity
急性毒素 acute toxin
急速冷卻 quick chilling
急速冷凍乾燥 accelerated freeze-drying (AFD)
急速冷凍乾燥機 accelerated freeze dryer
急驟加熱 flash heating
急驟冷卻 flash cooling
急驟冷凍 flash freezing
急驟汽化 flash vapourization
急驟乾燥 flash drying
急驟脫氣 flash deaeration
急驟蒸發 flash evaporation
恢復自然 renaturation
恢復活力 rejuvenation; resurgence
恢復健康的 restorative
恆重 constant weight
恆溫的 thermostatic oven
恆溫器 thermostat
恆濕器 humidistat; hygrostat
扁小的食物 pattie; patty
扁平的餐具 flatware
扁平洋菇 flat mushroom
扁帆鰭魚 pacific sailfish
扁形南瓜 turban squash
扁豆 *Dolichos lablab*; haricot bean; haricot; lablab bean
扁油酸甘油酯 glyceryl behenate
扁桃 almond; amygdala
扁桃形結構 amygdala
扁桃腺 tonsil
扁桃酸酯 amygdalate
扁桃酸鹽 amygdalate
扁舵鰹 frigate mackerel
扁魚 flatfish; flounder
扁蒲 bottle gourd; calabash gourd; white flowered gourd
扁嘴旗魚 swordfish
扁頭鰹 bulletmackerel
扁麵條 linguine
扁櫻桃 Surinam cherry
扁罐 flat can; flat tin
挖〔果〕心刀 coring knife
挖核器 corer
按規格切割 precut
按部位分割 cut up
按量分割 cut up
按蓋 snap-on cap
拼盤 hors-doeuvre
拭子 swab
拭抹 swab
拭紙 tissue
持水力 water binding capacity; water holding capacity
持水性 liquid-binding power
拮抗〔作用〕 antagonism
拮抗物 antagonist
指天椒 tabasco pepper
指示劑 indicator
指示器 indicator
指南 manual
指套形通心粉 ditali
指標培養基 indicator medium
指標細菌 indicator bacteria
指標微生物 indicator organism
指環粉 anelli siciliani
括下的肉 fleshings
括板篩 drag screen

括約肌 sphincter
挑選 pick(ing)
施氏赤刀魚 Japanese bandfish
施食處 soup kitchen
施湯處 soup kitchen
施糞肥 dung
春大麥 spring barley
春小麥 spring wheat
春化處理 jarovization
春天 spring
春季 spring
春季無花果 profichi
春捲 spring roll
春黃菊 camomile
春黃菊油 chamomile oil; camomile oil
春綠 spring green
春鮭 pacific salmon
春糧 spring grain
星王〔蘋果〕 starking
星狀結晶〔易溶〕食鹽 dendritic salt
星貂鮫 dog shark
星雞魚 banded grunter
星鯊 tope
星蘋果 cainito; star apple
星鰻 common Japanese conger
柿子 dateplum; kaki; Japanese persimmon; persimmon
柿乾 dried persimmon
柿餅 persimmon cake
柿糕 persimmon cake
柿澀 persimmon astringent
柿澀單寧 shibuol
柿澀醇 shibuol
染色 dye; staining; tincture
染色劑 stainer; staining reagent
染色體 chromosome
染料 dye; dyestuff; stain
柱狀田頭菇 poplar field cap
柱層析法 column chromatography
柔化 ageing; aging
柔化的肉 aged meat
柔和風味 mild flavo(u)r
柔美棕 wine plam
柔弱香蘑（蕈類） tawny funnel cap
柔軟多汁的 pulpy
柔軟性 softness; tenderness
柔軟的 soft biscuit
柔軟麵包 crumby bread
柔韌性 pliability
柔韌性包裝 flexible package
柔嫩度 tenderness
架〔子〕 rack
架子 shelf
架醃〔法〕 shelf curing
枯草〔芽孢〕桿菌 *Bacillus subtilis*
枯草桿菌素 bacillin
枯草菌素 subtilin
枯茗 cumin; cummin
枯茗油 cumin oil
枯茗籽 cumin seed; cummin seed
枯茗酸 cuminic acid
枯茗醛 cumaldehyde; cumin aldehyde
枯萎 damp off; scorch; shrivel; wilt; wither
柯衣定 chrysoidine
柯林斯雞尾酒 collins
柄 stalk; stem
柑 orange
柑瓜 mango melon
柑果 aurantium; hesperidium
柑香酒 curacao
柑桔 orange
柑桔剝皮機 orange peeler; orange peeling machine
柑桔溫室 orangery
柑橘 hesperidium; mandarin(e); orange
柑橘分瓣機 orange segmenting machine
柑橘去皮機 orange peeler; orange peeling machine

柑橘白蘭地 citrus brandy
柑橘皮絲果凍 marmalade
柑橘皮類胡蘿蔔素 citranaxanthin
柑橘果皮 citrus peel
柑橘枝 citrus juice
柑橘香料 citrus flavo(u)ring
柑橘香精 citrus flavo(u)ring
柑橘酒 citrus wine
柑橘球蛋白 pomelin
柑橘渣 citrus pulp
柑橘溫室 orangery
柑橘糖蜜 citrus molasses
柑橘壓榨機 citro-press
柑橘類水果 citrus fruit
柑橘類植物 *Citrus*
柑橘苷 citrin(e)
柚 *Citrus grandis*
柚〔子〕 pomelo; pompelmous; pummelo; shaddock
柚皮油 shaddock oil
柚皮苷 naringin
柚皮苷酶 naringinase
柚橘 ugli
柚苷 naringin
柚苷酶 naringinase
查布利白葡萄酒 Chablis
查爾特勒士甜酒（法國） chartreuse
查羅萊斯牛（法國） charolais
查證 verification
枸杞 boxthorn; Chinese wolfberry; lycium; matrimony vine
枸櫞 citron; cedrat
柏木油 cedar wood oil
柏拉圖糖度計 Plato saccharometer
柏林香腸 Berliner sausage
柏根迪葡萄酒 Burgundy
柏根迪酵母 Burgundy yeast
柳丁 sweet orange
柳豆 pigeon pea
柳眼 rib eye roll
柳葉菜 willow herb
柳酸 salicylic acid
柳酸乙酯 ethyl salicylate; salicylic ether
柳酸甲酯 methyl salicylate
柳酸異丁酯 isobutyl propionate
柳酸異戊酯 isoamyl salicylate
柳酸苄酯 benzyl salicylate
柳橙 orange; sweet orange
柳橙汁 orange juice
柳薄荷 hyssop
柳蘭 kaporic; willow herb
柳蘭茶 iwan tea; kaporic
毒力 virulence
毒固醇 toxisterol
毒性 toxicity; virulence
毒性評估 toxicological evaluation
毒性試驗 toxicity test
毒物 intoxicant; poison; toxic substance
毒物學 toxicology
毒物學評估 toxicological evaluation
毒的 toxic
毒朗 virulence
毒素 poison; toxicant; toxin
毒素型食物中毒 food intoxication; toxin-type food poisoning
毒液 venom
毒理學 toxicology
毒蛋白質 toxoprotein
毒魚 poisonous fish
毒麥 darnel
毒傘蕈 amanita
毒菌 toadstool
毒菇 poisonous mushroom
毒蕈 toadstool
毒藥 poison; toxicant
氟 fluorine
氟中毒 fluorosis
氟化〔作用〕 fluorination

氟化物 fluoride
氟化處理 fluoridization
氟化鈉 sodium fluoride
泉 spring
泉水 fountain; spring water
洋刀豆 jack bean
洋艾〔草〕 wormwood
洋艾草油 wormwood oil
洋艾酒 absinth(e)
洋李 common plum; plum
洋杉油 cedar oil
洋芋 potato
洋芋片 potato chip; potato crisp
洋芫茜 parsley
洋芫荽 parsley
洋芫荽籽油 parsley seed oil
洋芫荽莖油 parsley herb oil
洋紅 carmine; fuchsin; magenta; nacarat
洋茉莉醛 piperonal
洋香菜 parsley
洋香藜 epazote
洋梨 Bartlett (pear)
洋傘菇 parasol mushroom
洋紫蘇 basil; sweet basil
洋菜 agar; agar-agar
洋菜二糖 agarobiose
洋菜平板 agar plate
洋菜平板〔培養〕法 agar plate method
洋菜平板菌數 agar plate count
洋菜穿刺培養 agar stab
洋菜凍 agar jelly
洋菜原料藻 agar weed; agarophyte
洋菜培養基 agar medium
洋菜培養試管 agar tube
洋菜斜面培養 agar slant; agar slope
洋菜硫〔酸〕糖 agaropectin
洋菜畫線培養 agar streak
洋菜凝膠 agar gel
洋菜糖 agarose; gelose
洋菜擴散法 agar diffusion method
洋菇 champignon; cultivated mushroom; mushroom
洋菇菌絲體 mushroom mycelium
洋菇菌種 mushroom spawn
洋菇糖 mushroom sugar
洋菇醬 mushroom paste
洋菇醬油 mushroom sauce; mushroom soy
洋蔥 onion; *Allium cepa*
洋蔥奶油調味醬 soubise sauce
洋蔥皮 onion skin
洋蔥油 onion oil
洋蔥頭 onion bulb
洋蔥鱗莖 onion bulb
洋槐 black locust
洋薑 Jerusalem artichoke
洋薊 artichoke; globe artichoke
洋蘑菇 meadow mushroom
洋鱸 ocean perch
流口水 drool; slobber
流化〔作用〕 fluidization
流化床 fluidized bed
流化床乾燥 fluid bed drying; fluidized bed drying
流化床乾燥機 fluid bed dryer; fluidized bed dryer
流化床造粒 fluidized bed granulation
流化床造粒機 fluidized bed granulator
流出 efflux
流出物 exudate; efflux
流出的 effluent
流出液 effluent
流行病學 epidemiology
流度 fluidity
流度計 fluidimeter
流動性 fluidity; flowability; fluidness; liquidity; runnability
流動相 mobile phase

流動層 fluidized bed
流動層乾燥 fluid bed drying; fluidized bed drying
流動層乾燥機 fluid bed dryer; fluidized bed dryer
流動層造粒 fluidized bed granulation
流動層造粒機 fluidized bed granulator
流動廚房 rolling kitchen; soup kitchen
流動餐館膳食 mobile catering
流動攤販 vender; vendor
流涎 slobber
流理槽 scullery
流速計 tachometer
流程圖 flow chart; flow diagram; process chart
流線形的 fusiform
流質食物 liquid food; spoon meal; slop
流質滋補食物 caudle
流質膳食 fluid diet; liquid diet
流變計 rheometer
流變儀 rheometer
流變學 rheology
流體 fluid; liquid
流體化 fluidization
流體化蒸餾〔法〕 fluidized distillation
流體食物 fluid food
流體薄膜 fluid film
津白 Chinese cabbage
津津有味 relish
津貼 subsidy
洗 wash
洗衣機 washing machine
洗果機 fruit washer
洗指碗 finger bowl
洗氣器 scrubber
洗淨糖 affination sugar
洗瓶機 bottle washing machine
洗脫 elution
洗脫液 eluant; eluent
洗提 elution
洗提液 eluant; eluent
洗煉 affination
洗煉糖 affination sugar
洗煉離心機 affination centrifuge
洗碗工人 dishwasher
洗碗水 dishwater
洗碗槽 scullery
洗碗機 dishwasher
洗腸器 casing flusher
洗腸機 slimer
洗滌 scour; wash; washing
洗滌者 washer
洗滌粉 washing powder
洗滌液 washing liquid
洗滌瓶 washing bottle
洗滌處 wash house; washery
洗滌間 wash house; washery
洗滌廢液 washings
洗滌槽 sink; washing tank
洗滌劑 cleaning agent; strippant; washer; washing agent
洗滌器 scrubber
洗滌機 scourer; washer; washing machine
洗滌鹼 washing soda
洗潔劑 abstergent
洗盤機 tray washer
洗糖 affination
洗糖離心機 affination centrifuge
洗罐機 can washer; can washing machine
活化劑 activator; activating agent
活火 direct fire
活底烤蛋糕箍 cake hoop
活底烤蛋糕模子 cake hoop
活性 activity
活性〔作用〕 activation; actification
活性表面 active surface
活性染色 vital staining

活性氧氣法 active oxygen method (AOM)
活性氧測定法 active oxygen method (AOM)
活性能 activation energy
活性乾酵母 active dry yeast
活性部位 active site
活性鈣 active calcium
活性漂白土 activated Fuller's earth
活性碳 absorbent carbon; activated carbon; activated charcoal; active carbon
活性碳過濾器 activated carbon purifier
活性酵母 activated yeast
活性劑 activator
活性黏土 activated clay
活性離子鈣 active calcium
活重 fresh weight; green weight; live weight
活栓 stopcock; tap
活畜重量 live weight
活動性 activity
活細胞 living cell; viable cell
活菌數 viable count
活塞式裝填機 piston filler
活酵母 living yeast
活體中 in vivo
活體染色 vital staining
派 pie
派皮 pie crust
派餡 pie filling
洛氏硬度 Rockwell hardness
洛西〔葡萄〕酒 rose wine
洛克福〔耳〕羊奶芝士 Roquefort (cheese)
洛克福〔耳〕羊乳乾酪 Roquefort (cheese)
洛根莓 loganberry
洛神茶 roselle drink
洛神葵 roselle
泄放閥 bleeder
炭 charcoal; char
炭化 charring
炭疽〔芽孢〕桿菌 *Bacillus anthracis*
炭疽病 anthrax
炭疽熱 anthrax
炭疽熱桿菌 *anthrax bacillus*
炭過濾 char filtration
炸 friture
炸〔馬鈴〕薯片 chip; potato chip; potato crisp
炸丸子 croquette
炸土豆片 potato chip; potato crisp
炸肉片 schnitzel
炸肉排 cutlet
炸肉餅 croquette
炸肉餡餅 rissole
炸豆腐泡 fried bean curd (puff); fried tofu
炸的 fried
炸春捲 spring roll
炸洋葱圈 onion ring
炸馬鈴薯片 crisp
炸馬鈴薯條 French fired potato
炸馬鈴薯絲 shoestring potato
炸馬鈴薯餅 potato cake
炸甜圈 doughnut
炸魚條 fried fish stick
炸魚糕 fried fishcake
炸豬腿排 pork leg schnitzels
炸薯片 fried potato chip
炸薯條 French fired potato; French fries; French fry
炸雞 fried chicken
炸麵圈 doughnut; ring doughnut
牲畜 livestock
牲畜飼料 stock feed
狩獵臘腸 jagdwurst
珊瑚狀猴頭菇 coral tooth fungus
珊瑚礁魚類 coral fish
玻璃 glass

玻璃杯 glass
玻璃紙 cellophane; glassine (paper)
玻璃瓶裝食品 glassed food
玻璃器皿 glassware
玲瓏籽 annatto
珍 gizzard
珍味 regale
珍珠 pearl
珍珠母 nacre; pearl
珍珠貝 pearl shell
珍珠岩 perlite
珍珠粒大麥 pearl barley
珍珠麥 pearl barley
珍珠筍 baby corn
珀爾梅散乾酪 Parmesan (cheese)
玳玳花油 dai-dai flower oil
玳玳葉油 petitgrain dai-dai oil
玳瑁 turtle
玳瑁殼 turtle shell; tortoise shell
畏食〔症〕 sitophobia
界面活性 interfacial activity
界面活性物質 surface active substance
界面活性劑 interfacial agent; surface active agent; surfactant
界面區 boundary zone
界面帶 boundary zone
界面張力 interfacial tension
界面層 boundary layer
疤痕 scar
癸二烯酸 decadienoic acid
癸二酸 sebacic acid
癸二酸二乙酯 diethyl sebacate
癸二酸二丁酯 dibutyl sebacate
癸酸 capric acid; decanoic acid
癸酸乙酯 ethyl decanoate
癸酸甘油酯 caprin
癸醇 decanol; decyl alcohol
癸醛 decanal
癸醛二甲縮醛 decanal dimethyl acetal
皇冠羊肋排烤肉 crown roast
皇帝豆 lima bean
盆 pot; tub
相 phase
相〔轉變〕圖 phase diagram
相互擴散 counter diffusion
相加作用 additive action
相平衡 phase equilibrium
相界面 phase interface
相差顯微鏡 phase contrast microscope
相對密度 relative density
相對濕度 relative humidity (RH)
相轉換 phase inversion
相變換 phase inversion
眉豆 catjang; hyacinth bean
砂浴 sand bath
砂粒感 sandiness
砂質 sandiness
砂糖 crystal sugar; granulated sugar; sand sugar
砂糖容器 sugar basin
砂糖塊 knob
砂糖精製器 cartridge clarifier
砂鍋 earthen casserole; earthen pot; marmite
砂鍋菜 casserole
砂鍋燉肉什錦 casserole dish
砂濾 sand filtering; sand leach
砂濾沉澱器 permutite precipitator
砂濾器 sand filter
砂礫 grit
砂礫濾器 gravel filter
砂囊 gizzard
研究 research
研杵 pestle
研細 porphyrization
研缽 mortar
研棒 pestle
研碎 trituration

研磨 grind; grinding milling; pestle mill
研磨成粉末 pulverize; pulverization
研磨器 grinder; triturator
研磨機 grinding mill; grinding machine; muller
砍甘蔗刀 machete
科技 technology
科萊鱈 coley
科涅克〔白蘭地〕酒 cognac
秋〔熟〕作物 autumn crop
秋刀魚 mackerel pike; pacific saury; saury; saury pike; saira
秋刀魚毒素 saurine
秋水仙鹼 colchicine
秋作 fall crop
秋姑魚 red mullet
秋季長肥 autumn fattening
秋季南瓜 autumn squash
秋葵 gumbo; okra; ochro
秋播作物 winter crop
秋鯡 autumn herring
穿孔 puncture; perforation
穿孔的 perforated
穿孔罐 pin hole can
穿成串的 speated
穿刺（用烤肉叉） spit
穿刺培養 stab culture
穿流 percolation
穿透 penetration
穿透性 permeability
穿透度計 penetrometer
突伸小側耳（蕈類） angel's wings
突厥酮 damascone
突嘴頜針魚 needlefish
突變 mutation
突變性 mutagenicity
突變株 mutant
突變種 mutant
突變體 mutant

籽 seed
籽晶 inoculating crystal
紅三葉草 red clover
紅大麻哈魚 Alaska redfish; blueback salmon; red salmon; redfish; sockeye (salmon)
紅小麥 red wheat
紅五爪甜蘋果 red delicious
紅仁 red kernel
紅毛丹〔果〕 rambutan
紅外〔線〕輻射解凍 infrared radiation thawing
紅外線 infrared (ray)
紅外線〔加熱〕乾燥 infrared dehydration; infrared desiccation; infrared drying
紅外線〔加熱〕乾燥器 infrared dryer
紅外線分光光度計 infrared spectrophotometer
紅外線水分分析儀 infrared moisture analyzer
紅外線水分計 infrared moisture meter
紅外線水分測定儀 infrared moisture meter
紅外線加熱 infrared heating
紅外線去皮〔法〕 infrared peeling
紅玉蘋果 Jonathan
紅皮小麥 red wheat
紅皮軟質小麥 red winter wheat
紅皮硬質小麥 red spring wheat
紅目鱸 barramunda; barramundi
紅米 red rice
紅肉 dark meat; red meat
紅肉垂 wattle
紅肉旗魚 striped marlin
紅肌肉 red muscle
紅色2G，單偶氮 red 2G, azogeramine
紅色肉 red meat
紅色油 red oil
紅色根菜 red vegetable

紅色素 red colo(u)rs
紅色高麗菜 red cabbage
紅色無角牛 red poll
紅色雄松雞 gorcock
紅色酵母 red yeast
紅血球 erythrocyte; red blood cell
紅血球素 erythrin
紅血球凝集素 haemagglutinin; hemagglutinin
紅衣蘿蔔 red radish
紅尾蝦 red tail prawn
紅李 apricot plum
紅杜林小麥 red durum wheat
紅杜蘭小麥 red durum wheat
紅角葵 roselle
紅豆 adsuki bean; adzuki bean; azuki bean; *Phaseolus angularis*; red bean
紅豆腐乳 red preserved beancurd
紅松雞 moorfowl
紅花 safflower
紅花〔菜〕豆 white Dutch runner bean
紅花油 cardy oil; carthamus oil; safflower oil
紅花籽油 safflower seed oil
紅花紅色素 carthamus red
紅花素 carthamin
紅花菜豆 multiflora bean; scarlet runner bean
紅花黃 safflower yellow
紅花黃〔色素〕 carthamus yellow
紅花覆盆子 salmonberry
紅長鰭鱈 red hake
紅帥蘋果 red delicious
紅柄金錢菇 red stemmed tough shank
紅苜蓿 meadow clover
紅核 red kernel
紅桑果 red mulberry
紅桑葚 red mulberry
紅海鱈 red hake
紅海鱸 red perch
紅茶 black tea; darjeeling
紅酒 red wine
紅馬頭魚 branquillos
紅豇豆 red cowpea
紅球甘藍色素 red cabbage pigment
紅甜蘋果 red delicious
紅眼魚 redeye; rudd
紅笛鯛 sea snapper
紅莓 cranberry
紅頂枝瑚菌（蕈類） pink tipped coral fungus
紅魚子 red caviar
紅棗 red date
紅紫葡萄皮的花色素 oenin
紅菜豆 red kidney bean
紅菜花豆 runner bean; string bean
紅菜頭 beet (root)
紅黃褶孔菌（蕈類） gilled bolete
紅菇 russule
紅葱頭 eschalot; shallot
紅椰菜 red cabbage
紅腹雅羅魚 redbellied dace; redbelly
紅腹鮭 torgoch
紅腹鯉魚 redbellied dace
紅萱 dwarf yellow daylily
紅葡萄酒 Cabernet (wine); claret; red wine; vin rouge
紅瑪莉酒 Bloody Mary
紅腐乳 red preserved beancurd
紅翁菜 gynura's deux couleurs
紅蜘蛛蟹 red crab
紅辣椒 chil(l)i; chile; hot red pepper; red pepper
紅辣椒〔粉〕 paprika
紅辣椒豬肉乾香腸 pepperoni (sausage)
紅酵母屬 *Rhodotorula*
紅鳳菜 gynura's deux couleurs
紅潮 red tide

紅糊精 erythrodextrin(e)
紅線米 red streak rice
紅蔗糖（含6—9%蔗糖漿） demerara sugar
紅蝦 red shrimp
紅豌豆 red pea
紅醋栗 garnet berry; red currant
紅醋菌 *Acetobacter roseum*
紅橙 blood orange
紅橘 *Citrus tangerina*; tangerine
紅橘油 tangerine oil
紅燒肉 stewed meat
紅燒的 braised; teriyaki
紅燒鳥肉 salmi
紅糖 black sugar; brown sugar; dark crystal; muscovado (sugar)
紅頭菜 chard
紅頭鮭 char
紅穀粒 red kernel
紅芸豆 red kidney bean
紅擬庸鰈 flathead flounder
紅薯 batata
紅螺 oyster drill
紅鮪 silver snapper
紅鮭〔魚〕 redfish; red salmon; blueback salmon; sockeye (salmon)
紅點石斑魚 garrupa
紅點鮭 charr; torgoch
紅鮨 grouper
紅雞油菌（蕈類） cinnabar chanterelle
紅蟳 mud crab
紅羅非魚 red tilapia
紅蟹 mask crab
紅麴 anka; red rice starter
紅麴〔霉〕色素 monascus colo(u)r
紅麴米 monascus colo(u)r; red koji rice
紅麴紅素 monascorubin
紅麴紅胺 monascorubramine
紅麴紫 monascorubramine
紅麴菌 monascus
紅麴黃素 ankaflavin
紅麴霉 monascus
紅鯔 red mullet
紅藻 laver; nori; pink laver; porphyra; red algae; red laver; red seaweed
紅藻膠 carrageenan; carrageenin; carragheenin; Danish agar; furcellaran
紅藻澱粉 floridian starch
紅麵包黴 neurospora
紅鰈 rusty dab
紅櫻桃 red cherry
紅鰭淡水魚 redfin
紅鱈 cusk
紅蘿蔔 carrot
紅變 pinking
紅黴素 erythromycin
紅鱒 rainbow trout; specked trout
約旦杏仁 Jordan almond
約克夏豬 Yorkshire
約翰內斯堡白葡萄酒 Johannisberger
約翰尼餅 Johnny cake
美人蛤 ocean quahog
美人蕉 canna
美人蕉澱粉 touslesmois
美多克紅葡萄酒 Medoc
美式咖啡 americano; caffe americano
美杜克紅葡萄酒 Medoc
美味 delicate flavo(u)r; tastiness
美味牛肝菌 penny bun
美味性 savoriness; savouriness
美味的 dulcet; dainty; palatable; savory; savoury; tasteful
美味的食物 ambrosia
美味食品 delicacy; delicate food
美拉德反應 aminocarbonyl reaction; carbonyl-amine reaction; Maillard reaction
美洲山核桃 pecan
美洲白鮭 tullibee
美洲西鯡 American shad; shad

美洲沙瑙魚 pacific sardine; pacific pilchard
美洲牡蠣 American oyster
美洲防風 parsnip
美洲狗魚 banded pickerel; redfin pickerel
美洲狐臭葡萄 fox grape
美洲花楸 rowan
美洲青魚 bluefish
美洲紅點鮭 brook trout
美洲胡瓜魚 American smalt; sparling
美洲狼鱸 white perch
美洲茶 Jersey tea
美洲梧桐樹 sycamore
美洲野牛 buffalo
美洲黑櫟 quercitron
美洲楊梅 gale
美洲酸果蔓果實 American cranberry
美洲箭齒鰈 arrow-toothed flounder
美洲薄荷 horsemint
美洲薊 American artichoke
美洲鯡 blue herring
美洲鰻鱺 American eel
美洲鱘 white sturgeon
美紅菇 yellow staining russule
美食 fleshpots; goody
美食主義 epicurism; gourmandism
美食法 gastronomy
美食家 epicure; gastronome(r); gastronomist; gourmet
美食學 gastronomy
美食〔學〕家 gastrologer
美食餐館 gastronomic emporium
美國大豆協會 American Soybean Association (ASA)
美國公認分析化學〔師協〕會 Association of Official Analytical Chemists (AOAC)
美國生物分析委員會 American Board of Bioanalysis (ABBA)
美國冷凍食品研究所 American Frozen Food Institute (AFFI)
美國乳品工業協會 American Dairy Association (ADA)
美國乳粉學會 American Dry Milk Institute (ADMI)
美國乳業科學協會 American Dairy Science Association (ADSA)
美國油脂化學會 American Oil Chemists Society (AOCS)
美國食用化學法典 Food Chemicals Codex (FCC)
美國食品科技學會 Institute of Food Technologists (IFT)
美國香辛料協會 American Spices Tvade Association (ASTA)
美國專利 United Stated Patent (USP)
美國菌種保存中心 American Type Culture Collection (ATCC)
美國飲食學會 American Dietetic Association
美國黑莓 black raspberry; blackcap
美國微生物學會 American Association of Microbiology (AAM); American Society for Microbiology (ASM)
美國農業部 United States Department of Agriculture (USDA)
美國標準菌庫 American Type Culture Collection (ATCC)
美國穀物化學師協會 American Association of Cereal Chemists (AACC)
美國穀類化學會 American Association of Cereal Chemists (AACC)
美國磚形芝士 brick cheese
美國磚形乾酪 brick cheese
美國膳食營養學會 American Dietetic Association
美國應用營養研究院 American Academy of Applied Nutrition (AANN)

美國營養學會 American Institute of Nutrition (AIN)
美國藥典 United States Pharmacopoeia (USP)
美國蘋果白蘭地〔酒〕 apple jack; calvados
美國罐頭協會 National Canners Association (NCA)
美國釀造化學師學會 American Society of Brewing Chemists (ASBC)
美藍 methylene blue
耐久存放的食品 lasting food
耐久性 durability; water resistance
耐水的 waterproof
耐火材料 refractory material
耐火的 refractory
耐冬寒性 winter hardiness
耐光性 photostability
耐冷性 cold endurance
耐冷試驗 cold test
耐油包裝〔材料〕 greaseproof wrapping
耐油的 greaseproof; oil-proof; oil-resistant
耐油紙 pergament
耐凍油 winter oil
耐氧細菌 aerotolerant bacteria
耐氧厭氧菌 aerotolerant anaerobe
耐真菌性 fungus resistance
耐寒性 cold endurance; cold resistance; freezing resistance
耐寒蔬菜 hardy vegetable
耐揉性 crumpling resistance
耐損 wearproof
耐溫性 temperature tolerance
耐滲壓細菌 osmophilic bacteria
耐滲壓微生物 osmophile
耐滲壓酵母 osmophilic yeast
耐熔的 refractory
耐酸性 acid endurance; acid resistance
耐酸的 acid-proof; acid resistant; aciduric
耐酸染色 acid-fast stain
耐酸細菌 acid-fast bacteria; aciduric bacteria
耐熱〔度〕 heat resistance
耐熱材料 refractory material
耐熱性 thermal tolerance; thermostability
耐熱的 heat proof; heat resistant; heat stable; heat-fast; heat tolerant; refractory; thermostable; thermoduric
耐熱培養〔物〕 thermoduric culture
耐熱細菌 heat resistant bacteria; thermoduric bacteria; thermotolerant bacteria
耐熱器皿 flame ware
耐皺性 crumpling resistance
耐糖性 sugar tolerance
耐儲藏的 long-keeping; long-life
耐壓性 pressure resistant
耐壓容器 pressure vessel
耐壓蒸餾器 pressurized still
耐藥量 tolerance
耐嚼的 chewy
耐鹽性 salt tolerance
耐鹽細菌 halotolerant bacteria; salt tolerant bacteria
耐鹽酵母 salt tolerant yeast
耐鹽黴菌 halotolerant mo(u)ld
耶仆蘭胡椒 Jaborandi pepper
胖罐 puffer
胚乳 albumen; endosperm
胚芽 embryo; germ; pangen; plumule
胚芽分離機 germ separator
胚芽米 germ rice
胚芽乳〔酸〕桿菌 *Lactobacillus plantarum*
胚芽油 germ oil
胚芽粉 germ meal
胚芽粉麵包 germ bread
胚芽麵包 hovis; wheat-germ bread
胚芽麵粉烘製品 hovis
胚胎 embryo
胚胎營養 embryotrophy

胚根鞘 coleorhiza
胚層 germ layer
胚囊 embryo sac
胃 gorge; maw; stomach; ventriculus; venter
胃〔病〕學 gastrology
胃口 stomach; tooth
胃炎 gastritis
胃的 gastric; peptic
胃病 gastrosis
胃液 gastric juice
胃蛋白酶 pepsin; pepsinum
胃蛋白酶的 peptic
胃蛋白酶原 pepsinogen
胃痛 stomachache
胃腸炎 gastroenteritis
胃腸的 gastrointestinal
胃腸病 gastroenteropathy
胃腸道 gastrointestinal tract
胃腸道脂肪酶 gastrointestinal lipase
胃腸激素 enterogastrone
胃酸 gastric acid
胃酸缺乏症 achlorhydria; anacidity
胃酸過多症 hyperchlorhydria
胃藥 digestant
背肋部 backribs
背肌 dorsal muscle
背面 backside
背脂 back fat
背脊肉 back pork; chine (bone); loin; short loin
背開冷燻鯡 newcastle kipper
背腰肉 loin
背腰肉火腿 loin ham
背膘 back fat
背膘去皮機 back fat skinner
背點棘赤刀魚 band fish
背鰭 back fin; dorsal fin
胡瓜 cucumber
胡瓜魚 smelt
胡桃 juglans; pecan; walnut
胡桃〔仁〕糊 walnut cream
胡桃巧克力小方餅 brownie
胡桃夾 cracker
胡桃油 walnut oil
胡桃蛋糕 walnut cake
胡桃鉗 nutcracker
胡麻 benne; til
胡荽 coriander
胡荽籽 coriander seed
胡荽籽油 coriander oil
胡椒 pepper; *Piper nigrum*
胡椒〔味〕的 peppery
胡椒〔粉〕瓶 pepper pot; pepperbox
胡椒子 peppercorn
胡椒油 pepper oil
胡椒粉 pepper
胡椒基丁醚 piperonyl butcxide
胡椒粒 peppercorn
胡椒啶 piperidine
胡椒酸 piperic acid
胡椒醇 piperonyl alcohol
胡椒薄荷 peppermint
胡椒薄荷油 peppermint oil
胡椒醛 piperonal
胡椒鹼 piperine
胡蜂 wasp
胡蒜 giant garlic; sand leek; Spanish garlic
胡薄荷油 pennyroyal oil
胡蘿蔔 carrot
胡蘿蔔籽油 carrot seed oil
胡蘿蔔素 carotene; carotin
胡蘿蔔素血症 xanthemia
胡蘿蔔素醛 carotenal
胡蘿蔔醇 carotenol; carotol; xanthophyll
胛心肉 shoulder picnic
胎生魚 live-bearing fish
胎球蛋白 fetuin

胞〔嘧啶核〕苷 cytidine
胞〔漿溶〕解 endolysis
胞內的 endocellular; intracellular
胞內液 intracellular fluid
胞內漿 intracytoplasm
胞內酶 endocellular enzyme; endoenzyme; intracellular enzyme
胞外脂酶 lyolipase
胞外酵素 extracellular enzyme; exocellular enzyme
胞外酶 extracellular enzyme; exoenzyme; lyoenzyme
胞芽 gemma
胞解素 endolysin
胞嘧啶 cytosine
胞漿素 plasmin
胞糖脂 cytolipin
胞囊 cyst
致死曲線 destruction curve
致死的 lethal
致死率 lethality
致死劑量 lethal dose (LD)
致冷 refrigeration
致命性 lethality
致命的 lethal
致病生物 pathogenic organism
致病性 pathogenicity
致病微生物 pathogenic microorganism
致腸病的 enteropathogenic
致醉物 intoxicant
致醉酒類 intoxicating beverage
致癌 cancerogenesis
致癌作用 carcinogenesis
致癌性 carcinogenicity
致癌劑 carcinogen
范氏植物牛油 vanaspati ghee
茅台酒 Mou Tai (Chiew)
苛性鈉 caustic soda
苛性鹼液 caustic lye
苦〔薄荷〕糖果 horehound
苦山核桃 pignut
苦木〔科植物〕 quassia
苦木抽提物 bitter wood extract; quassis extract
苦瓜 balsam pear; bitter gourd; bitter melon
苦艾 absinthium; wormwood
苦艾油 absinthic oil; wormwood oil
苦艾素 absinthin
苦艾酒 absinth(e); vermouth
苦艾腦 absinthol
苦艾醇 absinthol
苦杏仁 bitter almond
苦杏仁油 almond oil, bitter; bitter almond oil
苦杏仁酒 ratafia
苦杏仁酸酯 amygdalate
苦杏仁酸鹽 amygdalate
苦杏仁糖 amygdalose
苦杏仁苷 amygdalin
苦杏素 amarine
苦杏酸 amaric acid
苦杏質 amaroid
苦味 acerbity; bitterness; bitter
苦味物 bitters
苦味酒 bitters; quassia cup
苦味酸 bitter acid; picric acid
苦味質 amaroid
苦的 bitter
苦青草 bitter herb
苦哇啤酒 kvass
苦苣 endive
苦香〔樹〕皮 sweetwood bark
苦根 bitter cassava
苦根馬齒莧 bitterroot
苦草 bitter herb; eelgrass; wild celery
苦配巴油 copaiba oil
苦參子 khosam

苦啤酒 bitters
苦滷 bittern
苦精 angostura bitters
苦橙 bitter orange
苦橙油 bitter orange oil; neroli bigarade oil
苦橙苷 aurantiamarin
苦樹薯 bitter cassava
苦蕎麥 duck wheat; tartarian buckwheat; tartary buckwheat
苦薄荷 horehound
苦櫻桃 bitter cherry
苦肽 bitter peptide
茄子 aubergine; begoon; brinja(u)l; eggplant; garden-egg; guinea squash; mad apple; vegetable egg
茄酸 solanic acid
茄靈 solanine
茄鹼 solanine
若丹明 rhodamine
茂谷桔橙 muscott honey orange
茉莉 jasmine
茉莉〔花〕茶 jasmine tea
茉莉〔精〕油 jasmine oil
茉莉花 Arabian jasmine
茉莉浸膏 jasmine concrete
茉莉淨油 jasmine absolute
茉莉酮 jasmone
茉莉酯 jasmonyl
茉莉醛 jasmine aldehyde
苗 sprout
苗圃 nursery
苗條膳食 slender diet
苗種 offspring
英式的茶點小蛋糕 scone
英式麥餅 muffin
英式鬆餅 English muffin
英國紅人小麥 red man wheat
英國黃啤酒 pale ale (beer)
英國黑啤酒 porter
英國熱量單位 British Thermal Unit (BTU)
茁霉多糖 pullulan; pullulane
苜蓿 alfalfa; clover; lucern(e); medic
苜蓿芽 alfalfa sprout
苔 moss
苔色素 orchil; orselle
苔紅素 orcein
苔桃 cranberry; foxberry
苔素 zearin
苔蘚 lichen
苔蘚紅素 orchil; orselle
苔蘚食品 moss
苞片 bract
苞葉 bract; husk
苯 benzene
苯〔基〕丙胺酸 phenylalanine
苯乙烯 styrene; styrolene
苯乙酮 acetophenone; acetyl benzene
苯乙酸 phenylacetic acid
苯乙酸乙酯 ethyl phenylacetate
苯乙酸丁酯 butyl phenylacetate
苯乙酸二甲縮醛 phenylacetaldehyde dimethyl acetal
苯乙酸甲酯 methyl phenylacetate
苯乙酸苯乙酯 phenethyl phenylacetate
苯乙酸香葉酯 geranyl phenylacetate
苯乙酸烯丙酯 allyl phenylacetate
苯乙酸異丁酯 isobutyl phenylacetate
苯乙酸異戊酯 isoamyl phenylacetate
苯乙酸對甲酚酯 *p*-cresyl phenylacetate
苯乙酸苄酯 benzyl phenyl acetate
苯乙醇 phenethyl alcohol; styracitol(e)
苯乙醛 phenylacetaldehyde
苯二酸 phthalic acid
苯丙烯醇 cinnamyl alcohol
苯丙醇 phenylpropyl alcohol
苯代環氧丙酸乙酯 ethyl phenylglycidate

苯甲酸 benzoic acid
苯甲酸乙酯 ethyl benzoate
苯甲酸丙酯 propyl benzoate
苯甲酸甲酯 methyl benzoate
苯甲酸芳樟酯 linalyl benzoate
苯甲酸香葉酯 geranyl benzoate
苯甲酸異戊酯 isoamyl benzoate
苯甲酸鈣 calcium benzoate
苯甲酸鈉 sodium benzoate
苯甲酸鉀 potassium benzoate
苯甲酸酯 benzoate
苯甲酸鹽 benzoate
苯甲酸芐酯 benzyl benzoate
苯甲醇 benzyl alcohol
苯甲醚 anisole
苯甲醛 benzaldehyde; benzoic aldehyde
苯甲醛丙二醇縮醛 benzaldehyde propylene glycol acetal
苯甲醛甘油縮醛 benzaldehyde glyceryl acetal
苯甲酰化 benzoylation
苯甲酰甘胺酸 benzoylglycine; hippuric acid
苯并噻唑 benzothiazole
苯咪唑 thiabendazole
苯胺染料 aniline dye
苯胺紅 aniline red
苯胺黑 nigrosine
苯基縮水甘油酸乙酯 ethyl phenylglycidate
苯酚 carbolic acid; phenol
苯粽 phenylhydrazone
苯酮尿症 phenylketonuria (PKU)
苯醌 quinone
苷化 glycosidation
虹吸 siphon; syphon
虹吸管 siphon; syphon
虹鱒 rainbow trout
虹鱸 rainbow perch
衍生物 derivative
衍生脂質 derived lipid
衍生澱粉 derivatized starch
衍酪蛋白 paracasein
衍膽酸 apocholic acid
計時器 timer
計量杯 jigger
計量送料機 batcher
計量配料臺 batching table
計量裝填機 measuring filler
軍中小販 sutler
軍用口糧 rations
軍用食品 military foodstuff
軍用飲盒 messtin
軍用麵包 army bread; rations bread
軍需品 stores
軍糧 combat food; military foodstuff; operational ration
軍糧船 victual(l)er
郁金 turmeric
郁金粉 turmeric yellow
酊劑 tincture
重〔鹽〕醃 heavy salting
重牙鯛 pinfish
重金屬 heavy metal
重要控制點 critical control point (CCP)
重組肉類 restructured meat
重組乳 recombined milk
重蛋糕 high ratio cake
重煮 recoction
重量 weight
重量不足 underweight
重量分析〔法〕 gravimetric analysis
重新包裝 repack
重新燒熱的 warmed-over
重新曬乾 redry(ing)
重碳酸鹽 bicarbonate
重蒸煮 recooking
重鉻酸鹽 dichromate
重瓣胃 omasum

重鹽漬的 heavily salted
重鹽醃 hard salting
重鹽醃的 heavily salted
限制因素 limiting factor
限制流質膳食 restricted liquid diet
限制胺基酸 limiting amino acid
限制營養素 limiting nutrient
限量供應 ration
限鈉膳食 sodium restricted diet
降二氫癒瘡木酸 conidendrin
降速乾燥 falling rate drying
降解〔作用〕 degradation
降解產物 degradation product
降膜式蒸發器 falling film evaporator
降膜式蒸餾器 falling film still
革生瓦賽菌（蕈類） meadow puffball
革魨 leather jacket
韭菜 yellow chive
韭菜花 *Allium tuberosum*
音波〔振動〕處理 sonication
音波〔振動〕器 sonicator
頁 leaf
風〔動〕磨 pneumatic mill
風力分級 pneumatic classification
風力分級機 pneumatic classifier
風力分選機 pneumatic separator; wind sifter
風化 efflorescence
風穴儲藏 wind cave storage
風車磨坊 wind mill
風味 flavo(u)r; relish; sapor; sapour; savor; savour
風味不良的 mawkish
風味不純 unclean flavo(u)r
風味化合物 flavo(u)r compound
風味增效劑 flavo(u)r potentiator
風味增強劑 flavo(u)r potentiator; flavo(u)r enhancer; intensifier
風味變壞 off flavo(u)r
風信海扇〔貝〕 weathervane scallop
風乾 air dehydration; air drying; atmospheric drying; loft drying; open-air drying
風乾牛肉條 jerked beef
風乾肉 air dried meat
風乾肉條 jerky; jerked meat
風乾的 air dried; wind dried
風乾葡萄乾 cured raisin
風速計 air meter; airometer
風落果 windfall
風選 air classification; winnow
風選機 air separator; air classifier; pneumatic separator; winnower; winnowing machine
飛沫 spray
飛魚 flying fish; skipjack
飛塵 fly ash
飛機餐 inflight meal
食土癖 geophagy
食火雞 cassowary
食用大紅 fast red E
食用大黃 garden rhubarb, pieplant; tart rhubarb
食用牛 stocker cattle
食用牛脂 oleo stock
食用成熟度 eating maturity
食用羊脂 oleo stock
食用色素 edible pigment; food colo(u)r; food dye
食用作物 food crops
食用油 edible oil
食用油渣 greaves
食用的牲畜肺臟 lights
食用的屠體肺臟 lights
食用前需煮熟 cook-before-eating
食用紅色六號 cochineal red A
食用香料 flavo(u)ring substance; food flavo(u)ring; flavo(u)ring

食用香草 champignon
食用香精 flavo(u)ring essence; food flavo(u)r
食用真菌 edible fungus
食用脂肪 edible fat
食用草本植物 edible greens
食用混合油 edible blended oil
食用莧 edible amaranth
食用植物油 salad oil
食用紫紅藻 dulse
食用黃色4號 tartrazine
食用黃色4號鋁色澱 tartrazine aluminum lake
食用腸衣 edible casing
食用萱草 common yellow daylily
食用綠S green S
食用綠色3號 fast green FCF
食用綠色3號鋁色澱 fast green FCF aluminum lake
食用酵母 food yeast; torula (yeast)
食用蝸牛 escargot
食用豬 porker
食用糖 table sugar
食用薄膜 edible film
食用薊 cardoon
食用鮮肉 flesh
食用藍色1號 brilliant blue FCF
食用藍色1號色澱 brilliant blue FCF lake
食用藍色2號 indigo carmine
食用鹽 edible salt; sodium chloride
食肉中毒 kreotoxism
食肉的 zoophagous
食肉動物 zoophagan
食果主義者 fruitarian
食物 aliment; comestibles; eatable; fare; food; foodstuff; ingesta; muckamuck; nourishment; nutriment; pabulum; scran; sustenance; toke; viand; victual; vivers
食物〔媒介〕感染 food-borne infection
食物一份 helping; helping
食物上加裝飾 topping
食物中毒 alimentary toxicosis; food poisoning; food intoxication; sitotoxism
食物切碎機 food chopper
食物代用品 food equivalent; substitution food
食物均衡 food balance
食物供應者 victual(l)er
食物卷 food stamp
食物泥 puree
食物的 alimental; alimentary
食物室 buttery
食物毒素 sitotoxin
食物毒素型中毒 food-borne intoxication
食物要素 dietary factor
食物屑 scran
食物疾病 food-borne disease
食物產品 food product
食物循環 food cycle
食物殘渣 slop
食物貯藏室 garde-manger; pantry; still room; spense
食物傳染性中毒 food-borne intoxication
食物感染 food infection
食物過敏 food allergy
食物過量 surfeit
食物網 food web
食物製品 food product
食物學 bromatology
食物療法 alimentotherapy; regimen
食物櫥 safe
食物鏈 food chain
食物攝取量 food intake
食者 trencherman
食品 eatable; food; provisions; scoff; vivers
食品工程〔學〕 food engineering

食品工業 food industry
食品工藝師 food technologist
食品工藝學 food technology
食品工藝學家學會 Institute of Food Technologists (IFT)
食品分析 food analysis
食品化學 food chemistry
食品化學家 food chemist
食品及藥物管理局(美國) Food and Drug Adminstration (FDA)
食品及藥物管理規則 FDA regulation
食品加工 food processing
食品加工助劑 food processing aids
食品加工廠 packing house; packing plant
食品包裝 food packaging
食品用化學品藥典 Codex of Food Chemicals
食品安全 food safety
食品成分 food composition; food ingredient
食品污染物 food contaminant
食品色素 food colo(u)r
食品技術學 food technology
食品技術學家 food technologist
食品防腐劑 food preservative
食品供應 food supply
食品法規 Codex Alimentarius; food law
食品的 pabular
食品保藏 food preservation
食品室 larder; pantry
食品毒物學 food toxicology
食品科技學會 Institute of Food Technologists (IFT)
食品科學 food science
食品科學家 food scientist
食品紅14號 erythrosine
食品紅7號 ponceau 4R
食品紅色7號 erythrosine BS
食品紅色7號鋁色澱 erythrosine aluminium lake
食品配料成分 food ingredient
食品強化 food enrichment; food fortification
食品添加物 food additive
食品添加物法規 Food Chemicals Codex (FCC)
食品添加劑 food additive
食品添加劑及污染物委員會 Food Additives and Contaminants Committee (FACC)
食品添加劑法規委員會 Codex Committee on Food Additive
食品添加劑規范 Food Additive Regulation (FAR)
食品添加劑聯合專家委員會 Joint FAO/WHO Expert Committee on Food and Additives (JECFA)
食品貯藏室 buttery
食品微生物學 food microbiology
食品照射 food irradiation
食品摻假 food adulteration
食品腐敗 food spoilage
食品製備 food preparation
食品輔料 accessory food
食品價 food value
食品增補劑 food supplement
食品標準法規 Food Standards Code
食品標誌 food labelling
食品標籤 food labelling; food labels
食品衛生 food sanitation
食品衛生〔學〕 food hygiene
食品衛生法規委員會 Codex Committee on Food Hygiene
食品學 bromatology
食品膳食學 bromatology
食品檢驗 food inspection

食品營養強化 food enrichment; food fortification
食品營養學 sitology
食品雜貨 grocery
食品雜貨自助售貨店 groceteria
食品雜貨店 grocery store; grocery shop; grocery
食品櫥 ambry; larder; mess rack; meat safe
食品藥物管理局(美國) Food and Drug Adminstration (FDA)
食品鑑賞家 gourmand
食品罐頭 canned food; canned goods; tinned food; tinned goods
食客 eater
食真菌的 fungivorous
食草的 herbivorous
食草蟹 shore crab
食堂 dining hall; eatery; messhall; messhouse; refectory
食蛋小匙 egg spoon
食量大的人 good trencherman
食道 gullet; oesophagus; weasand
食管 weasand
食慾 orexis; stomach
食慾不振 depraved appetite; inappetence
食慾抑制劑 anorectic agent
食慾缺乏症 anorexia
食慾減退劑 anorexigenic drug
食慾過盛 bulimia; hyperorexia
食糖 sugar
食療 medicated diet
食療法 dietary cure; dietetics
食療的 dietetic
食療食物 dietetic food
食療飲料 dietetic beverage
食糜 chyme; chymus
食糧 food; pabulum
食蟲植物 carnivore
食蟲蝦製的菜餚 scampi
食櫥 cupboard
食譜 cookbook; recipe
食蘭粉 salep
食鹽 common salt; edible salt; salt; sodium chloride; table salt
食鹽水 brine
香〔辛〕味 spicery
香口膠 chewing gum
香子蘭 vanilla
香子蘭豆 vanilla bean
香子蘭香精 vanilla flavo(u)r
香丹參 clary sage
香丹參油 clary oil; clary sage oil
香化酒 aromatized wine
香木橼 citron
香水 perfume
香水樹 ilangilang; ylangylang
香水樹油 cananga oil
香水薄荷 balm-mint
香片 jasmine tea
香瓜 cantaloup(e); melon; muskmelon
香艾菊 estragon; tarragon
香艾菊油 estragon oil; tarragon oil
香旱芹油 ajowan oil
香旱芹酒 kummel
香豆 tonka bean
香豆素 coumarin
香豆酮 coumarone
香豆酸 coumaric acid
香辛小麵包 wig
香辛味強烈的 high seasoned
香辛性 spiciness
香辛的 spiced
香辛料 spicery; spice
香辛料包 bouquet garni
香辛料店 spicery
香辛料油樹脂 spice oleoresin
香辛料架 spice rack

香辛料調味 bespicing
香辛料罐 spice jar
香辛乾酪 spiced cheese
香辛醃漬 spice curing
香味 aroma; flavo(u)r; scent
香味化合物 flavo(u)r compound
香味輔佐料 flavo(u)ring adjunct
香杯傘（蕈類） blue green funnel cap
香油 balm; balsam; benne oil; gingelly oil; jinjili oil; sesame oil
香芹 parsley
香芹油 caraway oil
香芹籽 caraway seed
香芹籽油 parsley seed oil
香芹莖油 parsley herb oil
香芹酚 carvacrol
香芹菜小麵包 wig
香芹酮 carvone
香芹醇 carveol
香附〔子〕油 cyperus oil
香附子 nut grass
香紅茶 pekoe (tea)
香茅 citronella (grass); lemon grass
香茅油 citronella oil; lemon grass oil
香茅草油 lemongrass oil
香茅醇 citronellol; rhodinol
香茅醛 citronellal
香苦木油 cascarilla bark oil; cascarilla oil
香料 aromatics; flavo(u)ring material; perfume; spice; scent
香料作物 perfumery crop
香料店 perfumery
香料嗅瓶 vinaigrette
香料廠 perfumery
香根油 vetiver oil
香桃木油 myrtol
香氣 aroma; fragrance; perfume
香氣足 flavo(u)rsome
香脂 balsam; balm
香脂木 balsam
香荊芥酚 carvacrol
香草 herb; vanilla
香草〔砂〕糖 vanilla sugar
香草冰淇淋 vanilla ice-cream
香草豆 vanilla bean
香草油香精 vanilla oily flavo(u)r
香草的 herbal
香草香精 vanilla flavo(u)r
香草浸液 vanilla extract
香草紮 bouquet garni
香草萃 vanilla extract
香草酸 vanillic acid
香草醛 vanillin
香液 balsam; balm
香甜熱酒 mulled wine
香甜熱飲料 mull
香莢蘭豆 vanilla bean
香莢蘭豆酊 vanilla bean tincture
香莢蘭豆浸膏 vanilla bean concrete; vanilla extract
香莓 thimbleberry
香袋 scent bag
香魚 argentine; ayu (fish); smelt; sweetfish
香堇 florist violet; garden violet; sweet violet
香焦串 stem
香紫蘇 clary sage
香紫蘇油 clary (sage) oil; clary oil; clary sage oil
香菜 coriander
香菜心 asparagus lettuce
香菜芹 turnip-rooted chervil
香菇 *Lentinula edodes*; pasania fungus; shiitake
香菇多糖 lentinan
香菇香精 lenthionine
香酢漿草 Jamaica sorrel

香楊梅 Scotch gale; sweet gale
香豌豆 sweet pea
香腸 sausage; wurst
香腸內膠凍塊 jelly pocket
香腸內凝膠塊 jelly pocket
香腸因數(瘦肉的比率) sausage factor
香腸肉餡 sausage meat
香腸段 link
香腸捲 sausage roll
香腸紮節 link; linking
香腸結紮機 linger
香腸腸衣 sausage casing
香腸製品 sausage goods
香腸餡 sausage dough
香腸餡料 sausage mix(ture)
香腸麵包 sausage roll
香腸灌紮機 filler linker
香葉 bay leaf; laurel leaf
香葉油 geranium (rose) oil; geranium oil; myrcia oil
香葉烯 myrcene
香葉烯醇 myrcenol
香葉酸 geranic acid
香葉醇 geraniol
香葉醛 geranial
香精 essence
香精油 attar
香膏 balsam; balm
香膠木 balsam
香豬殃殃 woodruff
香樹油 amyris oil
香樹素 amyrin
香樹膠 balsam; balm
香糖酒 bumbo
香蕉 banana
香蕉瓜 banana squash
香蕉油 banana oil
香蕉油香精 banana oily flavo(u)r
香蕉香精 banana flavo(u)r
香蕉粉 bananase
香蕉乾(去皮) banana fig
香薄荷 summer savory; savory; savoury
香薄荷油 savory oil; savoury oil
香薑草油 gingergrass oil
香螺 neptune
香檳〔酒〕 bubbly; champagne
香檳在生產期間沉澱物從瓶壁轉移到瓶塞 remuage
香檳酒所用配料 tirage
香檳酒杯 champagne glass
香檳酒瓶的金屬絲網罩 wire hood
香檳酒發酵 tirage
香檳酒釀造 champagnization
香檳桶 wine cooler
香檳雞尾酒 champagne cocktail
香檳蘋果酒 champagne cider
香檸檬 bergamot
香檸檬油 bergamot oil
香檸檬薄荷油 bergamot mint oil
香櫞 cedrat; citron
香櫞瓜 chayote; vegetable pear
香蘭素 vanillin
香囊 scent bag
厘司(黏度單位) centistoke
厘史(黏度單位) centistoke
厘泊(黏度單位) centipoise (cp)
研光 polishing
胍 guanidine
恆溫動物 homoiotherm; homotherm
恆溫箱 incubator
籼米 non-glutinous rice; non-sticky rice; non-sweet rice; non-waxy rice

十畫

倍半碳酸鈉 sodium sesquicarbonate
倍西飲料 basi
倍單寧 gallotannin
值 value
倒捻子 mangosteen
個別包裝 item packaging; individual packaging
個別快速冷凍 individual quick freezing
個別儲藏 individual storage
個體速凍 individual quick freezing
修清肉眼 rib eye roll
修清臀肉 D-Rump
修補匠 tinker
修飾 trim
修飾劑 modifying agent; modifier
修飾澱粉 modified starch
修整 trim
倭瓜 cushaw
倉庫 depository; depot; garner; stock room; store; stores; storehouse; storage; storage room; warehouse
倉鼠 hamster
倉醃 kench curing; kench salting
倉燜穀物 bin-burned grain
倉儲費 storage
兼性的 facultative
兼性嗜熱細菌 facultative thermophile; facultative thermophilic bacteria
兼性厭氣細菌 facultative anaerobe; facultative anaerobe
凍灼 freezer burn
凍害 freezing injury
凍乾 freeze dehydration
凍乾法 freeze drying; lyophilization
凍結 congelation; freeze
凍結〔的〕 freezing
凍結室 freezing room
凍結隧道 freezing tunnel
凍飲 cold beverage; cold drink
凍傷 cold damage; freezing injury; frost damage; freezer burn
凍膠 gel; jell
凍凝 congelation
凍燒 freezer burn
凍藏 freezing storage; frozen storage; refrigerated storage
凋萎 wilt
凋謝 wither
剖開而燒烤 spitchcock
剖醃 split cure
剔出的殘果 cull fruit
剔除〔物品〕 cullage
剔骨肉 boneless meat
剔骨的 boned
剔骨的肉 boned meat
剔骨機 bone cleaner
剛毛 hog bristle; seta
剛生下的蛋 new-laid egg
剛果咖啡樹 Congo coffee
剛長大的雄家禽 stag

剛宰的胴體 hot carcass
剛學步行時期 toddler
剛斷乳嬰孩 weanling
剝去 strip
剝皮 decortication; dehide; dehiding; excorticate; hide pulling; peel; peeling; skinning; stripping; split; splitting
剝皮刀 flaying knife; peeling knife
剝皮工〔人〕 scalper; skinner
剝皮魚 skinned fish
剝皮機 decorticator; hide stripper; peeler; peeling machine; skinner; skinning machine; splitter; splitting machine; stripper; stripping machine
剝牲畜背皮的工人 backer
剝落 scale
原子吸〔收分〕光光度計 atomic absorption spectrophotometer
原子核 nuclei; nucleus
原水 raw water
原台位清洗〔法〕 clean in table
原汁 absolute juice; natural juice; primary juice; raw juice; single strength juice; straight juice
原汁牛肉湯 beef stock
原生動物 protozoa
原生質體 protoplast
原肌球蛋白 tropomyosin
原肌凝蛋白 tropomyosin
原色的 unbleached
原位清洗 in-place cleaning
原兒茶酸 protocatechuic acid
原兒茶酸乙酯 ethyl protocatechuate
原味 primary taste; raw taste
原味烹調 flat cooking
原始營養細菌 prototrophic bacteria
原果汁回添〔法〕 cut back
原果膠 pectose; protopectin
原果膠酶 pectosase; pectosinase; protopectinase
原油 crude oil; raw oil
原型 prototype
原料 raw material; stuff
原料肉 raw meat
原料乳 raw milk
原株 wild type
原脂 raw fat
原酒 brut
原培養物 stock culture
原宿群落 indigenous flora
原條醃魚製品 balbakwa
原液 stock solution
原棉 cotton wool
原湯 aujus; stock
原湯罐 stock pot
原紫質 protoporphyrin
原葉大黃 tart rhubarb
原葉綠素 protochlorophyll
原種 original breed
原維生素 provitamin
原彈性蛋白 tropoelastin
原膠原〔蛋白〕 tropocollagen
原養型細菌 prototrophic bacteria
原糖 dark crystal; gur; raw sugar
原膽酸 apocholic acid
原纖維 fibril
原鹽 crude salt
唐蒿苣 Swiss chard
哥打芝士 Gouda cheese
哥地膠 ghatti gum; gum ghatti
哥達乾酪 Gouda cheese
哲羅魚 taimen
哺乳的母羊 chilver; ewe lamb
哺乳的嬰兒 suckling; sucker
哺乳動物 mammal
哺乳瓶 feeding bottle
哺乳期 lactation

埃（10^{-8}cm） Angstrom (Å)
埃及土豆 ful medames
埃及小麥 mummy wheat; poulard wheat
埃希氏大腸桿菌 *Escherichia coli* (*E. coli*)
埃京（漂白麵粉用的三氯化氮） Agene
埃京作業法 Agene process
埃淨 Agene
埃淨作業法 Agene process
埃淨漂白過的麵粉 agenized flour
埃塞俄比亞香蕉 abyssinian banana
埃爾夏乳牛 Ayrshire (cow)
夏丁格糊精 Schardinger dextrin
夏丁格酶 Schardinger enzyme
夏布里白葡萄酒（法國） Chablis
夏作 summer crop
夏季西葫蘆 summer squash
夏季作物 summer crop
夏季南瓜 summer squash
夏季香腸 summer sausage
夏拉萊牛 charolais
夏花山楂 May hawthorn
夏威夷果 Australian nut; macadamia nut; Queensland nut
夏洛來牛 charolais
夏香風輪菜 summer savory
夏香薄荷 savory; savoury; summer savory
夏塊菌（蕈類） summer truffle
夏葡萄 pigeon grape
夏爾特甜酒 chartreuse
夏橙 natsudaidai; summer orange
套管式熱交換器 double tube heat exchanger
套管冷凝器 double pipe condenser
套管的 double pipe
套蓋包裝 blister pack
娟姍乳牛 Jersey dairy cattle
姬茴香 caraway
姬茴香油 cumin oil
娥眉豆 bonavist bean
宰前檢查 antemortem inspection
宰後立即烤食的雞 spatchcock
宰後排酸 postmortem ag(e)ing
宰後嫩化 postmortem ag(e)ing
宰後僵直 postmortem rigidity
宰後檢驗 necropsy; postmortem inspection
宰後變化 postmortem change
宰馬場 knackery
宰殺台 killing bed
害蟲 pest; vermin
害獸 vermin
家牛 domestic cattle
家用冰箱 home freezer
家用的 household
家用的膳食 at-home meal
家用香腸 domestic sausage
家用凍箱 home freezer
家用麵粉 consumer flour; family flour
家兔 cony
家政 domestic economy
家政學 domestic science
家庭用〔量〕包裝 family-size package
家庭用具 utensil
家庭式製造 home making
家庭自製〔型〕香腸 country style sausage; homemade sausage
家庭自製的 homemade
家庭自製麵包 homemade bread
家庭自製麵條 homemade noodle
家庭的 household
家庭烘焙〔的〕 home baking
家庭製造香腸 domestic sausage
家畜 cattle; domestic animal; livestock; stock
家畜屠宰場 knackery
家常便飯 homely meal
家禽 domestic bird; domestic fowl; fowl; poultry

家禽內臟 abat(t)is
家禽流感 roup
家禽場 poultry yard
家禽飼養人 poultryman
家蠅 house fly
家釀酒 home brew
宴客 treat
宴席甜點 sweet course
宴席菜 soup and fish
宴會 dinner party; dinner; feast; spread
宴會廳 saloon
容忍 tolerance
容易腐爛的 perishable
容限 allowance
容許限量 tolerance limit
容許殘留量 permissible residue level
容許量 allowance; tolerance
容許劑量 permissible dose
容量 volume
容量〔滴定〕分析 volumetric analysis
容量分析法 volumetry
容量分析的 volumetrical
容量吸管 volumetric pipet(te)
容量計 volumeter
容量測定法 volumetry
容器 vessel
容器重 tare
容積 volume
容積計 volumeter
射出成型 injection mo(u)lding
射出成型機 injection mo(u)lding machine
射解作用 radiolysis
射頻巴氏殺菌 radiofrequency pasteurization
射頻加熱 radiofrequency heating; radioheating
展示〔櫥〕櫃 display cabinet
展示冷藏櫃 display cooler
展示櫃 showcase
展延性 ductility
峨螺 whelk
峰肉 hump; hunch
島青霉毒素 islanditoxin
島黴毒素 islanditoxin
席位的餐具墊布或墊紙 place mat
庫存不足 understock
庫存烈啤酒 stock beer
庫存過剩 overstock
庫拉索酒 curacao
庫頁島鱘 green sturgeon
庫恩菇 two toned wood tuft
庫斯庫斯 couscous
庭園蔬菜 garden sass
座頭鯨 nobhead
弱化毒素 toxone
弱氧化醋酸桿菌 *Acetobacter suboxydans*
弱酸 weak acid
弱酸性 subacidity
弱鹼 weak base
恐食症 phagophobia
恩氏黏度 Engler degree; Engler number; Engler viscosity
恩氏黏度計 Engler viscosimeter
恩格勒黏度 Engler degree; Engler number; Engler viscosity
恩格勒黏度計 Engler viscosimeter
恩納杜道夫醣解途徑 Entner-Doudoroff pathway
扇〔形〕片 tidbit; titbit
扇貝 escallop; scallop
扇貝〔殼〕 fan shell
扇貝肉 scallop muscle
扇貝南瓜 custard squash
扇椰子 palmyra (palm)
扇葉樹頭櫚 palmyra (palm)
拳參 bistort
拿破崙〔白蘭地酒〕 Napoleon
拿鐵咖啡 caffe latte; latte

振動 shake; vibration
振動〔式〕輸送帶 shaking conveyor
振動〔粗〕濾器 vibrating strainer
振動式洗菜機 rinsing shaker
振動式輸送機 jigging conveyer
振動進料機 vibrating feeder
振動器 oscillator; vibrator
振動篩 impact screen; oscillating sieve; shaker screen; shaking screen; sieve shaker; vibrating screen; vibrating sieve
振動篩分 vibratory screening
振動篩選 vibratory screening
振動輸送機 vibrating conveyor; vibration conveyor
振盪培養〔物〕 shake culture
振盪器 oscillator; shaker
振盪篩 rocker screen; shaking screen
捕砂器 sand trap
捕塵器 dust arrester
捕蟲草 flycatcher
捕蠅紙 flypaper
捕蠅器 flycatcher; flytrap
捕鯨〔業〕 whale fishery; whaling
捕鯨人 whaler
捕鯨加工船 whale factory ship
捕鯨船 spouter; whaler
捆 pack
捆包 tie
捆帶機 banding machine
捆紮 tie
捆紮機 binding machine; strapping machine; strapper; tier
捆綁 truss
捏和 knead(ing)
捏和桿 kneading arm
捏和槽 kneading trough
捏和機 kneader; kneading machine; kneading mixer
捏揉 knead(ing)
捏揉不足 underkneading
捏揉機 kneader; kneading machine
捏製 knead(ing)
挽畜 draught cattle
挪威幼鯡魚 mussa
挪威乳清乾酪 mysost
挪威海螯蝦 Norway lobster; scampi
挪威醃鱈魚乾 klipfish
挪威龍蝦屬 scampi
挪威燻鮭魚 rokelax
挪威鱈 Norway pout
挨餓者 starveling
效用 value
料酒 cooking wine
時刻表 time table
時計果 passionfruit; parchita
時新風味 brisk taste
朗姆酒 rum
朗姆醚 rum ether
校準 calibration
核 kernel; nuclei; nucleus
核〔白〕蛋白 nucleoalbumin
核〔酸〕磷酸酶 nucleophosphatase
核仁 nucleolus
核仁巧克力餅 brownie
核仁糊 kernel paste
核果 drupaceous fruit; drupe (fruit); pyrenocarp; stone fruit; tryma
核果糖 nougat
核油 kernel oil; stone oil
核原生質 nucleoplasm
核桃 juglans; walnut
核桃〔仁〕糊 walnut cream
核桃油 walnut oil
核桃蛋糕 walnut cake
核蛋白〔質〕 nucleoprotein

核黃素 hepatoflavin; lactochrome; lactoflavin; ovoflavin; riboflavin(e); verdoflavin; vitamin B2
核黃素-5'-磷酸鈉 riboflavin-5'-phosphate sodium
核黃素四丁酸脂 riboflavin(e) tetrabutyrate
核黃素缺乏症 ariboflavinosis
核酮糖 ribulose
核窩 pit cavity
核酸 nucleic acid
核酸調味料 nucleic acid condiment; nucleic acid seasoning
核酸酶 nuclease
核漿 nucleoplasm
核糖 ribose
核糖〔核〕苷 riboside
核糖核蛋白〔質〕 ribonucleoprotein
核糖核酸 ribonucleic acid (RNA)
核糖核酸酶 ribonuclease
核糖核苷酸 ribonucleotide; ribotide
核糖核苷酸酶 ribonucleotidase
核糖酸 ribonic acid
核糖醇 ribitol
核糖體 ribosome
核糖苷玉米素 ribosylzeatin
核黏桃 clingstone peach
核苷 nucleoside
核苷酸 nucleotide
核苷酸酶 nucleotidase
核苷酶 nucleosidase
框式壓濾板 frame plate
根 root
根毛 fibril
根用香芹 turnip-rooted parsley
根皮紅 phloxine
根皮素 phloretin
根皮苷 phloridzin
根狀球莖 rhizocorm
根狀莖 rhizome
根狀菌絲體 rhizomycelium
根芹菜 celeriac
根冠 tip cap
根砧木 rootstock
根啤酒 root beer
根甜菜 garden beet
根莖 rhizome; rootstock
根部的芽 sucker
根菜 root vegetable
根菜類〔蔬菜〕 earth vegetable
根瘤 root knot; root nodule
根瘤〔細〕菌 nodule bacteria
根瘤菌屬 *Rhizobium*
根黴糖化澱粉法 amylo process
根黴屬 *Rhizopus*
桂〔皮〕酸甲酯 methyl cinnamate
桂皮 bay bark; cassia bark; cinnamon bark; cassia; Chinese cinnamon; cinnamon
桂皮油 cassia oil; Chinese cinnamon oil; cinnamic oil; cinnamon bark oil; cinnamon oil
桂皮扁魚 Japanese tamper
桂皮斑鮃 Japanese tamper
桂皮酸 cinnamic acid
桂皮酸乙酯 ethyl cinnamate
桂皮酸桂皮酯 styracin
桂皮酸鈉 sodium cinnamate
桂皮醇 cinnamic alcohol
桂皮醛 cinnamic aldehyde
桂花 osmanthus
桂花油香精 osmanthus oily flavo(u)r
桂花浸膏 osmanthus concrete
桂圓 dried longan; dragon eye; longan; lungan
桂葉油 cinnamon leaf oil
桂酸戊酯 amyl cinnamate
桂酸苯乙酯 phenethyl cinnamate
桂酸苯甲酯 benzyl cinnamate
桂酸桂酯 cinnamyl cinnamate

桂酸烯丙酯 allyl cinnamate
桂酸異丁酯 isobutyl cinnamate
桂酸苄酯 benzyl cinnamate
桂醇 cinnamyl alcohol
桂醛 cinnamaldehyde; cinnamal
桂醛乙二醇縮醛 cinnamaldehyde ethylene glycol acetal
桂櫻 baycherry
桔 mandarin(e)
桔子酒 orange wine
桔皮 mandarin peel
桔黃擬蠟傘(蕈類) false chanterelle
桔瓣 orange section
栗子 chestnut
栗子粉 Farina dolce
栗色河魨 blowfish
栗泥 chestnut paste
桌子 table
桌布 tablecloth
桑 morus
桑加里酒 sangaree
桑果 sorosis
桑椹 melberry; mulberry
桑椹油香精 mulberry oily flavo(u)r
桑椹香精 mulberry flavo(u)r
桑葚 mulberry
桑葚蜂蜜汁 morat
桑葚蜂蜜酒 morat
桑橙 Osage orange
桑樹 mulberry
桑蠶 silkworm
栽培 cultivation
栽培〔品〕種 cultivar
栽培者 grower
柴魚 dried bonito
桐油 tung oil; wood oil
格氏〔乳脂〕測驗法 Gerber test
格氏乳脂〔計〕離心機 Gerber centrifuge
格氏乳脂計 Gerber test bottle
格瓦斯淡啤酒 kvass
格伯〔氏〕乳脂計 Gerber test bottle
格伯〔氏〕乳脂測定法 Gerber test
格恩西牛 Guernsey cattle
格辣燉肉 goulash
格魯糖 glutose
格蘭〔氏〕可變細菌 Gram variable bacteria
格蘭〔氏〕染色 Gram's stain
格蘭〔氏〕染色不定細菌 Gram variable bacteria
格蘭〔氏〕陰性細菌 Gram negative bacteria
格蘭〔氏〕陽性細菌 Gram positive bacteria
桃〔子〕 peach
桃子去核機 peach pitter
桃子白蘭地酒 peach brandy
桃子果醬 peach jam
桃子香精 peach flavo(u)r
桃子醬 peach butter
桃仁油 peach kernel oil
桃金娘 myrtle
桃金娘油 myrtol; myrtle oil
桃紅葡萄酒 pink wine
桃脯 peach
桃駁李 nectarine
桃醛 peach aldehyde; γ-undecalactone
栓塞 stopcock
氣化 gasify
氣化〔作用〕 aerification; gasification
氣升發酵槽 air-lift fermenter; gas-lift fermenter
氣升攪拌 air-lift agitation
氣升攪拌器 air-lift agitator
氣孔 air pore; pore
氣孔(乾酪或芝士) gas hole
氣生植物 aerophyte
氣生菌絲 aerial hyphae
氣生菌絲體 aerial mycelium
氣冷 air cooling
氣冷式冷凍 air freezing

氣冷的 air cooled
氣味 odor; odour; scent; smell
氣味計 odorimeter; odorometer
氣味強度 odo(u)r intensity
氣味測定法 odorimetry
氣味測定計 olfactometer
氣味劑 odorant
氣味濃度 odo(u)r intensity
氣泡 air bladder; bladder
氣泡黏度計 bubble viscometer
氣孢桿菌素 aerosporin
氣流 air flow; air stream
氣流分級 air classification
氣流分級機 air classifier
氣流冷凍器 blast freezer
氣流乾燥器 blast drier
氣流乾燥機 pneumatic conveying dryer; pneumatic dryer
氣流輸送器 air conveyor
氣相色譜法 gas chromatography (GC); vapo(u)r chromatography
氣相層析法 gas chromatography (GC); vapo(u)r chromatography
氣浴 air bath
氣動控制 pneumatic control
氣動輸送機 air conveyor; pneumatic conveyor
氣動選別 pneumatic classification
氣動選別機 pneumatic classifier
氣動擊昏器 pneumatic stunner
氣密的 airtight; gas tight; gas proof
氣控包裝 controlled atmosphere packaging
氣掃 air sweep
氣桿菌屬 *Aerobacter*
氣液〔相〕色譜法 gas-liquid chromatography (GLC)
氣液〔相〕層析法 gas-liquid chromatography (GLC)
氣溶膠 aerosol
氣溶膠罐 aerosol can
氣溶膠罐包裝 pressurized package
氣溶膠罐裝食品 pressurized food
氣態流化〔作用〕 gaseous fluidization
氣管 trachea
氣調包裝 modified atmosphere packaging
氣調貯藏 controlled atmosphere storage (CA storage); gas storage
氣調儲藏 modified atmosphere storage (MA storage)
氣壓計 barometer
氣簾 air curtain
氣霧劑 aerosol
氣體 gas
氣體昏厥〔法〕 gas stunning
氣體殺法 carbon dioxide stunning
氣體滅菌 gaseous sterilization
氣體滅菌劑 gaseous sterilant
氣罐 gassy can
氧 oxygen
氧化〔作用〕 oxidation; oxidization; oxygenation
氧化一氮 nitric oxide
氧化力 oxidizing power
氧化分解 oxidative decomposition; oxygenolysis
氧化安全性 oxidative stability
氧化抑制劑 oxidation retarder; oxidation inhibitor
氧化亞麻仁油 linoxyn
氧化亞氮 nitrous oxide
氧化味 oxidized flavo(u)r
氧化油膜 linoxyn
氧化物 oxide
氧化玫瑰 rose oxide
氧化的 oxidative
氧化阻滯劑 oxidation retarder
氧化促進物質 prooxidant
氧化促進劑 oxidation accelerator

氧化型血紅蛋白 methemoglobin
氧化降解 oxidative degradation
氧化風味 oxidized flavo(u)r
氧化值 oxidation number
氧化原血紅素 haematin; hematin
氧化氣味 oxidized flavo(u)r
氧化脂肪 oxidized fat
氧化強化劑 prooxidant
氧化產物 oxidative product
氧化魚油 degras
氧化氮血紅素 nitric oxide h(a)emoglobin
氧化氮血紅蛋白 nitric oxide h(a)emoglobin
氧化發酵 oxidative fermentation
氧化硬脂精 oxystearin
氧化鈣 calcium oxide
氧化過度 overoxidation
氧化酸敗 oxidative rancidity
氧化數 oxidation number
氧化醋菌 *Acetobacter oxydans*
氧化鋅 zinc oxide
氧化鋁 alumina
氧化劑 oxidant; oxidizer; oxidizing agent; oxygenant
氧化澱粉 oxidized starch
氧化還原反應 redox reaction
氧化還原作用 redox
氧化還原酶 oxidation-reduction enzyme; oxidoreductase
氧化鎂 magnesium oxide
氧化鐵 iron oxide
氧化鐵紅 iron oxide red
氧化鐵黃 iron oxide yellow
氧化鐵黑 iron oxide black
氧化纖維素 oxycellulose
氧化酶 oxidase; oxydase
氧代戊二酮酸 ketoglutaric acid
氧卡計 oxycalorimeter
氧合〔作用〕 oxygenation
氧合肌紅素 oxygenated myoglobin; oxymyoglobin
氧合肌紅蛋白 metmyoglobin; oxygenated myoglobin; oxymyoglobin
氧合血紅素 oxyh(a)emoglobin
氧合血紅蛋白 oxyh(a)emoglobin
氧合酶 oxygenase
氧含量 oxygen tension
氧量熱計 oxycalorimeter
氧彈卡計 oxygen bomb calorimeter
氧彈量熱計 oxygen bomb calorimeter
氧壓 oxygen tension
氨 ammonia
氨乙基亞磺酸 hypotaurine
氨化〔作用〕 ammonification
氨化甘草甜素 ammoniated glycyrrhizin
氨化合物 ammoniate
氨化細菌 ammonifying bacteria
氨合 ammonate
氨合〔作用〕 ammonation
氨合物 ammonate
氨系發粉 ammonia system baking powder
氨味 ammoniacal odor
氨氣味 ammoniacal odor
氨基乙基化澱粉 aminoethylated starch
氨基化合物 amide
氨基琥珀酸 aspartic acid
氨基酸 amino acid
氨基酸溶液 amino acid solution
氨解 aminolysis
氨解作用 ammonolysis
氨態氮 ammonia nitrogen
氦 helium
泰威糖 tyvelose
泰國蝦 giant river prawn
泰黴素 tylosin
泰樂菌素 tylosin
浪費 waste; wastage
涕必靈 thiabendazole

消化 digest; digestion
消化力不足 oligopepsia
消化不良 cacochylia; digestive problem; dyspepsia; indigestion; maldigestion
消化系數 digestibility coefficient
消化性 digestibility
消化的 digestive; peptic
消化液 digestive fluid; digestive juice
消化率 digestibility
消化率系數 digestibility coefficient
消化腺 digestive gland
消化道 alimentary canal; alimentary tract; digestive tract
消化道感染 alimentary infection
消化槽 digester
消化器官 alimentary organ
消化酶 digestive enzyme
消石灰 calcium hydroxide; slaked lime
消泡 antifoam; defoam
消泡裝置 defoamer
消泡劑 antifoaming agent; antifoamer; defoamer; defoaming agent; kilfoam
消泡劑GP polyoxypropylene glycerol ether
消泡劑GPE polyoxypropylene oxyethylene glycerol ether
消泡器 antifoamer; foam breaker; foam destroyer; foaming killer; froth breaker
消泡機 foaming killer
消毒 antisepsis; sanitize; sterilize; disinfection; sanitization; sterilization
消毒〔過〕的 sterile
消毒牛乳 city milk; market milk
消毒的牛乳(添加過氧化氫) buddeized milk
消毒紙 aseptic paper
消毒劑 antiseptic; disinfectant; disinfecting agent; sanitizer; sanitizing compound
消毒器 disinfector
消旋果糖 acrose
消閑食品 leisure food
消脹〔作用〕 deswelling
消費者 consumer
消煙器 smoke eliminator
消過毒的 antiseptic
消電離作用 deionization
消瘦 marasmus
浸 dip; steep
浸入 immersion
浸出 extraction; infuse; infusion
浸出肉汁 infusion broth
浸出法 infusion process
浸出物 extract; extractive
浸出液 infusion
浸出麥芽汁 infusion wort
浸出器 extractor; infuser
浸汁 bree
浸沉培養〔物〕 submerged culture
浸沉發酵 submerged fermentation
浸沒冷卻〔法〕 immersion cooling
浸沒冷凍〔法〕 immersion freezing
浸沒解凍〔法〕 immersion defrosting
浸泡 dunk; soak(ing); steeping
浸凍 dip freezing
浸掛糖衣 enrobe
浸掛糖衣機 enrober
浸軟 maceration
浸軟酶 macerating enzyme
浸透 impregnation; soakage; soak(ing); steep
浸透的 sodden
浸提 diffusion; lixiviation
浸提用水 diffusion water
浸提器 diffuser; diffusor
浸煮器 digester
浸解 retting

浸漬 dipping; immersion; maceration; soakage; soak(ing); sop; steep; steeping
浸漬冷凍 dip freezing
浸漬液 steep
浸漬槽 soaker; steeper
浸漬劑 soaker
浸漬器 immerser
浸漬罐 steeper
浸滲 impregnation
浸製麥芽汁 infusion wort
浸酸 pickling
浸劑 infusion
浸濕 wet
浸濕的 sodden
浸濾 lixiviation
浸蠟 waxdip
海〔鹽〕水 sea brine
海人〔草〕酸 digenic acid
海中微生物 brit(t)
海比目魚 ling
海水 sea water; salt water
海水魚 sea fish
海牛 manatee
海甘藍 cole; colewort; crambe; sea cabbage; sea kale
海生苔蘚 sea moss
海生植物 marine plant
海白魚 ocean whitefish
海白菜 sea cabbage; sea kale
海白鰻 conger eel
海地油 pennyroyal oil
海百合 crinoid; sea lily
海佛牛 Hereford (cattle)
海芋植物 arum
海貝 sea shell
海味 marine food; sea food
海底植物 phytobenthos
海松素 natamycin
海狗 sea dog
海芥蘭 seakale
海星毒素 asterotoxin
海洋小甲殼動物 brit(t)
海洋生物 halobios; marine organism
海洋生物學 marine biology
海洋食物 ocean food
海洋浮游生物 haloplankton
海洋細菌 marine bacteria
海洋魚類 marine fish
海洋漁業 marine fisheries
海砂 sea sand
海苔 laver; nori
海軍豆 navy bean
海扇 cockle; scallop
海扇貝 escallop
海草 sea grass
海豹 sea calf; seal
海豹油 sea oil
海參 sea cucumber; sea slug; trepang
海帶 kombu; laminaria; sea belt; sea girdle; tangle
海帶二糖 laminaribiose
海帶多糖 laminarin
海帶製的蜜餞 seatron
海帶酸 tangic acid
海帶醇 laminaritol
海帶糖 laminariose
海甜菜 Swiss chard
海產品 marine product
海產食物 sea food
海產食品 marine food
海產食品拼盤 seafood platter
海產動物油 train oil
海豚 dolphin; porpoise
海豚肉 sea beef
海魚 marine fish
海鳥 sea fowl
海棠〔蘋果〕 crab apple
海棗 date; date palm

海棗油 date kernel oil
海蛤 sea clam
海蛤蝓 sea slug
海黃瓜 sea gherkin
海獅 sea lion
海葵 sea anenone
海蜇 jellyfish; medusa
海蜇傘膜 umbrella
海福特牛 Hereford
海綿 sponge
海綿狀 spongy; sponginess
海綿狀肉 sponge meat
海綿狀物 sponge
海綿蛋糕 sponge cake; spongy cake; trifle
海蒔蘿 samphire
海蝸牛 oyster drill
海魴 dory; John Dory
海龜 sea turtle; turtle; terrapin
海鮒 Japanese sea-perch
海膽 echinus; sea chestnut; sea hedgehog; sea urchin; urchin
海膽類長腕幼蟲 pluteus
海螺 conch; sea whelk; sea snail; top shell; whelk
海鮮 marine food; sea fresh; sea food
海鮮拼盤 seafood platter
海鮮養殖 mariculture
海鮮濃湯 bisk; bisque; chowder
海鮭 grey trout
海藍色素 ultramarine
海鯽魚 ditrema; marine carp
海蟹 marine crab
海鯛 sea bream
海鯰 sea catfish
海鯡魚 English herring
海藻 alga(e); marine algae; seaweed; varec(h)
海藻灰 kelp; varec(h)
海藻抽提粉 algae extract power
海藻的 algal
海藻粉 kelp
海藻素 algin
海藻酸 alginic acid
海藻酸丙二醇酯 propylene glycol alginate
海藻酸鈣 calcium alginate
海藻酸鈉 sodium alginate
海藻酸鉀 potassium alginate
海藻酸銨 ammonium alginate
海藻酸鹽 alginate
海藻糖 fucose; mycose; trehalose
海藻糖酶 trehalase
海鰻 conger eel; marine eel; muraenesox; pike eel; sharp-toothed eel
海鹽 sea salt
海灣無鬚鱈 gulf hake
海鱸〔魚〕 orange rock fish; sea bass
海葱 scilla; squill
海葱〔聚〕糖 sinistrin
浮肉 scrap fat
浮沫 skimmings; spume
浮油 floated oil
浮油起泡性 creaming property
浮洗機 flotation washer
浮動洗滌機 flotation washer
浮游生物 plankton
浮游動物 zooplankton
浮游植物 phytoplankton
浮渣 scruff; scum; skimmings
浮渣般的 scummy
浮萍 duckweed
浮腫 dropsy
浮膜酵母 scum yeast
浮濁澄清 deemulsification
浮濁澄清劑 deemulsifying agent
浮選劑 flotation agent
浮雕 embossment
烘 toast; torrefaction; torrefy
烘烤 bake; baking; firing; roasting; stoving

烘烤不足的 underbaked
烘烤食品 bakery goods
烘烤食品用起酥油 bakery shortening
烘烤食品廠 bakery
烘烤製品 bakery product
烘烤膨脹 dough spring; oven kick; oven spring
烘烤爐 roaster
烘乾 parch
烘乾麥芽 kilned malt
烘焙 bake; baking; roasting
烘焙用油脂 baking grease
烘焙食品 baked food; baked good; baking product
烘焙食品工業 baking industry
烘焙過度 overbake
烘焙爐 roaster
烘箱 baking oven; drying oven; gas oven; oven
烘箱乾燥 oven drying
烘雜燴 casserole dish; covered dish
烘爐 baker; oast
烘爐乾燥 kiln dry(ing); oven drying
烤 broil; broiling; grilling; roast; toast; torrefy; torrefaction
烤〔餅〕盤 tin pan
烤牛肉 roast beef
烤用豬 roasting pig
烤肉 barbecue; barbeque; barbecued meat; broiled meat; buccan; carbonad; fricasse; fried meat; grill; grilled meat; roast meat; roti; seared meat
烤肉上塗油 baste
烤肉叉 roast meat fork; skewer; spit
烤肉用的細長木籤 skewer
烤肉串 brochette; kabob; kebab
烤肉店 grill; grill room; rotisserie
烤肉的長鐵針 skewer
烤肉架 barbecue; barbeque
烤肉野餐 barbecue; barbeque
烤肉塗醬 baste
烤肉調味醬 barbecue sauce
烤肉鐵叉 broach
烤肝片 fagot
烤乳豬 roasting pig
烤明蝦 roasted prawn
烤板 salamander
烤炙的肉片 carbonad
烤架 grill
烤乾 parch
烤焙用具 broiler
烤焦 sear
烤碎豬雜 faggot; fagot
烤過頭變乾硬的 hard-baked
烤餅 crumpet; flapjack; grigger; griddlecake; scone; slapjack
烤餅鍋 pattypan
烤盤 piepan; roasting pan; salamander
烤盤油 pan grease
烤盤塗油 pan greasing
烤盤醒麵 pan proof
烤箱 baking oven; gas oven; oven; roaster
烤豬的脆皮 crackling
烤鴨 roast duck
烤雞 roast chicken
烤爐 baker; baking oven; oven; roaster
烤爐燒烤 oven grilling
烤麵包 baked bread; baking load
烤麵包片 toast; toasted bread
烤麵包機 toaster
烤鰻 broiled eel; spitchcock
烙 ustion
烙肉 seared meat
烙餅 flapjack; griddlecake; hot cake; pancake; paratha; slapjack; scone
烚蛋 hard boiled egg
烟草 nicotiana
烟酸 vitamin PP

烟酰胺 nicotinic acid amide
烟酰胺核苷酸 nicotinamide nucleotide
烟酰胺單核苷酸 nicotinamide mononucleotide
烈味 sharp taste
烈性酒 heavy wine; lace; lush; toddy
烈性酒精飲料 strong liquor
烈性啤酒 stingo
烈性麥芽酒 bub
烈性飲料 strong drink
烈性黑啤酒 stout (beer)
烈性葡萄酒 fortified wine
烈酒 ardent spirit; firewater; grog; hard liquor; liquor; rich wine; spirit; spirits; schnap(p)s; tipple
烏仔魚 black mullet
烏豆 black bean
烏茲卡餃子（波蘭） uszka
烏骨雞 silky bird
烏梅 ebony
烏魚 black mullet; grey mullet; mullet
烏魚子 mullet roe
烏斯特〔調味〕醬 worcester(shire) sauce
烏斯特醋栗 worcesterberry
烏斯達醬 worcester(shire) sauce
烏賊 cuttle; cuttlefish; inkfish; sleeve fish; squid; sepia
烏賊卵 sea grape
烏賊骨 sepium
烏賊黑色素 sepia melanin
烏賊墨汁 ink; sepia
烏頭 black mullet
烏頭酸 achilleic acid
烏龍茶 oolong (tea)
烏鯧 black butterfish; black pomfret
烏鱧 snake head
烏桕樹 tallow tree
特有風味 race
特希紅菇 iodoform scented russule
特拉卡甘溶膠 tragacanthin
特拉卡甘膠 gum dragon; gum tragacanth; tragacanth gum
特拉西蝦醬 trassi
特氟隆 teflon
特殊用途食品 special food
特殊膳食 special diet
特殊營養品 special dietary product
特殊營養食物 special nutrition foodstuff
特殊營養素 specific nutrient
特級〔葡萄〕酒 supernaculum
特級牛乳 accredited milk
特級的 fancy; superfine
特級產品 specialty product
特乾葡萄酒 extra dry wine
特產 specialty product
特陳年白蘭地酒 extra old (X.O.)
特製食品 designer food
特濃義大利咖啡 doppio; double espresso
特濃糖漿 extra heavy syrup
特選的 fancy selected
特選級 U.S. Choice
狼吞虎咽 raven; scoff
狼吞虎咽地吃 wolf down
狼吞虎咽的 voracious
狼魚 sea wolf; wolffish
狹溫性細菌 stenothermic bacteria
狹葉柳味茶 iwan tea
狹鱗庸鰈 pacific halibut
班伯里果料烤餅 banbury cake
班伯里果料塔餅 banbury tart
班伯里餅乾 banbury (biscuit)
班戟 hot cake; pancake
班戟粉 pancake mix
班鳩 turtle dove
琉璃苣 borage
珠〔綠〕茶 gunpowder tea
珠母貝 black shell; nacre
珠瓜 yellow flowered gourd

珠珍雞 guinea fowl
珠茶 tea leaf ball
珠雞 guinea fowl; pearl hen
珠葱 eschalot; pearl onion; shallot
畜皮 hide
畜牧場 animal farm; livestock farm
畜牧業 animal husbandry; graziery; live farming
畜牧業者 stock breeder
畜牧學 animal husbandry
畜舍 barn; stall; stable
畜骨下腳料 bone tankage
畜產品 animal product
畜棚 stall; stable
畜群 herd
畜欄 stockyard
留蘭香 garden mint; *Mentha viridis*; spearmint
留蘭香油 spearmint oil
疾病 affection
病人 patient
病人食品 invalid food
病人膳食 invalid diet
病後療法 after-treatment
病毒 virus
病毒性疾病 virosis
病毒感染 viral infection
病毒學 virology
病原〔細〕菌 pathogenic bacteria
病原生物 pathogenic organism
病原性 pathogenicity
病原性大腸桿菌 pathogenic *Escherichia coli*
病原菌 pathogenic germ
病原微生物 pathogenic microorganism
病原體 pathogen
病理學 pathology
病媒 vector
病菌 pathogen
病徵 symptom
症狀 symptom
疲倦毒素 kenotoxin
益蟲 useful insect
盎司 ounce (oz.)
盎斯 ounce (oz.)
真正的 genuine
真皮 corium
真果 true fruit
真空 vacuo; vacuum
真空分餾塔 vacuum column
真空巴氏殺菌 vacuum pasteurization
真空巴氏殺菌器 vacuum pasteurizer
真空充填 vacuum filling
真空加熱 vacuo heating
真空加熱〔巴氏〕殺菌器 vacu-therm pasteurizer
真空包裝 vacuum pack; vacuum package; vacuum packaging; vacuumize
真空包裝機 vacuum packaging machine
真空冷卻 vacuum cooling
真空冷凍法 vac-ice process
真空冷凍乾燥 vacuum freeze drying
真空咖啡沖泡器 vacuum coffee maker
真空泡沫乾燥 foam vacuum drying; foam-mat drying; vacuum foam drying; vacuum puff drying
真空封密 vacuum seal(ing)
真空封瓶機 vacuum closing machine
真空封蓋機 vacuum capping machine
真空封罐 vacuum seaming
真空封罐機 vacuum closing machine; vacuum can sealing machine; vacuum seamer; vacuum seaming machine
真空計 vacuum gauge
真空捏揉機 vacuum kneader; vacuum kneading machine
真空烘焙〔法〕 vacuum baking

真空高溫瞬間滅菌機 vacu-therm instant sterilizer
真空乾燥 vacuum drying; vacuum dehydration
真空乾燥器 vacuum desiccator; vacuum dryer
真空乾燥爐 vacuum drying oven; vacuum oven
真空旋轉乾燥機 vacuum rotary dryer
真空殺菌脱臭 vacreation
真空殺菌器 vacreator
真空清潔器 vacuum cleaner
真空混肉機 vacuum meat mixer
真空脱水 vacuum dehydration
真空脱氧 vacuum degassing
真空脱氧器 vacuum deaerator
真空脱臭處理 vacreation
真空脱臭器 vacreator
真空揉混機 vacuum kneader; vacuum kneading machine
真空提取器 vacuum extractor
真空裝填機 vacuum filler
真空過濾 vacuum filtration
真空滾筒乾燥機 vacuum drum dryer
真空蒸煮鍋 vacuum cooker
真空蒸發 vacuum evaporation
真空蒸發器 vacuum evaporator
真空蒸餾 vacuum distillation
真空蒸餾塔 vacuum column
真空蒸餾器 vacuum still
真空濃縮 vacuum concentration
真空濃縮器 vacuum concentrator
真空儲藏 vacuum storage
真空壓力 underpressure
真空鍋 vacuum pan
真空燻煙 vacuum fumigation
真空灌裝 vacuum filling
真空灌裝機 vacuum filler
真空罐 vacuum pan
真胃 abomasum; maw; true stomach
真消化率 true digestibility
真細菌 eubacteria
真菌 eumycete; fungi; fungus
真菌孢子 fungal spore
真菌毒素 mycotoxin
真菌素病 mycotoxicosis
真菌蛋白質 mycoprotein
真菌蛋白酶 fungal proteinase
真菌學 mycology
真菌澱粉酶 fungal amylase
真菌酶 fungal enzyme
真溶液 true solution
真酵母 true yeast
真酸度 true acidity
真鯛 crimson sea bream; genuine porgy; porgy; red porgy
真鱒 salmon trout
真鱸 common sea bass; Japanese sea-perch; perch; sea perch
真鰺 horse mackerel; Japanese horse mackerel
砧板 cutting board; chopping board; trencher
砧排 topside steak
砧排牛肉 topside (round)
砝碼 weight
破布子 sebastan plum cordia
破冰鐵鑿 ice pick
破乳劑 deemulsifying agent
破粒 brokens
破蛋器 egg breaker
破碎 crush; crushing; grind; grinding
破碎小麥粒 cracked wheat
破碎殼粒 cracked grain
破碎葡萄 grape mass
破碎機 cracking mill; crusher; grinder; grinding mill; pounder
破壞曲線 destruction curve

破壞穩定〔作用〕 destabilization
破壞穩定劑 destabilizer
砷 arsenic
神奇果 miracle berry
神秘果 miracle berry
神經元 neurone
神經甲殼質 neurochitin
神經角蛋白 neurokeratin
神經性食慾缺乏症 anorexia nervosa
神經性厭食症 anorexia nervosa
神經毒素 neurotoxin
神經細胞 neurone
神經蛋白〔質〕 neuroprotein
神經幾丁質 neurochitin
神經殼糖 neurochitin
神經酸 nervonic acid; selacholeic acid
神經鞘脂類 sphingolipid
神經鞘胺醇 sphingosine
神經鞘髓磷脂 sphingomyelin
神經營養素 trophon
神經纖維 nerve fiber
神經鹼 neurine
祝酒者 wassailer
祝福酒 Benedictine
秤量 scaling
秤量員 scaler
秤盤 scale pan
秧苗 seedling
秧雞 baby chick
秦椒 xanthoxylum
秘魯香膏油 Balsam Peru oil
秘魯香膠油 Peru balsam oil
秘魯酸漿 Peruvian ground cherry
秘魯鯷 Peruvian anchovy
窄條 strip
窄溫性細菌 stenothermic bacteria
窄鱗庸鰈 pacific halibut
笑氣 nitrous oxide
粉 powder
粉〔碎〕化 atomization
粉化 efflorescence
粉心〔麵〕粉 patent flour
粉末 powder
粉末牛肝菌 blackening bolete
粉末油脂 powdered oil
粉末狀的 powdered
粉末食品 powdered food
粉末香料 powdered flavo(u)r
粉末飲料 powdered drink
粉光參 American ginseng
粉性 mealiness
粉狀 mealiness
粉狀的 floury; mealy; powdery
粉紅葡萄酒 pink wine
粉絲 bean noodle; bean thread (vermicelli); bean vermicelli; green bean vermicelli; starch noodle; vermicelli
粉碎 crumble; crush; pulverize; shiver; reduction; comminution; disintegration; porphyrization; pulverization
粉碎芝士 comminuted cheese
粉碎乾酪 comminuted cheese; grated cheese
粉碎過度 over grinding
粉碎器 atomizer
粉碎機 comminutor; disintegrator; desintegrator; mill; muller; powdering machine pulverizer; pulverizing mill; pulverator
粉葛 kuza
粉質甘薯 big stem jersey
粉質的 farinose; floury; mealy
粉糖 confectioner's sugar; icing sugar; lawn sugar; powder sugar; pulverized sugar
粉糖機 icing machine
粉蕉 dwarf banana
紡紗 spinning

紡絲大豆蛋白 spinning soy protein
紡絲蛋白 spun protein
紡錘鰤 blue-striped runner
紗布 cheese cloth
紋理 vein
紊流 turbulent flow
素巧克力 plain chocolate
素食 maigre food; plant food; vegetable food; vegetarian diet; vegetable diet; vegetarian food
素食〔主義〕者 vegetarian
素食主義 grahamism; vegetarianism; veganism
素食的 maigre
素食餐館 vegetarian restaurant
素料餅乾 galette
素乾魚 unsalted dried fish
素燒板 porous sheet
素燒瓷體 porous body
素餐 vegetarian diet
素馨 jasmine
素馨淨油 jasmine absolute
素麵條 plain noodle
索 chord(a); cord; rope
索氏萃取法 Soxhlet extraction method
索氏萃取器 Soxhlet extractor
索托烈酒 sotol
索馬甜 thaumatin
純化 depuration; purification
純化器 purifier
純天然果汁 pure natural juice
純水 pure water
純巧克力 plain chocolate
純正的 genuine; unadulterated
純正食品 unadulterated food
純汁 absolute juice
純白蘭地〔酒〕 neat brandy
純咖啡 black coffee
純果汁 straight juice
純的 neat; straight; unmingled
純威士忌〔酒〕 straight whisk(e)y; whisky straight
純度 pureness; purity
純素食主義 veganism
純素食者 vegan
純茶 genuine tea
純淨 pureness; purity
純淨酵母 net yeast
純蛋白質 true protein
純菌培養 sterile culture
純種 pure culture
純種牛 pedigree cattle
純種培養 pure culture
純潔 pureness; purity
純橄欖油 florence oil
純麵粉 clean flour
紐約克 striploin
紐約客牛排 striploin steak
紐紋粉 gemelli
紐結狀椒鹽脆餅 pretzel
級 sort
納豆 natto
納豆〔芽孢〕桿菌 *Bacillus natto*
納斯果 naseberry
紙上電泳 paper electrophoresis
紙色層分析法 paper chromatography
紙色譜法 paper chromatography
紙身罐 composite can; composite paper can
紙杯 paper cup
紙杯蛋糕 cup cake
紙板 carton board
紙盒製造機 cartoning machine
紙袋 paper bag; satchel
紙漿 pulp
紙箱 carton
紙罐 paper can
缺口 notch

缺乏 dearth; deficiency; shortage
缺乏風味的 savorless; savourless
缺乏營養的 jejune
缺乏營養膳食 jejune diet
缺氧生活 anaerobiosis
缺氧呼吸 anaerobic respiration
缺氧症 anoxia
缺脂肪蛋白 apolipoprotein
缺鈣症 acalcerosis
缺輔基蛋白 apoprotein
缺輔基酶 apoenzyme
缺鐵性貧血 hypoferric anemia; iron deficiency anemia
羔羊 kid; lamb
羔羊肉 lamb; lamb meat; lambkin
翅 ala
翅尖 wing tip
翅板 mid-joint; middle section; wing flat
翅根 drummette
翅膀 wing
翅鯊 tope
耕地白糖 plantation white (sugar)
耕地蔗糖 plantation sugar
耕作 cultivation
耕作季 cropping season
耕作期 cropping season
耕種 cultivation; farm
耕種者 tiller
耕種栽培 crop husbandry
耙式輸送機 rake conveyor
耙式攪拌器 rake stirrer
耗熱量 heat consumption
耗糖度 attenuation degree
胱氨酸 cystine
胱胺酸 cystine
脂 grease
脂母細胞 lipoblast
脂白蛋白 lipoalbumin
脂多糖 lipopolysaccharide
脂色素 lipochrome
脂血症 lipidemia
脂油 fatty oil
脂肪 fat
脂肪〔族〕化合物 fatty compound
脂肪〔族〕醇 fatty alcohol
脂肪分解 adipolysis; fat splitting; lipoclasis; lipodieresis; lipolysis; steatolysis
脂肪分解酸敗 lipolytic rancidity
脂肪水解 lipolytic hydrolysis
脂肪代謝 lipometabolism
脂肪代謝障礙 lipodystrophia
脂肪加氧酶 lipoxgenase
脂肪皮 bacon rind
脂肪形成 adipogenesis
脂肪乳濁液 fat emulsion
脂肪性食品 fatty food
脂肪的 adipic; adipose; sebaceous
脂肪氧化的臭味 tallowiness
脂肪氧化酶 lipoxidase
脂肪族的 aliphatic
脂肪族酸 aliphatic acid
脂肪族醇 aliphatic alcohol
脂肪球 fat globule; oil globule
脂肪細胞 adipocyte; adipose cell; fat cell; lipocyte
脂肪組織 adipose tissue; fat tissue; fatty tissue
脂肪過多症 adiposis
脂肪蓄積 lipopexia
脂肪酸 fatty acid
脂肪酸山梨醇酐酯 sorbitan fatty acid ester (Span)
脂肪酸丙二醇酯 propylene glycol esters of fatty acids
脂肪酸甘油酯 fatty glyceride
脂肪酸甘油酯類 glycerol esters of fatty acids
脂肪酸乳酰酯 lactylic esters of fatty acids

脂肪酸組成 fatty acid composition
脂肪酸聚甘油酯 polyglycerol esters of fatty acid (PGFE)
脂肪酸蔗糖酯 sucrose esters of fatty acids (SE)
脂肪酸糖酯 sucrose fatty acid ester
脂肪酸鹽 salts of fatty acids
脂肪熬煉 fat rendering
脂肪質的 aliphatic
脂肪凝固過程 graining process
脂肪體 fat body
脂肪酶 lipase
脂氧化酶 lipoxidase
脂族酸 aliphatic acid
脂族醇 aliphatic alcohol
脂紅質 liporhodin
脂香菊 costmary
脂清蛋白 lipoalbumin
脂眼凹肩鰺 big-eye scad
脂眼鯡 pacific round herring; round herring
脂蛋白 lipoprotein
脂蛋白脂肪酶 lipoprotein lipase
脂嗜熱肪〔芽孢〕桿菌 *Bacillus stearothermophilus*
脂溶〔性〕的 liposoluble
脂溶性維他命 fat-soluble vitamin; lipid-soluble vitamin
脂溶性維生素 fat-soluble vitamin; lipid-soluble vitamin; lipovitamin
脂解 adipolysis
脂酮酸 liponic acid
脂褐素 lipofuscin
脂褐質 lipofuscin
脂質 lipid(e); lipin
脂質的 lipidic
脂質消化率 lipid digestibility
脂質營養不良 lipodystrophia
脂質體 liposome
脂類 lipid(e); lipin
脂蠟 ceride
脂肽 lipopeptide
脂酶 lipase
胰島素 insulin
胰脂酶 pancreatic lipase; steapsin
胰液 pancreatic juice
胰蛋白酶 pancreatic protease; trypsase; trypsin; tryptic enzyme
胰蛋白酶抑制劑 trypsin inhibitor
胰蛋白酶活性 tryptic activity
胰蛋白酶原 trypsinogen
胰腺 pancreatic gland; pancreas
胰增血糖素 glucagon
胰凝乳蛋白酶 chymotrypsin
胰凝乳蛋白酶元 chymotrypsinogen
胰凝乳蛋白酶原 chymotrypsinogen
胰澱粉酶 amylopsin; pancreatic amylase
胰臟 pancreas
胰臟（牛或羊） sweetbread
胰酶 pancreatin
胭脂 kermes
胭脂籽 annatto
胭脂紅 cochineal; carmine; kermes; nacarat; ponceau 4R
胭脂紅A cochineal red A
胭脂紅酸 carminic acid
胭脂紅鋁色澱 ponceau 4R aluminium lake
胭脂魚 catostomid fish; longnosed sucker
胭脂樹 achiote
胭脂樹籽 achiote
胭脂樹紅色素 bixin
胭脂樹橙 annatto; bixin
胭脂樹橙〔紅〕色素 annotta; arnotta; annatto
胭脂蟲紅 cochineal
胭脂蟲紅〔鋁〕 carmine
胭脂蟲紅萃取液 cochineal extract
胴肉 drum meat

胴腔 visceral cavity
胴體 carcass
胴體死後僵直 carcass setting
胴體肉 carcass meat
胴體沖洗裝置 carcass washer
胴體後背 backside
胴體修整 carcass dressing
胴體腹腔沖洗 inside washing
脆化 embrittlement
脆化劑 crisping agent
脆心巧克力 chocolate brittle
脆皮酥盒 croustade
脆性 brittleness; crispness; friability
脆物 crisp
脆的 crumbly
脆度 brittleness; crispness
脆核果糖 short nougat
脆烤圓餅 crumpet
脆甜豌豆 mangetout
脆硬甜麵包 rusk
脆煎肉餅 cracknel
脆餅 cookie
脆餅烤模 waffle iron
脆餅乾 cracker
脆質凝膠 short gel
脆質麵筋 short gluten
脆質麵糊 short paste
脆凝膠 short gel
脆燕麥餅 crisppock
脆薄吐司 melba toast
脆薄空心鬆餅 popover
脆薑餅乾 ginger nut; gingersnap
脆麵包丁 crouton
脆麵包片 croustade
脆麵筋 short gluten
胸肉 brisket; breast; breast cut
胸肉尖 brisket end
胸骨 sternum
胸排 breast cut
胸部 breast
胸腔 thorax
胸腹肉 brisket navel end; short plate
胸腺 thymus
胸腺(牛或羊) sweetbread
胸膛 thorax
脈動篩 pulsating screen
脈衝 pulse
胺 amine
胺〔基〕多肽酶 aminopolypeptidase
胺〔基〕嘌呤 aminopurine
胺〔基〕肽酶 aminopeptidase
胺乙酸 aminoacetic acid
胺化〔作用〕 amination
胺化酶 aminase
胺比西林 ampicillin
胺卡青黴素 ampicillin
胺甲酸乙酯泡沫 urethane foam
胺脂質 aminolipid
胺基 amino group
胺基乙酸 collagen sugar; glycine
胺基乙磺酸 taurine
胺基化合物 amino compound
胺基甲酸酯 carbamate
胺基甲酸鹽 carbamate
胺基葡〔萄〕糖 glucosamine
胺基態氮 amino nitrogen
胺基酸 amino acid
胺基酸分析儀 amino acid analyzer
胺基酸去羧酶 amino acid decarboxylase
胺基酸均衡 amino acid balance
胺基酸尿症 aminoaciduria
胺基酸定序 amino acid sequence
胺基酸定序儀 amino acid sequenator; amino acid sequencer
胺基酸剖析 amino acid profile
胺基酸氧化酶 amino acid oxidase
胺基酸側鏈 amino acid side chain
胺基酸強化 amino acid fortification

胺基酸組成 amino acid composition
胺基酸脱羧酶 amino acid decarboxylase
胺基酸殘基 amino acid residue
胺基酸發酵 amino acid fermentation
胺基酸評分 amino acid score
胺基酸順序 amino acid sequence
胺基酸順序分析儀 amino acid sequenator; amino acid sequencer
胺基酸態氮 amino acid nitrogen
胺基酸價 amino acid score
胺基酸範式 amino acid pattern
胺基酸調味料 amino acid seasoning
胺基酸輪廓 amino acid profile
胺基酸轉乙酰酶 amino acid transacetylase
胺基酸轉胺作用 amino acid transamination
胺基酸酰化酶 amino acid acylase
胺基酸酰基轉移酶 amino acid acylase
胺基醋酸 aminoacetic acid; collagen sugar; glycine; glycocoll
胺基糖類 aminosaccharide
胺喋呤 aminopterin
胺端 amino-terminal
胺醇 amino alcohol
胺羰反應 aminocarbonyl reaction; carbonylamine reaction; Maillard reaction
胺糖 amino sugar
胺醣 glycosamine
能母（1克母乳之營養值單位） nem
能值 energy value
能量 energy
能量攝取量 energy intake
能源 energy source
脊骨 spine
脊椎骨 chine (bone); vertebra(e)
脊椎動物 backboned animal; vertebrate
脊髓 medulla; spinal cord; spinal marrow
脊髓液 spinal fluid
臭味 nasty smell; rotten flavo(u)r
臭氣 effluvium; mephitis; niff; off odo(u)r; stench
臭氣難聞的 rank
臭氧 ozone (O_3)
臭氧化〔作用〕 ozonization; ozonation
臭氧計 ozonometer
臭氧處理 ozonization
臭氧處理水 ozonized water
臭氧發生器 ozonizer
臭粉 ammonia soda
臭腺 scent bag
舐食 suck
航空膳食 air catering
航空餐 inflight meal
荔枝 leechee; lichee; lichi; litchi; lychee
荔枝油香精 lychee oily flavo(u)r
荔枝香精 lychee flavo(u)r
荔枝螺 oyster drill
茚三酮 ninhydrin
茚三酮反應 ninhydrin reaction
荊樹皮 wattle
草乙酸 oxaloacetic acid
草本 vegetation
草本的 herbaceous
草本香料 odoriferous herb
草本植物 herb
草本調味香料 kitchen herb
草地濕傘（蕈類） buff wax cap
草味 grassy odo(u)r
草味的 grassy
草食〔性〕的 herbivorous
草食〔性〕動物 herbivore
草食動物 grazer
草香苦酮 humulone
草料 forage
草根沙士 root beer
草烯 humulene

草笠竹 morel
草莓 strawberry
草莓〔果〕醬 strawberry jam
草莓油香精 strawberry oily flavo(u)r
草莓香精 strawberry flavo(u)r
草莓番石榴 cattley guava; strawberry guava
草莓酸 2-methyl-2-pentenoic acid
草莓醛 ethyl methylphenylglycidate
草魚 Chinese ide; grass carp
草菇 Chinese mushroom; paddy straw mushroom; padi-straw mushroom; straw mushroom
草蜀黍 grass sorghum
草對蝦 grass prawn
草蒿腦 estragole
草酸 oxalic acid
草酸鉀 potassium oxalate
草酸鹽 oxalate
草蝦 grass prawn; tiger prawn
草藥 herb
草藥的 herbal
草藥茶 herb tea
茵陳蒿 estragon; tarragon
茵陳蒿油 estragon oil; tarragon oil
茵絲 spawn
茴芹 anise; anise seed; aniseed; Chinese aniseed; chervil; star anise
茴芹油 anise oil
茴香 anise; anise seed; aniseed; fennel
茴香伏特加酒 anise vodka
茴香利口酒 anesone
茴香油 anise oil; fennel oil
茴香籽 fennel seed
茴香酒 anisette
茴香基丙酮 anisyl acetone
茴香烯 anethene
茴香腦 anethole
茴香酮 anisyl acetone
茴香酸 anisic acid
茴香酸乙酯 ethyl *p*-anisate
茴香醇 anisyl alcohol
茴香醚 anisole
茴魚 grayling
茶 tea
茶几 coffee table; teapoy; tea table
茶包 tea bag
茶卡凝乳(印度) chakka
茶末 broken tea; fannings; tea fannings
茶汁 tea liquor
茶多酚 tea polyphenols
茶尖 tea shoot
茶尖第四片葉 souchong
茶托 saucer
茶皂苷 thea saponin
茶具 tea equipage
茶具抹布 tea cloth
茶具組 tea service; tea set
茶杯 tea cup
茶室 tea room; tea shop; tea house
茶籽油 tea seed oil
茶屑 fannings; tea fannings
茶桌 tea table
茶素 yerbine
茶胺 theamin
茶胺酸 teanine; theanine
茶匙 teaspoon
茶袋 tea bag
茶單寧 tea tannin
茶壺 teakettle; teapot; urn
茶壺〔保溫〕套 tea cosy; tea cozy
茶渣 tea leaves
茶黃素 theaflavin
茶園 tea garden; tea plantation
茶會 tea fight; tea party
茶腸 teewurst
茶葉 tea; tea leaf
茶葉末 fannings; tea fannings

茶葉兒茶素 tea catechin
茶葉兒茶酸 tea catechin
茶葉香精油 theol
茶葉揉捻 tea rolling
茶葉碎屑 broken tea
茶葉嫩芽 tea shoot
茶葉種植業者 tea grower
茶葉箱 tea chest
茶葉罐 caddy; tea caddy
茶葉鹼 theophylline
茶道 tea ceremony
茶碟 saucer
茶精 caffeine
茶盤 tea board; tea plate; tea tray
茶盤蓋布 tea cloth
茶樹 tea
茶樹嫩芽 flush
茶樹叢 tea bush
茶磚 brick tea; cake tea; compressed tea; tea brick; tile tea
茶館 tea room; tea shop; tea house
茶點 eating between meal; refection; refreshment
茶點〔甜〕麵包 tea bread
茶點的小甜點 tea cake
茶點室 refreshment room
茶點時間 teatime
茶點推車 tea trolley
茶濾 tea strainer
茶鹼 theine
茉莉〔薄荷〕甜酒 julep
茭白 Indian rice; water rice; water bamboo; wild rice
茭白筍 water bamboo
茭筍 water rice
荖藤 betel
茜紅 alizarin red
茜素 alizarin
茜黃 alizarin yellow
茼蒿 garden chrysanthemum; garland chrysanthemum
茯苓 tuckahoe
茯苓聚糖 pachyman
茯苓酸 pachymic acid
茯苓糖 pachymose
蚌蛤 clam
衰老 senility
衰老的 senile
衰弱 marasmus
衰變 decay
記錄溫度計 recording thermometer
貢丸 minced pork ball
貢蒿萜酮 carvone
貢蒿萜醇 carveol
起子 starter
起子培養物 starter culture
起司 cheese
起司色素 cheese colo(u)r
起司糕 cheese cake
起泡 effervescence; foaming; yeastiness
起泡〔沫〕 froth
起泡作用 barbotage
起泡性 frothiness; whipping quality
起泡沫 spumescence
起泡沫的 yeasty
起泡的 effervescent; frothy
起泡能力 whipping ability
起泡發酵 foam fermentation
起泡劑 creaming agent; foamer; frother; frothing agent; foaming agent
起氣泡 blister
起晶點 crystallizing point
起酥油 shortening
起酥計 shortometer
起皺 shrivel
送牛奶車 dandy cart
送奶〔工〕人 milkman
送飯菜升降機 dumbwaiter

逆反應 reversion
逆河產卵的鮭 grilse
逆流 counter-current; counterflow; reverse flow
逆流多段萃取 counter-current multistage extraction
逆流式乾燥機 contraflow dryer
逆流式噴霧乾燥機 counter-current spray dryer; counterflow spray dryer
逆流式熱交換器 upstream heat exchanger; counterflow heat exchanger
逆流式隧道乾燥機 counterflow tunnel dryer
逆流冷凝器 counterflow condenser
逆流流動 counter-current flow
逆流乾燥 counterflow drying
逆流提取 counter-current extraction
逆流萃取 counter-current extraction
逆滲透 reverse osmosis
逆擴散 counter diffusion
逆轉現象 reversion
迷迭香 rosemary
迷迭香油 rosemary oil
退化 degeneration
退火 annealing
退色 colo(u)r fading
迴腸 ileum
迴腸炎 ileitis
迴轉式食品架 dumbwaiter
追船魚 rudderfish
追熟 after-ripening
酒 drink; jig water; joy water; rack; squirrel dew; vin; vino; wet; wet goods; wine
酒化酶 zymase
酒心 liqueur center
酒仙 wine bag; winebibber; wine bowl
酒母 seed mash
酒石 cream of tartar; potassium acid tartrate; potassium bitartrate; potassium hydrogen tartrate; tartar; wine stone
酒石黃 tartrazine
酒石酸 tartaric acid
酒石酸二乙酯 diethyl tartrate
酒石酸二甲 dipotassium tartrate
酒石酸氫鉀 cream of tartar; potassium acid tartrate; potassium bitartrate; potassium hydrogen tartrate; tartar
酒石酸氫膽鹼 choline bitartrate
酒石酸硬脂酰酯 stearyl tartrate
酒石酸鈣 calcium tartrate
酒石酸鈉 sodium tartrate
酒石酸鉀 potassium tartrate
酒石酸鉀鈉 Rochelle salt; sodium potassium tartrate
酒石酸酯 tartrate
酒石酸鹽 tartrate
酒曲 distillers' yeast; distillery yeast
酒色 vinosity
酒吧 bar; groggery; public house; pub; taphouse; taproom
酒吧間 saloon
酒尾 afterrun
酒味 vinosity
酒味果汁 cordial punch
酒味強烈的 nappy
酒杯 cup; stoup; wine glass
酒泥 lees; wine lees
酒的 vinic
酒的酸敗 acescence
酒花油 hops oil
酒垢 crust
酒度計 oenometer
酒後的無酒精清淡飲料 chaser
酒香 bouquet
酒香的 oenanthic
酒宴 drinking bout; wassail

酒席 feast
酒徒 tippler; toper; winebibber
酒鬼 guzzler; sot; wino
酒商 vintner
酒桶 barile; tierce; tun
酒桶內容物 cuvee
酒瓶 wine bottle
酒瓶架 cellaret; tantalus
酒瓶軟木塞 wine cork
酒粕 distillers' grains
酒堡 chateau
酒壺 flagon; stoup
酒廊 cocktail lounge
酒渣 wine lees
酒窖 buttery; cellar; vault; vintry; wine cellar
酒量大的人 soaker; tippler
酒椰 raffia; raphia; wine raffiapalm
酒碗 wine bowl
酒精 alcohol; ethanol; ethyl alcohol; grain alcohol; spirit; spirit of wine
酒精不可溶的 alcohol-insoluble
酒精中毒 alcoholism
酒精比重計 alcohol hydrometer; alcohol gauge; alcoholimeter; alcoholmeter; alcoholometer
酒精加強啤酒 spiked beer
酒精可溶的 alcohol-soluble
酒精含量 alcoholicity
酒精含量高的 nappy
酒精含量測定法 alcoholometry
酒精含量超過標準的 overproof
酒精性飲料 alcoholic beverage; alcoholic drink
酒精性興奮劑 alcoholic cordial
酒精抽提 alcoholic extraction
酒精的 alcoholic; spirituous
酒精表 alcohol gauge
酒精度 alcoholicity; proof
酒精度低的 low-proof
酒精計 alcohol gauge; alcoholimeter; alcoholmeter; alcoholometer
酒精強度 alcohol strength
酒精測定法 alcoholometry
酒精發酵 alcohol fermentation; alcoholic fermentation
酒精發酵的廢液 alcohol slop
酒精萃取 alcoholic extraction
酒精飲料 intoxicating beverage; spirits; spirituous liquor
酒精蒸餾 alcohol distillation
酒精酵母 alcohol yeast
酒精儀 alkograph
酒精廢液 alcohol slop
酒精廠 alcohol plant; grain distillery
酒精醋 spirituous vinegar
酒精凝固試驗 alcohol coagulation test
酒精凝集裝置(釀酒用) vinificator
酒精濃度 alcohol strength
酒精濃度低於標準的 underproof
酒精燈煮水器 etna
酒增味劑 dosage; dose
酒槽 vinasse
酒膠糖 wine gum
酒質 vinosity
酒醉 ebriety; intoxication
酒劑 vinum
酒糖心 liqueur center
酒頭 first cut
酒館 café; grogshop; pub; taphouse
酒燴肉卷 braciola
酒糟 distillers' grains; distillery residue; lees; pot ale; spent brewer's grain; spent grain; wine lees
酒糟水 spent wash
酒餾殘液 spent wash
酒醪 distillers' s wort
酒類 spirituous liquor

酒類飲料 potation
酒體 wine body
酒櫥 cellaret
配方 formula; recipe
配料 ingredient
配料器 proportioner
配備 outfit
配給 ration
配給口糧 tommy
配給量 ration
配給麵包 rations bread
配菜 side dish; side plate; trimmings
配種 hybridization
配製牛乳 mock milk
配製食品 formulated food
配製茶 tea blend
配劑 elixir
配糖物 glycoside
配糖基 aglycon
配糖體 glucoside
配聯糖 coupling sugar
釘箱機 stitcher
針入度計 penetrometer
針孔偵測器 pin hole detector
針樅啤酒 spruce beer
針劑(一次用量) ampoule
針頭魚 pinfish
針點狀菌落 pin point colony
釜 kettle; saucepot
釜餾物 stillage
閃光鱘魚子醬 sevruga
閃急蒸發 flash evaporation
閃凍 flash freezing
閃蒸冷卻 flash cooling
除毛 epilation
除去內臟 disembowel
除去咖啡因 decaffeination
除石機 stoner; stoning machine
除冰裝置 deicer
除芒 hummel
除芒器 hummeller
除沫器 foam destroyer
除泡沫 despumation
除沫劑 antifoaming agent; defoaming agent; kilfoam
除垢器 scaler
除砂槽 sand trap
除胚芽機 disgerminator
除香檳酒泥機 disgorger
除氣 degassing
除氣〔作用〕 degasification
除氧劑 deoxidant
除臭 deodorization
除臭劑 deodorant; deodorizer
除草 weed
除草劑 herbicide; phytocide; weed control chemicals; weed killer
除草鏟 spud
除酒泥機 disgorger
除酒渣 disgorgement
除莠劑 phytocide; weed control chemicals; weed killer
除雪器 flanger
除渣杓 scummer
除絲器 silker
除菌 degerm
除溶劑 desolventization; desolventize
除溶劑器 desolventizer
除塵 dedust
除塵器 dust separator; duster
除碳酸 decarbonation
除酵母泥渣 disgorgement
除穀物的芒 hummel
除蔓機 viner
除濕〔作用〕 dehumidification
除濕劑 exsiccator
除濕機 dehumidifier
除鍋垢器 scale breaker

除霜 defrost; defrostation
除霜裝置 defroster
除霜機 defrosting machine
除蟲 disinfestation; pest control
除蟲菊 pyrethrum
除蟲菊素 pyrethroid
除蟲菊酯 pyrethrin
除蟲劑 disinfestant; pesticide
除蟲藥 pest control chemical
除雜草 weed control
除鐵機 metal eliminator
除鱗器 scale breaker
飢餓 starvation
飢餓者 starveling
飢餓療法 starvation therapy
馬 equine; horse
馬丁尼酒 martini
馬丁香 palmarosa
馬口鐵皮 can plate; sheet tin; tin; tinplate
馬口鐵罐 tin can
馬氏珍珠貝 pearl oyster
馬加鰆 Japanese mackerel
馬奶酒 koumiss; kumiss; kumys
馬西亞魚 mahseer
馬尿酸 benzoylglycine; hippuric acid
馬尾蟹 horse hair crab
馬尾藻 sargasso; sea grape
馬沙拉白葡萄酒 marsala
馬沙拉甜點 marsala
馬乳酒 koumiss; kumiss; kumys
馬來乳膠 gutta-percha
馬來菜豆 malay bean
馬來蓮霧 jambo
馬來樹膠 gutta-percha
馬來雞 Malay fowl
馬奇雅圖咖啡 caffe macchiato; espresso macchiato; macchiato
馬拉巴栗 malabar chestnut
馬拉里氏殺菌 mallorizing
馬拉里氏殺菌器 mallorizer
馬拉斯加櫻桃 marasca (cherry); maraschino cherry
馬拉斯加櫻桃酒 maraschino
馬拉斯金櫻桃 marasca (cherry); maraschino cherry
馬林魚 marlin
馬勃〔菌〕 puffball
馬看山核桃 big-bud hickory; black hickory
馬郁蘭 marjoram; sweet marjoram
馬郁蘭油 marjoram oil; sweet marjoram oil
馬提尼酒 martini
馬菇 horse mushroom
馬葛爾魚 meagre
馬鈴薯 murphy; potato; spud; white potato
馬鈴薯片 potato chip; potato crisp
馬鈴薯去皮機 potato peeler
馬鈴薯泥 mashed potato; mushy potato (es); potato puree
馬鈴薯泥搗碎器 potato masher
馬鈴薯威士忌〔酒〕 potato whisky
馬鈴薯粉 potato flour
馬鈴薯條 French fries; French fry
馬鈴薯球蛋白 tuberin
馬鈴薯雪片 potato flake
馬鈴薯塊莖 potato tuber
馬鈴薯葡萄糖洋菜〔培養基〕potato dextrose agar
馬鈴薯葡萄糖瓊脂〔培養基〕potato dextrose agar
馬鈴薯餅 potato pancake
馬鈴薯澱粉 farina; potato starch
馬鈴薯麵包 boxty bread
馬爾托環醇 maltol
馬爾薩拉葡萄酒 marsala
馬德拉海棉蛋糕 maderia cake
馬德拉甜酒 maderia

馬齒玉米 dent corn
馬蹄 water chestnut
馬蹄形餐桌 wine table
馬錢子鹼 strychnine
馬頭魚 tilefish
馬糞紙 strawboard
馬薯 water chestnut
馬黛茶 mate; yerba mate
馬檳榔 mabinlang
馬鞭草 vervain
馬鞭草油 verbena oil
馬鞭草屬植物 verbena
馬蘇大麻哈魚 cherry salmon
骨 bone
骨化醇 calciferol; ergocalciferol; vitamin D_2
骨灰 bone ash
骨肉粉 bone tankage
骨肉雜 tankage
骨架 skeleton
骨盆 pelvis
骨粉 bone dust; bone meal
骨鈣流失或缺乏 halisteresis
骨膠原 ossein
骨質疏鬆症 osteoporosis
骨質磷酸鹽 bone phosphate
骨骼 skeleton
骨骼肌 cross striated muscle; skeletal muscle; striated muscle
骨骼肌纖維 skeletal muscle fiber
骨髓 bone marrow; marrow; medulla
高三酸甘油酯血〔症〕 hypertriglyceridemia
高大環柄菇 parasol mushroom
高山士多啤梨 alpine strawberry
高山色拉米腸 alpine salami
高山草莓 alpine strawberry
高山茶 high grown tea
高山薩拉米香腸 alpine salami
高不飽和脂肪酸 highly unsaturated fatty acid
高分子 macromolecule
高分子電解質 polyelectrolyte
高支鏈澱粉 high amylopectin (starch)
高水分食品 high moisture food
高加索乳〔酸〕桿菌 *Lactobacillus caucasicus*
高加索酸凝乳 Caucasian milk
高半胱胺酸 homocysteine
高甘油〔起〕酥油 high ratio shortening
高甘油脂肪 high ratio fat
高甘油酯血〔症〕 hyperglyceridemia
高甲氧基果膠 high methoxyl pectin
高成分蛋糕 high ratio cake
高羊肚菌 black morel
高血糖〔症〕 hyperglycemia
高血壓 hypertension
高含糖量的 high-sugar
高尿酸血〔症〕 hyperuric(a)emia
高身罐 tall can; tall tin
高果糖玉米糖漿 high fructose corn syrup (HFCS)
高果糖葡萄糖漿 high fructose glucose syrup (HFGS)
高果糖漿 high fructose corn syrup (HFCS)
高果糖糖漿 high fructose syrup
高杯拉雞尾酒 cobbler
高杯雞尾酒 highball
高油脂曲奇餅 shortbread cookie
高直鏈澱粉玉米 amylomaize
高空輸送機 overhead conveyor
高度不飽和脂肪酸 highly unsaturated fatty acid
高峰〔氏〕澱粉酶 taka-amylase; takadiastase
高效液相色譜法 high performance liquid chromatography (HPLC)
高效液相層析法 high performance liquid chromatography (HPLC)
高效碾磨 high milling

高純度酒精 silent spirit
高級酒 fine wine
高級商品 prime goods
高級糖蜜 high test molasses
高胱胺酸 homocystine
高脂〔質〕脂蛋白 high lipid lipoprotein
高脂血〔症〕 hyperlip(a)emia
高脂肪食品 fatty food
高脂〔肪〕飲食 high fat diet
高脂稀奶油 concentrated cream; heavy cream; thick cream
高脂鮮奶油 heavy cream
高脂鮮乳油 double cream
高能量小吃 gorp
高酒度飲料 high-proof beverage
高酒精飲料 strong drink
高酒精蒸餾酒 high wine
高密度脂蛋白 high density lipoprotein (HDL)
高密度脂蛋白膽固醇 high density lipoprotein cholesterol
高密度營養 nutrient-dense
高密度營養食物 nutrient-dense food
高粱 grass sorghum
高產率 high yield
高蛋白〔質〕飲食 high protein diet
高蛋白〔質〕膳食 high protein diet
高蛋白〔質〕麵包 high protein bread
高蛋白乳 protein milk
高蛋白假絲酵母 Candida utilis
高速離心 supercentrifugation
高速離心機 high speed centrifuge; supercentrifuge
高麥芽糖〔糖〕漿 high maltose syrup
高單甘油酯〔起〕酥油 high ratio shortening
高單甘油酯脂肪 high ratio fat
高湯 soup stock; stock
高筋麵粉 aleuronat; gluten flour; high strength flour; macaroni flour
高絲胺酸 homoserine
高鈉飲食 high sodium diet
高溫分解 pyrolysis
高溫巴氏殺菌 high temperature pasteurization
高溫巴斯德殺菌 high temperature pasteurization
高溫肉 semi-condemned meat
高溫計 pyrometer
高溫烘烤 high bake
高溫殺菌的食品 sterile food
高溫短時 high-temperature short-time (HTST)
高溫短時〔間〕滅菌 high-short sterilization
高溫短時巴氏殺菌 high-temperature short-time pasteurization (HTST pasteurization)
高溫短時巴斯德殺菌 high-temperature short-time pasteurization (HTST pasteurization)
高溫短時殺菌法 high-temperature short-time process (HTST process)
高溫短時滅菌 high-temperature short-time sterilization (HTST sterilization)
高溫溫室 hothouse
高溫熬製 high boiled
高粱 durra; grain sorghum; jowar; kafir; kafir corn; kaoliang; milo; panic grass; sorghum
高粱米 sorghum rice
高粱色素 kaoliang colo(u)r
高粱紅 sorghum red
高粱啤酒 kafir beer
高粱酸 kafiroic acid
高粱醇溶蛋白 kafirin
高粱糖蜜 sorghum molasses

高粱糖漿 sorghum molasses; sorghum syrup; sorgo syrup
高粱屬 *Andropogon*
高腳酒杯 chalice; goblet
高酯化果膠 high ester pectin
高滲〔壓〕的 hypertonic
高滲〔壓〕溶液 hypertonic solution
高熔點脂肪 solid fat
高碳水化合物飲食 high carbohydrate diet
高精胺酸 homoarginine
高酸性食品 high acid food
高層發酵 high fermentation
高熱量食品 high calorie food
高調味的 high seasoned
高濃度 heavy concentration
高盧瓜（以色列） gallia
高糖用麵粉 high ratio flour
高糖高水分的蛋糕 high ratio cake
高糖高水分蛋糕用的麵粉 high ratio flour
高糖量轉化糖蜜 high test molasses; invert molasses
高錳酸鉀 potassium permanganate
高頻加熱 diathermic heating; dielectric heating; high frequency heating
高頻加熱乾燥 dielectric drying
高頻加熱滅菌〔法〕 electronic sterilization
高頻加熱解凍 dielectric defrosting; dielectric thawing
高頻乾燥 high frequency drying
高頻滅菌 high frequency sterilization
高壓〔殺菌〕釜 retort
高壓〔蒸汽〕滅菌器 autoclave
高壓滅菌釜 retort sterilizer
高壓蒸汽去皮〔法〕 high pressure steam peeling
高壓蒸汽處理 autoclaving
高壓蒸汽滅菌 autoclaving; autoclave sterilization
高壓蒸汽滅菌〔法〕 high pressure steam sterilization
高壓蒸汽滅菌器 autoclave sterilizer
高壓蒸汽噴射滅菌法 uperization
高壓蒸煮鍋 digester
高嶺土 kaolin
高嶺石 kaolinite
高膽固醇血〔症〕 hypercholesterolemia
高醣飲食 high carbohydrate diet
高黏度澱粉 high viscosity starch
高檸檬酸 homocitric acid
高麗馬加鰆 Korean mackerel
高麗參 ginseng; sang
高麗菜 cabbage; white cabbage
高麗菜捲 cabbage roll
高麵筋麵粉 glutinous flour
高鐵肌紅蛋白 metmyoglcbin
高鐵血色原 hemichrome
高鐵血紅蛋白 methemoglobin
高纖〔維〕飲食 high fiber diet
高體若鰺 kingfish
高罐 tall can; tall tin
高鹽度 high salinity
高鹽飲食 high salt diet
鬼頭刀 dolphin fish
桉油精 eucalyptol
桉腦 eucalyptol
桉葉〔油〕素 eucalyptol
桉葉油 eucalyptus oil
桉葉素 cineole
桉葉腦 eucalyptol
桉葉醇 eucalyptol
桉膠 eucalyptus gum
桉樹 eucalyptus
桉樹油 eucalyptus oil
桉樹烯 eucalyptene
桉樹腦 cineole; eucalyptol
桉樹膠 eucalyptus gum
砵酒 port

豚肉 D-Rump; rump

蚕豆 tick bean

豇豆 asparagus bean; black bean; black-eyed bean; black-eyed pea; catjang; China bean; cowpea; frijole; yardlong bean

豇豆球蛋白 vignin

豇豆屬 *Vigna*

酐 anhydride

酐烙酸 3-methyl pentanoic acid

陟釐屬 *Batrachospermum*

飢餓 abrosia; famine; hunger

飢餓療法 hunger cure; limotherapy; nestotherapy; peinotherapy

十一畫

乾〔式〕冷卻 dry cooling
乾〔製〕豆腐 dried beancurd
乾牛肉條 jerky; jerked beef
乾冰 dry ice
乾式研磨〔法〕 dry grinding
乾式鹼法去皮 dry caustic peeling
乾灰化 dry ashing
乾灰法 dry ashing
乾肉粉 ammonite
乾肉條 biltong; jerk; jerked meat; jerky; tasajo
乾肉瓶 pemmican
乾肉棒 meat bar
乾李子 dried plum
乾性油 drying oil
乾果 dry fruit
乾果皮 mandarin peel
乾果李 prune
乾果李汁 prune juice
乾法亞硫酸處理 dry sulfitation; dry sulphitation
乾法煉〔畜〕油 dry rendering
乾物〔質〕含量 solid content
乾物質 dry matter
乾的 dry; moisture-free
乾的帶殼咖啡豆 parchment coffee
乾孢子 xerospore
乾重 dry weight
乾食〔法〕 xerophagia
乾香腸 dry sausage
乾香檳酒 dry champagne
乾扇貝 dried scallop
乾草（飼料） hay
乾混〔法〕 dry mixing
乾球溫度計 dry bulb thermometer
乾貨 sun dried food
乾貨貯藏 dry storage
乾透空氣 bone dry air
乾魚 dried fish
乾渣機 pulp dryer
乾硬 hard drying
乾硬香腸 hard dry sausage
乾椰子肉 copra
乾椰子瓤 copa
乾葡萄酒 dry wine
乾裝 dry pack
乾酪 cheese
乾酪上的黴 cheese mould
乾酪孔 cheese eyeing
乾酪片 sliced cheese
乾酪包布 cheese cloth
乾酪皮 cheese rind
乾酪吐司 rarebit; Welsh rabbit; Welsh rarebit
乾酪成熟 cheese ripening
乾酪肉糕 cheese and meat loaf
乾酪肉羅夫 cheese and meat loaf
乾酪色素 cheese colo(u)r
乾酪刨 cheese grinder
乾酪乳〔酸〕桿菌 *Lactobacillus casei*

乾酪乳皮 cheese whey
乾酪乳清 cheese whey
乾酪取樣器 cheese trier
乾酪味的 cheesy
乾酪的 cheesy
乾酪毒素 tyrotoxin
乾酪洋葱雞肉煨飯 risotto
乾酪研磨機 cheese grinder
乾酪風味 cheese flavo(u)r; cheesy flavo(u)r
乾酪食品 cheese food
乾酪粉 cheese powder
乾酪素 casein
乾酪桶 chesset; chisset; chessel; cheese vat
乾酪發酵劑 cheese starter
乾酪菌元 cheese starter
乾酪榨乾機 cheese press
乾酪餅 cheese cake; ramekin; ramequin
乾酪箍 cheese hoop
乾酪槽 cheese vat
乾酪模 cheese mould
乾酪模子 cheese hoop; chessel
乾酪熟成 cheese curing; cheese ripening
乾酪凝乳切碎機 curd mill
乾酪凝塊 cheese coagulum; cheese curd
乾酪壓榨機 cheese press
乾酪壓模機 cheese mould
乾酪蠟 cheese wax
乾蒸 dry cook
乾蒸汽 dry steam
乾製品 dried product
乾酵母 dried yeast; dry yeast
乾熟香腸 dry sausage
乾熱烹調法 dry heat cooking
乾熱滅菌〔法〕 dry air sterilization; dry heat sterilization; hot-air sterilization
乾熱滅菌器 hot-air sterilizer
乾豌豆 parched peas
乾醃〔法〕 dry curing; dry salting
乾醃肉 dry cured meat; dry packed meat
乾醃的 dry salted
乾凝膠 xerogel
乾橘皮 dried tangerine peel
乾磨 dry milling
乾濕計 psychrometer
乾濕球溫度計 wet and dry bulb thermometer
乾濕球濕度計 dry and wet bulb thermometer; psychrometer; physical hygrometer
乾燥 dry; drying; desiccate; siccation
乾燥〔作用〕 desiccation; exsiccation
乾燥包裝 dry pack
乾燥失重 drying shrinkage
乾燥助劑 drying aid
乾燥果實 dried fruit
乾燥物 desiccate
乾燥的 desiccative; unwatered
乾燥室 drying chamber; hothouse
乾燥度 dryness
乾燥食品 dried food
乾燥殺菁 dry blanch
乾燥減重 drying loss
乾燥損失 drying loss
乾燥箱 cabinet dryer; drying cabinet; drying oven
乾燥蔬菜 dried vegetable
乾燥劑 desiccant; desiccating agent; dehydrater; dehydrator; dehydrant; desiccative; drying agent; drier; dryer; exsiccant; siccative
乾燥器 desiccator; drier; dryer; exsiccator
乾燥機 drier; dryer
乾燥爐 drying oven
乾縮 drying shrinkage
乾薄鬆餅 flatbread
乾燻〔法〕 dry smoking

乾燻硫〔法〕 dry sulfitation; dry sulphitation
乾藏 dry storage
乾蘋果粒 apple nugget
乾麵包 toke
乾鹼去皮 dry caustic peeling
乾鹽腌法 dry salt cure
乾鹽腌製 country cure
偽虎鯨 false-killer whale
偽造 falsification
偽裝香蘑（蕈類） blue legs
假羊皮紙 pergament
假果 accessory fruit; false fruit; pseudocarp
假果糖 pseudofructose
假封 false seam
假封機 clincher machine; clincher
假根 rhizoid
假捲封機 can clincher
假球蛋白 pseudoglobulin
假單孢菌屬 *Pseudomonas*
假晶體 pseudocrystal
假絲酵母菌屬 *Candida*
假菌絲體 pseudomycelium
假菌落 pseudocolony
假種皮 aril
假酵母 false yeast
假酸 pseudoacid
假黴菌 pseudomycete
做麵包 breadmaking
健康 whole; health
健康的 sound
健康食品 health food
健康檢查 medical inspection
偶合糖 coupling sugar
偶氮〔二〕甲酰胺 azodicarbonamide
偶氮化合物 azo compound
偶氮玉紅 azorubin(e); carmoisine
偶氮胭脂紅 azocarmine
偶氮蛋白 azoprotein
偶氮藍 azo blue
偵察器 detector
側耳屬（食用菌） *Pleurotus*
側肌 lateral muscle
側枝芽 sucker
側金盞戊糖 adonose
側柏葉油 cedar leaf oil; thuja oil
側基 side group
側穗燕麥 banner oat
側鏈 side chain
偏二氯乙烯 vinylidene chloride
偏光鏡 polarizer; polariscope
偏亞硫酸鈉 sodium metabisulfite
偏亞硫酸鉀 potassium metabisulfite
偏食 unbalanced diet; monophagia
偏差 deviation
偏酒石酸 metatartaric acid
偏硅酸鈉 sodium metasilicate
偏酸式亞硫酸鹽 metabisulfite
偏磷酸 metaphosphoric acid
偏磷酸鈉 sodium metaphosphate
偏磷酸鉀 potassium metaphosphate
偏磷酸鹽 metaphosphate
剪力 shear
剪切 shear
剪切力 shear force
剪切機 snipper
剪切黏度 shear viscosity
剪支酶 debranching enzyme
剪毛機 scraper
剪床 shears
剪斷 snip
副大腸桿菌 paracolon bacilli
副甲狀腺 parathyroid gland
副白蛋白 paralbumin
副肌凝蛋白 paramyosin
副作用 side effect
副乳酸 paralactic acid
副食品 non-staple food; subsidiary food

副球蛋白 paraglobulin
副產物 coproduct
副產品 afterproduct; by-product; coproduct; residual product; secondary product; second product
副蛋白質 paraprotein
副發酵〔作用〕 side fermentation
副菜 entrée
副傷寒沙門氏菌 *Salmonella paratyphi*
副傷寒沙門桿菌 *Salmonella paratyphi*
副酪蛋白 paracasein
副樣 subsample
副凝乳酶 parachymosin
動物〔性〕蛋白 animal protein
動物內臟 umbles
動物化學 zoochemistry
動物色素 animal pigment
動物固醇 zoosterol
動物性食品 animal food
動物明膠 animal gelatin(e)
動物油 animal oil
動物油脂 animal grease
動物甾醇 zoosterol
動物毒素 scaritoxin; zootoxin
動物脂 adeps; tallow
動物脂的 tallowy
動物脂肪 adipose; animal fat; animal tallow
動物寄生菌 zoobiotic
動物產品 animal product
動物細胞 zooblast
動物蛋白水解物 hydrolyzed animal protein (HAP)
動物腸衣 animal casing
動物膠 gelatin(e)
動物學 zoology
動物澱粉 animal starch; glycogen; liver starch
動物蠟 animal wax
動脈 artery
動脈〔粥狀〕硬化症 atherosclerosis
動脈硬化〔症〕 arteriosclerosis
動能 kinetic energy
動態性〔質〕 dynamic property
動黏度 dynamic viscosity
匙 spoon
匙用食物 spoon meal
匙用麵包 spoon bread
匙吻鱘 paddlefish
區 zone
區分 fractionation
區帶電泳〔法〕 zone electrophoresis
區帶離心〔法〕 zonal centrifugation
参比樣品 reference sample
参比標準 reference standard
參考樣品 reference sample
參考標準 reference standard
參茨 bitter cassava
曼蜜蘋果 mammee; mammey
商用級 U.S. Commercial
商用菌株 commercial strain
商店 store
商品 commercial goods
商品展銷室 wareroom
商品萬用條碼 universal product code (UPC)
商品壽命 marketable life
商品儲藏室 wareroom
商業魚類 commercial fishes
商業滅菌 appertizing
商業滅菌法 appertization; commercial sterilization
商標 trademark; label
商標名稱 brand name
啤酒 barley-bree; barley-broo; barley broth; beer
啤酒〔靴形〕杯 beer boot
啤酒火腿腸 bierschinken

啤酒用大麥 brewers' barley
啤酒用大麥芽 brewers' barley malt
啤酒用粗碾穀粒 brewers' grit
啤酒用麥芽 brewers' malt
啤酒用麥芽汁 brewers' wort
啤酒用碎穀 brewers' grit
啤酒味 beeriness
啤酒店 beer hall; beerhouse; porterhouse; brasserie
啤酒店女主人 alewife
啤酒杯 beer mug; beer tumbler; mug; stein
啤酒泡沫持久度 sigma value
啤酒泡沫穩定度 sigma value
啤酒花 hop(s); *Humulus lupulus*
啤酒花油 hop oil
啤酒花的 hoppy
啤酒花苦味物質 hop bitter substance
啤酒花苦味素 lupulin
啤酒花香的 hoppy
啤酒花浸取槽 hop-back
啤酒花浸膏 hop extract
啤酒花提取物 hop extract
啤酒花萃 hop extract
啤酒花樹脂 hop resin
啤酒屋 brasserie
啤酒後發酵 lager fermentation
啤酒苦味劑 quassia
啤酒香腸 bierwurst
啤酒師 brewer
啤酒桶 beer barrel; tun
啤酒粕 brewers' grain
啤酒麥芽糟 brewers' grain
啤酒發酵桶 tun
啤酒過濾器 beer filter
啤酒酵母 beer yeast; brewers' yeast
啤酒酵母〔菌〕 *Saccharomyces cerevisiae*
啤酒廢麥糟 spent brewer' grain; spent grain
啤酒廠 brewery; brewhouse
啤酒醋 alegar
啤酒罐 beer can; growler
啤酒釀造師 brewmaster
售奶〔工〕人 milkman
售後服務 after service
啜 sup
啜食 slurp
啜飲 sip; sup
圈狀麵包 wreath
國立衛生研究所(美國) National Institute of Health (NIH)
國光蘋果 Ralls
國家科學院 National Academy of Sciences
國家標準 National Standard
國際化學聯合會 International Union of Chemistry (IUC)
國際生物化學聯合會 International Union of Biochemistry (IUB)
國際仲裁標準 International Reference Standard
國際冷凍學會 International Institute of Refrigeration (IIR)
國際乳品聯合會 International Dairy Federation (IDF)
國際乳業聯盟 International Dairy Federation (IDF)
國際果汁聯合會 International Fruit Juice Union (IFJU)
國際果汁聯盟 International Fruit Juice Union (IFJU)
國際油脂研究學會 International Society for Fat Research (ISFR)
國際食品法典 Codex Alimentarius
國際食品法典委員會 Codex Alimentarius Commission
國際食品科技聯盟 International Union of Food Science and Technology (IUFoST)

國際食品科學技術協會 International Union of Food Science and Technology (IUFoST)
國際食品添加物法典委員會 Codex Committee on Food Additive
國際食品資訊服務中心 International Food Information Service
國際食品製造協會 International Food Manufacture Association (IFMA)
國際食品標準 International Food standard
國際食品衛生法典委員會 Codex Committee on Food Hygiene
國際容許〔殘留〕量 international tolerance
國際消費者同盟 International Office of Consumers Union (IOCU)
國際單位 international unit (IU)
國際對照標準 International Reference Standard
國際製冷技術學會 International Institute of Refrigeration (IIR)
國際標準 International Standard
國際標準〔化〕組織 International Organization for Standardization (ISO)
國際穀物化學協會 Association Internationalede Chimie Cerealiene
國際稻米研究所 International Rice Research Institute (IRRI)
國際營養師協會 International Dietetic Association (IDA)
國際藥典 International Pharmacopoeia (IP)
堅牢紅 fast red E; fast red
堅牢紅色澱 fast red lake
堅牢黃 fast yellow
堅牢綠 fast green; fast green FCF
堅牢綠鋁色澱 fast green FCF aluminum lake
堅果 hard nut; nut; nut fruit
堅果仁 nut meat
堅果仁糊 nut paste
堅果巧克力糕 brownie
堅果油 nut oil
堅果核仁 nutmeat
堅果殼 nutshell
堅果鉗 nutcracker
堅果樹 nuttree
堅果糖 praline
堅硬性 solidness
堅硬度 firmness
堅韌 tenacity
堅韌質地的乾酪 leathery cheese
堅實 firmness
堅靭 toughness
堆填區 strategic landfill
堆醃 kench salting
堆積發酵 heap fermentation
堆鹽漬 kench; kench curing
基本 fundament
基本作物 basic crop
基本食物 basic foodstuff
基本原料 base stock
基本配料 basic ingredient
基本培養基 minimum medium
基本組份 base stock
基本農作物 basic; basic cro
基本農產品 basic crop
基因 gene
基因位點 gene locus
基因治療 gene therapy
基因表現 gene expression
基因〔改造〕食物 gene hcally modified (GM)food
基因型 genotype
基因突變 gene mutation
基因重組 gene recombination
基因組 genome

基因圖譜 gene map; genetic map
基因選殖 gene cloning
基因療法 gene therapy
基料 base stock
基普（453.59 kg或1000 lb） kip
基普斯燻鯡魚 kipper
基質 substrate
基質蛋白質 stroma protein
基礎 base
基礎代謝 basal metabolism
基礎代謝率 basal metabolic rate (BMR)
基礎的 basic
基礎培養基 basal medium
棠梨 birchleaf pear
培根 bacon
培根肉片 sliced bacon
培根肉整形壓製機 bacon press(er)
培養 cultivation; culture; incubation
培養皿 cultured dish; petri dish; petri plate
培養物 culture
培養基 cultured medium; growth medium; medium
培養液 culture broth
培養瓶 culture bottle
培養試管 culture tube
培養酵母 culture yeast
培養箱 incubator
婀娜多籽 achiote
婀娜多紅色素 annotta; annatto; arnotta
婀娜多樹 achiote
婆羅門參 oyster plant; salsify; vegetable oyster
寄生 parasitism
寄生牛肝菌 parasitic bolete
寄生性營養 paratrophy
寄生蟲 parasite; vermin
寄生蟲病 parasitism; parasitosis
寄生蟲感染 parasitic infection
寄生麴菌 *Aspergillus parasiticus*
寄居蟹 hermit crab; soldier crab
密生西葫蘆 bush pumpkin
密封 airproof; seal; seal off; sealing
密封材料 packing material; sealer; sealing material
密封性 sealability
密封的 airtight
密封圈 packing ring
密封蓋 joint cap
密封劑 sealant
密度 density
密度計 densimeter
密閉式發酵罐 closed fermenter
密閉容器 tight container
密閉循環乾燥機 closed circuit dryer
密著包裝 skin packaging; skin tight packaging
密穗小麥 club wheat; cluster wheat
專用食品 specialpurpose food
專利 patent
專利藍V patent blue V
專利權 patent
專性嗜冷菌 obligate psychrophile
專性嗜熱細菌 obligate thermophilic bacteria
專性厭氧菌 obligate anaerobe
專性腐敗菌 obligate saprophyte
專性需氧菌 obligate aerobe
專家 specialist
屠刀 butcher knife
屠夫 butcher; flesher
屠前飼養 preslaughter care
屠前檢查 antemortem inspection
屠前檢驗 ante-mortem inspection
屠前護養 preslaughter care
屠後陳化 postmortem ag(e)ing
屠後僵直 postmortem rigidity
屠後熟成 postmortem ag(e)ing
屠後檢驗 postmortem inspection

屠後變化 postmortem change
屠宰 slaughter
屠宰工 butcher; slaughterer
屠宰台 killing bed
屠宰前飼養 preslaughter care
屠宰前護養 preslaughter care
屠宰設備 slaughtering facilities
屠宰場 abattoir; butchery; shambles; slaughterhouse; slaughtering plant
屠宰業 butchery
屠場 butchery
屠體 carcass; dressed carcass
屠體切割 dressing
屠體切割台 dressing table
屠體收率 carcass yield
屠體重 dressed weight; slaughter weight
屠體重量 carcass weight
屠體剖洗 carcass dressing
巢菜 vetch
巢蜜 comb honey; section honey
常吃水果的人 fruitarian
常見乳菇 pickle milk cap
常規檢驗 routine analysis
常量分析 macroanalysis
常量營養素 macronutrient
常溫 normal temperature; ordinary temperature
常綠果樹 evergreen fruit tree
常綠的 evergreen
常綠梨 evergreen pear
常綠植物 evergreen
常壓 ordinary pressure
常壓冷凍乾燥 atmospheric freeze drying
常壓烹煮 atmospheric cooking
常壓蒸發器 atmospheric evaporator
帶 chord(a); cord; zone
帶子 scallop muscle
帶冰結晶的冰淇淋 soggy ice cream
帶式烤爐 band oven
帶式乾燥機 band dryer; belt dryer; conveyor dryer
帶式輸送機 band conveyor; belt conveyor
帶肉果汁 comminuted juice
帶肉果汁加糖的不含酒精飲料 squash
帶肉豬肋骨 sparerib
帶肉蘋果汁 liquid apple
帶果肉果汁 fruit squash
帶果肉果汁飲料 fruit nectar
帶肩胛骨的前腿肉 rough picnic
帶柄水罐 pitcher
帶柄雪糕 ice lolly
帶原汁的 aujus
帶核水果 unpitted fruit
帶紋狗魚 banded pickerel; barred pickerel
帶翅胸肉 wing breast
帶骨火腿 bone-in ham; regular ham
帶骨的 bone-in
帶骨豬腰肉排 pork loin chop
帶域電泳〔法〕 zone electrophoresis
帶域離心〔法〕 zonal centrifugation
帶梗茶葉 stalky tea
帶甜味的 sweetish
帶莢咖啡豆 bullhead coffee
帶魚 belt fish; cutlass fish; frostfish; hairtail; silver belt fish
帶晶粒蜂蜜 granulated honey
帶殼水煮蛋 boiled egg
帶殼蛋 shell egg
帶殼穀粒 grout
帶微酸的 acidulous
帶碘化合物 iodophor
帶麩粗粉 middlings
帶麩粗粒小麥粉 adhering semolina
帶麩粗麵粉 middling flour; middlings
帶穗玉米 cob corn
康乃克油 cognac oil
康氏烟管魚 comet fish; cornetfish; flutemouth

康尼希雞 Cornish
康吉鰻 conger; conger eel
康考特葡萄 Concord grape
康科特葡萄 Concord grape
康培娜咖啡 caffe con montata; caffe con panna
康富力 comfrey
康雅〔白蘭地〕酒 cognac
康黴素 kanamycin
庸鰈 halibut
張 piece
張力 tensile force; tensile strength; tension
張力計 tensiometer
強力清潔劑 heavy duty detergent
強化〔營養素〕 enrichment
強化〔營養素〕米 enriched rice
強化〔營養素〕食品 enriched food
強化牛乳 fortified milk
強化食品 fortified food
強化培養基 enrichment medium
強化嬰兒奶粉 arlac
強化麵包 fortified bread
強壯劑 corroborant; invigorator; roborant
強制循環蒸發器 forced circulation evaporator
強制進料 positive feed
強制餵食 positive feed
強風冷卻 air-blast cooling
強風冷凍 air-blast freezing
強風冷凍機 air-blast freezer
強風解凍 air-blast defrosting
強風解凍機 air-blast thawer
強烈氣味 sharp odo(u)r
強菌培養基 enriched medium
強飼法 superalimentation
強酸 strong acid
強鹼 strong base
強鹼水 strong lye
彩印罐 lithographed can
彩虹香蘑（蕈類） strong scented blewit
彩帶麵 festonati
彩斑狀乳牛肝菌 variegated bolete
患者 patient
情人羊排 valentine chop
情人果 green mango
惕各酸乙酯 ethyl tiglate
惕各酸烯丙酯 allyl tiglate
惕各酸苄酯 benzyl tiglate
惕各醛 tiglaldehyde
戚風蛋糕 chiffon cake
掠食性動物 predator
控制 control
控制大氣貯藏 controlled atmosphere storage (CA storage); gas storage
控制氣體 controlled atmosphere
控氣包裝 controlled atmosphere packaging
控氣保藏 controlled atmosphere storage (CA storage); gas storage
捲心菜紅色素 red cabbage pigment
捲肋脊肉 rolled rib
捲形布丁甜點 roly-poly
捲狀物 roll
捲狀食品 rolled food
捲肩腿肉 rolled shoulder
捲封 seaming
捲封鈎 seam hook
捲封滾輪 seaming roll
捲封機 seamer
捲條脆餅 pretzel
捲筒粉 garganelli
捲酥 turnover
捲開罐 key opening can
捲醃燻肉 rolled bacon
捲餡餅 roll
捲縮 shrivel
捲邊滾輪 seaming roll
捲邊機 curling machine

探針 probe
探測器 detector
接合 conjugation
接合子 zygote
接合孢子 zygospore
接合菌 zygomycete
接收槽 receiving tank
接受性 acceptability
接枝澱粉 grafted starch
接骨木〔醬果〕 elderberry
接骨木果油 elderberry oil
接骨木果茶 elderberry tea
接菌 inoculate
接菌器 inoculator
接種 inoculate; inoculation
接種針 inoculating needle
接種廚 inoculating hood
接種箱 inoculating hood
接種器 inoculator
接種環 inoculating loop
接種體 inoculum; inocula
接縫 seam
接觸冰藏 contact icing
捷徑 short cut
掛肉鉤架 gambrel
掛糖衣 icing; panning
掛糖衣機 covering machine; icing machine
推進器 propeller
推薦每日營養素供給量 recommended daily allowance (RDA)
推薦營養素日供給量 recommended daily nutrient allowance (RDNA)
採收者 picker
採果器 picker
採茶 pluck tea
採摘 strip; pick(ing); pluck
採摘工人 plucker
採摘者 picker
排水 drain; draining
排水口 outfall
排水系統 drainage system; sewerage
排水渠 offtake
排水溝 drain; gutter
排水閥 drain valve
排出 efflux
排列 rank
排肋 costa rib
排放 drain; tapping
排放閥 drain valve
排放噴霧器 discharge atomizer
排泄 eccrisis; egestion; excrete
排泄物 excrement; excretion; excreta; faeces; feces
排泄物的 faecal; fecal
排便 egestion
排氣 vent
排氣風扇 discharge fan; exhaust fan
排骨 breast cut; chop; rib; sparerib
排骨肉 rib cut
排管式蒸發器 calandria evaporator
捻角山羊 markhor
救急口糧 emergency ration; iron ration
救急食品 emergency food
敗菌素 yurotin
敗壞 deterioration
敏豆 snap bean; kidney bean; navy bean; wax bean
敏感性 sensitivity; susceptibility
敏感症 allergy
斜面洋菜〔培養基〕 slant agar
斜面培養 slant; slant culture
斜面培養物 slant culture
斜面培養基 slant
斜面瓊脂〔培養基〕 slant agar
斜管〔式〕蒸發器 inclined tube evaporator
斜齒鯊酸 scoliodonic acid
斬拌機 chopper; chopping machine; cut mixer

族 race
旋毛蟲 trichina; *Trichinella spiralis*
旋毛蟲病 trichinosis
旋光性 optical activity; optical rotation
旋光度 optical activity; optical rotation
旋光計 polariscope
旋風分離器 cyclone; cyclone separator
旋風收集器 cyclone collector
旋風除塵器 cyclone scrubber
旋風器 cyclone
旋動篩 revolving screen; revolving sifter
旋渦混合器 vortex mixer
旋渦槽 whirlpool tank
旋渦攪拌器 vortex stirrer
旋鈕 knob
旋塞 cock; plug cock; tap
旋管加熱器 spiral heater
旋管式熱交換器 coil-type heat exchanger
旋管冷凝器 coil condenser
旋管的 spiral
旋管蒸發器 coil evaporator
旋轉 rotation; spinning
旋轉式冷凍機 rotary refrigerator
旋轉式洗滌機 rotary washer
旋轉式乾燥 rotary drying
旋轉式乾燥機 rotary dryer
旋轉式乾燥爐 rotary kiln
旋轉式混合槽 rotary drum mixer
旋轉式混合器 rotating mixer
旋轉式發酵槽 rotary drum fermenter
旋轉式萃取器 rotary extractor
旋轉式搖床 rotary shaker
旋轉式過濾器 rotary filter
旋轉式翻邊機 rotary flanging machine
旋轉式罐身雙頭翻邊機 rotary double-end flanger
旋轉板乾燥機 rotary shelf dryer
旋轉軋碎機 gyratory breaker
旋轉研磨機 revolving mill
旋轉乾燥機 revolving dryer; revolving rotary dryer
旋轉瓶蓋 roll on closure
旋轉蒸發器 rotary evaporator
旋轉窯 rotary kiln
旋轉篩 gyratory sifter; rotary screen
旋轉黏度計 rotational visco(si)meter
晚生斑褶菇 olive oyster
晚香玉浸膏 tuberose concrete
晚香玉淨油 tuberose absolute
晚崙西亞橙 Valencia orange
晚熟品種 late variety
晚稻 late rice
晚餐 supper
晚霞黃 sunset yellow; sunset yellow FCF
晨乳 morning milk
粱粟 foxtail millet
梯形罐 tapered shape can
桿 culm
桿狀體 rod
桿菌 bacilli; bacillus; rod
桿菌血症 bacillemia
桿菌病 bacillosis
桿菌素 bacillin
桿菌溶素 bacilysin
桶 barrel; cask; pail; runlet; tank; tub
桶(容量單位) hogshead
桶口 bunghole
桶內發酵 cask fermentation
桶內葡萄酒 cuvee
桶孔 bunghole
桶式乳油攪拌機 barrel churn
桶身木板 cask stave
桶柑 tankan
桶裝果汁 barrelled juice
桶裝啤酒 cask beer; keg beer
桶裝葡萄酒 cask wine
桶蓋 can cover
桶醃 vat salting

梗 stalk; stick; stem
梗破碎機 stem crusher
梭〔孢桿〕菌屬 *Clostridium*
梭子魚 pike
梭狀的 fusiform
梭狀芽孢桿菌屬 *Clostridium*
梭狀孢子 fusiform spore
梭魚 mullet
梭菌芽孢 clostridial spore
梭瑪丁 thaumatin
梭鏈孢酸 fusidic acid
梭麵包 koppe
梅 plum; Japanese apricot; mume; ume
梅扎尼〔花式〕通心粉 mezzani
梅扎尼里格體〔波紋狀〕通心粉 mezzani rigati
梅扎奈里〔花式〕通心粉 mezzanelli
梅花鱸 treefish
梅納反應 aminocarbonyl reaction; carbonyl-amine reaction; Maillard reaction
梅納汀 melanoidin
梅納環丁 melanoidin
梅乾 dried plum; prune
梅勒〔發酵〕法 Melle process
梅頭肉 collar butt
梔子 gardenia
梔子花浸膏 gardenia concrete
梔子紅色素 gardenia red
梔子黃 gardenia yellow
梔子綠色素 gardenia green
梔子藍色素 gardenia blue
條 bar
條例 act; regulation
條板箱 crate
條紋東方魨 yellowfin puffer
條紋槍魚 striped marlin
條紋鱸 striped perch
條帶細包 striped package
條蕈毒素 amanitatoxin
條麵 termette
條鱈 pout
梨〔子〕醬 pear butter
梨子 pear
梨子油香精 pear oily flavo(u)r
梨子香精 pear flavo(u)r
梨山楂 pear hawthorn
梨瓜 chayote; vegetable pear
梨形馬勃（蕈類） stump truffle
梨形罐 pear shaped can
梨果 pomaceous fruit; pome; pome fruit
梨籽油 pear seed oil
梨酒 perry
梨樹 pear
梨醬 pear sauce
殺生物劑 biocide
殺念珠菌素 candicidin
殺芽孢劑 sporicide
殺孢子〔能〕力 sporicidal activity
殺孢子劑 sporicide
殺病毒的 viricidal; virucidal
殺病毒劑 viricide; virucide
殺真菌素 fungicidin
殺真菌劑 fungicidal agent; fungicide; mycocide
殺草劑 herbicide; weed control chemicals; weed killer
殺假絲菌素 candicidin
殺桿菌劑 bacillicide
殺細菌的 bactericidal
殺細菌活性 bactericidal activity
殺細菌效應 bactericidal effect
殺細菌劑 bactericidal agent; bactericide
殺菁 blanch(ing)
殺菁劑 blancher
殺菌 disinfection; sterilize
殺菌不足 underpasteurization; understerilization
殺菌作用 germicidal action

殺菌-冷卻-充填法 heat-cool-fill method
殺菌性能 sterility
殺菌的 antiseptic; germicidal
殺菌活性 germicidal activity
殺菌食品 retort food
殺菌釜 autoclave; autoclave sterilizer; retort
殺菌釜滅菌 retort sterilization
殺菌袋 retort pouch
殺菌袋裝食品 retort pouch food
殺菌過度 overpasteurization
殺菌劑 bactericidal agent; bactericide; disinfectant; disinfecting agent; germicidal agent; germicide; microbicide; sanitizer
殺菌燈 germicidal lamp
殺菌籃 retort basket
殺菌籃〔吊架〕 carrier cradle
殺菌籃筐 retort crate
殺微生物劑 microbicide
殺鼠機 rodenticide
殺鼠藥 raticide
殺酵母素 yeastcidin
殺線蟲劑 nematocide
殺蟲 pest control
殺蟲粉 insect powder
殺蟲滅〔真〕菌劑 insectofungicide
殺蟲劑 disinfestant; helminthic; insect killer; insecticide; pesticide
殺蟲藥 pest control chemical; vermifuge
殺藻劑 algaecide; algicide
殺蠕蟲藥 vermicide
殺蟎劑 acaricide; miticide
毫升 milliliter (ml)
毫米 millimeter (mm)
毫克 milligram (mg)
毫微米 nanometer (nm)
毫微克 nanogram (ng)
氫 hydrogen
氫己二烯酸 hydrosorbic acid
氫化 hardening
氫化〔作用〕 hydrogenation
氫化松香甘油酯 hydrogenated ester gum
氫化油 hydrogenated oil; hardened oil
氫化脂 hydrogenated fat
氫化〔起〕酥油 hydrogenated shortening
氫化葡萄糖醬 hydrogenated glucose syrup
氫化器 hydrogenator
氫化澱粉水解物 hydrogenated starch hydrolysate (HSH)
氫化罐 hydrogenator
氫化酶 hydrogenase
氫氧化物 hydroxide
氫氧化鈣 calcium hydroxide; slaked lime
氫氧化鈉 sodium hydroxide
氫氧化鈉溶液 soda lye
氫氧化鉀 caustic soda; potassium hydroxide
氫氧化銨 ammonium hydroxide
氫氧化鎂 magnesium hydroxide
氫氧基 hydroxyl group
氫鹵化物 hydrogen halide
氫氯化噻胺 thiamine hydrochloride
氫氯酸 hydrochloric acid
氫過氧化物 hydroperoxide
氫膽固醇 hydrocholesterol
氫鍵 hydrogen bond
氫離子濃度測定計 pH meter
涎酸 sialic acid
涎澱粉酶 ptyalase; ptyalin
涎糖 ptyalose
涼瓜 balsam pear; bitter gourd; bitter melon
涼拌蔬果 macedoine
涼拌卷心菜〔絲〕 coleslaw; slaw
涼拌菜 salad
涼爽 coolness
涼薯 jicama; wayaka yam bean; yam bean

液 succus
液化 deliquescence
液化〔作用〕 liquefaction
液化蛋白〔質〕 liquefied protein
液化劑 liquefacient; liquefying agent; liquifier
液化器 liquefier
液化澱粉酶 liquefying amylase
液化酶 liquefying enzyme; liquifying enzyme
液固提取 liquid solid extraction
液固萃取 liquid solid extraction
液泡 vacuole
液封 liquid seal(ing)
液相 liquid phase
液相色譜法 liquid chromatography
液面培養〔物〕 surface culture
液盎司 fluid ounce (fl. oz.)
液乾酪羹 fondue
液液提取 liquid-liquid extraction
液液萃取 liquid-liquid extraction
液蛋 liquid egg
液啢 fluid ounce (fl. oz.)
液氮 liquid nitrogen
液氮冷凍 liquefrozen; liquid nitrogen freezing
液態牛油 oleo oil
液態食品 liquid food
液態香辛料 liquid spice
液態脂肪 liquid fat
液態〔起〕酥油 liquid shortening
液態培養基 liquid medium
液態蛋 liquid egg
液態氮 liquid nitrogen
液態氮冷凍 liquid nitrogen freezing
液態酥油 fluid shortening
液態豬油 lard olein
液態糖 liquid sugar
液態膳食 liquid diet
液膜 liquid film
液醃法 pickle curing
液凝膠 lyogel
液糖 liquid sugar
液燻 artificial smoking; liquid smoking
液麴 liquid koji
液蠟 fluid wax
液體 aqua; liquor; liquid
液體比重計 areometer; hydrometer
液體充填機 liquid filler
液體冷卻 liquid cooling
液體冷凍 liquid freezing
液體吸氣計 absorptiometer
液體物質 wet goods
液體培養 liquid culture
液體培養基 fluid medium
液體萃取 liquid extraction
液體層析法 liquid chromatography
液體燻製劑 liquid smoke
液體麴 liquid koji
液體灌裝機 liquid filler
淡化 freshen
淡化水 product water
淡水 freshwater
淡水貝 naiad
淡水的 limnetic
淡水魚 freshwater fish; limnetic fish
淡水藻 freshwater algae
淡水鱈 burbot
淡奶 evaporated milk
淡而無味 blank taste; wishy-washiness
淡而無味的 milk-and-water; wishy-washy
淡而無味的流質食物 slop
淡而無味的湯 wash
淡而無味的飲料 wish-wash
淡而無味食物 slip-slop
淡色肉 light meat
淡色啤酒 ale

淡色啤酒酵母 lager yeast
淡色麥芽 light malt
淡艾爾啤酒 pale ale (beer); small ale
淡味 flat taste; thin taste
淡味小烤餅 crule
淡季 off season
淡紅 azorubin(e); carmoisine
淡紅葡萄酒 clairet
淡烤酥餅 popover
淡烈酒 light liquor
淡茶 catlap
淡酒 tiff; wish-wash
淡酒再餾器 spirit still
淡酒飲料 wine cooler
淡乾魚 unsalted dried fish
淡啤酒 light beer; near beer; sissy beer; small beer
淡菜 dried mussel; mussel
淡飲料 catlap
淡煉乳 evaporated milk
淡葡萄酒 light wine
淡褐色粗糖 demerara sugar
淡醃 pale cure
淡醃的 lightly salted; slack salted
淡糖漿 light syrup
淡麵包 plain bread
淡麵粉 light flour
淡鹽醃食品 pale cure
淤泥 slush
淤渣 sludge
添加〔作用〕 addition
添加物 adjunct; additive; addition
添加香辛料 bespicing
添加維生素食品 vitaminized foodstuff
添加劑 additive; additive agent
淺皿 flatware
淺色卵黃 white yolk
淺色拉格啤酒 pale lager beer
淺色啤酒 bright beer; pale beer
淺色蛋黃 white yolk
淺色麥芽 pale malt
淺色麥酒 ale
淺色貯藏啤酒 pale lager beer
淺色糖蜜 second molasses
淺沖拔罐 shallow drawn can
淺沖壓罐 shallow drawn can
淺底闊籃 trug
淺炸 shallow fry(ing)
淺盆形罐 tapered shape can
淺紅色釘菇 pine spike cap
淺海魚〔類〕 epipelagic fish; neritic fish
淺蜊 baby-neck clam
淺盤 shallow pan; salver; saucer; tray
淺盤乾燥 pan drying
淺盤乾燥器 pan dryer
淺橢圓罐 shallow oval can
淺鍋 shallow pan
淺灘 shoal
淺灘捕魚 bank fishing
清肉湯 consomme
清果汁 clear juice
清洗 scour; wash
清酒 Japanese wine
清酒酵母〔菌〕 *Saccharomyces sake*
清酒醪 sake mash
清酒麴 sake koji
清除 scavenging
清除劑 scavenger
清涼飲料 belly wash; cold drink; refreshment beverage; refreshment drink; soft drink
清涼劑 freshener
清淡風味 bland flavo(u)r
清淡膳食 bland diet
清淨器 defecator
清蛋白 albumin; thin white
清袋機 sack cleaner
清湯 bree; clear soup; light soup; stock

清道夫 scavenger
清澈的 limpid
清管器 tube cleaner
清蒸的 steamed
清蒸魚 steamed fish
清潔 depuration
清潔劑 cleaning agent; cleanser; detergent; detersive
清穀篩 paddy cleaner
清燉肉 boiled meat
清篩刷 sieve brush
淋巴 lymph
淋巴系統 lymphatic system
淋巴管 lymphatic; lymphatic vessel
淋式濾床 trickling filter
淋洗 spray washing
混入氣體 aeration
混合 admix; admixture; blending; concoction; interfusion; intermingle; mix; mixing
混合不均 undermixing
混合包裝 mixed packing
混合打漿機 mixer-beater
混合床 mixed bed
混合乳 mixed milk
混合果汁 blended juice; cocktail juice; fruit juice cocktail
混合油 miscella
混合油脂 compound fat
混合油精煉 miscella refining
混合物 admix; admixture; concoction; mixture; maslin; mix
混合室 mixing chamber
混合香料 potpourri
混合時間 mixing time
混合茶 tea blend
混合培養〔物〕 mixed culture
混合氫化油 vanaspati ghee
混合造粒 mixing granulation
混合發酵 mixed fermentation; mixed culture fermentation
混合菌種 mixed culture
混合飲料 twist; mixed drink
混合填充 mixed packing
混合裝置 mixing device
混合酵母 compound barm
混合醃 mixed curing
混合醃劑 curing mix
混合器 mixer; mixing device
混合機 melangeur; mingler
混合糖漿 mixed syrup
混合膳食 mixed diet
混合攪拌機 mixing agitator
混合體 mixture
混合鹽醃製 mixed curing
混肉機 meat mixer
混和 blend; blending
混和不足 undermixing
混和機 mingler
混料 mix
混料罐 mixing bowl
混酸三甘油酯 mixed triglyceride
混酸甘油酯 mixed glyceride
混穀麵包 maslin
混濁 feculence; turbidity; turbidness
混濁果汁 cloudy juice
混濁的 turbid
混濁度 cloudiness; opacity; turbidity
混濁溫度試驗 cloud test
混濁劑 clouding agent
混糖 muscovado (sugar)
混雜物 mishmash
淘汰果 cull fruit
淘汰品 cullage
深水魚 bathypelagic fish
深水藻 abyssal algae
深色肉 dark meat
深色肌肉 dark muscle

深色麥芽 dark malt; Munich malt
深色劑 saddening agent
深色醬油 dark soy (sauce)
深色麵粉 dark flour
深沖壓罐 deep drawn can
深拉拔罐 deep drawn can
深度冷凍 deep freezing
深度冷凍櫃 deep freezer
深度凍結食品 deep frozen product
深紅甜葡萄酒 tent
深紅越橘 Japanese blueberry
深海底的魚 benthic fish
深海魚 abyssal fish; deep sea fish
深海魚業 deep sea fishery
深暗緊密和乾燥肉 dark, firm, dry muscle (DFD muscle)
深綠色厚壁玻璃瓶 junk bottle
深層牛乳培養基 deep milk medium
深層洋菜〔培養〕法 deep agar method
深層培養 deep culture
深層培養〔物〕 submerged culture
深層發酵 submerged fermentation; sub-surface fermentation
深層菌落 deep colony
淨化 clarification; decontamination; defecation; depuration; fining; purification; rectification; scavenging
淨化水 purified water
淨化汁 purified juice
淨化劑 decontaminant; decontaminating agent; depurant; depurator; fining agent
淨化器 cleanser; depurant; depurator; purifier
淨化糖蜜 clarified molasses
淨水 pure water
淨水處理 water purification
淨含量 net content
淨乳機 milk clarifier
淨重 net weight; suttle (weight)
淨容量 net capacity; net volume
淨產物 net product
淨蛋白質利用率 net protein utilization (NPU)
淨袋機 sack cleaner
淨豬肉 clear pork
淨鹽 straight salt
烹香 nidor
烹煮收縮率 cooked shrinkage rate
烹煮味 cooked taste
烹煮風味 cook flavo(u)r
烹煮損耗 cooking loss
烹飪 cooking; cuisine
烹飪法 cookery
烹飪學 gastrology
烹飪學家 gastrologer
烹調 cook; cooking
烹調用之香料紗布袋 fagot
烹調用油 cooking oil
烹調用料 kitchen stuff
烹調用脂肪 cooking fat
烹調用混合胡椒粉 kitchen pepper
烹調技術 culinary art
烹調法 recipe
烹調風格 cuisine
烹調時間 cooking time
烹調酒 cooking wine
烹調術 cookery
烹調設備 cooking equipment
烹調損失 cooking loss
烹調裹紙 papillote
烹調器 cooker
烹調鍋 cooking pan
烹燒時之香味 nidor
烯丙基澱粉 allyl starch
烯丙醇 allyl alcohol
烯烴 olefin
烯醇 enol

烯醇化〔作用〕 enolization
烯醇鹽 enolate
烯醇酶 enolase
爽口味 smooth flavo(u)r
牽牛花素 petunidin
牽絲牛乳 stringy milk
猛鮭 ferox
球化劑 nodulizer
球形物 bulb
球形厚大蛤 tiger lucina
球〔形〕磨機 pot mill
球果 cone
球狀蛋白質 globular protein
球狀朝鮮薊 globe artichoke
球芽甘藍 baby cabbage; brussel sprout
球根 bulb; tuber
球莖 corm
球莖甘藍 even stem cabbage; kohlrabi; turnip cabbage; turnip-rooted cabbage
球蛋白 globin; globulin
球菌 cocci; coccus
球磨碎機 ball crusher
球磨機 ball mill; globe mill
球薑 zerumbet ginger
球囊 sacculus
現成可用的 ready-to-use
現成的 ready made
現成食物 delicatessen
現成食品 ready-to-eat food; ready-to-serve food
現成菜餚 delicatessen
現宰烤雞 spatchcock
現場試驗 site test
現象莓 phenomenal berry
現榨果汁 fresh squeezed juice
瓠瓜 calabash gourd; pepo
瓶 pot
瓶口加環機 discer
瓶口套 capsule
瓶內巴斯德殺菌 bottle pasteurization
瓶托 bottle pedestal
瓶帽 capsule
瓶塞 bottle block; stopper
瓶塞機 bottle corking machine
瓶裝巴氏殺菌 in-bottle pasteurization
瓶裝牛奶 bottled milk
瓶裝乳 bottled milk
瓶裝果汁 bottled juice
瓶裝食物 bottle food
瓶裝滅菌 in-bottle sterilization
瓶蓋 bottle cap
瓶蓋機 bottle capping machine
瓶頸 neck
瓶頸商標 neck label
瓶頸標籤 neck label
瓷 porcelain
瓷土 kaolin; porcelain clay
瓷坩堝 porcelain crucible
瓷器 porcelain
甜小茴香油 sweet fennel oil
甜小茴香籽 fennel seed
甜化 edulcoration
甜巴旦杏仁 sweet almond
甜木薯 aipi; sweet cassava
甜巧克力 sweet chocolate
甜玉米 sweet corn; sweet maize; sugar corn
甜玉蜀黍 sweet corn; sweet maize
甜瓜 melon; muskmelon; sweet melon
甜瓜醛 melonal
甜佐料 sweetener
甜杜松子酒 mahogany
甜豆 sugar pea
甜味 sweet flavo(u)r; sweet taste; sugariness; sweetness
甜味料 sweetening; sweetener; sweetening agent
甜味素 aspartame

甜味草 sweet grass
甜味醃漬〔品〕 sweet pickle
甜味劑 edulcorant; sweetener; sweetening agent
甜性奶油 sweet butter; sweet cream
甜性酪乳 sweet buttermilk
甜炒玉米粉 pinole
甜的 luscious; melliferous; sweet; sugary
甜品 dessert; sweets
甜度 sugariness; sweetness
甜柿 sweet persimmon
甜美化 dulcification
甜食 sweet stuff; sweetmeat; sweet goods; tuck
甜料 sweetening
甜烈酒 créme
甜脆餅乾 rusk
甜茴香油 roman fennel oil
甜酒 sweet wine; sweets
甜馬郁蘭 sweet marjoram
甜高粱 saccharine sorghum; sorgho; sorgo; sweet sorghum; sugar sorghum
甜捲麵包 sweet roll
甜甜圈 doughnut; ring doughnut
甜麥芽汁 sweet wort
甜椒 bell pepper; bullnose pepper; Jamaica pepper; non-pungent pepper; pimento; sweet pepper
甜筒 ice cream cone
甜菊 stevia
甜菊糖 stevioside
甜菊糖苷 stevioside
甜菜 beet (root); sugar beet
甜菜片 beet chip; beet cossette; cossette
甜菜去頂器 beet topper
甜菜汁 beet juice
甜菜收穫機 beet harvester
甜菜肉 beet marc
甜菜挖掘機 beet puller
甜菜紅 beet red; beet root red
甜菜根 beetroot; beet (root)
甜菜根紅 beet red; beet root red
甜菜〔根〕糖 teutlose
甜菜素 oxyneurine
甜菜渣 beet pulp
甜菜絲 beet chip; beet cossette; cossette
甜菜榨渣 beet marc
甜菜廢絲 beet pulp; exhausted cossette
甜菜糖 beet sugar
甜菜糖膏 beet massecuite
甜菜糖蜜 beet molasses
甜菜糖廠 beet mill
甜菜糖漿 beet syrup
甜菜〔頭〕 red beet
甜菜鹼 betain(e); lycine; oxyneurine
甜菜苷 betanin
甜酥餅 cookie
甜煉乳 sweetened condensed milk
甜葉菊苷 stevioside
甜葡萄酒 sweet wine
甜旗草 sweet flag; sweet root
甜精 dulcin(e); sucrol; valzin
甜蜜的 candied; treacly
甜蜜的食物 sugar candy
甜蜜素 sodium cyclamate
甜辣椒〔粉〕 paprika
甜酸泡菜 sweet pickle
甜酸的 sweet and sour
甜酸醬 sweet and sour sauce
甜餅乾 sweet biscuit
甜豌豆 honey pea
甜醇 dulcite; dulcitol; dulcose
甜凝乳 junket
甜橙 *Citrus sinensis*; sweet orange
甜橙油 orange oil; sweet orange oil
甜樹薯 sweet cassava
甜橄欖 sweet olive

甜膩的 sweetish; syrupy
甜環餅 jumbal; jumble
甜點 dessert; sweet; sweet goods
甜點餐具 dessert service
甜蘋果 sweeting
甜麵包 pikelet
甜麵包乾 rusk
甜麵醬 sweet bean paste
甜櫻桃 whiteheart cherry
甜露酒 liqueur
產生糖的 sacchariferous
產色作用 chromogenesis
產色細菌 chromogenic bacteria
產卵 spawn
產卵後的〔廢〕魚 spent fish
產卵後鮭魚 black fish
產卵魚 layer; spawner
產卵期魚 seed fish
產卵禽 laying fowl
產乳的 milch
產乳酸細菌 lactic acid forming bacteria
產乳酸酶 lactalase
產物 offspring; product
產芽孢桿菌 spore-forming bacteria
產芽孢酵母 sporogenous yeast
產孢子酵母 sporogenous yeast
產孢的 spore-bearing
產孢細菌 spore-bearing bacteria
產孢菌 spore-former; spore-forming bacteria
產前營養 prematernal nutrition
產品 produce; product; production
產品檢驗 product inspection
產毒性 toxigenicity
產毒的 toxicogenic; toxigenic
產香物 aroma former
產香菌 aroma former
產氣 gassing
產氣性 gassing property
產氣桿菌 gas bacillus
產氣桿菌屬 *Aerobacter*
產氣細菌 aerogenic bacteria; gas forming bacteria; gas producing bacteria
產氣莢膜桿菌 *Clostridium perfringens*
產氣莢膜梭〔孢桿〕菌 *Clostridium perfringens*
產氣莢膜梭菌的食物中毒 perfringens poisoning
產氣發酵 gassy fermentation
產堅果的 nutty
產蛋雞 layer
產麥芽糖澱粉酶 maltogenic amylase
產量 yield
產酮〔體〕胺基酸 ketogenic amino acid
產酮食物 ketogenic diet
產酮膳食 ketogenic diet
產蜜的 melliferous
產酸作用 acid production
產酸細菌 acid-forming bacteria
產酸酵母〔菌〕 *Saccharomyces acidifaciens*
產熱食物 calorigenic food
產熱細菌 thermogenic bacteria
產膜酵母 film-forming yeast; film yeast
產〔澱〕粉的 farinose
產糖原的 glycogenic
產鹼食品 basic food
產鹼桿菌屬 *Alcaligenes*
產鹼細菌 alkali producing bacteria; alkaligenous bacteria
略帶酸味 acescence
略帶酸味的 acescent
畢卡〔麵粉粒度及色澤〕試驗 Pekar test
異〔葎〕草酮 isohumulone
異丁子香酚 isoeugenol
異丁香酚 isoeugenol
異丁香酚甲醚 methyl isoeugenol
異丁香酚苄醚 benzyl isoeugenyl ether
異丁烷 isobutane
異丁酸 isobutyric acid

異丁酸-2-苯氧基乙酯 2-phenoxyethyl isobutyrate
異丁酸乙酯 ethyl isobutyrate
異丁酸丁酯 butyl isobutyrate
異丁酸己酯 hexyl isobutyrate
異丁酸甲酯 methyl isobutyrate
異丁酸辛酯 octyl isobutyrate
異丁酸庚酯 heptyl isobutyrate
異丁酸芳樟酯 linalyl isobutyrate
異丁酸苯乙酯 phenethyl isobutyrate
異丁酸香茅酯 citronellyl isobutyrate
異丁酸香葉酯 geranyl isobutyrate
異丁酸桂酯 cinnamyl isobutyrate
異丁酸異丙酯 isopropyl isobutyrate
異丁酸異戊酯 isoamyl isobutyrate
異丁酸對甲苯酯 *p*-tolyl isobutyrate
異丁酸橙花酯 neryl isobutyrate
異丁酸苄酯 benzyl isobutyrate
異丁醇 isobutanol; isobutyl alcohol
異丁橡膠 butyl rubber
異丁醛 isobutyraldehyde
異己酮 hexone
異己糖酸 isosaccharinic acid
異元多糖 heteroglycan; heteropolysaccharide
異元聚己醣 heterohexosan
異化作用 catabolism
異丙醇 isopropanol; isopropyl alcohol
異戊二烯 isoprene
異戊酸 isovaleric acid
異戊酸-2-甲基丁酯 2-methylbutyl isovalerate
異戊酸乙酯 ethyl isovalerate
異戊酸丁酯 butyl isovalerate
異戊酸己酯 hexyl isovalerate
異戊酸壬酯 nonyl isovalerate
異戊酸苯乙酯 phenethyl isovalerate
異戊酸香葉酯 geranyl isovalerate
異戊酸桂酯 cinnamyl isovalerate
異戊酸烯丙酯 allyl isovalerate
異戊酸異戊酯 isoamyl isovalerate
異戊酸苄酯 benzyl isovalerate
異戊醇 isopentanol; isoamyl alcohol; isopentyl alcohol
異戊醛 isovaleraldehyde
異白胺酸 isoleucine
異冰片 isoborneol
異色 off colo(u)r
異位效應 allosteric effect
異形罐封罐機 Adriance seamer
異抗壞血酸 erythorbic acid; isoascorbic acid
異抗壞血酸鈉 sodium erythorbate
異沒食子酸戊酯 isoamyl gallate
異赤藻糖 threose
異味 foreign flavo(u)r; taint; off flavo(u)r; off odo(u)r
異松油烯 terpinolene
異油酸 isooleic acid; vaccenic acid
異物 foreign substance
異亮胺酸 isoleucine
異哇啉 isoquinoline
異型孢子 anisospore
異型發酵 heterofermentation
異型發酵乳酸桿菌 heterofermentative lactobacilli
異型發酵細菌 heterofermentative bacteria
異型罐 irregular (shaped) can
異胡薄荷醇 isopulegol
異食症 pica
異食癖 allotriophagy; pica
異核黃素 isoriboflavin
異常乳 abnormal milk
異常風味 abnormal flavo(u)r
異常蛋 abnormal egg
異常發酵 abnormal fermentation
異常飢餓 limosis
異甜菜苷 isobetanin

異硫氰酸 isothiocyanic acid
異硫氰酸烯丙酯 allyl isothiocyanate
異硫氰酸鹽 isothiocyanate
異蛇麻葫酮 isohumulone
異麥芽酮糖 isomaltulose; palatinose
異麥芽糖 isomaltose
異麥芽糖醇 isomaltitol
異硬脂酸 isostearic acid
異絲胺酸 isoserine
異黃酮 isoflavone
異黃樟油素 isosafrole
異黑鯛 porgy
異鼠李糖 isorhamnose
異羥丁胺酸 isothreonine
異構化〔作用〕 isomerization
異構化乳糖液 isomerized lactose syrup
異構化糖漿 isomerized syrup
異構性 isomerism
異構物 isomer
異構現象 isomerism
異構催化〔作用〕 isocatalysis
異構糖 isomerose
異構體 isomer
異構酶 isomerase
異滲溶液 anisotonic solution
異維生素C isovitamin C
異蒲勒醇 isopulegol
異質同晶〔現象〕 allomerism
異養的 heterotrophic
異養細菌 heterotrophic bacteria
異麩醯胺酸 isoglutamine
異澱粉酶 isoamylase
異糖二酸 isosaccharic acid
異糖林酸 isosaccharinic acid
異糖質酸 isosaccharic acid
異龍腦 isoborneol
異營〔作用〕 heterotrophy
異營的 heterotrophic
異營細菌 heterotrophic bacteria
異檸檬酸 isocitric acid
異蘇胺酸 isothreonine
疏水性 hydrophobicity
疏水物 hydrophobe
疏水物質 hydrophobic substance
疏水的 hydrophobic
疏水溶膠 hydrophobic sol
疏水膠體 hydrophobic colloid
疏油的 oleophobic
疏脂的 lipophobic
疏液物 lyophobe
疏液的 lyophobic
疏液膠體 lyophobic colloid; lyophobe
疏鬆的結構或質地（乾酪） short texture
疏鬆機 fluffer
痕量 trace
痕量元素 trace element
痕量金屬 trace metal
痕量級 trace level
盛肉汁船形器皿 gravy boat
盛宴 regale; royal feast; spread
盛菜用具 server
盛餐 full meal
眾香子 allspice; Jamaica pepper; pimento
眾香子油 allspice oil; pimento oil
眾香葉油 allspice leaf oil
眼孔 eye hole
眼鏡魚 moonfish
硫 sulfur; sulphur
硫化 vulcanization; vulcanize
硫化〔作用〕 sulfuration
硫化油酸丁酯 sulfated butyl oleate
硫化物 sulfide
硫化物黑斑污染 sulfide staining
硫化氫 hydrogen sulfide
硫化黑斑 sulfide staining
硫化黑變 sulfide blackening; sulfur blackening
硫化鐵黑變 iron sulfide blackening

硫化鐵變色 iron sulfide discoloration
硫代乙酸糠酯 furfuryl thioacetate
硫代二丙酸 thiodipropionic acid
硫代二丙酸二月桂酯 dilauryl thiodipropionate
硫〔代〕巴比妥酸 thiobarbituric acid (TBA)
硫〔代〕巴比妥酸值 thiobarbituric acid value (TBA value)
硫代丙酸糠酯 furfuryl thiopropionate
硫代硫酸鈉 sodium thiosulfate
硫代葡萄糖苷 glucosinolate
硫甘油 thioglycerin
硫辛酸 thioctic acid
硫脂質 sulfolipid
硫胺 aneurin; oryzanin; thiamine; vitamin B1
硫胺素 aneurin; thiamine; vitamin B1
硫胺素二月桂基硫酸鹽 thiamine dilaurylsulfate
硫胺素十六烷基硫酸鹽 thiamine dicetylsulfate
硫胺素硫氰酸鹽 thiamine thiocyanate
硫胺素焦磷酸鹽 thiamine pyrophosphate (TPP)
硫胺素硝酸鹽 thiamine (mono)nitrate
硫胺素鹽酸鹽 thiamine hydrochloride
硫胺素萘-1,5-二磺酸鹽 thiamine naphthalene-1,5-disulfonate
硫胺素萘-2,6-磺酸鹽 thiamine naphthalene-2,6-disulfonate
硫胺素酶 aneurinase; thiaminase
硫桿菌屬 *Thiobacillus*
硫細菌 thiobacteria
硫氰酸 thiocyanic acid
硫氰酸鈉 sodium thiocyanate
硫葡萄糖 thioglucose
硫葡萄糖苷酶 thioglucosidase
硫酯 thioester
硫酸 sulfuric acid; sulphuric acid
硫酸二銨 ammonium phosphate, dibasic
硫酸化劑 sulfating agent
硫酸亞鐵 ferrous sulfate
硫酸奎寧 quinine sulfate
硫酸脂 sulfolipid
硫酸氫二銨 ammonium phosphate, dibasic
硫酸氫鈉 sodium bisulfate
硫酸氫鉀 potassium bisulfate
硫酸蛋氨酸鋅 zinc methionine sulfate
硫酸軟骨素鈉 sodium chondroitin sulfate
硫酸鈣 calcium sulfate; gypsum
硫酸鈉 sodium sulfate
硫酸腦苷脂 sulfatide
硫酸鈷胺素 sulfatocobalamin
硫酸鈷維生素 sulfatocobalamin
硫酸鉀 potassium sulfate
硫酸酯 sulfate; sulphate
硫酸銅 copper sulfate; cupric sulfate
硫酸銨 ammonium sulfate
硫酸銨鋁 aluminium alum, alumin(i)um ammonium sulfate
硫酸鋅 zinc sulfate
硫酸鋁 alum; alumin(i)um sulfate
硫酸鋁鈉 sodium alum; aluminium sodium sulfate
硫酸鋁鉀 potassium alum; aluminium potassium sulfate
硫酸錳 manganese sulfate
硫酸鎂 magnesium sulfate
硫酸鐵 ferric sulfate
硫酸鹽 sulfate; sulphate; vitriol
硫醇 mercaptan; thiol
硫醇胺基酸 mercaptoamino acid
硫糖 thiosugar
硫磺 sulfur; sulphur
硫磺菌（蕈類） chicken of the woods
硫磺濕傘（蕈類） golden wax cap

硫黏蛋白 sulfomucin
硫醚 thioether
硫燻 sulfuring; sulphuring
硫燻果 sulfured fruit
移動爐 travelling furnace
移液管 pipet(te)
移植 transplantation
笛豆 flageolet bean
笛鯛 snapper
第一效蒸發器 first evaporator
第一級胺 primary amine
第一級醇 primary alcohol
第二次碳酸飽和 second carbonation
第二效蒸發器 second evaporator
第二胺 secondary amine
第二醇 secondary alcohol
第三丁氫醌 tertiarybutylhydroquinone (TBHQ)
第三級結構 tertiary structure
第三級廢料處理 tertiary waste treatment
第四級結構 quaternary structure
粒 granule
粒子 particle
粒化 graining; granulation
粒性 granularity
粒狀物 saccharoid
粒狀的 granular
粒狀麵食製品 farfal; farfel
粒度 graininess; granularity
粒線體 mitochondria
粒質糖 granulose
粗小麥粉 semolina
粗切 coarse cut
粗切肉塊 primal cut
粗孔篩 coarse mesh
粗加工 roughing
粗玉米粉 grits
粗米糠 brown shorts
粗油 crude oil; raw oil
粗的 raw
粗軋碎 coarse break(ing)
粗砂糖 sanding sugar
粗食 low-diet
粗海鹽 bay salt
粗粉 cribble; grout; meal; middlings; sizings
粗脂 raw fat
粗脂肪 crude fat
粗茶淡飯 simple diet
粗酒石 argol
粗高粱粉 milo grit
粗粒 coarse grain; grit
粗粒〔穀〕粉 coarse grit
粗粒小麥粉 coarse semolina; farina; semolina
粗粒分散膠體 macrodispersoid
粗粒杜林小麥粉 durum semolina
粗粒杜蘭小麥粉 durum semolina
粗粒的 hard-grained
粗粒度 coarseness
粗粒砂糖 coarse granular sugar; coarse granulated sugar
粗粒硬質小麥粉 durum semolina
粗粒糖 barley sugar
粗粒麵粉 coarse sizings; semolina
粗粒鹽 coarse salt
粗蛋白質 crude protein
粗通心粉 fovantini; macaroncelli
粗通心麵 macaroncelli
粗陶器 stoneware
粗麥粉 semolina
粗麥麩 brown shorts
粗晶粒 macrograin
粗晶粒硬糖 hard sugar
粗晶糖 hard-grains
粗殼粉 farina
粗硬麵包 rations bread
粗絞 coarse cut

粗絞〔餡〕香腸 coarse chopped sausage; coarse cut sausage
粗絞肉機 meat shredder
粗萃 crude extract
粗黑麥酸麵包 pumpernickel
粗碎 kibble
粗碎米 second head
粗碎肉機 meat shredder
粗碎麥芽 grist
粗碎穀粒 kibbled grain
粗碎機 kibbler; primary crusher
粗飼料 roughage
粗製油 raw oil
粗製砂糖 crude sugar; intermediate sugar
粗製脂 raw fat
粗製麵粉 alcomeal
粗碾 coarse grinding
粗穀粉 cribble; grist; kibbled grain
粗麩 broad bran; coarse bran
粗燕麥粉 grits
粗磨 coarse grinding; kibble
粗磨全麥麵粉 chisel
粗磨製麵粉 coarse flour
粗磨穀粉 grits
粗磨機 kibbler
粗磨麵粉 crude flour
粗篩 coarse mesh; cribble; coarse screen; coarse sieve; riddle
粗篩屑 coarse screenings
粗篩餘物 coarse screenings
粗糖 raw sugar; unrefined sugar
粗壓碎 coarse break(ing)
粗糠 bran coat; broad bran; coarse bran; rice hull; rice husk
粗糙〔型〕菌落 rough colony; R-colony
粗糙肉齒菌(蕈類) blue footed scaly tooth
粗糙性 harshness
粗糙度 coarseness
粗糙食物的纖維質 roughage
粗糙環柄菇 sharp scaled parasol
粗濾 rough filtration; strain
粗濾果泥 strained fruit
粗濾食品 strained food
粗濾產物 colature
粗濾器 cullender
粗糧 rough food
粗麵 linguine
粗麵粉 atta; chisel; middling flour; middlings; weatings
粗麵粉製的 wholemeal
粗纖維 crude fiber
粗鱗大環柄菇 shaggy parasol
粗鹽 coarse salt; crude salt
粕 cake
粕餅 expeller cake
統計學 statistics
統粉 full straight; straight flour
紮腸機 linking machine
紹興酒 shao-hsing wine
細 thinness
細小鬆脆餅 lady's finger
細切 fine cut
細切煙草 fine cut
細毛尖香紅茶 pekoe souchong
細玉米粉 golden meal
細白砂糖 berry sugar; castor sugar
細米粉絲(中國) locksoy
細枝 twig
細長葡萄 lady's finger
細咬 nibble
細流 trickle
細胞化學 cytochemistry
細胞外酶 lyoenzyme
細胞生物學 cytobiology
細胞自溶〔作用〕 autocytolysis
細胞自溶酶 cathepsin
細胞色素 cytochrome; histohaematin

細胞色素氧化酶 cytochrome oxidase
細胞色素還原酶 cytochrome reductase
細胞毒素 cytotoxin
細胞核 nuclei; nucleus
細胞培養〔物〕 cell culture
細胞液 cellular fluid
細胞組織 cell tissue
細胞間液 intercellular fluid
細胞間質 intercellular substance
細胞黃素 cytoflavin
細胞溶素 cytolysin; lysin
細胞溶素生成的 lysogenic
細胞溶解 cytolysis
細胞溶解酶 cytase; cytolytic enzyme
細胞增代時間 doubling time; generation time
細胞漿 enchylema
細胞質 cell plasm; cytoplasm
細胞壁 cell wall
細胞糖苷脂 cytolipin
細胞壞死 necrobiosis
細香葱 chive
細根 rootlet
細粉 fine grit; fine powder
細粉條 vermicelli
細粉磨 pulverizing mill; regrind mill
細粒小麥粉 fine semolina
細粒禾穀 small grain
細粒狀的 acinose
細粒砂糖 fine-grained sugar; fine granulated sugar
細粒麵粉 fine semolina
細通心粉 small macaroni
細晶糖 fine sugar
細絲 filament
細菌 bacteria; bacterium; germ; microphyte
細菌分解 bacterial decomposition
細菌平板菌落計數 bacterial plate count
細菌污染 bacterial contamination
細菌含量 bacterial content
細菌性食物中毒 bacterial food poisoning
細菌性疾病 bacteriosis
細菌性腐敗 bacterial spoilage
細菌的 bacterial
細菌芽孢 bacterial spore
細菌孢子 bacterial spore
細菌毒素 bacterial toxin; bacteriotoxin
細菌活性 bacterial activity
細菌降解 bacterial degradation
細菌區系 bacterial flora
細菌培養 bacterial cultivation
細菌發酵 bacterial fermentation
細菌菌落 bacterial colony
細菌感染 bacterial infection
細菌群落 bacterial flora
細菌群體 bacterial population
細菌腐壞味 bacterial taint
細菌酵素 bacterial enzyme
細菌數 bacterial count
細菌數〔目〕 bacterial population
細菌凝聚團 zoogloea
細菌學 bacteriology
細菌學的 bacteriological
細菌學家 bacteriologist
細菌澱粉酶 bacterial amylase
細碎肉 fine meat
細腰蜂 wasp
細葉香芹 beaked parsley; chervil
細管 tubule
細管狀通心粉 fovantini; macaroncelli
細質乳菇 mild milk cap
細麩 dunst; red short; weatings
細麩皮 bran shorts; pollen
細齒紫魚 flower flute porgy
細磨 fine grinding
細磨機 super grinder
細磨麵粉 pollen

細頸瓶 ampoul
細壓碎機 fine crusher
細麵粉 cake flour
細麵條 fedelini; vermicelli
細鱗大麻哈魚 humpback salmon; pink salmon
細鱗紫茵鯛 flower flute porgy
細鹽 fine salt
組分 constituent
組合件 unit
組合食品 fabricated food
組合罐 built-up tin
組氨酸鹽酸鹽 histidine monohydrochloride
組胺 ergamine; histamine
組胺酸 histidine
組蛋白 histone
組裝紙盒 set-up box
組織 tissue
組織化大豆蛋白〔質〕texturized soybean protein
組織化蛋白〔質〕 texturized protein
組織化植物蛋白〔質〕textured vegetable protein; texturized vegetable protein
組織脂肪 tissue fat
組織胺 ergamine; histamine
組織培養〔物〕 tissue culture
組織液 enchyma; tissue fluid
組織細胞 histiocyte
組織蛋白 histone; tissue protein
組織蛋白酶 cathepsin
組織學 histology
終止 termination
終產物 end product; final product
終發酵 end fermentation
缽 bowl
羚羊 antelope
羚羊奶油 goa butter
羚羊乳酪 goa butter
脯胺酸 proline
脖子 neck
脱下的玉米粒 dredge corn
脱支酶 debranching enzyme
脱毛 defeather; dehair; depilate; depilation; epilation; unhairing; unwooling
脱毛的 depilatory
脱毛膠料 depilating compound
脱毛劑 depilatory; epilator
脱毛機 dehairing machine; dehairer; depilator
脱水 dewater
脱水〔作用〕 anhydration; deaquation; dehydration; exsiccation
脱水奶油 cream powder
脱水白脱 anhydrous butter
脱水收縮〔作用〕 syneresis
脱水肉 dehydrated meat; dried meat
脱水冷凍〔法〕 dehydrofreezing
脱水冷凍的 dehydrofrozen
脱水乳酪 anhydrous butter
脱水食品 dehydrated food
脱水湯粉 dehydrated soup
脱水製罐 dehydrocanning
脱水蔬菜 dehydrated vegetable
脱水劑 dehydrant; dehydrater; dehydrator
脱水器 anhydrator; dehydrater; dehydrator
脱水機 hydro-extractor
脱水糖 anhydro sugar
脱水酶 dehydrase
脱去 strip
脱去氯〔作用〕 dechlorination
脱皮 decortication; dehull; shed
脱皮玉米粒 corn rice; maize rice; mealie rice
脱皮機 decorticator; dehuller

脱色 bleaching; depigmentation; discolo(u)ring; discolo(u)ration
脱色〔用黏〕土 bleaching clay
脱色〔作用〕 decolo(u)ration; decolo(u)rization
脱色土 bleaching earth
脱色油 bleached oil
脱色脂肪 bleached fat
脱色醋 crystal vinegar
脱色劑 decolo(u)rant; decolo(u)rizer; decolo(u)rizing agent
脱乳糖乳 delactosed milk
脱乳糖乳清 delactosed whey
脱果膠 depectinization
脱果膠果汁 depectinized juice
脱油裝置 deoiler
脱附〔作用〕 desorption
脱胚 degerm
脱胚芽機 germ separator
脱苦味 debittering; debitterize
脱苦味酵母 debittering yeast
脱氣 deflation; degas; degassing; outgassing
脱氣〔作用〕 deaeration; degasification
脱氣不足 underexhaustion
脱氣果汁 deaerated juice
脱氣裝置 deaerator
脱氧〔作用〕 deoxidation; deoxidization; disoxidation
脱氧核糖 deoxyribose; desoxyribose
脱氧核糖核酸 deoxyribonucleic acid (DNA); desoxyribonucleic acid
脱氧核糖核酸酶 deoxyribonuclease
脱氧劑 deoxidant; oxygen absorber; oxygen scavenger; oxygen scavenging agent
脱氧糖 deoxysugar; desoxysugar
脱氧膽酸 deoxycholic acid; desoxycholic acid
脱氧酶 deoxygenase
脱脂 degrease; oil off
脱脂〔淡〕煉乳 evaporated skim(med) milk
脱脂大豆粉 defatted soy flour
脱脂奶 nonfat milk; skim milk
脱脂奶粉 nonfat dry milk; powdered skim milk; skim milk powder
脱脂乳 nonfat milk; skim milk
脱脂乳粉 nonfat dry milk; powdered skim milk; skim milk powder
脱脂的 fat-free; nonfat
脱脂芝士 skimmed cheese
脱脂花生 defatted peanut
脱脂食品 defatted food
脱脂乾酪 lean cheese; skim milk cheese; skimmed cheese
脱脂棉 cotton wool
脱脂煉乳 condensed skim milk
脱脂劑 degreaser
脱臭 deodorization
脱臭劑 deodorant; deodorizer
脱臭器 deodorizer
脱胺〔基〕作用 deamination; desamination
脱胺〔基〕酶 aminase; deaminase; desaminase
脱氫〔作用〕 dehydrogenation
脱氫抗壞血酸 dehydroascorbic acid
脱氫視黃醇 vitamin A_2
脱氫醋酸 dehydroacetic acid (DHA)
脱氫醋酸鈉 sodium dehydroacetate
脱氫膽固醇 dehydrocholesterol
脱氫酶 dehydrogenase
脱硫〔作用〕 desulfurization
脱粒 threshing
脱粒機 thrasher; thresher; maize sheller
脱蛋白質 deproteinize
脱蛋白質〔作用〕 deproteinization
脱殼 dehull; dehusk; hulling; husking; stripping

脱殼玉米粗粉 hominy grits
脱殼稻米 hulled rice; husked rice
脱殼蝦米〔乾〕 dry-naked shrimp
脱殼機 dehuller; huller; hulling mill; husker; stripping machine
脱氮〔作用〕 denitrification
脱氮細菌 denitrifying bacteria
脱鈣 decalcify
脱鈣作用 decalcification
脱溶劑器 desolventizer
脱落 shed
脱落素 abscisin
脱落酸 abscisic acid
脱葉劑 defoliant
脱羧基〔作用〕 decarboxylation
脱羧酶 decarboxylase
脱酯〔作用〕 deesterification
脱酯粕 defatted meal
脱碳〔作用〕 decarbonization
脱輔基蛋白 apoprotein
脱輔基酶蛋白 apoenzyme
脱酸〔作用〕 deacidification
脱模劑 remover
脱漿 destarch
脱穀 threshing
脱穀機 paddy machine; paddy separator
脱膜劑 parting agent; release agent
脱膠 degum; degumming
脱膠果汁 depectinized juice
脱膠劑 degumming agent
脱澱粉 destarch
脱糖 desugar
脱糖乳 delactosed milk
脱糖乳清 delactosed whey
脱錫 detinning
脱澀味 deastringent
脱鎂葉綠素 phaeophytin; pheophytin
脱鎂葉綠酸 phaeophorbide; pheophorbide
脱礦質〔作用〕 demineralization
脱蠟 dewax
脱鹽 desalination; desalting
脱鹽的 desalted
脱酰胺〔基〕作用 deamidization
舵手橘蘋果 cox's orange pippin
船上〔的〕廚房 caboose; cookhouse
船上廚師 sea cook
船糧 sea stock; sea stores
莎草油 cyperus oil
莞豆 navy bean
荸薺 chufa; water chestnut
莢 cod; hull; husk; pod; shuck
莢果 seedpod
莢膜 capsule
莖 culm; stalk; stem
莖用菜 stem vegetable
莖桿 haulm
莖菜 stem vegetable
莖椰菜 asparagus broccoli; calabrese
莖葉 haulm
莖藍 kohlrabi
莽吉柿 mangosteen
莫內林〔天然蛋白甜味劑〕 monellin
莫氏乳品試驗器 Mojonnier milk test
莫氏乳脂試驗 Mojonnier fat test
莫氏提香精油機 Mojonnier deoiler
莫氏管 Mojonnier tube
莫氏黏度計 Mojonnier visco(si)meter
莫耳 mole
莫耳溶液 molar solution
莫耳濃度 molar concentration; molarity
莫利洛黑櫻桃 morello (cherry)
莫拉石斑魚 kelp bass
莫塞爾白葡萄酒 Moselle
莫瓊尼爾乳品測定器 Mojonnier milk test
莫瓊尼爾乳脂肪測定法 Mojonnier fat test
莫瓊尼爾提香精油機 Mojonnier deoiler

莫瓊尼爾管 Mojonnier tube
莫瓊尼爾黏度計 Mojonnier visco(si)meter
莫櫻桃 morello (cherry)
荷花 water lily
荷荷巴豆 jojoba
荷荷巴油 jojoba oil
荷葉 lotus leaf
荷葉離褶傘（蕈類） clustered grey gill
荷爾蒙 hormone
荷蘭艾丹姆乾酪 Edam cheese
荷蘭杜松子酒 schnap(p)s; Holland gin; Hollands
荷蘭沙司 hollandaise sauce
荷蘭豆 honey bean; sugar pea
荷蘭乳牛 Holstein dairy cattle
荷蘭芝士 Dutch cheese
荷蘭芹 parsley
荷蘭珍甘藍 white cabbage
荷蘭乾酪 Dutch cheese
荷蘭琴酒 Holland gin
荷蘭黑白花牛 Friesian; Holstein (-Friesian)
荷蘭愛丹乾酪 Edam cheese
荷蘭銀箔甜酒 silver liqueur
荷蘭調味醬 hollandaise sauce
荷蘭薄荷 spearmint; *Mentha viridis*
荼末 dust tea
莧菜 amaranth; edible amaranth; Ganges amaranth
莧菜紅 amaranth
莧菜紅鋁色澱 amaranth alum lake; amaranth aluminium lake
處女蜂巢 virgin comb
處女蜂巢取出的蜂蜜 virgin honey
處理 processing; treat
處理室 handling room
蛇 snake
蛇瓜 club gourd; serpent gourd; snake gourd; snake squash
蛇皮果 salak
蛇甜瓜 snake melon
蛇莓 Indian strawberry; mock strawberry
蛇麻 lupulus
蛇麻花 hop
蛇麻花油 hop oil
蛇麻素 lupulin
蛇麻葫烯 humulene
蛇麻葫酮 humulone
蛇麻〔蘆〕酮 lupulone
蛇羹 snake soup
蛇鰻 snake eel
蛀孔 wormhole
蛀蟲 moth
蛆 maggot
蛋 egg
蛋切片器 egg slicer
蛋卡士達 egg custard
蛋奶甜羹 custard
蛋奶酥 soufflé
蛋奶羹 custard cream
蛋白 albumen; egg albumen; egg white; glair
蛋白水解素 proteolysin
蛋白多糖 proteoglycan
蛋白色素 proteinochrome
蛋白尿症 proteinuria
蛋白杏仁甜餅 macaroon
蛋白杏仁糖果 marchpane
蛋白乳 albumin milk
蛋白的 proteic
蛋白甜酥餅 kiss
蛋白溶素 proteolysin
蛋白聚糖 proteoglycan
蛋白質 protein
蛋白質〔態〕氮 protein nitrogen
蛋白質分散指數 protein dispersibility index
蛋白質分解 proteolysis

蛋白質分解的 proteolytic
蛋白質分解活性 proteolytic activity
蛋白質分解細菌 proteolytic bacteria
蛋白質分解酶 protein degrading enzyme; proteolytic enzyme
蛋白質分離物 protein isolate
蛋白質日糧 protein ration
蛋白質水解 proteolysis
蛋白質水解物 hydrolyzed protein; protein hydrolysate; protein hydrolyzate
蛋白質水解的 proteolytic
蛋白質功效比〔值〕 protein efficiency ratio (PER)
蛋白質生物合成 protein biosynthesis
蛋白質合成 protein synthesis
蛋白質的 proteinaceous
蛋白質保留效率 protein retention efficiency
蛋白質係數 protein factor
蛋白質食品 protein food
蛋白質容許量 protein allowance
蛋白質效率比〔值〕 protein efficiency ratio (PER)
蛋白質消化率 protein digestibility
蛋白質脂 proteolipid
蛋白質能量比 protein energy ratio
蛋白質評分 protein score
蛋白質當量 protein equivalent
蛋白質構造 protein structure
蛋白質滴定度 protein titre
蛋白質維持需要量 protein maintenance requirement
蛋白質需要量 protein requirement
蛋白質價 protein score; protein value
蛋白質質量指數 protein quality index
蛋白質凝膠 protein gel
蛋白質濃縮物 protein concentrate
蛋白質營養缺乏症 kwashiorkor
蛋白質療法 protein therapy
蛋白質轉移 protein shift
蛋白質轉換 protein shift
蛋白質變性 protein denaturation
蛋白糖飾 meringue
蛋白攪泡器 albumen whisk
蛋白鹽 proteinate
蛋白腖 peptone
蛋白腖水 peptone water
蛋白酶 protease; proteinase
蛋乳酒 nog
蛋卷冰淇淋 ice cream cone
蛋杯 eggcup
蛋粉 egg powder
蛋胺酸 methionine
蛋匙 egg spoon
蛋捲 egg roll; waffle roll
蛋球蛋白 egg globulin
蛋殼 egg shell
蛋殼內的卵膜 putament
蛋殼色素 protoporphyrin
蛋殼角皮層 shelled cuticle
蛋殼角質膜 putament
蛋殼膜 shell membrane
蛋殼護膜 shelled cuticle
蛋黃 egg yolk; yolk
蛋黃色 yolk colo(u)r
蛋黃沙司 custard sauce
蛋黃沙律醬 mayonnaise
蛋黃粉 yolk powder
蛋黃醬 mayonnaise
蛋禽 laying fowl
蛋製品 egg product
蛋糕 cake; castilla
蛋糕用混料粉 cake mix
蛋糕捲 Swiss roll
蛋糕裝飾用花邊墊紙 lacepaper edging
蛋糕裝飾師 confectioner
蛋糕飾花機 lidding machine
蛋糕麵包 cake bread

蛋糖霜 meringue
蛋雞尾酒 coffee cocktail
蛋麵〔條〕 egg noodle; noodle
蚯蚓 earthworm; worm
被包裝的 packed
被判定為不適宜食用的 condemned
被吸附物 adsorbate
被感染的 infected
被蒸餾物 distilland
被膜酵母 scum yeast
被膜劑 coating agent
袖珍罐 pocket (type) can
袋 sac; sack; pouch
袋（容量單位） sackful
袋內解凍食品 thaw-in-bag food; thaw-in-pouch food
袋包裝 pouch packaging
袋烤食品 bake-in-bag (food)
袋煮食品 boil-in-bag food
袋萃取器 bag extractor
袋裝 bagging
袋裝的 in-sack
袋裝食品 pouch food
袋鼠 kangaroo
袋鼠肉 kangaroo meat
袋濾器 bag filter
袋濾機 bag filter
規定度加侖 proof gallon
規定食物 dietary
規定溶液 normal solution
規定濃度 normality (N)
規定濃度之烈度酒 proof spirit
規則 regulation
規度 normality (N)
規度溶液 normal solution
規章 regulation
規模放大 scale up
訣竅 know-how
許可限度 tolerance limit
設宴 repast
設備 device; equipment
豚脂 adeps
豚魚 filefish
貫葉連翹 St. John's wort
貨架 shelf
貨棧 storehouse
貪吃者 esurient; guttler
貪食 engorgement; esurient; gluttony; gorge; polyphagia; polyphagy; voracity
貪食的 voracious
貪食者 glutton
貪食症 bulimia; sitomania
貪食癖 phagomania
貧乏培養基 minimum medium
貧血 anaemia; anemia; spanemia
趾 toe
軟化 mellow
軟化〔處理〕 softening
軟化水 demineralized water
軟化脂 plastic fat
軟化乾酪 liquefied cheese
軟化劑 emollient; mollifier; softener
軟化器 mollifier; softener
軟心豆粒糖 jelly bean
軟木 cork; suber
軟木脂 suberin
軟木塞 cork; cork stopper
軟木塞的 corky
軟木墊 cork liner
軟木墊圈 cork washer
軟木橡樹 cork oak
軟水 soft water
軟水劑 water softener
軟牛脂 soft tallow
軟包裝 flexible package
軟包裝袋 flexible pouch
軟奶油 salvy butter
軟冰淇淋 soft ice cream

軟肋骨 rib cartilage
軟血餅 soft clot
軟乳酪 salvy butter
軟性〔樹〕膠 gum elastic
軟果 mushy fruit; pulpy fruit
軟果肉 soft flesh
軟的 soft biscuit
軟紅〔冬〕小麥 soft red winter (wheat)
軟紅冬麥 soft red winter (wheat)
軟食 pap
軟脂 glyceryl tripalmitate; palmitin; tripalmitin
軟脂肪 soft fat
軟脂酸 palmitic acid
軟骨 cartilage; gristle; soft bone
軟骨的 cartilaginous
軟骨症 halisteresis
軟骨素 chondroitin
軟骨蛋白 cartilagin; chondroprotein
軟骨膠 chondrin
軟殼蟹 soft crab
軟殼蟹 soft shelled crab
軟乾香腸 soft dry sausage
軟乾酪 soft cheese
軟雪糕 soft ice cream
軟殼蛋 soft shelled egg
軟殼蟹 buckram; papershell crab
軟飲料 soft drink
軟填料 wad material
軟質〔小麥〕麵粉 pastry flour
軟質小麥 miller wheat; soft wheat
軟質小麥粉 soft wheat flour
軟質包裝 flexible package
軟質奶油 salvy butter
軟質玉米 soft corn
軟質乳酪 salvy butter
軟質食品 spread
軟質乾酪 soft cheese
軟質蛋奶麵包 spoon bread
軟質糖霜泥 soft texture fondant
軟質膳食 soft diet
軟質麵粉 smooth flour; soft wheat flour
軟凝塊 soft clot
軟凝塊乳 soft curd milk
軟樹脂 soft resin
軟糖 gum drops; jelly bean; jelly candy; soft jelly; soft sweets; soft texture fondant
軟糖〔粒〕 jelly drops
軟糖料 fondant
軟麵糰 mobile dough; soft dough
軟體動物 mollusc; mollusk
軟罐 flexible can
軟罐頭 soft can
軟罐頭食品 retort pouch food
通心粉 chifferi; macaroni
通心粉用麵粉 macaroni flour
通心粉的模心 pin
通心菜 water convolvulus; water spinach
通心麵 favantini; macaroni
通用封罐機 universal seamer
通用產品記碼 universal product code (UPC)
通用緩衝液 universal buffer
通用麵粉 all purpose flour
通便性食物 stiparolytic food
通便的 purgative
通便藥 aperient
通風 aeration; ventilation
通風口 vent
通風孔 vent
通風的 ventilated container
通風乾燥 aeration drying
通風乾燥器 air dryer
通風乾燥機 through circulation dryer; through flow dryer
通風設備 ventilator
通風槽 aeration tank

通風器 ventilator
通風儲藏 ventilator storage
通風櫥 fume hood
通氣 aeration
通氣烘箱 air oven
通氣烤箱 air oven
通氣乾燥 aeration drying
通氣槽 aeration tank
通氣器 aerator
通氣攪拌器 aeration agitator
通脊肉 back pork
通草 akebia
通菜 water convolvulus; water spinach
連四硫酸鹽培養基 tetrathionate broth
連苯三酚 pyrogallic acid; pyrogallol
連接酶 ligase
連袋解凍食品 thaw-in-bag food; thaw-in-pouch food
連袋蒸煮食品 cook-in-pouch food
連鎖反應 chain reaction
連鎖商店 chain store
連鎖終止反應 chain termination
連鎖開始反應 chain initiation
連鎖繁衍反應 chain propagation
連續〔高壓〕殺菌釜 continuous retort
連續式巴氏殺菌器 continuous pasteurizer
連續式烤爐 travelling oven
連續式滅菌器 continuous sterilizer
連續板式烤爐 travelling plate oven
連續流動離心分離 continuous flow centrifugation
連續效應 serial effect
連續稀釋 serial dilution
連續萃取 continuous extraction
連續過濾 continuous filtration
連續盤式烤爐 travelling tray oven
連續黏度圖 amylogram; viscoamylogram
連續黏度儀 amylograph
連續離心分離機 continuous centrifuge
速度計 velocimeter
速食 fast food
速食化 instantizing
速食食品 instant food
速食甜點 instant dessert
速食湯 instant soup
速食飯 instant rice; pregelatinized rice
速食癖 tachyphagia
速食麵 instant noodle
速凍 quick freezing; rapid freezing; sharp freezing
速凍室 quick freezing room
速凍食品 quick frozen food
速凍乾燥機 accelerated freeze dryer
速凍機 quick freezer
速烤 sear
速烤肉 seared meat
速乾 quick drying
速煮 fast cooking
速煮食品 quick cooking food
速發麵包 quick bread
速溶化 instantizing
速溶化設備 instantizer
速溶奶粉 instant milk powder
速溶咖啡 instant coffee
速溶茶晶 crystalline tea
速溶湯 instant soup
速製麵包 no-time bread
速製麵糰 no-time dough
速醃 quick cure; rapid curing
速醃〔法〕 quick curing
速凝 quick setting; rapid set; short-set (ting)
速濾劑 accelerator
逐別包裝 item packaging
逐步水解 progressive hydrolysis
逐步冷凍〔法〕 progressive freezing
逐步乾燥〔法〕 progressive drying
逐步乾燥器 progressive dryer

造血〔作用〕 hemopoiesis
造型食品 architectured food
造粒 agglomeration; granulation; pelletization; pelletizing
造粒機 agglomerator; granulator; granulating machine; pelletizer
透水汽性 moisture permeability
透汽性 vapo(u)r permeability
透明包裝 transparent package; vision package
透明冰衣 clear ice
透明性 transparency; transmittance
透明的 transparent
透明度 transparency
透明酒櫃 tantalus
透明塑膠條門簾(冷庫用) plastic strip door
透明薄荷硬糖 clear mint
透析 dialysis
透析液 dialysate
透析管 dialysis tube
透析蒸餾 perdistillation
透析器 dialyzer
透射比 transmittance
透射係數 transmittance
透氣性 air permeability; gas permeability
透過液 permeate
透濕性 moisture permeability
透濾 infiltration
部分不孕 partial sterility
部分不完全蛋白質 partially incomplete protein
部分水解 partial hydrolysis
部分冷凍 partial freezing
部分滅菌 partial sterility
部件 unit
部份 section
部份二聚松香甘油酯 glycerol ester of partially dimerized rosin
都柏林灣斑節蝦 Dublin bay prawn
都漢短角肉牛 Durham
酗酒 bibulosity; insobriety; winebibbing
酗酒者 alcoholic; sot; winebibber
野山椒 rod chil(l)i
野公豬 wild boar
野木瓜 stauntonvine
野外乾燥 field drying
野外鮮蘑菇 meadow mushroom
野生型 wild type
野生動植物 wildlife
野生無花果 caprifig
野生菊苣 wild succory
野生黃〔草〕莓 cloudberry
野生酵母 wild yeast
野生稻米 wild rice
野生蘋果 codling
野米 wild rice
野芋 eddo
野兔 hare
野味 game; venison
野味肉 game meat
野味的風味 gaminess; gamy
野油菜 rape
野芥菜 charlock
野洋菇 field mushroom
野苣 lamb's lettuce; white potherb
野核桃 Chinese walnut
野草 weed; wild grass
野茶 faham
野梅 sloe
野梅酒 sloe gin
野菜 potherb
野豌豆 tare
野禽 feather game; game bird; game fowl; jungle fowl; wild fowl
野禽飼養 aviculture
野鼠尾草籽 chia seed
野辣椒 bird pepper; bush pepper

野豌豆 chickling; vetch
野豬 boar; wild boar
野燕麥 *Avena fatus*; wild oat
野餐 picnic
野餐〔小〕香腸 hunting sausage
野鴨 wild duck
野鴨〔肉〕 mallard
野薔薇果〔實〕 hep
野雞 jungle fowl; pheasant; praire fowl; red grouse
野蘑菇 horse mushroom; true mushroom
野蘿蔔 wild radish
野葱 wild leek
鈕子洋菇 button mushroom
閉式烤爐 blind roaster
閉果 indehiscent fruit
閉殼肌 adductor (muscle); occlusor
陳〔腐〕味 storage flavo(u)r
陳化 ageing; aging; ripening
陳化的 mellow
陳化劑 maturing agent
陳皮 dried tangerine peel; tangerine peel
陳列〔櫥〕櫃 display cabinet
陳列室 stock room
陳列櫃 showcase
陳年白蘭地酒(15-20年酒齡) very special old (V.S.O.)
陳年老白蘭地酒(25-30年酒齡) very special old pale (V.S.O.P.)
陳年酒 mature wine
陳年葡萄酒液面的薄膜 beeswing
陳酒 aged wine; ripe wine
陳酒庫 ageing cellar; aging cellar
陳貯啤酒 lager (beer)
陳貯啤酒酵母 lager yeast
陳腐 staleness; staling
陳腐味 stale flavo(u)r
陳腐的 stale
陳腐氣味 stale odo(u)r
陳戰味 stale taste
陳舊 staleness
陳舊的 stale
陳舊風味 old flavo(u)r
陳釀風味 aged flavo(u)r
陳釀酒窖 ageing cellar; aging cellar
陳釀葡萄酒 aged wine
陸生芹菜 land cress
陸生動物 terrestrial animal
陸地蟹 land crab
陰性染色法 negative staining
陰柳 flank steak
陰乾 shade drying
陰離子 anion
陰離子交換 anion exchange
陰離子交換樹脂 anion exchange resin
陰離子表面活性劑 anionic surface-active agent; anionic surfactant
陰離子界面活性劑 anionic surface-active agent; anionic surfactant
陰離子清潔劑 anionic detergent
陰離子澱粉 anionic starch
陶土 pottery clay
陶石器 stoneware
陶瓷器皿 ceramic ware
陶瓷濾器 porcelain filter
陶器 ceramic; earthenware; pottery
陶器的 ceramic
陶罐食品 terrine
雀麥 cheat
雪利〔葡萄〕酒 sherry
雪利酒蛋糕 trifle
雪豆 sieva bean
雪果霜 sherbet
雪松〔堅〕果 cedar nut
雪松果 pinon
雪松油 cedar oil
雪松葉油 cedar leaf oil; thuja oil
雪泥 water ice

雪花蟹 queen crab
雪封保藏 contact icing
雪茄毒素 ciguatoxin; ciguatera toxin
雪菜 leaf mustard; potherb mustard
雪裡紅 leaf mustard
雪糕 ice cream bar
雪櫃 refrigerator; refrigeratory
章魚 devilfish; octopus
章魚肉鹼 octopine
頂 tip
頂芽 tippy leaf
頂部空間 headspace
頂部修整 topping
頂替色譜法 displacement chromatography
頂葉 tippy leaf
頂飾 topping
頂蓋 top lid
頂隙 headspace
頂層發酵 top fermentation
頂層酵母 top yeast
魚 fish; ikan
魚丸 fish ball
魚叉 harpoon; spear
魚子 fish berry; roe; spawn
魚子醬 caviar; carvier
魚中毒 ichthyism; ichthyotoxism
魚片 filet; fillet; fish fillet; fish steak
魚汁 fish gravy
魚生 sashimi
魚白 soft roe
魚皮 fish skin
魚池 piscina
魚肉 fish; fish flesh
魚肉中毒 ciguatera poisoning; ichthyosarcotoxism
魚肉毒素 ciguatoxin; ciguatera toxin; ichthyosarcotoxin
魚肉香菇餡餅 volauvent
魚肉香腸 fish sausage
魚肉糜 minced fish (meat)
魚卵 fish berry; fish roe; roe; spawn
魚卵毒素 ichthyootoxin
魚卵磷蛋白 ichthulin
魚肝油 fish liver oil
魚肚 fish maw
魚貝中毒 fish and shellfish poisoning
魚味的 fishy
魚油 fish oil; gurry oil
魚的 fishy; piscine
魚的細刺骨 pin bone
魚柳 fish fillet
魚毒 fish poison
魚毒〔素〕 ichthyotoxin
魚背肉 fish loin
魚苗 advanced fry; fry
魚風味 fishy flavo(u)r
魚香腸 fish sausage
魚苗 fish fry
魚凍 aspic jelly
魚秧 advanced fry
魚粉 fish meal
魚翅 dried shark fin; fin; shark fin
魚翅瓜 noodle squash; spaghetti squash; vegetable spaghetti
魚脂 fish fat; fish tallow
魚臭 fish odo(u)r
魚乾 dried fish; stockfish
魚排 fish fillet; fish steak; steak
魚條 fish bar; fish stick
魚產品 fish product
魚蛋白濃縮物 fish protein concentrate
魚棒 fish bar
魚湯 bouillabaisse; fish soup
魚絲 shredded fish
魚塘 pound; warren
魚塭 fish farm
魚塊 fish block
魚溶漿 fish soluble

魚群 shoal
魚腥味 fishy odo(u)r; fishy flavo(u)r; fishiness; fishy smell
魚腥味的 fishy
魚腥臭 fish odo(u)r
魚腹黑膜 belly nape
魚精 soft roe
魚精蛋白 protamin(e)
魚精蛋白酶 protaminase
魚製調味料 fish sauce
魚酸 clupanodonic acid
魚餅 fish cake
魚漿 minced fish (meat); surimi
魚膠 fish gelatin; fish glue; fish isinglass; ichthyocolla; ichthyocol(l); isinglass
魚膠片 leaf isinglass
魚膠粉 gelatin(e)
魚糕 fish cake; kamaboko
魚艙 fish hold
魚糜 minced fish (meat)
魚醬 prahoc
魚醬油 fermented fish sauce; fish sauce
魚醬油（印尼） petis
魚醬油（地中海沿岸） garum
魚鬆 shredded fish
魚羹 fish soup
魚類學 ichthyology
魚觸鬚 barbel; wattle
魚露 fermented fish sauce; fish gravy; fish sauce
魚鰾 air bladder; fish maw; gas bladder; maw; swim bladder
魚體廢棄物 fish waste
魚鱗 fish scale; scale
魚鱗粉 pearl white
魚罐頭 canned fish
鳥〔核〕苷酸 guanosine nucleotide
鳥〔核〕苷酸二鈉 disodium guanylate
鳥胺酸 ornithine
鳥椒 bird pepper; bush pepper; tabasco
鳥結糖 nougat
鳥蛤 cockle
鳥葡萄 bird grape
鳥嘌呤 guanine
鳥嘌呤核苷 guanosine
鳥嘌呤核苷酸 guanylic acid
鳥嘴莓 bilberry; whortleberry
鳥頭酸 aconitic acid
鳥鷸 woodcock
鳥苷 guanosine
鳥苷一磷酸 guanosine monophosphate
鳥苷酸 guanylic acid
鹵化物 halide; halogenide
鹵素 halogen
鹵素消毒劑 halogen disinfectant
鹿 deer
鹿內臟餡餅 humble pie
鹿肉 venison
鹿角 hartshorn
鹿角菜 carrageen; carragheen; *Chondrus crispus*; Irish moss
鹿角菜雙糖 carrabiose
鹿角菜苷 carrageen; carragheen; Irish moss
鹿茸 hartshorn
鹿蹄草 wintergreen
鹿蹄草油 wintergreen oil
麥心 middlings
麥心粉 middling flour
麥氏培養基 MacConkey's medium
麥氏瓊脂〔培養基〕 MacConkey's agar
麥片湯 brose
麥片粥 brochan; brose; flummery; porridge; skilly; stirabout
麥片糊 brochan
麥仙翁 cockle
麥白蛋白 leucosin
麥皮 bran coat

麥角 ergot
麥角中毒〔症〕 ergotism
麥角固醇 ergosterin; ergosterol
麥角甾醇 ergosterin; ergosterol
麥角紅素 sclererythrin
麥角鈣化固醇 calciferol; ergocalciferol; vitamin D_2
麥角酸 ergotic acid; lysergic acid
麥乳粥 frumenty; furmety
麥乳精 malted milk
麥芽 barley malt; byne; malt; malted barley; wheat germ
麥芽三糖 maltotriose
麥芽小根 malt culm
麥芽片 malt flake
麥芽汁 beer must; bub; cereal mash; malt wort; must; wort
麥芽汁浸出槽 malt extract bath
麥芽汁發酵時的泡沫 krausen
麥芽汁飲料 beer
麥芽汁濾渣 wort sludge
麥芽汁醪糟 wort sludge
麥芽乳 malted milk
麥芽乳粉 malted milk powder
麥芽味 malty flavo(u)r
麥芽味的 malty
麥芽的 malty
麥芽威士忌酒 malt whisky
麥芽紅色素 sclererythrin
麥芽風味 malty flavo(u)r
麥芽氧化酶 spermase
麥芽浸出液 sweet wort
麥芽浸膏 malt extract
麥芽粉 malt dust; malt flour; malt meal
麥芽粉碎機 malt mill
麥芽酒 malt liquor
麥芽粗粉 malt meal
麥芽酚 maltol
麥芽渣 malt dust
麥芽萃 malt extract
麥芽飲料 malt beverage
麥芽黃素 maltoflavin
麥芽精 malt extract
麥芽膏 malt extract
麥芽製造者 maltster
麥芽廠 malt house
麥芽漿 cereal mash
麥芽碾機 malt mill
麥芽糊精 maltodextrin
麥芽醇溶蛋白 bynin(e)
麥芽醋 malt vinegar
麥芽澱粉酶 malt amylase; malt diastase
麥芽糖 barley sugar; malt sugar; maltobiose; maltose
麥芽糖化醪 malt mash
麥芽糖飴 malt sugar
麥芽糖漿 malt syrup; maltose syrup
麥芽糖糊精 amyloin; maltodextrin
麥芽糖醇 maltitol
麥芽糖醇糖漿 maltitol syrup
麥芽糖苷 maltoside
麥芽糖酶 malt carbohydrase; maltase
麥芽醪 mash
麥芽麵包 malt bread
麥芽釀酒 malt brewing
麥金塔蘋果 McIntosh
麥胚油 wheat-germ oil
麥胚麵包 wheat-germ bread
麥粉布丁 hasty pudding
麥酒醋 alegar
麥康凱洋菜〔培養基〕 MacConkey' agar
麥康凱培養基 MacConkey's medium
麥桿 haulm
麥清蛋白 leucosin
麥蛋白 aleuronat
麥斯卡爾酒 mescal
麥渣 coarse semolina
麥菜肉濃湯(蘇格蘭) Scotch broth

麥飯石 wheat ball stone
麥黃酮 tricin
麥精 malt extract
麥製薄餅 mill bannock
麥餅 bannock
麥穀蛋白 glutenin
麥糊 caudle; kasha
麥醇溶蛋白 gliadin
麥麩 bran; wheat bran
麥蕈 truffle
麥稭 straw
麻仁球蛋白 edestin
麻布袋 gunny; packing canvas; sack
麻芋 leopard palm
麻油 sesame oil
麻花 twist
麻花狀麵包圈 braided ring
麻花麵包 twister
麻迪菊油 madia oil
麻袋 jute bag
麻袋布 sackcloth; sacking
麻雀 sparrow
麻黃素 ephedrine
麻痺性貝毒素 saxitoxin
麻痺性貝類中毒 paralytic shellfish poisoning
麻醉 anaesthesia
麻醉劑 anaesthetic; drug
麻醉藥 narcotic
唡 ounce (oz.)
堇紫絲膜菌（蕈類） violet web cap
烷自由基 alkyl radical
烷基 alkyl group
烷基化劑 alkylating agent
焗 baking
焗爐 oven
焗蘋果酥 apple strudel
烴 hydrocarbon
焓 enthalpy
硒 selenium
硅 silicon
硅油 silicone oil
硅酮 silicone
硅酮樹脂 silicone resin
硅酸鈣 calcium silicate
硅酸鈉 sodium silicate
硅酸鋁 kaolin
硅酸鋁鈣 aluminium calcium silicate; calcium aluminium silicate
硅酸鎂 magnesium silicate
硅酸鹽 silicate
硅膠 silica gel; kiese gel
硅鋁酸鈉 sodium aluminosilicate; sodium silicoaluminate
硅鋁酸鈉鎂 sodium magnesium aluminosilicate
硅樹脂 silicone
硅藻 diatom
硅藻土 diatom earth; diatomaceous earth; diatomite; infusorial earth; kieselguhr; randanite; tripolite
硅藻石 tripolite
脛 shank; shin
脛骨 shin; tibia
莙薘菜 chard; leaf beet; spinach beet; Swiss chard
莪述 zedoary
莪述根 zedoary root
莰烯 camphene
莰醇 borneol; bornyl alcohol
酚化合物 phenolic compound
酚紅 phenol red
酚氧化酶 phenol oxidase
酚-硫酸呈色法 phenol-sulfuric acid method
酚單位（煙燻食品） phenol unit
酚樹脂 phenol resin
酚醛樹脂 phenolic resin

酚鹽 phenolate
酚酶 phenolase
酚酞 phenolphthalein
啫喱 jelly

脲 urea
脲基甲酸 allophanic acid
脲酶 urease

十二畫

傍晚奶 afternoon milk
備烤的 ready-to-bake
備煮的 ready-to-cook
備餐室 buttery
傘形多孔菌（蕈類） umbrella polypore
傘形毒蕈 amanita
傘菌 agaric; toadstool
傘菌氨酸 agaritine
傘菌屬 *Agaricus*
最上等的酒 supernaculum
最大日糧 maximum ration
最大可能數 most probable number (MPN)
最大安全量 maximum safety level
最大耐量 maximum tolerated dose
最低日需〔要〕量 minimum daily requirement
最低抑制濃度 minimum inhibitory concentration (MIC)
最低熱殺菌條件 minimum thermal process
最低營養素培養基 minimum medium
最佳化 optimization
最後一次發麵 final proof
最後產品 end product
最後糖膏 after-massecuite
最高級的酒（法國） surfine
最確數 most probable number (MPN)
最適〔宜〕的 optimum
最適pH optimum pH
最適化 optimization
最適酸鹼度 optimum pH
最適膳食 optimum diet
凱氏〔測定氮〕法 Kjeldahl method
凱氏氮 Kjeldahl nitrogen
凱氏燒瓶 Kjeldahl flask
凱門鱷 caiman
凱斯特納蒸發器 Kestner evaporator
凱斯特勒值 Koestler number
凱撒卷 kaiser roll
剩餘乳 surplus milk
乘餘的 residual
乘餘部分 residual fraction
勞丹油 ladanum oil
勞立法 Lowry method
博伊森樹莓 boysenberry
博多奧香腸 bobbio (sausage)
博克香腸 bockwurst sausage
博克啤酒 bock (beer)
博爾門萃取器 Bollman extractor
博爾曼抽提器 Bollman extractor
博爾德葡萄酒 vintage port
博羅特豆 borlotto bean
喝 drink
喝酒 winebibbing
喝醉 tipsiness
喝醉的 sloshed; tipsy
喜馬拉雅山長形米 basmati rice
喜筵 wedding feast
喇叭菌（蕈類） horn of plenty

單一維生素缺乏症 monovitaminosis
單乙醇胺 ethanolamine
單不飽和脂肪酸 monounsaturated fatty acid
單元操作 unit operation
單月桂酸山梨醇酐酯 sorbitan monolaurate (Span 20)
單片罐 one-piece container
單件包裝 individual packaging
單件儲藏 individual storage
單式顯微鏡 simple microscope
單位 unit
單果 simple fruit
單油酸山梨醇酐酯 sorbitan monooleate (Span 80)
單孢子 monospore
單柄高酒杯 tankard
單柄深鍋 saucepan
單流的 uniflow
單相 uniphase
單籽山楂 singleseed hawthorn
單胃動物 monogastric animal
單面煎的 sunnyside up
單食癖 monophagia
單效發粉 single baking powder
單效蒸發器 single-stage evaporator; single-effect evaporator
單純三甘油酯 simple triglyceride
單純多醣 simple polysaccharide
單純脂質 simple lipid
單純蛋白質 simple protein
單級均質機 single-stage homogenizer
單級萃取 single-stage extraction
單級蒸發器 single-stage evaporator
單針頭肌肉注射醃漬法 single-needle stitch pumping
單胺基酸 monoamino acid
單烯脂肪酸 monoenoic fatty acid
單烯酸 monoenoic acid
單細胞 single cell; unicell
單細胞生物 monoplast
單細胞培養 monocell culture
單細胞蛋白〔質〕single cell protein (SCP)
單酚氧化酶 monophenol oxidase
單棕櫚酸山梨醇酐酯 sorbitan monopalmitate (Span 40)
單棘魨 filefish
單發酵啤酒 gail; gyle
單硬脂酸山梨醇酐酯 sorbitan monostearate (Span 60)
單硬脂酸木糖醇酐酯 xylitan monostearate
單硬脂酸丙二醇酯 propylene glycol monostearate
單硬脂酸甘油酯 glycerin monostearate; glycerol monostearate; glyceryl monostearate
單程乾燥機 single-pass dryer
單筒顯微鏡 monocular microscope
單軸擠壓機 single-screw extruder
單煎一面的 sunnyside up
單寧 tannin
單寧酸 gallotannin; tannic acid; tannin
單寧酸鹽 tannate
單寧酸酶 tannase
單滾筒乾燥機 single-drum dryer
單種維生素缺乏症 monovitaminosis
單酸甘油酯 monoglyceryl ester
單層 monolayer; unilayer
單層的 single-layer
單層袋 single ply bag
單輪碾碎機 kek mill
單輥壓碎機 single-roll crusher
單澱粉磷酸酯 monostarch phosphate
單糖 glycose; monose; monosaccharide; monosaccharose; simple sugar
單頭封罐式封罐機 one-spindle seamer
單頭牲畜的圍欄 stall

單螺杆擠壓機 single-screw extruder
單鍵 single bond
單雙甘油酯 mono-and diglycerides
單雙鍵脂肪酸 monoenoic fatty acid
單雙鍵酸 monoenoic acid
單鰭多線魚 Atka fish
單鰭鱈 cusk
單體 monomer
單峯牛 zebu
單苷 monoside
唾液 saliva; salivary fluid; spit
唾液酸 sialic acid
唾液酸酶 sialidase
唾液澱粉酶 ptyalase; ptyalin; salivary amylase
喉 guttur; larynx
喉片 pastille
喉錠 troche
喉嚨 throat; weasand
圍巾 pinny
圍裙 apron
圍嘴兒 pinny
堪帶蠟 candelilla wax
堪察加蟹 Japanese crab
壺 jug; kettle; pot
媒介 vector
富奇軟糖 fudge
富馬酸 fumaric acid
富馬酸一鈉 monosodium fumarate
富馬酸亞鐵 ferrous fumarate
富馬酸鈉 sodium fumarate
富硒酵母 yeasts, riched selenium
富貴蝦 mantis shrimp
富雷斯糖 fourres
就地清洗 in-place cleaning
就地清洗〔法〕 cleaning in place (CIP)
就地清洗系統 cleaning in place system (CIP system)
就餐 repast
就餐時間 repast
帽貝 limpet
帽狀通心粉 cappelletti
幾丁胺糖 chitosamine
幾丁聚糖 chitosan
幾丁糊精 chitodextrin
幾丁質 chitin
幾丁質酶 chitinase
幾丁糖 chitose
幾內亞胡椒 Guinea pepper
幾內亞高粱 guinea corn
廂式乾燥器 compartment dryer
廄肥 manure
復水 rehydration; reconstitution; water regain
復水助劑 rehydration aid
復水食品 reconstituted food
復壯 rejuvenation
復染色 counterstain
復活節彩色蛋 Easter egg
復原 reconstitution; recovery; restoration; reversion
復原乳 reconstituted milk
復原性 renaturation
復原果汁 reconstituted juice
復原食品 reconstituted food
復煮 recoction
復甦 anabiosis
復溶 remelt
復蓋捲開罐 key opening reclosure
復蒸餾 redistillation
復鮮 freshelize
復蘇 anabiosis
循序稀釋 serial dilution
循環水 circulating water
惡性貧血 pernicious anemia
惡臭 effluvium; fetor; malodo(u)r; mephitis; niff; offensive odo(u)r; stench; stink

惡臭物 malodorant
惡臭的 foul; malodorous; rank
悶燒 smolder
掌狀紅皮藻 dulse
掌玫 palmarosa
掌玫油 palmarosa oil
揀選 sorting
揀選台 cull eliminator
揉合不足 underkneading
揉和 knead(ing); malaxation
揉茶 tea panning
揉混儀 mixograph
揉圓機 baller
揉緊麵糰 knock back
揉麵 dough kneading
揉麵桿 kneading arm
揉麵槽 kneading trough
揉麵機 developer; dough kneader
揉麵糰 dough up
揉麵糰機 kneader; kneading machine
揉捏 malaxation
提子乾 raisin
提取 abstract; extraction
提取物 extract; extractive
提取劑 extractant; extracting agent
提取器 extractor
提神 freshen
提神飲料 freshener; refresher; refreshment
提秤 steelyard
提桶 pail
提糖 desaccharify; desugar
提糖〔作用〕 desaccharification
揮發 volatilize
揮發〔作用〕 volatilization
揮發性的 volatile
揮發性 volatility
揮發性芳香物質 volatile flavo(u)r
揮發性氨基態氮 volatile basic nitrogen (VBN)
揮發性脂肪酸 volatile fatty acid
揮發性組分 volatile component
揮發性溶劑 volatile solvent
揮發性鹼氮 volatile basic nitrogen (VBN)
揮發油 essential oil; ethereal oil; volatile oil
揮發物 volatile
揮發度 volatility
揮發度計 vapo(u)rimeter
揮發胺 volatile amine
揮發酸 volatile acid
捶擊機 thrasher; thresher
換算係數 conversion factor
換熱器 heat exchanger
揚穀器 winnower
揚穀機 grain blower
敞口式發酵罐 open fermenter
敞口發酵 open fermentation
散葉萵苣 leaf lettuce; loosehead lettuce
散裝 bulk pack(ing)
散裝〔填充〕 dump packing
散裝物品 bulk goods
散熱 thermolysis
散熱器 radiator
斑豆 pinto bean
斑紋 mottle
斑節對蝦 giant tiger prawn; tiger prawn
斑節蝦 banded shrimp; kuruma prawn; prawn
斑點 mottle; speck
斑點奶油 mottled butter
斑點乳酪 specked butter
斑點鮭魚 specked trout
斑蝥黃〔質〕 canthaxanthin
斑雞 partridge
斑雞魚 banded grunt

斑鱒 brown trout
斐林溶液 Fehling's solution
斐林試驗 Fehling's test
斯巴達〔蘋果〕 spartan
斯卡珀農葡萄 scuppernong
斯可左那拉參 scorzonera
斯托克〔超高溫〕滅菌法 Stork process
斯李白蘭地 slivovitz
斯里蘭卡肉桂 Srilanka cinnamon
斯佩爾特小麥 dinkel (wheat)
斯卑爾脱小麥 spelt
斯林酒 sling
斯馬什雞尾酒 smash
斯馬喜雞尾酒 smash
斯脱克〔超高溫〕滅菌法 Stork process
斯普雷格乳牛肝菌 painted bolete
斯潘20 sorbitan monolaurate (Span 20)
普桑雞 poussin
普通小麥 miller wheat
普通豌豆 field pea
普通膳食 regular diet
普通麵條 plain noodle
普澤餅 pretzel
普澤餅棒 pretzel stick
晶核 nuclei; nucleus
晶粒 grain
晶粒砂糖 granulated sugar
晶種 inoculating crystal; seed
晶種母粒 seed grain
晶糖 crystal sugar; crystallized sugar
晶體 crystal; crystalloid
晶體蛋白 crystallin
晶鹼 washing soda
智利辣椒 chile; chile pepper; chil(l)i; chil(l)i pepper
晾乾 atmospheric drying
替代品 substitute
朝天辣椒 red cluster pepper
朝鮮泡菜 kimchi
朝鮮馬鮫 Korean mackerel
朝鮮薊 artichoke; globe artichoke
棕灰口蘑(蕈類) grey knight cap
棕色FK brown FK
棕色HT brown HT
棕色油炒麵粉 brown roux
棕色著色劑 brown colo(u)r
棕色調味醬 brown sauce
棕櫚 maripa; palm
棕櫚仁 palm kernel
棕櫚仁油 palm kernel oil
棕櫚心 palm heart
棕櫚汁 toddy
棕櫚油 palm butter; palm grease; palm oil
棕櫚油酸 palmitoleic acid; zoomaric acid
棕櫚核 palm nut
棕櫚脂 palm butter
棕櫚酒 palm wine; toddy
棕櫚葉 palm
棕櫚精 glyceryl tripalmitate; palmitin; tripalmitin
棕櫚酸 palmitic acid
棕櫚燒酒 beno; bino
棕櫚糖 palm sugar
棕麴菌 *Aspergillus orchraceus*
棕麴霉素 ochratoxin
棘皮類動物 echinoderm
棘茄魚 red batfish
棘鯛 black bream
棘鰆 king mackerel
棗 Chinese date; common jujube; date; jujube
棗子酊 jujube tincture; red date tincture
棗泥 date paste
棗椰子 date palm
棗椰果 date
棗椰樹 date palm
棒 bar; rod; stick

棒狀包裝 stick packaging
棒狀物 stick
棒狀桿菌的 coryneform
棒狀桿菌屬 *Corynebacterium*
棒柄杯傘（蕈類） club footed funnel cap
棒棒糖 bonbon; candy stick; lollipop; sucker
棒磨機 rod mill
棒糖 stick candy; sugarloaf
棒鱸 smare
棲息地 habitat
棍 stick
植固醇 sitosterin; sitoterol
植物 plant; vegetation; vegetal
植物化合物 phytochemicals
植物化學 phytochemistry
植物化學的 phytochemical
植物水份計 succulometer
植物牛油 butterine; margarine; oleomargarine
植物汁液 plant sap
植物甲萘醌 phytonadione; vitamin K_1
植物白蛋白 phytalbumin
植物光敏色素 phytochrome
植物自營 phytotrophy
植物色素 vegetable pigment
植物色蛋白 phytochrome
植物乳〔酸〕桿菌 *Lactobacillus plantarum*
植物固醇 phytosterin; phytosterol
植物性〔起〕酥油 vegetable shortening
植物性色素 vegetable colo(u)r
植物性色素著色 vegetable colo(u)ring
植物性油脂 vegetable fat and oil
植物性的 vegetative
植物性芝士 vegetable cheese
植物性毒素 plant toxin
植物性食品 plant food; vegetable food
植物性乾酪 vegetable cheese
植物性乾酪素 vegetable casein
植物性產品 plant product
植物性蛋白質 vegetable protein
植物性硬脂 vegetable tallow
植物性著色劑 vegetable colo(u)ring
植物性酪蛋白 vegetable casein
植物性調味香料 mirepoix
植物性凝乳酶 plant rennet
植物油 vegetable oil
植物油脂 vegetable fat and oil
植物油脂製的冰淇淋 mellorine
植物的 vegetal; vegetative
植物毒素 phytotoxin
植物相 flora
植物脂 green butter; vegetable butter; vegetable fat
植物脂多醣 phytolipopolysaccharide
植物球蛋白 phytoglobulin
植物組織 plant tissue
植物蛋白〔質〕 plant protein
植物蛋白水解物 hydrolyzed plant protein; hydrolyzed vegetable protein (HVP)
植物蛋白質 vegetable protein
植物蛋白腖 phytone
植物黃油 butterine; margarine; oleomargarine
植物黑素 phytomelane
植物滲出物 plant exudates
植物酸 plant acid
植物膠 gum; vegetable glue; vegetable gum
植物養料 plant food
植物凝集素 lectin; phytagglutinin; plant lectin
植物學 botany; phytology
植物激素 plant hormone
植物營養 plant nutrition; phytotrophy
植物黏膠 mucilaginous gum
植物纖維 vegetable fiber

植物鹼 plant alkaloid
植物鹼基 alkaloid
植酸 phytic acid
植酸鈣鎂 phytin
植酸酯 phytate
植酸鹽 phytate
植酸酶 phytase
植醇 phytol
椒類蔬菜 peppers
椒鹽脆餅棒 pretzel stick
棉子酚 gossypol
棉子糖 gossypose; melitriose; raffinose
棉子糖酶 raffinase
棉豆 butter bean; lima bean; sieva bean
棉花 cotton wool
棉花〔軟〕糖 marshmallow
棉花殼 cottonseed hull
棉花塞 cotton plug
棉花糖 candy floss; cotton candy; floss candy; spun sugar
棉拭檢驗法 swab method
棉籽 cottonseed
棉籽油 cottonseed oil
棉籽粕 cottonseed meal
款待 treat
欽美酒 kummel
殘存曲線 survival curve
殘油 resid; residuum
殘料 waste
殘留物 residue; residuum
殘留的 residual
殘留殺蟲劑 residual pesticide
殘留農藥 pesticide residue; residual pesticide
殘茶盆 slop basin
殘渣 debris; residue; residuum; tailings
殘渣〔淺〕碟 slop basin
殘廢牲畜 cripple
殘餘〔物〕 residue
殘餘的 residual
殘餘液 raffinate
殘餘產物 residual product
殘餘餾分 residual fraction
殘羹 scran; scrap
殘鬚去除機 resilker
殖培養基 enrichment medium
殼 cod; shell; shuck
殼菜 mussel
殼管式熱交換器 shell and tube heat exchanger
殼聚糖 chitosan
殼糊精 chitodextrin
殼糖 chitose
殼糖胺 chitosamine
氮 nitrogen
氮化作用 nitrogenization
氮氣儲藏 nitrogen storage
氮氧化物 nitrogen oxide
氮〔素〕代謝 nitrogen metabolism
氮素計 azotometer; nitrometer
氮〔素〕溶出率 nitrogen solubility index
氮蛋白質轉換因數 nitrogen-protein conversion factor
氮蛋白質轉換係數 nitrogen-protein conversion factor
氮量計 azotometer; nitrometer
氮轉換因數 nitrogen conversion factor
氮轉換係數 nitrogen conversion factor
氯 chlorine
氯乙烯 vinyl chloride
氯化〔處理〕 chlorination
氯化亞乙烯 vinylidene chloride
氯化亞汞 calomel
氯化亞錫 stannous chloride
氯化物 chloride
氯化鈣 calcium chloride
氯化鈉 sodium chloride
氯化鉀 potassium chloride

氯化銨 ammonium chloride
氯化鋁 aluminium chloride
氯化錳 manganese chloride
氯化磷酸三鈉 sodium phosphate, chlorinated
氯化膽鹼 choline chloride
氯化鎂 magnesium chloride
氯化鐵 ferric chloride; iron chloride
氯仿 chloroform; trichloromethane
氯含量 chlorinity
氯酸鹽 chlorate
氬 argon
港口 port
游走孢子 zoospore
游霉素 natamycin
游離水 free water
游離水分 free moisture
游離脂肪酸 free fatty acid
游離胺基酸 free amino acid
游離基 free radical
游離滴液 free drip
游離酸 free acid
游離鹼 free alkali
渣分離機 pulp separator
渣滓 draff; dreg; faeces; feces; feculence; foots; offscouring; recrement; residue
渣滓油 foot oil
減少 reduction
減少鹹味 freshen
減肥食物 diet food
減肥膳食 reducing diet
減肥餐 weight loss diet
減肥藥 obesity drug
減胖 slimming
減重 slimming
減率乾燥 falling rate drying
減濃器 deconcentrator
減壓加熱 vacuo heating
減壓濃縮 vacuum concentration
減濕〔作用〕 dehumidification
湖〔泊〕魚 lacustrine fish
湖泊的 lacustrine
湖南稗子 barnyard millet
湖紅點鮭 char; lake trout; namaycush
湖魚 laker
湖鮭 grey trout
湖鯡 lake herring
湖鱒 lake trout
渦輪型混合器 turbine type mixer
渦輪型攪拌機 turbine type mixer; turbine (type) agitator
湯 bree; soup
湯丸 glutinous dumpling
湯尼汽水 tonic water
湯姆生無籽葡萄 Thompson seedless
湯杯 soup cup
湯料 soup base; stock
湯料濃縮方塊 stock cube
湯粉 dried soup; powdered soup; soup mix; soup powder
湯匙 tablespoon
湯碗 sour tureen
湯盤 soup plate
湯鍋 stock pot
渴 thirst
測定 determination
測脂器 butyrometer
測試對象 testee
測濕計 humidistat; hygrostat
滋味 savor; sapidity; sapor; sapour; savour; tastiness; zest
滋味物質 sapid substance
滋補 invigoration
滋補的 tonic
滋補品 nourishment
滋補食品 corroborant food; tonic food
滋補劑 restorative; tonic
滋補膳食 nourishing diet

滋養 nutriture; nourishment
滋養物 nutriment
滋養的 nutritious; nourishing
滋養品 aliment
滋養食物 alible food
焙 broil; roast; torrefaction
焙炒咖啡 roast coffee
焙根包 bard
焙烤食品 bake; baked food; baked good
焙烤食品用油 baker's oil
焙烤製品 baked product
焙烤酵母聚糖 bakers yeast glycan
焙粉 baking powder
焙茶 tea firing
焙茶〔平〕鍋 kuo
焙茶機 tea firing machine
焚化 incineration
焦〔風〕味 burnt flavo(u)r
焦五倍子酸 pyrogallic acid; pyrogallol
焦化 carbonizing; charring; scorch
焦化麥芽 crystal malt
焦化糖 burnt sugar; caramelized sugar
焦木酸 pyroligneous acid; wood vinegar
焦耳 Joule
焦谷胺酸 pyroglutamic acid
焦亞硫酸鈉 sodium pyrosulfite
焦亞硫酸鉀 potassium metabisulfite
焦兒茶酚 pyrocatechol
焦味 burned taste; scorched flavo(u)r
焦柑 tangor
焦炭 coke
焦臭〔味〕 empyreuma
焦乾碎小麥 bulgur
焦軟糖 fudge
焦麥芽 brown malt; caramel malt; melanoidin malt; roasted malt
焦碳酸二乙酯 diethyl pyrocarbonate
焦碳酸二甲酯 dimethyl dicarbonate (DMDC)
焦糊精 pyrodextrin
焦麩胺酸 pyroglutamic acid
焦糖 black jack; burnt sugar; caramel; caramelized sugar
焦糖化〔作用〕 caramelization
焦糖色 burnt sugar colo(u)r; caramel colo(u)r; sugar colo(u)ring
焦糖風味 caramelized flavo(u)r
焦磷酸 pyrophosphoric acid
焦磷酸二氫二鈉 disodium (dihydrogen) pyrophosphate
焦磷酸二氫鈣 calcium dihydrogen pyrophsophate
焦磷酸二鈉 disodium (dihydrogen) pyrophosphate
焦磷酸四鈉 tetrasodium pyrophosphate
焦磷酸四鉀 tetrapotassium pyrophosphate
焦磷酸亞鐵 ferrous pyrophosphate
焦磷酸鈣 calcium pyrophsophate; dicalcium pyrophosphate
焦磷酸鈉 sodium acid pyrophosphate; sodium pyrophosphate
焦磷酸鉀 potassium pyrophosphate
焦磷酸酯 pyrophosphate
焦磷酸鐵 ferric pyrophosphate
焦磷酸鐵鈉 sodium ferric pyrophosphate
焦磷酸鹽 pyrophosphate
焦磷酸酶 pyrophosphatase
無力密封鍋 waterless cooker
無子白葡萄 sultana
無子西瓜 aspermous watermelon
無子柑橘 aspermous orange
無孔性 imporosity
無孔眼的乾酪 blind
無支架番茄 upright tomato
無水甘露醇 mannitan
無水的 anhydrous; unhydrous; waterless

無水酒精 absolute alcohol; anhydrous alcohol; unwatered alcohol
無水硫酸銨鋁 burnt ammonium alum
無水硫酸鋁鉀 aluminium potassium sulfate, anhydrous
無水碳酸鈉 soda ash
無水醋酸鈉 sodium acetate, anhydride
無火烹調 fireless cooking
無火蒸煮 haybox cooking
無牛油麵包 dry bread
無汁的 juiceless
無用副產品 waste product
無皮的 hulless; rindless; skinless
無皮芝士 rindless cheese
無皮乾酪 crustless cheese; rindless cheese
無皮腿肉 skinless ham
無灰的 ashless
無灰濾紙 ashless filter paper
無肉的 fleshless; maigre
無色 leuco
無色亞甲藍 leucomethylene blue
無色的 colo(u)rless
無色花青素 leucoanthocyanin
無色花青苷 leucoanthocyanin
無色美藍 leucomethylene blue
無色核黃素 leucoflavin; leucoriboflavin
無色桿菌屬 *Achromobacter*
無色糊精 achrondextrin
無衣香腸 skinless sausage
無伸縮性 inelasticity
無角的 polled
無角紅毛牛 red poll
無乳油冰淇淋 mellorine
無乳症 agalactosis
無乳酪麵包 dry bread
無乳糖膳食 lactose-free diet
無刺激膳食 bland diet
無味 sans taste; scentless; unsavo(u)riness
無味的 flavo(u)rless; insipid; milk-and-water; tasteless
無味食物 insipid food
無味覺症 ageusia
無咖啡因咖啡 decaffeinated coffee
無咖啡因的 caffeine-free
無咖啡鹼的 caffeine-free
無定形 amorphism
無定形的 amorphous
無定形糖 amorphous sugar
無性生殖 asexual reproduction
無性孢子 asexual spore
無果仁的 pipless
無法回收再用的 unrenewable
無法凝固的 incoagulable
無油油炸〔法〕 dry frying
無芽孢細菌 asporous bacteria; non-spore (forming) bacteria
無花果 fig
無花果片 pulled fig
無花果乾 pulled fig
無花果蛋白酶 ficin
無花果葉瓜 melabar gourd
無花果樹 sycamore
無柄大口杯 beaker
無毒性 avirulence
無毒物質 innocuous substance; non toxic substance
無毒的 innocuous; non toxic
無突緣罐 unflanged can
無籽西瓜 seedless watermelon
無籽的 pipless; seedless
無籽品種 seedless variety
無籽葡萄乾 seedless raisin
無香味的 flavo(u)rless; unscented
無害物質 innocuous substance
無害的 innocuous
無效食品 inert food
無核小葡萄乾 currant

無核西瓜 seedless watermelon
無核的 seedless; stoneless
無核細胞 akaryote
無核葡萄 raisin grape
無核葡萄乾 dried currant; seedless raisin
無氣味的 inodorous; odo(u)rless; scentless
無氣葡萄酒 still wine
無脂固形物 nonfat solids
無脂的 fat-free
無脂肪的 nonfat
無脊椎動物 invertebrate
無臭的 inodorous; odo(u)rless
無酒精的 alcohol-free
無酒精飲料 non-alcoholic beverage; teetotal drink; temperance drink
無骨 clear bone
無骨火腿 boneless ham
無骨的 boneless
無添加食品 pure food
無甜度香檳酒 dry champagne
無甜度葡萄酒 dry wine
無蛋白質小麥澱粉 protein-free wheat starch
無蛋白質的 protein-free
無蛋白質膳食 protein-free diet
無魚卵的鯡魚 empty herring
無晶形 amorphism
無殼大麥 hulless barley
無殼的 hulless
無殼燕麥 hulless oat
無氮化合物 non-nitrogenous compound
無氮膳食 non-nitrogenous diet
無渣油 sludgeless oil
無滋味 vapidity
無發酵乳酪 sweet butter
無發酵的甜蘋果汁 sweet cider
無發酵麵包 matzo
無發酵麵糰 sweet dough
無硬皮芝士 crustless cheese
無硬殼的 shell-less
無筋力的麵筋 soft gluten
無筋菜豆 stringless bean
無菌 asepsis; sterility
無菌〔操作〕法 aseptic method
無菌水 sterile water
無菌牛奶 germ-free milk
無菌加工 aseptic process
無菌包裝 aseptic package
無菌技術 aseptic technic
無菌乳 germ-free milk; sterile milk
無菌性 aseptic
無菌法 asepsis
無菌的 abacterial; aseptic; axenic; germ-free; sterile
無菌室 aseptic room
無菌食品 sterile food
無菌紙 aseptic paper
無菌動物 axenic animal
無菌培養 sterile culture
無菌桶 aseptic tank
無菌處理 asepsis
無菌填充 sterile filling
無菌裝填 aseptic filling
無菌裝罐 aseptic canning
無菌過濾 aseptic filtration; sterile filtration
無菌製作 aseptic process
無菌製程 aseptic process
無菌槽 aseptic tank
無菌箱 aseptic box
無菌鋁箔包 tetra-pack; tetra-pak
無菌灌裝 aseptic filling; sterile filling
無菌罐 aseptic tank
無鈉〔的〕 sodium free
無鈉〔食〕鹽 sodium free salt
無鈉膳食 sodium free diet
無韌性麵筋 short gluten
無嗅覺症 anosmia

無塗漆罐 plain can
無意食品添加劑 unintentional food additive
無溶質水 freestone water
無煙燻製 artificial smoking; smokeless smoking
無照酒館 shebeen
無精蛋 clear egg; infertile egg
無輔料餅乾 water biscuit
無酵餅 matzo
無酸的 acid-free; acidless
無彈性 inelasticity
無齒鰺 kingfish
無擋板發酵槽 unbaffled fermenter
無機化合物 inorganic compound
無機鹽 mineral salts
無糖的 sugarless
無糖煉奶 plain condensed milk
無糖煉乳 evaporated milk
無錫〔鐵〕罐 tin free can; tinless can
無頭的 headless
無餡大餅 blind
無營養的 inalimental; non-nutritive
無營養甜味劑 non-nutritive sweetener
無縫罐 seamless can; seamless tin
無縫罐身 seamless body
無黏性 non-sticking
無黏性的肉 short meat
無鞭毛細菌 aflagellar bacteria; aflagellate bacteria
無攙雜的 unadulterated
無攙雜食品 unadulterated food
無麵筋 gluten-free
無麵筋食品 gluten-free food
無續熱烹調 fireless cooking; haybox cooking
無襯墊瓶蓋 linerless cap
無鬚鱈 hake
無鰾石首魚 kingfish
無鱗鯉魚 leather carp
無鹽的 salt-free; saltless; unsalted
無鹽魚乾 unsalted dried fish
無鹽膳食 salt-free diet
無鑰匙罐 keyless can
無萜 terpeneless
無萜油 terpeneless oil
無萜烯油 terpeneless oil
煮合子 ravioli
煮老的 hard boiled
煮老蛋 hard boiled egg
煮肉鍋 fleshpot
煮咖啡 coffee brew
煮咖啡壺 coffee urn
煮咖啡漏壺 coffee percolator
煮咖啡機 coffee maker
煮沸 boil; seethe
煮沸過度 overboiled
煮得嫩的 underdone
煮得鬆散的肉 short meat
煮蛋 boiled egg
煮透的 done
煮湯用小型通心粉 pastina
煮硬的 hard boiled
煮硬蛋 hard boiled egg
煮飯 rice cooking
煮解 digestion
煮解器 digester
煮過久的 overcooked; overdone
煮過度 overcook
煮嫩蛋 rate egg
煮熟味 cooked taste
煮熟的 fully cooked
煮熟風味 cook flavo(u)r
煮熟澱粉 boiling starch
煮熬場 boilery
煮熬器 boilery
煮濃 boil down; boiling down; heavily boiling
牌子 brand name

猶太律法食品 kasher; kosher
猶太教食品 kasher
猴胡桃 paradise nut; sapucaya nut
猴頭菇 monkey head mushroom
猴麵包樹 baobab
琺瑯 enamel
琥珀 amber
琥珀色 amber
琥珀色〔焦〕麥芽 amber malt
琥珀魚 amberjack
琥珀酸 amber acid; succinic acid
琥珀酸一鈉 monosodium succinate
琥珀酸二乙酯 diethyl succinate
琥珀酸二鈉 disodium succinate
琥珀酸去氫酶 succinic dehydrogenase
琥珀酸生育酚 tocopheryl acid succinate
琥珀酸酐 succinic anhydride
琥珀酸脱氫酶 succinic dehydrogenase
琥珀酸鈉 sodium succinate
琥珀酸酯 succinate
琥珀酸檸檬酸鐵鈉 sodium ferric succinic citrate
琥珀酸鹽 succinate
琥珀酰單甘油酯 succinylated monoglyceride
琵琶桶 barrel
琵琶魚 anglerfish; frogfish; monkfish
琴柱草 salvia
琴酒 gin; jacky
琴酒飲料 mahogany
琴雞尾酒 Tom Collins
畫眉草（埃塞俄比亞） teff
畫眉笛鯛 black striped snapper
畫線分離〔培養〕 streaking
畫線培養〔物〕 streak culture
畫線接種 streaking
番木瓜 carica papaya; melon tree; papaw; pawpaw
番木瓜汁 papayotin
番木瓜屬 *Carica*
番木鼈鹼 strychnine
番石榴 guava
番石榴酸 piscidic acid
番杏 New Zealand spinach
番紅定性素 safranol
番紅花 crocus; saffron; safran
番紅花色素 saffron colo(u)r
番紅花油 saffron oil
番紅花紅 safranine
番紅花醇 safraninol
番紅花醛 saffranal
番紅花苷 crocin
番紅精 safranine
番茄 love apple; tomato
番茄打漿機 tomato pulper
番茄汁 tomato juice
番茄泥 tomato puree
番茄素 licopersicin; tomatine
番茄粕 tomato pomace
番茄渣 tomato pomace
番茄漿 strained tomato; tomato pulp
番茄調味醬 tomato catchup; tomato catsup; tomato ketchup; tomato sauce
番茄醬 catchup; catsup; ketchup; tomato catchup; tomato catsup; tomato ketchup; tomato sauce
番茄苷 tomatine
番椒 cayenne pepper; hot pepper
番椒粉 cayenne pepper
番辣椒 allspice
番薯 batata; sweet potato
番薯片 sweet potato chip
番瀉〔樹〕葉 senna leaf
番瀉樹 senna
番荔枝 custard apple; ilama; sugar apple; sweetsop
痢疾 dysentery
痢疾志賀桿菌 *Shigella dysenteriae*
痛飲 swig; swill

發光真菌 luminous fungus
發光細菌 luminous bacteria
發光蜜環菌(蕈類) ringless honey fungus
發光燈籠魚 lantern fish
發汗損害 sweat damage
發汗箱 sweat box
發色細菌 chromogenic bacteria
發色劑 chromogenic reagent; colo(u)r fixative
發育 development
發育不全 aplasia
發育不良 dysontogenesis
發泡 foaming; lather
發泡〔葡萄〕酒 sparkling wine
發泡力 lathering power
發泡及嘶嘶響的飲料 fizz
發泡作用 barbotage; zeiosis
發泡性 whipping quality
發泡的 frothy
發泡飲料 effervescent beverage; sparkling beverage
發泡葡萄酒 spumante wine
發泡劑 blowing agent; foamer; foaming agent; frother; frothing agent; inflating agent; whipping agent
發泡體 foamy body
發炎 inflammation
發芽 budding; germination; sprout
發芽大麥 malted barley
發芽小麥粉的烘製品 hovis
發芽〔生殖〕 gemmation
發芽測定器 germinator
發芽器 germinator
發亮劑 glazing agent
發展 development
發粉 baking agent; baking powder; raising powder; saleratus
發臭 stink
發臭的 rotten; stinking
發麥芽 malting
發黑 darkening; nigrescence
發暖劑 calefacient
發煙性 smokiness
發煙劑 smoke agent
發煙器 smoke generator
發滿泡沫的 barmy
發酵 ferment; leavening; zymosis
發酵〔大〕桶 fermentation vat
發酵〔作用〕 fermentation
發酵工藝 zymotechnics; zymotechnique
發酵不足的 underfermented
發酵奶油 ripen cream
發酵玉米粥(南非) mahewu
發酵用麥芽汁 gail; gyle
發酵成熟的芝士 cured cheese
發酵成熟的酒 completely fermented wine
發酵池 fermentation vat
發酵助劑 fermentation adjuvant
發酵完的麵糰 ripe dough
發酵完畢的麵糰 matured dough
發酵技術 zymotechnics; zymotechnique
發酵抑制劑 antizymotic agent
發酵乳 acidified milk; cultured milk; fermented milk
發酵乳〔酸〕桿菌 *Lactobacillus fermentum*
發酵乳飲料 fermented milk drink
發酵乳酪 ripen cream; sour cream butter
發酵果汁 fermented juice
發酵狀態 yeastiness
發酵的 barmy; yeasty
發酵室 fermentation room
發酵度 attenuation degree
發酵計 zymometer; zymoscope; zymosimeter
發酵食品 fermented food; fermented goods
發酵香腸 fermented sausage

發酵粉 baking powder; saleratus
發酵能力測定器 zymometer; zymoscope; zymosimeter
發酵臭 ferment odor
發酵茶 fermented tea
發酵培養基 fermentation medium
發酵桶 fermentation cask
發酵液 fermentation broth
發酵產物 fermentation product
發酵產品 fermented product; fermentation product
發酵細菌 zymophyte
發酵魚醬 bagoong
發酵麥芽汁 krausen
發酵程度 attenuation degree
發酵窖 fermentation cellar
發酵輔佐物 fermentation adjuvant
發酵過久的麵糰 long dough
發酵過度 everfermentation
發酵酪乳 cultured buttermilk
發酵綿羊乳清 skuta
發酵製品 fermented goods
發酵槽 fermenter; fermentor; fermentation tank
發酵蝦醬 blacan; blachan
發酵醋 fermented vinegar
發酵劑 barm; starter
發酵劑培養物 starter culture
發酵劑餾出液 starter distillate
發酵學 fermentology; zymology
發酵學家 zymologist
發酵甕 keeve
發酵醪 distillers' wort; fermented mash; yeast mash
發酵罐 fermenter; fermentor
發酵酶 zymase
發嘶嘶聲 fizz
發熱器 heater
發霉 staleness
發霉的 mildewed; musty; rotten; stale
發黏 sliming; sliminess
發黏肉 slimy meat
發黏的 slimy
發黏凝塊 stringy curd
發黏麵包 clammy bread
發黏麵糰 sticky dough
發麵底子 sourdough
發麵粉 raising powder
發麵起子 flour barm
發麵過度後的回縮 knock back
發麵酵母 flour barm
發麵槽 dough box; dough trough
發麵糰 sponge; sponge dough
短小果 nubbin
短小蛋 dwarf egg
短小穗 nubbin
短切 short cut
短肋肉 short ribs
短吻舌鰨 tongue sole
短角牛(英國) shorthorn
短乳〔酸〕桿菌 *Lactobacillus brevis*
短油 short oil
短桿菌素 tyrothricin
短桿菌屬 *Brevibacterium*
短粒米 short grain rice
短期保藏 semi-preservation
短期保藏食品 semi-preserved food
短溝對蝦 bear prawn
短管通心粉 ziti
短鏈脂肪酸 short-chain fatty acid
短鰭舵魚 lembus rudderfish
硝化〔作用〕 nitration; nitrification
硝化細菌 nitrifier; nitrifying bacteria; nitrobacteria
硝化細菌屬 *Nitrobacter*
硝化糊精 nitrodextrin
硝化蔗糖 nitrosaccharose
硝化澱粉 nitrostarch; starch nitrate

硝化纖維素 nitrocellulose
硝石 niter; nitre; saltpeter; saltpetre
硝基化合物 nitro-compound
硝蛋白 nitroprotein
硝態氮 nitrate nitrogen
硝酸 aqua fortis; nitric acid
硝酸根 nitrate (NO_3)
硝酸鈉 sodium nitrate
硝酸鉀 niter; nitre; potassium nitrate; saltpeter; saltpetre
硝酸糊精 nitrodextrin
硝酸蔗糖 nitrosaccharose
硝酸澱粉 nitrostarch; starch nitrate
硝酸纖維素 nitrocellulose
硝酸鹽 nitrate (NO_3)
硝酸鹽肉湯〔培養基〕 nitrate broth
硝酸鹽還原細菌 nitrate reducing bacteria
硝酸鹽還原酶 nitrate reductase
硬小麥 durum (wheat); flint wheat; hard wheat
硬小麥粉 hard wheat flour
硬小麵包 hard roll
硬化 hardening; sclerosis; vulcanization; vulcanize
硬化油 hardened oil; hydrogenated fat; hydrogenated oil
硬化植物脂 vegetable tallow
硬化劑 firming agent; hardener; hardening agent; stiffener; stiffening agent
硬化機 hardener
硬木質果 xylocarp
硬水 hard water
硬牛軋糖 short nougat
硬皮桔 Bengal quince
硬皮甜瓜 cantaloup(e); rock melon
硬皮橘 bael
硬皮麵包 crusty bread
硬米 keng rice
硬性 hardness; solidness
硬芝士 hard cheese
硬度 hardness; solidness
硬度計 durometer; sclerometer
硬柄小皮傘（蕈類） fairy ring champignon
硬紅冬麥 hard red winter (wheat)
硬紅春麥 hard red spring(wheat)
硬紙板 carton board; strawboard
硬紙盒 carton
硬脂 glyceryl tristearate; stearin; tristearin
硬脂精 glyceryl tristearate; stearin; tristearin
硬脂酸 octadecanoic acid; stearic acid
硬脂酸丁酯 butyl stearate
硬脂酸山梨糖醇酯 sorbitan monostearate
硬脂酸鈣 calcium stearate
硬脂酸鈉 sodium stearate
硬脂酸鉀 potassium stearate
硬脂酸酯 stearate
硬脂酸鎂 magnesium stearate
硬脂酸鹽 stearate
硬脂酰乳酸鈣 calcium stearoyl lactylate (CSL)
硬脂酰乳酸鈉 sodium stearoyl lactate (SSL)
硬脂酰酒石酸酯 stearyl tartrate
硬脂酰富馬酸鈉 sodium stearyl fumarate
硬脆的麥芽 vitreous malt
硬脆餅乾 cracknel
硬骨魚 teleost
硬乾酪 hard cheese
硬盒裝 hard pack
硬粒玉米 flint corn; flint maize
硬粒的 hard-grained
硬粒麥芽 vitreous malt
硬蛋白 albuminoid
硬蛋白質 scleroprotein
硬殼 crust; incrustation; quahog
硬殼蛤 hard clam; hard-shelled clam; quahaug

硬殼蟹 hard crab; hard-shelled crab
硬菌絲團 sclerotium
硬餅乾 hardtack; sea biscuit
硬質小麥 durum (wheat); flint wheat; hard wheat
硬質小麥粉 durum flour; hard wheat flour
硬質牛油 firm butter
硬質包裝 rigid package
硬質芝士 hard cheese
硬質紅小麥 red durum wheat
硬質紅皮冬小麥 hard red winter (wheat)
硬質紅皮春小麥 hard red spring(wheat)
硬質乾酪 hard cheese
硬質麵粉 hard wheat flour; strong flour
硬樹脂 hard resin
硬糖〔果〕 barley candy; barley sugar; boiled sweet; hard candy; hard sugar; high boiled sweet; rock; rock candy
硬膨罐 hard swell (can)
硬頭鱒 rainbow trout
硬黏蛋白 scleromucin
硬麵包 hardtack
硬麵粉 hard wheat flour
硬體 hardware
硬鱗 ganoid scale
硬鱗的 ganoid
硬鱗魚 ganoid
稀〔鮮〕乳油 single cream
稀化 thinning
稀奶油 cream; single cream
稀奶油粉 dried cream
稀奶油乾酪 cream cheese; single cream cheese
稀奶油層 cream layer
稀汁 thin juice
稀白蘭地酒 mahogany
稀有胺基酸 rare amino acid
稀卵白 thin white
稀乳油乾酪 single cream cheese
稀泥 slurry
稀的 thin; wishy-washy
稀粥 gruel
稀粥飲食 gruel diet
稀糊 thin paste
稀糊澱粉 thin boiling starch
稀糖汁 thin juice
稀糖漿 light syrup
稀薄飲料 catlap
稀釋 cut back; dilution; thinning
稀釋〔的〕 dilute
稀釋物 dilution
稀釋液 diluent
稀釋劑 diluent; extender; liquifier; reducer; thinner; thinning agent
稀麵糰 mobile dough
窖藏 cellar storage
窖藏酒 cellar
童子雞 poussin; spring chicken
等 sort
等卡〔熱〕飲食 equicaloric diet
等外品 offgrade product
等級 rank; rating
等溫吸收線 absorption isctherm
等溫吸附線 adsorption isctherm
等溫吸濕曲線 moisture absorption isotherm; moisture sorption isotherm; water sorption isotherm
等溫的 isothermal
等溫脫濕曲線 moisture desorption isotherm; water desorption isotherm
等溫線 isotherm
等電的 isoelectric
等電蛋白質 isoelectric protein
等電點 isoelectric point (pI)
等滲〔壓〕的 iso-osmotic; isosmotic; isotonic
等滲〔壓〕稀釋液 isotonic diluent

等滲〔壓〕溶液 iso-osmotic solution; isotonic solution
等滲性 isosmoticity; isotonicity
等滲壓 iso-osmotic pressure; isosmoticity
等熱食品 isodynamic food
等熱量飲食 equicaloric diet
等熱量膳食 equicaloric diet
等壓吸收線 absorption isobar
等壓的 isobarometric
等壓裝瓶機 isobarometric filler
等壓灌裝機 isobarometric filler
筆尖形通心粉 penne
筆管粉 penne
筒式乾燥機 cylinder dryer; cylinder drying machine
筒式換熱器 barrel heat exchanger
筒式過濾器 cartridge filter; drum filter
筒式熱交換器 barrel heat exchanger
筒倉 silo
筒管式香腸 bobbio (sausage)
筍瓜 banana squash; courgette; hubbard; winter squash; zucchini
筍尖 spear
筋 tendon
筋力低的麵粉 soft wheat flour
筋力弱的麵粉 weak flour
筋力強的麵粉 strong flour
筋力強的麵筋 strong gluten
筋肌 sinew
筋膜 fascia
粟 German millet; Italian millet; millet; proso millet
粟米 corn; maize
粟米油 corn oil
粟粉 cornstarch
粟粉粥 polenta
粥 congee; conjee; porridge; rice congee; water gruel
粥湯 caudle
粥碗 porringer
絞 wring
絞〔細〕肉 ground meat
絞肉 cutting
絞肉午餐肉 minced luncheon meat
絞肉火腿 minced ham
絞肉機 mangler; meat chopper; meat cutter; meat grinder; mincer; mincing machine
絞細牛肉 ground beef
絞碎 mince; mincing
絞碎牛肉 hamburg(er); minced beef
絞碎肉餡 mince collops
絞碎物 mince
結 node
結皮 skin formation
結合 conjugation
結合水 bound water
結成粒狀 graining
結衣 skin formation
結果實 fructification; fruitage
結疤 scar
結核〔芽孢〕桿菌 *Bacillus tuberculosis*
結核桿菌 *Bacillus tuberculosis*
結婚蛋糕 bride cake; wedding cake
結球甘藍 ball cabbage; common head cabbage; head cabbage; headed cabbage; heading cabbage
結球白菜 celery cabbage; Chinese cabbage; Chinese heading cabbage
結球芥菜 cabbage mustard
結球高麗菜 ball cabbage
結球椰菜 ball cabbage
結球萵苣 cabbage lettuce; crisphead lettuce; head lettuce
結粒機 agglomerator
結紮 ligating; tie
結紮機 clipper; linking machine; tier
結晶 crystal; crystallizing; crystallization; rime

結晶不好的糖 damaged sugar
結晶〔助〕劑 crystallizing agent
結晶砂糖 crystallized sugar
結晶溫度 crystallizing point
結晶葡萄糖 crystalline dextrose
結晶蜂蜜 set honey; solid honey
結晶蒸發器 crystallizing evaporator
結晶點 crystallizing point
結晶纖維素 cellulose gel; crystal cellulose; crystalline cellulose
結殼 incrustation
結塊 agglomeration; caking; lump
結塊奶油 curdled cream
結塊性 lumpiness
結塊的糖 lump sugar
結塊蔗糖 caked sugar
結腸 colon; middle gut
結腸炎 colitis
結構蛋白質 structural protein
結網 netting; skin formation
結締組織 connective tissue
結頭菜 kohlrabi
結霜 frost; frosting; rime
絨毛 villi; villus
絕育 sterilization
絕食者 starver
絕乾重量 absolute dry weight; oven dry weight
絕對 absolute (abs.)
絕對好氧菌 obligate aerobe
絕對低限 absolute threshold
絕對戒酒的 teetotal
絕對無菌性 absolute sterility
絕對嗜冷菌 obligate psychrophile
絕對嗜氧菌 obligate aerobe; strict aerobe
絕對嗜熱細菌 obligate thermophilic bacteria
絕對嫌氧菌 obligate anaerobe
絕對溫度 absolute temperature
絕對禁酒主義 teetotalism
絕對零度 absolute zero
絕對厭氧菌 obligate anaerobe; strict anaerobe
絕對腐生菌 obligate saprophyte
絕對閾值 absolute threshold
絕對壓力 absolute pressure
絕對濕度 absolute humidity
絕熱 thermal insulation
絕熱杯套 insulating sleeve
絕熱泡沫材料 insulating foam
絕熱量熱器 adiabatic calorimeter
絕熱熱量計 adiabatic calorimeter
絕熱箱 insulated cabinet
絕熱體 heat insulator
絕緣 insulation
絕緣材料 insulating material
絕緣物 insulant
絕緣體 insulator
紫丁香 lilac
紫丁香蘑（蕈類） wood blewit
紫外光 ultraviolet light
紫外線 ultraviolet (UV)
紫外線分光光度計 ultraviolet spectrophotometer
紫外線分光光度測定〔法〕 ultraviolet spectrophotometry
紫外線水消毒器滅菌 ultraviolet water sterilizer
紫外線偵測器 ultraviolet detector; UV detector
紫外線殺菌處理 rentschlerizing
紫外線滅菌 ultraviolet sterilization
紫外線照射 ultraviolet irradiation
紫外線檢測器 ultraviolet detector; UV detector
紫玉米色素 mazi morado colo(u)r
紫色甘藍 red cabbage
紫花苜蓿 alfalfa; lucern(e)

紫花豌豆 canada field; field pea
紫紅黃紅菇 yellow swamp russule
紫紅膽汁洋菜培養基 violet red bile agar (VRBA)
紫紅麴菌 *Monascus purpureus*
紫紅麴霉 *Monascus purpureus*
紫茉莉 jalap
紫苑油 madia oil
紫苑苷 asterin
紫珠 beauty berry
紫草 alkanet
紫草科植物 comfrey
紫草紅 alkannin; gromwell pigment
紫草根 alkanet
紫草素 alkannin
紫草寧 shikonin
紫草屬植物 gromwell
紫帶蘑菇 porphyry mushroom
紫莖樹莓 purple raspberry
紫晶蠟蘑（蕈類） amethyst deceiver
紫菜 laver; nori; purple laver
紫菜頭 beet (root)
紫菜屬 porphyra
紫菜鹼 porphyrine
紫黑醬果 blueberry
紫萵苣 radicchio
紫膠 lac; shellac
紫膠紅 lac dye; lac red
紫膠酸 laccaic acid
紫膠蠟酸 lacceroic acid
紫質 porphyrin
紫霉素 viomycin
紫錐花 echinacea root
紫羅酮 ionone
紫羅蘭 violet
紫羅蘭花 stock
紫羅蘭葉浸膏 violet leaves concrete
紫羅蘭酮 ionone
紫蘇 perilla; shiso
紫蘇子 perilla seed
紫蘇色素 perilla colo(u)r
紫蘇油 perilla herb oil; perilla oil
紫蘇素 shisonin
紫蘇精油 perillatin
紫蘇糖 perillartine
紫蘇醛 perilla aldehyde
紫鰤 amberjack
絮狀物 floc
絮狀酵母 flocculent yeast
絮凝 flocculate
絮凝〔作用〕 flocculation
絮凝物 floc; flocculate
絮凝劑 flocculant; flocculating agent
絲狀藻類 filamentous algae
絲 silk
絲心蛋白 fibroin
絲瓜 dishcloth gourd; loofah; luffa; rag gourd; sponge cucumber; sponge gourd; towel gourd; vegetable sponge
絲瓜鹼 luffanine
絲光硬糖 pulled hard candy
絲狀真菌 filamentous fungus; filamentous mould
絲狀菌 hyphomycete
絲狀酵母 filamentous yeast
絲狀黏質 rope
絲狀黴菌 filamentous fungus; filamentous mould
絲胺酸 serine
絲胺酸磷脂 phosphatidylserine
絲蛋白 fibroin
絲絨狀垂齒菌（蕈類） weeping widow
絲綢 silk
絲蓋口蘑（蕈類） deceiving knight cap
絲蓋小包腳菇 silky volvar
絲蘭 yucca
絲纖維蛋白 fibroin
絲黴菌 hyphomycete

絡合物 complex
給水 water supply
給水裝置 water works
給料 supply
絳珠籽 annatto
腋芽 sucker
腎〔臟〕 kidney
腎上腺素 epinephrine
腎上腺酮 corticosterone
腎區板油 kidney suet
脹罐 can buckling; gassy can
脾臟 spleen
腓利牛排 fillet steak
舒夫里〔甜點〕 souffle
菩提樹 linden
萃取 extraction
萃取物 extract; extractive
萃取劑 extractant; extracting agent
萃取器 extractor
菸鹼 nicotine
菸鹼酸 niacin; nicotinic acid; pellagramin; vitamin PP
菸鹼醯胺 niacinamide; nicotinamide; nicotinic acid amide
菸鹼醯胺核苷酸 nicotinamide nucleotide
菸鹼醯胺單核苷酸 nicotinamide mononucleotide
菸鹼醯胺腺二核苷酸 nicotinamide adenine dinucleotide (NAD)
菸鹼醯胺腺二核苷酸磷酸 nicotinamide adenine dinucleotide phosphate (NADP)
菠菜 spinach; spinage
菠菜固醇 spinasterol
菠菜泥 pureed spinach
菠菜甾醇 spinasterol
菠菜素 spinacin
菠蘿 ananas; pineapple
菠蘿切片機 pineapple slicer
菠蘿包 pineapple bread
菠蘿汁 pineapple juice
菠蘿刮肉機 eradicator
菠蘿油香精 pineapple oily flavo(u)r
菠蘿香精 pineapple flavo(u)r
菠蘿扇片 pineapple tidbit
菠蘿旋片 pineapple spiral slice; spiral slice
菠蘿蛋白酶 bromelain; bromelin
菠蘿圓 pinery
菠蘿圓片 pineapple whole slice
菠蘿酯 allyl 3-cyclohexanepropionate
菠蘿醛 allyl hexanoate
華夫餅乾 waffle
華夫餅乾烤模 waffle iron
華臍魚 anglerfish
華蘭西橙 Valencia orange
菱平魚 brill
菱〔角〕 caltrop; singharanut; water caltrop
菱角澱粉 caltrop starch
菱紅菇 bare toothed russule
菱粉 caltrop starch
菱體兔牙鯛 pinfish
著色 colo(u)ring; pigmentation; staining
著色劑 colo(u)r; colo(u)rant; colo(u)ring; colo(u)ring agent; dyestuff; food colo(u)r; stain; stainer; staining reagent
萊豆 lima bean
萊姆 lime
萊姆汽水 limeade
萊姆果汁 lime juice
萊姆油 lime oil; limette oil
萊姆金柑 limequat
萊姆素 citroptin; limettin
萊姆莓 limeberry
萊姆飲料 limeade
萊姆酸橙 lime
萊姆酸橙油 lime oil; limette oil
萊茵〔河區〕白葡萄酒 hock; Rhenish; Rhine wine

萊蘭鰤 amberfish
菰 water rice; wild rice
菰〔米〕 water bamboo
菖蒲油 calmus oil; calamus oil
菇 mushroom
菇傘 mushroom cap
菇類開傘 open veil
菾菜 garden beet
萌芽 flush
菌元 mother starter; starter
菌母 mother culture
菌托 volva
菌血症 bacteremia
菌狀真菌 mycelium fungus
菌核 sclerotium
菌根 mycorrhiza
菌株 strain
菌蛋白 mycoprotein
菌傘 pileus; umbrella
菌絲 hypha; hyphae; mycelial filament
菌絲〔體〕 mycelia; mycelium
菌絲的 hyphal
菌絲塊 sclerotium
菌絲體 hyphal body; mushroom spawn
菌絲體的 mycelial
菌塊 clump
菌落 clump; colony
菌落形成單位 colony forming unit (CFU)
菌落形態 colonial morphology
菌落計數計 colony counter
菌落數 colony count
菌種 culture; inocula; inoculum; spawn
菌腐味 bacterial taint
菌蓋 pileus
菌膜 mycoderm; veil; velum
菌膠團 zoogloea
菌環 annulus
菌褶 gill
菲力 tenderloin steak
菲律賓發酵魚〔蝦〕醬 bagoong
菲律賓蛤仔 short necked clam
菲滋酒 fizz
菊牛蒡 goboazami
菊色素 asterin
菊芋 girasole; Jerusalem artichoke
菊花 chrysanthemum
菊花浸膏 chrysanthemum concrete
菊花茶 chrysanthemum tea
菊花晶 chrysanthemum tea
菊花黃浸膏 coreopsis yellow extract
菊苣 chicory; endive; succory
菊苣浸膏 chicory concrete
菊粉 alantin; inulin
菊粉酶 inulase; inulinase
菊胺 chrysanthemin
菊黃酮 chrysanthemin
菊黃質 chrysanthemaxanthin
菊糖 alantin; inulin; synanthrin
菊糖酶 inulase; inulinase
菊薊 American artichoke
萎葉 betel
萎葉酚 chavicol
萎縮果 rivelled fruit
菜〔子〕油 rubsen oil
菜子〔白〕蛋白 plantose
菜子固醇 brassicasterol
菜牛 butcher cow
菜瓜 dishcloth gourd; sponge cucumber; sponge gourd
菜用甜菜 garden beet
菜肉餡餅 potpie
菜夾 kitchen tongs
菜豆 asparagus bean; common bean; field bean; frijole; garden bean; green bean; haricot; haricot bean; kidney bean; *Phaseolus vulgaris*; potato bean; string bean; snap bean; yardlong bean
菜豆亭 phaseolunatin

菜豆球蛋白 phaseolin
菜豆鹼 phaseoline
菜豆苷 phaseolunatin
菜泥 mashed vegetable; puree; vegetable puree
菜泥壓榨器 ricer
菜油 coleseed oil; rape oil
菜油固醇 campesterol
菜籽 colza; rapeseed
菜籽油 coleseed oil; colza oil; rape oil; rapeseed oil
菜籽酸 rapic acid
菜籽餅 rape(seed) cake
菜乾 dried vegetable
菜販 greengrocer
菜單 bill of fare; menu
菜棕 palm cabbage
菜湯 vegetable soup
菜園 garden; kitchen garden
菜農 trucker
菜盤 server
菜燉牛肉 goulash
菜頭 radish
菜餚 cuisine; dish; mess
菜餚的飾菜或飾物 garnish; garniture
菜櫥 meat safe
菜譜 menu
菜麵濃湯 minestrone
萘 naphthalene
萘丸 moth ball
萘酚紅 naphthol red
萘酚黃 naphthol yellow
萜品油烯 terpinolene
萜品烯 terpinene
萜品醇 terpineol
萜烯 terpene
萜烯樹脂 terpene resin
萜〔烴〕 terpene
蛙 frog
蛔蟲 ascarid; eelworm
蛤 clam
蛤肉 clam meat
蛤肉雜燴 clam chowder
蛤蜊 clam
蛤濃汁 clam nectar
蛤餡餅 clam patty
街市 street market
裂合酶 lyase
裂肌球蛋白 meromyosin
裂果 dehiscent fruit
裂紋疣柄牛肝菌 yellow pored scaber stalk
裂紋蛋 cracked egg
裂紋穀粒 cracked grain
裂莢豌豆 split pea
裂解酶 lyase
裂緣 knockdown flange
裂縫 split; splitting
視消化率 apparent digestibility
視測濁度計 scopometer
視窗紙盒 window carton
視窗袋 window bag
視紫紅〔質〕 rhodopsin
視紫質 rhodopsin
視黃素 xanthophane
視黃質 xanthopsin
視黃醇 retinol; vitamin A; vitamin A alcohol
視黃醛 vitamin A aldehyde
視網酸 retinoic acid
視網膜黃素 xanthophane
視網醇 retinol
視網醛 retinene
評味 tasting
評味員 assayer; panel; panelist; taster
評酒 wine tasting
評價 rating
証明 proof; proofing

象牙果 ivory nut
象牙棕櫚之果實 ivory nut
象鼻蟲 weevil
貯存 storage; store
貯存〔期〕濃稠化 age thickening
貯槽 stock tank
貯藏室 conservatory
貯藏限期 limited storage
貯藏窖 pit
貯罐 stock tank
貼上標籤的罐頭 labelled can
貼商標 labelling
貼標籤 labelling
貽貝 mussel; sea mussel
貽貝肌醇 mytilite; mytilitol
貽貝毒素 mytilotoxin
貽貝乾 dried mussel
貽貝酚 mytilite; mytilitol
貽貝糖醇 mytilite; mytilitol
貽貝屬 *Mytilus*
費林溶液 Fehling's solution
費林試驗 Fehling's test
費約果 feijoa; pineapple guava
費喬亞果 feijoa
費斯透那提麵 festonati
越冬作物 winter crop
越瓜 oriental pickling melon
越年果 biennial fruit
越桔 cowberry
越橘 cowberry; mountain cowberry; mountain cranberry; red huckleberry; tangleberry
越橘紅 cowberry red
越橘莓 bilberry
超出量 overrun
超市 supermarket
超甘油化的 superglycerinated
超甘油化脂肪 superglycerinated fat
超低溫冷凍 deep freezing
超低溫冷凍肉 deep frozen meat
超低溫冷凍庫 deep freezer
超低溫冷凍櫃 deep freezer
超低溫凍結食品 deep frozen product
超重 overweight
超音波均質機 supersonic homogenizer; ultrasonic homogenizer
超音波的 ultrasonic
超音波清潔器 ultrasonic cleaner
超音波滅菌〔法〕 ultrasonic sterilization
超音波碎解器 supersonic disintegrator
超氧化物 superoxide
超純化 ultrapurification
超純的 ultrapure
超級市場 supermarket
超脂稀奶油 solid cream
超高速離心分離 ultracentrifugation
超高速離心機 ultracentrifuge
超高溫 ultrahigh temperature (UHT)
超高溫殺菌 ultrahigh temperature sterilization (UHT sterilization)
超高溫滅菌 ultrahigh temperature sterilization (UHT sterilization)
超高溫滅菌乳 UHT milk
超細研磨機 super grinder; ultrafine grinding mill
超細粉碎機 ultrafine pulverizer
超細微粒 ultramicron
超細磨粉機 superfine pulverizer
超提純 ultrapurification
超量營養法 superalimentation
超量餵食法 superalimentation
超微〔細〕粒 ultramicron
超微生物 ultramicrobe; ultramicroorganism
超微粒體 ultramicrosome
超微組織 ultrastructure
超微結晶 ultramicrocrystal
超微結構 ultrastructure
超微量分析 ultramicroanalysis

超微量天平 ultramicro balance
超微濾〔法〕 hyperfiltration
超溫 overtemperature
超溫加熱 superheating
超載 overload
超載的 overladen
超過劑量 overdose
超熱蒸汽 superheated steam; superheated vapo(u)r
超熱器 superheater
超濃縮果汁 superconcentrated juice
超聲波 ultrasonic
超聲波均質機 supersonic homogenizer; ultrasonic homogenizer
超聲波粉碎機 supersonic disintegrator
超聲波清潔器 ultrasonic cleaner
超聲波滅菌〔法〕 ultrasonic sterilization
超臨界流體萃取 supercritical fluid extraction (SFE)
超薄切片機 ultramicrotome
超薄的 ultrathin
超薄馬口鐵 double reduced tinplate
超薄膜 ultrathin film
超濾 hyperfiltration
超濾〔作用〕 ultrafiltration (UF)
超濾〔法〕 ultrafiltration (UF)
超濾液 ultrafiltrate
超濾膜 ultrafiltration membrane
超濾器 ultrafilter
超顯微鏡 ultramicroscope
週期表 periodic table
進料 feeding
進料斗 loading hopper
進料器 feeder
進貨過多 overstock
進餐 dine
進餐時間 mealtime
鄉村式香腸 country style sausage
鄉村乾酪 cottage cheese
鄉村燻腸 katenrauchwurst
酥皮 puff pastry; short crust
酥油 ghee; ghi; shortening
酥油餅 paratha
酥的 crisp
酥炸 frizzle
酥炸麵包塊 croustade
酥餅 flake pastry
酥糖 brittle; light textured confection
酥鬆性測定計 shortometer
量 quantity
量杯 graduated glass; measuring cup
量油計 oil gauge
量氣計 air meter; airometer
量酒杯 jigger
量匙 measuring spoon
量桿 measuring rod
量瓶 measuring flask; volumetric flask
量筒 graduated cylinder; measuring cylinder; volumetric cylinder
量熱法 calorimetry
量熱彈 calorimetric bomb
鈣 calcium
鈣化 calcification
鈣化固醇 viosterol; vitamin D_2
鈣化醇 calciferol; ergocalciferol; vitamin D_2
鈉 sodium
鈉明礬 aluminium sodium sulfate; sodium alum
鈉酪蛋白 nutrose
鈍化〔作用〕 deactivation; inactivation; passivation
鈍化劑 deactivator; inactivator; passivating agent; passivator
鈍葉鼠曲草 balsamweed
開口 abroach
開口鍋 open kettle; open pan
開孔 abroach

開心果 pistachio nut
開水 boiling water
開貝器 oyster dredge
開放發酵 open fermentation
開花 efflorescence
開封 unsealing
開胃 piquancy
開胃〔葡萄〕酒 aperitif (wine); appetizer wine
開胃小菜 relish
開胃物 whet
開胃的 piquant
開胃食品 appetizer; cocktail
開胃酒 table wine
開胃菜 appetizer; hors-doeuvre
開胃菜餚 savory; savoury
開胃醬 piquant sauce
開背冷燻魚 kipper; kippered fish
開瓶 decap
開瓶匙 bottle opener
開瓶費 corkage
開瓶器 bottle opener; decrowner; decapper; lid flipper
開頂式二重捲封機 open-top can
開頂容器 open-top container
開頂貨櫃 open-top container
開傘 open veil
開窗孔紙盒 window carton
開窗袋 window bag
開塞鑽 corkscrew
開蓋 uncork
開蓋器 cap opener
開箱裝置 uncaser
開膛〔去內臟〕 viscerate
開膛取臟 disembowel
開罐刀 tin opener
開罐試驗 cut-out test
開罐器 can opener; tin opener
開罐糖度 cut-out Brix
間接火乾燥機 indirect fired dryer
間接加熱 indirect heating
間接冷凍 indirect freezing
間接食品添加物 indirect food additive
間接乾燥 indirect drying
間接熱量測定法 indirect calorimetry
間菜 entrée
間歇巴斯德殺菌法 discontinuous pasteurization
間歇式巴氏滅菌器 batch pasteurizer
間歇式乾燥〔法〕 intermittent drying
間歇式乾燥機 intermittent dryer
間歇式殺菌器 batch sterilizer
間歇式離心機 batch centrifuge
間歇滅菌〔法〕 fractional sterilization; intermittent sterilization;tyndallization; discontinuous sterilization
階段 phase
階級 rank
階梯式凍結裝置 multistage freezer
陽性染色〔法〕 positive staining
陽離子 cation
陽離子交換 cation exchange
陽離子表面活性劑 cationic surfactant
陽離子清潔劑 cationic detergent
隆起菌落 raised colony
雁 wild goose
雅芝竹 artichoke; globe artichoke
雄火雞 bubbly-jock; gobbler; turkey cock
雄赤松雞 gorcock
雄性動物 tom
雄松雞 wood grouse
雄紅松雞 moorcock
雄鹿 buck; hart; stag
雄鹿角 staghorn
雄禽 rooster
雄蕊 stamen
雄鮭〔魚〕 redfish; kipper
雄鵝 gander

雄蘇格蘭雷鳥 moorcock
集中器 concentrator
集合體 aggregate
集粉器 powder collector
集散市場 terminal market
集塵袋 dust bag
集塵器 dust collector; dust trap
集體〔性食物〕中毒 mass poisoning
雲木香油 costus root oil
雲吞 wonton
雲杉啤酒 spruce beer
雲豆 haricot; haricot bean
雲紋石斑魚 kelp bass
雲莓 cloudberry
韌皮部 phloem
韌度 tenacity
韌帶 ligament
順式-3-己烯異戊酸酯 *cis*-3-hexenyl isovalerate
順式-4-辛烯酸乙酯 ethyl *cis*-4-octenate
順式-4-庚烯醛 *cis*-4-hepten-1-al
順式-4-癸烯醛 *cis*-4-decen-1-al
順序效應 serial effect
順流式乾燥機 cocurrent dryer; concurrent flow dryer
順流式噴霧乾燥機 cocurrent spray dryer
順流乾燥機 parallel flow dryer
順流蒸發機 parallel flow evaporator
飯 rice
飯豆 rice bean
飯店 restaurant
飯前酒 aperitif (wine); appetizer wine
飯後甜〔葡萄〕酒 sweet dessert wine
飯桶 dinner bucket
飯票 meal ticket
飯焦 rice crust
飯煲 rice cooker
飯鍋 rice cooker
飯鮹 ocellated octopus
飲水感染 water-borne infection
飲水器 fountain
飲水機 drinking fountain
飲用水 drinking water; potable water
飲乳療法 galactopathy
飲食 diet; meat and drink; repast
飲食店 eating house; victual(l)ing house
飲食治療 dietary therapy
飲食物 ingesta
飲食的 dietetic; dietary
飲食過度 surfeit
飲食過量 overeat
飲食學 bromatology; dietics
飲食學家 dietician; dietitian
飲食營養學 sitology
飲食療法 dietotherapy; sitotherapy; trophotherapy
飲宴 wassail; wine and dine
飲料 beverage; drink; potable; refreshment
飲料充氣機 gasogen(e); gazogene
飲料粉 drink powder
飲料酒〔精〕 potable spirit
飲烈酒 tipple
飲酒 potation
飲盒 messtin
黃刀豆 wax bean
黃片糖 brown slab sugar; brown sugar slab
黃牛 burmese; kine
黃牛肝菌 butter mushroom
黃玉米 yellow corn; yellow maize
黃瓜 cucumber; cuke
黃瓜子油 cucumber oil; cucumber seed oil
黃瓜條 chuck tender
黃瓜醇 cucumber alcohol
黃皮 wampee
黃皮小胡瓜 yellow zucchini
黃皮筍瓜 yellow zucchini
黃石鰭 yellow bass
黃肉甜〔甘〕薯 orange flesh sweet potato

黃色根菜 yellow vegetable
黃色素 flavochrome
黃色素血症 xanthemia
黃色酸櫻桃 pale sour cherry
黃色糊精 canary dextrin
黃夾竹桃糖 thevetose
黃豆 soy; soja (bean); soya (bean); soybean
黃豆皂苷 soy saponin; soya saponin
黃豆油 soy oil; soya oil; soybean oil
黃豆芽 soy sprout; soybean sprout
黃豆粉 soy flour; soybean flour
黃豆粕 soy cake; soybean cake
黃豆蛋白 soy protein; soybean protein
黃豆蛋白食品 soy protein food; soybean protein food
黃姑魚 croaker
黃油 butter; grease; oleo oil
黃花 citron day-lily
黃花南芥菜 rocket
黃花魚 yellow croaker
黃花菜 common yellow daylily; grassleaf daylily; lemon daylily; lemon lily; long yellow daylily; yellow daylily
黃花萱草 lemon daylily
黃芥末 yellow mustard
黃芥菜 yellow mustard
黃金菊 camomile
黃金菊油 camomile oil
黃金銀耳 yellow brain fungus
黃孢紅菇 crab russule
黃扁豆 yellow lentil
黃秋葵 bamies; gumbo; okra
黃原酸酯 xanthate
黃原酸鹽 xanthate
黃原膠 xanthan gum
黃核苷 xanthosine
黃桃 yellow peach
黃素 flavin
黃素蛋白 flavoprotein
黃耆 tragacanth
黃耆溶膠 tragacanthin
黃耆膠 gum dragon; gum tragacanth; tragacanth gum
黃耆質 tragacanthin
黃耆糖 tragacanthin
黃脂色素 lipoxanthine
黃酒 yellow wine
黃啤〔酒〕 pale ale (beer)
黃梔子 gardenia yellow
黃蛋白〔質〕 xanthoprotein
黃魚 yellow croaker
黃麻 jute
黃烷酮 flavanone
黃烷醇 flavanol
黃喋呤 xanthopterin
黃番茄 yellow tomato
黃貂魚 stingray
黃黃質 flavoxanthin
黃楊栗 jojoba
黃葵 musk mallow
黃葵籽油 ambrette seed oil
黃蜂 wasp
黃酮 anthoxanthin; flavone
黃酮醇 flavonol
黃酮類化合物 flavonoid
黃綠藻 yellow green algae
黃蓋鰈 dab; mud dab
黃嘌呤 xanthine
黃嘌呤氧化酶 xanthine oxidase
黃蓍膠 bassora gum
黃蓍膠糖 bassorin
黃蓍膠黏素 bassorin
黃樟油 sassafras oil
黃樟素 safrole
黃樟烯 safrene
黃樟腦 safrole
黃熟 gold ripe(ness)
黃糊精 canary dextrin; yellow dextrin

黃豌豆 yellow pea
黃質氧化酶 Schardinger enzyme
黃糖 brown sugar; yellow sugar
黃錫鯛 silver bream
黃麴 yellow koji
黃麴毒素 alfatoxin
黃麴毒素中毒 aflatoxicosis
黃麴菌 *Aspergillus flavus*
黃麴黴 *Aspergillus flavus*
黃麴黴〔毒〕素 alfatoxin
黃麴黴素中毒 aflatoxicosis
黃藻 yellow green algae
黃蠟 yellow wax
黃鰭油鯡 yellowfin menhaden
黃鰭金槍魚 allison tuna; pacific yellowfin tuna; yellowfin; yellowfin albacore; yellowfin tuna
黃鰭魚 yellowfin
黃鰭鮪〔魚〕 allison tuna; yellowfin tuna
黃鰭鯛 yellowfin sea bream
黃變米 yellow damaged rice
黃變米〔霉〕毒素 luteoskyrin
黃體素 lutein
黃鱔〔魚〕 rice field eel; mud eel
黃黴素 xanthomycin
黃芪膠 tragacanth gum
黃苷 xanthosine
黃酶 yellow enzyme
黍 proso millet
黍鯡 brisling
黑口 black mouth croaker
黑子南瓜 melabar gourd
黑小麥 triticale
黑木耳 black fungus; jew's ear
黑毛黴菌 chetomin
黑加侖子 blackcurrant
黑加侖紅 black currant red
黑卡布奇諾咖啡 cappuccino scuro; dark cappuccino
黑卡布基諾咖啡 cappuccino scuro; dark cappuccino
黑巨頭鯨 black fish
黑巧克力 dark chocolate
黑布丁 black pudding; black sausage
黑皮波羅門參 black salsify; oyster plant; salsify; vegetable oyster
黑皮旗魚 black spearfish
黑安格斯牛 Aberdeen Angus
黑朱古力 dark chocolate
黑色素 brilliant black; brilliant black BN; melanin(e)
黑色雄松雞 blackcock
黑色覆盆子 black raspberry; blackcap
黑豆 black bean
黑豆紅 black bean red
黑刺李 blackthorn; sloe
黑刺李琴酒 sloe gin
黑刺莓 blackcap raspberry
黑咖啡 black coffee
黑姑魚 black croaker
黑果 huckleberry
黑河鱸 black perch
黑芝麻 black sesame (seed)
黑芥子 black mustard
黑芥菜 black mustard
黑芥酸鉀 sinigrin
黑孢塊菌（蕈類） perigord truffle
黑度計 nigrometer
黑砂糖 muscovado (sugar)
黑胡椒 black pepper
黑胡椒油 black pepper oil
黑面魨 black scraper
黑根黴 *Rhizopus nigericans*
黑桑 black mulberry
黑海帶 arame
黑海鯛 oldwife
黑茶藨子 blackcurrant
黑豇豆 black cowpea

黑啤酒 black beer; dark beer; porter
黑啤酒（英國） entire
黑眼豆 black eyed bean
黑莓 blackberry; dewberry
黑魚子醬 black caviar
黑麥 rye
黑麥〔小麥〕麵包 brown bread
黑麥芽 black malt; rye malt
黑麥威士忌〔酒〕 rye whisky
黑麥粉 rye flour
黑麥草 darnel
黑麥粗粉 rye meal
黑麥醇溶蛋白 secalin
黑麥糖 secalose
黑貂 sable
黑黍 Indian millet
黑雷雞 blackcock
黑旗魚 black marlin
黑槍魚 black marlin; black spearfish; blue marlin
黑綠豆 black gram; urd (bean)
黑線銀鮫 oriental chimaera
黑線鱈 haddie; haddock
黑醋栗 blackcurrant
黑醋菌 *Acetobacter melarogenum*
黑樹莓 black raspberry; blackcap; blackcap raspberry
黑濁啤酒 stout (beer)
黑糖 black sugar; brown sugar; nigerose
黑鮑 black abalone
黑檀樹 ebony
黑邊石斑魚 red grouper
黑邊吳郭魚 tilapia
黑鯧 black pomfret
黑鯛 black porgy; black sea bream
黑麴 black koji
黑麴菌 *Aspergillus niger*; black aspergillus
黑麴黴 black mould
黑麴黴糖 nigerose
黑糯米 black glutinous rice
黑麵包 brown bread; rye bread
黑鰔 black mouth croaker
黑櫻桃酒 maraschino
黑鰭髭鯛 grunt
黑鰭鯛 black bream
黑鰱 bighead carp
黑蘿蔔 black radish
黑變 nigrescence
黑變肉 black meat
黑黴 black mould
黑鱸 blackbass; largemouth bass
黑鮗 ocellated plesiope
椪柑 ponkan
氰化物 cyanide
氰鈷胺素 cyanocobalamin; vitamin B_{12}
氰鈷維生素 cyanocobalamin; vitamin B_{12}
猢猻麵樹 baobab
酢漿草 oxalis; shamrock; wood sorrel
鈦 titanium
韌性 toughness
韌度 toughness
腖 peptone
腖化作用 peptonization
腖化乳 peptonized milk
腖水 peptone water
喼汁 worcester(shire) sauce
葱 cibo; cibo(u)l; scallion; spring onion; stone leek; Welsh onion
葱油 scallion oil
葱蒜 *Allium*
葱蒜〔性〕調味料 alliaceous condiment
葱蒜味的 alliaceous
葱蒜氣味 alliaceous odor
葱屬 *Allium*
韮菜 Chinese chive; Chinese leek; leek; scallion
韮葱 leek

十三畫

傳染 infect; infection
傳染病 infectious disease
傳送吊軌 conveyor rail
傳送式烤爐 travelling oven
傳送式乾燥機 travelling dryer
傳送板式烤爐 travelling plate oven
傳送帶 conveyor belt
傳送盤式烤爐 travelling tray oven
傳送機 conveyor
傳送鏈式烤爐 travelling chain oven
傳統食品 traditional food
傳統乾醃法 conventional dry curing
傳統發酵的麵包 salt-rising bread
傳統醃製 conventional curing
傳統濕醃法 conventional pickle curing
傳感器 sensor
傳熱性 thermal conductivity
傳導 conduction
傳導加熱 conduction heating
傾卸桶 tipping tank
傾卸槽 dump tank; tipping tank
傾析 decantation
傾析〔式〕離心機 decanter
傾析用的玻璃瓶 decanter
傾析槽 decanting tank
傾析器 decanter
傾注平板〔培養基〕 pour plate
催化 catalyze
催化〔作用〕 catalysis
催化反應 catalytic reaction
催化劑 catalyst; catalytic agent
催吐劑 emetic
催乳激素 lactogen
催涎的 sialagogue
催涎劑 sialagogue
催熟 after-ripening
催熟劑 maturing agent; ripener
傷口感染 wound infection
傷痕 scar
傷寒沙門氏菌 *Salmonella typhi; Salmonella typhosa*
傷寒沙門桿菌 *Salmonella typhi; Salmonella typhosa*
傷寒熱 typhoid fever
嗎哪 manna
嗎啡 morphine
嗎啉 morpholine
嗎啉脂肪酸鹽 morpholine fatty acid salt
嗎啉脂肪酸鹽果蠟 morpholine fatty acid salt fruit wax
嗜好 addictive
嗜好〔性〕試驗 preference test
嗜好食品 fancy food
嗜血細菌 hemophilic bacteria
嗜冷生物 psychrophile
嗜冷的 psychrophilic
嗜冷細菌 cryophilic bacteria; psychrotrophic bacteria; psychrophilic bacteria; psychrophile
嗜冷微生物 cryophile

嗜旱生物 xerophile
嗜旱的 xerophilic
嗜旱細菌 xerophilic bacteria
嗜食家 gourmand
嗜食症 hyperphagia
嗜氣性細菌 aerophilic bacteria
嗜氧培養 aerobic culture
嗜氧細菌 aerobic bacteria
嗜氧發酵 aerobic fermentation
嗜氧菌 aerobe; aerobium; aeromicrobe
嗜氧菌種 aerobic culture
嗜氧微生物 aerobic microorganism
嗜酒狂 alcoholomania; dipsomania; potomania
嗜酒癖 dipsomania
嗜高滲〔壓〕微生物 osmophile
嗜高滲細菌 osmophilic bacteria
嗜高滲酵母 osmophilic yeast
嗜甜食者 sweet tooth
嗜異食症 cissa
嗜異癖 allotriophagy; cissa
嗜硫細菌 thiophilic bacteria
嗜寒細菌 cryophilic bacteria
嗜寒微生物 cryophile
嗜飲水者 water drinker
嗜溫生物 mesophile
嗜溫的 mesophilic
嗜溫細菌 mesophilic bacteria
嗜滲壓細菌 osmophilic bacteria
嗜滲壓微生物 osmophile
嗜滲壓酵母 osmophilic yeast
嗜酸乳〔酸〕桿菌 *Lactobacillus acidophilus*
嗜酸乳桿菌 *Bacillus acidophilus*
嗜酸的 acidophilous; acidophilic
嗜酸細胞 acidophile
嗜酸細菌 acidophilic bacteria; acidophile
嗜酸菌乳 acidophilus milk
嗜熱生物 thermophile
嗜熱乳〔酸〕桿菌 *Lactobacillus thermophilus*
嗜熱細菌 thermophilic bacteria; thermophile
嗜熱菌蛋白酶 thermolysin
嗜熱澱粉分解〔芽孢〕桿菌 *Bacillus thermoamylolyticus*
嗜熱鏈球菌 *Streptococcus thermophilus*
嗜壓細菌 barophilic bacteria
嗜壓微生物 barophile
嗜鹼細菌 alkalophilic bacteria; basophilic bacteria
嗜鹽生物 halobiont
嗜鹽細菌 halophilic bacteria; halobacteria; halophilous bacteria
嗜鹽植物 halophyte
嗜鹽菌 obligate halophile
嗜鹽〔微〕生物 halophile
嗅味學 osmics
嗅學 osmology
嗅覺 olfaction; osmesthesia; smell; scent
嗅覺正常 euosmia
嗅覺低限 odo(u)r threshold; olfactory threshold
嗅覺的 olfactory
嗅覺計 olfactometer
嗅覺缺失症 anosmia
嗅覺評分 nasal rating
嗅覺障礙 dysosmia
嗅覺學 osmology
嗅覺閾值 odo(u)r threshold; olfactory threshold
嗉囊 maw
園藝 horticulture
園藝學 horticulture
園蘿蔔 turnip radish
圓刀切片機 disc slicer
圓大麥米 pearl barley
圓片狀通心粉 disc; disk

圓白鯧 orbfish
圓尾頜針魚 garfish
圓狀豬腰肉排 pork loin medallion steak
圓花鰹 frigate mackerel
圓金柑 marumi kumquat; round kumquat
圓金橘 round kumquat
圓柚酮 nootkatone
圓珠茶 gunpowder tea
圓荳蔻 round cardamon
圓球狀的小〔白蘭地酒〕杯 pony
圓球糖成形機 ball machine
圓舵鰹 bulletmackerel
圓頂麵包 open top bread; round top bread
圓筒形殺菌釜 cylindrical retort
圓筒形換熱器 cylindrical heat exchanger
圓筒形熱交換器 cylindrical heat exchanger
圓蛤類 quahaug; quahog
圓滑番荔枝 pond apple
圓當歸籽油 angelica seed oil
圓當歸根油 angelica root oil
圓腹雅羅魚 orfe
圓腹鯡 round herring; sprat
圓葉當歸 lovage
圓葉當歸油 lovage oil
圓葉葡萄 muscadine grape; muscat grape; scuppernong
圓葉錦葵 running mallow
圓葉薄荷 *Mentha rotundifolia*
圓蓋機 lid curler
圓酵母 torula (yeast)
圓酵母紅素 torularhodin
圓酵母素 torulin; vitamin T
圓盤 disc; disk
圓盤切片機 disc slicer
圓穀倉 silo
圓篩 bolting reel; reel
圓錐小麥 cone wheat; duckbill wheat
圓錐形 cone
圓錐形冰淇淋杯 ice cream cone
圓錐鐘菌（蕈類） thimble cap
圓頭甘藍 roundhead cabbage
圓鰭魚 lumpfish
圓鰭雅羅魚 chub
圓體魚類 round fish
圓罐 round can; round tin
圓鰺 amberfish
圓稨牛肉 knuckle
圓稨牛排 knuckle steak
塞子 bung; stopper
塞子發酵罐（製啤酒用） plug ferment
塞卡雞尾酒 sidecar
塞克爾梨 seckel
塞瓶工人 corker
塞瓶機 plugging machine
塞〔爾〕維拉〔特〕燻香腸（德式） cervelas; cervelat
塑化 plasticization
塑化劑 plasticizer; plasticizing agent; plastifier
塑性 plasticity
塑性奶油 plastic cream
塑性的 plastic
塑性計 plastometer
塑性脂肪 plastic fat
塑性膜 plastic film
塑性凝膠 plastigel
塑度計 plastometer
塑料溶膠 plastisol
塑料製品 plastic
塑料薄膜 plastic film
塑膠 plastic
塑膠容器 plastic container
塑膠瓶 plastic bottle
塑膠盤 plastic tray
塑變記錄儀 plastograph
塘鵝 gannet; solan

塗上熔化乾酪〔及啤酒〕的烤麵包 rarebit; Welsh rabbit; Welsh rarebit
塗片培養〔物〕 smearing culture
塗抹食品 spread
塗抹培養〔物〕 smearing culture
塗抹蛋白 glair
塗抹器 spreader
塗油 inunction
塗料（罐頭） enamel
塗漆罐 lacquered can
塗層劑 coating agent
塗敷 spread
塗蠟 waxing
塔巴斯哥酸辣醬 tabasco sauce
塔式乾燥器 tower dryer
塔式蒸發器 tower evaporator
塔式蒸餾器 tower still
塔板 plate
塔格糖 tagatose
塔曼里羅果 tamarillo
塔菲亞酒 tafia
塔塔沙司 tartare sauce
塔塔粉 cream of tartar; potassium acid tartrate; potassium bitartrate; potassium hydrogen tartrate
塔糖 sugarloaf
塔羅糖 talose
塔羅糖醇 talitol
填充 pack
填充床過濾器 packed bed filter
填充封口機 filling seal machine
填充柱 packed column
填充料 packing material; stuffing
填充塔 packed tower
填充劑 bulking agent
填充器 applier
填充機 filler; filling machine; stuffer; stuffing machine
填肥肉餡 larding
填料 filling; filler; wad material
填塞 ram; stuff
填塞物 stuffing
填裝 filling
填裝機 filler
填餡青椒 stuffed green pepper
填餡雞 stuffed chicken
填鴨 crammed duck
塌麵糰 flowy dough
塊 lump; piece
塊冰 block ice
塊狀肉乾 meat bar
塊狀物 cake
塊狀酵母 pressed yeast
塊根 root
塊根芹 celeriac
塊根農作物 root crop
塊根澱粉 root starch
塊根類蔬菜 root vegetable
塊莖 tuber
塊莖作物 tuber crop
塊莖形多孔菌（蕈類） tuberous polypore
塊莖澱粉 tuber starch
塊菌 truffle
塊裝朱古力 block chocolate
塊糖 caked sugar; loaf sugar
塊鱗灰鵝膏（蕈類） stout agaric
塊鹽 block salt
奧士瓦〔毛細管〕黏度計 Ostwald viscosimeter
奧地利牛奶軟芝士 abgesottener cheese
奧地利牛奶軟乾酪 abgesottener cheese
奧地利蘋果餡餅 strudel
奧利亞吳郭魚 blue tilapia
奧斯瓦爾特〔毛細管〕黏度計 Ostwald viscosimeter
嫌氣〔性〕的 anaerobic; anerobic
嫌氣性生化處理 anaerobic biochemical treatment

嫌氣性發酵 anaerobic fermentation
嫌氣培養 anaerobic culture
嫌氣細菌 anaerobic bacteria
嫌氣菌 anaerobe; anaerobic formers
幹 stem
廉價小酒店 barrelhouse
廉價出售 undersell
廉價威士忌酒 red eye
廉價食品 low cost food
廉價魚 offal fish
廉價餐館 beanery
微升 microliter (μ l)
微孔〔過〕濾器 millipore filter
微火 low fire
微卡計 micro calorimeter
微生物 germ; microbe; microbion; microgerm; microorganism
微生物污染 microbial contamination
微生物性感染 microbiosis
微生物的 microbial
微生物毒素 microbial toxin
微生物降解 microbial degradation
微生物區系 microbial flora; microflora
微生物培養基 growth medium
微生物族群 microbial population
微生物產物 microbiological product
微生物產品 microbiological product
微生物測定法 microbial assay; microbiological assay
微生物菌叢 microbial flora; microflora
微生物群落 flora; microbial flora; microflora
微生物群體 microbial population
微生物製劑 microbiological product
微生物數 microbial population
微生物學 microbiology
微生物學的 microbiological
微生物學家 microbiologist
微米 micrometer (μ m); micron
微克 microgram (μ g)
微妙的味道 subtle taste
微波 microwave
微波加熱 microwave heating
微波食品 microwavable food
微波乾燥 microwave drying
微波烹調 microwave cooking
微波輻射 microwave irradiation
微波爐 microwave oven
微粉末 impalpable powder
微球菌〔屬〕 *Micrococcus*
微粒 particle
微粒分散膠體 microdispersoid
微粒體 microsome
微晶 crystallite; microcrystal
微晶砂糖 ultrafine granulated sugar
微晶糖膏 fondant
微晶纖維素 cellulose gel; crystal cellulose
微植物 microphyte
微發酸 acescence
微結構 microstructure
微絨毛 microvilli
微絲 microfilament
微絲血管 capillary
微菌落 microcolony
微量 scruple; soupcon; toothful; trace
微量元素 microelement; trace element
微量分析 microanalysis
微量化學分析 microchemical analysis
微量天平 microbalance
微量成分 micro-constituent
微量克耳達法 micro Kjeldahl method
微量法 micromethod
微量金屬 trace metal
微量原素 minor element
微量凱氏〔測定氮〕法 micro Kjeldahl method
微量過濾 microfiltration
微量滴定 microtitration

微量熱量計　micro calorimeter
微量營養素　micronutrient
微嗜氧細菌 microaerophile; microaerobe; microaerophilic bacteria
微溫　tepidity
微溫的　tepid
微過濾　microfiltration
微團　micella(e); micelle
微管　microtubule
微酸味　acescency
微酸性　subacidity
微酸的　acescent; sourish
微需氧性細菌　microaerophile
微熱　tepidity
微熱的　tepid
微膠粒　micella(e); micelle
微膠粒化　micellisation
微膠囊　microcapsule
微膠囊化　microencapsulation
微膠囊化酵素 microencapsulated enzyme
微醉　tipsiness
微醉的　tipsy
微澀的　subastringent
微鹹的　brackish; saltish
微纖維　microfibril
微鹼性的　alkalescent
意大利巴多利諾紅葡萄酒　bardolino
意大利水牛芝士　mozzarella
意大利生鮮香腸　Italian fresh sausage
意大利〔式〕直麵　spaghetti
意大利式水果蛋糕　panettone
意大利式多色冰淇淋　spumoni
意大利式咖啡杯　espresso cup
意大利式脫脂硬乾酪 Parmesan (cheese)
意大利米　Italian rice; risotto rice
意大利色拉米香腸　salami (sausage)
意大利西昂蒂紅葡萄酒　Chianti
意大利李　gage; greengage
意大利果仁味羊奶〔硬〕乾酪　fontina
意大利直麵南瓜　spaghetti squash
意大利〔芝士洋葱雞肉〕調味飯　risotto
意大利阿本尼香腸　Appennio sausage
意大利春捲　cannelloni
意大利紅洋葱　Italian red onion
意大利胡椒香腸　cappicola sausage
意大利粉　spaghetti
意大利酒量單位　anfora
意大利細寬麵（加雞蛋）　tagliarini
意大利細麵　capellini
意大利雲吞　gnocchi; tortellini
意大利黑醋　Italian black vinegar
意大利圓細麵　fedeli
意大利蒸餾咖啡　espresso
意大利辣肉腸　pepperoni (sausage)
意大利餃子　gnocchi; ravioli; tortellini
意大利寬麵（麵球）　tagliatelle
意大利濃縮咖啡　espresso
意大利薄餅　pizza
意大利薩拉米香腸　Italian salami
意大利麵　spaghetti
意式咖啡杯　espresso cup
意式咖啡機　espresso machine
意式苦艾酒　Italian vermouth
意式餃子　ravioli
意式濃咖啡　espresso
慈菇　arrowhead; oldworld arrowhead
感光分解　photodecomposition
感光降解　photodegradation
感受性　susceptibility
感官性質　organoleptic property; sensory property
感官評價 organoleptic evaluation; sensory evaluation
感官試驗　organoleptic test; sensory test
感官閾值　organoleptic threshold
感官檢驗　organoleptic examination
感染　infect; infection
感染性疾病　infectious disease

感染型食物中毒 food infection; infection type food poisoning; infectious food poisoning
感測器 sensor
感應〔加〕熱滅菌 induction heating sterilization
感應加熱 induction heating
感覺器官 sense organ
愛玉 ficus awkeotsang
愛玉子 awkeotsang; jelly-fig
愛吃甜食 sweet tooth
愛克曼試驗 Eijkman test
愛爾卡酶 alcalase
愛爾夏乳牛 Ayrshire (cow)
愛爾蘭〔混合〕甜酒 Irish mist
愛爾蘭咖啡 Irish coffee
愛爾蘭威士忌〔酒〕 Irish whisky
愛爾蘭馬鈴薯菜糊 colcannon
愛爾蘭薯 Irish cobbler
愛歐那葡萄 Iona
癒創木油 guaiac wood oil
癒創木酚 guaiacol
癒創樹脂 guaiacum (resin)
搓草形通心粉 gramigna
搓碎 grate
搓碎乾酪 grated cheese
搪瓷 enamel; porcelain enamel
搭鎖蓋 snap-on cap
損耗 waste; wastage
損耗量 wastage
損壞 spoilage
搖烤食品 shake and bake food
搖烤〔調理〕 shake and bake
搖動 shake; vibration
搖動器 shaker
搖動篩 swing sieve; swinging screen
搖蜜機 honey extractor
搖篩機 sieve shaker
搖盪篩 rocking sieve
搗杵 pounder; stamp pestle
搗棒 tamper
搗碎 crumb; mashing; pounding; quassation
搗碎的 mashed
搗碎器 blender; blendor
搗碎機 hacker; hacking machine; masher; stamping mill; triturator
搗樁機 stamping mill
搗擊機 stamper
搗錘 pounder; tamper
搗錘磨 pestle mill
搗鎚研磨機 stamp mortar
搗爛 pound
搗爛機 masher
敬酒 toast
新月形小麵包 crescent
新月菌屬 *Fusarium*
新生 regenerate
新生嬰兒 neonate
新地 sundae
新品酸性紅 erythrosine
新型 neotype
新洋菜二糖 neoagarobiose
新紅 new red
新紅鋁色澱 new red aluminium lake
新胡蘿蔔素 neocarotene
新桔皮糖苷 neohesperidoside
新桔皮苷 neohesperidin
新桔皮苷二氫查耳酮 neohesperidin dihydrochalone
新胭脂紅 cochineal red A; new coccine
新酒 must
新啤酒 new beer
新陳代謝 metabolism
新麥角固醇 neoergosterol
新番茄紅素 neolycopene
新菸鹼 neonicotine
新黃質 neoxanthin
新煙鹼 neonicotine

新葉黃素 neoxanthin
新葡萄酒 must
新漂木酸 neochlorogenic acid
新綠原酸 neochlorogenic acid
新維生素 neovitamin
新維生素A neovitamin A
新橙皮苷 neohesperidin
新橘皮糖苷 neohesperidoside
新橘皮苷 neohesperidin
新橘皮苷二氫查酮 neohesperidin dihydrochalone
新鮮 greenness; verdure
新鮮切割的肉塊 butcher' s cut
新鮮的 green; unsalted; young
新鮮度 freshness
新鮮蔬菜 fresh vegetable; green goods
新瓊脂雙糖 neoagarobiose
新黴素 neomycin
新釀〔葡萄〕酒 new wine
暗反應 dark reaction
暗色肉 dark meat
暗紅色肉 dark muscle
暗紅色的蘋果（英國） biffin
暗乾肉 dark, firm, dry muscle (DFD muscle)
暗鱗口蘑（蕈類） dark scaled knight cap
暖〔水〕壺 thermos (bottle); thermos flask
會餐 mess
椰色木漿果 langsat
業務用包裝 institutional pack
極低密度脂蛋白 very low density lipoprotein (VLDL)
極低溫冷凍 cryogenic freezing
極冷化學 cryochemistry
極冷保藏 cryopreservation
極冷溫度 cryogenic temperature
極冷劑 cryogen
極佳級 U.S. Prime
極性脂肪 polar lipid
極性胺基酸 polar amino acid
極性溶劑 polar solvent
極度飢餓 limosis
極限糊精 limit dextrin
極限糊精酶 limit dextrinase
極高密度脂蛋白 very high density lipoprotein (VHDL)
極乾香檳酒 brut
椰子 coconut
椰子汁 coconut milk; coconut water
椰子油 coconut oil; copra oil
椰子油香精 cocoanut oily flavo(u)r
椰子油餅 coconut cake
椰子脂 coconut butter
椰子粕 coconut meal; coconut meal; copra meal
椰子漿 coconut cream
椰子樹 coco; coco plam; coconut tree
椰子糖 jaggery
椰子醛 coconut aldehyde
椰子蟹 coconut crab
椰汁 coconut milk; coconut water
椰李 coco plum
椰粕 poonac
椰棗 date
椰菜 broccoli; cabbage; palm cabbage
椰菜花 cauliflower
椰菜絲沙律 coleslaw; slaw
椰糖 gula kelapa
楊桃 carambola; star fruit
楊梅 arbutus; babysberry; strawberry tree; wax myrtle; waxberry; yangmei
楊梅果醬 arbutus jam
楊梅油 myrica oil
楊梅酸 myricinic acid
楊梅醛 ethyl methylphenylglycidate
楓茅油 lemongrass oil
楓香浸膏 liquidambar concrete
楓樹 maple

楓樹液 maple sap
楓樹軟糖料 maple fondant
楓樹糖 maple sugar
楓樹糖蜜 maple honey
楓樹糖漿 maple syrup
楓糖 maple sugar; tree sugar
楓糖初漿 first running
楓糖蜜 maple honey; tree molasses
楓糖廠 sugarhouse; sugar camp
楓糖漿 maple syrup
楓糖霜泥 maple fondant
溢出 overflow; spillage; slop
溢出量 spillage
溢流 defluxion; effusion; overflow
溢流管 run down pipe
溶化鍋 liquefier
溶血〔作用〕 haemocytolysis; haemolysis; hemocytolysis; hemolysis
溶血卵磷脂 lysolecithin
溶血菌毒肽 phallin
溶性油 soluble oil
溶析 elution
溶析液 eluant; eluent
溶胞〔作用〕 lysis
溶胞酶 cytase
溶原性的 lysogenic
溶液 aqua; liquor; solution
溶雪水 meltwater
溶媒 solvent
溶菌〔作用〕 lysis; bacteriolysis
溶菌素 bacteriolysin
溶菌產物 lysate
溶菌酶 lysozyme
溶解 dissolution; diffluence; solubilize
溶解〔化〕 solubilization
溶解性 solubleness; solubility
溶解物 solute; soluble
溶解度 solubility
溶解〔能〕力 solvency
溶解產物 lysate
溶解劑 solubilizer
溶解總固體 total dissolved solid (TDS)
溶解酶 lysozyme
溶製乾酪 fondue
溶膠 sol
溶膠化〔作用〕 solation
溶質 solute
溶劑 dissolvent; solvent
溶劑混合油 miscella
溶劑混合油精煉 miscella refining
溶劑發酵 solvent fermentation
溶劑萃取 solvent extraction
溶劑萃〔取物〕 solvent extract
溶酶體 lysosome
滅活〔作用〕 inactivation
滅活劑 inactivator
滅菌 sterilization; sterilize; sterilizing
滅菌〔牛〕乳 sterilized milk
滅菌不足 understerilization
滅菌乳 sterile milk
滅菌室 sterilizing chamber
滅菌值 F value; sterilizing value
滅菌效果 sterilizing effect
滅菌率 sterilizing rate
滅菌率曲線 sterility curve
滅菌過度 oversterilization
滅菌箱 sterilizing chamber
滅菌劑 sterilant; sterilizing agent
滅菌器 sterilizer
滅鼠機 rodenticide
滅蟲 disinfestation
滅蟲劑 disinfestant
溫水 lukewarm water
溫州蜜柑 mandarin(e)
溫肉 warm meat
溫血動物 endotherm; homoiotherm; homotherm; warm-blooded animal
溫和的酒 velvet spirit

溫和消毒劑 mild disinfectant
溫和膳食 bland diet
溫室 conservatory; greenery; greenhouse
溫度 temperature (temp.)
溫度計 thermometer
溫度過高 overtemperature
溫度調節作用 attemperation
溫泉 spa
溫洲蜜柑 satsuma orange
溫屠體 hot carcass; warm carcass
溫標 temperature scale
溫體 warm carcass
溫體肉 warm meat
溫鹽漬 warm salting
滑石〔粉〕 talc; talcum
滑頂薄殼烏蛤 torigai
滑溜 sliminess
溜滑的魚 butterfish
溜麴菌 *Aspergillus tamarii*
溪紅點鮭 brook trout
溴化〔植物〕油 brominated vegetable oil
溴化麵粉 brominated flour
溴甲酚紫 bromocresol purple
溴甲酚綠 bromocresol green
溴酚紅 bromophenol red
溴酚藍 bromophenol blue (BPB)
溴酸鈣 calcium bromate
溴酸鉀 potassium bromate
溴酸鹽 bromate
煎 friture; pan frying; pan-searing; saute-(ing); shallow fry(ing); wok-searing
煎小牛肉片 schnitzel
煎牛肉餅 beefburger
煎肉餡餅 rissole
煎炸 frizzle
煎炸油 frying oil
煎脆 frizzle
煎蛋 fried egg
煎蛋包 omelet(te)
煎蛋〔餅〕 omelet(te)
煎煮 decoct; decoction
煎餅 flapjack; griddlecake; slapjack
煎熬 decoction
煎熬出的藥汁 decoction
煎劑 decoction
煎燉 braise; braising; braize
煎燉的 braised
煎鍋 frier; fryer; frying pan
煎藥 decoct
煎鏟 turner
煎麵包 cronton
煙 fume; smoke
煙斗 pipe
煙肉 bacon
煙色垂幕菇 conifer tuft
煙囪 stovepipe
煙窗 funnel
煙雲杯傘（蕈類） clouded funnel cap
煙道式乾燥器 flue dryer
煙酸 niacin; nicotinic acid
煙燻 besmoking; fume; fume smoking; smoke; smoking
煙燻火腿 gammon; smoked ham
煙燻材料 smoking wood
煙燻味 tangy flavo(u)r
煙燻味的 smoky
煙燻的 flue cured
煙燻室 black house; smoke house; smoker; smokery; smoking room
煙燻架 curing frame
煙燻風味 smoked flavo(u)r; smoky flavo-(u)r
煙燻食品 smoked food
煙燻香料 smoke flavo(u)ring; smoked flavo(u)r
煙燻液 fumeol
煙燻脱水 smoke dry
煙燻程度 smokiness

煙燻溶液 smoke solution
煙燻過度的 oversmoked
煙燻劑 smoke agent
煙燻爐 smoke oven
煙麴黴酸 helvolic acid
煙酰胺 niacinamide; nicotinamide
煤氣 coal gas; gas
煤氣烤爐 gas oven
煤氣燒毛器 gas singer
煤氣燎毛器 gas singer
煤苔 coal tar
煤苔色素 coal tar colo(u)r
煤焦油 coal tar
煤焦油色素 coal tar colo(u)r
煉合肉 meat emulsion
煉好的油脂 rendered fat
煉乳 condensed milk; lactein
煉乳（無蔗糖） concentrated milk
煉乳廠 condensery
煉油 render(ing)
煉脂 tallow rendering
照射 irradiation; radiation
照射巴氏殺菌〔法〕 irradiation pasteurization; radiation pasteurization; radiopasteurization
照射巴斯德殺菌 irradiation pasteurization
照射肉 irradiated meat
照射食品 irradiated food
照射殺菌〔法〕 radiosterilization; radurization
照射處理 radiation treatment
照射量測定計 dose meter; dosimeter
照射滅菌〔法〕 cold sterilization; irradiation sterilization; radappertization; radiation sterilization
照燒 teriyaki
瑞士牛肉乾 bunderfleish
瑞士衣曼塔乾酪 Emment(h)al cheese
瑞士乳〔酸〕桿菌 *Lactobacillus helveticus*
瑞士埃曼塔爾硬乾酪 Emment(h)al cheese
瑞士恭菜 Swiss chard
瑞士〔蛋糕〕捲 Swiss roll
瑞士〔硬〕芝士 Emmentaler; Swiss cheese
瑞士〔硬〕乾酪 Emmentaler; Swiss cheese
瑞典式自助餐 smorgasbord
瑞典蕪菁 rutabaga; swede
瑞典藍瑞斯豬 Swedish landrace
瑞典蘭德瑞斯豬 Swedish landrace
瑞香油 daphne oil
瑞香素 daphnetin
瑞香草 thyme
瑞香草油 thyme oil
瑞香草酚 thymol
當地品種 local variety
當量 equivalent weight
當量溶液 normal solution
當量濃度 normality (N)
當歸 angelica
當歸油 angelica oil
矮小而發育不良的動物 runt
矮生西葫蘆 summer squash
矮生果樹 dwarf
矮生南瓜 bush pumpkin
矮生茄子 dwarf eggplant
矮生香蕉 dwarf banana
矮生菜豆 bush bean
矮松子 pinon
矮型罐 pocket (type) can
矮扁桃 Russian almond
矮腳雞 bantam
碎大豆 soybean grit
碎小麥 bulgur; burghul
碎小麥粒 wheat grits
碎火腿 chopped ham
碎片 shiver; shred; snip
碎牛肉 minced beef

碎牛肉〔製牛〕排 stuffed beef steak
碎玉米 corn grit; maize grit
碎玉米羹 cream style corn
碎石 rubble
碎冰 cracked ice; rushed ice
碎冰冷藏 dry icing
碎冰甜雞尾酒 swizzle
碎冰機 ice chopper; ice crusher
碎冰錐 ice pick
碎米 brokens; mealie rice
碎米薺 bitter cress
碎肉 chopped meat; comminuted meat; ground meat; mince collops; minced meat; mincemeat
碎肉夾餅 burger
碎肉做的肉排 mock cutlet
碎肉機 meat mincing machine; mincer; mincing machine
碎咖啡豆 triage
碎果肉 eradicator meat
碎屑 brokens; debris; offal; scrap; shred residue
碎茶 fannings; tea fannings
碎高粱 milo grit
碎堅果仁 nutmeat
碎塊 crush
碎溶技術 melt-flake method
碎麩皮 pollard
碎斷 fragmentation
碎蘋果肉 apple pulp
碎鹹牛肉 cornbeef
碎麵包 crumb
碎麵包皮 chippy bread
碎麵條 pastina
碗 bowl
碗布 dishcloth
碗菌 cup fungus; cup mushroom
碗碟洗滌處 scullery
碗磨 bowl mill
碗櫃 cupboard
碗櫥 cupboard
碘 iodine
碘化亞銅 cuprous iodide
碘化物 iodide
碘化食鹽 iodized salt
碘化酪蛋白 iodinated casein
碘化鉀 potassium iodide
碘值 iodine number; iodine value (IV)
碘蛋白 iodoprotein
碘量滴定法 iodimetry; iodometry; iodometric method
碘試驗 iodine test
碘滴定法 iodimetry; iodometry
碘酸鈣 calcium iodate
碘酸鉀 potassium iodate
碘酸鹽 iodate
碘價 iodine number; iodine value (IV)
硼 boron
硼矽玻璃 boron-silicate glass
硼砂 borax
硼酸 boric acid
硼酸鈉 sodium borate
硼酸鹽 borate
禁食 abrosia; abstinence; apastia; fast
禁食日 fast day
禁食〔的〕 fasting
禁食療法 fasting cure; hunger cure
禁酒 temperance
萬壽菊 marigold; tagetes
萬壽菊油 tagetes oil
禽肉 fowl
禽尾〔椎〕部 rump
禽流感 avian flu; avian influenza
禽畜內臟 viscera
禽腿去毛機 leg finisher
禽類內臟或雜碎 giblet
稜紋貝通心粉 conchiglie rigate
稜紋管通心粉 pipe rigate

稠化 thickening
稠化劑 gelling agent; gelatinizer; gelatinizing agent; jelling agent
稠李 bird cherry; European bird cherry
稠沸澱粉 thick boiling starch
稠的 thick
稠厚凝膠 short gel
稠度 thickness
稠度曲線記錄儀 plastograph
稠度計 consistometer
稠度測定儀 texturometer
稠漿 heavy paste; thick slurry
稠糊 heavy paste
稠糊澱粉 thick boiling starch
稠凝膠 short gel
稠變乳 ropy milk
稠變酒 ropy wine
筷子 chopsticks
節 node
節瓜 fuzzy melon; hairy squash; jointed wax gourd
節路頓膠 jelutong
粳稻 keng rice
經口腔感染 per oral infection
經常庫存 running stock
經銷商 distributor
經濟魚類 commercial fishes
經濟餐館 beanery; hash house
絹光紅藻屬 *Caloglossa*
絹篩 tammy
綁 tie
絛蟲 tapeworm
置換培養基 replacement medium
置換層析法 displacement chromatography
置蓋機 clincher
群 cluster
群生金錢菇 tufted tough shank
群青 ultramarine
群青色素 ultramarine
聖代冰淇淋 sundae
聖餅 wafer
聖餐 sacrament
聖餐杯 chalice
聖餐葡萄酒 sacramental wine
腱 chord(a); cord; sinew; tendon; thew
腱黏蛋白 tendomucoid
腰 loin
腰上部肉 sirloin
腰子 kidney
腰內肉 back fillet; tenderloin
腰內肉片 filet mignon; fillet mignon
腰肉 loin
腰肌 psoas
腰果 cashew nut
腰脊牛排 sirloin steak
腰眼骨肉 shortloin
腰部 haunch; loin
腰部上肉 topside (round)
腰椎里脊肉 short fillet
腰短骨肉小圓排 noisette
腰裏脊肉 tenderloin
腰臀 haunch
腰臀的 ischiatic
腸 bowel; gut; intestine
腸內的 intestinal
腸內病原細菌 enteropathogenic bacteria
腸內桿菌 enteric bacilli
腸內細菌 enteric bacteria; enterobacteria; intestinal bacteria
腸內菌叢 intestinal flora
腸衣 casing; hank; sausage skin; ausage casing
腸衣包裝 casing packaging
腸衣去黏膜機 gut stripper
腸衣穿串 casing threading
腸衣剝刮 casing stripping
腸衣剝刮機 casing stripper
腸抑胃素 enterogastrone

腸肝循環 enterohepatic circulation
腸油 casing fat; intestinal fat
腸炎 enteritis
腸炎沙門氏菌 *Salmonella enterica*
腸炎沙門桿菌 *Salmonella enterica*
腸炎弧菌 *Vibrio parahaemolyticus*
腸的 enteral; enteric; intestinal
腸毒素 enterotoxin
腸胃炎 gastroenteritis
腸胃的 gastrointestinal
腸胃氣脹 flatulence
腸病毒 enterovirus
腸脂 casing fat; intestinal fat; mesenteric fat
腸液 intestinal juice; succus entericus
腸球菌 enterococci; enterococcus
腸蛋白酶 erepsin; ereptase
腸道細菌 enteric bacteria; enterobacteria
腸線 catgut
腸膜狀牛耳（蕈類） tripe fungus
腸壁 intestinal wall
腸蟲 helminth
腸繫膜 mesentery
腸繫膜油 ruffle fat
腸肽酶 erepsin; ereptase
腥臭 fishiness
腳爪 paw
腳趾 toe
腳踩 tread
腳鐐 shackle
腫脹 swelling
腹白米 white belly rice
腹肉 belly
腹脂 caul fat; caul web fat
腹脅 flank
腹脅肉 flank; flank and knuckle
腹脅肉排 flank steak
腹部 belly; venter
腹腔 peritoneal cavity; visceral cavity
腹脹 flatulence
腹膜 peritoneum
腹壁 belly flap
腹瀉 diarrhea; lax
腹鰭 ventral fin
腹脇肉 belly; belly meat; side bacon
腹脇肉輥展機 belly roller
腺嘌呤 adenine
腺嘌呤去胺酶 adenine nucleotide
腺苷 adenosine
腺苷一磷酸 adenosine phosphate; adenosine monophosphate (AMP); adenylic acid
腺苷二磷酸 adenosine diphosphate (ADP)
腺苷三磷酸 adenosine triphosphate (ATP)
腺苷酸 adenosine phosphate; adenosine monophosphate (AMP); adenylic acid
腦 brain
腦硫脂 sulfolipid
腦糖 cerebose
腦磷脂 cephalin; kephalin; phosphatidylethanolamine
葷食 meat diet
落果 drop; fruit drop(ping)
落花生 earthnut; goober; ground pea; groundnut; monkey nut; peanut
落花生屬 *Arachis*
落球黏度計 falling ball viscosimeter
落葵 Ceylon spinach
落葵紅 basella rubra red
落葉 defoliation
落葉松蕈 agaric; *Agaricus*; white agaric
落葉松蕈素 agaricin
落葉劑 defoliant
落塵 fallout
萱花 yellow daylily
萱草 yellow daylily
葵花 sunflower

葵花籽 sunflower seed
葵花籽油 sunflower oil; sunflower seed oil
葫蘆 bottle gourd; calabash gourd; cucurbit; gourd
葫蘆巴 fenugreek
葫蘆巴籽 fenugreek seed
葫蘆屬 gourd
葉 leaf
葉子 leafage
葉甘素 phyllodulcin
葉白素 leukophyll
葉肉 mesophyll
葉狀植物 thallophyte
葉狀濾器 leaf filter
葉柄 leaf stalk; scape
葉紅素 carotene; carotin; erythrophyll; phylloerythrin
葉脈 vein
葉帶酸味的植物 sorrel
葉甜素 phyllodulcin
葉甜菜 spinach beet
葉甜精 phyllodulcin
葉蛋白 leaf protein
葉紫質 phylloporphyrin
葉菜 leaf vegetable
葉黃素 lutein; phylloxanthin; phytoxanthin; xanthophyll
葉黃質 phylloxanthin; phytoxanthin
葉綠素 chlorophyll
葉綠〔素〕蛋白 phyllochlorin
葉綠素銅 copper chlorophyll
葉綠素銅鈉鹽 sodium copper chlorophyllin
葉綠素銅鉀鹽 potassium copper chlorophyllin
葉綠素鎂 magnesium chlorophyll
葉綠酸 chlorophyllin
葉綠酸鐵鈉鹽 sodium iron chlorophyllin
葉綠醇 phytol
葉綠體 autoplast; chloroplast
葉綠醌 phylloquinone
葉酸 folic acid; folacin; pteroylglutamic acid; vitamin Bc; vitamin M
葉酸鹽 folate
葉輪 impeller
葉醇 leaf alcohol
葉醛 leaf aldehyde
葉醌 phylloquinone
葛粉 arrowroot starch
葛〔澱〕粉 maranta
葛縷子 caraway; *Carum carvi*
葛縷子硬餅乾 abernethy (biscuit)
葛縷籽 caraway seed
葛鬱金 arrowroot
萼片 sepal
萵苣 lettuce
萵苣筍 asparagus lettuce
葡〔萄〕糖二酸 glucaric acid; saccharic acid
葡〔萄〕糖生成 glucogenesis
葡〔萄〕糖型抗壞血酸 glucoascorbic acid
葡〔萄〕糖胺 glucosamine
葡〔萄〕糖基化 glucosylation
葡〔萄〕糖腦苷酯 glucocerebroside
葡〔萄〕糖酸 gluconic acid
葡〔萄〕糖酸內酯 gluconolactone
葡〔萄〕糖酸亞鐵 ferrous gluconate
葡〔萄〕糖酸發酵 gluconic acid fermentation
葡〔萄〕糖酸鈣 calcium gluconate
葡〔萄〕糖酸鈉 sodium gluconate
葡〔萄〕糖酸銅 copper gluconate
葡〔萄〕糖酸鹽 gluconate
葡〔萄〕糖酸-δ-內酯 glucono-delta-lactone; glucono-δ-lactone
葡〔萄〕糖醇 glucitol; sorbitol
葡〔萄〕糖澱粉酶 glucoamylase

葡〔萄〕糖激酶 glucokinase
葡〔萄〕糖糖化酶 glucoamylase
葡〔萄〕糖磷酸變位酶 phosphoglucomutase
葡〔萄〕糖醛酮 glucosone
葡〔萄〕糖醛酸 glucuronic acid
葡〔萄〕糖醛酸內酯 glucuronolactone
葡〔萄〕糖醛酸鹽 glucuronate
葡〔萄〕糖醛酸苷 glucuronide
葡〔萄〕糖醛酸苷酶 glucuronidase
葡〔萄〕糖苷 glucoside
葡〔萄〕糖苷酶 glucosidase
葡甘露聚糖 glucomannan
葡基轉移作用 transglycosylation
葡萄 grape; vitis
葡萄子 grape seed; grapestone
葡萄汁 grape juice
葡萄白腐病 white rot
葡萄皮色素 grape skin extract
葡萄皮紅 grape skin extract
葡萄皮提取物 grape skin extract
葡萄皮渣 grape pomace
葡萄皮萃 grape skin extract
葡萄收成量 vintage
葡萄收獲期 vintage
葡萄果皮色素萃 enocianina
葡萄油香精 grape oily flavo(u)r
葡萄狀白黴菌 *Mucor racemosus*
葡萄花青素 enocianina; oenin
葡萄柚 *citrus paradisi*; grapefruit; natsudaidai
葡萄柚油 grapefruit oil
葡萄籽油 raisin seed oil
葡萄香精 grape flavo(u)r
葡萄香檳酒 vintage champagne
葡萄栽培 vine culture; viniculture; viticulture
葡萄栽培兼釀酒者 winegrower; winegrowing
葡萄烈酒 spirit of wine
葡萄酒 vin; vino; vinho; wine
葡萄酒色素 oenolin
葡萄酒杯 wine glass
葡萄酒的 vinic; vinaceous; vinous
葡萄酒的酒石沉澱 beeswing
葡萄酒的釀造〔季節〕 vintage
葡萄酒的釀造年 vintage year
葡萄酒花色素 oenocyanin
葡萄酒酒母 wine starter
葡萄酒酒度計 oenometer
葡萄酒酒度測定法 oenometry
葡萄酒酒精比重計 vinometer
葡萄酒商 vintner
葡萄酒桶 wine barrel
葡萄酒發酵 vinous fermentation
葡萄酒窖 wine vault
葡萄酒菌元 wine starter
葡萄酒飲料 wine punch
葡萄酒愛好者 wino
葡萄酒榨渣 wine marc
葡萄酒酵母 wine yeast
葡萄酒醋 wine vinegar
葡萄酒醋菌 *Acetobacter vini acetati*
葡萄酒釀造 wine making
葡萄酒〔釀造〕廠 winery
葡萄酒釀造學 enology; oenology
葡萄酒釀造學家 enologist; oenologist
葡萄酒釀製〔法〕 vinification
葡萄胺糖 chitosamine; glucosamine
葡萄乾 raisin
葡萄乾比思克 Bath biscuit
葡萄乾布丁 duff; plum pudding; spotted dog
葡萄乾夾心餅 banbury (biscuit)
葡萄乾甜捲麵包 Bath bun
葡萄乾蛋糕 plum cake
葡萄乾葡萄 raisin grape
葡萄乾餡〔酥〕餅（英式） Eccles (cake)

葡萄乾麵包 currant bread; raisin bread; sultana bread
葡萄乾〔釀造〕酒 raisin wine
葡萄基麥芽糖 panose
葡萄球菌中毒 staphylococcal poisoning
葡萄球菌毒素 staphylococcin
葡萄球菌食物中毒 staphylococcal food poisoning
葡萄球菌產的腸毒素 staphylococcal enterotoxin
葡萄球菌溶血素 staphylolysin
葡萄球菌屬 *Staphylococcus*
葡萄渣 grape pomace; pressed grape
葡萄渣油 wine yeast oil
葡萄園 grapery; vinery; vineland; vineyard
葡萄溫室 vinery
葡萄葉 vine
葡萄圖譜 ampeliograph
葡萄榨汁裝置 wine press
葡萄榨汁機 grape press
葡萄榨白蘭地酒 Marc brandy
葡萄榨渣 grape pomace
葡萄種植園 chateau
葡萄醋 grape vinegar
葡萄學 ampeliology
葡萄樹 vine
葡萄樹的 viny
葡萄糖 amylaceum; dextrose; glucose; grape sugar; starch sugar
葡萄糖二酸鹽 saccharate
葡萄糖去氫酶 glucose dehydrogenase
葡萄糖尿症 glucosuria
葡萄糖值 dextrose equivalent (DE); dextrose equivalent (value)
葡萄糖氧化酶 glucose oxidase; notatin
葡萄糖異構酶 glucose isomerase
葡萄糖當量 dextrose equivalent (DE); dextrose equivalent (value)
葡萄糖酸桿菌 gluconobacter
葡萄糖漿 dextrose syrup; glucose syrup
葡萄糖磷酸 glucose phosphate
葡萄糖瓊脂〔培養基〕 glucose agar
葡萄糖苷 heteroside
葡萄糖苷化 glucosylation
葡萄糖苷移轉酶 transglucosidase
葡萄糖苷鍵 glucosidic bond
葡萄壓榨機 wine press
葡萄醛酮糖 glucosone
葡萄吡喃糖 glucopyranose
葡萄呋喃糖 glucofuranose
葡聚糖 dextran; glucan; glucosan
葡聚糖酶 dextranase; glucanase
葡糖胺 glycosamine
葡糖硫苷酶 thioglucosidase
葡糖酸鉀 potassium gluconate
葡糖酸鋅 zinc gluconate
葡糖酸錳 manganese gluconate
葡糖酸鎂 magnesium gluconate
葡糖酸鹽 glyconate
葡糖糖化酶 glucoamylase
葡糖苷 asterin
蜀葵屬植物 althea
蜀黍 durra; jowar; milo
蛾 moth
蜂王 queen bee
蜂王漿 royal jelly
蜂后 queen bee
蜂乳 royal jelly
蜂房 apiary
蜂花 melissa
蜂花素 myricoid
蜂花酸 melissic acid; myricyl acid
蜂巢 beehive; hive
蜂巢胃 honeycomb; honeycomb tripe; reticulum
蜂巢蜜 hive honey
蜂群 hive
蜂窩 beehive

蜂窩胃 reticulum
蜂窩組織 areolar tissue
蜂蜜 honey
蜂蜜果脯酥餅 lebkuchen
蜂蜜酒 honeymead; hydromel; mead; metheglin
蜂蜜菇 honey mushroom
蜂蜜酵母〔菌〕 *Saccharomyces mellis*
蜂蜜糖漿 honey syrup
蜂箱 hive
蜂蠟 beeswax; cera
蜂蠟的 cerotic
蜆 corb shell; corbicula
蜆肉 clam
蜆殼粉 abassina rigata; conchiglie; shell pasta
蜊仔 corbicula
裙帶豆 yardlong bean
裙帶菜 wakame
補充 replenish
補充食品 reprovision; supplement food
補充糧食 reprovision
補色 complementary colo(u)r
補助金 subsidy
補助劑 adjuvant
補足 replenish
補品 restorative; tonic
補酒 alcoholic cordial; tonic wine
補塵器 dust trap
補藥 corroborant; invigorator
補鱈人 banker
裝桶 barrelling
裝桶醃漬 tierce curing
裝桶機 barrel filler
裝瓶 bottling; potting
裝瓶後巴氏殺菌 in-bottle pasteurization
裝瓶後滅菌 in-bottle sterilization
裝瓶機 bottler; bottle filler; bottle filling machine; bottling machine
裝袋 sacking
裝袋工人 bagger
裝袋漏斗 bagging hopper
裝袋機 bagger; bag filling machine; bagging machine; sack packer; sacker
裝袋儲藏 sack storage
裝備 outfit
裝填 potting
裝填不足 underfill
裝填過量 overfill
裝填機 packing machine; packer
裝置 device; equipment
裝載過多的 overladen
裝飾 trim
裝飾包裝 ornamental package
裝飾品 trimmings
裝飾蛋糕 decorated cake
裝箱 casing
裝箱醃製 box curing
裝罐 can filling; potting
裝罐不足 slack fill
裝罐機 can filler; can stuffer
裡肌肉 tenderloin
解百衲葡萄酒 Cabernet (wine)
解吸附〔作用〕 desorption
解毒 detoxi(fi)cation
解毒劑 antidote; detoxicant; toxicide
解毒藥 alexipharmic; theriac
解凍 defrost; defrostation; defreeze; thaw (ing); unfreeze
解凍肉 defrosted meat
解凍房 thawing house
解凍室 defrosting room; thawing room
解凍損失 thawing loss
解凍溫度 thaw point
解凍裝置 defroster; thawer
解凍滴液 thawing drip
解凍滴液〔現象〕 thaw dripping
解凍滲出液 thaw exudates

解凍僵直 thaw rigor
解凍器 thawer
解凍機 defrosting machine
解凍點 thaw point
解氨酶 ammonialyase
解脂〔作用〕 lipolysis
解脂性酸敗 lipolytic rancidity
解脂的 lipolytic
解脂細菌 lipolytic bacteria
解脂酶 lipolytic enzyme
解除飢渴 refection
解渴 quench(ing)
解渴的 thirstquenching
解渴飲料 quencher
解酮〔作用〕 ketolysis
解聚〔合〕作用 depolymerization
解聚〔作用〕 disaggregation
解聚酶 depolymerase
解僵 off rigor; resolution of rigor; rigor off; thaw rigor
解膠作用 peptization
解膠劑 peptizator
解糖的 sucroclastic
解糖酶 sucroclastic enzyme
解離 dissociation
解離常數 dissociation constant
解離酶 lyase
解饞食物 junk food
詳細審查 sift
試用 trial
試車 trial run
試味 tasting
試味者 taster
試紙 indicator paper
試管 test tube
試管內 in vitro
試管培養〔物〕 tube culture
試製 trial produce; trial manufacture
試製品 trial produce
試樣 sample
試劑 reagent
試驗 experiment; test; trial
試驗工場 pilot plant
試驗工場規模 pilot plant scale
試驗工場試產 pilot plant trial
試驗者 tester
試驗食物 test meal
試驗裝置 tester
試驗運轉 trial run
試驗器 tester
試驗膳食 test diet
試驗餐 test meal
詹巴什錦飯 jambalaya
跟腱 Achilles tendon
路特鱈魚乾 lutefish
路邊小吃店 transport café
跳汰機 jigging machine
跳汰篩 jigging screen
跳封 jumped seam
跳動運輸帶 jigging conveyer
跳動篩 jigging screen
載玻片 microscope slide
載奧辛 dioxin
載體蛋白 carrier protein
農夫 tiller
農用化學品 agricultural chemicals
農作物 crop; farm crop
農家式火腿 farm-style ham
農家式香腸 farmer sausage
農家自製麵包 cottage bread
農產品 farm product; produce
農產品集散市場 terminal market
農場 farm
農業 agriculture
農業工程學 agricultural engineering
農業化學 agricultural chemistry
農業法規 agricultural code
農業的 agricultural

農業品 agricultural product
農業產品 agricultural product
農學 agronomy
農藝 agriculture
農藝的 agricultural
農藝學 agronomy
農藥 agricultural chemicals; pesticide; pest control chemical
農藥殘留 pesticide residue
農藥增效醚 butoxide
運水果船 fruiter
運果箱 lug box
運動飲料 sport drink
運載家畜之火車車廂 stock car
道納 doughnut
達拉謨短角肉牛 Durham
達馬樹脂 dammar resin
達馬樹膠 dammar resin
達爾斯海藻 dulse
達蘭麵粉 daren flour
過乙酸 peracetic acid
過大 oversize
過多包裝 over packaging
過冷 subcool
過冷的 undercooled
過冷〔卻〕 overcooling; superchilling; supercooling; undercooling
過冷〔現象〕 undercooling
過冷器 subcooler
過度水合 overhydration
過度充氣 overgassing
過度充氣果汁 overcarbonated juice; overgassed juice
過度加工 overprocessing
過度加工風味 overprocessing flavo(u)r
過度加熱 overheat; overprocessing
過度冷卻 overcooling; supercooling
過度冷卻的 overcooled
過度受熱 superheat
過度拌合 overmix
過度氧化 overoxidation
過度粉碎 over grinding
過度乾燥 overdry
過度混合 overmix
過度烹煮 overcooking
過度發酵 over fermentation
過度發酵麵糰 old dough; overripe dough; over-aged dough
過度精製 overrefining
過度碾磨 overmilling
過度醃漬 overcuring
過度醃漬的 overcured
過度攝取 over intake
過重 overbalance; overweight
過氧乙酸 peroxyacetic acid
過氧化反應 peroxidation
過氧化丙酮 acetone peroxide
過氧化物 hyperoxide; peroxide; superoxide
過氧化物酶 peroxidase
過氧化物酶體 peroxisome
過氧化苯〔甲〕酰 benzoyl peroxide
過氧化值 peroxide value (POV)
過氧化氫 hydrogen peroxide
過氧化氫物 hydroperoxide
過氧化氫酶 catalase; hydrogen peroxidase
過氧化鈣 calcium peroxide
過氧化醋菌 *Acetobacter peroxydans*
過氧化體 peroxisome
過烘 overbake
過乾的 overdried
過敏元 anaphylactogen
過敏性反應 anaphylaxis
過敏毒素 anaphylatoxin
過敏原 allergen; anaphylactogen
過敏〔原〕食物 allergenic food
過敏症 allergy

過硫酸鉀 potassium persulfate
過硫酸銨 ammonium persulfate
過剩乳 surplus milk
過氯酸 hyperchloric acid
過渡期 transitional period
過渡溫度 transition temperature
過量進料 overfeed
過量裝填 overfill
過量餵食 overfeed
過溝菜蕨 bracken
過載 overload
過飽 fed-upness; repletion; satiety
過飽充汁 overcarbonated juice; overgassed juice
過飽和 oversaturation; supersaturation
過飽和的 supersaturated
過漂 overbleach
過酸化〔作用〕 superacidulation
過酸性 peracidity
過熟 hypermaturity
過熟的 overmature; overripe
過熟乾酪 old cheese
過熟麵糰 old dough; over-aged dough
過熱 overheat; overtemperature; superheat (ing)
過熱蒸汽 overheated steam; superheated steam; superheated vapo(u)r
過熱蒸汽去皮 retort peeling
過熱器 superheater
過醋酸 peracetic acid
過燒的 overburn; overburnt
過篩物 undersize
過篩部分 screen-throughs
過醒麵糰 overproofed dough
過濕的 overwet
過濾 colation; filtration; filter; lauter; sift; sieve
過濾介質 filter medium
過濾咖啡用紙 coffee filter paper
過濾袋 filter bag
過濾麥芽汁 lauter
過濾滅菌〔法〕 sterile filtration; filtering sterilization
過濾滅菌器 filter sterilizer
過濾網 screen
過濾槽 lauter tank
過濾器 filter
過濾襯墊 filter pad
過鹹的 oversalted
過鹼性 peralkaline
酪乳 buttermilk
酪朊酸鈣 calcium caseinate
酪胺 tyramine
酪胺酸 tyrosine
酪胺酸代謝〔紊亂〕症 tyrosinosis
酪胺酸酶 tyrosinase
酪胺酶 tyraminase
酪梨 aguacate; alligator pear; avocado; butter-fruit
酪蛋白 casein
酪蛋白元 caseinogen
酪蛋白水解物 casein hydrolyzate
酪蛋白乳清 casein whey
酪蛋白乳〔酸〕桿菌 *Lactobacillus casei*
酪蛋白洋菜〔培養基〕 casein agar
酪蛋白原 caseinogen
酪蛋白鈣 calcium caseinate
酪蛋白鈉 nutrose
酪蛋白酸鈉 sodium caseinate
酪蛋白鐵 iron caseinate
酪蛋白鹽 caseinate
酪蛋白酶 casease
酪農芝士 cottage cheese; pot cheese
酪農乾酪 cottage cheese; pot cheese
酪酸 butyric acid
酪酸乳〔酸〕桿菌 *Lactobacillus pastorianus*
酪醇 tyrosol
鈷 cobalt

鈷胺酸 cobamic acid
鈷維生素 cobalamin
鉗 claw
鉗子 pincers; tongs
鉗合〔作用〕 chelation; sequestration
鉗合劑 chelating agent; sequestering agent; sequester; sequestrant
鉀 potassium
鉀〔明〕礬 potassium alum; aluminium potassium sulfate
鉀硝 saltpeter; saltpetre
鉛 lead
鉛中毒 plumbism; lead poisoning; saturnism
鉛色灰球菌（蕈類） lead grey bovist
隔水燉鍋 bain-marie
隔年結果果樹 alternate bearing; biennial bearing
隔柵式轉篩 rod washer
隔間乾燥器 compartment dryer
隔熱 thermal insulation
隔熱杯套 insulating sleeve
隔熱板 thermal baffle
隔膜〔肉〕 skirt
隔離 isolation; quarantine
隔離品評法 closed panel method
隔離屠宰 isolation slaughter
雉 pheasant
雉鳥 pheasant
雷魚 standfish
雷鳥 capercaillie; capercailzie
雷諾數 Reynold's number
電子 electron
電子滅菌〔法〕 electronic sterilization
電子顯微鏡 electron(ic) microscope
電化學 electrochemistry
電昏 electrical stunning
電昏器（屠宰用） applicator
電泳 electrophoresis
電容器 condenser
電烘爐 electric oven
電烤爐 electric oven
電殺法 electrical stunning
電焊罐 welded seam can
電視餐 TV dinner
電飯煲 electric rice cooker
電飯鍋 electric rice cooker
電極 electrode
電溶膠 electrosol
電葫蘆 trolley hoist
電解 electrolysis
電解液 electrolyte; electrolytic solution
電解質 electrolyte
電解鐵 iron, electrolytic
電熔爐 electric furnace
電磁盤 electromagnetic plate
電磁爐 electromagnetic plate
電磁攪拌器 electromagnetic stirrer; magnetic stirrer
電蒸餾器 electric still
電熱板 electric hot plate
電熱器 electric heater
電熱鍋爐 electric boiler
電擊 shock
電擊昏 electrical stunning
電擊昏器 electrical stunner
電擊器 electric tongs
電鍍馬口鐵皮 electric tinplate; electrolytic plate; electrolytic tinplate
電鍍錫 electrolytic tinning
電鍍錫板 electric tinplate; electrolytic plate; electrolytic tinplate
電鍋 electric rice cooker
電離〔作用〕 ionization; electrolytic dissociation
電鯰 electric cat-fish; raad
電爐 electric furnace; electric stove
電鰻 electric eel; raad
零下溫度 subzero temperature

零食 sock; snack
零售 retail
零售分割肉 retail cut meat
零售包裝 consumer pack(age); retail packaging
零售市場 retail market
零售酒商 cooper
零售商 retail dealer
零售商店 retail store
零售商品 consumer goods
零陵香豆 tonka bean
零點 zero point
零點調整 zeroing
靴口蘑（蕈類） brown matsutake
預〔先磨〕碎 foregrinding
預切 precut
預水解 prehydrolysis
預加熱 preliminary heating
預包裝 prepack; prepackage
預份化 prefractionation
預份化器 prefractionator
預先加工 preparation
預先包裝 prepack(age)
預先乾燥 pre-drying
預先煮好的 ready-cooked
預先塑造成型 preform
預先塗助濾劑的過濾器 precoat filter
預回潮 prehumidification
預冷〔卻〕 prechill; precool(ing)
預冷卻室 prechill room
預冷室 precooling room
預冷凍 prefreeze; prefreezing
預冷凝器 precondenser
預冷器 precooler
預沖洗 prerinse
預保溫 preincubation
預封機 can clincher; clincher; clincher machine
預洗 preliminary washing; prewash
預洗機 prewasher
預炸 prefrying
預凍 prefreeze
預凍結 prefreezing
預消化 predigestion
預浸〔泡〕 presoak(ing)
預浸漬 preimpregnation
預破碎機 prebreaker
預粉碎 foregrinding
預粉碎的 preground
預乾 predrying
預培養 preincubation
預清洗 prerinse
預混合 premix
預混合料 premix
預處理 preconditioning; pretreatment
預備 preparation
預煮 blanch(ing); parboil(ing); parcook-(ing); precook(ing)
預煮的冷凍食品 precooked frozen food
預煮食品 precooked food
預煮器 precooker
預煮機 blancher
預萃取 pre-extraction
預搗碎 premash
預滅菌 presterilizarion
預稠化 prethickening
預解凍 prethawing
預過濾 prefiltration; prefilter
預蒸煮 parboil(ing); parcook(ing)
預蒸發 pre-evaporation
預蒸發器 pre-evaporator
預製冷凍食品 prepared frozen food
預製食品 prepared food
預增稠 prethickening
預澄清 predefecation
預熱 forewarm; preheat(ing)
預熱室 preheating chamber
預熱槽 preheating tank

預熱箱 preheating chamber
預熱器 preheater
預糊化 pregelatinization
預糊化米 pregelatinized rice
預糊化的 pregelatinized
預糊化食品 pregelatinized food
預糊化澱粉 pregelatinized starch
預膠凝 pregelatinization
預醃〔製〕 presalting
預凝〔固〕 presetting
預濃縮器 preconcentrator
預磨 pregrinding
預磨碎的 preground
預壓〔榨〕 prepress
預濕 prewet
預濕潤 prehumidification
預濾器 prefilter
預覆濾器 precoat filter
預餾 predistillation
頑強便秘 obstipation
飼用粟 German millet
飼料 feed; feed stuff; fodder; provender
飼料玉米 field corn
飼料作物 forage crop
飼料酵母 nutrient yeast
飼料穀物 fodder grain
飼槽 feeder
飼養 alimentation; feed(ing)
飼養者 feeder; grower
飴糖 caramel; malt sugar; malt syrup
飽充罐 saturator
飽和 saturation
飽和水蒸汽 moist steam
飽和曲線 saturation curve
飽和度 saturation level
飽和值 saturation value
飽和脂肪 saturated fat
飽和脂肪酸 saturated fatty acid
飽和等溫線 saturation isotherm
飽和溶液 saturated solution
飽和蒸汽 saturated steam; saturated vapo(u)r
飽和劑 saturant
飽和器 saturator
飽和濕度 saturated humidity
飽和點 saturation point
飽食 satiety; satiation; stuff
飽滿度 plumpness
飽餐 repletion
飽餐食品 belly-full food
馴鹿 reindeer
鼓式剝皮機 drum barker
鼓式過濾器 drum filter
豉油 light soy sauce; soy sauce; soya sauce; soybean sauce
鼓泡塑料襯墊包裝 blister pack
鼓風凍結 air-blast freezing
鼓風凍結室 air-blast freezer
鼓風乾燥 forced air drying
鼓風機 air blower
鼠 rat
鼠色小孔菌(蕈類) cone cap
鼠尾草 clary; clary sage; sage; sage clary; salvia; *Salvia officinalis*
鼠尾草油 clary oil; clary sage oil; saga oil; sage oil
鼠尾草蜜 sage honey
鼠尾草醇 carnosol
鼠李糖 rhamnose
鼠李糖苷 rhamnoside
鼠李〔糖〕醇 rhamnitol
鼠李酶 rhamnase
鼠海豚 porpoise
鼠傷寒沙門氏菌 *Salmonella typhimurium*
鼠傷寒沙門桿菌 *Salmonella typhimurium*
巰〔基〕乙酸鈉 sodium thioglycollate
巰基胺基酸 mercaptoamino acid
摃肉鎯頭 steak hammer

搥打 pounding
搥肉器 meat beater
搥薄 battering
稗子 barnyard millet
羥乙〔基〕澱粉 hydroxy ethyl starch
羥乙基纖維素乙基醚 ethyl hydroxyethyl cellulose
羥乙酸 glycolic acid; hydroxyacetic acid
羥丁胺酸 threonine
羥化酶 hydroxylase
羥丙基二澱粉甘油酯 hydroxypropyl distarch glycerol
羥丙基二澱粉磷酸酯 hydroxypropyl distarch phosphate
羥丙基甘油雙澱粉 hydroxypropyl distarch glycerol
羥丙基甲基纖維素 hydroxypropyl methylcellulose
羥丙基澱粉 hydroxypropyl starch
羥丙基磷酸雙澱粉 hydroxypropyl distarch phosphate
羥丙基纖維素 hydroxypropyl cellulose (HPC)
羥丙澱粉 hydroxypropyl starch
羥甲基纖維素 hydroxymethyl cellulose (HMC)
羥谷胺酸 hydroxyglutamic acid
羥高鐵血紅素 haematin; hematin
羥胺 hydroxylamine
羥胺基酸 hydroxy amino acid
羥基 hydroxyl group
羥基乙酸 glycolic acid; hydroxyacetic acid
羥基化卵磷脂 hydroxylated lecithin
羥基香茅醇 hydroxycitronellol
羥基香茅醛 hydroxycitronellal
羥基香茅醛二甲縮醛 hydroxycitronellal dimethyl acetal
羥基胺基酸 hydroxy amino acid
羥基脯胺酸 hydroxyproline; oxyproline
羥基醋酸 glycolic acid; hydroxyacetic acid
羥基賴胺酸 hydroxylysine
羥基離胺酸 hydroxylysine
羥脯胺酸 hydroxyproline; oxyproline
羥鈷胺 hydroxocobalamine hydrochloride; vitamin B_{12a}
羥鈷維生素 hydroxycobalamin
羥維生素D_3 hydroxyvitamin D_3
羥醋酸 glycolic acid; hydroxyacetic acid
羥賴胺酸 hydroxylysine
羥膽固醇 hydroxycholesterol; oxycholesterol
羥離胺酸 hydroxylysine
羧〔基〕酶 carboxylase
羧化〔反應〕 carboxylation
羧甲基澱粉 carboxymethyl starch
羧甲基澱粉鈉 sodium carboxymethyl starch
羧甲基纖維素 carboxymethyl cellulose (CMC); cellulose gum
羧甲基纖維素鈣 calcium carboxymethylcellulose
羧甲基纖維素鈉 sodium carboxymethylcellulose
羧基 carboxyl group
羧酸 carboxylic acid
腩肉 belly
腩條 intercostal meat; rib fingers
葑醇 2-ethyl fenchol; fenchol; fenchyl alcohol
葎草烯 humulene
葎草酮 humulone
蒎烯 pinene
酮 ketone
酮戊二酸 ketoglutaric acid
酮血〔症〕 ketonemia
酮尿〔症〕 acetonuria; ketonuria

酮症 ketosis
酮葡〔萄〕糖酸 ketogluconic acid
酮解〔作用〕 ketolysis
酮酸 keto acid; ketonic acid
酮糖 keto sugar; ketose
酮〔糖〕苷 ketoside
酮體 acetone body; ketone body
酯 ester
酯〔化〕值 ester number; ester value
酯化〔作用〕 esterification
酯交換〔作用〕 interesterification
酯交換〔油〕脂 interesterified lard
酯基轉移 transesterification
酯膠 ester gum
酯蠟 ester wax
酯酶 esterase
鉬 molybdenum
裹白樹莓 wineberry
裹襯包裝 lined wrapping
裹襯袋 lined bag

十四畫

廄 stall; stable
像糖漿的 syrupy
厭食症 anorexia; phagophobia
厭氣生活 anaerobiosis
厭氣呼吸 anaerobic respiration
厭氣微生物 anaerobia; anaerobic microorganism; anaerobion; anaerobium
厭氧去氫酶 anaerobic dehydrogenase
厭氧生活 anoxybiosis; anaerobiosis
厭氧生活細菌 anoxybiotic bacteria
厭氧性生化處理 anaerobic biochemical treatment
厭氧性發酵 anaerobic fermentation
厭氧的 anaerobic; anerobic
厭氧培養 anaerobic culture
厭氧脱氫酶 anaerobic dehydrogenase
厭氧菌 anaerobe; anaerobia; anaerobic bacteria; anaerobic formers; anaerobion; anaerobium
厭氧微生物 anaerobe; anaerobia; anaerobic microorganism
厭鹽的 halophobic
嘗〔味〕 degust(ate); degustation; gustation
嘗味的 gustatory
嘗味試驗 taste test
嘔吐 emesis; vomiting
嘉魚 charr
嘉鱲 red porgy
團 lump
團子 dumpling
團塊 lumpiness
團體食品 institutional food
圖林根香腸 thuringer
墊布 undercloth
墊板 pallet; trencher
墊料 wad material
墊圈 gasket; packing ring; washer
壽司 sushi
壽魚 banjofish
壽喜燒 sukiyaki
嫩化 tenderization
嫩化劑 tenderizer; tenderizing agent
嫩化器 tenderizer
嫩牛肉 tender beef
嫩玉米 green corn
嫩肉 tender meat
嫩肉劑 meat tenderizer
嫩里肌肉 tenderloin
嫩性 tenderness
嫩枝 sprig; tiller; twig
嫩度計 tenderometer
嫩莖花椰菜 sprouting broccoli
嫩莖萵苣 asparagus lettuce
嫩煮蛋 soft boiled egg
嫩葉 leaflet
嫩豌豆 mangetout
嫩鴨 duckling
嫩雞 roaster
孵化 incubation

孵化場 hatchery
孵化期 incubation period
孵化器 hatcher
孵卵所 hatchery
孵卵室 incubator house
孵卵器 hatcher; incubator
寧希德林 ninhydrin
寧希德林反應 ninhydrin reaction
寧海準 ninhydrin
寧海準反應 ninhydrin reaction
寧堡乾酪 Limburg(er) cheese
寡果糖 fructooligosaccharide
寡聚糖 oligose; oligosaccharide
寡糖 oligosaccharide
寡糖酶 oligosaccharidase
寡肽 oligopeptide
實物工資制 tommy
實驗 experiment
實驗室規模 laboratory scale
實驗動物 laboratory animal
實體鏡 stereoscope
實罐 filled can
察氏肉汁〔培養基〕 Czapek bouillon
察氏培養基 Czapek Dox agar; Czapek's medium
對乙氧苯脲 dulcin(e); sucrol; valzin
對乙基苯酚 *p*-ethylphenol
對二甲氧〔基〕苯 dimethoxybenzene
對比品評試驗 duo-trio test
對比染色 counterstain
對丙基苯甲醚 *p*-propyl anisole
對甲氧基苯乙酮 acetanisole
對甲氧基苯甲醇 anisyl alcohol
對甲氧基苯甲醛 *p*-methoxybenzaldehyde
對甲基大茴香醚 *p*-methyl anisole
對甲基苯乙酮 4-methyl acetophenone
對甲基苯甲醚 *p*-methyl anisole
對甲基苯甲醛 *p*-methyl benzaldehyde
對甲酚甲醚 *p*-methyl anisole
對百里香素 *p*-cymene
對流 convection
對流〔式〕烘箱 convection oven
對流式乾燥機 contraflow dryer; convection dryer
對胺苯基胂酸 arsanilic acid
對開片 halve
對照 blank; control
對照實驗 control experiment
對羥苯甲酸甲酯 methyl paraben
對羥苯基乙醇 tyrosol
對羥基苯甲酸乙酯 ethyl-*p*-hydroxybenzoate; ethylparaben
對羥基苯甲酸丁酯 butyl *p*-hydroxybenzoate; butylparaben
對羥基苯甲酸丙酯 propyl-*p*-hydroxybenzoate; propylparaben
對羥基苯甲酸甲酯 methylparaben
對羥基苯甲酸庚酯 heptylparaben
對羥基苯甲酸異丁酯 isobutylparaben
對羥基苯甲酸異丙酯 isopropylparaben
對羥基苯甲酸酯 paraben; parahydroxybenzoic acid ester
對數〔生長〕期 log phase; logarithmic phase
對數生長期 logarithmic growth phase
對數增殖期 logarithmic growth phase
對蝦 prawn
對蝦屬 Penaeus
幕 veil
慢性中毒 chronic intoxication
慢性毒性 chronic toxicity
慢性毒素 toxone
慢性胃炎 chronic gastritis
慢性鉛中毒 saturnism
慢性磷中毒 phosphorism
慢速冷卻 slow cooling
慢速冷凍 slow freezing
慢速除霜 slow defrosting

慢速解凍 slow defrosting
慢發酵劑 slow starter
慢凝果膠 slow set pectin
撇去〔油沫〕 skimming
撇去〔浮沫〕 skim
撇去〔浮沫〕 skimming
撇去〔浮油〕 skim
撇沫器 strainer
撇浮渣 scum off; scumming
撇浮渣杓 scummer
撇除奶油 cream off
撇渣器 skimmer
摘要 abstract
摘蒂機 stemmer
摺疊紙盒 set-up box
催乾劑 siccative
摻入 admixture
摻水 watering
摻冰〔蘇打〕水威士忌飲料 highball
摻和物 admix; admixture
摻油麵粉糊 roux
摻酒精威士忌 spirit whisk(e)y
摻假 adulteration; sophisticate
摻假的酒 rotgut
摻假貨 adulteration
摻雜 adulteration; sophisticate
摻雜物 adulterant; dope
摻雜食品 adulterated food
摻雜麵粉 adulterated flour
敲 knocking
敲打 beating
敲肉木槌 meat pounder
旗魚 marlin; swordfish
暢飲 potation; quaff
榨 wring
榨出汁 press juice; press liquor
榨奶 milking
榨汁 juicing; juice extraction; pressed dripping; ream; squeeze juice
榨汁器 juicer; squeezer
榨汁器(錐形) reamer; reaming extractor
榨汁機 juice extractor; juicer; squeezer; wringer
榨取 rack; ream
榨油 expelling
榨油廠 oilery
榨油機 oil press
榨〔後〕油渣 pressed scrap
榨〔乾的〕甜菜渣 pressed beet pulp; pressed cossette
榨渣 press residue
榛子 filbert; hazel; hazelnut
榛果 hazelnut
榛樹 hazel
榛雞 hazel hen; hazel grouse
榴蓮 durian
槐二糖 sophorose
槐豆膠 locust bean gum
槍烏賊 calamary
槍烏鰂 squid
槌磨 sledge mill
滾動篩 revolving sieve; revolving sifter
滾筒 reel; roll; roller; rotary drum; tumbler
滾筒分離機 roll separator
滾筒式分級機 assorting cylinder
滾筒式熱燙機 roller scalder
滾筒式輸送機 roller conveyor
滾筒乳粉 roller milk
滾筒乾燥 roller drying
滾筒乾燥乳粉 roller dried milk
滾筒乾燥機 cylinder dryer; cylinder drying machine; roll drying machine; roller dryer; tumbler dryer
滾筒混合機 tumbler mixer
滾筒造粒 tumbling granulation
滾筒萃取器 roller extractor
滾筒碾磨機 roller mill

滾筒篩 rotary sieve
滾碾機 chaser mill
滾輪式提升機 trolley hoist
滾糖衣機 dragee machine; revolving silvering pan
滾壓線紋 knuring
滾鹽 rouse
滴 drop
滴汁 drip
滴定 titration
滴定分析〔法〕 titrimetric analysis; titrimetry; volumetric analysis
滴定測量〔法〕 titrimetric analysis; titrimetry
滴流 trickle
滴酒不沾的 teetotal
滴乾 draining
滴乾重 drained weight
滴液 drip; drop
滴液〔現象〕 dripping
滴淋冷卻器 trickling cooler
滴管 dropper
滴濾池 trickling filter
滴濾器 trickling filter
滴瀝咖啡 drip coffee
滴瀝咖啡壺 Dripolator
滴瀝咖啡機 drip coffee maker; drip maker
漏斗 funnel
漏出 seep(ing); transudation
漂木酸 chlorogenic acid
漂白 bleaching; blanch(ing)
漂白土 bleaching earth; Fuller's earth
漂白〔用黏〕土 bleaching clay
漂白〔作用〕 decolo(u)ration; decolo(u)-rization
漂白粉 bleaching powder; chlorinated lime
漂白粉消毒 javellization
漂白液 bleaching liquor
漂白紫膠 white shellac
漂白裝置 bleacher
漂白過度 overbleach
漂白蔗糖 clayed sugar
漂白劑 bleacher; bleaching agent; decolo-(u)rant; decolo(u)rizer; decolo(u)rizing agent; whitener
漂白澱粉 bleached starch
漂白麵粉 bleached flour
漂洗 rinse; rinsing
漂洗槽 potcher
漂浮分級機 flotation grader
漂粉精 calcium hypochlorite
漢堡 burger
漢堡牛排 hamburger steak
漢堡包 hamburg(er)
漢堡西洋芹 hamburg parsley
漢堡煎牛肉餅 hamburg(er)
漢堡麵包 hamburger bun
漢普夏〔爾〕豬 Hampshire
漢森酵母 Hansenula yeast
漢遜〔氏〕酵母 Hansenula yeast
滿口袋 bagful
滿杯 cupful
滿浮渣的 scummy
滿袋 sackful
滿盤 trayful
滯後期 lag phase
滯留水 stagnant water
滯留時間 retention time
滯留體積 retention volume
漆蠟蘑（蕈類） common deceiver
漱口 gargle
漱口〔藥〕水 gargle
漁產品 fishery product
漁場 fishery
漁業 fishery; fishery industry
滲入 infiltration; interfusion

滲出 diffusion; exudation; seep(ing); seepage; transudation; weep
滲出汁 diffusion juice
滲出物 exudate; transudate
滲出液 diffusate; transudate; weep
滲出現象 seepage
滲出量 seepage
滲出器 diffuser; diffusor
滲析 dialysis
滲析液 dialysate
滲析管 dialysis tube
滲析膜 dialyzer
滲析器 dialyzer
滲透 infiltration; penetration; permeate
滲透平衡 osomotic equilibrium
滲透〔作用〕 osmosis; permeation
滲透性 osomotic permeability; permeability
滲透物 permeant
滲透膜 osomotic membrane; permeable membrane
滲透學 osmology
滲透選擇性 permeaselectivity
滲透壓 osomotic pressure
滲透壓力計 osmometer
滲漏 percolation; seep(ing); seepage; transudation
滲壓計 osmometer
滲濾 percolate; percolation
滲濾液 percolate
滲濾器 percolation filter; percolator
滌氣器 scrubber
滷汁 marinade; pickle
滷液 marinade
熔化的 melted
熔化鍋 melter
熔接罐 welded seam can
熔點 melting point
熔爐 furnace
熙春茶 hyson
熊蝦 bear prawn
熊貓 panda
熄滅 quench(ing)
熒光桃紅 phloxine
瑤柱 conpoy
瑪拉加白葡萄酒（西班牙） Malaga
瑪珈琳 butterine; margarine; oleomargarine
瑪珈琳製造器 complector
瘋牛症 bovine spongiform encephalopathy (BSE); mad cow disease
睡前飲料 nightcap
睡菜 buckbean
睡蓮 water lily
磁力捕集機 magnetic interceptor
磁力除鐵機 magnetic interceptor
磁力攪拌器 magnetic stirrer
碟 dish
碟子 saucer
碟式離心機 disc bowl centrifuge
碳 carbon
碳化 carbonizing
碳水化合物 carbohydrate
碳化〔作用〕 carbonization
碳烤 char grilling
碳氫化合物 hydrocarbon
碳黑 carbon black
碳酸 carbonic acid
碳酸化 carbonating
碳酸化作用 carbonatation; carbonation
碳酸化器 carbonator; carbonizer
碳酸水 carbonated water; soda
碳酸泉水 carbonated spring
碳酸酒 carbonated wine
碳酸酒精飲料 alcoholic carbonated drink
碳酸氫三鈉 sodium sesquicarbonate
碳酸氫鈉 baking soda; bicarbonate of soda; sodium bicarbonate; sodium hydrogen carbonate

碳酸氫鉀 potassium bicarbonate
碳酸氫銨 ammonium bicarbonate; ammonium hydrogen carbonate; baker's salt
碳酸氫鹽 bicarbonate
碳酸鈣 calcium carbonate
碳酸鈉 sodium carbonate
碳酸飲料 brause; carbonated beverage; carbonated drink; soda pop
碳酸鉀 potassium carbonate
碳酸飽充糖汁 saturation juice
碳酸飽和 carbonatation; carbonation
碳酸飽和汁 saturation juice
碳酸酯 carbonate
碳酸銨 ammonium carbonate
碳酸鎂 magnesium carbonate
碳酸鹽 carbonate
碳酸鹽法 carbonatation; carbonation
福〔爾〕馬林 formalin
福美達美斯豆 ful medames
福壽魚 Nile tilapia
福壽螺 apple snail
福瑪喬山羊奶乾酪（意大利） formaggio cheese
種 race; species
種子 pip; seed
種〔子〕仁油 seed kernel oil
種子油 seed oil
種子培養物 seed culture
種子植物 seed plant
種子酵母 seeding yeast
種母牛 breeding cow
種母豬 sow
種用酵母 mother starter
種皮 episperm; seed coat
種馬 stallion
種莢 seedpod
種植園 plantation
種菌 culture; mother culture
種菌發酵槽 seed fermenter
種酵母 pitching yeast; seed yeast
種醋 mother vinegar
種臍 hilum; scar
種醪 seed mash
種類 sort; species; variety
種麴 mold starter; mould starter; mother culture
稱重 weigh(ing)
稱重瓶 weighing bottle; weight bottle
稱量紙 weighing paper
稱量瓶 weighing bottle; weight bottle
端菜者 server
管 duct; pipe; tube
管式加熱器 tubular heater
管式冷卻器 pipe cooler
管式冷凝器 tubular condenser
管式乾燥器 tube dryer
管式蒸發器 tubular evaporator
管式鍋爐 tubular boiler
管形雞油菌（蕈類） trumpet chanterelle
管狀腸衣 tubular casing
管狀噴霧器 tubular atomizer
管柱層析法 column chromatography
管殼式熱交換器 shell and tube heat exchanger
管道式速凍機 bazooka freezer
管導式乾燥機 pipe dryer
箕作氏槍魚 blue marlin; pacific marlin; spearfish
筵席 feast
箔 foil
精子 sperm
精白米 polished rice
精白麵粉 fine flour
精米機 rice pearling machine; rice polisher; rice polishing machine; rice whitening machine; rice sweep machine; rice sweeper

精油 essential oil; ethereal oil
精油樹脂 oleoresin
精研磨機 refiner
精美食品 delicacy
精美菜餚 kickshaw
精氨酸 arginine
精氨酸鹽酸鹽 arginine monohydrochloride
精神食糧 manna
精胺酸 arginine
精胺酸酶 arginase
精密天平 analytical balance; precision balance
精細的 superfine
精細砝碼 precision weight
精萃 quintessence
精華 distillate; quintessence
精煉 depuration; refine; refinement
精煉不足 underrefining
精煉油 purified oil; refined oil
精煉煙燻料 fumeol
精煉廠 refinery
精煉豬油 refined lard
精煉機 refiner
精煉糖廠 sugar refinery
精過濾 secondary filter
精製 affination; refine; refining; refinement
精製油 refined oil
精製食鹽 table salt
精製酒 fine wine
精製酒精 neutral spirit
精製〔產〕品 refinement
精製廠 refinery
精製豬油 prepared lard; refined lard
精製劑 refining agent
精製機 refiner
精製澱粉 refined starch
精製糖 refined sugar
精製糖蜜 refinery molasses
精製糖漿 refinery syrup
精製麵粉 purified starch
精製鹽 refined salt
精碾 fine grinding; glazing
精碾不足的米 undermilled rice
精碾米 glazed rice
精緻食物 delicate food
精糖 fine sugar; refined sugar
精鍊 refining
精鍊劑 refining agent
精鍊糖蜜 refinery molasses
精鍊糖漿 refinery syrup
精餾 rectification; rectify
精餾酒精 rectified alcohol; rectified spirit
精餾塔 rectifier
精餾器 rectifier
精鹽 fine salt; pan salt; pure salt; refined salt; stoved salt; straight salt
綜合果汁 mixed juice
綜合果汁飲料 mixed juice drink
綜合病症 syndrome
綜合維生素 multivitamin
綠皮小胡瓜 green zucchini
綠皮西葫蘆 green zucchini
綠皮蘋果 greening
綠色 greenness
綠色植物 verdure
綠豆 golden gram; green gram; mung bean
綠豆芽 green bean sprout; mung bean sprout; silver sprout
綠的 green
綠花椰菜 broccoli
綠扁豆 green lentil
綠原酸 chlorogenic acid
綠真鰺 jack mackerel
綠茶 green tea
綠茶粉 powdered green tea

綠葉菜 greens
綠熟果 ripe green fruit
綠霉 green mo(u)ld
綠薄荷 *Mentha viridis*; spearmint
綠薄荷油 spearmint oil
綠藻 chlorella; green algae
綠藻類 chlorophyceae
綠蘆筍 green asparagus
綠鰭馬面魨 black scraper
綠鱈 saithe
綠變肉 green meat
綠變鮪魚 green tuna
綠黴 green mo(u)ld
緊密包裝 compact package
緊密性 tightness
緊密容器 tight container
緊貼包裝 skin tight packaging
緊縮包裝 compact package; skin packaging
綴合蛋白質 conjugated protein
網包裝 net package
網油 caul fat; caul web fat; intestinal fat; omental fat; ruffle fat; veil fat; webfat
網狀牛肝菌 summer bolete
網狀組織 reticular tissue
網狀蛋白 reticulin
網狀結構 network structure; reticulum
網狀結締組織 areolar tissue
網狀濾器 gauze strainer; mesh filter
網狀纖維 reticular fiber; reticulin fiber
網紋瓜 netted melon
網紋馬勃(蕈類) common puffball
網紋甜瓜 nutmeg melon
網硬蛋白 reticulin
網膜 omentum
網膜脂 caul fat; caul web fat; veil fat
網質纖維 recticulin fiber
網篩 mesh screen; mesh sieve; net sieve; silk sieve
綿羊 jumbuck; sheep
綿羊肉 mutton
綿羊脂 mutton fat
綿羊排 mutton chop
綿羊腸衣 sheep casing
綿棗兒 scilla
綿糖 soft sugar
維也納〔式〕小麵包 Vienna roll
維也納〔式〕麵包 Vienna bread
維也納香腸 Vienna sausage; Wiener (sausage); wienerwurst
維他命 nutramin(e); vitamin; vitamine
維生食物 maintenance food; vitagen
維生素 nutramin(e); vitamin; vitamine
維生素A retinol; vitamin A
維生素A_1 vitamin A_1
維生素A_2 vitamin A_2
維生素A醇 vitamin A alcohol
維生素B_1 aneurin; oryzanin; thiamine; vitamin B_1
維生素B_{12} cobalamin; cyanocobalamin; hydroxocobalamine hydrochloride; vitamin B_{12}
維生素B_{12a} vitamin B_{12a}
維生素B_{13} orotic acid; vitamin B_{13}
維生素B_{15} vitamin B_{15}
維生素B_2 lactochrome; lactoflavin; riboflavin(e); verdoflavin; vitamin B_2
維生素B_2缺乏症 ariboflavinosis
維生素B_3 pantothenic acid; vitamin B_3
維生素B_6 pyridoxine; vitamin B_6
維生素Bc folic acid; vitamin Bc
維生素B_t L-carnitine; vitamin B_t
維生素B群 vitamin B group
維生素C ascorbic acid; cevitamic acid; vitamin C
維生素D vitamin D
維生素D_2 calciferol; ergocalciferol; vitamin D_2; viosterol

維生素D_3 cholecalciferol; vitamin D_3
維生素D_4 vitamin D_4
維生素D_5 vitamin D_5
維生素E tocopherol; vitamin E
維生素F vitamin F
維生素G vitamin G
維生素H antiavidin; biotin; vitamin H
維生素J vitamin J
維生素K menadione; vitamin K
維生素K_1 phytonadione; phylloquinone; vitamin K_1
維生素K_2 vitamin K_2
維生素K_3 menadione sodium bisulfite; vitamin K_3
維生素K_4 vitamin K_4
維生素K_5 vitamin K_5
維生素K_6 vitamin K_6
維生素K_7 vitamin K_7
維生素L vitamin L
維生素M vitamin M
維生素P citrin(e); methyl hesperidin; vitamin P
維生素PP niacin; nicotinic acid; vitamin PP
維生素R vitamin R
維生素S vitamin S
維生素T vitamin T
維生素U vitamin U
維生素V vitamin V
維生素X vitamin X
維生素元 previtamin
維生素化作用 vitaminization
維生素先質 vitamin precursor
維生素油劑 oleovitamin
維生素的 vitaminous
維生素前驅物 vitamin precursor
維生素食物 vitagen
維生素原 primary vitamin; previtamin; provitamin; vitamin precursor
維生素缺乏症 avitaminosis; hypovitaminosis; subvitaminosis; vitamin deficiency
維生素強化 vitamin fortification; vitaminization
維生素強化乳 vitaminized milk
維生素強化果汁 vitaminized juice
維生素強化的 vitaminous
維生素強化食品 vitamin-enriched food; vitamin fortified food; vitaminized foodstuff
維生素添加 vitaminization
維生素過多症 hypervitaminosis; supervitaminosis
維生素增補 vitamin fortification; vitamin supplementation
維生素學 vitaminology
維生最低飲食 subsistence diet
維生飲食 subsistence diet
維吉尼亞火腿 Virginia ham
維多酚 sodium sulfocarbonate
維長粒米 long grain rice
維持生命的食物 maintenance food
維持培養基 maintenance medium
維持體力的食品 sustenance
維持體重的飲食量 aequum
維珍尼亞火腿 Virginia ham
維特契莓 veitchberry
維管束 vascular bundle
維管束組織 vascular tissue
維管系統 vascular system
維護 maintenance
翡翠貽貝 green mussel
聚乙二醇 polyethylene glycol; polythylene glycol
聚乙烯 polyethylene (PE); polythene
聚乙烯收縮膜 polyethylene shrink film
聚乙烯塑膠 vinyl plastic

聚乙烯聚吡咯烷酮 polyvinylpolypyrrolidone (PVPP)
聚乙烯樹脂 polyvinyl resin
聚乙烯薄膜 polyethylene film
聚乙烯〔薄膜〕袋 polybag; polythene bag
聚乙烯吡咯烷酮 polyvinylpyrrolidone (PVP)
聚乙酸乙烯酯 polyvinyl acetate
聚丁烯 polybutene
聚二甲基硅氧烷 polydimethylsiloxane
聚二甲基硅醚 polydimethylsiloxane
聚二氯亞乙烯 polyvinylidene chloride (PVDC)
聚二氯亞乙烯腸衣 polyvinylidene chloride casing
聚三葡萄糖 pullulan
聚山梨〔糖〕醇酯 polysorbate
聚山梨酸酯(乳化劑) polysorbate
聚山梨酸酯 20 polysorbate 20
聚山梨酸酯40 polysorbate 40
聚山梨酸酯60 polysorbate 60
聚山梨酸酯65 polysorbate 65
聚山梨酸酯80 polysorbate 80
聚己糖 hexosan
聚六碳糖 hexosan
聚木糖 xylan
聚木糖酶 xylanase
聚丙烯 polypropylene (PP)
聚丙烯乙二醇 polypropylene glycol
聚丙烯酸鈉 sodium polyacrylate
聚丙烯酰胺凝膠 polyacrylamide gel
聚丙烯酰胺 polyacrylamide
聚半乳糖 galactan; galactosan
聚半乳糖醛酸 polygalacturonic acid
聚半乳糖醛酸酶 polygalacturonase
聚四氟乙烯 teflon
聚左旋糖 fructosan; levan; levulosan
聚戊糖 pentosan
聚甘油酯 polyglycerol ester
聚甘胺酯 polyglycine
聚甘露糖 mannan; mannosan
聚甘露糖酶 mannase
聚甲基硅氧 methyl polysilicone
聚甲殼糖 chitosan
聚〔合〕偏磷酸鹽 kraul salt; polymetaphosphate
聚合〔作用〕 aggregation; polymerization
聚合果 aggregate fruit; syncarp
聚合油 polymerized oil
聚合物 polymer
聚合〔高分子〕電解質 polyelectrolyte
聚合現象 polymerism
聚合體 polymer
聚合酶 polymerase
聚果糖 fructosan; fructan; levan; levulosan
聚果糖蔗糖酶 levansucrase
聚果呋喃糖 fructofuranose
聚矽氧 silicone
聚矽氧油 silicone oil
聚花果 multiple fruit; syncarp
聚阿拉伯糖 araban; arabinan
聚苯乙烯 polystyrene (PS)
聚苯乙烯收縮膜 polystyrene shrink film
聚氧乙烯(40)硬脂酸酯 polyoxyethylene (40)stearate
聚氧乙烯(8)硬脂酸酯 polyoxyethylene (8)stearate
聚氧乙烯木糖醇酐單硬脂酸酯 polyoxyethylene xylitan monostearate
聚氧乙烯脱水山梨醇單油酸酯 polyoxyethylene sorbitan monooleate; Tween 80
聚氧乙烷高級脂肪醇 polyoxyethylene higher aliphatic alcohol
聚氧丙烯甘油醚 polyoxypropylene glycerol ether
聚氧丙烯氧化乙烯甘油醚 polyoxypropylene oxyethylene glycerol ether

聚海藻糖 fucosan
聚胺 polyamine
聚胺酯 polyurethane
聚胺酯樹脂 polyurethane resin
聚偏二氯乙烯 polyvinylidene chloride (PVDC)
聚偏二氯乙烯腸衣 polyvinylidene chloride casing
聚偏磷酸鉀 potassium polymetaphosphate
聚異丁烯 polyisobutylene
聚氯乙烯 polyvinyl chloride (PVC)
聚氯乙烯收縮膜 polyvinyl chloride shrink film
聚氯乙烯薄膜 polyvinyl chloride film
聚結 coalescence
聚集〔作用〕 aggregation
聚集體 aggregate
聚葡甘露糖 glucomannan
聚葡萄胺糖 chitosan
聚葡萄糖 glucan; glucosan; polydextrose
聚葡萄糖酶 glucanase
聚酯 polyester; polyethylene terephthalate (PET)
聚酯收縮膜 polyester shrink film
聚酯薄膜 polyester film
聚對苯二甲酸乙二醇酯 polyethylene terephthalate (PET)
聚碳酸酯 polycarbonate
聚糊精 polydextrose
聚醋酸乙烯酯 polyvinyl acetate
聚糖 glycan; osan; polyose; polysaccharide; polysaccharose
聚糖醛酸 polyuronic acid
聚磷酸鈣 calcium polyphosphate
聚磷酸鈉 sodium polyphosphate
聚磷酸鉀 potassium polyphosphate
聚磷酸銨 ammonium polyphosphate
聚磷酸鹽 polyphosphate
聚酰胺 polyamide
腐化 pythogenesis
腐生生物 saprobe
腐生細菌 metatrophic bacteria; saprogenic bacteria; saprophytic bacteria
腐生植物 saprophyte
腐生菌 saprobe
腐生營養 metatrophy
腐皮 soy skin; soya skin; soybean milk film; yuba
腐朽 decay
腐竹 dried bean skim; soy skin; soya skin; soybean milk film; yuba
腐肉中毒 kreotoxism
腐肉毒素 kreotoxin
腐肉植物 necrophyte
腐乳 bean curd cheese; Chinese cheese; fermented tofu; preserved bean curd; soy cheese; soybean cheese; sufu
腐物寄生細菌 saprophytic bacteria
腐物寄生植物 saprophyte
腐臭 putrefaction; putrid odo(u)r; rancidness
腐臭物 putrilage
腐臭細菌 putrefactive bacteria
腐臭發酵 putrefactive fermentation
腐臭腐敗 putrefactive spoilage
腐胺 putrescine; pythogenesis; putrefaction; putrescence; rot; spoilage; taint
腐敗〔芽孢〕桿菌 *Bacillus putrificus*
腐敗生物 spoilage organism
腐敗味 rotten flavo(u)r
腐敗性 saprogenicity
腐敗物 spoilage
腐敗的 putrefactive; putrid; rotten
腐敗氣味 spoiled odo(u)r
腐敗〔細〕菌 putrefactive bacteria; spoilage bacteria
腐敗發酵 putrefactive fermentation
腐敗程度 degree of spoilage

腐敗菌 spoilage organism
腐敗變黏的酒 ropy wine
腐植酸 humic acid
腐植質 humic substance; humus; muck
腐爛 blet; decay; foul; putrefactive spoilage; putrescence; rot
腐爛物 putrilage
腐爛的 putrefactive; putrid; rotten
膀胱 bladder; urinary bladder
膏 paste; spread
膏狀藥 poultice
腿 gam
腿肉 thigh meat
臺上攪拌器 bench mixer
臺式混合器 bench mixer
臺灣胡桃 Chinese walnut
與空氣化合 aerification
蒿腦 estragole
蓄水池 cistern; reservoir; water cistern
蓄熱器 regenerator
蒙古牛 Mongolian cattle
蒙汗酒 Mickey Finn
蒙那靈〔天然蛋白甜味劑〕 monellin
蒙蒂拉葡萄酒 amontillado
蒲公英 dandelion; taraxacum
蒲公英乾根莖（藥用） taraxacum
蒲公英黃質 taraxanthin
蒲公英醇 taraxol
蒲芹蘿蔔 parsnip
蒲桃 jambo; rose apple
蒜抑菌素 allistatin
蒜制菌素 allistatin
蒜味小皮傘（蕈類） wood garlic mummy cap
蒜味的 alliaceous
蒜味香腸 knoblanch; knoblauchwurst
蒜味調味料 allylic condiment
蒜味麵包 garlic bread
蒜泥 garlic spread
蒜泥蛋黃醬 aioli
蒜油 garlic oil
蒜苗 leek
蒜粉 garlic powder
蒜素 allicin
蒜蓉〔麵〕包 garlic bread
蒜頭 garlic
蒜瓣 clove; garlic hop
蒜鹽 garlic salt
蒜苷 allin
蓋子 lid
蓋印 seal
蓋板 coverplate
蓋斯卡乾酪 geska
蓋鈎 end hook
蓋緣機 can end curling machine
蓋邊 curl
蒸 steam
蒸成半熟 parboil(ing); parcook(ing)
蒸成半熟的肉 parboiled meat
蒸米餅 idla
蒸汽 steam; vapo(u)r
蒸汽化 vapo(u)rization
蒸汽加熱乾燥 steam drying
蒸汽加熱乾燥機 steam dryer; steam heated dryer
蒸汽加熱器 steam heater
蒸汽加熱攪拌 steam bubbling
蒸汽去皮 steam peeling
蒸汽回收 vapo(u)r recovery
蒸汽〔冷〕凝液 steam condensate
蒸汽夾套 steam jacket
蒸汽夾套鍋 steam jacket(ed) kettle
蒸汽夾層鍋 steam jacket(ed) kettle
蒸汽直空封口機 vapo(u)r vacuum sealing machine
蒸汽保溫檯 steam table
蒸汽室 steam chamber; vapo(u)r chamber
蒸汽套管 steam jacket

蒸汽浴 steam bath
蒸汽乾燥 steam drying
蒸汽乾燥機 steam dryer
蒸汽真空法 steam-vac method
蒸汽密封 vapo(u)r seal
蒸汽張力 vapo(u)r tension
蒸汽脱氣 steam exhausting
蒸汽脱脂 vapo(u)r degreasing
蒸汽發生器 steam generator
蒸汽滅菌 steam sterilization
蒸汽滅菌器 steam sterilizer
蒸汽煉油 steam rendering
蒸汽過熱器 steam superheater; vapo(u)r superheater
蒸汽蒸煮 steam boiling; steam cooking; vapo(u)r cooking
蒸汽蒸煮器 steam cooker
蒸汽蒸餾 steam distillation
蒸汽蒸餾器 steam still
蒸汽噴射加熱器 steam-jet heater
蒸汽噴射排氣封罐 steam flow closure; steam flow sealing
蒸汽噴射滅菌法 uperization
蒸汽噴射滅菌器 uperizer
蒸汽噴射製冷機 steam-jet refrigerating machine
蒸汽噴霧器 steam-jet atomizer
蒸汽熬油 steam rendering
蒸汽熬製的優質豬油 prime steam lard
蒸汽箱 steam box
蒸汽壓 vapo(u)r tension; vapo(u)r pressure
蒸汽壓力計 vapo(u)rimeter
蒸汽櫃 steam box; steam chest
蒸的 steamed
蒸氣攪拌鍋 blow-up pan
蒸散作用 transpiration
蒸煮 cooking; steam boiling; steam cooking; steaming; vapo(u)r cooking
蒸煮工人 pansman
蒸煮時間 cooking time
蒸煮袋 boiling bag; retort pouch
蒸煮袋食品 boil-in-bag food
蒸煮袋裝食品 retort pouch food
蒸煮過度 overcook; overcooking
蒸煮器 boiler; cooker
蒸煮鍋 boiling pan; cooking pan; steam cooker; saucepot
蒸發 vapo(u)rize
蒸發〔作用〕 evaporation; volatilization; vapo(u)rization
蒸發池 evaporation pond
蒸發乳 evaporated milk
蒸發脱脂乳 evaporated skim(med) milk
蒸發場 evaporating field
蒸發結晶器 crystallizing evaporator
蒸發盤組 pan bench
蒸發器 evaporator; vapo(u)rizer
蒸發濃縮 boil down
蒸發鍋 evaporating pan
蒸發罐 strike pan
蒸飯 steamed rice
蒸濃 inspissate
蒸濃〔作用〕 inspissation
蒸濃器 inspissator
蒸鍋 steamer
蒸餾 distil(l); distillation; still
蒸餾水 distilled water
蒸餾物 distillate; distillation
蒸餾室 distillery; grain distillery; still room
蒸餾酒 akvavit; aquavit; alcoholic liquor; distilled liquor; distilled spirit; spirit
蒸餾酒〔飲料〕 distilled beverage
蒸餾酒精 silent spirit
蒸餾酒廠 distiller
蒸餾釜 distillating still; still kettle
蒸餾得來的 distilled
蒸餾殘渣 distillation residue

蒸餾塔 distillation column
蒸餾廢液 distillery stillage
蒸餾廢醪 stillage
蒸餾醋 distilled vinegar
蒸餾器 distiller; distillating still; distillator; still; vapo(u)rizer
蒸騰作用 transpiration
蒸籠 steamer
蓓蕾 bud
蒼白軟質汁液滲出肉 pale, soft, exudative muscle (PSE muscle)
蒼白鬆軟和滲水肉 pale, soft, exudative muscle (PSE muscle)
蒼术油 atractylis oil
蒼术〔硬〕脂 atractylis concrete
蒼蠅 fly
蜜二糖 melibiose
蜜三糖 melitriose; raffinose
蜜三糖酶 raffinase
蜜瓜 honeydew; honeydew melon
蜜李 damson (plant)
蜜里糖 melitose
蜜思葡萄酒 mispelie
蜜柑蛋白 pomelin
蜜香瓜 honeydew melon
蜜栗 marron glace
蜜酒飲料 mead
蜜莓 honey berry
蜜棗 honey date; preserved date jujube
蜜源植物 nectariferous plant; nectar plant
蜜腺 nectary
蜜蜂 bee; honeybee
蜜蜂花 balm; melissa
蜜蜂食料 beebread
蜜蜂巢 honeycomb
蜜蜂學 apiology
蜜糖 honey
蜜糖豆 honey bean
蜜餞 candy; candied fruit; confection; conserved fruit; comfit; confiture; fruit preserve; preserve; preserved food; preserved fruit; succade; sweet stuff; sweetmeat; sweets; sweety
蜜餞水果 crystallized fruit
蜜餞夾心冰淇淋 auffait
蜜餞果皮 candy peel
蜜餞的 candied
蜜餞栗子 marron glace
蜜環菌 shoestring fungus
蜜環蕈 honey fungus
蜜環蕈屬 *Armillaria*
蜜露 honeydew
蜘蛛蟹 spider crab
裹上糖衣的 sugar coated
裹包式紙箱 wraparound case
裹包的 wraparound
裹漿 battering
裹麵包屑 breading
裹麵包屑後烹煮 breaded
裹麵包屑製品 breaded product
裹麵〔烤〕肉 toad-in-the-hole
裸大麥 hulless barley
裸麥 rye
裸麥芽 rye malt
裸麥威士忌〔酒〕 rye whisky
裸麥粉 rye flour
裸麥粗粉 rye meal
裸麥酸麵包 pumpernickel
裸麥麵包 rye bread
裸蓋魚 beshow; black cod; sablefish
裸燕麥 hulless oat; naked oat
製丸 pelletization; pelletizing
製丸機 pelletizer
製冰廠 ice plant
製冰器 ice maker
製冰機 ice generator; ice machine; ice maker; ice making machine

製成品 finished goods
製成蜜餞 preserve
製曲 koji-making
製曲機 koji-making machine
製冷的 refrigeratory
製冷裝置 refrigerating equipment; refrigeration equipment
製冷劑 refrigerant
製皂用脂 soap fat
製乳場 dairy
製品 preparation
製洋菜的海藻 tengusa
製粉〔用〕穀物 grist
製粉廠 miller
製紙盒機 cartoning machine
製酒用酵母 pitching yeast
製酒殘渣 distillery residue
製啤酒的麥芽汁 beer wort
製桶工人 cooper
製桶工場 cooperage
製桶業 cooperage
製梨酒用梨 perry fruit
製粒機 granulator; nodulizer
製袋機 bag making machine
製造日期 pack date
製造麥芽 malting
製造麥芽的工人 maltster
製造罐頭食品 tinning
製麥芽室 malt house
製備 preparation
製蓋機 lid maker
製糖 sugaring
製糖廠 sugarhouse
製醪 mashing
製瓊脂的海藻 tengusa
製藥 pharmaceutical
製麴 koji-making
製麴機 koji-making machine
製罐 can-making
製罐級 U.S. Canner
製罐廠 can factory
製罐機 can body making machine
製鹽廠 salt works
誤差 deviation
誘突變作用 mutagenesis
誘突變性 mutagenicity
誘突變劑 mutagen; mutagenic agent
誘發劑 inducing agent
誘導劑 inducing agent
誘導酶 induced enzyme
誘癌作用 carcinogenesis
誘癌性 carcinogenicity
誘癌物質 carcinogen
誘癌劑 carcinogen
誘變性 mutagenicity
誘變劑 mutagen; mutagenic agent
豪華酒店 gin palace
豪華酒館 gin palace
豪華餐廳 fleshpots
豪飲 winebibbing
豪飲者 toper; wassailer; wine bag
豪豬 porcupine
賓治飲料 punch
赫利福德牛 Hereford (cattle)
赫里福牛 Hereford (cattle)
輔助色素 accessory pigment
輔助泵 donkey pump
輔助食品 accessory food
輔料 adjuvant; adjunct
輔基 prosthetic group
輔蛇麻蘆酮 colupulone
輔酶 coenzyme; coferment
輔酶I nicotinamide adenine dinucleotide (NAD)
輔酶II nicotinamide adenine dinucleotide phosphate (NADP)
輔酶Q ubiquinone
輕工業 light industry

輕包裝 light packaging
輕收斂的 subastringent
輕沸 thin boiling
輕沸澱粉 thin boiling starch
輕便飲料 gasogen(e)
輕便電爐 hot plate
輕度焙烤 light roast
輕度凝結的牛乳 soft curd milk
輕度膨罐 flipper
輕食食品 snack food
輕蛋糕 low ratio cake
輕煮 soft cook
輕煮蛋 soft boiled egg
輕榨優質橄欖油 virgin olive oil
輕醃 mild cure
輕醃的 corning; lightly salted; slack salted
輕瀉劑 laxative
輕燻 mild smoke
辣沙司 hot sauce; hot spice sauce
辣乳菇 peppery milk cap
辣味 acridity; hot taste
辣味的 devilled; peppery
辣味濃湯 pepper pot
辣的 hot; pungent
辣根 horseradish
辣根菜類 scurvy weed
辣椒 capsicum; cayenne; cayenne pepper; chile; chile pepper; chili; chilli; chil(l)i pepper; hot pepper
辣椒〔粉〕 red pepper
辣椒汁 tabasco sauce
辣椒色素 paprika red; red paprika
辣椒油 capsicum oil; pepper oil
辣椒油樹脂 capsicum oleoresin; paprika oleoresin
辣椒紅 paprika red; red paprika
辣椒紅素 capsanthin
辣椒粉 cayenne pepper; chil(l)i powder
辣椒素 capsaicin
辣椒紫紅素 capsorubin
辣椒質 capsanthin
辣椒醇 capsanthol
辣椒橙 chilli orange; paprika orange
辣椒醬 chil(l)i sauce
辣椒屬 capsicum
辣椒鹼 capsicine
辣醬 hot sauce; hot spice sauce; piquant sauce; pungent sauce; piquant sauce
遠東多線魚 Atka mackerel
遠東沙丁魚 oriental sardine
遠東寬突鱈 navaga
遠洋魚 pelagic fish
遠洋魚業 deep sea fishery; pelagic fishery
遠藤〔氏〕培養基 Endo's agar; Endo's medium
酵母 barm; bulgood; sourdough; yeast
酵母水解物 yeast hydrolyzate
酵母自溶物 yeast autolysate
酵母乳 yeast cream
酵母固醇 zymosterol
酵母抽提物 yeast extract
酵母泥 yeast paste
酵母的 barmy; yeasty
酵母甾醇 zymosterol
酵母活化劑 yeast food
酵母風味 yeasty flavo(u)r
酵母浸膏 yeastrel; yeast extract
酵母粉 powdery yeast; yeast powder
酵母培養〔物〕 yeast culture
酵母發酵 yeast fermentation
酵母發酵醪 yeast wash
酵母發麵能力測定計 zymotachegraph
酵母萃 yeast extract
酵母萃培養基 yeast extract medium
酵母菌 microzyme; yeast
酵母菌元 yeast starter
酵母菌分解 saccharomycetolysis

酵母菌母 yeast starter
酵母菌株 yeast strain
酵母塊 yeast cake
酵母聚糖 zymosan
酵母餅 yeast cake
酵母調味品 marmite
酵母養料 yeast food
酵母壓塊 yeast cake
酵母醪 yeast wash; yeast cream
酵母屬 *Saccharomyces*
酵性的 fermentative
酵素 enzyme; ferment
酵素分析法 enzyme assay
酵素分解〔作用〕 enzymolysis
酵素反應 enzymatic reaction; enzyme reaction
酵素水解 enzymatic hydrolysis
酵素作用 enzyme action
酵素抑制劑 enzyme inhibitor
酵素受質複合物 enzyme-substrate complex
酵素性劣變 enzymatic deterioration
酵素性降解 enzymatic degradation
酵素性氧化 enzymatic oxidation
酵素性氧化褐變 enzymatic oxidative browning
酵素性敗壞 enzymatic deterioration
酵素性腐敗 enzymatic spoilage
酵素性褐變 enzymatic browning
酵素的 emzymic
酵素活性 enzymatic activity; enzyme activity
酵素特異性 enzyme specificity
酵素專一性 enzyme specificity
酵素處理澱粉 enzyme treated starch
酵素學 enzymology
酵頭 sourdough
酸 acid
酸不溶灰分 acid insoluble ash
酸分解 acid decomposition
酸化 acidation; acidulation; acidification; ripening; souring
酸化乳 acidified milk
酸化果汁 acidified juice
酸化食物 acidified food
酸化劑 acidulant; acidulating agent; acidifier
酸化器 acidifier
酸水解 acid decomposition; acid hydrolysis
酸牛奶 clabber; lapper milk; yoghourt; yoghurt; yogurt
酸牛乳酒 kefir; kefyr; kephir
酸牛乳酒麴 kefir grain
酸去皮〔法〕 acid peeling
酸奶 acidophilus milk; yoghourt; yoghurt; yogurt
酸奶油 sour cream
酸奶製品 leben
酸式鹽 acid salt
酸灰質食物 acid ash food
酸血症 acidosis; acidemia
酸忌廉 sour cream
酸改性澱粉 acid modified starch
酸豆 caper; tamarind
酸乳 cultured milk; sour milk
酸乳油 sour cream
酸乳清 acid whey
酸乳飲料 sour milk beverage
酸乳酪 yoghourt; yoghurt; yogurt
酸味 acid flavo(u)r; acid taste; sour taste; sourness
酸味水果棒糖 acid stick
酸味的 sour; vinaigrous
酸味軟糖料 acid fondant
酸味飲料 sour
酸味劑 acidifier
酸性 acidity; sourness
酸性白土 acid clay; acid earth

酸性品紅 acid fuchsin; acid magenta
酸性紅 acid red
酸性食品 acid food
酸性酒精 acid alcohol
酸性胺基酸 acid amino acid
酸性培養基 acid medium
酸性蛋白酶 acid proteinase
酸性焦磷酸鈣 calcium dihydrogen pyrophsophate
酸性焦磷酸鈉 disodium dihydrogen pyrophosphate
酸性紫 acid violet
酸性黃 acid yellow; curcumin; turmeric yellow
酸性煤焦色素 acidic coal tar dye
酸性酪乳 cultured buttermilk
酸性調味料 acid condiment
酸性糖化醪 sour mash
酸性靛藍 indigo carmine
酸性磷酸鈣 acid calcium phosphate; calcium acid phosphate
酸性磷酸〔酯〕酶 acid phosphatase
酸性磷酸鋁鈉 sodium aluminium phosphate, acidic (SALP)
酸性黏土 acid clay
酸性罐頭食品 acid canned food
酸果 tart fruit
酸果汁 verjuice
酸果蔓 cranberry
酸泡菜 sour pickle
酸的 acidic; acetous; sour; tart
酸度 acid degree; acidity; sourness
酸度計 acidimeter; acidometer; acidity meter
酸度常數 acidity constant
酸度測定 acid test
酸度調節劑 acidity regulator
酸度檢驗 acidity test
酸指數 acid index
酸毒症 acidosis
酸洗 acid cleaning; acid wash
酸洗不足 underpickling
酸致凝結 acid coagulation
酸值 acid number; acid value (A.V.)
酸修飾澱粉 acid modified starch
酸氣味 acid smell
酸消化 acid digestion
酸臭 acidic odor
酸臭氣 acid smell
酸酒酵母〔菌〕 *Saccharomyces aceti*
酸酐 acid anhydride
酸敗 acidification; rancidness; rancidification; rancidify; rancidity; sour rot
酸敗味 rancid flavo(u)r
酸敗的 rancid
酸液比重計 acidimeter; acidometer
酸處理澱粉 acid treated starch
酸煮解 acid digestion
酸番荔枝 guanabana; soursop
酸菜 pickle
酸越橘 huckleberry
酸量滴定法 acidimetry
酸黃瓜 pickled cucumber; sour cucumber; sour pickle
酸煉脫膠 acid degumming
酸解〔作用〕 acidolysis
酸過多的 superacid
酸酪乳奶油 sour cream butter
酸漬 acid pickling
酸漬菜 kraut; pickled vegetable; sauerkraut; sourcrout; sourkrout
酸漬黃瓜 cucumber pickle
酸腐 sour rot
酸腐的 sour
酸辣芹菜魚湯 court bouillon
酸辣醬油 vinaigrette sauce
酸僵直 acid rigor
酸價 acid number; acid value (A.V.)

酸模〔屬植物〕 sorrel
酸漿果 alkekengi
酸漿果紅素 physalin
酸漿莓 ground cherry
酸凝〔作用〕 acid coagulation
酸凝乳 clotted milk
酸凝乳〔塊〕 acid curd
酸凝試驗 acid coagulation test
酸凝酪 yoghourt; yoghurt; yogurt
酸橙 bigarade; *Citrus aurantium*; lime; seville orange; sour orange
酸橙汁 lime juice
酸橙飲料 limeade
酸橙樹 bigaradier
酸糖化 acid saccharification
酸糖比 acid-sugar ratio
酸澀水果 acerbic fruit
酸澀味 acerbity
酸澀性 harshness
酸轉化微晶糖膏 acid fondant
酸轉化澱粉 acid-converted starch
酸轉化糖漿 acid-converted syrup
酸醪 sour mash
酸麵糰 sour dough
酸櫻桃 amarelle; sour cherry
酸變性 acid denaturation
酸變性澱粉 acid treated starch
酸鹼度 pH
酸鹼度指示劑 pH indicator
酸鹼度計 pH meter
酸鹼度試紙 pH paper
酸鹼值 pH scale; pH value
酸鹼滴定 acid-base titration
酸—酶法〔澱粉〕糖漿 acid-enzyme syrup
銀 silver
銀大眼鯧 batfish; butterfish; kitefish
銀大麻哈魚 coho (salmon); jack salmon; river salmon; silver salmon
銀耳 jelly fungus; *Tremella fuciformis*
銀肉荳蔻 macassar nutmeg
銀杏 gingko; ginkgo; maidenhair tree fruit
銀杏黃素 ginkgetin
銀杏黃酮 ginkgetin
銀杏醇 ginnol
銀杏樹 maidenhair tree
銀芽 green bean sprout; silver sprout
銀栲皮樹 wattle
銀栲皮樹膠 wattle gum
銀魚 whitebait
銀製食具 flat silver
銀製餐具 silverware
銀鳶鯧 kitefish
銀盤 silver plate
銀器 silverware
銀器皿 silver plate
銀薄荷 *Mentha silvestris*
銀鮫 ratfish
銀鮭 coho (salmon); river salmon; silver salmon
銀邊肉 silverside
銀鯧 butterfish; pomfret; white pomfret
銀鰻 silver eel
銀鱈 silver cod
銅 copper
銅中毒 copper poisoning
銅葉綠素 copper chlorophyll
銅綠紅菇 verdigris russule
鉻 chromium
鉻酸鹽 chromate
銨化合物 ammonium compound
銨明礬 aluminium alum; aluminium ammonium sulfate
雌紅松雞 moorhen
雌魚 hen
雌禽 hen
雌綿羊 chilver
雌蘇格蘭雷鳥 moorhen
需氧細菌 aerophilic bacteria

需氧生活 oxybiosis
需氧的 aerobic
需氧培養 aerobic culture
需氧細菌 aerobic bacteria; oxybiotic bacteria
需氧發酵 aerobic fermentation
需氧菌 aerobe; aerobium; aeromicrobe
需氧菌種 aerobic culture
需氧微生物 aerobic microorganism
需氧醣〔酵〕解 aerobic glycolysis
領試工廠 pilot plant
領試工廠規模 pilot plant scale
領試工廠試產 pilot plant trial
餃子 Chinese dumpling; dumpling
餃形通心粉 tortellini
餅 cake
餅皮 crust
餅乾 biscuit; cooky
餅乾烤爐 biscuit oven
餅乾麵糰 biscuit dough; biscuit paste
餅乾罐 biscuit tin
餅餡 pie filling
骯髒的麵包 squalid loaf
鳶尾〔草〕 orris
鳶尾草根 orrisroot
鳶尾草根油 orrisroot oil
鳶尾淨油 orris absolute
鳶尾酮 irone
鳶尾凝脂 orris oil; orrisroot oil
鳳尾魚 anchovy; long tail anchovy
鳳尾魚油 anchovy oil
鳳尾魚醬 anchovy paste
鳳尾菇 phoenix mushroom
鳳果 mangosteen
鳳梨 ananas; pineapple
鳳梨切片機 pineapple slicer
鳳梨去皮芯機 ginaca
鳳梨去皮通心機 pineapple peeler-corer
鳳梨汁 pineapple juice
鳳梨果汁 pineapple juice
鳳梨扇片 pineapple tidbit
鳳梨酒 pineapple wine
鳳梨旋片 pineapple spiral slice
鳳梨甜酒 pineapple wine
鳳梨處理機 ginaca
鳳梨蛋白酶 bromelain; bromelin
鳳梨番石榴 feijoa
鳳梨圓片 pineapple whole slice
鳳梨溫室 pinery
鳳梨精油 pineapple oil
鳳梨螺片 spiral slice
鳳眼果 ping pong
鳳榴 pineapple guava
鼻腔 nasal cavity
鼻聞分析 sniff analysis
齊全膳食 full diet
嘌呤 purine
攢奶油 whipped cream
熏衣草 lavender
熏衣草油 lavender oil
熏衣草花油 lavender flower oil
箍桶 cooper
緋紅濕傘(蕈類) scarlet wax cap
蒟蒻 elephant foot; giantarum; konjac; konnyaku
蒟蒻甘露聚糖 konjac mannan
蒟蒻粉 konjac flour
蒟蒻聚甘露糖 konjac mannan
蒴果 capsule
蓴菜 water shield
蒔蘿 dill
蒔蘿利口酒 kummel
蒔蘿籽 dill seed
蒔蘿籽油 dill seed oil
蒔蘿烯 anethene
蒔蘿葉 dill weed
蒔蘿醛 cumaldehyde
蓖麻 castor; ricinus

蓖麻油 castor oil
蓖麻毒素 recin
蓖麻毒蛋白 ricin
蓖麻籽 castor bean; castor oil bean; castor seed; ricinus
蓖麻蛋白 ricin
蓖麻酸 ricinic acid; ricinoleic acid
蓖麻鹼 ricinine
銫 cesium
魟魚 skate
鵻鳩 turtle dove
蓴菜 water shield
酶 enzyme; ferment; zyme
酶元 proenzyme; zymogen
酶改性乳脂〔肪〕 enzyme-modified milkfat
酶的 emzymic
酶活化劑 enzyme activator
酶原 pre-enzyme; proferment; proenzyme; zymogen
酶處理澱粉 enzyme treated starch
酶解〔作用〕 enzyme digestion; enzymc-lysis; zymohydrolysis
酶解肌球蛋白 meromyosin
酶製劑 enzyme preparations; enzyme' s product
酶凝乳作用 renneting
酶凝乳塊 rennet curd
酶凝乾酪素 rennet casein
酶凝酪蛋白 rennet casein
酶學 enzymology
睪丸 testes; testicles
睪丸激素 testosterone
睪固酮 testosterone
榲桲 quince

十五畫

儀器 instrument
僵直 mortal rigor; rigor; rigor mortis
僵直的 stiff
僵直期 rigor phase
僵直熱 heat of rigor
僵硬的 stiff
價值 value
劇烈發酵 boiling fermentation
劈開 split; splitting
劈開兩半的魚 split fish
劈開機 splitter; splitting machine
劈頭機 heading cutter
劍豆 sword bean
劍魚 dolphin fish
劍嘴鯊 goblin shark
嘴 mouth
噴水池 spray pond
噴水器 sprinkling machine
噴水頭 spray ball; sprinkler head
噴汽封罐 steam flow closure; steam flow sealing
噴泉 fountain
噴洗 spray washing
噴紅乳菇 rufous milk cap
噴風式碾米機 pneumatic rice pearler
噴射式食物 aerosol food
噴射式熱風乾燥機 jet dryer
噴射型攪拌器 ejector type agitator
噴射洗滌機 spray washer
噴射混合器 jet mixer
噴射裝置 sprayer
噴射器 ejector
噴射攪拌器 jet agitator
噴粉器 duster
噴乾乳粉 spray-dried milk
噴淋式洗滌機 spray washer
噴淋池 spray pond
噴淋冷卻 shower cooling; spray cooling
噴淋冷卻器 spray cooler
噴絲 spinning
噴絲大豆蛋白 spinning soy protein
噴槍 spray gun
噴嘴 nozzle; spray head; spray nozzle; spraying nozzle
噴頭 nozzle; spray ball; spray head
噴霧 mist spray; spray
噴霧包〔冰〕衣 spray glazing
噴霧包覆 spray glazing
噴霧式巴氏殺菌器法 biorisation
噴霧〔作用〕 nebulization
噴霧冷凍 spray freezing
噴霧室 atomizing chamber; spray chamber
噴霧乾燥〔法〕 spray dehydration; spray drying
噴霧乾燥乳粉 spray-dried milk
噴霧乾燥的 spray-dried
噴霧乾燥造粒 spray drying granulation
噴霧乾燥塔 spray drying tower
噴霧乾燥機 spray drier; spray dryer

噴霧殺菌器 biorisator
噴霧造粒 spray granulation
噴霧嘴 atomizing nozzle
噴霧盤 atomizing disk; spray disc
噴霧劑 spray
噴霧器 atomizer; nebulizer; sprayer; spray atomizer; spray jet; spray
噴霧濕潤器 atomizing damper
噴霧罐 aerosol can
噴灑 sprinkle
噴灑器 sparger
增加物 additament
增色劑 colo(u)ring adjunct; saddening agent; toner
增味劑 flavo(u)r enhancer
增泡器 lather booster
增重 weight gain
增香乳 flavo(u)red milk
增香劑 flavo(u)ring agent
增效作用 potentiation
增效劑 potentiating agent; synergist
增強劑 enhancer; intensifier
增殖 proliferation
增殖培養 enriched culture
增殖培養基 enriched medium
增殖率 growth rate
增殖期 growth phase
增殖體 vegetation
增菌培養 enriched culture
增量劑 extender
增塑 plasticization
增塑劑 elasticizer; plasticizer; plasticizing agent; plastifier
增溶〔作用〕 solubilization
增溶溶解 solubilize
增溶劑 hydrotropic agent; solubilizer
增碘劑 iodizing compound
增稠 thickening
增稠劑 bodying agent; stiffener; stiffening agent; thickener; thickening agent; texturizing agent
增補劑 extender
增補礦物質 mineral supplementation
增積劑 bulking agent; bulking filler
增壓 pressurization
增濕 humectation; humidification; ordering
增濕房 ordering room
增濕劑 moistener
增濕器 humidor; humidifier; moistener
增黏劑 tackifier; tackifying agent
增體劑 bulking agent
寬突鱈 navaga
寬紋虎鯊 cat shark
寬條麵 fettuccine
寬葉薰衣草 aspic
寬鼻白鮭 broad whitefish
寬鰭鱲 Oikama minnow
寬鰭鱲魚 common minnow
層 layer
層板 ply
層板〔巴氏〕殺菌器 plate pasteurizer
層板加熱器 plate heater
層板冷卻器 plate cooler
層板冷凍〔法〕 plate freezing
層板冷凍機 plate freezer
層板乾燥器 plate dryer
層板接觸冷凍機 plate contact freezer
層板蒸發器 plate evaporator
層板熱交換器 plate (type) heat exchanger
層醃 kench curing; kench salting
層濾機 metafilter
履帶果汁萃取器 caterpillar juice extractor
廢水 effluent; spent water; sewage water; tail water; waste water
廢水池 waste lagoon
廢水氧化池 waste lagoon
廢水處理 waste water treatment

廢水管 waste pipe
廢汽 spent steam; waste steam
廢物 draff; garbage; offscouring; recrement; rubbish; sewerage; trash; waste
廢物利用 reclamation
廢物處理 waste disposal; waste treatment
廢品 reject; rejected product
廢料 garbage; reject; scrap; trash; tailings; waste; waste product
廢料回收再用 recovery
廢氣 discharge gas
廢棄物 offal
廢棄〔麵粉〕原料 dead meal
廢液 exhausted liquid; spent liquor
廢產物 waste product
廢渣 off scum; solid waste
廢蒸氣 exhaust steam
廢蒸氣加熱 exhaust steam heating
廢蜜紅〔葡萄酒〕 blackstrap
廢熱 waste heat
廢糖蜜 blackstrap molasses; black molasses; black treacle; exhausted molasses; final molasses; molasses; waste molasses
廢糖蜜漿 molasses mash
廚灶 cooking stove
廚具 kitchenware
廚房 cookroom; galley; kitchen
廚房必備食品 staple
廚房用具 kitchenware
廚房廢料或油垢 kitchen stuff
廚房器具 kitchen utensil
廚師 chef; cook; hasher; kitchener
廚櫃 spense
廣口瓶 jar
廣木香根油 costus root oil
廣用捲封機 universal seamer
廣用緩衝液 universal buffer
廣藿香 patchouli
廣藿香油 patchouli oil
廠內水源加氯 in-plant chlorination
廠內用水加氯消毒 in-plant chlorination
廠內的 in-plant
彈力 elasticity
彈力計 elastometer
彈式熱量計 bomb calorimeter
彈性 elasticity; springiness
彈性的 elastic
彈性計 elastometer
彈性〔強度〕 resiliency
彈性蛋白 elastin
彈性蛋白原 tropoelastin
彈性增進劑 elasticizer
彈性凝膠 elastogel
彈性纖維 elastic fiber
彈性變形 elastic deformation
彈塗魚 mud skipper
彈簧 spring
彈簧秤 spring balance
彈簧蓋廣口瓶 spring-top jar
德氏乳〔酸〕桿菌 *Lactobacillus delbruckii*
德式酸泡菜 kraut; sauerkraut; sourcrout; sourkrout
德洛克豬 Duroc (Jersey)
德國布氏肝腸 braunschweiger
德國布勞恩施魏克香腸 braunschweiger
德國白葡萄酒 hock
德國式油煎小香腸 bratwurst (sausage)
德國式燻肝香腸 braunschweiger
德國色查礦泉水 seltzer water
德國粗香腸 bockwurst sausage
德國葡萄酒 German wine
德國賽爾脫茲礦泉水 seltzer water
徵候群 syndrome
慕尼黑啤酒 munchen(er) beer; Munich beer
慕尼黑麥芽 Munich malt
慕思(泡沫狀奶油甜點) mousse

憔悴 emaciation
摩卡咖啡 caffe mocha; mocha
摩西果醬 mossberry
摩姆啤酒 mum
摩泰臺拉香肚(生燻) mortadella
摩爾溶液 molar solution
摩爾濃度 molar concentration; molarity
摩澤爾白葡萄酒 Moselle
摩擦去皮〔法〕 friction peeling
摩擦去皮機 friction peeler
撞擊破碎機 impact breaker
撞擊粉碎機 impact crusher; impact grinder
撞擊攪乳器 dash churn
撞擠罐 impact extruded can
撞錘 ram
撈圈罐 beaded can
撕開蓋 tear-off lid
撕開罐 tear open can
撕碎 shredding
撕碎機 shredder; shredding machine
撒 sprinkle
撒〔上麵〕粉 flour
撒粉 dust; dusting
撒粉器 duster; sifter
撒能山羊 Saanen
撒糖粉 dust; dusting
撒糖餅乾 galette
撒鹽 rouse
撒鹽器 salt shaker
播種 seedling
播種者 seeder
播種機 seeder
撬開〔式〕蓋 pry-off cap
撬蓋器 lid flipper
敷粉 dusting powder
數量 quantity
暴食 glut; gluttony; gorge
暴食者 glutton; swallower
暴氣槽 aerator
暴飲暴食 engorgement; gormandise; gormandize; gourmandise; gourmandize
樣本 specimen
樣品 sample; specimen
樟腦 camphor
標本 sample; specimen
標貼 sticker
標準 standard
標準化 standardization
標準以下的 substandard
標準包裝 standard pack
標準平板計數 standard plate count
標準曲線 standard curve
標準狀況 standard condition
標準的 standard
標準烈度酒 proof spirit
標準烈酒加侖 proof gallon
標準偏差 standard deviation (SD)
標準培養〔物〕 type culture
標準條件 standard condition
標準規格 standard specification
標準菜單 standard fare
標準飲食 standard diet
標準溶液 standard solution
標準試樣 standard sample
標準誤差 standard error (SE)
標準樣品 standard sample
標準篩 standard screen
標準篩制 standard screen scale
標準篩網尺度 standard screen scale
標誌 tag
標籤 label; labelling; tab; tag
槽 tank
槽的容量 tankage
槽紋面中粗〔管狀〕通心粉(1/2英吋粗) zitoni rigati
模子 shaper
模具 mold; mould

模烤長方形麵包 pan loaf
模烤麵包 pan bread
模製餡餅 timbale
模製麵包 tin bread
模壓火腿 pressed ham
模擬食品 simulated food
槳式混合器 blade mixer; paddle mixer
槳式攪拌器 blade agitator; blade stirrer; paddle mixer; paddle (type) agitator; paddle stirrer
樂園堅果 paradise nut
槭樹 maple
槭樹液 maple sap
槭〔樹〕糖 maple sugar
槭〔樹〕糖蜜 maple honey
槭〔樹〕糖漿 maple syrup
歐〔陸〕式早餐 Continental breakfast
歐芹 parsley
歐芹籽油 parsley seed oil
歐芹莖油 parsley herb oil
歐南瓜 marrow
歐洲〔大〕楓樹 sycamore
歐洲山楂 medlar
歐洲李 common plum
歐洲防風草 parsnip
歐洲花楸 rowan
歐洲紅莓 red raspberry
歐洲栗子 European chestnut
歐洲海蓬子 samphire
歐洲淡水鱸魚 barfish
歐洲甜櫻桃 mazzard cherry; sweet cherry
歐洲細辛 European wildginger
歐洲野生薑 European wildginger
歐洲越桔 bogberry
歐洲越橘 whortleberry
歐洲黑莓 black berry
歐洲榛 nut bush; nuttree
歐洲榛子 cob nut; European filbert
歐洲酸櫻桃 marasca (cherry); maraschino cherry; pie cherry
歐洲龍蝦 langouste
歐洲薄荷 summer savory
歐洲薄荷油 savory oil; savoury oil
歐洲鯽魚 crucian (carp)
歐洲鯷 sardelle
歐洲鰻鱺 European eel
歐洲鹽乾鱈 klipfish
歐蒔蘿 cumin; cummin
歐根香瓜 ogen
歐斯塔魚子醬 osetr caviar
歐渣〔果〕 medlar
歐蒔蘿籽 cummin seed; cumin seed
歐撒橘 Osage orange
歐蝶魚 American plaice
歐鱘 Oregon sturgeon
漿 brei; paste; plasma
漿果 bacca; berry; sap fruit; small fruit
漿果派 berry pie
漿狀的 pasty
漿〔料〕 pulp
漿料 size
漿液 sera; serum
漿糊 starch paste
漿糊狀 starchiness
漿麵糊 panada
澄粉 protein-free wheat starch
澄清 clarification; defecation; fining; lauter; settle
澄清汁 subsider juice
澄清池 settling pond
澄清果汁 clarified juice; clear juice
澄清的 settling
澄清槽 clarifying tank; defecator; separating tank
澄清劑 clarificant; clarifier; clarifying agent; clearing agent; fining; fining agent
澄清器 clarifier; defecator; settler

澄清糖蜜 clarified molasses
澄麵 tang flour
潔白濕傘(蕈類) snowy wax cap
潔冷膠 gellan gum
潛水者 ducker
潛伏 incubation
潛伏期 incubation period; latent period
潛伏感染 latent infection
潛熱 latent heat
潮州柑 ponkan
潮解 air slake; air slaking; deliquescence
潮霉素 hygromycin
潮濕 dampness; wateriness
潮濕的 madid; moist; moisty
潮濕醃製 moist curing
潰瘍 ulcer; ulceration
潰爛 ulceration
潤滑 lubrication
潤滑油 lubricating oil
潤滑脂 grease; lubricating grease
潤滑劑 emollient; lubricant
潤滑器 lubricator
潤濕 humectation; madefaction; ordering; remoisten
潘〔諾〕糖 panose
潘胺酸 pangamic acid
潘趣飲料 bumbo; punch
熟化槽 aging vat
熟化劑 maturing agent
熟火腿 boiled ham; cooked ham
熟可可粒 cacao nib; cocoa nib
熟石灰 calcium hydroxide; slaked lime
熟成 ageing; aging; conditioning; cure; curing; ripening
熟成牛肉 aged beef
熟成肉 aged meat
熟成芝士 aged cheese; cured cheese
熟成室 curing room
熟成風味 aged flavo(u)r
熟成乾酪 aged cheese; cured cheese
熟成期 curing period
熟羊後腿〔肉〕 gigot
熟肉 cooked meat
熟肉製品 cooked meat product
熟油 boiled oil
熟食 delicatessen
熟食店 delicatessen (shop); food shop
熟食的葉用菜 potherb
熟食品 ready-to-eat food; ready-to-serve food
熟香腸 cooked sausage
熟啤酒 pasteurized beer
熟嫩程度 degree of doneness
熟麴 matured koji
熬 decoct; decoction
熬出物 decoction
熬汁 extractive
熬油 render(ing)
熬煮硬糖 boiled sweet
熬糖鍋 boiling pan; sugar boiler; sugar pan
熬鹽場 boilery
熱 heat
熱力學 thermodynamics
熱不穩定性 thermolability
熱不穩定的 thermolabile
熱分解〔作用〕 thermal decomposition; thermolysis
熱水瓶 thermos (bottle); thermos flask; vacuum bottle; vacuum flask
熱水燙毛缸 scalder
熱牛奶咖啡 cappuccino
熱充填 hot filling
熱加工 heat processing
熱包裝 hot pack(ing)
熱去皮 thermopeeling
熱平衡 heat balance
熱生活 heating inactivation

熱交換 heat exchange
熱交換器 heat exchanger
熱休克 heat shock; thermal shock
熱回收換熱器 regeneration heat exchanger
熱收縮模 heat-shrinkage film
熱老化 heat ag(e)ing
熱含量 heat content
熱沖擊 thermal shock
熱函 enthalpy
熱固性樹脂 thermosetting resin
熱固的 thermoset
熱析法 calorimetry
熱狗 hot dog
熱狗小腸 hot dog
熱的 hot
熱空氣烤箱 air oven
熱阻 thermal resistance
熱封 heat seal(ing); hot seal
熱封口的袋子 heat sealable bag
熱封機 heat sealer
熱穿透 heat penetration
熱致死時間 thermal death time (TDT)
熱致死時間曲線 thermal death time curve (TDT curve)
熱致死率 thermal death rate (TDR)
熱致死速度 thermal death rate (TDR)
熱致死點 thermal death point (TDP)
熱重 hot weight
熱風烘烤 air cooking
熱風烘箱 hot-air oven
熱風烤爐 hot-air oven
熱風乾燥〔法〕 hot-air drying
熱風乾燥機 hot-air dryer
熱風烹煮 air cooking
熱風滅菌〔法〕 hot-air sterilization
熱風滅菌器 hot-air sterilizer
熱風機 air heater
熱值 caloric value; calorie value; calorific value; energy value
熱容量 heat capacity; thermal capacity
熱氧化 heat oxidation
熱氧化大豆油 thermally oxidized soya bean oil (TOS)
熱烤去皮 thermopeeling
熱脈衝封口 impulse seal(ing)
熱能 heat energy; thermal energy
熱能回收器 recuperator
熱能食品 energy food
熱帶水果 tropical fruit
熱帶作物 tropical crop
熱情果 granadilla; parchita; passionfruit
熱敏性物質 thermosensitive substance; thermosensitive food
熱敏的 thermosensitive
熱敏感的 heat sensitive
熱殺菌 heat processing
熱處理 heat treatment; heat processing; thermal treatment
熱蛋啤酒 egghot
熱陳化 heat ag(e)ing
熱單位 caloric unit
熱絕緣 heat insulation
熱脹〔現象〕 hot swelling
熱脹罐 hot swell can; hot swelling
熱量平衡 heat balance
熱量所需量 aequum
熱量計 calorimeter
熱量計彈筒 calorimetric bomb
熱量產生 heat production
熱量單位 caloric unit
熱量測定 calorimetry
熱鈍化 heating inactivation
熱傳 heat transfer
熱傳係數 thermal conductivity
熱傳導 heat conduction; thermal conduction
熱塑性 thermal plasticity
熱塑性的 thermoplastic
熱損傷 heat damage

熱損壞 heat damage
熱滅率 thermal death rate (TDR)
熱滅點 thermal death point (TDP)
熱裝 hot filling; hot pack(ing)
熱裝〔罐〕 heat pack
熱裝填 warm filling
熱解〔作用〕 pyrolysis; thermolysis
熱電偶 thermocouple
熱電偶溫度計 thermocouple thermometer
熱對流 heat convection; thermal convection
熱熔接合袋 heat sealable bag
熱聚合 heat polymerization
熱熟成 heat ag(e)ing
熱蔗糖鹽 hot saccharate
熱醋醃 hot marinating
熱醃〔法〕 hot curing; thermal cure; ther-mocuring
熱震 thermal shock
熱凝固〔作用〕 thermal coagulation; thermocoagulation
熱凝固的 thermoset
熱凝結 heat coagulation
熱凝聚〔作用〕 thermocoagulation
熱凝膠 thermal gel
熱燙 blanch(ing); scald; scalding
熱燙去皮機 scalder-peeler
熱燙拔毛 scald picking
熱燙槽 scalding tank
熱燙機 scalder
熱膨脹 thermal expansion
熱輻射 thermal radiation
熱壓榨 hot pressing
熱擴散 thermal diffusion
熱燻〔法〕 hot smoking
熱燻製〔法〕 hot smoking
熱雞蛋酒 egg-flip; egghot
熱穩定化〔作用〕 thermostabilization
熱穩定性 thermal stability; thermostability
熱穩定的 heat-fast; thermostable
熱變形 thermal degradation; heat denaturation
熱變性 thermal denaturation
熱鹽醃〔法〕 hot salt curing
犛牛 yak
瘤牛 zebu
瘤胃 rumen; tripe
瘦 leanness; thinness
瘦肉 lean; lean meat; muscle
瘦肉型 meat type
瘦肌肉 lean muscle
瘦果 achene; akene
瘦的 fleshless; lean; rawboned; thin
瘦弱 emaciation
瘦魚 logey
瘦穀粒 hollow kernel
皺皮瓜 rock melon
皺皮果 rivelled fruit
皺皮種青豌豆 wrinkled pea
皺皮豌豆 wrinkled pea
皺胃 abomasum; runnet; true stomach
皺紋鮑〔魚〕 pink abalone
皺馬鞍菌（蕈類） common white saddle
皺葉包心菜 savoy cabbage
皺葉甘藍 savoy cabbage
皺葉卷心菜 savoy cabbage
皺葉芥菜 curled mustard
皺葉萵苣 curled lettuce
皺環球蓋菇 burgundy slime head
皺縮 shrivel
皺褶羅鱗傘（蕈類） gypsy
皺鰓鯊 frill shark
盤 dish; plate; salver; tray
盤尼西林 penicillin
盤打酥油 panda
盤式切碎機 disc shredder
盤式比色計 disc colorimeter
盤式烤爐 tray oven

盤式粉碎機 disc pulverizer
盤式乾燥 pan drying
盤式乾燥器 disc dryer; pan dryer; tray dryer
盤式噴霧 driving spray
盤式噴霧器 disc atomizer
盤式磨 disc mill
盤式篩選機 plansifter
盤式輸送機 tray conveyer
盤式離心蒸發器 disc evaporator
盤托 pan arrest
盤架乾燥器 shelf dryer
盤烤麵包 hearth bread
盤菜 dish; tray meal
盤菇 rubber brush
盤碗離心機 disc bowl centrifuge
盤裝菜 dish
盤管式熱交換器 coil-type heat exchanger
盤管冷凝器 coil condenser
盤管蒸發器 coil evaporator
盤製乳油 pan-set cream
盤磨 pan milling
盤磨機 pan mill
磅 pound
磅蛋糕 pound cake
磅餅 pound cake
確認 verification
碾白大麥 milled barley
碾米 rice pearling
碾米機 polishing machine; rice mill; rice milling machine; rice polisher; rice polishing machine; rice sweep machine; rice sweeper
碾米機的砂輥 abrasive roller
碾麥機 scourer; scouring mill
碾碎 crush; pan milling
碾碎的乾小麥 bulgur
碾碎機 chaser mill; crushing mill; pan mill
碾槌 pestle
碾穀機 grain mill; pearling mill; rice huller; rice hulling machine; rice husker
碾磨 mill milling; trituration
碾磨面 mill surface
碾磨業〔者〕 miller
碾壓 roller compaction
穀仁油 seed kernel oil
穀乳 cereal milk
穀刺 awn
穀固醇 sitosterin; sitoterol
穀物 grain
穀物加工品 cereal product
穀物毒素 sitotoxin
穀物粉末 pulverized grain
穀物乾燥機 seed drier; seed dryer
穀物脫粒機 grain thresher
穀物飼料 grain feed
穀物澱粉 cereal starch
穀甾醇 sitosterin; sitoterol
穀倉 barn; bin; garner; grain bin; grain elevator; granary
穀氨酸 glutamic acid
穀氨酰胺 glutamine
穀粉粥 groats
穀粒 cereal grain; grain; rice grain
穀蛋白 glutelin
穀殼 chaff
穀殼吸風分離器 hull aspirator
穀殼機 chaffer
穀蜀黍 grain sorghum
穀種 seed corn
穀漿 cereal milk
穀膠蛋白 gliadin; mucedin
穀糠 bran coat; chaff
穀類 cereal; grain
穀類作物 cereal crops; grain crop
穀類食品 cereal; cereal food

穀類製品 cereal product
穀蠹 lesser grain borer; weevil
稷 millet; proso millet
稻 paddy; rice
稻田 paddy field
稻米脱殼機 rice scourer
稻草 haulm; straw
稻粒 rice grain
稻殼 husk; paddy; paddy rice; rice hull; rice husk; rough rice
稻穀脱殼機 rice huller; rice hulling machine; rice husker
稻霉素 orymycin
稻屬 *Oryza*
窰 kiln
窰式乾燥 kiln dry(ing)
窰式乾燥機 kiln dryer
窰乾 kiln dry(ing)
窰乾麥芽 kilned malt
箭魚 broadbill swordfish; swordfish
箭葉黃體芋 yautia
箭豬 porcupine
箱 tank
箱型乾燥機 bin dryer; box dryer; cabinet dryer; chamber dryer; chest drying machine
箱型壓榨機 box press
糊 paste; spread
糊化 gelatinate; gelatinize
糊化〔作用〕 gelation; gelatinization
糊化程度 gelatinization degree
糊化溫度 gelatinized point
糊化澱粉 gelatinization starch; gelling starch
糊化點 gelatinized point
糊化麵粉 scalded flour
糊狀物 paste; pasty mass
糊狀的 pasty
糊狀食物 mush; paste (like) food
糊粉 aleurone
糊粉粒 aleurone grain
糊粉細胞 aleurone cell
糊粉測定儀 aleurometer
糊粉層 aleurone layer
糊精 dextrin(e); starch gum
糊精化〔作用〕 dextrinization
糊精膠 dextrin(e) gum
糊精酶 dextrinase
糊劑 poultice
緬甸墨吉對蝦 banana prawn
緣膜 velum
線紋通心粉 rigatoni
線粒體 mitochondria
線蟲 eelworm; nematode
緩沖劑 buffer
緩發酵元 slow starter
緩慢發酵 sluggish fermentation
緩慢凝結 long setting
緩衝板 baffle (plate)
緩衝能力 buffer capacity
緩衝液 buffer; buffer solution
緩衝劑 buffer
緩衝器 buffer
緩凝 long setting; slow set
緩凝乳〔凝〕塊 long-set curd
緩凝果膠 slow set pectin
緩凝劑 retarder
膜 membrane
膜通透性 membrane permeability
膜過濾 membrane filtration
膜滲透性 membrane permeability
膜濾器 membrane filter
膠 gum
膠〔水〕 glue
膠化 gelatinize; gelatinate; gelatinization; gelation
膠化形成 micellisation

膠化劑 gelling agent; gelatinizer; gelatinizing agent; jelling agent
膠木膠 gutta-percha
膠合物 agglutinant
膠束 micella(e); micelle
膠乳 rubber latex
膠姆糖基礎劑 chewing gum base components
膠狀的 jelly-like
膠度計 jelly tester
膠凍 jell; jello; jelly
膠凍狀的 jelly-like
膠凍粉 jelly powder
膠凍強度測定儀 jelly tester
膠凍軟糖 jelly
膠原蛋白 collagen
膠原蛋白酶 collagenase
膠原腸衣 collagen casing
膠原〔質〕 collagen
膠原纖維 collagen fiber
膠料 size; sizings
膠圈 gasket
膠強度 jelly strength
膠粒 micella(e); micelle
膠韌質地 rubbery texture
膠溶〔作用〕 solation; peptization
膠溶劑 peptizator
膠溶體 peptizate
膠態離子 micella(e); micelle
膠質的 colloidal; gelatinous
膠凝 gel
膠凝力 gelling power; jellying power
膠凝化 gelatinate; gelatinize
膠凝〔作用〕 gelatinization; gelation; gelling effect; jellification; pectisation; pectization
膠凝性〔能〕 gel forming property; gelling property
膠凝〔狀〕物質 gelatinous substance
膠凝食品 gelled food(stuff)
膠凝效應 gelling effect
膠凝強度 gelling strength
膠凝強度計 gelometer
膠凝溫度 gel point; gelling point
膠凝劑 gelling agent; gelatinizer; gelatinizing agent; jelling agent
膠凝點 gel point; gelling point; jellying point
膠樹脂 gum resin
膠磨機 colloid mill
膠糖 gum sugar
膠黏乳 stringy milk
膠黏性 gummosity; tackiness
膠黏的 gummy
膠黏脱毛化合物 adhesive depilating compound
膠黏劑 adhesive; tackifier; tackifying agent
膠囊 capsule
膠囊化 encapsulation
膠囊〔包裹〕法 capsulating process
膠體 colloid
膠體的 colloidal
膠體溶液 colloidal solution
膠體滲透壓 oncotic pressure
膠體磨 colloid mill
膠體懸浮液 colloidal suspension
蔗汁 cane juice; sugar cane juice
蔗汁比例 java ratio
蔗汁比率 java ratio
蔗果三糖 kestose
蔗渣 bagasse; cane trash; megasse
蔗渣板 bagasse board; celotex
蔗渣屑 cushcush
蔗節 cane joint
蔗糖 cane sugar; colonial sugar; saccharobiose; saccharose; sucrose; saccharum; sugar

蔗糖乙酸異丁酸酯 sucrose acetate isobutyrate (SAIB)
蔗糖二硬脂酸酯 sucrose distearate
蔗糖不耐症 sucrose intolerance
蔗糖化合物 sucrate
蔗糖化學 sucrochemistry
蔗糖甘油酯 sucroglyceride
蔗糖脂肪酸酯 sucrose esters of fatty acids (SE); sucrose fatty acid ester
蔗糖單硬脂酸酯 sucrose monostearate
蔗糖酯 sucrose ester
蔗糖酸鈣 calcium saccharate
蔗糖廠 cane mill
蔗糖漿 cane syrup
蔗糖磷酸化酶 sucrose phosphorylase
蔗糖酶 invertase; saccharase; sucrase
蓮 lotus
蓮子 lotus nut; lotus seed
蓮根 lotus root
蓮蓉 lotus nut paste; lotus seed paste
蓮蓬頭 sprinkler head
蓮藕 lotus root; lotus rhizome; lotus tuber
蓮霧 wax apple; wax jambo
蔬果泥 mashed vegetable
蔬果保鮮室 vegetable compartment
蔬菜 garden sass; green; green stuff; green vegetable; greengrocery; vegetable; vegetal; verdure
蔬菜切絲機 vegetable shredder
蔬菜水果店 greengrocery
蔬菜水果販 greengrocer
蔬菜汁 vegetable juice
蔬菜皮 peel
蔬菜作物 truck crop
蔬菜沙拉 vegetable salad
蔬菜沙律 vegetable salad
蔬菜性食品 vegetable food
蔬菜凍 macedoine
蔬菜栽培〔學〕 olericulture
蔬菜脫水機 vegetable dehydrator
蔬菜通心粉 vegetable dyed macaroni
蔬菜通心粉湯 minestrone
蔬菜湯 vegetable soup
蔬菜絲 julienne
蔬菜絲肉湯 julienne
蔬菜菜餚 vegetable dish; vegetable plate
蔬菜園 vegetable garden
蔬菜園藝 olericulture
蔬菜盤 vegetable server
蔬菜燉羊肉 haricot
蔬菜燴肉 mulligan
蔬菜〔類〕 garden stuff
蔬菜罐頭 canned vegetable
蔓 runner; vine
蔓生作物 vine crop
蔓越桔 bogberry
蔓越橘 cranberry; mossberry
蓬莪蒁 zedoary root
蓬萊米 japonica (type rice)
蓬萊蕉 ceriman
蝶酰谷胺酸 folic acid; pteroylglutamic acid
蝴蝶形通心粉 bow tie pasta; farfalla; farfalle
蝴蝶鰩 butterfly ray
蝶酰麩胺酸 folic acid; pteroylglutamic acid
蝦子 shrimp
蝦仁 peeled shrimp; shrimp meat
蝦片 prawn crisp
蝦白脫 shrimp butter
蝦夷扇貝 comb shell
蝦夷葱 chive
蝦米 dried shrimp
蝦〔米〕乾 dried shrimp
蝦肉 shrimp meat
蝦乳酪 shrimp butter

蝦青素 astaxanthin
蝦紅素 astacin
蝦粉 shrimp meal
蝦蛄 mantis shrimp; squilla
蝦黃質 astaxanthin
蝦腸 vein
蝦膏 blacan; blachan; shrimp paste
蝦濃湯 shrimp bisque
蝦醬 blacan; blachan; shrimp paste
蝸牛 escargot; snail
蝸牛通心粉 lumache medie
蝸形管 worm pipe
衛生 hygiene; sanitation
衛生安全 wholesomeness
衛生的 hygienic; sanitary; wholesome
衛生指標細菌 sanitary indicator bacteria
衛生條件 sanitary condition
衛生細菌學 sanitary bacteriology
衛生處理 sanitization
衛生設備 sanitation
衛生稽查員 sanitary inspector
衛生學 hygiene; hygienics; sanitary science
衛生學的 hygienic
衛生學家 hygienist
衛生檢驗 sanitary inspection
衛生環境 sanitary condition
衛生罐 open-top can; sanitary can
衛矛〔己六〕醇 melampyrin
衛矛醇 dulcin(e)
衛矛糖 dulcitose
衝壓機 ram
褐色貝蒂布丁 brown betty
褐色炒麵粉 brown roux
褐色海草 brown seaweed
褐色海藻 brown algae; brown seaweed
褐色著色劑 brown colo(u)r
褐疣柄牛肝菌 brown birch scaber stalk
褐梅鯛 caesio
褐魚粉 brown fish meal
褐斑點 scald
褐絨蓋牛肝菌 bay bolete
褐圓孔牛肝菌 chestnut bolete
褐腐病 brown rot
褐蝦 sand shrimp
褐環乳牛肝菌 slippery jack
褐藻 brown seaweed; brown algae
褐藻〔門〕 phaeophyta
褐藻酸 algin; alginic acid
褐藻酸丙二醇酯 propylene glycol alginate
褐藻酸鈣 calcium alginate
褐藻酸鈉 sodium alginate
褐藻酸鉀 potassium alginate
褐藻酸銨 ammonium alginate
褐藻酸鹽 alginate
褐藻酸酶 alginase
褐藻膠 algin
褐藻類製的蜜餞 seatron
褐變 browning
褐變反應 browning reaction
褐鱒 brown trout
複合物 complex
複合食品 composite food
複合脂質 complex lipid; conjugative lipid
複合培養基 complex medium
複合蛋白質 conjugated protein
複合維生素 vitamin complex
複合維生素B B complex; vitamin B complex
複合調味料 complex seasoning
複合凝膠 plural gel
複合薄膜 laminated film
複合麵粉 composite flour
複式包裝 multipack; multiple package
複式包裝機 multipackaging machine
複式混合器 multimixer
複果 multiple fruit
複查 reinspection
複製〔法〕 replica

複製品 replica
複層袋 multiwall bag
複檢 reinspection
複驗 reinspection
複體 complex
諸葛菜 root mustard
課程 course
調合 blend; blending; concoction
調合果汁 blended juice
調合物 blend
調合的調味料 blended seasoning
調合糖漿 syrup blend
調色玉米澱粉 custard powder
調色劑 toner
調味 season; seasoning
調味午餐肉 spiced luncheon meat
調味汁 dip
調味用草本香料 kitchen herb
調味肉泥 pate
調味乳 flavo(u)red milk
調味品 condiment; relish; seasoner; spice
調味品瓶 crewet; cruet; muffineer; seasoning jar
調味品樽 crewet; cruet
調味食品 flavo(u)r food
調味料 condiment; flavo(u)ring; flavo(u)ring agent; flavo(u)ring material; flavo(u)ring substance; food flavo(u)ring; sapid substance; sass; sauce; seasoner; seasoning
調味料浸漬 marinate
調味料瓶 castor
調味料瓶架 caster
調味瓶 castor; caster
調味蛋黃漿 remoulade
調味番茄醬 tomato catchup; tomato catsup; tomato ketchup; tomato sauce
調味飲料 flavo(u)red drink
調味碎肉 forcemeat
調味辣椒醬 chil(l)i sauce
調味醋醬 vinaigrette sauce
調味劑 flavo(u)ring agent
調味醬 dressing; sass; sauce
調和 tempering
調查 research
調氣包裝 modified atmosphere packaging
調氣保藏 modified atmosphere storage (MA storage)
調脂〔水〕乳 toned milk
調酒〔長柄〕匙 bar-spoon
調酒師 bartender
調控大氣 controlled atmosphere
調理 preparation
調理食品 prepared food
調溫 attemperation; tempering
調溫室 tempering room
調溫裝置 attemperator
調溫槽 attemperating tank
調溫器 attemperator
調節 conditioning
調節劑 modifier
調節器 regulator; temperer
調製 concoction
調製乳 modified milk; recombined milk
調漿機 paste mixture
調質劑 texturizer; texturizing agent
調整乳 toned milk
調整膳食 modified diet
調壓閥 pressure regulating valve
調壓器 pressure regulator
調濕 tempering
調濕小麥 tempered wheat
調濕室 tempering room
調濕穀倉 tempering bin
調濕機 damping machine
調羹 spoon
調麵機 dough mill; doughing machine
豌豆 garden pea; *Pisum sativum*

豌豆〔仁〕 green pea
豌豆去莢機 pea huller; pea podder
豌豆粉 pea flour
豌豆素 pisatin
豌豆脱粒機 pea huller
豌豆脱殼機 pea huller
豌豆莢 pea pod
豌豆 pea; pease
豌豆球蛋白 vicilin
豌豆澱粉 pea starch
豎管〔式〕蒸發器 vertical-tube (type) evaporator
豎鰭 vertical fin
豬 pig; swine
豬〔後〕臀肉 ham butt
豬內臟 harslet; haslet
豬皮 hog skin; pork skin
豬耳 hog ears
豬肉 pig; pork
豬肉香腸 pork sausage
豬肉貢丸 minced pork ball
豬肉絲 pork floss
豬肉條蟲 pork tapeworm
豬肉餡餅 pork pie
豬肋肉 side pork
豬舌 pork tongue
豬血腸 boudin
豬肝肉丸 faggot
豬肘 hock
豬肚 pork stomach
豬裏脊肉 griskin; loin butt
豬油 lard; pork fat
豬油果 lard fruit
豬油的 lardy
豬油的固態部分 lard stearin
豬油的液態部分 lard olein
豬油粕 pork scrap
豬油渣 crackling; cracknel; crisp fat; lard dreg
豬油壓濾棧 lard filter press
豬的 porcine
豬直腸 bung; hog bung
豬肩〔上〕端肉 shoulder butt (pork)
豬肩肉 picnic; picnic shoulder
豬肩背肉 Boston butt
豬肩胛肉 Boston butt
豬前腿肉 hock
豬前蹄 fore feet
豬後腿剝皮 ham skinning
豬柳 pork fillet
豬胃 pork stomach
豬食 hogwash; swill
豬脂 lard
豬〔脂〕油 lard oil
豬脂〔肪〕 hog fat
豬胸肉 side pork
豬胸肋肉 rib belly
豬屠體 hog carcass; pork carcass
豬排 griskin; pork chop
豬結腸 hog middle
豬群 swinery
豬腰肉排 pork loin steak
豬腰部瘦肉 griskin
豬腸衣 hog casing
豬腳 pettitoes
豬腹部肉 belly; belly meat
豬腿排 pork leg steak
豬嘴 snout
豬蹄 pettitoes
豬頭皮 hog face
豬頭皮凍 brawn
豬頭肉凍 head cheese
豬頭雜凍 head cheese
豬臀肉排 pork rump steak
豬雜碎 harslet; haslet
豬顎 chaps
豬鬃 hog bristle
豬欄 swinery

賦形劑 forming agent
賣主 vender; vendor
質粒 plasmid
質量 quality
質量低 undergrade
質量保證 quality assurance (QA)
質量控制 quality control (QC)
質感化 bodying
質感化劑 bodying agent
質體 plasmid
赭麴霉素 ochratoxin
踐踏 tread
踝 ankle
輪作 crop rotation
輪枝小皮傘(蕈類) common wheel mu-mmy cap
輪狀薄糖果 jumbal; jumble
輪耕 crop rotation
輪碾機 edge mill; edge runner; wheel mill
輥式軋碎機 roller crusher
輥式粉碎機 roller breaker
輥式榨油機 Scotch hands
輥式輸送機 roller conveyor
輥軋機 roller mill; rolling mill
輥筒 roll
輥碎機 roll crusher
輥輪 roller
輥壓機 roller press
適冷細菌 psychrotrophic bacteria
適旱生物 xerophile; xerophyte
適旱的 xerophilic
適旱細菌 xerophilic bacteria
適於飲用的 potable
適當膳食 adequate diet
適酸的 acidophilous
適壓細菌 barophilic bacteria
適壓微生物 barophile
適應酶 adaptive enzyme
適鹽生物 halobiont
遮目魚 milkfish
鄰甲氧基苯酚 guaiacol
鄰甲氨基苯甲酸甲酯 dimethyl anthranilate
鄰甲硫基苯酚 *o*-methylthiophenol
鄰苯二酚 pyrocatechol
鄰苯基苯酚 *o*-phenyl phenol (OPP)
鄰苯基苯酚鈉 sodium *o*-phenylphenolate
鄰氨基苯甲酸乙酯 ethyl anthranilate
鄰氨基苯甲酸甲酯 methyl anthranilate
鄰氨基苯甲酸苯乙酯 phenethyl anthranilate
鄰氨基苯甲酸桂酯 cinnamyl anthranilate
醇化 alcoholization; mellow
醇血症 alcoholemia
醇酒 ripe wine; velvet spirit
醇脫氫酶 alcohol dehydrogenase
醇溶的 alcohol-soluble
醇溶〔穀〕蛋白 prolamin(e)
醇溶膠 alcohol-sol; alcosol
醇解 alcoholysis
醇酸 alcoholic acid
醇酸樹脂 alkyd
醇酸澱粉 alkyd starch
醇凝膠 alcogel
醇酶 alcoholase
醉瓜 nutmeg melon
醉蛋糕 tipsy cake
醋 vinegar
醋化 acetify
醋化作用 acetification
醋化器 acetifier; acetator
醋母 mother vinegar; vinegar mother
醋泡菜 vinegar pickle
醋的 acetic; vinegary
醋栗 gooseberry; goosegog; Indian gooseberry; ribes

醋栗蕃茄 currant tomato
醋酐 acetic anhydride
醋蛆 vinegar eel
醋粟 emblic
醋漬 pickle; pickling; vinegar cure; vinegar pickling
醋漬食品 acetic acid preserves; pickle; pickled food; vinegar pickled product
醋漬液 marinade; vinegar pickle
醋漬菜 pickled vegetable
醋漬槽 pickling tank
醋精 acetin; glyceryl monoacetate
醋酸 acetic acid; ethanoic acid
醋酸乙酯 acetic ether; ethyl acetate
醋酸比重計 acetimeter; acetometer
醋酸生育酚 tocopheryl acetate
醋酸甲酯 methyl acetate
醋酸含量測定法 acetimetry
醋酸和脂肪酸甘油酯 acetic and fatty acid esters of glycerol
醋酸的 acetous; acetic
醋酸芳樟酯 linalyl acetate
醋酸癸酯 decyl acetate
醋酸計 acetimeter; acetometer
醋酸浸漬食品 acetic acid preserves
醋酸桿菌 *Bacillus aceticus*
醋酸桿菌屬 *Acetobacter*
醋酸細菌 vinegar bacteria
醋酸測定法 acetimetry; acetometry
醋酸發酵 acetic acid fermentation
醋酸短桿菌 *Brevibacterium acetylicum*
醋酸菌 acetic acid bacteria
醋酸鈣 calcium acetate
醋酸鈉 sodium acetate
醋酸鉀 potassium acetate
醋酸鉛 lead acetate
醋酸酯 acetate; acetic ester
醋酸醋菌 *Acetobacter aceti*
醋酸澱粉〔酯〕 starch acetate
醋酸環己酯 cyclohexyl acetate
醋酸纖維素 acetyl cellulose; cellulose acetate
醋酸纖維薄膜 acetate film
醋酸鹽 acetate
醋線蟲 vinegar eel
醋醛 acetaldehyde; ethanal
醃什錦菜 chowchow
醃牛肉 bully (beef); corned beef
醃肉 cured meat; meat curing; salted meat
醃肉切片機 bacon slicer
醃肉片 sliced bacon
醃肉用豬 bacon pig; baconer
醃肉用閹豬 bacon hog
醃肉色 cured meat colo(u)r
醃肉香味 cured flavo(u)r
醃肉壓型機 bacon press(er)
醃肉雜碎 bacon weed
醃缸 curing tank
醃料 curing agent; marinade
醃乾魚 cured fish
醃液注射法 stitch pumping
醃液注射醃漬法 pickle injection method
醃液動脈注射法 artery pumping
醃菜 pickle; pickled vegetable
醃黃瓜 cucumber pickle; pickled cucumber
醃漬 curing; marinate; pickle; pickling
醃漬小黃瓜 gherkin
醃漬不足 undercuring; underpickling
醃漬水果 pickled fruit
醃漬用鹽水 souse
醃漬用鹽汁 curing pickle
醃漬沙丁 marinated sardine
醃漬洋葱 pickling onion
醃漬食品 pickle; pickled food; souse
醃漬液 pickle
醃漬過久 overcuring
醃漬過久的 overcured
醃漬槽 pickling tank

醃漬箱 kench
醃製 cure; curing; souse
醃製工廠 saltery
醃製火腿 cured ham
醃製用混合鹽 curing mix
醃製用鹽 curing salt
醃製肉食 cured meat
醃製的野豬肉 brawn
醃製室 curing room
醃製產品 cured product
醃製期 curing period
醃製槽 curing tank; salting tank
醃製劑 curing agent
醃槽 curing tank
醃豬肋肉 flitch
醃豬腹肉用箱 belly box
醃豬頭肉 bath chap
醃鮭魚子醬 red caviar
醃燻牛肉 beef bacon
醃燻肉捲 rolled bacon
醃燻魚 kipper; kippered fish
醃燻豬五花肉 bacon
醃燻鯖魚 bloater
醃燻鯡魚 bloater
醃藏食品 cured product
醃鱈〔肉〕塊 codfish brick
醃鱈魚 cape hake; haberdine
鋅 zinc
銻 stibium
銷售期限 pull date
鋁 aluminium; aluminum
鋁化作用 calorization
鋁色澱 aluminium lake; aluminum lake
鋁粉 aluminium powder
鋁箔 aluminum foil
鋁箔包裝材料 foil wrapper
鋁箔紙 metallic paper
鋁箔與膠膜複合之包裝材料 multifoil; polyfoil
鋁箔複合薄膜 foil laminated film
鋁箔積層膜 foil laminated film
鋁蓋封機 aluminum capping machine
鋁製拉環式易開蓋 aluminum ring pull end
鋁酸鹽 aluminate
鋁罐 aluminum can
銳首擬石斑 barramundi cod
鋇 barium
震昏 stun(ning)
震動 shock
震動輸送帶 tapping conveyor
震凝現象 rheopexy
霉 mildew
霉白黴菌 *Mucor mucedo*
霉味 musty odor; stale flavo(u)r; stale taste
霉狀小奧德菇 porcelain fungus
霉臭 musty smell; stale odo(u)r
霉臭〔味〕 fustiness
霉臭的 fusty; musty
霉菌 mold; mould
頜 jaw
頜式破碎機 jaw breaker
頜式壓碎機 jaw crusher
頜油 jaw oil
頜針魚 garfish; needlefish
頜圓鰺 horse mackerel
頜鬚鮈 minnow
養分 nutrient
養牛場 cattle farm
養生法 regimen
養育 nurture
養兔場 rabbitry; warren
養兔業 rabbitry
養料 aliment
養魚池塘 piscina
養魚的 piscine
養魚場 fish farm; fishery

養魚〔業〕 fish farming
養殖場 farm
養禽業 aviculture
養禽學 aviculture
養蜂人 apiarist
養蜂家 apiarist
養蜂場 apiary
養蜂〔業〕 apiculture
養豬者 swineherd
養豬場 hoggery
養鴨者 ducker
養鴨場 duckery
養雞場 hennery
養蠔場 oyster bank; oyster bed; oyster ground
餓死 starvation
餓虛 inanition
餓瘦的人 starveling
餘味 after-taste
餘柑子 emblic
餘苦味 lingering bitterness
餘渣 debris
餘熱 waste heat
餘熱利用 recuperation
駝牛 zebu
駝背笛鯛 paddle tail
駝背鮭魚 dog salmon; humpback salmon; keta
駝背鯨 humpback whale
駝峰 hump
駒 foal
髮菜 black moss; long thread moss
魷魚 sea arrow; squid
魷魚肝油 squid liver oil
魷魚乾 flattened dried squid
魯冰豆 lupine
魯酵母〔菌〕 *Saccharomyces rouxii*
鴉片 opium
鴉葱 scorzonera
麩 bran; chisel
麩皮 chaff; ship stuff; shorts
麩皮飼料 bran shorts
麩〔皮〕麴 mold bran; mould bran
麩皮麵包 bran bread
麩氨酸鈉 monosodium glutamate (MSG); sodium glutamate
麩粉 shorts; weatings
麩粉飼料 seconds
麩胱甘肽 glutathione
麩胺酸 glutamic acid
麩胺酸〔一〕鈉 aginomoto; monosodium glutamate (MSG)
麩胺酸〔一〕鉀 monopotassium glutamate (MPG)
麩胺酸鈉 sodium glutamate
麩胺酸酯 glutamate
麩胺酸鹽 glutamate
麩醯胺 glutamine
麩醯胺酸酶 glutaminase
墨水 ink
墨汁 ink
墨西哥玉米薄餅 tortilla
墨西哥油炸玉米粉餅 flour quesadilla
墨西哥粗糖 panocha
墨西哥塔可包 taco
墨西哥塔可醬 taco sauce
墨西哥塔瑪玉米粉肉粽子 tamale
墨西哥鼠毛草 Mexican sage
墨西哥薄餅 flour quesadilla
墨角藻 black wrack; sea oak
墨紅花浸膏 rose crimson glory flower concrete
墨紅花淨油 rose crimson glory absolute
墨魚 cuttle; cuttlefish; inkfish; sepia
墨魚骨 cuttle bone
墨腺（墨囊） ink gland; ink bag; ink sac
墨鰂 cuttle; cuttlefish
蔹鰆 pacific bonito

齒齦 gum
噁心 nausea
噁心的 nauseous
噁心食物 nauseating food
槲皮粉 quercitron
槲皮素 quercetin
槲皮酮 quercetin
槲皮醇 acorn sugar; quercitol
槲皮苷 quercitrin
槲實 acorn
槲實糖 acorn sugar; quercitol
潲水 swill
熵 entropy
獠牙 tusk
羰胺反應 aminocarbonyl reaction; carbonyl-amine reaction; Maillard reaction
羰基 carbonyl
羰基化合物 carbonyl compound
羰基鐵 iron, carbonyl
堇菜黃質 violaxanthin
堇菜苷 violanin
蓽澄茄油 cubeb oil
魴〔魚〕 John Dory; gurnard
魨魚 bellows fish
魬鯛 golden skinned porgy
銹斑 rusty stain
踪跡 trace
醌 quinone

十六畫

皺粒豌豆 marrow fat pea
凝血維生素 vitamin K
凝血酶 thrombin
凝血酶原 prothrombin
凝乳 curd; lapper milk; milk curd; renneting
凝乳狀的 curdy
凝乳甜點 junket
凝乳粒 curd particle
凝乳塊 clotted milk
凝乳槽 curd vat
凝乳劑 milk coagulant
凝乳餡餅 curd tart
凝乳酶 chymase; chymosin; milk clotting enzyme; milk coagulating enzyme; pexin; rennase; rennet; rennin
凝乳酶元 chymosinogen
凝乳酶原 chymosinogen; pexinogen; prorennin; renninogen; rennogen
凝乳酶凝乳〔作用〕 rennet clotting
凝固 solidify; setting; congelation; solidification
凝固乳 coagulated milk
凝固的 curdled
凝固素 coagulin
凝固緩慢 slow set
凝固劑 solidification reagent
凝固點 freezing point
凝固酶 coagulase
凝結 coagulate; coalescence; clot; setting; coagulation
凝結成塊的 curdy
凝結物 coagulate; coagulum
凝結的 clotting; curdled
凝結〔芽孢〕桿菌 *Bacillus coagulans*
凝結塊 coagulum
凝結酵素 clotting enzyme
凝結酸乳 clabber
凝結劑 coagulant; coagulating agent; coagulator; setting agent
凝結點 setting point
凝結酶 clotting enzyme; coagulase
凝集〔作用〕 agglutination
凝集素 agglutinin
凝集劑 agglutinant; agglomerant
凝塊 clot; coagulate; coagulum; curd
凝塊乳 peptonized milk
凝塊鮮乳油 curdled cream
凝酪槽 chesset; chisset
凝酪鍋 chesset; chisset
凝聚 agglomeration; coacervation; coagulation; condensation
凝聚素 coagulin
凝聚層 coacervate
凝聚膠 coagel
凝聚劑 coagulator
凝膠 gelatin(e); jell
凝膠色譜法 gel chromatography
凝膠物 jello

凝膠狀的 gelatinous
凝膠食品 gelled food(stuff)
凝膠剛度 gel rigidity
凝膠堅度計 gelometer
凝膠強度 jelly strength
凝膠過濾 gel filtration
凝膠電泳 gel electrophoresis
凝膠層析法 gel chromatography
凝膠纖維 gelatinous fiber
凝膠體 gel
劑量 dosage; dose
劑量不足 underdose
劑量計 dose meter; dosimeter
劑量過度 overdose
噸 ton
器皿 utensil; vessel; ware
器材 equipment
器具 utensil
器官 organ
噬菌作用 phagocytosis
噬菌細胞 phagocyte
噬菌體 bacteriophage; phage
噻苯咪唑 thiabendazole
噻唑 thiazole
噻胺 aneurin; oryzanin; thiamine; vitamin B_1
噻胺焦磷酸鹽 thiamine pyrophosphate (TPP)
噻胺酶 thiaminase
學前時期 preschooler
學校膳食服務 school meal service
學校餐廳 lunchroom
學童的食品 tuck
導流板 baffle (plate)
導管 duct; trachea; vessel
導熱系數 thermal conductivity
導熱性 thermal conductivity
擋板 baffle (plate)
擂潰 mincing
擂潰機 mincer; mincing machine
撿蛋 egging
擔(一百市斤) picul
擔子果 basidiocarp
擔子菌〔類〕 basidiomycetes
擔孢子 basidiospore
整分部分 aliquot
整米 head rice
整個 whole
整套的餐具 dinner service; dinner set
整隻火雞 whole turkey
整條麵包 bread loaf
整粒 whole grain
整粒玉米 whole grain corn; whole kernel (style) corn
整粒米 whole grain rice
整魚 round fish
整魚鹽醃 round cure
整等分液量 aliquot
整齊菌落 entire colony
整穀粒 whole grain
整穗玉米 corn-and-cob; corn on the cob
整雞 whole chicken
樺子 birch seed
樺木汽水 birch beer
樺木油 sweet birch oil
樺焦油 birch tar oil
樺樹汽水 birch beer
橙 orange
橙〔皮〕油 orange oil
橙汁 orange juice
橙汁汽水 orange soda
橙汁飲料 orange beverage; orange drink; orangeade
橙皮 orange peel
橙皮油 orange peel oil
橙皮素 hesperetin
橙皮烯 hesperidene
橙皮酸 hesperitinic acid

橙皮苷 aurantiamarin; hesperidin
橙皮苷酶 hesperidinase
橙色〔色素〕 orange colo(u)r
橙泥 orange puree
橙花叔醇 nerolidol
橙花油 neroli bigarade oil; neroli oil; orange flower oil
橙花醇 nerol
橙花醛 neral
橙金桔 orangequats
橙扁豆 orange lentil
橙黃白毫紅茶 orange pekoe
橙黃網孢盤菌（蕈類）orange peel fungus
橙葉油 petitgrain oil
橙膏 orange butter
橫杆式脱粒滾筒 bar-type mill
橫紋肌 cross striated muscle; striated muscle
橫帶髭鯛 grunt
橫膈膜 diaphragm; midriff
橘子 mandarin orange; orange; tangerine
橘子水 orange squash; orangeade
橘子汁 orange juice
橘子皮 zest
橘子汽水 orange soda; orangeade
橘子油 mandarin oil
橘子油香精 orange oily flavo(u)r
橘子香精 orange flavo(u)r
橘子酒 orange wine
橘皮果醬 marmalade
橘皮油 citrus oil
橘皮烯 hesperidene
橘皮蜜餞 cedra
橘皮酸 hesperitinic acid
橘皮苷 hesperidin
橘皮苷素 hesperetin
橘肉果汁 orange squash
橘味烈酒 curacao
橘味葡萄酒 bishop
橘柚 tangelo
橘紅 chrysoidine
橘紅2號 citrus red No.2
橘香紅茶 orange pekoe
橘黃質 citroxanthin
橘黃鵝膏（蕈類）orange grisette
橘園 orangery
橘瓣 orange section
橘黴素 citrinin
樹汁 sap
樹皮 bark
樹豆 pigeon pea
樹枝蟲膠 stick lac
樹茄 tree tomato
樹苗 sapling
樹苔浸膏 tree moss concrete
樹脂 resin; rosin
樹脂〔質〕交換劑 resinous exchanger
樹脂床 resin bed
樹脂物質 resinous substance
樹脂酸鋅 zinc resinate
樹梅 strawberry tree
樹液 sap
樹莓 raspberry
樹蛇狀葡萄 pepper vine
樹番茄 tamarillo; tree tomato
樹〔膠〕脂 gum resin
樹膠醛糖 arabinose
樹薯 cassava; manihot; manioc
樹蘭〔花〕油 aglaia odorata flower oil
橄欖 olive
橄欖仁油 olive kernel oil
橄欖油 extra virgin olive oil; olive oil; sweet oil
橄欖的苦澀物質 oleuropein
橄欖粕 sansa
橢圓〔形〕罐 oblong can; oval can; oval shaped can; oval tin
橢圓酵母〔菌〕*Saccharomyces ellipsoideus*

橡子 acorn
橡木酒桶 oak oval
橡木桶 oak cask; oak wood
橡皮 rubber
橡皮糖 gum drops
橡苔浸膏 oak moss concrete
橡實南瓜 acorn squash
橡膠 rubber
橡膠基質 isoprene
橡樹 oak
橡樹果 acorn
機能性 functional property
機械分級 mechanical sorting
機械去皮 mechanical peeling
機械選別 mechanical sorting
機械壓榨機 mechanical press
機械擊昏 mechanical stunning
機器擠奶 inchurn milking
澱粉 amylum; starch
澱粉分解 amylolysis
澱粉分解發酵 amylolytic fermentation
澱粉分解酶 amylolytic enzyme; diastatic enzyme
澱粉水解 amylolysis
澱粉水解酶 amylolytic enzyme
澱粉回凝 retrogradation; starch retrogradation
澱粉老化 retrogradation; starch retrogradation
澱粉作物 starch crop
澱粉改性劑 starch modifying agent
澱粉沉降 starch settling
澱粉味 starchy taste
澱粉和碘的藍色試驗 starch-iodine blue test
澱粉性的 amyliferous
澱粉的 starchy
澱粉食物 amyloid
澱粉核 pyrenoid
澱粉消化不良 amylodyspepsia
澱粉剪支酶 isoamylase
澱粉液化酶 alpha-amylase; α-amylase
澱粉粒 starch grain
澱粉粒質 granulose
澱粉粒臍點 hilum
澱粉—碘反應 starch-iodine reaction
澱粉—碘試驗 starch-iodine blue test
澱粉葡萄糖苷酶 amyloglucosidase
澱粉酯 starch ester
澱粉製成的食品 fecula
澱粉製品 starch product
澱粉廠 starch mill
澱粉漿 starch slurry
澱粉糊 starch paste
澱粉糊化 starch pasting
澱粉糊精 amylodextrin; starch dextrin
澱粉糊精化 starch dextrinization
澱粉糊黏度圖示儀 amylograph
澱粉糊黏度圖譜 amylogram
澱粉膠 starch gum
澱粉質 starchiness
澱粉質的 amyloid
澱粉質原料 starchy material
澱粉質蔬菜 starchy vegetable
澱粉質體 amyloplast
澱粉凝沉 retrogradation; starch retrogradation
澱粉凝固酶 amylocoagulase
澱粉凝膠 starch gel
澱粉糖 corn sweetener; starch sugar
澱粉糖化發酵 amylolytic fermentation
澱粉糖化酶 amylolytic enzyme; beta-amylase; β-amylase; diastase; saccharogenic amylase; starch converting enzyme
澱粉糖化酶活性 diastatic activity
澱粉糖漿 corn syrup; maize syrup; starch syrup

澱粉磷酸鈉 sodium starch phosphate
澱粉磷酸酯鈉 sodium phosphate starch
澱粉薄膜 oblate
澱粉顆粒 starch granule
澱粉黏焙力曲線圖 amylogram
澱粉黏焙力測量器 amylograph
澱粉醚 starch ether
澱粉瀉 amylorrhea
澱粉臍 hilum
澱粉轉化糖漿 starch conversion syrup
澱粉纖維素 amylocellulose
澱粉體 amyloplast
澱粉酶 amylase; amylomyces; diastase; saccharogenic amylase
澱粉酶活性 diastatic activity
濃〔厚〕味 brisk taste
濃奶昔 frappe
濃汁 heavy gravy; thick juice
濃汁白燴肉丁 fricassee
濃汁炖肉 salmi
濃味 strong taste; thick taste
濃味魚肉湯 bouillabaisse
濃味魚羹 bouillabaisse
濃味蔬菜燉肉 ragout
濃咖啡 coffee robusta; robust coffee; Robusta
濃固型酸乳酪 thick yoghurt
濃的 thick
濃度 consistence; consistency; concentration
濃郁香氣 full aroma
濃烈的 stiff
濃酒 rich wine
濃液 strong liquor
濃麥芽汁 heavy wort
濃湯 bisk; bisque; cream style soup; hoosh; potage; pottage; puree; robust soup; thick soup
濃湯粉 bouillon powder
濃稠肉汁〔醬〕 heavy gravy
濃稠食物 stodge
濃稠調味醬 thick sauce
濃稠體 heavy body
濃葡萄汁 thickened must
濃糖漿 heavy syrup
濃膩的 stodgy
濃膩食物 stodge
濃縮 concentrate; concentration; inspissate; inspissation; thickening
濃縮大豆蛋白 soy protein concentrate; SPC; soybean protein concentrate
濃縮奶油(加糖) condensed cream
濃縮白脱乳 condensed buttermilk
濃縮乳 condensed milk; concentrated milk; plain condensed milk
濃縮乳清 concentrated whey
濃縮果汁 concentrated juice; condensed juice; fruit juice concentrate; inspissated juice
濃縮物 concentrate
濃縮液 concentrated solution
濃縮蛋白質 protein concentrate
濃縮魚汁 fish soluble
濃縮魚蛋白 fish protein concentrate
濃縮魚湯 liquid fish
濃縮湯 condensed soup
濃縮葡萄汁 reduced must
濃縮酪乳 condensed buttermilk
濃縮維生素E vegol
濃縮調味料 condensed seasoning
濃縮器 concentrator; inspissator
濃縮鮮奶油 concentrated cream
濃縮罐 strike pan
濃鮮乳油 thick cream
濃醪 thick mash
濃鹽水 strong brine
濁度 cloudiness; turbidity; turbidness
濁度計 nephelometer; turbidimeter

濁度測定法 nephelometry
澳洲及新西蘭食物管理局 Australia New Zealand Food Authority (ANZFA)
澳洲肺魚 barramunda; barramundi
澳洲胡桃 Australian nut; macadamia nut; Queensland nut
澳洲鮭 newfish
澳斯特沙司 worcester(shire) sauce
激冷 snap-chilling
激勃素 gibberellin
激活 activation
激素 hormone
激酶 kinase
燉〔煮〕 stew; stewing
燉什錦 mulligan
燉牛肉 beef stew; stewed beef
燉牛蹄筋 cowheel
燉肉 meat stew; pot roast; simmered meat; stew; stewed meat
燉壺 stew pot
燉菜 stew
燉鍋 casserole; covered dish; marmite; stew pan; stew pot
燉雜燴 casserole
燉雞濃湯 chicken stew
燐蝦 euphausiid shrimp
燐蝦目 euphausia
燐蝦目浮游甲殼動物 euphausiid
燐蝦〔類〕 euphausiid
燒 firing
燒毛 singe; singeing
燒毛器 singeing torch; singer
燒毛機 singeing machine
燒水壺 teakettle
燒肉 barbecued meat
燒灼 ustion
燒灼滅菌〔法〕 incineration
燒明礬 aluminium potassium sulfate, anhydrous; dry alum
燒杯 beaker
燒烤 broiling; grilling
燒烤用具 broiler
燒烤爐 barbecue grill
燒酒 alcoholic liquor; aguardiente; ardent spirit; rack
燒焦 scorch
燒焦氣味 burnt flavo(u)r
燒結 caking
燒結玻璃 sintered glass
燒結玻璃濾器 fritted glass filter; sintered glass filter
燒銨明礬 burnt ammonium alum
燈魚 lantern fish
燈籠果 cape gooseberry
燈籠椒 bell pepper; sweet pepper
燕子 swallow
燕仔魴 butterfly ray
燕麥 *Avena sativa*; haver; oat
燕麥大麥混合飼料 dredge corn
燕麥片 farina avena; oat flake; oat meal; rolled oats
燕麥布丁 hasty pudding
燕麥粉 farina avena; gerty; oat flour; oat meal
燕麥粉製糕餅 mill bannock
燕麥蛋白 avenin
燕麥蛋糕 farl(e)
燕麥渣 oat middlings
燕麥粥 brose; grouts; hasty pudding; oat meal; oat meal porridge
燕麥餅 haver cake
燕麥餅乾 jannock; oatcake
燕麥穀蛋白 avenin
燕麥薄餅 clap bread; farl(e)
燕麥麵包 jannock
燕窩 bird’s nest; swallow nest
燎毛 singe; singeing
燎毛器 singeing torch; singer

燎毛爐 singeing machine
燙毛缸 scalding tank
燙洗 scald; scalding
燙洗槽 scalder; scalding tank
燙洗機 scalder
燙滾 poaching
燙漂 blanch(ing)
燙麵 scalded flour
燜 simmer(ing); stew; stewing
燜煮 simmer(ing)
燜燒肉 pot roast
燃料 fuel
燃燒 inflammation
燃點 ignition point
獨行菜 garden cress; upland cress
獨活草 lovage
獨活草油 lovage oil
獨特風味 raciness
盧杷醇 luparol
磨 grind
磨刀石 whetstone
磨刀床 knife sharpener
磨刀器 knife sharpener; knife grinder; sharpener
磨刀機 knife machine
磨刀鋼條 sharpening steel
磨石 grinding stone; grindstone; millstone; rubbing stone
磨石碾磨機 millstone mill
磨光 polishing
磨光機 polishing machine
磨成粉 powder
磨成粉的 powdered
磨坊 gristmill; mill
磨床 grinding machine
磨砂玻璃 frosted glass
磨粉 pulverize; pulverization
磨粉分離器 millerator
磨粉用套篩 plansifter
磨粉廠 gristmill
磨粉機 flour mill; gristmill; mill; miller; milling machine; pulverizer; pulverizing mill
磨粉篩 milling separator
磨菇木耳 agaric
磨菇素 agaricin
磨菇酸 agaric(ic) acid; agaricinic acid
磨碎 grind; grinding; mill; milling; trituration
磨碎咖啡〔粉〕 ground coffee
磨碎的 mashed
磨碎器 grater
磨碎機 attrition grinder; attrition mill; desintegrator; grinder; grinding machine; grinding mill; pulper; pulping machine; rasper; rubbing machine
磨漿機 paste mill
磨盤 abrasive disc; millstone
磨擦去皮 abrasion peeling
磨擦式去皮機 abrasive peeler
磨擦式砂輥 abrasive roller
磚乳 block milk
磚紅雞油菌（蕈類） smooth chanterelle
磚茶 brick tea; cake tea; compressed tea; tea brick; tile tea
積層薄膜 laminated film
穎果 caryopsis
穆哈咖啡 caffe mocha; mocha
篩 colander; screen; sieve; sifter
篩下料 riddling; scalpings; sieve residue
篩上料 overs
篩子 bolter; shifter
篩分 sift; sizing; sieving; scalp(ing); screening
篩分器 duster
篩分機 classificator; sieving machine; sifting machine; screening machine

篩孔 mesh; screen aperture; screen opening; sieve aperture; sieve opening
篩孔〔大小〕 mesh size
篩孔目數 mesh number
篩孔效應 sieve effect
篩尺寸 sieve size
篩出物 siftings
篩去麩皮 dusting
篩布 bolting cloth
篩目 mesh number; screen mesh; sieve mesh
篩目孔徑 sieve size
篩目號 sieve number
篩板 sieve plate
篩板塔 perforated plate column; sieve plate column
篩析 screen analysis; sieve analysis
篩面 screening surface
篩屑機 jigger
篩格 sieve grating
篩粉機 bolter; bolting machine; flour dresser
篩除物 overs
篩理 dressing
篩眼 mesh
篩渣 screenings; sieve residue; tailover
篩筒 screen drum
篩絹 screen cloth
篩號 mesh number; screen size
篩網 sieve mesh; sifting screen
篩網過濾器 mesh filter
篩盤 sieve tray
篩盤塔 sieve tray column
篩餘物 scalpings; screenings
篩餘料 riddling; triage
篩磨機 bolting mill
篩選 riddle; scalp(ing); screening; sieving; sift
篩選試驗 screening test
篩選機 bolting machine; chaffer; scalper; screening machine; sieving machine; sifting machine; sizer
篩濾 strain
篩濾果泥 strained fruit
篩濾食品 strained food
篩濾番茄泥 strained tomato
篩濾蜂蜜 strained honey
篩濾器 screen filter
篩濾機 finisher
篦條篩 grate
篦麻油 ricinus oil
糕餅 pastry
糕餅用盒裝現成材料 cake mix
糕餅用麵粉 cake flour
糕餅的花飾糖果或蜜餞 hundreds and thousands
糕餅表面裝飾機 lidding machine
糕餅模具 cake mo(u)ld
糕點 gateau
糕點用油 bake oil
糕點的裝飾糖衣 piping sugar
糕點的糖製花飾 piping sugar
糖 saccharide; saccharum; sugar
糖二酸 saccharic acid
糖二酸鹽 saccharate
糖化 malting; mashing
糖化〔作用〕 saccharification
糖化原料 mash goods
糖化桶 mash tun
糖化〔細〕菌 saccharolytic bacteria
糖化酵母〔菌〕 *Saccharomyces diastaticus*
糖化酵素 saccharifying enzyme
糖化槽 saccharifying tank
糖化澱粉 converted starch
糖化澱粉酶 saccharifying amylase
糖化鍋 mash tub
糖化醪 mash; sweet mash
糖化酶 diastatic enzyme

糖水 sweet water
糖代用品 sugar substitute
糖包 sugar sachet
糖汁 sirup; sugar juice; syrup
糖汁加石灰 juice liming
糖汁脱〔石〕灰 juice deliming
糖汁澄清 juice liming
糖血症 glucemia; glycemia
糖衣 glace; glaze; icing; sugar coating
糖衣雪糕 ice lolly
糖衣蜜餞水果 glaze fruit
糖衣機 dragee machine; enrober
糖衣錠 dragee
糖作物 sugar crop
糖尿〔症〕 glycosuria
糖尿病 diabetes; diabetes mellitus
糖尿病人食品 diabetic food
糖豆 jelly bean
糖果 candy; comfit; confection; confectionary; confectionery; confetti; goody; sweet; sweet stuff; sweetie; sweetmeat; sweets; sweety; sugar candy
糖果片 tablet candy
糖果包裝 candy wrapping
糖果〔包裝〕紙 candy paper
糖果包裝機 sweet wrapping machine
糖果包覆 candy wrapping
糖果夾心 candy center
糖果店 confectionary; confectioner; confectionery; sweet shop; tuckshop
糖果拉製 candy pulling
糖果拉製機 puller
糖果的包衣 confectioner' s coating
糖果棒 candy stick
糖果絲光機 puller
糖果業 confectionery
糖果製造者 confectioner
糖果壓形機 candy press
糖果壓塊機 candy press
糖的 saccharine
糖花〔白〕 sugar bloom
糖度折光儀 sugar refractometer
糖度計 saccharimeter; saccharometer
糖耐量 sugar tolerance
糖食 confection; confectionery
糖食的 confectionary
糖原 animal starch; glycogen
糖原分解 glycogenolysis
糖原生成 glycogenesis
糖原酶 glycogenase
糖料作物 sugar crop
糖料拉條機 pulling machine
糖粉 castor sugar; confectioner' s sugar; icing sugar; powder sugar; powdered sugar; pulverized sugar
糖納子 doughnut
糖脂 glycolipid; glycolipin
糖高粱 saccharine sorghum; sugar sorghum
糖液 sugar liquor
糖瓶 sugar castor
糖莢豌豆 sugar pea
糖蛋白〔質〕 glycoprotein
糖晶 crystallox
糖晶水果蜜餞 crystallized fruit
糖晶粒固結 grain hardening
糖晶漿 magma
糖棕 sugar palm
糖煮水果 compote
糖結晶 sugar crystal
糖結晶抑制劑 candy doctor; sugar doctor
糖量計 saccharimeter; saccharometer
糖椰 sugar palm
糖楓 sugar maple
糖楓林 sugar bush
糖解 glycolysis
糖解酶 glycolytic enzyme
糖飾 confectioner' s coating

糖酮酸 keturonic acid
糖酯 sugar ester
糖漬 sugaring; syrup curing
糖漬水果 glaze fruit
糖漬水果片 compote
糖漬的 candied; glace; glaze; sugar cured; sugared
糖漬食品 chowchow
糖精 benzoic sulfimide; benzosulfimide; gluside; saccharin; saccharine
糖精鈣 calcium saccharin
糖精鈉 crystallox; saccharin sodium; sodium saccharin; soluble saccharin
糖精鉀 potassium saccharin
糖精酸 saccharinic acid
糖精銨 ammonium saccharin
糖膏 fillmass; fullmasse; massecuite; magma; melada; sirup; sugar mass
糖蜜 gur; molasses; treacle
糖蜜的 treacly
糖蜜酒 rum
糖蜜塊 molasses sugar
糖蜜醪 molasses mash
糖酵解 glycolysis
糖酸 aldonic acid; sugar acid
糖酸比 acid-sugar ratio; Brix acid ratio
糖酸鹽 glyconate; saccharate
糖廠 sugar mill
糖槭 rock maple; sugar maple
糖槭樹林 sugar bush
糖漿 liquid sugar; sirup; syrup; sugar liquor; sugar syrup; thick juice; treacle
糖糊 magma
糖膠樹膠 chicle (gum); gum chicle
糖醇 glycitol; sugar alcohol
糖醋肉 sweet and sour pork
糖醋的 sweet and sour
糖醋醬 sweet and sour sauce
糖錠 tablet sugar
糖磷脂 glycophospholipin
糖薑 sugared ginger
糖霜 frost; frosting; icing sugar
糖霜泥 fondant
糖醛酸 alduronic acid; uronic acid
糖醛酸酯 uronate
糖醛酸鹽 uronate
糖類 carbohydrate; saccharide
糖罐 sugar basin
糖苷 glycoside
糖苷化 glycosidation
糖苷配基 aglycon
糖苷酶 glycosidase
糖肽 glycopeptide
糖酶 carbohydrase; saccharidase
翮羽 quill
膳食 diet; meal
膳食治療 dietary therapy
膳食表 dietary
膳食要素 dietary factor
膳食計劃 meal planning
膳食規定 dietary
膳食費用 board
膳食管理員 steward
膳食製備 catering
膳食增補劑 dietary supplement
膳食學 dietetics
膳食營養缺乏 dietary deficiency
膳食營養學 sitology
膳食營養攝取量 dietary allowance
膳食療法 dietotherapy
膳食療養法 dietary cure
膳食療養食品 diet food
膳食纖維 dietary fiber
膨化 puffing
膨化食品 puffed food
膨化乾燥 puff drying
膨化穀物 puffed grain
膨化穀類食品 puffed cereal

膨化器 puffer
膨脹計 dilatometer
膨發 leavening; rising
膨發得不鬆的麵包 heavy bread
膨發劑 leaven; leavening agent
膨脹 inflation; puffing; swell; swelling
膨脹不足 underexpansion
膨脹度 swelling capacity
膨脹度計 expansimeter
膨脹率 overrun
膨脹率測定器 overrun tester
膨脹劑 inflating agent; plumping agent; swelling agent
膨脹壓 oncotic pressure
膨脹鬆化 puff
膨潤 swelling
膨潤土 bentonite
膨潤度 swelling capacity
膨潤劑 swelling agent
膨潤顆粒 swollen granule
膨鬆 leavening
膨鬆劑 leaven; leavening agent
膨鬆麵糰 leaven dough
膨罐 bloat can; blown can; blown tin; can buckling; swell; swelled can
興奮性食品 excitant food
興奮性飲料 bracer; cordial
興奮劑 cordial; incitant; stimulant
艙口 port
蕈 mushroom
蕈托 volva
蕈幕 veil
蕈環 annulus
蕈褶 gill
蕨菜 bracken; fiddlehead fern
蕃茄紅素 lycopene
蕃茄紫素 lycophyll
蕃茄黃色素 lycoxanthin
蕃茄黃質 lycoxanthin
蕃茄膏 tomato paste
蕃茄糊 tomato paste
蕃荽 parsley
蕃椒 red pepper
蕃荔枝 custard apple
蕪菁 green turnip; turnip
蕪菁甘藍 rutabaga; swede
螃蟹 crab
熒光 fluorescence
熒光素 fluorescein
熒光假單孢菌 *Pseudomonas fluorescens*
熒光黃 fluorescein
螢烏賊 firefly squid
融化 thaw(ing); unfreeze
融化乳酪 process(ed) cheese
融化的 melted
融製奶油 drawn butter
融製乾酪 crustless cheese
融霜 defrost; defrostation
褪色 discolo(u)ring; discolo(u)ration
親水性 hydrophilicity
親水性膠體 hydrocolloid
親水物 hydrophile
親水物質 hydrophilic substance
親水的 hydrophilic
親水溶膠 hydrophilic sol
親水膠體 hydrocolloid; hydrophilic colloid
親水劑 hydrophile
親水親油平衡 hydrophilic-lipophilic balance; HLB; hydrophile-lipophile balance
親水親脂均衡 hydrophilic-lipophilic balance; HLB; hydrophile-lipophile balance
親合性 affinity
親和吸收劑 affinity absorber
親和性 affinity
親和層析法 affinity chromatography
親油的 oleophilic; oleophylic

親氣細菌 aerophilic bacteria
親脂性 lipotropy
親脂的 lipophilic
親液物 lyophile
親液的 lyophilic
親液膠體 lyophilic colloid; lyophile
親鹽細菌 halophilic bacteria
諧麗〔葡萄〕酒 sherry
諾金（液量單位） noggin
諾格蛋黃〔乳〕酒 nog
諾格濃啤酒 nog
賴白當浸膏 labdamum concrete
賴百當油 labdanum oil
賴氨酸鹽酸鹽 lysine monohydrochloride
賴胺酸 lysine
蹄 hoof; trotters
輻射 irradiation; radiation
輻射巴氏殺菌〔法〕 radiation pasteuriz-ation; radiopasteurization
輻射加熱器 radiant heater
輻射保藏〔法〕 radiation preservation
輻射毒性 radiotoxicity
輻射能 radiant energy
輻射高溫計 radiation pyrometer
輻射殺菌〔法〕 radiosterilization; radici-dation; radurization
輻射處理 radiation treatment
輻射滅菌〔法〕 radappertization; radiation sterilization
輻射滅蟲 radiation disinfestation
輻射熱 radiant heat
輻射器 radiator
輻照食品 irradiated food
輻照滅菌 irradiation sterilization
輸送台 conveyor table
輸送帶 band conveyor; belt conveyor; conveyor; conveyor belt
辦公室包飯 office catering
選分篩 separating screen
選別性培養基 selective and differential medium
選別機 classifier; grading machine
選擇性 selectivity
選擇性平皿培養〔法〕 selective plating
選擇性抑菌作用 selective bacteriostasis
選擇性培養〔物〕 selective culture
選擇性培養基 selective medium
選擇性增菌培養 selective enrichment
選擇滲透性 permeaselectivity; selective permeability
遲滯期 lag phase
遲緩期 lag phase
遺傳工程〔學〕 genetic engineering
遺傳因子 genetic factor
遺傳密碼 genetic code
遺傳學 genetics
醒麵 proof; proofing
醒麵室 proofer; proofing room
醒麵過度的麵糰 overproofed dough
醒麵箱 proof box; proofer
錠〔劑〕 pastille; tablet
鋸大眼鯛 big-eye porgy
鋸峰齒鮫 blue shark
鋸緣青蟹 mud crab
鋸齒形菌落 erose colony
錳 manganese
錯合物 complex
錯合劑 complexing agent
錯流式乾燥機 cross flow dryer
鋼 steel
鋼絲攪打器 wire whipper
鋼鐵 steel
錫 tin
錫味 tinny taste
錫硫化物 tin sulfide
錫箔 tin foil
錫器 tinware
錫蘭肉桂 Ceylon cinnamon

錫蘭紅茶 Ceylon tea
錫蘭桂皮 Ceylon cinnamon
錫蘭茶 Ceylon tea
錫蘭醋栗 kitembilla
錐形冰淇淋杯 cone
錐形破碎機 conical breaker
錐形紙袋 cornucopia
錐形乾燥機 conical dryer
錐形混合機 conical blender; conical mixer
錐形球磨 conical ball mill
錐形榨汁機 cone crusher
錐形磨 cone mill
錐形壓碎機 cone crusher; cone mill
錐形罐 tapered shape can
錦葵色素 malvidin
錦葵色素苷 malvin
錦葵定 malvidin
錦鱗蜥魚 brush-toothed lizard fish
隧道式凍結機 tunnel freezer
隧道式烤爐 tunnel oven
隧道式乾燥機 canal drier; canal dryer; tunnel dryer
隨軍小販 sutler
隨意肌 skeletal muscle; voluntary muscle
隨機採樣 random sampling
雕紋口蘑（蕈類） yellow staining knight cap
霍克白葡萄酒 hock
霍亂 cholera
霍亂弧菌 Cholera vibrio; Comma bacillus
霍亂菌 *Vibrio cholera*
靛 indigo
靛青 indigotine
靛紅 isatin
靛酚 indophenol
靛酚氧化酶 indophenol oxidase
靛藍 indigo; indigo blue; indigotine
靛藍鋁色澱 indigotine alum lake
靛苷 indican
靜止 still
靜止代謝 resting metabolism
靜止〔式〕殺菌釜 still retort
靜止空氣冷凍法 still-air freezing
靜止相 stationary phase
靜止培養〔物〕 stationary culture; still culture
靜水 dead water
靜水壓殺菌釜 hydrostatic retort
靜水壓滅菌〔法〕 hydrostatic sterilization
靜水壓滅菌器 hydrostatic sterilizer
靜水壓蒸煮器 hydrostatic cooker
靜脈 vein
靜脈輸入高營養液 hyperalimentation
靜脈營養補給 intravenous alimentation
靜菌作用 bacteriostasis
靜置 setting
靜置法製的奶油 pan-set cream
鞘脂類 sphingolipid
鞘胺醇 sphingosine
鞘磷脂 phosphosphingolipid
鞘髓磷脂 sphingomyelin
頰肉 cheek; cheek meat; jowl
頰骨 cheek bone
頸 neck
頸肉 neck
頸背 scruff
頸部 neck
頭刀 topside (round)
頭奶 foremilk
頭骨 skull
頭罩 helmet
頭榨汁 mother juice
頭榨蔗汁 primary cane juice
頭髮 hair
餐 repast
餐刀 table knife
餐叉 fork; table fork

餐巾 napkin; serviette; table napkin
餐巾紙 napkin
餐用比司克 dinner biscuit
餐用砂糖 table sugar
餐用食醋 table vinegar
餐用稀奶油 table cream
餐用餅乾 dinner biscuit
餐用糖漿 table syrup
餐用鮮乳油 table cream
餐用麵包 dinner roll
餐車 diner; dining car; trolley
餐車業務 mobile catering
餐具 cutlery; mess gear; mess kit; service; table plate; tableware
餐具室 pantry; spense
餐具架 plate rack
餐具組 table set
餐具櫃 buffet; sideboard
餐券 meal ticket
餐後血糖 postprandial blood sugar
餐後低血糖症 postprandial hypoglycemia
餐後的 postprandial
餐後的含白蘭地酒咖啡 pousse café
餐後的雞尾酒 pousse café
餐後酒 afterdinner drink; dessert wine
餐後混合甜酒 pousse café
餐後甜酒 dessert wine
餐後甜點心 entremets
餐後葡萄酒 dessert wine
餐宴 feast; repast
餐桌 board; dining table; table
餐桌布 table linen
餐桌布置 table setting
餐桌用小墊子 table mat
餐桌用巾 napery
餐桌用布 table linen
餐桌用台布 tablecloth
餐桌用油 table oil
餐桌轉盤 lazy susan
餐桌鹽 table salt
餐酒 dinner wine; table wine
餐盒 dinner bucket; dinner pail
餐飲部 catering department
餐飲業 food service
餐館 café; eatery; messhall; messhouse; porterhouse; restaurant; victual(l)ing house
餐館主 victual(l)er
餐館用包裝 institutional pack
餐館老板 restauranteur
餐館〔營〕業者 restauranteur
餐廳 dining hall; dining room; lunchroom; refectory; restaurant
餞別酒 stirrup cup
餛飩 Chinese dumpling; ravioli; wonton
餡 stuffing
餡角 turnover
餡料 filling
餡料夾層的甜點心 buncuit
餡餅 pasty; pie; pate
餡餅皮 pastry; pie crust
餡餅烤盤 piepan
駱駝 camel
鮑魚 abalone; ear shell; haliotis; ormer; Venus ear
鮑魚中毒 abalone poisoning
鮑魚菇 abalone mushroom; *Pleurotus cystidiosus*; *Pleurotus ostreatus*
鮑爾溫蘋果 Baldwin
鴨 duck
鴨肉 duck; duck meat
鴨肫 duck gizzard
鴨蛋 duck egg
龍舌蘭〔發酵飲料〕 agave
龍舌蘭酒 pulque; tequila
龍宮菜 Ceylon spinach
龍涎香 ambergris
龍眼 dragon eye; longan; lungan

龍眼乾 dried longan
龍腦 borneol; bornyl alcohol
龍葵 morel
龍蒿 estragon; tarragon
龍蒿油 estragon oil; tarragon oil
龍蝦 lobster; langouste; sea crawfish; sea crayfish; spiny lobster
龍頭 stopcock
龍頭魚 bombay duck
龍膽花 genitan
龍膽根 genitan; gentian root
龍鬚蛋麵 fine egg noodle
龍鬚菜 gracilaria
龜 tortoise
龜甲 tortoise shell; turtle shell
龜肉 turtle meat
龜板 tortoise shell
龜背肉 calipash
龜殼 turtle shell
龜腹肉 calipee
穀粒 kernel
穀粒重(千粒重) kernel weight
穀渣 spent brewer's grain; spent grain
芸豆 kidney bean
蕓香油 rue oil
蕓香浸膏 liquidambar concrete
蕓香素 rutin
蕓香糖 rutinose
蕓香糖苷 rutinoside
蕓香苷 rutin; rutoside
蕁麻 nettle
蕎麥 brank; buckwheat
蕎麥中毒 fagopyrism
蕎麥片 kasha
蕎麥粉 buckwheat flour
蕎麥粥 kasha
蕎麥鹼 fagopyrin
蕎頭 rakkyo; scallion
閾限 threshold
閾限值 threshold limit value
閾值 threshold; threshold value
閹公牛 ox; steer
閹公羊 wether
閹公豬 barrow
閹牛 beef steer; bullock; gelded ox
閹母雞 poulard
閹割 castrate; emasculation
閹豬 gelded pig
閹雞 capon

十七畫

優良產品 quality goods
優良豬油 neutral lard
優良營養 eutrophia; eutrophy
優格 yoghourt; yoghurt; yogurt
優酪乳 yoghourt; yoghurt; yogurt
優質咖啡 sound coffee
儲水槽 reservoir
儲存性 storability
儲存室 store room
儲存脂肪 deposited fat; depot fat
儲存處 reservoir
儲存期 storage period
儲存期限 shelflife
儲存槽 storage tank
儲倉 storage bin
儲備 stock
儲備培養 stock culture
儲備液 stock solution
儲藏 preservation; storage; store
儲藏地窖 storage cellar
儲藏室 depository; storage room; store room; stores
儲藏〔性〕蛋白質 reserve protein
儲藏限度 preservation limit
儲藏容器 storage container
儲藏庫 stock room; storage bunker
儲藏脂肪 deposited fat; depot fat
儲藏條件 storage condition
儲藏期 storage period
儲藏期限 marketable life; shelf life; storage life
儲藏〔期〕濃稠化 age thickening
儲藏溫度 storage temperature
儲藏槽 reserve tank
儲藏澱粉 reserve starch
儲藏穩定食品 shelf-stable food
壓力 pressure
壓力不足 underpressure
壓力式噴霧乾燥機 pressure spray dryer
壓力計 compressometer; manometer; piezometer; pressure gauge
壓力透析 pressure dialysis
壓力噴霧 pressure atomization
壓力調節閥 pressure regulating valve
壓力調節器 pressure regulator
壓力鍋 pressure cooker, pressure pan; pressure saucepan
壓方塊機 cuber
壓片 pelletization; pelletizing
壓片機 flaking mill; pelleting press; pelletizer; sheeter; tablet machine; tablet press; tabletting machine; tabletting press; waferer
壓汁 juicing
壓成塊的食品 compressed food
壓制劑 repressor
壓型火腿 pressed ham
壓型沙丁魚 pressed sardine
壓型酵母 pressed yeast

壓封式瓶蓋 press on cap
壓封蓋 squeeze cap
壓扁 squash
壓旋蓋 press-twist cap (PT cap)
壓軟木塞機 cork press
壓渣器 pulp press
壓菜泥器 ricer
壓塞機 corker
壓塊奶粉 block milk
壓塊茶磚 baled tea
壓塊機 block moulding machine
壓碎 crush; crushing; quassation; squash
壓碎汁 crushed juice
壓碎機 crusher; crushing mill
壓榨 expelling; press; pressing; squeeze
壓榨出的果汁 expressed juice
壓榨出的液滴 pressed dripping
壓榨板 press plate
壓榨乾燥〔法〕 expressive drying
壓榨殘渣 press residue
壓榨筒 press cylinder
壓榨葡萄汁 press must
壓榨酵母 compressed yeast; pressed yeast
壓榨酵母餅 yeast cake; yeast press cake
壓榨器 presser
壓榨機 expeller; expressive machine; press; presser; squeezing machine
壓蓋機 capper; capping machine; crowning machine
壓製食品 compressed food
壓製脂 oleostearin(e)
壓製麵條 press noodle
壓模 stamper
壓縮火腿 pressed ham
壓縮包裝 compression packing
壓縮計 compressometer
壓縮軍糧 compact ration
壓縮乾燥食品 compressed dehydrated food
壓縮乾糧 compact ration
壓縮脱水食品 compressed dehydrated food
壓縮造粒〔法〕 compression granulation
壓縮機 compressor
壓薄片機 flaking mill
壓薄拉深罐 ironed can
壓薄的麵糰 abaisse
壓濾 compression filtration; pressing; pressure filtration
壓濾油 pressed oil
壓濾殘渣 press residue
壓濾餅 press cake
壓濾器 blotter press; leaf filter; press filter; pressure filter
壓濾機 filter press; press filter; pressure filter
壓麵機 sheeter
嬰兒 infant
嬰兒乳 infant formula
嬰兒食品 baby food; infant food
嬰兒配方食品 infant formula
嬰兒期 infancy
應急口糧 emergency ration
應急食物 emergency food
戴白氏乳〔酸〕桿菌 *Lactobacillus delbruckii*
擊昏 knocking; stun(ning)
擊昏法 captive-bolt stunning
擊昏棒 captive-bolt gun
擊昏槍 captive-bolt gun; captive bolt pistol; power actuated stunner
擠 wring
擠出 extrusion
擠出器 ejector
擠出機 extruder; extruding machine
擠奶 milking

擠奶的女工 milkmaid
擠奶機 mechanical milking; milking machine
擠乳 milking
擠乳者 milker
擠乳站 milk depot
擠乳器 milker
擠乳機 milking machine
擠花袋 paste bag
擠扁罐 collapsed can
擠製 extrusion
擠製機 extruder; extruding machine
擠製糖果 extruded candy
擠壓 extrusion; squeeze; squash
擠壓食品 extrusion food
擠壓烹煮 extrusion cooking
擠壓造粒〔法〕 extruding granulation
擠壓熟化 extrusion cooking
擠壓機 extruder; extruding machine; squeezing machine; wringer
擠檸檬器 lemon squeezer
擦 scrape
擦光 polishing
擦抹取樣法 swab method
擦細器 grater
擦菜板 grater; rasp; rasper; rubbing board
擦碎 rasp
擦碎機 attrition mill; rasp; rubbing machine
擬庸鰈 American plaice
擬清蛋白 paralbumin
擬維生素 vitamer
擱板 shelf
曙紅 eosin(e)
檀香 sandal; sandalwood
檀香〔木〕 santal
檀香油 sandalwood oil
檀香醇 santalol
檀香屬 santalum
檢定 verification
檢波器 detector
檢查 inspection
檢查員 inspector
檢疫 medical inspection; quarantine
檢疫員 quarantine inspector
檢驗 inspection; test
檢驗員 inspector
濱菜 sea cabbage; sea kale
濱螺 winkle
濱藜屬植物 orach(e)
澀味 astringency; astringent taste; astriction; harsh taste
澀味〔性〕 austerity
澀味的 austere
澀味劑 astringent
澀的 astringent
澀風味 astringent flavo(u)r
濕包裝 wet pack
濕式加工 wet process(ing)
濕式加熱 wet heat
濕式冷卻〔法〕 wet cooling
濕式研磨〔法〕 wet grinding
濕式乾燥 humidity drying
濕式清潔 wet cleaning
濕式處理 wet process(ing)
濕式煉油 wet rendering
濕式熬油 wet rendering
濕式篩分〔法〕 wet screening
濕式壓榨 wet pressing
濕灰化 wet ashing
濕乳菇 distant gilled milk cap
濕法去皮 wet dehulling
濕法醃漬 brine salting; pickle salting
濕法鹽醃 pickle curing
濕的 humid; wet
濕空氣 moist air
濕度 humidity
濕度平衡 moisture equilibrium

濕度計 hygrometer; hygroscope; succulometer
濕度記錄儀 hygrograph
濕度儀 hygroscope
濕度調節器 humidistat; hygrostat
濕重 wet weight
濕氣 dampness; moisture
濕氣箱 moist cabinet
濕球溫度計 wet-bulb thermometer
濕貨 wet goods
濕碎〔法〕 wet crushing
濕蒸汽 moist steam; wet steam; wet vapo(u)r
濕潤 madefaction; watering
濕潤性 wettability
濕潤的 madid; moist
濕潤度 wettability
濕潤腐爛 damp off
濕潤箱 sweat box
濕潤劑 humectant; moisture-retaining agent; wetting agent
濕潤器 humidifier
濕熱 moist heat; wet heat
濕熱殺菌〔法〕 wet pasteurization
濕熱處理 heat-moisture treatment
濕熱滅菌〔法〕 moist heat sterilization; wet heat sterilization
濕醃〔法〕 wet curing; wet salting
濕磨〔法〕 wet milling
濕麵筋 wet gluten
濕黴素 hygromycin
濕鹽 wet salt
營業額 turnover
營養 alimentation; nutriture; nutrition; threpsis
營養〔成分〕標簽 nutritional labeling
營養〔體〕的 vegetative
營養口糧 nutritive ration
營養不良 cacotrophia; cacotrophy; dystrophy; innutrition; kakotrophy; malnutrition; paratrophy; undernutrition
營養不足 denutrition; hypotrophy; inanition; nutrient deficiency; nutritional deficiency; oligotrophy; subalimentation; subnutrition; undernourishment; underfeed; undernutrition
營養不足的膳食 inadequate diet
營養分析 trophic analysis
營養代謝 trophism
營養充足的 well-fed
營養失調 dystrophy; nutritional disorder
營養失調症 trophonosis
營養平衡 nutrient balance
營養成分 nutritional ingredient
營養成分表 nutrition board
營養成分標籤 nutrition labeling
營養含量聲明 nutrient content claims
營養抑制劑 nutritional inhibitor
營養佳良 eutrophia; eutrophy
營養供給量 nutrient allowance
營養委員會 nutrition board
營養性水腫 nutritional edema; trophedema
營養性疾病 trophonosis
營養性衰竭 abiotrophy
營養性甜味劑 nutritive sweetener
營養性貧血 alimentary anemia; nutritional anemia
營養性糖尿 alimentary glucosuria
營養治療 nutritional therapy; therapeutic nutrition
營養物 nutrient; nutriment
營養狀況 nutritional status; nutriture; nourishment
營養的 alimentary; nutritional; nutritive; nutrient; trophic
營養品 nourishment; nurture; sustenance
營養品質 nutritional quality

營養品質指數 index of nutritional quality (INQ)
營養指南 nutritional guidance
營養指數 nutrition index; nutritive index
營養指標 nutrition index; nutritive index
營養洋菜〔培養基〕 nutrient agar
營養科學 nutritional science
營養要素 nutritive factor
營養食品 restorative
營養值 culinary value
營養師 dietician; dietitian
營養素 nutrient; pabulum
營養素強化 ennoblement
營養素增補 ennoblement
營養缺乏 denutrition; nutrient deficiency; nutritional deficiency
營養缺乏的 jejune
營養缺乏症 athrepsy; deficiency disease
營養缺乏膳食 jejune diet
營養培養液 nutrient broth
營養密度 nutrient density
營養強化 fortification; nutrition fortification
營養強化大麥 enriched barley
營養強化牛乳 fortified milk
營養強化米 enriched rice
營養強化食品 enriched food; fortified food
營養強化劑 nutrition enhancer; nutritional enriching agent; nutritional fortification substance
營養強化麵包 enriched bread; fortified bread
營養教育 nutrition education
營養液 enchyma
營養異常 anomalotrophy; metatrophy
營養細胞 vegetative cell
營養菌絲 vegetative hypha(e)
營養菌絲體 vegetative mycelium
營養補助食品 supplement food
營養補償食品 restored food
營養資料 nutrition facts
營養過度 hypernutrition; hypertrophy; overnutrition; polytrophia; supernu-trition
營養障礙 nutritional disturbance
營養需求〔量〕 nutritional requirement
營養價值 food value; nutritional value; nutritive value; trophic value
營養價值高的 nutritious
營養增富 nutrition enrichment
營養增補 nutrition restoration
營養增補劑 dietary supplement
營養寬麵條 tagliarini
營養標示 nutrition labeling
營養標準 nutritional standard
營養標籤 nutrition labeling
營養調查 nutritional survey
營養質量 nutritional quality
營養質量指標 index of nutritional quality (INQ)
營養學 alimentology; dietetics; dietics; nutriology; nutrition; threpsology; trophology
營養學家 dietician; dietitian; nutritionist
營養療法 alimentotherapy; sitotherapy; trophotherapy
營養瓊脂〔培養基〕 nutrient agar
環己六醇 inose; inosite; inositol; myo-inositol
環己基乙酸烯丙酯 allyl cyclohexanea-cetate
環己基丁酸烯丙酯 allyl cyclohexane-butyrate
環己基氨基磺酸 cyclohexyl sulfamic acid
環己基氨基磺酸鈣 calcium cyclamate
環己基氨基磺酸鈉 sodium cyclamate
環己烷 cyclohexane
環己烷基磺酸鹽（人工甜味劑） cyclamate
環己醇 cyclohexanol

環多醇 cyclitol
環肌 circular muscle
環形小甜薄餅 jumbal; jumble
環狀通心粉 anelli
環狀菌落 circular colony
環流式釀醋器 recirculating type vinegar generator
環氧乙烷 ethylene oxide
環氧乙烷高級脂肪醇 oxyethylene higher aliphatic alcohol
環氧化物 epoxide
環烴 cyclic hydrocarbon
環開鋁蓋 aluminum ring pull end
環葡聚糖 cycloglucan
環境保護 environmental protection
環境保護局 Environmental Protection Agency (EPA)
環境保護署 Environmental Protection Agency (EPA)
環境溫度 ambient temperature
環糊精 cycloglucan; cyclodextrin; Schardinger dextrin
環輥碾磨機 ring roll mill
環輥壓碎機 ring roll crusher
環醇 cyclitol
環糖醇 cyclitol
環肽 cyclic peptide
療效性食品 therapeutical food
療效食品 curative food; dietetic food
療效飲料 dietetic beverage
療效膳食 therapeutical diet
瞬間巴氏殺菌 flash pasteurization; mallorizing; shocking pasteurization
瞬間巴氏殺菌器 mallorizer
瞬間封口 impulse seal(ing)
瞬間溫度 transient temperature
磷 phosphorus
磷中毒 phosphorism
磷肌酸 phosphagen
磷肌醇甘油酯 phosphoinositide
磷苯二甲酸 phthalic acid
磷神經脂質 phosphosphingolipid
磷脂 phosphatide; phospholipid
磷脂蛋白 phospholipoprotein
磷脂酸 phosphatidic acid
磷脂酸銨鹽 ammonium phosphatide
磷脂酰乙醇胺 phosphatidylethanolamine
磷脂酰甘油 phosphatidyl glycerol
磷脂酰絲胺酸 phosphatidylserine
磷脂酰糖 phosphatidyl sugar
磷脂酰膽鹼 lecithin(e); phosphatidylcholine
磷脂酶 phosphatidase; phospholipase
磷蛋白質 phosphoprotein
磷絲胺酸 phosphoserine
磷葡萄糖異構酶 phosphoglucoisomerase
磷葡萄糖酸 phosphogluconic acid
磷葡萄糖變位酶 phosphoglucomutase
磷酸 phosphoric acid
磷酸一鈣 calcium dihydrogen phosphate; monocalcium phosphate
磷酸一鈉 sodium phosphate monobasic
磷酸二氫鈣 calcium dihydrogen phosphate
磷酸二氫鈉 sodium phosphate, monobasic
磷酸二氫鉀 potassium dihydrogen phosphate
磷酸二氫銨 ammonium dihydrogen phosphate
磷酸二鈣 dicalcium phosphate
磷酸二鈉 disodium phosphate; sodium phosphate dibasic
磷酸三鈣 tricalcium phosphate
磷酸三鈉 sodium phosphate tribasic; trisodium phosphate
磷酸三鉀 tripotassium phosphate
磷酸化二澱粉磷酸酯 phosphated distarch phosphate
磷酸化〔作用〕 phosphorylation

磷酸化酶 phosphorylase
磷酸丙醣 triose phosphate
磷酸甘油 phosphoglycerol; glycerophosphate; glycerophosphoric acid
磷酸甘油酯 phosphoglyceride
磷酸肌酸 creatine phosphate; phosphocreatine
磷酸肌醇 phosphoinositide
磷酸原 phosphagen
磷酸核糖苷 phosphoriboside
磷酸氫二鈉 disodium hydrogen phosphate
磷酸氫二鉀 dipotassium (hydrogen) phosphate
磷酸氫鈣 acid calcium phosphate; calcium monophosphate; calcium phosphate, dibasic
磷酸氫鎂 magnesium hydrogen phosphate
磷酸球蛋白 phosphoglobulin
磷酸單澱粉酯 monostarch phosphate
磷酸鈣 calcium phosphate
磷酸鈉 sodium phosphate
磷酸葡〔萄〕糖異構酶 phosphoglucoisomerase
磷酸葡糖酸 phosphogluconic acid
磷酸解酶 phosphorylase
磷酸鉀 potassium phosphate; tripotassium orthophosphate
磷酸酯 phosphate
磷酸酯酶 phosphoesterase
磷酸銨 ammonium phosphate
磷酸澱粉鈉 sodium starch phosphate; sodium phosphate starch
磷酸糖蛋白 phosphoglucoprotein
磷酸膽鹼 choline phosphate
磷酸鎂 magnesium phosphate, tribasic
磷酸鎂銨 ammonium magnesium phosphate
磷酸雙澱粉 distarch phosphate
磷酸鐵 ferric phosphate
磷酸鹽 phosphate
磷酸吡哆醛 pyridoxal phosphate
磷酸酶 phosphatase
磷蝦 krill
磷膽鹼 phosphocholine
磺化〔作用〕 sulfonation
磺基丙胺酸 cysteic acid
磺基鈉代丁二酸二辛酯 dioctyl sodium sulfosuccinate
磺酸 sulfonic acid
磯沙蠶 palolo (worm)
穗 spike
穗狀山楂 spike hawthorn
穗狀花 spike
穗醋栗 currant
穗薰衣草油 spike lavender oil
簇 cluster
簇生辣椒 cluster pepper
糠 bran; chisel; husk; hull
糠屑 pollard
糠基乙酸異丁酯 isobutyl furfuryl acetate
糠基硫醇 furfuryl mercaptan
糠基吡咯 furfuryl pyrrole
糠醛 furfural
糜 porridge
糞肥 dung; muck; manure
糞便 dung; excrement; faeces; feces; intestinal discharge; stercus; stool
糞便大腸桿菌群 fecal coliform
糞便的 faecal; fecal
糞便細菌 fecal bacteria
糞便鏈球菌 fecal streptococcus
糞後膽色素 stercobilin
糞膽色素 stercobilin
糞鏈球菌 Streptococcus faecalis
糟粕 dreg
糙皮側耳（蕈類） common oyster mushroom; *Pleurotus ostreatus*
糙皮甜瓜 cantaloup(e)

糙米 brown rice; cargo rice; hulled rice; husked rice; paddy rice; rough rice; unpolished rice
糙米碎屑 brown grits
糙面精整 matte finish; stone finish
糙莓 blackcap raspberry; black raspberry; blackcap; thimbleberry
糙葉山楂 niagara hawthorn
縮二胺酸 dipeptide
縮二脲 biuret
縮二脲反應 biuret reaction
縮水甘油 glycidol
縮合 condensation
縮胺酸 peptide
縮短 shortening
縮聚〔作用〕 polycondensation
縮聚物 polycondensate
縮醛 acetal
縮醛樹脂 acetal resin
縮醛磷脂 acetal phosphatide
縫合 seam
總水分 total moisture
總可消化養分 total digestible nutrients (TDN)
總可溶固形物 total soluble solid
總可溶物 total solubles
總固形物 total solid
總活菌數 total viable count
總重〔量〕 gross weight; total weight
總脂肪物質 total lipid
總產量 total yield
總氮 total nitrogen
總菌數 total count
總菌數洋菜培養基 plate count agar
總匯三明治 club sandwich
總匯軟芝士 club cheese
總匯軟乾酪 club cheese
總溶固形物 total dissolved solid (TDS)
總酸度 total acidity
總樣品 gross sample
總熱量 total heat
總糖〔量〕 total sugar
總醣 total carbohydrate
總鹼度 total alkalinity
縱肌 longitudinal muscle
縱帶笛鯛 black striped snapper
縱飲 drinking bout
繁星康吉鰻 common Japanese conger
繁星糯鰻 conger eel
繁殖 breed; proliferation; reproduction
繁殖力 fertility
繁殖孢子 propagative spore
繁殖菌絲體 reproductive mycelium
繁殖酵母 propagate yeast
繁殖槽 propagation tank
翼 ala
翼式攪拌器 blade agitator; blade stirrer
翼豆 winged bean
聲波〔振動〕處理 sonication
聲音 sound
聯乙酰 biacetyl; diacetyl
聯合包裝 multipack
聯合國食品法典委員會 Codex Alimentarius Commission
聯合國蛋白質顧問組 Protein Advisory Group (PAG)
聯合國糧食〔及〕農業組織 Food and Agriculture Organization (FAO); United Nations Food and Agriculture Organization (UNFAO)
聯合國糧農組織 Food and Agriculture Organization (FAO); United Nations Food and Agriculture Organization (UNFAO)
聯合國糧農組織(FAO)世界衛生組織(WHO)聯合食品法規委員會 Joint FAO/WHO Food Codex Alimentarius Commission; FAO/WHO

聯邦科學與工業研究組織 Commonwealth Scientific and Industrial Research Organization (CSIRO)
聯邦食品藥品及化粧品法〔案〕(美國) Federal Food, Drug and Cosmetic Act (FD&C)
聯苯 biphenyl; diphenyl; phenylbenzene
聯苯胺試驗 benzidine test
聯苯酚 diphenol; *o*-phenyl phenol (OPP)
聯苯酚鈉 sodium *o*-phenylphenolate
聯胺 hydrazine
聯結 link; linking
膺品 duff
臂式攪拌器 arm mixer
臂胸臀指數 ACH index
臀肉 dock; rump
臀肉牛排 rump steak
臀肌 gluteus
臀骨 hip bone
臀部 buttocks; fundament; hip; hind end; rump
臀部肉 buttocks; cushion
臀腿 thick flank
臀鰭 anal fin
膿毒病 sepsis
膽汁 bile; gall
膽汁培養基 bile medium
膽汁酸 bile acid
膽石 gallstone
膽色素 bile pigment
膽固醇 cholesterol; cholesterin
膽固醇酯 cholesterol ester
膽固醇酶 cholesterase
膽甾醇 cholesterin; cholesterol
膽甾醇酶 cholesterase
膽紅素 bilirubin
膽骨化醇 vitamin D_3
膽鈣化固醇 cholecalciferol; vitamin D_3
膽綠素 biliverdin
膽綠黃素 verdoflavin
膽酸 cholic acid
膽酸鈉 sodium choleate
膽褐素 bilifuscin
膽囊 bile cyst; gall bladder
膽鹼 choline; sinkaline
膽鹼磷脂 lecithin(e); phosphatidylcholine
膽鹽 bile salt
臨床營養〔學〕 clinical nutrition
臨界值 threshold limit value
臨界溫度 critical temperature
臨界點 critical point
臨產牛 springer cow
舉杯祝賀者 toaster
薄水果餡甜餅 strudel
薄片 chip; foil; leaf; slice
薄片冰 flake ice
薄牛排 minute steak
薄包裝材料 light wrapper
薄皮的 thin skinned
薄肉片 collop; escalope
薄肌原絲 thin filaments
薄卷餅 blintze
薄板濾器 plate filter
薄的 thin
薄度 thinness
薄烤餅 hot cake; pancake
薄烙餡餅 blintze
薄紙 tissue
薄脆餅 snap; wafer
薄脆鬆餅 gauffre
薄細木條 slat
薄荷 brandy mint; mint; spearmint
薄荷油 brandy mint; *Mentha arvensis* oil; mint oil; peppermint; pennyroyal oil; peppermint oil; spearmint oil
薄荷香料 mint flavo(u)ring
薄荷香精 mint essence
薄荷草 spearmint

薄荷草油 spearmint oil
薄荷茉莉酒 mint julep
薄荷腦 menthol
薄荷葉 mint leaves
薄荷酮 menthone
薄荷調味料 mint flavo(u)ring
薄荷醇 menthol; peppermint camphor
薄荷糖 brandy mint; humbug; mint; peppermint
薄荷醬 mint sauce
薄荷屬 *Mentha*
薄魚片 escalope
薄麥餅 flannel cake
薄硬餅乾 water biscuit
薄煎餅 flapjack; griddlecake; hot cake; pancake; slapjack
薄〔煎〕餅用麵粉 pancake flour
薄煎餅混料粉 pancake mix
薄餅 grigger; pizza
薄餅捲 wafer roll
薄層色譜法 thin layer chromatography (TLC)
薄層乾燥 thin layer drying
薄層層析法 thin layer chromatography (TLC)
薄糊 gruel
薄膜 thin film
薄膜分離 membrane separation
薄膜包裝 film packaging
薄膜式巴氏殺菌法 stassanization
薄膜式冷凝 film type condensation
薄膜式脱氣機 film type deaerator
薄膜乾燥 film drying
薄膜乾燥機 film dryer
薄膜蒸發器 film evaporator; thin film evaporator
薄薑餅 parliament cake
薄鬆餅 wafer
薑 ginger; zingiber; ginger root
薑汁汽水 ginger ale; ginger beer; ginger pop
薑汁啤酒 ginger ale; ginger beer; shandygaff
薑汁餅乾 ginger nut; gingerbread; gingersnap
薑汁麵包 cock
薑味小酥餅 ginger nut; gingersnap
薑味糕點 gingerbread
薑油 ginger oil
薑油香精 ginger oily flavo(u)r
薑油脂 gingerin
薑油酮 zingerone
薑油樹脂 ginger oleoresin
薑根 ginger root; race
薑烯酚 shogaol
薑黃 curcumin; curcuma; turmeric; turmeric yellow
薑黃油 curcuma oil; turmeric oil
薑黃油樹脂 turmeric oleoresin
薑黃素 curcumin; turmeric yellow
薑黃酮 turmerone
薑黃酸 turmeric acid
薑黃醇 turmerol
薑餅 gingerbread; brandy snap
薔薇 rosa
薔薇果 hip; rose hip
薯仔 potato
薯芋 yam
薯泥 mashed potato
薯粉 potato flour
薯絲麵 cassava vermicelli
薯蓉 mushy potato(es); mashed potato
薯蕷 Chinese yam; cinnamon vine; yam
薊 thistle
蟑螂 roach
螳螂蝦 squilla
螺母 nut
螺紋 worm

螺紋保證蓋 securo cap
螺紋管 worm pipe
螺旋 spiral
螺旋口瓶 screw top jar
螺旋片 spiral slice
螺旋充填機 auger filler
螺旋打包機 auger packer
螺旋式熱交換器 votator
螺旋〔式〕輸送機 screw conveyor; spiral conveyor; worm conveyor
螺旋冷凍機 spiral freezer
螺旋形布卡提通心粉 fusilli bucati
螺旋形的 spiral
螺旋豆 mesquite
螺旋拔塞器 corkscrew
螺旋狀細菌 helical bacteria
螺旋泵 screw pump
螺旋粉 fusilli; spiral pasta; spirali
螺旋粉碎機 screw crusher
螺旋紋通心粉 elicoidali
螺旋送料機 screw feeder
螺旋通心粉 fusilli
螺旋喂送器 auger feeder
螺旋菌屬 *Spirillum*
螺旋進料器 auger feeder; spiral feeder
螺旋塞 screw top
螺旋榨汁機 pressure screw extractor; screw extractor
螺旋蓋 screw top; screw cap
螺旋蓋封蓋機 screw capper
螺旋蓋擰蓋機 screw cap tightener
螺旋槳 propeller
螺旋槳混合器 propeller mixer
螺旋槳葉片攪拌器 ribbon mixer
螺旋槳攪拌器 propeller agitator; propeller stirrer
螺旋壓榨機 expeller; jack screw press; pressure screw extractor; screw extractor; screw (type) press; worm press
螺旋擠出機 screw expeller press
螺旋擠壓機 plodder; screw (type) extruder
螺旋藻 spirulina
螺旋攪拌器 crutcher; spiral agitator; spiral stirrer
螺帽 screw cap
螺絲〔形〕通心粉 cavatappi; fusilli
螺絲母 nut
螺絲豆 mesquite
螺絲粉 fusilli; gemelli; spiral pasta; spirali
螺絲起子 screwdriver
螺運機 screw conveyor
講究飲食 gormandise; gormandize; gourmandise; gourmandize
講究飲食的人 gastronome(r)
賽尼可 Xenical
賽克拉美 cyclamate
賽克拉美鈉 sodium cyclamate
賽茨過濾器 Seitz filter
賽璐凡 cellophane
賽璐玢 cellophane
賽璐珞 celluloid
購物袋 shopping bag
趨脂性 lipotropy
還原 reduction; reconstitution
還原力 reducing power
還原角蛋白 keratein
還原乳 reconstituted milk
還原型肌紅蛋白 reduced myoglobin
還原食品 reconstituted food
還原酮 reductone
還原蝦紅素 astaxanthin
還原劑 reducer; reducing agent; reductant
還原糖 reducing sugar
還原鐵 iron, reduced
還原酶 reductase
醣蛋白 glucoprotein
醣解途徑 Entner-Doudoroff pathway
醣醛酸 glycuronic acid

醜橘 ugli
鍍 plating
鍍金口蘑（蕈類） sandy knight cap
鍍銀食具 flat silver
鍍錫 tin; tinning
鍍錫容器 tinware
鍍錫層 tin coating
鍍錫薄板 sheet tin
鍍錫鐵皮 tin; tinplate
鍍錫鐵罐 tin can
鎂 magnesium
鍋 braiser; braizer; kettle; pot
鍋子 braiser; braizer
鍋巴 rice crust
鍋垢 incrustation; scale
鍋煮〔沸〕 pan boiling
鍋爐 boiler
鍋爐防垢劑 boiler compound
鍋爐房 boiler room
鍋鹽 pan salt; stoved salt
闊麵 lasagna
闊麵條 fettuccelle
闊鱗方頭魚 ocean whitefish
隱足魚 cutlass fish
隱黃質 cryptoxanthin
霜 frost
霜降肉 marbled meat
霜害 frost damage
霜淇淋 soft ice cream
韓國泡菜 kimchi
顆粒 grain; granula; granule
顆粒性 graininess
顆粒狀的 granular
餵食 feeding
餵養不足 underfeed
鮮奶油 sweet cream
鮮肉 fresh meat; green meat
鮮乳 city milk; fresh milk; market milk
鮮乳車 dandy cart
鮮乳油 sweet cream
鮮乳油巧克力 creamy chocolate
鮮乳油乳酪 sweet cream butter
鮮乳油派 cream pie
鮮味 delicate flavo(u)r; umami
鮮果 fresh fruit
鮮度 freshness
鮮重 fresh weight; green weight; hot weight; wet weight
鮮蛋 fresh egg
鮮魚 fresh fish
鮮銷 marketing fresh
鮮凝乳〔塊〕 sweet curd
鮮麵糰 sweet dough
鮫肝醇 chimyl alcohol
鮪 tuna; tunny; yaito bonito
鮪魚三明治 tuna sandwich
鮪魚罐頭 tuna fish
鮭魚 calico salmon; lax; salmon; trout; truff
鮭魚卵 salmon roe
鮭魚油 salmon oil
鴿子 dove; pigeon
鴿豌豆 pigeon pea
麋 elk
麋鹿 elk
黏牙的糖果 stickjaw
黏合 adhesion; cohesion; coherence; glue
黏合用麵粉 binder flour
黏合〔作用〕 agglutination
黏合劑 adhesive; binding agent; bonding agent; binder; cohesive
黏多糖 mucotin; mucopolysaccharide
黏多糖酶 mucinase
黏住 stick
黏乳 ropy milk
黏固〔作用〕 cementation
黏性 mucosity; glutinousness; glutinosity; gummosity; ropiness; sliminess; stickiness; tackiness; viscidity; viscousness

黏性物質 sticky matter
黏性的 gummy; viscous
黏性產品 viscous product
黏性膠體 viscoloid
黏性凝膠 viscogel
黏附力 adhesion
黏度 viscosity
黏度曲線儀 viscograph
黏度計 viscometer; viscosimeter
黏度測定法 viscometry; viscosimetry
黏度測定儀 viscoamylogram
黏度儀 viscograph
黏流 viscous flow
黏核桃 clingstone peach
黏病毒 myxovirus
黏液 mucus; mucilage; slime
黏液化 mucification
黏液外膜 mucous lining
黏液性 mucosity
黏液狀的 mucoid
黏液狀菌落 mucoid colony
黏液的 slimy
黏液病毒 myxovirus
黏液素 mucin; mucotin
黏液素元 mucigen; mucinogen
黏液菌 myxobacteria
黏液腺 mucous gland
黏液層 slime layer
黏球蛋白 mucoglobulin
黏球菌 myxococcus
黏蛋白 mucin; mucoprotein
黏蛋白元 mucigen; mucinogen
黏蛋白原 mucigen; mucinogen
黏蛋白酶 mucinase
黏魚類 eelpout
黏結 cohesion
黏結劑 agglomerant; binder
黏絲狀 ropiness
黏絲菌 myxobacteria
黏絲腐敗 rope
黏著 coherence
黏著劑 adhesive; agglutinant
黏貼 stick; sticking
黏稠元 viscogen
黏稠物 brei; dope; sticky matter
黏稠度 consistence; consistency
黏稠度計 consistometer
黏稠凝膠 viscogel
黏稠劑 pasting agent
黏葡聚糖 myxoglucosan
黏滯 viscousness
黏滯性 glutinousness; tenacity
黏滯的 viscous
黏滯流動 viscous flow
黏滯液體 viscous liquid
黏精 viscogen
黏聚 agglomeration
黏聚劑 agglomerant
黏酸 mucic acid
黏彈性 viscoelasticity
黏漿 mucilage
黏糊狀 doughiness
黏膜 mucosa; mucous lining; mucous membrane
黏膜下層 submucosa
黏膠 mucilage
黏膠〔液〕 viscose
黏膠腸衣 viscose casing
黏膠纖維腸衣 viscose casing
黏質 viscidity
黏質形成細菌 slime-forming bacteria
黏質物 slime
黏質真菌 slime fungus
黏質細菌 slime bacteria
黏質發酵 ropy fermentation; slime fermentation
黏質層 slime layer
黏質樹膠 mucilaginous gum

黏質黴菌 slime mol(u)d
黏澱粉 slime starch
黏頭小包腳菇 stubble field volvar
黏麵糰 sticky dough
黏着物 sticker
黏肽 mucopeptide
點心 cake; eating between meal; refection; refresher; refreshment; snack; snack food
點柄乳牛肝菌 dotted stalk bolete
點滴式咖啡壺 Dripolator
點滴培養 drop culture
點霉素 notatin
黛可莉〔雞尾〕酒 daiquiri
黛西雞尾酒 daisy
齋月 ramadan
齋戒 fast
齋戒日 fast day
齋戒的 lenten
齋啡 black coffee
檟如樹 *Anacardium occidentale*
獼猴桃香精 kiwi flavo(u)r
薏仁 adlay; jobstears
薏米 adlay; jobstears
薏苡 adlay; jobstears
薤 rakkyo
蕹菜 water convolvulus; water spinach
螯(蝦或蟹等) pincers
螯合〔作用〕 chelation
螯合物 chelate
螯合的 chelate
螯合劑 chelating agent; sequester; sequestering agent; sequestrant
醚化〔作用〕 etherification
醛 aldehyde
醛去氫酶 aldehyde dehydrogenase
醛氧化酶 aldehyde oxidase
醛脱氫酶 aldehyde dehydrogenase
醛醇 alditol
醛醇樹脂 aldol resin
醛醇縮酶 aldolase
醛糖 aldose
醛糖酸 aldonic acid
醛糖苷 aldoside
醛縮酶 aldolase; zymohexase
醛還原酶 Schardinger enzyme
醛醣二酸 aldaric acid
醛醣醇 alditol
醛酶 aldehydrase
鍶 strontium
蟎 acarus

十八畫

餬口 subsistence
叢 cluster
擴張器 spreader
擴散 diffusion
擴散系數 diffusion coefficient
擴散性 diffusibility
擴散梯度 diffusion gradient
擴散率 diffusivity
擴散器 diffuser; diffusor
擺動分級機 shuffle grader
斷奶 wean
斷奶小牛 weaner calf
斷奶幼畜 weaner
斷乳 ablactation; wean
斷乳幼畜 weanling; weaner
斷乳食品 weanling food
斷食 abrosia
斷食療法 nestotherapy
斷頭法 kosher method
檳榔 areca nut; betel nut; pinang
檳榔樹 betel nut palm
櫃式巴氏滅菌器 cabinet pasteurizer
櫃式烤爐 deck oven
檸黃質 citroxanthin
檸檬 *Citrus limon*; lemon
檸檬子油 lemon pip oil
檸檬水 lemonade
檸檬汁 lemon juice
檸檬汁汽水 lemon squash
檸檬汁威士忌〔酒〕 whisky sour
檸檬汁飲料 lemon squash; lemonade
檸檬皮 lemon peel; zest
檸檬皮碎粒 lemon grating
檸檬杏仁餡〔酪〕餅 maid of honor
檸檬油 citron oil; lemon oil
檸檬油香精 lemon oily flavo(u)r
檸檬派 lemon pie
檸檬苦素 limonin
檸檬香水薄荷 lemon balm
檸檬香精 lemon essence; lemon flavo(u)r
檸檬素 citrin(e)
檸檬草 citronella (grass); lemon grass
檸檬草油 lemongrass oil
檸檬桉油 citriodora oil
檸檬烯 limonene
檸檬飲料 lemon beverage
檸檬黃 tartrazine
檸檬黃鋁色澱 tartrazine aluminum lake
檸檬葉油 lemon leaf oil; petitgrain lemon oil
檸檬酪醬 lemon curd
檸檬精油 limonene
檸檬酸 citric acid
檸檬酸一鈉 sodium dihydrogen citrate
檸檬酸一鉀 potassium dihydrogen citrate
檸檬酸二氫鉀 potassium dihydrogen citrate
檸檬酸二鈉 disodium citrate
檸檬酸三乙酯 triethyl citrate
檸檬酸三鉀 potassium citrate
檸檬酸三銨 triammonium citrate

檸檬酸亞錫二鈉 disodium stannous citrate
檸檬酸亞鐵鈉 sodium ferrous citrate
檸檬酸和脂肪酸甘油脂 citroglyceride
檸檬酸異丙酯混合物 isopropyl citrate mixture
檸檬酸單甘油酯 monoglyceride citrate
檸檬酸硬脂酰單甘油酯 stearyl monoglyceridyl citrate
檸檬酸硬脂酰酯 stearyl citrate
檸檬酸菌素 citromycetin
檸檬酸鈣 calcium citrate
檸檬酸鈉 sodium citrate
檸檬酸鉀 potassium citrate
檸檬酸錳 manganese citrate
檸檬酸鐵 ferric citrate; iron citrate
檸檬酸鐵銨，棕色 ferric ammonium citrate, brown
檸檬酸鐵銨，綠色 ferric ammonium citrate, green
檸檬酸鹽 citrate
檸檬糖 lemon drop
檸檬餡餅 lemon pie
檸檬醛 citral
檸檬醛二乙縮醛 citral diethyl acetal
檸檬羹 lemon curd
檸檬鰨 lemon sole
檯布 napery; tablecloth
檯秤 bench scale
歸零 zeroing
瀉流 effusion
瀉劑 purgative
瀉藥 cathartic
瀉鹽 bitter salt; Epsom salt; magnesium sulfate
濾水砂 permutite
濾水器 water filter; water purifier; water strainer
濾出液 filtrate; percolate
濾布 cloth filter; filter cloth; straining cloth; tammy
濾石 filtros
濾床 filter bed
濾杓 strainer-dipper
濾板 filter plate
濾泥餅 mudcake
濾泥壓乾機 mud press
濾砂 permutite
濾紙 filter paper
濾紙筒 cellulose thimble
濾紙電泳 paper electrophoresis
濾紙層析法 paper chromatography
濾茶器 tea infuser
濾液 colature; filter liquor; filtrate; press juice
濾淨蜂蜜 strained honey
濾袋 filter bag
濾渣 filter residue; residue
濾渣器 pulp filter
濾筒 filtering cartridge
濾過性病毒 virus
濾過性微生物學 virology
濾鼓 drum filter
濾墊 filter bed
濾網 filter screen; sieve; strainer
濾餅 filter cake; filter mass; mudcake; press cake
濾膜 filter membrane
濾器 colander; strainer
濾篩 filter screen; strainer
濾鍋 colander; cullender
燻大麻哈魚 lox
燻肉 smoked meat
燻肉薄片 bard
燻材 smoking wood
燻豆 smoky bean
燻味 smoky flavo(u)r

燻房 smoke house; smokery; smoking room
燻肩肉 shoulder bacon
燻芝士 smoked cheese
燻香 incense
燻香料 incense
燻香腸 smoked sausage
燻乾 smoke dry
燻乾魚 hard smoked fish
燻乾酪 smoked cheese
燻乾臘腸 saveloy
燻培根 smoked bacon
燻液 liquid smoke; liquid smoke solution; smoke solution; smoking fluid
燻硫 sulfuring; sulphuring
燻魚 smoked fish
燻魚乾 hard smoked fish
燻蒸〔消毒〕 fumigation
燻蒸〔消毒〕劑 fumigant; fumigating agent
燻蒸〔消毒〕器 fumigator
燻製 smoke curing; smoking
燻製品 smoked product
燻製食品 smoked food; smoked provision
燻製香腸 smoked sausage
燻製業者 smoker
燻劑 fumigant
燻燒 smolder
燻鮭 smoked salmon
燻鮭魚 lox
燻臘腸 cervelas; cervelat
燻鯡魚 red herring
燻鹹鰊魚 glasgow magistrate
燻鱈魚 finnan haddock
獵物 game
獵捕海豹業 sealing
獵禽 game bird; game fowl
甕 urn; vat
甕漬 vat salting
癖 addictive
癒創〔木〕樹脂 guaiac gum; guaiac resin; gum guaiac; guaiacum (resin)
癒創酸 traumatic acid
簡易午餐 buffet lunch
簡易包裝 convenience package
簡便早餐 Continental breakfast
簡便食品 convenience food
簡便商店 convenience store
簡便商品 convenience goods
簡單蛋白質 simple protein
糧食 foodstuff; provisions; rations; sustenance; victual; vivers
糧食供應 victual(l)ing
糧食補給船 victual(l)er
糧食〔農〕作物 alimentary crop
糧食儲備 victual(l)ing
糧倉 grain bin; grain elevator; granary; garner
糧草 provender
糧農組織 Food and Agriculture Organization (FAO); United Nations Food and Agriculture Organization (UNFAO)
糧穀酒精 grain alcohol
織布魚 weaver
繡球菌(蕈類) cauliflower fungus
翻車魚 moonfish; sunfish
翻車魨 ocean sunfish
翻烤長麵包 somerset brick
翻煮 reboil
翻滾式殺菌 end-over-end (processing)
翻熱 reheat(ing)
臍 navel
臍橙 navel orange
舊貨 second hand
藏花 crocus; saffron; safran
藏花素 crocin
藏花酸 crocetin
藏花醇 safranol

藏紅花 crocus; saffron; safran
藏紅花色素 saffron colo(u)r
藏紅醇 safraninol
藏茴香籽 caraway seed
藏酒量 cellar
薩拉米香腸 Italian salami; salami (sausage)
薩能奶山羊 Saanen
藍光酸性紅 azorubin(e); carmoisine
藍芝士 bleu cheese; blue cheese
藍背紅點鮭 oquassa trout
藍桉葉油 eucalyptus oil
藍乾酪 bleu cheese; blue cheese
藍莓 blueberry
藍普卡魚 dolphin fish
藍紫洋菇 blewits
藍越橘 tangleberry
藍黃紅菇 charcoal burner
藍圓孔牛肝菌 cornflower bolete
藍圓鰺 round scad
藍溪藻黃質 myxoxanthin
藍瑞斯豬 Landrace (swine)
藍綠藻 blue-green algae
藍豌豆 blue pea
藍錠果紅 sweetberry honeysuckle red
藍錐齒鯊 blue shark
藍蟹 blue crab
藍鯨 blue whale
藍藻色素 spirulina blue colo(u)r
藍鰓魚 bluegill sunfish
藍鰭鮪 bluefin tuna
藍鱈魚 bastard cod
藍鰮鯨 blue whale
薰衣草油 aspic oil
薰草素 coumarin
薰草酮 coumarone
薰蒸消毒 gassing
蟯蟲 threadworm
蟯蟲病 enterobiasis
蟲 worm
蟲孔 wormhole
蟲害 insect damage
蟲蛀〔水〕果 scaly fruit; wormy fruit
蟲蛀的 wormy
蟲漆 shellac
蟲漆蠟酸 lacceroic acid
蟲膠 lac; shellac
蟲膠紅 lac dye; lac red
蟲膠酸 shellolic acid
蟲膠蠟 shellac wax
蟲繭蜜 trehala
蟠桃 flat peach; peento
覆盆子 raspberry; red raspberry
覆盆子酮 raspberry ketone
覆蓋物 integument
覆糖軟糖 Turkish delight; Turkish paste
覆糖霜 top icing
豐產 fertility
豐滿度 plumpness
轉乙酰作用 transacetylation
轉化澱粉 converted starch
轉化糖 invert sugar; inverted sugar; invertose
轉化糖漿 invert syrup
轉化酶 invertin; invertase; saccharase; sucrase
轉台 rotary table
轉甲基作用 transmethylation
轉果糖苷酶 transfructosidase
轉胺作用 transamination
轉胺酶 aminotransferase
轉速計 tachometer
轉筒 reel; tumbler
轉筒式分級機 mammoth grader
轉筒式乾燥機 revolving dryer
轉筒式混合器 barrel mixer; drum mixer
轉筒剝皮機 drum barker
轉筒乾燥〔法〕 drum drying

轉筒乾燥機 drum dryer
轉筒篩 bolting reel; drum screen; screening trommel; trommel
轉筒濾器 rotary filter
轉葡〔萄〕糖苷酶 transglucosidase
轉鼓 roller; rotary drum; tumbler
轉鼓式洗滌機 rotary washer
轉鼓式乾燥〔法〕 drum drying
轉鼓式乾燥機 drum dryer
轉鼓式混合機 rotary drum mixer
轉鼓式發酵槽 rotary drum fermenter
轉羧作用 transcarboxylation
轉酯作用 transesterification
轉管培養〔物〕 roll tube culture
轉臺 lazy susan
轉盤 rotary table; turntable
轉盤斬拌機 bowl cutter
轉篩 reel; revolving screen; revolving sieve
轉糖苷作用 transglycosylation
轉糖苷酶 transglycosidase
轉磷酸作用 transphosphorylation
轉鍋 revolving pan
轉酰〔基〕作用 transacylation
轉酰胺〔基〕作用 transamidation
轉爐 rotating oven
轉肽作用 transpeptidation
醫院飲食 hospital diet
醫院膳食 hospital diet
醫療食品 diet food
醬 spread
醬色 burnt sugar colo(u)r; caramel; caramel colo(u)r; sugar colo(u)ring
醬果酒 berry wine
醬果類 berry fruit
醬油 light soy sauce; shoyu; soy; soy sauce; soya sauce; soybean sauce
醬油壺 gravy boat
醬油膏 condensed soy
醬油酵母〔菌〕 *Saccharomyces soya*
醬油醪 shoyu mash; soy sauce mash
醬油麴 soy sauce koji
醬油麴菌 *Aspergillus sojae*
醬狀食品 pate
醬料 sass sauce
醬醪 shoyu mash; soy sauce mash
醬醪 bean paste
鎳 nickle
鎮靜劑 depressant
鎘 cadmium
鎚式粉碎機 hammer mill; hammer grinder
鎚碎機 hammer mill
鎚磨機 hammer grinder
鎗殺法 gun shooting
離子 ion
離子化 ionization
離子交換水 ion exchange water
離子交換色譜法 ion exchange chromatography
離子交換〔作用〕 ion exchange
離子交換基 ion exchange group
離子交換過濾 ion exchange filtration
離子交換〔管〕柱 ion exchange column
離子交換劑 ion exchanger
離子交換器 ion exchanger
離子交換樹脂 ion exchange resin
離子交換樹脂層 resin bed
離子性清潔劑 ionic detergent
離子強度 ionic strength
離子過濾層析法 ion filtration chromatography
離子鍵 ionic bond
離心〔作用〕 centrifugation
離心力除菌 bactofugation
離心分離物 centrifugate
離心分離蜂蜜 centrifugal honey
離心分離糖漿 centrifugal syrup
離心式乾燥機 whizzer

離心洗滌 centrifugal washing
離心粉碎機 centrifugal disintegrator; centrifugal mill
離心除菌機 bactofugeur
離心液 centrifugate
離心淨化 centrifugal purification
離心淨化器 centrifugal clarifier
離心球磨機 centrifugal ball mill
離心瓶 centrifuge bottle
離心脫水機 centrifugal dryer
離心提煉油脂 centrifugal rendering
離心管 centrifuge tube
離心蒸發機 centrifugal evaporator
離心噴霧 centrifugal atomization
離心噴霧乾燥機 centrifugal spray dryer
離心噴霧盤 atomizing disk
離心噴霧器 centrifugal atomizer
離心澄清 centrifugal purification
離心盤式噴霧器 centrifugal dish atomizer
離心機 centrifuge; fugal machine
離心薄膜蒸發機 centrifugal film evaporator
離心籃 centrifuge basket
離水〔現象〕 syneresis
離別酒 doch-an-dorris
離乳 wean
離乳小牛 weaner calf
離乳食品 weanling food
離果 schizocarp
離核桃 free-stone peach
離胺酸 lysine
離解 dissociation
離解常數 dissociation constant
離解酶 macerating enzyme
離層素 abscisin
離層酸 abscisic acid
離體 in vitro
雜己聚糖 heterohexosan
雜交 hybridization
雜交玉米 hybrid corn
雜交育種 cross breeding
雜交果 allocarpy
雜交品種 hybrid variety
雜交繁殖 cross breeding
雜多糖 heteropolysaccharide
雜色 mottle
雜色鮑 small abalone
雜味 unclean flavo(u)r
雜物 sundries
雜食性 polyphagia; polyphagy
雜食性動物 heterophagous animal; omnivore
雜草 weed
雜草防治 weed control
雜草臭 weed faint
雜貨 sundries
雜貨西藥店 drugstore
雜貨店 grocer; grocery; grocery store; variety shop; variety store
雜貨舖 variety shop; variety store
雜酚油 creosote
雜碎 fancy meat; numbles; pluck
雜碎肉 variety meat
雜碎肉凍 brawn
雜碎餡 harslet; haslet
雜聚糖 heteroglycan
雜酵母 wild yeast
雜熏衣草油 lavandin oil
雜質 impurity
雜醇油 fusel oil
雜糖苷 heteroside
雜燴 chowder; hodgepodge; hotchpotch; mishmash
雜燴文蛤 clam chowder
雜環的 heterocyclic
雜環胺基酸 heterocyclic amino acid
雙乙酸鈉 sodium (hydrogen) diacetate
雙乙酰 biacetyl; diacetyl
雙三醇 farnesol

雙叉乳酸桿菌 bifidus; *Lactobacillus bifidus*
雙叉乳酸桿菌屬 *Bifidobacterium*
雙甘油酯 diacetyl tartaric acid esters of mono-and diglycerides
雙岐乳〔酸〕桿菌 *Lactobacillus bifidus*
雙把燉鍋 stew pot
雙歧藻屬 *Scytonema*
雙孢磨菇 champignon
雙柄鍋 saucepot
雙重紙盒 box in box
雙重蒸鍋 bain-marie
雙面磨 duplex mill
雙香豆素 dicoumarin
雙效打漿機 double beater
雙效蒸發器 double effect evaporator
雙氧噁噻嗪 acesulfame
雙帶鱸魚 sea perch
雙烯酸 dienoic acid
雙球菌 diplococcus
雙球菌素 diplococcin
雙硫鍵 disulfide bond
雙晶性 dimorphism
雙晶的 dimorphous
雙軸擠壓機 twin screw extruder
雙羥香豆素 dicoumarol
雙羥薰草素 dicoumarol
雙滾筒式乾燥機 double-drum dryer
雙層打麥機 double scourer
雙層袋 duplex bag
雙層蒸氣鍋 jacketed steam kettle
雙層蒸發器 jacketed evaporator
雙層鍋 double kettle; jacketed kettle
雙糖 disaccharide; disaccharose
雙糖酶 disaccharidase
雙〔縮〕脲 biuret
雙〔縮〕脲反應 biuret reaction
雙螺桿擠壓機 twin screw extruder
雙鍵 double bond
雙醛澱粉 dialdehyde starch
雙轉筒乾燥機 double-drum dryer; twin cylinder dryer; twin-drum dryer
雙鏈 double bond
雛 poult
雛型 preform
雛菊 daisy
雛雞 chick; chickling; spring chicken
雞上腿 chicken thigh
雞大腿 chicken thigh
雞小腿 chicken drumstick
雞心螺 cone shell
雞爪 chicken paw
雞肉 chicken
雞肉糜 minced chicken
雞尾酒 cocktail
雞尾酒杯 cocktail glass
雞尾酒會 cocktail lounge; cocktail party
雞兒豆 urd (bean)
雞油 chicken fat
雞油菌（蕈類） chanterelle; common chanterelle
雞肫 chicken gizzard
雞冠 cockscomb
雞冠形通心粉 cresti di gallo
雞翅膀 chicken wing
雞屠體 dressed chicken
雞排 chicken cutlet
雞蛋 egg
雞蛋果 egg fruit; granadilla; parchita; passion fruit
雞蛋麵〔食〕 egg noodle; egg pasta
雞場 fowl run
雞湯 chicken broth; chicken soup
雞脾 chicken thigh
雞腳 chicken feet
雞腿〔下段〕 drumstick
雞蓉 minced chicken
雞瘟〔疫〕 roup

雞雜 chicken giblet
鞣酸 gallotannin; tannic acid; tannin
鞣酸鹽 tannate
鞣酸酶 tannase
鞣質 tannin
鞭毛蟲 flagellate
鞭毛藻 flagellate
鞭蟲 trichuris
鞭蟲病 trichuriasis
顏色 colo(u)r
顏色固定 colo(u)r fixation
顏色固定劑 colo(u)r fixative
顏色強度 colo(u)r density; colo(u)r intensity
顏料 pigment
顎 jaw
顎式破碎機 jaw breaker
顎式壓碎機 jaw crusher
顎肉 palate meat
顎油 jaw oil
餾出液 distillate
餾酒器 spirit still
餿水 draff; hogwash; swill; wash
餿水桶 garbage can
鬆弛 relaxation
鬆弛效應 relaxation effect
鬆粉器 detacher
鬆脆性 shortness
鬆脆的 crisp
鬆脆餅乾 beaten biscuit
鬆脆薄片糖 brittle
鬆蛋糕 light cake
鬆軟烤餅 flannel cake
鬆軟圓烤餅 scone
鬆軟麵糰 runny dough
鬆散 openness
鬆弛 slack
鬆餅 muffin; puff pastry
鬆糕 sponge cake
鯊肝醇 batyl alcohol
鯊油酸 selacholeic acid
鯊油醇 selachyl alcohol
鯊烯 squalene
鯊魚 shark
鯊魚肝油 shark liver oil
鯊魚鰭 shark fin
鯊烷 squalane
鯉科小魚 minnow
鯉科的 cyprinid
鯉科的魚類 cyprinid
鯉魚 carp
鯉魚管 eye round
鯽〔魚〕 crucian (carp); golden carp
鵝 goose
鵝口瘡熱 aphthous fever
鵝肉 goose
鵝肉乾（德國）spinkganz
鵝肝 foie gras
鵝肝醬 goose liver paste
鵝梅 European gooseberry
鵝腳藜 quinoa
繖房花越橘 highbush blueberry
蟳 crab; serrated crab
酰化 acylation
酰化蛋白 acylated protein
酰化澱粉 acylated starch
酰化酶 acylase
酰胺 acid amide; amide
酰胺化 amidation
酰胺〔化〕果膠 amidated pectin

酰胺態氮 amide nitrogen
酰胺酶 amidase
酰基 acyl group
酰基轉移酶 acylase
醪 mash; moromi
醪桶 mash tub; mash tun
醪糟 pot ale
髻葉芥 curled mustard
鯇〔魚〕 Chinese ide; grass carp
鯪魚 mud carp

十九畫

壞死 necrosis
壞血病 scurvy
懷孕 gestation
懷孕母牛 springer cow
懷孕期 gestation period
懷卵幼鯡魚 matie
攀鱸 climbing perch
櫥存期限 marketable life; shelf life
櫥櫃 ambry
瀝水篩 dewatering screen
瀝乾 draining
瀝乾物重 drained weight
瀝乾的〔水果〕蜜餞 drained fruit
瀝濾 leach
瀝濾液 leaching liquor
瀝濾器 leach
爆化乾燥 cannon drying
爆玉米花 corn puff; maize puff; popcorn
爆玉米花機 corn popper; maize popper; popper
爆米花 puffed rice; popcorn
爆米花機 grain puffer
爆穀 popcorn
犢牛肉 baby beef
獸氏毛黴菌 *Mucor rouxii*
獸皮 hide; rind; skin; skin hide
獸醫 veterinarian; veterinary
獸醫的 veterinary
獸欄 pound
瓊芝屬海藻 eucheuma seaweed
瓊脂 agar; agar-agar
瓊脂二糖 agarobiose
瓊脂平板 agar plate
瓊脂平板〔培養〕法 agar plate method
瓊脂平板菌數 agar plate count
瓊脂穿刺培養 agar stab
瓊脂培養試管 agar tube
瓊脂斜面培養 agar slant; agar slope
瓊脂畫綫培養 agar streak
瓊脂酸 agaric(ic) acid; agaricinic acid
瓊脂膠 agaropectin
瓊脂凝膠 agar gel
瓊脂糖 agarose; gelose
瓊膠 agar; agar-agar
癟〔穀〕粒 imperfect grain
穩化澱粉 inhibited starch
穩定作用 stabilization
穩定性 stability
穩定泡沫 solid foam
穩定的 stable
穩定期 stationary phase
穩定劑 stabilizer; stabilizing agent
穩定澱粉 inhibited starch
簾蛤 hard clam; hard-shelled clam; quahaug; quahog
簸動篩 jigging screen
簸篩機 jigger
簸選機 jigging machine
繫留場 resting pen; stockyard
繭狀通心粉 bozzoli

蘭蜜 trehala
蘭蜜糖 mycose; trehalose
蘭蜜糖酶 trehalase
繩 rope
羅夫 loaf
羅巴斯達咖啡 coffee robusta; Robusta
羅氏沼蝦 giant river prawn
羅甘莓 loganberry
羅宋湯 borsch
羅非魚 tilapia
羅奎福特羊奶芝士 Roquefort (cheese)
羅奎福特羊乳乾酪 Roquefort (cheese)
羅浮 oval kumquat
羅馬甜瓜 cantaloup(e)
羅馬萵苣 romaine (lettuce)
羅偉邦色度 Lovibond colo(u)r
羅偉邦色調計 Lovibond tintometer
羅勒 basil; sweet basil
羅勒百里香 basil thyme
羅勒油 basil oil
羅曼茴香油 roman fennel oil
羅曼酒 roman punch
羅望子 tamarind
羅望子多糖膠 tamarind gum
羅望子果 tamarind
羅望子油 tamarind seed oil
羅望子膠 tamarind gum
羅雪鹽 Rochelle salt; sodium potassium tartrate
羅漢果抽提液 Lo-Han-Kuo extract
羅維邦色度 Lovibond colo(u)r
羅維邦色調計 Lovibond tintometer
羅謝爾鹽 Rochelle salt; sodium potassium tartrate
羹 bouillon
羹湯 skilly
臘腸 sausage
藤 vine
藤山椒 climbing-prickly ash
藤本作物 vine crop
藤崖椒 climbing-prickly ash
藤梨 Chinese gooseberry; kiwi (fruit)
藥丸 pill
藥片 tabloid
藥用作物 medicinal crop
藥用的 pharmaceutical
藥用脂肪 axunge
藥用植物 pharmaceutical plant
藥用植物學 pharmaceutical botany
藥材商 druggist
藥典 pharmacopoeia
藥房 drugstore
藥物 drug; medicine; pharmacon
藥物療法 pharmacotherapy
藥品 pharmaceutical
藥草 herb; medicinal herb
藥草的 herbal
藥酒 elixir; medicated liquor; medicinal liquor; tincture; vinum
藥酒廠 tincture distillery
藥商 druggist
藥理學 pharmacology
藥棉 swab
藥蜀葵〔蜜餞〕 marshmallow
藥劑師 druggist
藥膳 medicated diet
蟻酸 formic acid
蠅拍 flyflap
蠅虎蕈 amanita
蠅虎蕈毒素 amanitatoxin
蟹 crab
蟹子 crab roe
蟹肉 crab meat
蟹黃 crab roe
蟹螯 claw
蟹罐頭〔加工〕船 crab floating cannery
蟾魚 toadfish
蟾蜍 toad

譜系牛 pedigree cattle
邊封 edge sealing; side seam
邊封機 side seamer
邊封罐 side seam can
邊緣 edge; side seam
鏡檢 microscopic examination
鏡鯉 mirror carp
鏟 scoop
鏟子 spade
鏈孢黴 neurospora
鏈動連續烤爐 travelling chain oven
鏈烯 alkene
鏈球菌中毒 streptococcal poisoning
鏈球菌感染 streptococcal infection
鏈球菌溶血素 streptolysin
鏈球菌屬 streptococci; streptococcus
鏈烷 alkane
鏈烷酸 alkanoic acid
鏈黃黴素 xanthomycin
鏈黑菌素 streptonigrin
鏈黴素 streptomycin
鏈黴菌屬 *Streptomyces*
鏈黴糖 streptose
鏢石鱸 banded grunter
關刀豆 jack bean; sword bean
關華豆膠 guar gum
難以捉摸的味道 subtle taste
難聞氣味 malodo(u)r
霧化 atomization; pulverize
霧化室 atomizing chamber
霧化器 atomizer; mistorizer
霧化機 pulverizer
霧氣 vapo(u)r
霧點試驗 cloud test
類白蛋白 albuminoid
類似物 analog(ue)
類似葉綠素 allochlorophyll
類固醇 steroid
類固醇激素 steroid hormone
類毒素 anatoxin; toxoid
類洋菜 agaroid
類胡蘿蔔素 carotenoid
類真菌 fungoid
類脂的 lipoid
類脂質 lipoid
類黑素 melanoid
類黑精 melanoidin
類維生素 vitamer
類澱粉 amyloid
類澱粉黏膠性物質 amyloid
類黏液素 mucinoid
類黏蛋白 mucoid; mucinoid
類黏蛋白物質 mucoid substance
類瓊脂 agaroid
饅頭 bun; steamed bread
鯨 cetacean
鯨肉 whalemeat
鯨肝油 whale liver oil
鯨肝醇 kitol
鯨油 train oil
鯨脂 blubber; speck; whale tallow
鯨魚 train fish; whale
鯨魚肉 sea beef
鯨魚油 train oil
鯨魚油渣 fenks
鯨魚背條肉 back fillet
鯨魚群 gam
鯨魚類動物 cetacean
鯨腦油 spermaceti
鯨醇 kitol
鯨鯊 whale shark
鯨蠟 spermaceti
鯨蠟醇 cetyl alcohol; cetanol; spermol
鯧魚 butterfish; pomfret
鯧鰺 pampano
鯖兔河魨 blowfish
鯖〔魚〕 mackerel
鯖魚科 scombroid

鯖魚屬 *Scomber*
鯖鱈 pollack; pollock
鯛 crimson sea bream
鵲豆 bonavist bean; hyacinth bean
鵪鶉 partridge; quail
鵪鶉蛋 quail egg
麒麟菜屬 eucheuma
麗春紅4R ponceau 4R
麴 koji
麴二糖 kojiboise
麴汁 koji extract
麴汁洋菜〔培養基〕 koji agar
麴汁瓊脂〔培養基〕 koji agar
麴床 koji bed
麴味 koji flavo(u)r
麴菌病 aspergillosis
麴菌素 aspergillin
麴菌酸 aspergillic acid
麴菌屬 *Aspergillus*
麴酸 kojic acid
麴盤 koji tray
麴澱粉酶 koji amylase
麴黴 koji mo(u)ld
麴黴病 aspergillosis
麴黴素 aspergillin
麴黴酸 aspergillic acid
麴黴屬 *Aspergillus*
櫟生疣柄牛肝菌 red oak bolete
櫟皮粉 quercitron
櫟皮酮 quercetin
櫟皮苷 quercitrin
櫟金錢菇 russet tough shank
櫟實 acorn
櫟實糖 acorn sugar; quercitol
櫟精 quercetin
櫟醇 acorn sugar; quercitol
櫟樹 oak
藨莓 raspberry
藜豆 chickpea; Japanese velvet bean
醭酵母屬 *Mycoderma*
鯪魚 dace; mud carp
鯰魚 catfish; Japanese catfish; mud fish
鯔魚 black mullet; mullet
鯔魚子 mullet roe
鯥魚 blue fish
鯡〔油〕酸 nisinic acid
鯡斗 cran
鯡魚 clupeoid; herring; menhaden; pogy
鯡魚卵 herring roe
鯡魚酸 clupanodonic acid
鯡魚類 *Clupea*
鯡鰣 hilsa

二十畫

嚴格的素食主義者 vegan
嚴格厭氧菌 strict anaerobe
嚴格需氧菌 strict aerobe
寶刀魚 wolf herring
懸吊輸送機 trolley conveyor
懸空輸送器 overhead conveyor
懸浮 suspension
懸浮固形物 suspended solid
懸浮果肉漿 suspended pulp
懸浮相 suspended phase
懸浮培養〔物〕 drop culture; suspension culture
懸浮液 suspension
懸浮粒子 suspended particle
懸浮膠體 suspended colloid; suspension colloid; suspensoid
懸浮劑 suspending agent
懸浮體 suspension
懸掛〔式〕離心機 suspended centrifuge
懸鈎子酮 raspberry ketone
懸滴培養〔物〕 drop culture; hanging drop culture
懸膠體 suspensoid
懸濁劑 suspending agent
懸鈎子 raspberry
攙酒義式濃咖啡 caffe corretto
爐 furnace; hearth; kiln; stove
爐內增重 oven jump
爐內膨脹 dough spring; oven kick; oven spring
爐床 siege; hearth
爐灶 stove
爐底 siege
爐烤直徑增大量 oven flow
爐烤持水增重 oven jump
爐溫計 oven thermometer
礦物油 mineral oil
礦物質 mineral; mineral salts
礦物質增補 mineral supplementation
礦泉 spa
礦泉水 mineral water; spa water
礦脂 petrolatum
礬 alum; vitriol
礬土 alumina
籃式巴氏滅菌器 basket pasteurizer
籃式離心機 basket centrifuge
糯玉米 amioca; waxy corn
糯米 amylopectin rice; glutinous rice; sticky rice; sweet rice; waxy rice
糯米粉 glutinous rice flour; sticky rice flour; sweet rice flour; waxy rice flour
糯米紙 oblate
糯性〔穀物〕澱粉 waxy starch
糰性測定儀 mixograph
糰物性測定儀 farinograph
糰物性圖 farinogramme
糰展性測定圖 alveogram
糰展性測定儀 alveograph
繼代培養〔物〕 subculture
繼發性腐敗 secondary spoilage

罌粟 opium; papaver; poppy
罌粟籽 poppy seed
藻色素 phycochrome
藻青素 phycocyanobilin
藻青蛋白 phycocyan
藻朊酸鈉 sodium alginate
藻紅素 phycoerythrobilin
藻紅蛋白 phycoerythrin
藻菌 phycomycete
藻菌植物 thallophyte
藻黃素 phycoxanthin
藻黃質 phycoxanthin
藻酸丙二醇酯 propylene glycol alginate
藻褐素 phycophaein
藻膽色素 phycobilin
藻藍素 phycocyanobilin
藻類 alga(e); phycophyte
藻類多醣 algal polysaccharide
藻類的 algal
藻類培養物 algal culture
藻類養殖 algal culture
藻類學 algology; phycology
蘑菇 button mushroom; field mushroom; mushroom
蘑菇圈 fairy ring
蘑菇傘 mushroom cap
蘑菇菌柱 mushroom spawn
蘑菇菌絲體 mushroom mycelium
蘑菇糖 mushroom sugar
蘑菇醬 mushroom paste
蘑菇醬油 mushroom sauce; mushroom soy
蘑菇屬 *Agaricus*
蘆木 calamite; eldrin; rutin; rutoside
蘆花雞 plymouth rock
蘆柑 *Citrus foonensis*
蘆筍 asparagus; sparrowgrass
蘆筍清湯 asparagus consomme
蘆粟 sorgho; sorgo; sweet sorghum
蘆粟糖漿 sorgo syrup
蘆葦 reed
蘆薈 aloe; aloe vera
蘆薈素 aloin
蘆薈苷 barbaloin
蘋果 apple
蘋果切片器 apple slicer
蘋果去心器 apple corer
蘋果去皮機 apple peeler
蘋果布丁 pandowdy
蘋果汁 apple cider; apple juice; cider; cyder
蘋果白蘭地〔酒〕 apple brandy; cider brandy
蘋果沙司 applesauce
蘋果果渣 apple pomace
蘋果果醬 apple jam
蘋果泥 apple butter
蘋果油 amyl valerate
蘋果油香精 apple oily flavo(u)r
蘋果挖核器 apple corer
蘋果派 apple pie
蘋果香精 apple essence; apple flavo(u)r
蘋果烤布丁 brown betty
蘋果酒 apple wine; cider; cyder; fermented cider
蘋果乾 biffin
蘋果淡酒 apple cider
蘋果渣 apple pomace
蘋果塊 apple nugget
蘋果酪 apple cream
蘋果榨汁器 cider press
蘋果酸 malic acid
蘋果酸一鈉 sodium hydrogen malate
蘋果酸乳酸發酵 malo-lactic fermentation
蘋果酸琥珀酸發酵 malo-succinic fermentation
蘋果酸鈣 calcium malate
蘋果酸鈉 sodium malate

蘋果酸鉀 potassium malate
蘋果酸酶 malic enzyme
蘋果餅 pandowdy
蘋果漿 apple cream; apple pomace; apple pulp
蘋果糊 applesauce
蘋果醋 apple vinegar; cider vinegar
蘋果餡餅 apple pie
蘋果壓榨機 apple press
蘋果醬 apple butter; apple jam; applesauce
蘋果類的 pomaceous
蘋婆 ping pong; sterculia
蘋婆酸 sterculic acid
蘋婆膠 sterculia gum
蘇丹紅 Sudan red
蘇丹娜葡萄乾 sultana
蘇木紅 haematein; hematein
蘇木精 haematoxylin; hematoxylin
蘇打 soda
蘇打水 carbonated water; club soda; soda; soda pop; soda water; sparkling water
蘇打水〔供應〕器 soda fountain
蘇打汁威士忌〔酒〕 whisky soda
蘇打灰 barilla; soda ash
蘇打餅乾 cracker; soda cracker
蘇打麵包 soda bread
蘇合香 storax; styrax
蘇合香油 storax oil
蘇合香脂 storax; styrax
蘇合香腦 styracol
蘇合香醇 styracitol(e)
蘇式魚糜製品冷盤 gefillte fish
蘇利南櫻桃 Surinam cherry
蘇眉 Napoleon fish
蘇格蘭甘藍 Scotch kale
蘇格蘭沙雞 moorfowl
蘇格蘭青鱈 Scotch hake
蘇格蘭威士忌〔酒〕 Scotch whisky
蘇格蘭雷鳥 moorfowl
蘇胺酸 threonine
蘇糖 threose
蠔 oyster
蠔油 oyster sauce
蠔豉 dried oyster
蠔場 oyster farm
蠔殼 oyster shell
蠔菇 common oyster mushroom; *Pleurotus obalorus*; ostreat; oyster fungus; oyster mushroom
蠕動泵 peristaltic pump
蠕蟲 helminth; vermis; worm
蠕蟲病 verminosis
蠕蟲類 vermes
觸變性 rheopexy
觸變稀性流體 thixotropic fluid
觸酶 catalase; catalyst; catalytic agent
觸酶〔作用〕 catalysis
觸酶反應 catalytic reaction
釋迦〔果〕 custard apple; sugar apple; sweetsop
鐘式巴氏滅菌器 bell pasteurizer
饑荒 dearth; famine
鰓 gill
鰓蓋 gill cover
鹹五花肉 streaky bacon
鹹水 salt water
鹹水魚 sea fish
鹹火腿片 rasher
鹹牛肉 bully (beef); corned beef
鹹牛肉乾 salt junk
鹹肉 bacon; cured meat; salted meat
鹹肉片 rasher
鹹味 saline taste; salt taste; salty flavo(u)r; salty taste
鹹味小菜 savory; savoury
鹹味的 salted; salty
鹹味劑 salt substance
鹹性 saltiness

鹹的 salted; salty; savory; savoury
鹹金槍魚乾 mojama
鹹度 brininess; saltiness
鹹乾肉 junk
鹹乾魚 cured fish
鹹蛋 preserved egg; salted duck egg
鹹魚 cured fish; salted fish
鹹魚乾 dried salted fish
鹹菜 brined vegetable; salted vegetable
鹹鴨蛋 salted duck egg
鹹鱈魚乾 bacalao; klipfish
麵 pasta
麵包 bread
麵包小麥 bread wheat
麵包工業 baking industry
麵包切片機 bread slicer
麵包孔隙度 grain of bread
麵包方條 loaf
麵包片 sliced bread; tartine
麵包用小麥 bread wheat
麵包用麵粉 bread flour
麵包皮 crust
麵包皮〔殼〕 bread shell
麵包改良劑 bread improver
麵包店 bakehouse; bakery; bakeshop
麵包房的廢物渣屑 bakery refuse
麵包果 breadfruit; monkey fruit; marang
麵包果樹 breadfruit tree; bread-nut tree
麵包〔芽孢〕桿菌 *Bacillus panis*
麵包保軟劑 crumb softener
麵包柔軟劑 crumb softener
麵包屑 breadcrumb; crumb; crumble; crumbly bread
麵包師 baker
麵包烘房 bakehouse; bakeshop
麵包烤模 bread mo(u)ld; bread tin; loaf form; tin
麵包乾 biscuit; dried bread; sea biscuit
麵包乾燥爐 rusk oven
麵包捲 twist
麵包疏鬆度 grain of bread
麵包棒 bread stick
麵包製作 loaf making
麵包製品 bread
麵包製造 breadmaking
麵包酵母 baker's yeast; bakery yeast; bread yeast
麵包盤 bread plate
麵包穀物 bread grain
麵包箱 bread box
麵包糊 panada
麵包霉 hair mould
麵包樹 bread-nut tree; breadfruit (tree)
麵包餡 loaf filler
麵包糠 breadcrumb
麵包麵粉 bread flour
麵包黴 bread mo(u)ld
麵拖 battering
麵疙瘩 gnocchi
麵食 alimentary paste; pasta
麵食〔製品〕 pasta product
麵粉 flour; wheat flour
麵粉改良劑 flour improver
麵粉的 floury
麵粉烙餅 wheat cake
麵粉強度 flour strength
麵粉強筋劑 II 號 flour enhancing resistance agent II
麵粉陳化 flour ag(e)ing
麵粉減筋劑 flour reducing resistance agent
麵粉筋力 flour strength
麵粉煎餅 wheat cake
麵粉試驗儀 farinograph
麵粉漂白 flour bleaching
麵粉漂白劑 flour bleaching agent
麵粉製糕點 pastry
麵粉熟成 flour ag(e)ing
麵粉糊 flour paste; flummery

麵粉調粉性圖譜 farinogramme
麵粉調粉性圖譜儀 farinograph
麵粉調混性記錄儀 mixograph
麵粉磨機 flour mill
麵粉篩分機 milling separator
麵粉〔營養〕強化 flour enrichment
麵粉蠹 lesser grain borer
麵茶 roasted wheat flour
麵條 noodle
麵條切割機 noodle cutter; noodle cutting machine
麵條製品 alimentary paste
麵球 nidi
麵粒 farfal; farfel
麵筋 clear dough; gluten; wheat gluten
麵筋拉力測定儀 alveograph
麵筋球 gluten ball
麵筋蛋白 mucedin
麵筋筋力儀 aleurometer
麵漿 batter
麵漿混料 batter mix
麵糊 batter; panada
麵糊混料 batter mix
麵線 fedelini; vermicelli
麵糰 dough
麵糰分切機 dough divider
麵糰成形 loaf making
麵糰成圓機 roll moulder; rounder; rounding machine
麵糰伸展儀 extensograph
麵糰吹泡張力曲線圖 alveogram
麵糰刮刀 dough scraper
麵糰泄氣 bleeding
麵糰油酥料 shortener
麵糰狀 doughiness
麵糰〔狀〕的 doughy
麵糰軋片機 dough sheeter
麵糰後發酵 proofing
麵糰後發酵室 proofing room
麵糰後發酵箱 proof box
麵糰球 dough ball
麵糰揉和 doughing
麵糰揉圓機 baller
麵糰發起 dough raising
麵糰發酵 panary fermentation
麵糰發酵時間 dough time
麵糰發酵槽 dough trough
麵糰發麵力 baking strength
麵糰筋力 dough strength
麵糰脹力計 aleurometer
麵糰搓圓機 hander-up
麵糰滾圓機 rounder
麵糰熟成劑 dough maturing agent
麵糰調混性圖譜 area mixogram
麵糰調整劑 dough conditioner
麵糰〔質量〕改進劑 dough conditioner
麵糰醒發成形 proof; proofing
麵糰醒發成形室 proofing room
麵糰鋸刀 jagger
獼猴桃 Chinese gooseberry; kiwi (fruit)
獼猴桃蛋白酶 actinidin(e)
獼猴桃酶 actinidase
臙脂蟲酸 carminic acid
蝶螺 top shell
鰈〔魚〕 flatfish; flounder; plaice
鯷魚 anchovy
鯷魚油 anchovy oil
鯷魚醬 anchovy paste
鰆 Spanish mackerel
鰉〔魚〕 huso; huso sturgeon
鱷梨 aguacate; alligator pear; avocado; butter-fruit
鰺〔類〕 jack
鰕虎魚 goby

二十一畫

屬 genus
攝入 intake
攝氏溫度計 Celsius thermometer
攝生法 regimen
攝取 intake; ingestion; uptake
攝食 feeding
攝食過量 hyperphagia
攝食學 phagology
櫻花鉤吻鮭魚 cherry salmon
櫻桃 cherry
櫻桃白蘭地酒 kirsch
櫻桃李 cherry plum
櫻桃油香精 cherry oily flavo(u)r
櫻桃紅 cherry red
櫻桃香精 cherry flavo(u)r
櫻桃核 cherry stone
櫻桃酒 kirsch
櫻桃膠 cherry gum
櫻桃蕃茄 cherry tomato
櫻桃鮭 cherry salmon
櫻草 primrose
櫻草糖 primverose
灌木 shrub
灌木水果 bush fruit
灌木懸鈎子 foxberry
灌注 perfusion
灌洗 swill
灌食法 superalimentation
灌香腸機 sausage filler; sausage stuffer; stuffer
灌瓶 bottling
灌瓶機 bottler
灌溉 watering
灌腸嘴 casing applier
灌腸劑 enema
灌腸器 applier
灌腸機 sausage filler; sausage stuffer; stuffer
灌裝不足 slack fill
灌裝機 filling machine; racking machine
灌餡機 stuffer; stuffing machine
爛泥 slush
癩皮病 pellagra
癩蛤蟆 toad
蘭金溫度 Rankine temperature
蘭勃舵魚 bluefish; chub; lembus rudderfish; rudderfish
蘭酒 bumbo; rum
蘭溪藻黃質 myxoxanthin
蘭蜜酒 blackstrap
蘭德列斯豬 Landrace (swine)
蘭德瑞斯豬 Landrace (swine)
蘭撒漿昊 langsat
蘚 moss
蠟 cere; wax
蠟狀的 waxy
蠟封 wax-sealed
蠟素 cerin
蠟紙 wax paper; waxed paper
蠟酸 cerotic acid

蠟膜 cere
蠟質的 waxy
護士 nurse
護色劑 colo(u)r fixative
鐮刀 sickle
鐮刀菌屬 *Fusarium*
鐮狀細胞貧血症 sickle-cell anaemia
鐵 iron
鐵(III)肌紅蛋白 metmyoglobin
鐵(III)血紅素 haematin; hematin
鐵(III)血紅蛋白 methemoglobin
鐵叉 broach
鐵均衡 iron balance
鐵杉內酯 conidendrin
鐵氟龍 teflon
鐵混濁 iron turbidity
鐵蛋白 ferritin
鐵〔腥〕味 iron taste; metallic flavo(u)r
鐵酸鹽 ferrate
鐵質缺乏〔症〕 sideropenia
鐵銹 rust
霸王梨 prickly pear
露天倉庫 storage yard
露天乾燥 open-air drying
露天堆放 open stack
露珠莓 dewberry
露酒 cordial
露點 dew point
露點濕度計 dew point hygrometer
驅腸蟲劑 helminthic
驅蟲劑 helminthic; insect repellant; insectifuge; repellent
驅蟲藥 vermifuge
騾 mule
魔芋 elephant foot; giantarum; konjac; konnyaku;
魔芋甘露聚糖 konjac flour
魔芋粉 konjac flour
鰭 fin
鶴鱵魚 needlefish
麝香 cabardine; musk
麝香瓜 cantaloup(e)
麝香洋蔥 musk onion
麝香根油 musk root oil
麝香草 thyme
麝香草油 thyme oil
麝香草莓 hautboy
麝香草酚 thymol
麝香梨 ambrette
麝香梨子油 ambrette seed oil
麝香鹿 musk deer
麝香葡萄 muscadine grape; muscat grape
麝香葡萄酒 muscadine; muscat
麝香酮 musk ketone; muskone
麝香錦葵 musk mallow
麝香鴨 musk duck
麝香薄荷 horsemint
麝葵籽油 ambrette seed oil
礱穀 hulling; husking
礱穀機 huller; husker; rice huller; rice hulling machine; rice husker
纈草 valerian
纈草油 valerian oil
纈草酸 valeric acid
纈胺酸 valine
纈胺黴素 valinomycin
鰣〔魚〕 hils herring; hilsa; shad
鰨 common sole
鰩魚 skate

二十二畫

囊 pouch; sac; vesicle
囊果孢子 carpospore
囊蟲 cisticercus; cysticercus
囊蟲病 cysticercosis
彎刀 machete
彎曲層齒菌（蕈類） layered tooth fungus
彎管 elbow
彎頸南瓜 crookneck (pumpkin)
彎頸葫蘆 crookneck squash
歡樂食物（兒童生日會） blithemeat
灑 sprinkle
灑水 watering
灑水裝置 sprinkler
灑鹽 rouse
瓤瓣 section; segment
疊層蛋糕 double cake
籠式破碎機 cage mill
籠式榨油機 cage oil press
籠式磨 cage mill
籠式壓榨機 cage press
臟器 visceral organ
襯布 undercloth
襯裏 liner
襯裏袋 lined bag
襯墊 cushion; liner
襯墊紙 lining paper
鑑別培養基 differential medium
鑒定 verification
韃靼調味醬 tartare sauce
韃靼蕎麥 tartarian buckwheat; tartary buckwheat
鬚鯨 baleen whale
鱉 soft shelled turtle; terrapin
鰾 air bladder; maw; sound
鰾膠 ichthyocolla; ichthyocol(l)
鰻 eel
鰻草 eelgrass
鰻鱺 common eel; eel
鷓鴣 partridge
鷓鴣菜屬 *Caloglossa*
鑊 wok
鱈卵 cod roe
鱈油酸 gadoleic acid
鱈科的魚 gadoid (fish)
鱈〔魚〕 cod; codfish; gadoid (fish); hoki; pacific cod; pollack; pollock
鱈魚肉 codfish
鱈魚肝油 cod liver oil
鱈魚船 banker
鱈魚燻窯 finnan kiln
鱈場蟹 king crab
鰹〔魚〕 aku; bonito; oceanic bonito; oceanic skipjack; shoemaker; skipjack; striped tuna
鰹魚乾 dried bonito
鰹酸 bonitonic acid
鰳魚 Chinese herring; long-fin(ned) herring
鰽白 Chinese herring

二十三畫

攪切器 blender; blendor
攪打 beating; whipping
攪打助泡劑 whipping aid
攪打起泡力 whipping power
攪打起泡性 whipping property
攪打起泡能力 whipping ability
攪打混合機 mixer-beater
攪打疏鬆機 fluffer
攪打器 whip; whipper; whipping machine; whisk
攪打機 beating machine
攪乳 churning
攪乳器 churn
攪乳擋板 churn dasher
攪和 agitation; stir
攪和器 melangeur
攪拌 mixing; paddling; stir; stirring
攪拌式巴氏殺菌器 mixer pasteurizer
攪拌式結晶罐 agitating crystallizer
攪拌式瞬間脱水器 agitator flash evaporator
攪拌式瞬間蒸發器 agitator flash evaporator
攪拌式薄膜脱水器 agitating film evaporator
攪拌式薄膜蒸發器 agitating film evaporator
攪拌時間 mixing time
攪拌乾燥器 agitator dryer
攪拌桶 agitating vat
攪拌缽 mixing bowl
攪拌棒 stirring rod
攪拌裝置 agitating device; dasher
攪拌槽 stirred tank
攪拌槳葉 mixing paddle; stirrer arm
攪拌醃漬槽 agitated curing vat
攪拌器 agitator; beater; blender; blendor; mixer; rouser; stirrer; whisk
攪拌機 stirring machine
攪拌機槳葉 mixing arm; stirrer arm
攪拌鍋 agitator kettle
攪泡器 whip; whipper
攪起泡沫 whip
攪酒棒 swizzle stick
攪動 agitation; churning; stir; stirring
攪動加熱器 agitating heater
攪動式乾燥機 agitated dryer
攪動式殺菌斧 agitating retort
攪動式殺菌器 agitating sterilizer
攪動式蒸煮器 agitating cooker
攪動器 agitator; churn
攪動翼 churn dasher; dasher
攪蛋器 albumen whisk; egg whisk
攪凝性 rheopexy
曬乾 insolation; solar drying; sun cured; sun curing; sun dehydration; sun drying
曬乾的魚類 stockfish
曬乾食品 sun dried food
曬魚場 flake yard
曬斑 sunburn; sunscald

曬斑水果 sunburned fruit
曬焦 sunblister; sunburn
曬傷 sunscald
曬製 sun cured; sun curing
曬製水果 sun cured fruit
曬製鹽 solar salt
曬鹽 bay salt
纓穗 tassel
纖絲 filament
纖維 fiber; fibre
纖維二糖 cellobiose
纖維二糖酶 cellobiase
纖維三糖 cellotriose
纖維作物 fiber crop
纖維性食物 roughage
纖維狀蛋白 spun protein
纖維狀蛋白質 fibrillar protein; fibrous protein
纖維粉末 celluflour
纖維素 cellulose
纖維素人造腸衣 fibrous casing
纖維素分解酵素 cellulolytic enzyme
纖維素甲乙醚 methyl ethyl cellulose (MEC)
纖維素物質 cellulosic material
纖維素粉 celluflour; powdered cellulose
纖維素腸衣 cellulose casing
纖維素膠 cellulose gum
纖維素凝膠 cellulose gel
纖維素黏膠腸衣 viscose casing
纖維素酶 cellulase
纖維蛋白 spun protein
纖維蛋白溶解作用 fibrinolysis
纖維腸衣 fibrous casing
纖維聚糖 cellulosan
纖維糊精 cellodextrin
纖維膠 viscose
蘸 dip; dunk
蘿考籽 annatto
蘿梨 noni
蘿蔔 radish
蘿蔔紅 radish red
變化 variation
變色 discolo(u)ration; off colo(u)r
變形 deformation
變形〔芽孢〕桿菌 *Bacillus proteus*
變形麵包 squalid loaf
變味 flavo(u)r reversion; off flavo(u)r; off taste
變性〔作用〕 denaturation
變性毒素 anatoxin
變性蛋白質 denatured protein; metaprotein
變性劑 denaturant; denaturing agent
變性澱粉 modified starch
變紅 pinking
變風味 flavo(u)r reversion
變得平淡無味 pall
變異 variation
變稀的發酵麵糰 flowy dough
變黃雞油菌（蕈類） golden chanterelle
變黑 darkening
變暗藍裸蓋菇 blue rimmed liberty cap
變溫動物 allotherm
變構效應 allosteric effect
變種 variant; variety
變綠紅菇 green russule
變酸 sour
變酸性的 acidoid
變質 deterioration; perishability
變壞 perishability
顯色反應 colo(u)r reaction
顯色試驗 colo(u)r test
顯型 phenotype
顯著水準 significant level
顯著差異 significant difference
顯微結構 microstructure
顯微鏡 microscope

顯微鏡數菌法 microscopic count
顯微鏡檢查 microscopic examination
顯影 development
顯影劑 developing agent
顯熱 sensible heat
驗油比重計 elaeometer; elaiometer
驗酒員 wine conner
驗據 proof
髓骨 marrowbone
體內 in vivo
體內代謝中產酸的食品 acid residue food
體內寄生蟲 entozoa
體外 in vitro
體育型體質 mesomorph
體細胞 somatic cell
體腔洗滌 inside washing
體視鏡 stereoscope
體積 volume
鱔〔魚〕 common eel; eel; field eel; nineeyes
鱗片 scale; squama
鱗甲 scute
鱗多孔菌（蕈類） dryad's saddle
鱗狀的 scaly
鱗莖 bulb; scaly stem
鱗莖菜 scaly stem vegetable
鱗烟管魚 cornetfish; flutemouth
麟鳳蘭 yucca

黴 mildew; must
黴脂酸 mycolipenic acid
黴酚酸 mycophenolic acid
黴菌 mold; mould
黴菌元 mold starter; mould starter
黴菌多糖類 mold polysaccharide; mould polysaccharide
黴菌抑制劑 mold inhibitor; mould inhibitor
黴菌毒素 mycotoxin
黴菌毒素中毒 mycotoxicosis
黴菌培養〔物〕 mold culture
黴菌蛋白質 mycoprotein
黴菌蛋白酶 mold protease; mould protease
黴菌菌絲體 mold mycelia; mould mycelia
黴菌酸 mycolic acid
黴菌學 mycology
黴菌澱粉酶 mold amylase; mould amylase
黴菌糖化法 amylo process
鱒魚 trout; truff
鱘魚 sturgeon
鱘魚子 sturgeon roe
鱘魚子醬 carvier; caviar
鱘精蛋白 sturin(e)
鱘鰾膠 fish isinglass; ichthyocolla; ichthyocol(l)
鷸 snipe

二十四畫

罐 can; jar; kettle; pot; tank; tin
罐內腐蝕 internal corrosion
罐身 can body
罐身切角機 notching machine
罐身成形機 can body making machine; can body forming machine
罐身成圓機 can body forming machine; flexing machine
罐身拉舌 side strip
罐身板 can body blank
罐身捲緣 body flanged; can flange
罐身接縫 body joint
罐身脫鈎 knockdown flange
罐身開啟拉舌 side strip
罐身摺邊 edge
罐身滾壓淺紋 knuring
罐身撓曲機 flexing machine
罐身鐵皮 can body blank
罐身鈎丁 edge
罐底 factory end
罐底捲封 bottom seam
罐乾燥機 can dryer
罐脹 packing swelling
罐裝水果 canned fruit
罐裝火腿 canned ham; potted ham
罐裝牛油 tinned butter
罐裝肉 tinned meat
罐裝乳酪 tinned butter
罐裝果汁 canned fruit juice
罐裝食品 canned food; canned goods; tinned food; tinned goods
罐裝啤酒 canned beer
罐裝魚 canned fish
罐裝湯 canned soup
罐裝蔬菜 canned vegetable
罐蓋 can cover; can end; lid; packer's end
罐蓋二重捲封 top double seam
罐蓋打代號 embossment
罐蓋打印機 lid embossing machine
罐蓋沖壓機 end press
罐蓋封口膠 can end cement
罐蓋捲曲部 curl
罐蓋捲邊機 can end curling machine
罐蓋滅菌器 cover sterilizer
罐蓋壓印機 lid embossing machine
罐緣 body flanged; can flange; flange
罐燜燒肉及蔬菜 hot pot
罐頭 can; tin can
罐頭刀 tin opener
罐頭工業 canning industry
罐頭工廠 cannery
罐頭不產氣酸敗 flat sour
罐頭牛肉 bully (beef)
罐頭充填機 can filler
罐頭凹陷 denting
罐頭包裝法 tinning
罐頭肉 canned meat
罐頭的翻滾轉動 end-over-end agitation

罐頭食品 canned food; canned goods; can goods; preserved food; tinned food; tinned goods
罐頭食品中毒 can poisoning
罐頭食品廠 cannery
罐頭乾燥機 can dryer
罐頭捲邊不良 faulty closure
罐頭旋轉式滅菌 end-to-end sterilization
罐頭殺菌釜 cannery retort
罐頭殺菌鍋 canner
罐頭湯 canned soup
罐頭貼標機 tin labeller
罐頭塗料 can enamel
罐頭製品 canned product
罐頭製造 canning
罐頭製造商 canner
罐頭廠 packing house; packing plant; tinning factory
罐頭翻滾式殺菌 end-to-end sterilization
罐頭翻邊機 flanger
罐藏〔法〕 appertizing; canning
羈絆 shackle
蠶 silkworm
蠶豆 broad bean; field bean; horse bean
蠶豆嘧啶葡萄糖苷 vicine
艷紅鋁色澱 allura red AC aluminium lake
釀酒 brew; wine making
釀酒用葡萄 vinifera
釀酒技師 brewmaster
釀酒麥芽 distillers' malt
釀酒麥芽汁 distillers' wort
釀酒酵母 distillers' yeast; distillery yeast
釀酒廠 brewery; brewhouse; distillery; grain distillery; winery
釀造 brew; brewage; brewing
釀造工藝 zymotechnics; zymotechnique
釀造年份 vintage
釀造技術 zymotechnics; zymotechnique
釀造者 brewer; brewster
釀造後再貯藏陳化的啤酒 lager (beer)
釀造酒 brewage; brewer; brewster
釀造啤酒用之酵母 beer yeast; brewers' yeast
釀造業 brewer
釀造學 zymurgy
釀葡萄酒學 enology; oenology
釀葡萄酒學家 enologist; oenologist
釀製成香檳酒 champagnize
釀醋裝置 acetator
釀醋器 vinegar generator
釀酶 zymase
靈芝 *Ganoderma lucidum*
靈芝屬 *Ganoderma*
靈敏性 sensitivity
鷹爪豆 weavers broom
鷹嘴豆 Bengal gram; chickpea; gram
鹼 alkali; base; soda
鹼土金屬 alkaline-earth metal
鹼中毒 alkalosis
鹼化 alkalize
鹼化〔作用〕 alkalinization; alkalization; basification
鹼水解 alkaline hydrolysis
鹼去皮〔法〕 alkali peeling
鹼石灰 soda lime
鹼式碳酸鎂 magnesium hydroxide carbonate
鹼式鹽 basic salt
鹼灰 soda ash
鹼灰化 alkaline-ashing
鹼灰質飲食 alkaline-ash diet
鹼灰質膳食 alkaline-ash diet
鹼皂化 alkaline saponification
鹼味 alkaline taste
鹼性 alkalinty
鹼性水解 alkaline hydrolysis
鹼性的 alkalinous; basic

鹼性度 basicity
鹼性染料 basic colo(u)r; basic dye
鹼性泉水 alkali fountain
鹼性洗滌劑 alkalis
鹼性洗潔劑 alkali detergent
鹼性食品 base forming food; base residue food; basic food
鹼性胺基酸 basic amino acid
鹼性蛋白 basic protein
鹼性蛋白酶 alcalase
鹼性溶液 alkali solution
鹼性煤塔染料 basic coal tar dye
鹼性僵直 alkaline rigor
鹼性劑 alkalizer
鹼性磷酸鋁鈉 sodium aluminium phos-phate, basic
鹼的 alkaline; basic
鹼金屬 alkali metal
鹼度 alkalinty; basic capacity; basicity
鹼度計 alkalimeter; alkaline meter
鹼毒症 alkalosis
鹼洗 alkali wash; caustic washing; lye washing
鹼氣味 alkaline flavo(u)r
鹼臭 alkaline soapy odor
鹼液 buck; lye
鹼液去皮〔法〕caustic peeling; lye peeling
鹼脱皮麥粒 wurld wheat
鹼處理 alkaline treatment
鹼處理澱粉 alkaline treated starch
鹼量計 alkalimeter; alkaline meter
鹼量滴定 alkalimetric titration
鹼溶性的 alkali soluble
鹼溶纖維素 viscose
鹼煉 alkali refining; caustic refining
鹼滴定 alkalimetric titration
鹼澱粉 alkali starch
鹼濃度測定器 alkalimeter; alkaline meter
鹽 salt
鹽土植物 halophyte
鹽井 salt well
鹽化 salify(ing)
鹽化〔作用〕 salification
鹽水 brine; saline; salt brine; salt water
鹽水比重計 brine gauge; salimeter; salt gauge; salt meter
鹽水充填包裝 brine pack
鹽水注射針 pickle pumping needle
鹽水凍結 brine freezing
鹽水浮游生物 haloplankton
鹽水浮選〔法〕 brine flotation
鹽水漬 brine salting; brining
鹽水醃肉 brined meat
鹽水醃漬 brine cure; pickle salting; wet salting
鹽汁 pickle
鹽田 salina; saline; salt farm; salt field; salt pan; saltern
鹽冰 salt ice
鹽池 evaporation pond; salt pond; salina; saline
鹽灼 salt burn
鹽味的 salted
鹽析 salting out
鹽的 salted
鹽度 brininess; salineness; salinity
鹽度計 salimeter; salinimeter; salinometer; salometer; salt gauge; salt meter
鹽浸冷凍 brine freezing
鹽浴 salt bath
鹽粉 dust salt
鹽乾 salted drying
鹽乾食品 salted and dried food
鹽液 salt brine
鹽瓶 salt cellar
鹽場 evaporating field; salina; saline; salt farm; salt works; saltern
鹽斑 salt rust; salting stain

鹽湖 salina; saline
鹽量計 halometer; salinimeter; salinometer; salometer
鹽溶 salting in
鹽溶性 salt-soluble
鹽溶性蛋白質 salt-soluble protein
鹽溶的 salt-soluble
鹽漬 brine salting; salination; salting
鹽漬〔化〕 salinization
鹽漬肉類 brined meat
鹽漬的 brine cured
鹽漬食物 brined food
鹽漬食品 salted food
鹽漬桶 briner
鹽漬過久的 oversalted
鹽漬蔬菜 brined vegetable
鹽滷 bittern
鹽碟 salt cellar
鹽酸 hydrochloric acid
鹽酸奎寧 quinine hydrochloride
鹽酸硫胺〔素〕 thiamine hydrochloride
鹽酸膽鹼 choline hydrochloride
鹽醃 dry curing; dry salting; salinization; salination; salt curing; salt pickling; salting
鹽醃泡菜 salt stock pickle
鹽醃物 salt stock pickle
鹽醃的 corned; corning; dry salted; salt cured; salted
鹽醃液 brine pickle
鹽醃產品 salt preserved product
鹽磚 compressed salt
鹽鍋 salt pan
鹽霜 salt rust
鹽礦 salt mine
囓齒動物 rodent
鱧 snake head

二十五畫或以上

觀音莧 gynura' s deux couleurs
觀葉南瓜 yellow flowered gourd
鑲青椒 stuffed green pepper
鑲點石斑 black spotted grouper
鑰匙罐 key opening can
髖骨 hip bone
鑷子 tongs
驢 donkey
驢耳朵鮑魚 ass' s-ear

鱵 halfbeak
鱵魚 garfish
鱷魚 crocodile
鱸科魚類 bacalao
鱸魚 bass; Japanese sea bass; weever
鱸滑石斑 bullnose bass
鬱金 turmeric
鬱金菌 chanterelle
鸛 stork

非中文字元

1,1,2-三氯三氟乙烷 1,1,2-trichlorotrifluoroethane

1,2-二氯乙烷 ethylene dichloride

1,3-丁二醇 butan-1,3-diol

1,6-聚葡萄糖 dextran

1,6-聚葡萄糖酶 dextranase

10-十一烯酸乙酯 ethyl 10-undecenoate

2,4-十一碳二烯醛 2,4-undecadienal

2,6-二叔丁基對甲酚 butylated hydroxytoluene (BHT)

2-乙基-1-己醇 2-ethyl-1-hexanol

2-乙基-3-甲基吡嗪 2-ethyl-3-methylpyrazine

2-乙基小茴香醇 2-ethyl fenchol

2-乙酰基吡嗪 2-acetyl pyrazine

2-乙酰基噻唑 2-acetyl thiazole

2-乙酰基呋喃 2-acetylfuran

2-乙酰吡咯 2-acetylpyrrole

2-丁烯酸己酯 hexyl-2-butenoate

2-丁烯酸異丁酯 isobutyl-2-butenoate

2-丁酮 2-butanone

2-十三烯醛 2-tridecenal

2-十三酮 2-tridecanone

2-己烯酸甲酯 methyl 2-hexenoate

2-壬烯酸甲酯 methyl 2-nonenoate

2-甲氧基吡嗪 2-methoxypyrazine

2-甲基-2-戊烯酸 2-methyl-2-pentenoic acid

2-甲基-3-巯基呋喃 2-methyl-3-furanthiol

2-甲基丁酸 2-methylbutyric acid

2-甲基丁酸-2-苯乙酯 2-phenethyl-2-methylbutyrate

2-甲基丁酸乙酯 ethyl 2-methylbutyrate

2-甲基丁酸己酯 hexyl 2-methylbutyrate

2-甲基丁酸甲酯 methyl 2-methylbutyrate

2-甲基丁酸葉醇酯 cis-3-hexenyl 2-methylbutyrate

2-甲基丁醛 2-methylbutyradehyde

2-甲基戊酸 2-methylvaleric acid

2-甲基戊醛 2-methyl pentanal

2-甲基吡嗪 2-methylpyrazine

2-辛炔羧酸甲酯 methyl 2-octynoate

2-硝基丙烷 2-nitropropane

2-巯基丙酸 2-mercaptopropionic acid

3,4-二甲氧基苯甲醛 3,4-dimethoxybenzenecarbonal

3,7-二甲基黄嘌呤 3,7-dimethylxanthine; theobromine

3-乙酰基-2,5-二甲基呋喃 3-acetyl-2,5-dimethyl furan

3-乙酰吡啶 3-acetylpyridine

3-甲基戊酸 3-methyl pentanoic acid

3-甲硫基丁醛 3-methylthiobutanal

3-甲硫基丙酸乙酯 ethyl 3-methyl thiopropionate

3-甲硫基丙酸甲酯 methyl 3-methylthiopropionate

3-甲硫基丙醇 methionol

3-甲硫基丙醛 methional

3-苯丙酸乙酯 ethyl 3-phenyl propionate

3-巰基-2-丁酮 3-mercapto-2-butanone
3-羥基丁酸乙酯 ethyl 3-hydroxybutyrate
3-酮基己糖 glutose
3-環己基丙酸烯丙酯 allyl 3-cyclohexanepropionate
4-乙基癒創木酚 4-ethylguaiacol
4-庚烯醛 *cis*-4-hepten-1-al
4-苯丁酸乙酯 ethyl 4-phenylbutyrate
4-苯丁酸甲酯 methyl 4-phenylbutyrate
4-氯苯氧乙酸鈉 sodium 4-chlorophenoxyacetate
5-甲基糠醛 5-methylfurfural
5'-肉苷酸鈉 sodium 5'-inosinate (5'-IMP-Na)
5'-肌苷酸 inosine-5'-monophosphoric acid; 5'-inosinic acid
5'-肌苷酸〔二〕鈉 disodium 5'-inosinate
5'-肌苷酸二鉀 dipotassium 5'-inosinate
5'-肌苷酸鈣 calcium 5'-inosinate
5'-肌苷酸鈉 sodium 5'-inosinate (5'-IMP-Na)
5'-尿苷酸二鈉 disodium 5'-uridilate
5'-胞苷酸〔二〕鈉 disodium 5'-cytidylate
5'-核糖核苷酸〔二〕鈉 disodium 5'-ribonucleotide
5'-核糖核苷酸鈣 calcium 5'-ribonucleotide
5'-鳥苷酸 guanosine-5'-monophosphoric acid (GMP); 5'-guanylic acid
5'-鳥苷酸〔二〕鈉 disodium 5'-guanylate
5'-鳥苷酸二鉀 dipotassium 5'-guanylate
5'-鳥苷酸鈣 calcium 5'-guanylate
5'-鳥〔嘌呤核〕苷酸鈉 sodium 5'-guanylate (5'-GMP-Na)
6-甲基香豆素 6-methyl coumarin
ACH指數 ACH index
A帶 A band
DFD肉 dark, firm, dry muscle (DFD muscle)
D-泛醇 dexpanthenol
D值 decimal reduction value; D value
EDTA鐵鈉鹽 sodium iron(Ⅲ) ethylenediaminetetraacetate
F值 F value
I帶 I band
L-半胱氨酸鹽酸鹽 L-cysteine (mono) hydrochloride
L-肉鹼 L-carnitine
L-抗壞血酸鈉 sodium L-ascorbate
L-酒石酸鈉 sodium L-tartrate
pH指示劑 pH indicator
pH計 pH meter
pH值 pH value
pH試紙 pH paper
pH調節劑 pH adjusting agent
PP蓋 pilfer-proof cap (PP cap)
PSE肉 pale, soft, exudative muscle (PSE muscle)
TBA值 thiobarbituric acid value (TBA value)
TCA循環 tricarboxylic acid cycle (TCA cycle)
Tris緩衝液 Tris buffer
T-骨牛排 T-bone steak
V型混合機 V-type mixer
Z值 Z value
Z線 Z line
α-己基〔肉〕桂醛 α-hexyl cinnamaldehyde
α-戊基桂醇 α-amylcinnamyl alcohol
α-戊基桂醛 α-amylcinnamaldehyde
α-戊基桂醛二甲縮醛 α-amylcinnamaldehyde dimethylacetal
α生育酚 alpha-tocopherol; α-tocopherol
α-甲基〔肉〕桂醛 α-methyl cinnamaldehyde
α-甲基紫羅蘭酮 α-methyl ionone

α-烯丙基紫羅蘭酮 allyl α-ionone
α-異甲基紫羅蘭酮 iso-α-methyl ionone
α葡聚糖 alpha-glucan
α澱粉 alpha-starch; α-starch
α澱粉酶 alpha-amylase; α-amylase
α螺旋 alpha-helix
α纖維素 alpha-cellulose
β生育酚 beta-tocopherol; β-tocopherol
β-阿朴胡蘿蔔素醛 β-Apo-8' carotenal
β-阿朴胡蘿蔔酸乙酯 β-Apo-8' carotenoic acid ethyl ester
β胡蘿蔔素 beta-carotene; β-carotene
β澱粉 beta-starch; β-starch
β澱粉酶 beta-amylase; β-amylase
β-環糊精 β-cyclodextrin
γ-丁內酯 γ-butyrolactone
γ-十一碳烷內酯 peach aldehyde; γ-undecalactone
γ-十二內酯 γ-dodecalactone
γ-己內酯 γ-hexalactone
γ-戊內酯 γ-valerolactone
γ-辛內酯 γ-octalactone
γ-庚內酯 γ-heptalactone
γ-癸內酯 γ-decalactone
δ-十一內酯 δ-undecalactone
δ-十二內酯 δ-dodecalactone
δ-己內酯 δ-hexalactone
δ-壬內酯 δ-nonalactone
δ-辛內酯 δ-octalactone
δ-癸內酯 δ-decalactone